Commercial Flowers

— *Volume 2* —

THE EDITORS

Professor T.K. Bose obtained his M.Sc. (Ag.) and Ph.D. degree from Calcutta University, West Bengal, India. Trained at the Royal Botanic Garden, Kew, UK. Former Secretary of the Agri-Horticultural Society of India; Project Coordinator, All India Coordinated Floriculture Improvement Project (ICAR); Professor and Head, Department of Horticulture; Dean, Faculty of Agriculture; Director of Research, Bidhan Chandra Krishi Viswavidyalaya, West Bengal; Member, State Agriculture Commission, West Bengal; Member of Scientific Advisory Committees of ICAR, CSIR (Govt. of India) and Government of West Bengal. Awarded (D.Sc.) Honoris Causa from OUAT, Bhubaneshwar and UBKV, Coochbehar, India. Guided 45 Ph.D. students, published more than 250 research papers and authored 30 books on horticultural crops.

Dr. Chenna Reddy Aswath obtained his B.Sc. (Hort), M.Sc. (Hort.) in Floriculture from University of Agricultural Sciences, Bengaluru and Ph.D. degree in Biotechnology from Guwhati University, Assam. Serving as Principal Scientist in ICAR-IIHR, Bengaluru and has made significant contributions in the field of Floriculture and Biotechnology. Recipient of "Dr BP Pal Best Scientist Award for Agriculture" from NABS, Chennai, became FELLOW of the NABS, Chennai, ISOH, New Delhi and SAB, Dharwad, CHAI, New Delhi. Received the 2012 SAB Award of Excellence in Horticultural Biotechnology. Recipient of Brain pool Post-Doctoral Fellowship from South Korea Science Federation for the outstanding research contributions in Horticultural Biotechnology. His short term fellowship in Japan under INSA has recognized him among the Orchid tissue culture group. Published more than 110 research papers besides several books. Guided more than 10 post-graduate students.

Dr. Reeta Bhatia is working as senior scientist at the Division of Floriculture and Landscaping, ICAR- Indian Agricultural Research Institute, New Delhi. She has more than 13 years of research and teaching experience in the field of floriculture. She has obtained her Master's degree from Punjab Agricultural University, Ludhiana and doctoral degree from Indian Agricultural Research Institute, New Delhi. She has been awarded with the prestigious Endeavor Research Fellowship by the Government of Australia and pursued her postdoctoral research at the University of Western Australia. She has an excellent academic and research career and awarded with several medals. She has published more than 65 research papers in the reputed high impact scientific journals and written more than 30 review/popular/scientific articles. She is also serving as editor for the Journal of Ornamental Horticulture.

Dr. Koushik Dutta obtained his M.Sc. (Hort.) from CCS Haryana Agricultural University, Hisar, Haryana and Ph.D. from Uttar Banga Krishi Viswavidyalaya, West Bengal. Acting as Assistant Secretary of The Agri-Horticultural Society of India, Kolkata, West Bengal. Recipient of Inspire Fellowship from Department of Science and Technology, New Delhi. Published several scientific papers in national journals on different aspect of horticulture particularly in tissue culture and written book chapters on nursery management and jasmine. Associated with micropropagation of different of horticultural crops during last 7 years.

Dr. Rajiv Kumar obtained his B.Sc. (Ag and AH) and M.Sc. (Ag) Horticulture from C.S. Azad University of Agriculture and Technology, Kanpur, Uttar Pradesh, India and completed his Ph.D. (Horticulture) from I.A.R.I., New Delhi. Presently working as Principal Scientist at ICAR-IIHR, Bengaluru. Recipient of Young Scientist Associate Award by HTHS, U.P.; Young Scientist Award by SSDAT, U.P.; Gold Medal by HTHS, Meerut, and BR and IC Pvt. Ltd.; Fellow (Horticulture) by HAHS, Muzaffarnagar, U.P., AABS, Karnakata, and ISOH, New Delhi. Published 110 research articles, 25 popular articles, 6 edited books, 20 book chapters, 8 technical bulletins and 6 extension folders. Developed two varieties of China aster, associated in five and three varieties of Gerbera and gladiolus, respectively; registered 17 genetic stocks of China aster, Gerbera, Chrysanthemum and Gladiolus with ICAR-NBPGR, New Delhi. Guided two M.Sc. (Hort.) and two Ph.D. (Hort.) students in Floriculture.

Dr. Tarak Nath Saha graduated in Agriculture from Banaras Hindu University, Varanasi (Uttar Pradesh), obtained his Masters and Doctoral degree in Horticulture with specialization in Floriculture from India Agricultural Research Institute (IARI), New Delhi. Presently working as Senior Scientist at ICAR-Directorate of Floricultural Research, Pune. He has an experience of more than 12 year in research, teaching and extension, besides coordinating work of All India Coordinated Research Project (AICRP) on Floriculture. He has contributed 25 research papers, 41 technical articles, 5 book chapters, 20 technical bulletins, 6 folders, besides in the editorial team of 30 Annual Reports and 10 AICRP workshop proceedings. Developed one variety and registered one genetic stocks of tuberose with ICAR-NBPGR, New Delhi. He is actively involved in PG teaching; recognized as faculty of MPKV, Rahuri (also at IARI, New Delhi 2009-11) and guided more than 5 students in Floriculture as advisory.

Commercial Flowers

— Volume 2 —

— *Senior Editor* —

C. Aswath

Division of Flower and Medicinal Crops
ICAR-Indian Institute of Horticultural Research
Bengaluru 560089, Karnataka, India

— *Editors* —

T.K. Bose

Faculty of Horticulture
Bidhan Chandra Krishi Viswavidyalaya
West Bengal 741 252, India

Reeta Bhatia

Division of Floriculture and Landscaping,
ICAR- Indian Agricultural Research Institute, New Delhi 110012, India

Koushik Dutta

The Agri-Horticultural Society of India
1, Alipore Road, Kolkata 700027
West Bengal, India

Rajiv Kumar

Division of Flower and Medicinal Crops,
ICAR-Indian Institute of Horticultural Research,
Bengaluru 560089, Karnataka, India

Tarak Nath Saha

ICAR-Directorate of Floricultural Research, College of Agriculture Campus, Shivajinagar
Pune 411005, Maharashtra, India

2023

Daya Publishing House®

A Division of

Astral International Pvt. Ltd.

New Delhi – 110002

Third Revised & Illustrated edition 2023

Cataloging in Publication Data--DK

Courtesy: D.K. Agencies (P) Ltd. <docinfo@dkagencies.com>

Commercial flowers/senior editor, C. Aswath ; editors, T.K. Bose [and four others]. -- Third revised edition.

Volumes cm

Contributed articles.

Includes index.

ISBN 9789390435296 (HB)

1. Floriculture. 2. Flowers--Pictorial works. I. Aswath, C. (Principal scientist at the Indian Institute of Horticulture Research), editor. II. Bose, T. K., editor.

LCC SB405.C66 2021 | DDC 635.9 23

Published by : **Daya Publishing House®**
A Division of
Astral International Pvt. Ltd.
– ISO 9001:2015 Certified Company –
4736/23, Ansari Road, Darya Ganj
New Delhi-110 002
Ph. 011-43549197, 23278134
E-mail: info@astralint.com
Website: www.astralint.com

Digitally Printed at : **Replika Press Pvt. Ltd.**

From the Desk of Prof. T.K. Bose

The undersigned had initiated the publications entitled 'Fruits: Tropical and Sub-tropical', 'Vegetable Crops' and 'Commercial Flowers', one volume each in the 1980's in collaboration with co-editors, humbly expresses from the core of his heart that the updated revised editions of the above mentioned publications with colour illustrations in thirteen volumes which was planned in June, 2020, are being printed from January this year, a rare event in Horticultural Science has been possible by the blessing of Almighty God, with the cooperation of dedicated horticulturists as editors and contributors.

Kolkata

T.K. Bose
Editor Emeritus

Preface to the Third Edition

The first edition of the book entitled **'Commercial Flowers'** published in 1989, reprinted in 1993 and 1998 was greatly appreciated and proved valuable scientific review of commercially important flowers, to the students, teachers, researchers and extension specialists in horticulture in general and floriculture in particular, in different parts of the world. The book was revised in 2002 with the same title in two volumes due to its increasing demand, which contained updated scientific information of commercially grown flower crops and considered as valuable contribution on the subject. After about 18 years from the last revised publication in 2002, voluminous information pertaining to crop improvement, production, protection, postharvest management, biotechnology has been generated necessitating further revision of the existing chapters and inclusion of more ornamentals plants of commercial importance in four volumes, with relevant colour illustrations.

Second volume in this edition includes chapters on bulbous plants in ornamental horticulture and updated research review on all major aspects on lotus, tulip, lilium, gladiolus, tuberose, dahlia, alstroemeria, amaryllis and hippeastrum, peony, anemone, canna, ornithogalum, achimenes, torch ginger, Victoria lily and kniphofia with relevant colour illustrations, covering more than 800 pages.

Apart from own collection of photographs, editors have also taken some photographs which are in public domain from different websites, as far as possible. Therefore, we acknowledge with enormous thanks to all State Agricultural Universities and Research Institutes under Indian Council of Agricultural Research, New Delhi, India; Public and Private International Agricultural Institutes across the world for using photographs from their websites to make this book a valuable source of knowledge for the readers. We think images can be powerful teaching tools, as illustrations to in-class lectures, or for studying concepts outside of the classroom. We also think that identification of flower crops, their importance and uses, species and cultivars, methods of propagation, cultivation practices including greenhouse cultivation, diseases and pests, harvesting and postharvest technology and achievement of biotechnology, with the help of suitable illustrations in a consolidated manner will make a dent for the students, researchers and teachers of floriculture and ornamental horticulture.

We sincerely thank the contributors for preparing updated revised manuscripts. Our thanks are due to Shri Prateek Mittal, Director, Astral International (P) Ltd., New Delhi, for his keen interest in bringing out this publication.

Kolkata **Editors**

Contents

List of Contributors

Ahmed, Z.

Division of Floriculture and Landscape Architecture, Sher-e-Kasmir University of Agricultural Sciences and Technology of Kashmir, Srinagar 190019, Jammu and Kashmir, India

Anand, P.

Division of Floriculture and Landscaping, ICAR-Indian Agricultural Research Institute, New Delhi 110012, India

Aswath, C.

Division of Flower and Medicinal Crops, ICAR-Indian Institute of Horticultural Research, Bengaluru 560089, Karnataka, India

Bal, Solanki

Department of Vegetable Science, Faculty of Horticulture, Bidhan Chandra Krishi Viswavidyalaya, Mohanpur 741252, Nadia, West Bengal, India

Bharati, T.U.

Division of Flower and Medicinal Crops, ICAR-Indian Institute of Horticultural Research, Bengaluru 560 089, Karnataka, India

Bhattacharjee, S.K.

Division of Floriculture and Landscaping, ICAR-Indian Agricultural Research Institute, New Delhi - 110012, India

Bhargav, V.

Department of Horticulture, Multi-Technology Testing Centre and Vocational Training Centre, College of Horticulture and Forestry, CAU, Pasighat 791102, Arunachal Pradesh, India

Bhatia, R.

Division of Floriculture and Landscaping, ICAR-Indian Agricultural Research Institute, New Delhi 110012, India

Bose, T.K.

Department of Horticulture, Bidhan Chandra Krishi Vishwavidyalaya, Mohanpur 741252, West Bengal, India

Chatterjee, Subhrajyoti

Centurion University of Technology and Management, Department of Horticulture, Paralakhemundi 761 211, Gajapati, Odisha, India

Dubey, R.K.

Department of Floriculture and Landscaping, Punjab Agricultural University, Ludhiana, Punjab, India

Dutta, K.

The Agri-Horticultural Society of India, 1, Alipore Road, Kolkata 700027, West Bengal, India

Dohare, S.R.

Division of Floriculture and Landscaping, ICAR-Indian Agricultural Research Institute, New Delhi 110012, India

Jain, R.

Division of Floriculture and Landscaping, ICAR-Indian Agricultural Research Institute, New Delhi 110012, India

Kumar, Rajiv

Division of Flower and Medicinal Crops, ICAR-Indian Institute of Horticultural Research, Bengaluru 560089, Karnataka, India

Kumar, Sunil

Department of Floriculture and Landscape Architecture, College of Horticulture and Forestry, CAU, Pasighat 791102, Arunachal Pradesh, India

Maity, R.G.

Department of Horticulture, Bidhan Chandra Krishi Vishwavidyalaya, Mohanpur 741252, West Bengal, India

Misra, R.L.

Division of Floriculture and Landscaping, ICAR-Indian Agricultural Research Institute, New Delhi 110012, India

Mitra, M.

Department of Horticulture, Bidhan Chandra Krishi Vishwavidyalaya, Mohanpur 741252, West Bengal, India

Momin, K. Ch

Department of Floriculture and Landscape Architecture, College of Horticulture and Forestry, CAU, Pasighat 791102, Arunachal Pradesh, India

Mukherjee, Debmala

Department of Vegetable Science,Faculty of Horticulture, Bidhan Chandra Krishi Viswavidyalaya, Mohanpur 741252, Nadia, West Bengal, India

Nazki, I.T.

Division of Floriculture and Landscape Architecture, Sher-e-Kasmir University of Agricultural Sciences and Technology of Kashmir, Srinagar 190019, Jammu and Kashmir, India

Pal, P

Department of Horticulture, Bidhan Chandra Krishi Viswavidyalaya, Mohanpur - 741252 Nadia, West Bengal, India

Palai, S.K.

Regional Plant Resource Centre, Bhubaneswar 751015, Odisha, India

Parvathi, B.

Division of Floriculture and Medicinal Crops, IARI- Indian Institute of Horticultural Research, Hessaraghatta, Bengaluru 560089, Karnataka, India

Parthasarathy, V.A.

ICAR – Indian Institute of Spices Research, 32/482 C Narmada Nilayam, Bharathan Bazar, Chelavoor, Calicut 673571, Kerala, India

Patel, D.A.

Department of Floriculture and Landscape Architecture, S.D. Agricultural University, Jagudan, Mehsana 384 460, Gujarat, India

Roychoudhury, N.

Department of Horticulture, Bidhan Chandra Krishi Vishwavidyalaya, Mohanpur 741252, West Bengal, India

Safeena, S.

Division of Flower and Medicinal Crops, ICAR-Indian Institute of Horticultural Research, Bengaluru 560 089, Karnataka, India

Sangeetha Priya

Division of Flower and Medicinal Crops, ICAR-Indian Institute of Horticultural Research, Bengaluru - 560 089, Karnataka, India

Singh, Babita

Division of Floriculture and Landscaping, ICAR-Indian Agricultural Research Institute, New Delhi 110012, India

Singh, Brijendra

Division of Floriculture and Landscaping, ICAR-Indian Agricultural Research Institute, New Delhi 110012, India

Shreya

Department of Basic Science, College of Horticulture, S.D. Agricultural University, Jagudan, Mehsana 384460, Gujarat, India

Sumalatha, A.

Department of Floriculture and Landscape Architecture, Dr. Y.S.R. Horticultural University, Venkataramannagudem 534101, Andhra Pradesh, India

Swami Vinayananda

Ramakrishna Mission Ashram, Asansol 713 301, West Bengal, India

Tiwari, A.K.

Division of Floriculture and Landscaping, ICAR-Indian Agricultural Research Institute, New Delhi 110012, India

Varun, M. Hiremath

Division of Floriculture and Landscaping, ICAR-Indian Agricultural Research Institute, New Delhi 110012, India

Vanlalruati

Division of Floriculture and Landscaping, ICAR- Indian Agricultural Research Institute, New Delhi 110012, India

Yadav, L.P.

Department of Horticulture, Bidhan Chandra Krishi Vishwavidyalaya, Mohanpur 741252, West Bengal, India

Yatung, T.

Department of Horticulture, Multi-Technology Testing Centre and Vocational Training Centre, College of Horticulture and Forestry, CAU, Pasighat 791102, Arunachal Pradesh, India

1

BULBOUS PLANTS IN ORNAMENTAL HORTICULTURE

K. Dutta, T.K. Bose and C. Aswath

1.0 Importance and Uses

In horticulture, the word 'bulb' includes underground modified stems which are used for propagation, *e.g.*, bulb, corm, tuber and rhizome. Tuberous roots are also covered by this term. A large number of these plants producing attractive flowers and foliage are grown for their ornamental value. They are extensively grown in beds, borders, rockery, in home garden, parks and public places. Many species and cultivars of bulbous ornamentals are cultivated in large scale for commercial production of cut flowers and bulbs, outdoors and in greenhouses for domestic and export markets. Several flowering species are suitable for growing in semi-shade and marshy areas. Water lily, lotus and Victoria lily are important features in lily pools. Lotus is the national flower in India. It is the symbol of purity in Buddhist temple and used as a respectful offer to worship in Hinduism. The flower is so sacred in Japan that it is never used other than Buddhist services.

Potted bulbous ornamental plants are also widely used in different purposes. They can be kept well in lawn, along roads of the garden, balcony and roof, in window garden and also inside the room for indoor decoration. Pot grown plants are also used for decorations of conference halls, stages, meeting places and exhibited in flower shows.

Because of enormous variation in growth and flowering habit in different genera and species, wide range of biodiversity among the cultivars and suitability for adaptation in different climatic conditions, bulbous plants are considered as one of the most important group of ornamental plants bearing beautiful flowers in large numbers with variation of colours, shape, size and appearance. Cut flowers of *Lilium, Gladiolus, Alstroemeria, Polianthes, Strelitzia* (Bird of paradise), *Zantedeschia*, grown in mild topics and subtropics are important in domestic and global flower markets.

Hippaestrum, Haemanthus, Cyclamen, Ornithogalum, Lilium, Hyacinthus, Strelitzia etc. grow and flower well in containers, often used for decoration and sold as pot plants. *Nymphaea* and *Nelumbo* flower well in large containers in moist or waterlogged soil for many months except in winter, in addition to their spectacular display of colour in tanks, lakes and lily pools.

A bulbous plant has normally three phases of growth and development including flowering during a year; vegetative growth, flowering and dormancy. Vegetative growth may precede flowering as in *Gladiolus, tuberose* or follow it as in *Amaryllis* and *Haemanthus*. After growth and flowering, many species of bulbous plants, enter into a rest period, the duration of which varies with the type of plants and environmental conditions, like temperature, light and humidity. In some bulbous plants, dormancy is represented by temporary cessation of active growth, *e.g., Polianthes* (tuberose); Canna has no dormancy when grown in tropics. A large number of bulbous ornamental plants are grown in tropical and subtropical climate. The flowering group includes *Achimenes, Acorus, Alstroemeria, Amaryllis, Crinum, Dahlia, Etlingera, Eucharis, Gladiolus, Gloriosa, Haemanthus, Hedychium, Heliconia, Hemerocallis, Hippeastrum, Hymenocallis, Nelumbo, Nymphaea, Ornithogalum,*

Pancratium, Polianthes, Sinningia, (gloxinia), *Zantedeschia, Zephyranthes,* while *Alocasia, Asparagus, Caladium, Calathea, Kaemferia, Maranta, Sanseviera, etc.* belong to the foliage group. *Agapanthus, Cyclamen, Lilium* and *Ornithogalum* also grow and flower in mild climate (subtropical).

Bulbous flowering plants which grow and flower well in temperate climate are *Agapanthus, Allium, Anemone, Begonia tuberhybrida* (tuberous begonia), *Clivia, Crocus, Cyclamen, Freesia, Fritillaria, Iris, Ixia, Hyacinthus, Lilium, Moraea, Montbretia, Narcissus, Nerine, Paeonia, Kniphofia, Ornithogalum, Ranunculus, Sparaxis, Tulipa, Watsonia, etc.*, are grown outdoor in temperate climate, or in greenhouses when temperature is low. Some of the bulbous plants which are extensively grown in tropical gardens also flower very well in mild subtropics. Several species and cultivars of *Lilium, Ornithogalum, Agapanthus, Begonia tuberhybrida* grow and flower well in subtropical and warm temperate climate (Bose *et al.*, 2002).

Bulbs of many species grown as ornamental plants are well known for their therapeutic properties and have long been used for treatment of many diseases, around the globe.

The tuber of *Gloriosa superba* is a rich source of alkaloid colchicine which is used for the treatment of various diseases like anti-gout, anti-inflammatory, and anti-tumor activity. It is also a very effective chemical used for inducing polyploidy for improvement of plants.

Total area and production value under bulbous crops is maximum is in The Netherlands *i.e.* 20,577 ha and 570 millions euro, respectively (Bose *et al.*, 1999; 2015). Europe, the trade center of world floriculture shares about 77 per cent of the global market. The consumption per capita has increased strongly over the years.

2.0 Season of Flowering of Bulbous Plants

Soil and climate largely influence growth and development of plants including bulbous ornamentals. Many bulbous plants continue to grow and flower for many years at the same site of planting without separation of clumps, *viz., Canna, Crinum, Pancratium, Heliconia etc.*

2.1 Spring Flowering Bulbs

Spring is the most typical season for bulbs to flower (IFBC, 2014). Some examples are *Asphodelus, Camassia, Convallaria, Crocus, Cyclamen, Eranthis, Freesia, Fritillaria, Galanthus, Hyacinthus, Hippeastrum, Iris, Ixia, Leucojum, Muscari, Narcissus, Ornithogalum, Ranunculus, Scilla, Trillium, Tulipa* and *Zephyranthes*.

2.2 Summer Flowering Bulbs

Achimenes, Agapanthus,Alstroemeria, Amaryllis, Anemone, Begonia, Calochortus, Canna, Crinum, Crocosmia, Dahlia, Dierama, Eucomis, Galtonia, Gladiolus, Gloriosa, Haemanthus, Hymenocallis, Lilium,Oxalis, Pancratium, Paradisea, Polianthes, Sprekelia, Tritonia, Watsonia and *Zantedeschia*.

Importance of Bulbous Plants

Lotus in Tank

Lilium in Pot

Amaryllis in Pot

Tulip Bulbs

Gladiolus Corms

Tuberose Bulbs

Rhizomes of Iris

Tubers of Dahlia

Canna Rhizomes

***Begonia tuberhybrida* in Pot**

Amaryllis in Pot

Alstromeria in Flower Vase

Importance of Bulbous Plants

Cultivation of Tulip in Open

Greenhouse Cultivation of lilium

Tulip Bulbs in Market

Seeds of Bulbous Plants in Market

Cut Tulip in Market

Quality of Cut Gladiolus in Market

Cyclamen in Pots

Achimenes in Pot

Importance of Bulbous Plants

Dahlia in Pot

Gladiolus in Garden

Zantedeschia in Pot

Tulip Cut Flowers

Gloriosa superba in Garden

Haemanthus in Pot

Crinum (Spider lily)

Hedychium (Ginger Lily)

Etlingera (Torch ginger) in Garden

Ranunculus in Pot

Kniphofia in Bed

Importance of Bulbous Plants

Hippeastrum in Garden

Ornithogalum in Bloom

Hyacinthus in Bed

Paeonia in Bed

Iris in Bed

Hymenocallis

Nerine in Bed

Agapanthus in Garden

Calathia in Pots

Maranta leuconeura

Rhizomatous Foliage Plants

2.3 Autumn Flowering Bulbs

Crocus, Colchicum, Cyclamen, Nerine and *Sternbergia.*

2.4 Winter Flowering Bulbs

Some species from the following genera are produce flowers in winter months. *Galanthus, Crocus, Cyclamen* and *Eranthis.*

2.5 Shade Loving Species

Anemone blanda, Anemone nemorosa, Arum italicum, Convallaria majalis, Corydalis flexuosa, Cyclamen purpurascens, Disporum flavescens, Erythronium, Fritillaria pallidiflora, Galanthus, Iris douglasiana, Leucojum vernum, Lilium martagon, Ranunculus ficaria, Sanguinaria canadensis and *Smilacina racemosa.*

3.0 Bulbous Plants for Rock Garden

A rock garden is a garden which is made of a combination of rocks and small plants. Some of the genera of bulbous plants that are most suitable for rock gardens include

Allium, Anemone, Anthericum, Bulbocodium, Cyclamen, Eranthis, Erythronium, Galanthus, Muscari, Ornithogalum, Oxalis, Romulea, Rhodohypoxis and *Scilla.*

4.0 Research and Development on Bulbous Ornamental Plants

Because of the aesthetic and commercial value, large number of bulbous ornamental plants are grown in different climates throughout the world. Intensive research is being carried out on important bulbous plants for decades for the improvement of species and cultivars, methods of propagation, production of flowers and bulbs including greenhouse cultivation, protection from diseases and pests and postharvest management of flowers, bulbs, corms (planting materials).

As majority of bulbous ornamentals are commonly grown in sub-tropical and temperate regions for cut flower and bulb production, technology of greenhouse establishment, management, production and protection has been standardized through continuous research and trials on important bulbous plants. In order to facilitate production and reduction of management cost, automation in greenhouse technology has been adapted in the production of flowers and bulbs of good quality.

Many aspects of several bulbous plants of commercial importance grown in tropics, subtropics and temperate regions are reviewed and illustrated in volume 2 and 3.

5.0 References

Anonymous (2014) *International Bulb Centre Spring Blooming Bulbs.*

Bose, T. K., Maity, R.G. and Dhua, R.S. (1999) *Floriculture and Landscaping*, Naya Prakash, Kolkata.

Bose, T. K., Singh, L.J., Sadhu, M.K. and Maity, T.K. (2015) *Ornamental Plants and Garden Design in Tropics and Subtropics*, Astral International, New Delhi. Vols. 1 and 2.

Bose, T. K., Yadav, L.P., Pal, P., Das, P. and Parthasarathy, V.A. (2002) Commercial Flowers, Naya Prokash, Vols. 1 and 2.

2

LOTUS

V.A. Parthasarathy, C. Aswath and Solanki Bal

1.0 Introduction

Sacred lotus, so named because of its religious significance in both Buddhism and Hinduism, belongs to the small plant family Nelumbonaceae, with only one genus, Nelumbo, and two species: *Nelumbo nucifera* (Asia, Australia, Russia) and *N. lutea* (eastern and southern North America) (Shen - Miller, 2002). Lotus is truly unique. All parts of the plant – the roots, stems, leaves, seeds and flowers have been used for millennia as a food source and in traditional medicines. The lotus is revered in many cultures as a symbol of purity and perfection. It is not surprising that it is embedded in our psyche as a sacred plant symbolising divinity (Battaglia, 2019). It is also the National flower of India.

The lotus was extant 135 million years ago in aquatic habitats of the Northern Hemisphere (Mandal and Brar, 2013). Historically, the lotus has been grown in the far East for 5000–7000 years (Wong, 1987), with evidence of its cultivation more than 3000 years ago for use as food, medicine and for cultural and religious activities. According to Shen Millier 2002, in Buddhism, lotus symbolizes purity and is placed at the shrine to maintain serenity; more than 1000 years ago, Buddhist monks had grown the lotus extensively at a particular lake in China (Machuria). Rhizomes from *N. nucifera* are used as one of the major vegetables in China, Japan and Korea. In 2005, production of lotus in China was estimated at 300,000 ha resulting in approximately 6.75 million tons of lotus rhizomes per year based on an average yield of 22.5 ton/ha (Li *et al.*, 2005).

2.0 Importance and Uses

Apart from the ornamental importance of lotus flowers, the plant as a whole has got enough neutraceutical and medicinal value for curing ailments involving digestive, reproductive, circulatory and excretory systems. Almost each and every plant part *viz.*, leaves, petioles, flowers, seeds, rhizomes are used as food in various ways. Moreover, these are recommended for the treatment of spermatorrhoea, gonorrhoea, insomnia, metrorrhagia, dyspepsia, *etc.* (Nadkarni, 1954). *Nelumbo nucifera*, or sacred lotus, has been valuable for us to use as vegetable, functional food, and herb medicine for over 2,000 years. Chen *et al.* (2019) presented state-of-the-art overview on its traditional and modern uses in food industry and medicines, together with the comprehensive profiles of phytochemicals, and health promoting bioactivities of this valuable plant. Traditionally, lotus has been used to treat chronic dyspepsia, hematuria, insomnia, nervous disorders, cardiovascular diseases, and hyperlipidemia. Now, phytochemical investigations on different parts of *N. nucifera* have indicated a wide spectrum of at least 255 constituents belonging to different chemical groups, including proteins, amino acids, polysaccharides, starch, flavonoids, alkaloids, essential oils, triterpenoids, steroids, and glycosides. Meanwhile, its pharmacological activities, including anti-obesity, antioxidant, anti-inflammatory, cardiovascular, hepatoprotective, hypoglycemic, hypolipidemic, antitumor, memory-improving and antiviral activities, have been amply worked. Zhang *et al.* (2015) wrote a detailed review on the nutrition, composition, physiological functions and processing methods of lotus seeds, describing the impact of the latter on nutrient preservation. According to them, Lotus seeds, which are currently the oldest known plant seeds, contain many functional ingredients. They can be eaten raw or cooked, and are often added to foods as ingredients or supplements. Many naturally occurring ingredients isolated from lotus seeds are certified to be multiple functional compounds, such as polyphenols, protein, polysaccharides. Proteins and carbohydrates are the main nutrients of lotus seeds. Low fat content and good proportion of amino acids confer to lotus seeds unique nutritional values that have attracted increasing attention around the world. Multiple studies have assessed the functional components of lotus seeds. The bioactivity of ingredients from lotus seeds *in vitro* and *in vivo* include antioxidant activity, hypoglycemic, immunomodulatory, antibacterial, anti-inflammatory, analgesic effects as well as gastrointestinal regulation. Lotus seeds show prospective application in function food area and traditional medicine research. Furthermore, structure–activity relationship of functional compounds from lotus seeds will attracts much more interests in recent years.

Nelumbo nucifera has also been extensively used for nutritional and traditional medicinal purpose by people in many parts of the world. Mandal and Brar (2013) described the medicinal uses of lotus as follows. Therapeutically, flowers are used as astringents in diarrhoea, cholera, fever and recommended as cardiac tonic. Seeds are used to curb vomiting, and given to children as a diuretic, for skin diseases and leprosy, and are considered as an antidote to poison. Petioles are considered as astringent, useful for burning sensations and piles. The root in powder form is prescribed for curing piles, dysentery, and dyspepsia, and is used as a paste for skin infections and ringworm (Chopra *et al.*, 1986). Lotus honey is considered as good as *Amrita* in Ayurveda and believed to be valuable as it can cure many common ailments (Porterfield, 1941). Virtually, all parts of the lotus plant are used: the rhizome is used as food, seed as medicine, thalamus as fruit, leaves as plate (*thali*), stalks as pickle, petals for colour extraction, and tender leaves as food after being blended with vegetables (Mandal *et al.*, 2007). The pharmacological studies have shown tremendous potential of the plant against a wide range of diseases and infections. So the need of the hour is to further evaluate the medicinal importance of *Nelumbo nucifera*,in view of its large scale use in traditional medicine and recently identified pharmacological activities and also to develop the protocols for efficient extraction and validation of the active principles for their use to combat different human disease conditions (Sheikh, 2014). The phytochemical profile of the plant is different by part to part. There are numerous therapeutic profits of this plant for which different parts are used. The extracts of rhizomes, seeds flowers and leaves have been reported to have varied therapeutic potential (Buddhadev and Buddhadev, 2014). Leaves, carpels and rhizomes are medicinally important too as they possess different alkaloids. It is reported that the leaves contain three alkaloids namely nuciferine, roemerine and nornuciferine. Nelumbine, another alkaloid which acts as cardiac poison, has been isolated from petioles, pedicel and embryo (Anon, 1966). Besides, the juice of leaves and floral stalk, is used in diarrhoea. Some unusual major polyamines such as homospermidine, aminopropylhomospermidine, neospermidine, norspermine, thermospermine and caldopentamine (Hamana *et al.*, 1994).

The lotus root is used to add seasoning to food. Lotus root is a moderate calorie root vegetable (100 g of root-stem provides about 74 calories) and is composed of several vitamins, minerals, and nutrients: 83.80 per cent water, 0.11 per cent fat, 1.56 per cent reducing sugar, 0.41 per cent sucrose, 2.70 per cent crude protein, 9.25 per cent starch, 0.80 per cent fiber, 0.10 per cent ash and 0.06 per cent calcium.100 g of root provides 44 mg of vitamin C or 73 per cent of daily recommended values (RDA). The nutritional value of cooked roots are given in Table 1.

Table 1: Nutritional Value of Cooked Lotus Root

Nutritional value per 100 g (3.5 oz) of Lotus root (Cooked)	
Energy	278 kJ (66 kcal)
Carbohydrates	16.02 g
Sugars	0.50 g
Dietary fiber	3.1 g
Fat	0.07 g
Protein	1.58 g
Vitamins	Quantity per cent DV
Thiamine (B1)	11 per cent (0.127 mg)
Riboflavin (B2)	1 per cent (0.01 mg)
Niacin (B3)	2 per cent (0.3 mg)
Pantothenic acid (B5)	6 per cent (0.302 mg)
Vitamin B6	17 per cent (0.218 mg)
Folate (B9)	2 per cent (8 μg)
Choline	5 per cent (25.4 mg)
Vitamin C	33 per cent (27.4 mg)
Minerals	Quantity per cent DV
Calcium	3 per cent (26 mg)
Iron	7 per cent (0.9 mg)
Magnesium	6 per cent (22 mg0
Manganese	10 per cent (0.22 mg)
Phosphorus	11 per cent (78 mg)
Potassium	8 per cent (363 mg)
Sodium	3 per cent (45 mg)
Zinc	3 per cent (0.33 mg)
Other constituents	Quantity
Water	81.42 g

Units: μg = micrograms; mg = milligrams; IU = International units.

Source: USDA Nutrient Database

https://en.wikipedia.org/wiki/Nelumbo_nucifera Seeds are source of high quality protein and are rich in many essential amino acids including high contents of albumin (42 per cent) and globulin (27 per cent), they also contain unsaturated fatty acids, carbohydrates, vitamins, calcium, iron, zinc, phosphorus and ther trace element (Wang and Zhang, 2010).

Green fruiting torus containing carpels is a popular edible item in the rural and sub-urban areas in India. The carpels are eaten raw, roasted, boiled and sometimes grounded into flour which is a good source of cereal and nutritive too as evident from the following analysis (Table 2) of matured dried carpels (Anon, 1966).

Table 2: Nutritional Composition of Carpels*

Constituents	*Availability (per cent)*	*Constituents*	*Availability (mg/100g)*
Total carbohydrate	66.6	Calcium	136.0
Protein	17.2	Phosphorus	294.0
Water	10.0	Iron	2.3
Sucrose	4.1		
Ash	3.8		
Fibre	2.6		
Reducing Sugar	2.4		
Fat	2.4		

*Anon. (1966).

Lotus leaves and embryos are rich in flavonoids and other secondary metabolites, and they have been extensively studied for their antioxidant, antibacterial, anti-HIV and anti-obesity functions (Chen *et al.*, 2012). Flavonoid composition and concentration were investigated in 12 different tissues of 'Ti-1' lotus by high performance liquid chromatography equipped with photodiode array detection tandem electrospray ionization mass spectrometry (HPLC-DAD-ESI-MSn). A total of 20 flavonoids belonging to six groups (myricetin, quercetin, kaempferol, isohamnetin, diosmetin derivatives) were separated and identified (Chen *et al.*, 2012). Myricetin 3-O-galactoside, myricetin 3-O-glucuronide, isorhamnetin 3-O-glucuronide and free aglycone diometin (3,5,7-trihydroxy-4-methoxyflavone) were first reported in lotus. Flavonoid composition varied largely with tissue type, and diverse compounds (5–15) were found in leaf and flower stalks, flower pistils, seed coats and embryos. Flower tissues including flower petals, stamens, pistils, and, especially, reproductive tissue fruit coats had more flavonoid compounds (15–17) than leaves (12), while no flavonoids were detectable in seed kernels. The flavonoid content of seed embryos was high, 730.95 mg/100 g DW (dry weight). As regards the other tissues, mature leaf pulp (771.79 mg/100 g FW (fresh weight) and young leaves (650.67 mg 100/g FW) had higher total flavonoid amount than flower stamens (359.45 mg/100g FW) and flower petals (342.97 mg/100g FW), while leaf stalks, flower stalks and seed coats had much less total flavonoid (less than 40 mg/100g FW) (Chen *et al.*, 2012).

Fresh lotus flowers as such generally remain in good demand for the purpose of worship throughout India.

Uses of Lotus

In a Tank

In Public Garden

In Pot

Lotus in Pond

In Decoration

As Cut Flowers

In Flower Arrangement

3.0 Origin, History and Distribution

3.1.0 Origin and History

The genus *Nelumbo* is widely distributed in the tropical and sub-tropical regions of the world.It is found in lakes, ponds, quiet streams and estuaries of tropical and sub-tropical zones in Asian continent extending up to Caspian Sea and United States of America.Reading the History and distribution of Lotus is just like reading the history and spread of two oldest religions of the world, Hinduism and Buddhism. Here again, the wonderful piece of work by Shen Miller *et al.* (2013) is worth mentioning. According to them originating in India, Buddhism later spread into China and by 372 A.D. had been introduced into the Korean peninsula via China's Liaoning Province, the source of the centuries-old Nelumbo fruit where the plant had been cultivated in the large lake that is now the site of Xipaozi Village. Interestingly, the three Chinese characters for this village, Xi, Pao, and Zi, literally mean "Steeped Seeds of the West", in reference to a lake filled with seeds of an aquatic plant, Nelumbo, which had been introduced from western regions. As a result of centuries of historically recorded earthquakes in the Xipaozi region (Shen Miller *et al.*, 2002), the Nelumbo-populated lake ultimately drained into the Bo Hai Sea converting the former lakebed to arable land.

It is believed that lotus (*N. nucifera*) has at least 6000 years old association with Indian culture and religion. Lotus has been mentioned in several mythological legends, epics, Sanskrit literature and historical records. It has also been delineated in sculpture, temple carvings, architecture, cave murals, frescoes and in innumerable paintings during early Hindu-Buddhist period of Indian history. An examination of the mythic and artistic records of India and Southeast Asia indicates that the famous psychotropic of the ancient Aryans was the eastern lotus, *Nelumbo nucifera* (McDonald, 2004). Vedic epithets, metaphors, and myths that describe the physical and behavioural characteristics of the 'soma' plant as a sun, serpent, golden eagle, arrow, lightning bolt, cloud, phallic pillar, womb, chariot, and immortal navel, relate individually or as a whole to the eastern lotus. Since most Hindu and Buddhist gods and goddesses trace their origins from the Vedas and have always shared close symbolic associations with *Nelumbo*, there is reason to believe the divine status of this symbolic plant derives from India's prehistoric past (McDonald, 2004).

Possibly, the most interesting and authentic history of lotus was presented by Shen Miller *et al.* (2013). According to them, during the Sino-Japanese conflict of the 1920s, Japanese botanist Ichiro Ohga was presented a single-seeded fruit of *Nelumbo nucifera* var. China Antique collected by a local farmer from a dry lake bed in Northeast China (then, Manchuria). Ohga (1926) studied the fruit and published his findings. Years later, Shen Miller *et al.* (2013) tested the germination of Nelumbo fruit from the same locality. The oldest seed sprouted, having a germination time of ~3 days, was radiocarbon dated to be ~1300 years old. These cold- and drought-tolerant seeds exhibited shoot-before-root emergence and a primary green plumule capable of "dim-light" photosynthesis. Such traits and the notable long-term viability of the fruit spurred the interest of Ray Ming, University of Illinois that has now led to the sequencing of the Nelumbo genome. Analyses of this genome may provide insight into the biochemistry of Nelumbo on wax-biosynthesis genes, and application of aging-related thermostable proteins to the extension of seed-life and improvement of food quality of economic crops. They (Shen Miller *et al.*, 2013) reviewed the history of these long-lived Nelumbo fruit, and their occurrence, discovery, collection, propagation, and methods of seedling care. The robust impermeable wax and suberin covered pericarp is a major factor contributing to their remarkable longevity. New findings are presented on the modern and 459- and 464-year-old pericarp anatomy, impermeability to water, and whole fruit and pericarp mechanical properties, and comparison of the mode of fruit weight-gain during imbibitions and germination time relative to fruit maturity.

The first Indian fossilised record of *N. nucifera* was reported from the Pleistocene epoch of Kashmir. Further, evidence on the presence of *Nelumbo* during the Tertiary period of Assam was on the basis of impressions of leaves and rhizomes discovered in the collection from Eocene bed near Damalgiri (Lakhanpal, 1955). These reports confirm that *Nelumbo nucifera* was indigenous to India. This species is widely distributed into different phyto-geographical regions of India exhibiting great phynotepic variation and plasticity. Other countries where the *Nelumbo nucifera* and its cultivars are found – Sri Lanka, Indonesia, Korea, Thailand, Japan, China, Brazil, USA and Australia. *N. lutea* is distributed in North America and West Indies. It has been found naturally growing profusely along the delta of Mississippi river, southern Ontario to Florida and from Minnesota to Texas. It was speculated that through human activity, *N. nucifera* was introduced to tropical areas of South American, South and Southeast Asia and Australia (Borsch and Barthlott, 1994; Green, 2002). The historical reconstruction of the origin and dispersal of plant taxa in space and time facilitates a better understanding of their modern distribution patterns. However, most studies of paleobiogeography have focused on terrestrial plants, and the distribution changes of aquatic plants are less well understood. Hence, According to Wang and Zhang (2010), Lotus has at least 5,000 years of lotus growing history since the two carbonized lotus seeds were found on the sites of 'Yangshao Culture' (5,000-3,000 BC) in Zhengzhou, Henan of China. It has more than 2,500 years of history in cultivation since the construction of "Pond of Playing Flowers" (490 BC or so), which was built by Fu Chai, the King of Wu, for Xi Shi for enjoying lotus in Lingyan Hill of Suzhou. 'Hua Lian' (flower lotus), cultivated by the Chinese ancestors before Christ, is purely a few-petalled red cultivar of the "wild lotus". During centuries of cultivation, the genetic composition of flower lotus is gradually becoming complicated, and the polyploidy has transformed a few-petalled flower to semi-double, double, duplicate-petalled, thousand-petalled form. As a result, some excellent cultivars occurred in Chinese dynasties. However, due to the continuous wars in the Chinese feudal society, many lotus cultivars from either the royal parks or private gardens have not been handed down to the middle 1700s (Wang and Zhang, 2010).

3.2.0 Distribution

Li *et al.* (2014) studied the lotus plant *Nelumbo* (Nelumbonaceae), an aquatic perennial herb, with a disjunctive distribution across East, South and Southeast Asia-North Australia and North America. They described reproductive organs of *Nelumbo changchangensis* He et Jin. from the Eocene of Hainan, China as well. Analysis of the spatial and temporal distributions of Nelumbo in the geologic past indicates that the genus first occurs in mid-latitude area of Laurasia in the Early Cretaceous, then becomes widespread in North America and Eurasia and expands into South America during the Late Cretaceous, and reaches its maximum northern limit during the Eocene. The genus persists and thrives in North America and Eurasia until the Pliocene. The Pleistocene ice age causes the extinction of Nelumbo in Europe and central Asia, and its populations in North American and Asia are also restricted to refuges of lower latitude. Like the terrestrial plants Metasequoia (Cupressaceae) and Nordenskioeldia (Trochodendraceae), the fluctuations of Nelumbo distribution ranges are also linked to climatic changes in the Cenozoic. The cooling climate and increasing seasonality in the Eocene of East Asia may favour the origin of tubers and the differentiating of the ecotypes in lotus, which allow the deciduous type to survive in cold winters (Li *et al.*, 2014).

The other species, *N. lutea* (Yellow Lotus) is a native of North America (New York, Southern Ontairo, Florida, Texas); Central America, West Indies and Columbia (Niklitschek, 1991).

4.0 Species and Cultivars

4.1.0 Taxonomic Classification

Lotus classification took a long time because of its superficial similarities in the flowers and vegetative body with the waterlily (Lin *et al.*, 2019). Nelumbo used to be regarded as one genus of the Nymphaeaceae family in the old classification system. From the genetic point of view, both species of lotus are diploid with the number of chromosomes 2n = 16. The predicted size of the lotus genome is 929 Mb, which is based on the flow cytometry analysis (Diao *et al.*, 2006).

4.2.0 Species

Lotus (*Nelumbo* Adans.) belongs to the monogeneric family Nelumbonaceae and is represented by two species *Nelumbo nucifera* Gaertn. and *N. lutea* Pers. The flowers of *N. nucifera* races are more attractive as compared to the pale yellow flowers of *N. lutea*. The botanical descriptions of the two species are as follows.

4.2.1 Botany

N. nucifera (Sacred Lotus)

Leaves (Pal and Dey, 2012): Leaves are large, of both types, aerial as well as floating orbicular 20-90 cm. In diameter, abruptly acute to form a short tip, petiolate, entire glaucous, non-wettable, strong cupped in case of aerial leaves and flat in case of floating ones, radiantly nerved, the fresh leaves are leathery, bout on drying they are nearly membranous and brittle, there is more or less brownish red blotching on the lower surface, petioles of the aerial leaves are erect and stout white those of the floating ones are not strong enough. The usual length varies from 24 to 33cm in aerial leaves and 23 to 30 cm in case of floating, petioles are smooth, greenish or greenish brown in colour with small brown dots sometimes rough with very

Species of Lotus

Nelumbo nucifera

Nelumbo lutea

Cultivars of Lotus

'Alba Grandiflora'

'Charles Thomas'

'Empress'

'Ink Bowl'

'Nikki Gibson'

'Buddha Seat'

small, but distinct, odour is distinct, fracture is fibrous. When transversely cut, the petiole of leaf stalk always shows four distinct, large cavities in the centre and small cavities in the periphery.

Flowers: Flowers 8-25 (-30.5) cm across, rose-pink to white, fragrant; sepals 1.5-5.5 × 0.8-3.5 cm, ovate to elliptic, concave, green (in white flowers) or pinkish-green (in rose-pink flowers). Petals 16-24 (in single forms) or 116-160 (-198) in double forms; stamens up to 230, 2.2-5.0 cm long; outermost staminodial in double forms; receptacle 2.0-4.5 cm across, spongy, yellow during anthesis; stigma papillose (Sharma and Goel, 1997).

Seeds: Two types of dried lotus seeds can be found commercially; brown peel and white. The former is harvested when the seed head of the lotus is ripe or nearly ripe and the latter is harvested when the seed head is still fully green, but with almost fully developed seeds. White lotus seeds are de-shelled and de-membraned. The bitter tasting germ of the seed is also removed at the time of harvest using a hollow needle, though some may still remain in the seed due to production oversight. Brown peel lotus seeds are brown because the ripened seed has adhered to its membrane. These seeds are usually cracked in half in order to remove the germ since the seeds are hard enough to make the germs' removal by needle difficult. Seeds that are sold in packages or in bulk at many Asian markets must first be soaked in water overnight prior to use. They can then be added directly to soups and congee, or used in other dishes.

N. lutea (American Lotus, Water Chinquapin, Yanquapin)

Aquatic perennial; leaves peltate, usually raised out of the water, cup shaped or depressed at the centre, over the point of attachment of the petiole, 0.30-0.90 m diameter, bluish green above, light yellowish on lower surface at young stages; petioles 1.5-2.0 m long, papillate; flowers 12-25 cm across on long pedicles, 1.5-2.0 m, pale sulphur yellow in colour, like a double tulip; petals ovate-obtuse; anthers hook-appendaged; seeds acron-like, 1.5 × 1.30 cm (Bailey and Bailey, 1976; Everett, 1980). *Nelumbo lutea* was introduced into India from Missouri Botanical Garden, USA (Goel *et al.*, 2001).

4.3.0 Cultivars

A large number of lotus (*Nelumbo nuicfera*) cultivars are available in China, Japan and Thailand which have flowers in single or double forms in white to light pink and red colours. The important ones are 'Alba Grandiflora' (white), 'Alba Striata', 'Charls Thomas' (rose), 'Chawan Basu', 'Debbie Gibson' (cream), 'Empress' (white), 'Momo Botan' (rose), 'Nikki Gibson' (Pink with white centre). Other Chinese and Japanese races/cultivars are'Buddha's Seat', 'Children', 'Han lotus', 'Japanese', Semi-double', 'Nehru Lotus', 'Pink Bowl', 'Pink Double', 'Pink Jade', 'Pink Rose', 'Red Paeony', 'Single Pink, 'Sino-Japenese Friendship', 'Shiroman' (double, flowers creamy with greenish tine, becoming pure white, fragrant) (Graf, 1973), 'Welcoming Guest', 'White Cheny', 'White Paeony', 'White Sea', 'White Swan' and 'Yushan Pink'.

A selection of *N. nucifera* named 'Krishna' with large double flowers of rose colour, blooming from April to October (Sharma *et al.*, 1998).

5.0 Breeding

The genus *Nelumbo* is represented by two species *viz., N. nucifera* and *N. lutea.* In India, lotus (*N. nucifera*) widely occurs in the wetlands from temperate regions to the plains. There is a lot of phenotypic variation with regard to the flower colour, shape and size under different geographic locations. Though there are shades of only two main floral colours *i.e.,* white and pink to red intermediate shades are also seen besides variation in shape and size of the petals.

A detailed study on flowering, pollination, fertilization and seed formation of *N. nucifera* was done by Kasumi and Sakuma (1998) in Japan. They observed that the flowers fully opened at 6-7 h in the morning and closed by 12 h. Dehiscence of anthers occurred on second and third day while stigma was receptive on the first day. In China also considerable breeding and improvement work have been done. As a result, a good number of cultivars have been evolved such as 'Buddha's Seat', 'Han Lotus', 'Sino-Japanese Friendship', 'Yushan Pink' *etc.* (Nix, 1987; 1989). Works on breeding and improvement have been reported from Bonn Botanical Garden (Willams and Schneider 1993; Borsch and Wilde, 2000). The largest area under cultivation of lotus is located in the regions surrounding mid-down Yangtse River, including Hubei, Jiangsu, Zhejiang, Anhui, Jiangxi and Hunan provinces of China (Guo, 2009). According to different purposes or morphological differences, the Chinese lotus (*N. nucifera* ssp. *nucifera*) is usually classified into three types: rhizome lotus, seed lotus and flower lotus. Rhizome lotus is mainly cultivated in Hubei, Jiangsu, Anhui and Zhejiang provinces; Seed lotus in Jiangxi, Fujian and Hunan, and flower lotus in Wuhan, Hubei province, and Beijing. Up to the year 2002, a total of 572 lotus accessions (including landraces, cultivars and breeding lines) were conserved in National Garden of Aquatic Vegetable, Wuhan, Hubei province, including those collections from 153 counties in 18 provinces, and lines bred by breeders. Out of these accessions, 310 were rhizome lotus which contains 201 landraces and 109 breeding lines; 229 were flower lotus including 172 cultivars and 57 breeding lines; and the rest 33 were seed lotus with 18 cultivars (Guo, 2009).

Inspite of the importance Lotus is being given to Hindu tradition, research on lotus in India is scanty. Mitchel (2010) wrote the status of lotus research in India as follows. There are numerous areas of wild lotus growing throughout India. However, many of these of these are in a perilous predicament due to farming, pollution, industrialization, drought, and sometimes even floods. Often these varieties exist in very small ponds in the countryside and will soon disappear if not preserved. Lotuses are also grown around many temples, but this doesn't always save them from the ravages of nature and humanity. A Lotus Research Station in Lucknow in northern India boasted of having 37 indigenous varieties, but in fact they only have about eight left alive at the most and do not actually carry out any lotus research. They estimate that there are probably about one hundred native varieties. One variety near Samastipur grows in a lake whose water reputedly measures pH 9.5. India does have lotus and water lily growing enthusiasts, but there doesn't seem to be many; no organization exists for hobby growers. To most Indians there are only two different lotus colors, pink and white. However, a vast range exists within the pink varieties. This range extends from white with various amounts and shades of pink on the tips and/or petal margins (usually in singles) through to various light pinks, mid pinks and on into dark pinks/reds.

5.1.0 Breeding Methods

5.1.1 Selection

Thousand Petalled Lotus Variety

In China, where it went from India, Dr. Daike Tian discovered the long-lost Nelumbo 'Zhiun Qianban' ('Ultimate Thousand-petal' in English) in South China in 2009. It is the third cultivar found of the ancient thousand-petalled lotus group after 'Qianban Lian' and 'Yiliang Qianban'. According to Daike there is a group of 1000-petal type lotus based on classification of lotus. According the form of lotus flower, lotus can be divided into single flower/few-petalled, semi double, double, duplicate-petalled, and thousand-petalled type which is an extreme form of double lotus.

Wang and Zhang (2010) described the status of Lotus breeding in China. According to them, that there were 33 cultivars of pot- or jar-lotus in the southern regions of China according to the book of Hong He Pu (Pot Lotus) (1808) written by Zhongbao Yang in the Qing dynasty. Some cultivars can still be seen, like 'Hong Qian Ye', 'Bai Qian Ye', 'Chong Tai Lian, 'Qian Ban Lian', 'Bi Lian', 'Fo Zuo Lian', 'Da Sa Jin', 'Jiao Rong San Bian', 'Xiao Bi Lian', 'Wa Wa Lian', 'Zhuo Shang Lian', 'Jin Bian Lian' and 'Xiamen Wan Lian', and so on. These cultivars came from natural or artificial selection, not from the human's intentional breeding (Cited by Wang and Zhang, 2010).

5.1.2 Hybridization

According to Wang and Zhang (2010), the beginning of lotus breeding started in 1960 at Wuhan Botanical Garden. The first hybrid 'Zhong-Ri You Yi Lian' (Sino-Japanese Friendship Lotus) was bred by crossing the Chinese 'Gu Dai Lian' (ancient lotus) with the Japanese ancient lotus 'Oga Lotus'. Three years later, 41 lotus cultivars from across China were re-introduced to Wuhan. Together with the cultivars introduced from Japan and the American yellow lotus (*Nelumbo lutea* Pers.), the total number of cultivars became 57. These plant materials became the crucial genetic basis for breeding new lotus cultivars. From 1980 to 1985, there were 88 notable new lotus cultivars, including a rare natural triploid 'Yan Yang Tian' (Bright Sky).Other cultivars with high ornamental values produced were 'Man Jiang Hong', 'Ying Bin Fu Rong', and the Sino-American hybrids like 'Xiao Wu Hui', 'Fei Long', 'Mei Zhong Hong' and 'Xiang Nü Sang Hua'. During this period, the artificial direct selection and hybridisation were the major methods for lotus breeding. It is rich in germplasm and more than 2000 cultivars have been cultivated through hybridization and natural selection (Zhang *et al.*, 2019).

Wang and Zhang (2010) described some interesting varieties developed in China as follows. Besides an attractive yellow flowered type 'Ming Mei Lian'introduced from Japan, many hybrids have been developed such as the red double 'Chang Ban Da Hong' and pink double 'Fen Tuan' were bred by Beijing Normal University and the other institutions from 'Baiyangdian Hong Lian' through ion implantation technique. 'Guo Qing Hong', a large-sized cultivar with red double flowers and longer inflorescence, was bred by Chongqin Yameijia Aquatic Flower Co., Ltd. Corp. through crossbreeding of 'Dong Hua Hong', a tropical type lotus and 'Tai Kong Hong Qi', a cultivar of "space-lotus" lines. 'Muling Qiu Se', a large-medium sized cultivar with yellow flower was bred by NLG from the hybrids of the American yellow lotus and Chinese lotus. 'Mei Yan Jia Ren', a tropical type double flower cultivar and 'Jü Ren' with deep-water tolerance, were bred by SLW. 'Lü Zhu', a cultivar with ball-shape green bud and flower selected from the seedlings, and red double 'Donghu Qing Si' and double pink 'Nan Gui Yan' from 'Die Lian Hua' treated by radiation, were bred by CLRC. The pink double 'Quancheng Yu Nü' was bred by Jinan Daminghu Park. The red double 'Hong Lou Meng' and white double 'Xue Tao' were bred by Jiangsu Yangcheng Ailian Yuan. 'Laoshan Hong Ri' and 'Laoshan Hong Wan' were produced by Qing Dao Water lily World from continuous self-pollination of 'Mang Jiang Hong'.

Breeding of lotus work has been done in USA by Perry D. Slocum. Some of his hybrids are unique and stunning in their form and colour. 'Perry's Giant Sunburst' is a cross between *N. nucifera* 'Alba Plena' and *N. lutea* developed in 1987. This cultivar is a summer blooming one with soft yellow flowers, 25-30 cm in diameter and suitable for growing in large containers. Another remarkable cultivar is 'Carolina Queen' developed by him in 1984 which is a hybrid of *N. lutea* × *N. nucifera* 'Rosea Planea' with magnificent colour change effect from rose with yellow centre to pink and finally creamy yellow at the third day. 'Angel Wing' is another successful cross made in 1984 involving *N. nucifera* 'Pekimense Rubra' × N. 'Shirokunshi' – the tulip lotus. Flowers of this hybrid are delicately white and slightly fragrant, highly suitable for home garden and recurrent bloomer (Green, 2002; Slocum and Robinson, 1996). Induced mutation by irradiating seeds and rhizomes by gamma rays is another method for the improvement and developing new varieties of lotus.

An interesting story of the development of a yellow cultivar cv. Friendship Peony was narrated by Wang and Zhang (2010) in a special issue of The water Garden Journal According to them that in ancient books of the Ming and Qing dynasty it has been recorded about yellow cultivars of Chinese lotus, but none of these cultivars can be found now. The breeders have been anxious for breeding such cultivars for a long time (Guo, 1987). According to Guo (1987), he was invited to the United States by the Great Pacific Enterprise Inc. to carry on research work at the lotus garden in Modesto, California. In July of this year, 'Mrs.Perry D. Slocum', a famous lotus cultivar in America, being a distant hybrid of *N. nucifera* × *N. lutea* and with double and pink-colored flowers, was used as a parent and pollinated by selfing artificially. Some seeds were obtained and sown in August. Six young plants have been kept in and three of them bloomed in 1986. The first one is yellow in flower color and possesses huge double flowers; the second one is semi-double and with pink flowers; and the third is a single-flowered white one. The first one was selected and named as *Nelumbo* × *nucifera* cv. Friendship Peony by Professor Chen Junyu. It means that its flowers are alike to those of the tree peony mudan, symbolizing the friendship between the Chinese and the American. The main characteristics of this new cultivar are as follows. Flower's color is creamy yellow; flower diameter 17.5 ±1.2cm number of petals 237.3 ±25.6;the highness and width of leaves 52.6 ± 7.6cm and 30.2 ± 4.6cm respectively; carpel number 21.0±1.0 (Guo, 1987 cited by Wang and Zhang, 2010).

5.1.3 Mutation Breeding and Polyploidy

Using radiation breeding and polyploidy, improvement has been tried and already produced some preliminary results. Breeding efforts led to the development of 276 new cultivars, including the double-flower lotus from mutants of seed lotus after space radiation treatment, such as 'Chan Juan' with white petals, 'Jiao Yang Zui Wu' and 'Xing Kong Mu Dan' with pink petals, 'Feng Juan Hong Qi' with red petals and so on. Sanshui Lotus World (SLW) successfully bred the tropical-type lotus, such as 'Dong Hua Hong', 'Ao Shuang' and 'Xue Li Hong Hua', with the flowering duration lasting for eight months in the Pearl River Delta. By the radiation breeding technique, Nanjing Lotus Garden (NLG, http://www.yileen.com.cn/) has produced the lotus cultivars suitable for cut-flower production, such as 'Nü Mu Dan', 'Yü Ji Shan', 'Jin Tai Yang', 'Jü Wu Ba', 'Huan Qing' and 'Chun Hong', *etc.* The total number of Chinese lotus cultivars amounts to 608 including those in the books of China Lotus Flower Cultivars and China Lotus Flower Cultivars (Wang and Zhang, 2010).

5.2.0 Breeding Objectives for Future

Unfortunately, very few workers have been working on Lotus. Wang and Zhang (2010) have identified the following breeding objectives.

1. Large-sized, deep-water-tolerant cultivars: The short-term goal of lotus breeders should be focused on the deep-water tolerant cultivars.
2. Micro-shade-tolerant cultivars of bowl lotus: Bowl lotus refers to the small-sized ornamental lotus cultivars that can be planted in a bowl. It has a huge potential in the market of house plants.
3. Long-blooming cultivars: Flowering duration could be classified at the single-flower level and group (or population) level. The duration of a single lotus flower is generally three to four days. The group flowering period can last for three months in pond and less than two months in jars, but only one month or so for bowl lotus in small containers.
4. Cultivars with novel colors: There are only three basic colors in Chinese lotus, namely red, pink and white. According to the Chinese historical records, the single and double lotus cultivars with yellow flowers, and 'Sa Jin Lian' (a cultivar with sparse golden spots or strips on petals) occurred in the Ming Dynasty (1368-1644). 'Mi Bo' (honey pot) was included in Hong He Pu of the Qing Dynasty. Unfortunately, these cultivars have lost. Breeding for blue colour would be a great challenge.
5. Breeding more cultivars of thousand-petalled type: Thousand-petalled lotus is an extreme form of the double-flower cultivars in Chinese lotus. 'Qian Ban Lian', a pink cultivar and the only thousand-petalled lotus, has been showing its beauty over 1,000 years. Recently, two other cultivars

have been discovered, namely 'Yilian Qian Ban' and 'Zhi Zun Qian Ban', both with pink flowers.

6. To breed the thousand-petalled cultivar with novel colors, it is only achieved through one of the following treatments, such as the radiation of ^{60}Co at a strong dose, space radiation, and ion implantation.
7. Pest-resistant cultivars such as the chilli thrips (*Scirtotuthrips dorsalis* Hood).
8. The cultivars suitable for the European climate: The European Atlantic coast countries with mid-latitude (40°-60°), such as Britain, France, Germany, Italy, Spain, Portugal, Poland, Denmark, *etc*., have a temperate maritime climate, with a significant feature of warm winter and cool summer. It is year-round mild and rainy without a very cold winter. The average temperatures are below 22°C (72°F) in the hottest month (July) and only 17°C-28°C (63°F-82°F) for the hottest time in the late July.
9. The cultivars with dual uses: Lotus was originally an economic plant combining ornamental, edible and medicinal values. Later, the cultivar groups were bred and classified based on different special values and the purpose of utilization. For the majority of flower lotus, the underground rhizomes have no edible values, the stamens turn into petals and the pistils develop into bubble-like structures or petals. The plants bear few fruits, even no fruits. The seed lotus blooms first and bears fruits later. Therefore, it has both ornamental and edible uses. However, the seed lotus has not been popularly used as an ornamental due to its single or less petalled flowers for most cultivars. The majority of edible lotus (also called rhizome lotus) don't bloom or have few flowers. Therefore, they virtually have no ornamental value. Currently, wetland restoration and water gardens are the largest market for lotus. The most popular cultivars are the lotus which can produce both productive flowers and fruits. Thus, the breeders should have a deep insight into the market's trends and breed the dual-purpose cultivars.

6.0 Growth and Flowering

Many rooted water plants possess an interconnected lacunar system that facilitates gas exchange between their green aerial parts and the rhizomes and roots, which are buried in the mud and are potentially subject to a shortage of oxygen for respiration. In no other water plant this gas flux is apparent as in Nelumbo (Vogel, 2004). The floating as well as emergent peltate leaf blades of Nelumbo have a slight central depression with a centrally located pale-coloured, plate-like "umbilicus" or navel. This plate lies above the insertion of the petiole (Vogel, 2004). In an interesting narrative Vogel (2004) describes the work of Ohno (1910) who, while passing a temple pond crowded with sacred lotus near Hiroshima, noticed a surprising phenomenon: a constant stream of air bubbles arose from the sheet of water that had collected in the depression of the leaves after a rain shower. He identified the umbilical plate as the source of the air bubbles. He then conducted a series of experiments to explain the source of the gas pressure. Unknown to Ohno, natural gas emission by the Nelumbo leaves had been described as early as 1841 by RaffeneauDelile from plants cultivated in Montpellier. The gas emission took place in the sun, but sometimes also at night. When emission was strong, large bubbles rushed through a water cover placed upon the leaf centre and produced a real sprinkling; when emission was low, air gently crept below the water sheet to the outside via the silvery air film that separates the wax-covered, unwettable leaf epidermis (Vogel, 2004).

The thermo-osmotically driven air stream that temporarily ventilates the plant body of *Nelumbo nucifera*, supplying rhizome and roots with oxygen, takes its way through inter-connected canals and caverns. The course of this circulation, starting with the uptake of gas through leaf stomata and ending with its expulsion via the specialized navel (central plate) of the leaves, as well as organs regulating the circuit, were examined using anatomical, pneumatic, and silicone casting methods (Vogel, 2004). The aerenchyma of the peltate leaf, closely paralleling the nervature, is divided into four separate domains: two mirroring lateral halves including halves of the central plate, and each half again separated into an adaxial (basal) and abaxial (distal) sector. Gas absorbed by the adaxial sectors flows through a definite pair of pipes down the petiole into three lateral pairs of the eight main tubes of the horizontal shoot. There it passes several nodes with adjoining leaves, where homologous ducts, arranged in series along the shoot, contribute to the gas flux in its basipetal course. A reverse gas flow confined to another pair of cauline pipes and also arranged in series, enters in the nodes another pair of petiolar pipes that directly lead to the foliar central plates, where it is released through stomatal pores. These pores are three times as large as the laminal stomata and control the gas release by opening and closing. Each abaxial leaf sector is, apart from its ducts leading down the petiole, via a shortcut connected with the upstream pipes supplying the central plate, so that its air circuit, when active, is only intra laminar and does not join the entirety of the system. The question where, and how, the two own streaming currents merge into the two up streaming ones remains unresolved.

Sacred lotus is known to vent pressurized air in the leaf blade through stomata on the central plate of the leaf via the rhizome. Riya *et al.* (2020) measured diurnal CH_4 emissions from lotus-cultivated in tanks during July and August 2016. CH_4 emissions from the central plate of leaves of different age were also evaluated. Average daily CH_4 fluxes of lotus-cultivated tanks were 23.1±5.6 and 11.4±6.0 mg $cm^{-2}h^{-1}$ during July and August, respectively. Both the diurnal pattern of flow rate and the CH_4 concentration of gas emitted from the central plate were similar to those of lotus and other wetland plants. However, the diurnal pattern of CH_4 flux of younger leaves was dissimilar to that of tanks with cultivated lotus. Younger leaves exhibited a higher flow rate and greater concentration of CH_4 than older leaves. This implies that pressurized air in older leaves is emitted through younger leaves, which contrasts with other plants that emit gas from older leaves or columns via pressurization. They concluded that CH_4 fluxes from lotus-cultivated in tanks were comparable to values from other plants. Patterns of CH_4 fluxes from the tanks and from individual leaves were not identical. Younger leaves exhibited higher gas flow rate and CH_4 concentration than older leaves (Riya *et al.*, 2020). Contrary to earlier reports, well organized but fewer stomata develop on the lower surface of the leaves of *Nelumbo nucifera* Willd. during aerial growth. The stomata, however, become obliterated by the readjustment of neighboring epidermal cells. During initial stages of degeneration the guard cells show irregularly thickened walls, disintegrated nuclei, and highly vacuolated cytoplasm. Such abnormal features finally lead to the disappearance of stomata from the lower surface of leaves. The ontogeny, structure and distribution of stomata on leaves, perianth lobes, stamens, receptacles and carpels are described. The stomata are haplocheilic in development and are anomocytic (ranunculaceous) at maturity (Gupta *et al.*, 1968).

Biologically, lotus has not only the common aquatic plant features, but also certain unique features that distinguish it from other plant species (Lin *et al.*, 2019). These features include seed longevity, leaf ultrahydrophobicity and floral thermoregulation. Understanding of the mechanisms that lead to the formation of these unique properties is important, for not only the basic plant biology, but also their great usage potential through bionics. Lotus fruit is famous for its longevity. It was reported that lotus fruits buried underground over 1300 years in the Northeast of China could still be germinated (Shen Miller *et al.*, 1995). Understanding the underlying mechanism of lotus seed longevity may contribute to enhancing seed storage in agriculture, and even in the healthcare of human beings (Lin *et al.*, 2019).

Growth of *Nelumbo nucifera* in Lake Kasumigaura, Japan was investigated with special emphasis on leaf dynamics (Tsuchiya and Nohara, 1989). Floating leaves were found throughout the growing period, while emergent leaves were not found until July. However, emergent leaves contributed approximately 75 per cent of the total leaf area index at its seasonal maximum, 2.76 m^2/m. Mean leaf life span of emergent leaves with thick laminae and stiff petioles was much longer than that of floating leaves (44.5 *vs.* 17.1 days).

The floral development of *Nelumbo nucifera* was compared with that noted in previous studies of *Nelumbo*, Nymphaeaceae, and other basal angiosperms (Hayes *et al.*, 2000). Important features include developmental evidence of only two sepals, development of an androecial ring meristem, and an apocarpous gynoecium composed of ascidiate carpels that become embedded in an expanded receptacle. Secretory papillate trichomes cover the stigma and line the stylar canal. The unique apocarpous gynoecium, which lacks a conical residual floral apex, and a greatly expanded receptacle distinguish Nelumbonaceae from the Nymphaeaceae, as does the distinctive androecial ring. Nelumbo is characterized by polysymmetric floral development, with some organs originating spirally (petals) and some in simultaneous whorls (stamens and carpels). This pattern of floral development, as well as the pattern of carpel closure by secretion, are common in several paleo herbs and eudicots and indicate phylogenetic affinity between Nelumbonaceae and basal angiosperms. Their study does not support the conclusion that sepals vary in number and intergraded with the petals. Late in the development of the flower, the sepals and some of the outer petals no longer completely enclose the enlarging bud. These outermost petals may become photosynthetic and are intermediate in size, giving the appearance of intergradations of sepals and petals. The larger size of the sepal primordia and the auricle located at their outermost interface to the peduncle distinguish them from the petals during the early development of the flower. Because of its unique floral development and anatomy, Nelumbo appears to be an isolated member of the eudicot dade (Hayes *et al.*, 2000).

An interesting phenomena in the development of flower in lotus (*N. nucifera*), was observed by Seymour and Motel (1996). They found that significant amount of heat is generated during the sequence of anthesis. The ideal temperature for best and profuse flowering has been observed between 30 and 35°C. This thermo-regulating mechanism helps the development of flower besides providing a warm hospitable environment to the insect pollinators. Schneider and Buchanan (1980) were also of the same opinion that floral structures and pattern of anthesis were most suitable for beetle pollination.

Seymour and Motel (1998) also reported that the heat production in lotus flower depends on ambient temperature and there was no effect of light at any time of the day and concluded that the flowers are thermogenic and physiologically thermoregulatory. They suggested that temperature regulation apparently occurred at a cellular level as a result of some biochemical reactions. The ideal temperature forbest and profuse flowering has been observed between 30 and 35°C. This thermo-regulating mechanism helps the development of flower besides providing a warm hospitable environment to the insect pollinators. Schneider and Buchanan (1980) were also of the same opinion that floral structures and pattern of anthesis were most suitable for beetle pollination. Seymour and Motel (1998) also reported that the heat production in lotus flower depends on ambient temperature and there was no effect of light at any time of the day and concluded that the flowers are thermogenic and physiologically thermoregulatory. They suggested that temperature regulation apparently occurred at a cellular level as a result of some biochemical reactions.

Thermogenesis in *Nelumbo nucifera* (Gaertn.) has been known to scientists for many years; however, the extent of heating by different floral parts remains unclear. Watling *et al.* (2010) presented evidence that the receptacle, stamens and petals produce heat independently, and that the source of heating in these tissues is most likely the alternative oxidase (AOX). The temperatures of the receptacle, petals and stamens were significantly higher than non-thermogenic leaf tissue. After removal from the pedicel, the receptacle retained the most heat (8.1±1.9°C above non-thermogenic tissue temperature) and the petals the least (2.8±4.2°C), with the stamens intermediate. High AOX protein levels and flux through the AOX pathway (in all tissues) during the thermogenic period are consistent with AOX being the mechanism used for thermogenesis. Lipids and carbohydrates were investigated as possible substrates for thermogenesis. There was little change in total lipids during floral development; however, soluble carbohydrate levels decreased.

The lotus is a long-day plant. Its flowers open in the early morning and the life span of one single flower usually does not exceed four days. The lotus flower is solitary and hermaphrodite, and iscomposed of the perianth, androecium, gynoeceum,receptacle, and peduncle. The calyx isthe outermost whorl of the perianth, usually consisting of four sepals that are green, thick, tenacious,and similar to tetals in structure, except that they fall earlier. One flower contains twenty or more tepals that vary widely in size, shape, number and color from population to population and from cultivar to cultivar, which results in a diverse floral morphology. Usually, a lotus flower is self- incompatible because the multiple stigmas of a receptacle mature ahead of the stamens in the same flower (Shen Miller, 2002). Microsporogenesis and male gametogenesis in the anther are important for hybridization in flowering plants. However, little is known about the cytological events, especially related to the stamen, during the reproduction of the lotus. To better understand the mechanism controlling the male reproductive development of the lotus, Zhang *et al.* (2019) investigated the flower structure of the Asian lotus (*N. nucifera*). The cytological analysis of anther morphogenesis showed both the common and specialized cytological events as well as the formation of mature pollen grains via meiosis and mitosis during lotus anther development. Intriguingly, an anatomical difference in anther appendage structures was observed between the Asian lotus and the American lotus (*N. lutea*). To facilitate future study on lotus male reproduction, we categorized pollen development into 11 stages according to the characterized cytological events. This discovery expands our knowledge on the pollen and appendage development of the lotus as well as improving the understanding of the species differentiation of *N. nucifera* and *N. lutea* (Zhang *et al.*, 2019).

Zhang *et al.* (2019) described the morphologic characteristics of the anther and its appendage in *N. nucifera*. According to them that in the underwater flower bud, the stamens originate from the stamen primordium, which comes from the lateral edges of the growth cone inside the tepal primordium and appears around the base of the receptacle. Shortly after the formation of the stamen primordium, the pistil primordium forms at the top central domain of the growth cone. Meanwhile, the receptacle gradually develops and elevated incrementally. Before the anther sac initiates, the appendage could be observed at the top of the stamen. Along with stamen development, the anther and appendage gradually elongates and reaches 2.5-4.5 cm at full length. At the late stage of anther development, the filament continues to extend while the anther and appendage no longer elongated. Each stamen contains a filament and an anther with four pollen sacs linked to the filament by connective tissues. The anther wall consists of the epidermis, endothecium, middle layer, and tapetum. The sporogenic cells are enclosed in a fluid-filled locule surrounded by tapetal cells. When the flower fully opens, the fiber of the wall layers becomes thick and dehiscent, before the mature pollen grains are dispersed. The appendage of the Asian lotus stamen is long, oval shaped, and lay at the top of the connective. It is usually milky white or sometimes pink to rosy red in color for Asian lotus cultivars, while the appendage of the American lotus is bright yellow with a sickle-like shape. The appendage only had two layers: the epidermis and hypodermis parenchyma. The subepidermal parenchyma also contains a high density of granules that could be deeply stained with toluidine blue (Zhang *et al.*, 2019).

The macromolecular constituents of the sclerotic propagule wall of *Nelumbo nucifera* and seed coat of *Nymphaea caerulea* were studied using scanning electron- and light microscopy in combination with Curiepoint pyrolysis-gas chromatography-mass spectrometry (Van Bergen *et al.*, 1997). They also analysed *Nelumbo* material using solid state ^{13}C nuclear magnetic resonance and in-source pyrolysis-mass spectrometry. The sclerotic seed coat of the *Nymphaea caerulea* revealed the presence of angiosperm lignin-cellulose similar to that found in most sclerotic plant remains. In sharp contrast, the fruit wall plus seed coat of *Nelumbo* is believed to be composed of a complex of polysaccharides, based on primarily galactose and mannose units, and insoluble tannins, which are suggested to play the same structural role as the lignin-cellulose in the sclerotic seed coat. The distinctive nature of the chemical constituents present in the propagule wall of *Nelumbo,* supports the systematic distinction of this genus in the separate family Nelumbonaceae. The characteristic chemical composition of the propagule walls of *Nelumbo* could be an additional factor in favour of a prolonged longevity of these fruits. However, the distinctive composition of polysaccharides and tannins without the presence of lignin is considered to be the main reason for the absence of these propagules in the fossil record, despite their physical resistance (Van Bergen *et al.,* 1997).

7.0 Propagation

7.1.0 Seed Propagation

Lotus is generally propagated both sexually and asexually. Sexual propagation through seeds is the most common and simple method. However, the seeds are having hard and impermeable seed coat resulting in quite a long dormancy. When the seeds are kept in water, there is no sign of water absorption even after a long period of time (Stefferud, 1959; Slocum, 1985). Studies on the germination of the seeds under laboratory condition were done by Nohara and Tsuchiya (1990). They found that the optimum temperature for seedling growth was 30°C while it was minimum at 20°C or below. Seeds even germinated under high acidic (pH 4.0) and alkaline (pH 9.0) conditions but in very low percentage and poor growth of the seedlings. The optimum pH range was found to be 7.5 - 8.5. Besides, depth of seed sowing significantly affected germination. When the seeds were buried below 2 cm soil depth, the germination was reduced by 60 per cent.

American lotus can easily be propagated from seeds. Sayre (2004) details the propagation protocol for lotus as follows. First, it is crucial to use viable seeds. The easiest way to determine the viability of lotus seeds is to throw them in a cup of water. Retain the seeds that sink and discard the seeds that float; floating seeds have poorly developed or rotten embryos. The hard seed coats must be scarified to allow the seeds to imbibe water. The simplest approach is to scratch off a small area of seed coat using medium grade sandpaper. It is important to change the water daily as you wait for seeds to germinate. Once seeds have germinated and the root begins to emerge (but before the leaves are up), transplant them to a 10 cm pot filled with a loamy-clay. We place pots in 2.5 – 5 cm of warm water in holding ponds. A seedling is not likely to bloom the first year. As the seedlings develop, gradually increase the water depth but not so much that the emergent true leaves are submerged below the water surface. Generally, the fertility in the soil we use to fill pots is sufficient for growth. At the end of the growing season when leaves have died back, we remove tubers from the containers and store them surrounded by moistened sphagnum peat in a cool (not freezing), dark storage room. Seedlings can be held in their containers in storage ponds as long as the water is more than 1 m (3 ft) deep and the tops of the containers do not freeze (Sayre, 2004).

Wahdah and Ahmad (2004) studied the seed quality including physiological maturity. They tested the best germination medium for seeds (sands or rice papers) and seed age at physiological maturity. They found that there was no significant interaction between seeds germination medium with seed maturity. Seed germination percentage in rice papers medium was higher than sands medium. Seed age at physiological maturity was four weeks after blooming flower, and at that moment, dry matter seeds and germination seeds were maximum (Wahdah and Ahmad, 2004).

Im *et al.* (2012) determined the effects of scarification temperature, and sulfuric acid treatments on seed germination of white lotus collected from the Muan districts, Jeonnam in Korea. Without scarification, white lotus seeds did not germinate at all at 15 days after seeding. However, seeds sacrificed at basal parts showed 100 per cent germination rate at 25°C 6 days after seeding. All the seeds floated (0.90 g) and soaked (1.18 g) in the water were completely germinated. Especially, the lotus seeds soaked in the water at 25°C showed high germination rate. Seeds treated with 80 per cent sulfuric acid for 40-160 hours were germinated completely within 6 days after seeding. No difference in seed germination rate of white lotus stored up to 0, 12 and 24 months after harvest was observed.

Development of radical of no-primary-root type seedlings of *N. nucifera* was reported by Mao *et al.* (2002). The radical converted from meristem into storage tissue so that radical sieve element of the primary vascular system of the seedling was sieve tube element. On the basis of this, they opined that special structure of the cotyledon and the radical were related with life conditions and systematic evolution. In an investigation on the development of photosystem, Ji *et al.* (2001) reported that though no chlorophyll synthesis and photosystem biogenesis have been detected in dark-grown lotus (*N. nucifera*) seedlings, but they showed that both PS II and PS I could be formed in dark grown lotus seedlings.

Lotus seeds are known to possess the maximum period of viability. Some seeds collected by a Japanese paleobotanist from the peat deposits in Manchuria (China) were estimated to be 4000 years old but successfully germinated in 1924 (Ogha, 1926; Meyer, 1930). Morozova and Vasil'eva (1993) reported that mature seeds stored at room temperature for 1 or 8 years gave 100 per cent germination in water after scarification while unscarified seeds failed to germinate even after 2 years. Therefore, it was concluded that scarification facilitates germination process. Besides, they also detected the presence of abscissic acid or cytokinin activity after storage of the seeds.

In order to overcome the problem of long dormancy, sulphuric acid treatment for the lotus seeds was suggested. Chromic acid and potassium hydrate (conc.) have also been found useful in overcoming the seed coat permeability. Seeds are also scarified finely at both ends with the help of a sand paper just to cut the outer seed coat to hasten imbibition of water. The germination starts after 3-5 days at normal room temperature and the seedlings become ready for transplantation after 15 days.

7.2.0 Vegetative Propagation

Vegetative propagation by the division of rhizomes is also practised and advantageous to get true to type plants. Rhizomes with new sprouts are cut into small pieces containing 3 nodes and planted horizontally at the depth of 10 cm with eyes above the surface of the soil during February-March.

In Thailand and Asian countries, lotus growers usually propagate by cutting stolon segment since this technique allows cloning to supply uniform planting stock (Hongpakdee *et al.*, 2019). They however consider that only longer stolon cutting with many nodes and active buds would be suitable as propagules. To clarify this condition, the effect of vegetative propagation technique by node number on plant growth and flowering in 'Sattabutsaya' sacred lotus was conducted with 3 stolon cutting types *i.e.*, stolon cutting with 1, 2 and 3 nodes, respectively, using a completely randomized design. All active stolons were planted singly in 1.20×1.00 m circular concrete pots, filled with 1/3 clay soil and 2/3 tap water under natural conditions (May to August 2016). It was found that increasing node number of cutting material increased in total leaf number at 6 weeks after planting (WAP). None of the stolon cutting treatments affected total leaf area, new internode length, flower stalk length and flower length at 3 months after planting. Nevertheless, planting sacred lotus with single node stolon cuttings gave the lowest new stolon length, flower number, percentage of flowering and delay more 2-3 weeks of flowering. Triple nodes of propagated stolon cutting gave the best result in plant dry weight components (flower, pod, new stolon and fibrous root). Planting sacred lotus stolon cuttings with 2 or 3 nodes did not affect total leaf number, new stolon length and girth and visible flowering time. It could be concluded that stolon cuttings with at least two nodes for propagative material is sufficient for creating plant stock for flowering lotus production (Hongpakdee *et al.*, 2019).

Protocol for nursery production using rhizome was described for American lotus by Sayre (2004). The protocol is as follows. Propagating *N. lutea* from rhizomes requires several steps. For good results start with dormant tubers (before new runners form). Cut the tubers into pieces, ensuring each section contains at least 1 bud. It is from these buds, or eyes, that leaves will emerge. Plant the rhizome pieces in wide, round pots that lack drainage holes. Avoid square pots because the growing tip of the rhizome is tender and can break-off if it grows into a corner. Pot size depends on the size of the rhizome pieces. In general, expect to use a pot with a diameter ranging from 20 – 36 cm. Since *N. lutea* grows vigorously, no more than 1 piece should be planted in each pot. Use extra care so that the growing tip is not damaged during planting because at this sensitive stage, a damaged tip can result in rhizome death. The pot should be filled with a rich, loamy-clay soil and then topped with a thin layer of pea gravel. Correctly planted rhizome pieces have about 1.3 cm (0.5 in) of the growing tip visible above the pea gravel.

8.0 Soil and Climate

8.1.0 Soil

Lotus grows well in silt loam to clay loam soil with adequate organic matter. Since it grows in standing water, rapid percolation from the soil is not desirable. It

Propagation of Lotus

Seed Propagation of Lotus

Fruits Containing Seeds

Collection of Seeds

Seed

Germination of Seed in Water

Growth and Flowering

Cultivation of Lotus

Cultivation in Pots

Cultivation in Pond

is also important that rhizomes will not develop well if the soil is heavy clay and poor in organic matter.

It was reported by Sharma and Goel (1997) that lotus is grown well in natural lakes having polluted acidic and alkaline water. Moreover, it can absorb heavy metals from water and recommended for plantation in the lakes and ponds where liquid effluents are discharged by the industries.

8.2.0 Climate

8.2.1 Temperature

Nelumbium nucifera prefers tropical and subtropical climate. In the plains of India it grows and flowers between March and October and remains dormant in winter. *N. nucifera* was also tried as aquatic vegetables for cultivation in ponds and marshes in temperate climates of northern Europe along with other aquatic plants but needed some protection against low temperature (Fritz *et al.*, 1992). In another report from United Kingdom, it was revealed that lotus can be grown in greenhouses. However, plants may be grown in tubs and placed outdoor during summer but they should be shifted to greenhouse during winter (Anon., 1989). *N. lutea* tolerates cold climate. Lotus requires an open sunny to cloudy climate.

8.2.2 Light

As large areas are required for its cultivation (approximately 2 m^2 per plant), physiological research, such as into the effect of environmental factors on dormancy, has not been well studied until recently (Masuda *et al.*, 2006). They conducted studies using seedlings to examine environmental factors affecting dormancy induction. In a first experiment, seeds were sown from 6 April to 6 October at 2-month intervals, and cultivated for 2 months in an unheated greenhouse. In a second experiment, seeds were prepared for germination on 16 November and 16 May and the seedlings were grown at 25 or 30°C under natural daylength in phytotron growth rooms. After 1 month, the seedlings were cultivated at 20, 25 or 30°C for a further month. The number of leaves and rhizome branches on the main stem were counted, and growth of rhizomes on the main stem was calculated using a rhizome enlargement index (maximum internode diameter/internode length) after 2 months of culture in both experiments.

They found that rhizomes elongated without enlargement when the seeds were sown in April and June. Sowing the seeds in August and October resulted in rhizome enlargement from the tenth and fifth internodes, respectively. Rhizomes enlarged in the November-sowing but elongated in the May-sowing irrespective of temperature treatments under natural daylength in the phytotron rooms. The seedlings cultivated from May at 25-30°C for 2 months had more leaves, and more rhizome branches and nodes than those cultivated from November. The studies have proven that short days led to induced dormancy in lotus (Masuda *et al.*, 2006).

Later, Masuda *et al.* (2007) made first report on phytochrome-dependent morphogenesis of storage organs in rhizomous plants. They examined photoperiodic response of lotus rhizome morphogenesis (its transition to a storage organ) by using seed-derived plants. Rhizome enlargement (increase in girth) was brought under 8, 10 and 12 h photoperiods, whereas the rhizomes elongated under 13 and 14 h photoperiods. Rhizomes elongated under 14 h light regimes supplied as 8 h of natural light plus 6 h supplemental hours of white, yellow or red light, but similar treatments with supplemental blue, green or far red light, caused enlargement in girth of the rhizomes. A 2 h interruption of the night with white, yellow or red light, in plants entrained to 8 h photoperiod brought rhizome elongation, whereas 2 h-blue, green or far red light night breaks still resulted in rhizome increase in girth. The inhibitory effect of a red (R) light night break on rhizome increase in girth was reversed by a far-red (FR) light given immediately afterwards. Irradiation with R/FR/R inhibited the rhizome increase in girth. FR light irradiation following R/FR/R irradiation cancelled the effect of the last R light irradiation. It was demonstrated that the critical photoperiod for rhizome transition to storage organ is between 12 and 13 h photoperiod. It was also evident that the optimal light quality range for interruption of dark period (night break) is between yellow and red light and that a R/FR reversible reaction is observed. Masuda *et al.* (2007) proposed that phytochrome plays an important role in photoperiodic response of rhizome increase in girth in lotus. The effects of photoperiod on growth of *N. nucifera* were investigated under six day-length conditions *viz.*, 11 h for 2 months, 11 h for 1 month and then to 13 h for 1 month, 11 h for 1 month and then 15 h for 1 month, 13 h for 1 month and then 11 h for 1 month, 13 h for 2 months and 13 h for 1 month and then 15 h for 1 month. Each pot was planted with 90 stem cuttings, filled to 2/3 of the pot level with natural clay soil and then filled to 3/4 of the pot level with tap water after planting (Hongpakdee and Ruamrungsri, 2017). The stolons were grown under natural conditions (ambient temperature 30/18°C with an 11 h day length in mild winter from January-February. Night interruption after 6:00 pm was used to induce a long day-length condition (13 and 15 h). None of photoperiod conditions affected the total number of leaves. However, extending the photoperiod from 11 to 15 h and from 13 to 15 h and a constant photoperiod of 13 h produced the largest leaves. Shortened photoperiods from 13 to 11 h decreased the flowering percentage and number of flowers. Extended photoperiods from 11 to 13 h, 11 to 15 h and 13 to 15 h only increased the flowering percentage compared with that of constant photoperiods of 11 and 13 h. Nevertheless, the extended photoperiod delayed the date of visible flowering (Hongpakdee and Ruamrungsri, 2017).

8.2.3 Air

Aquatic vascular plants depend on an adequate oxygen supply in order to maintain growth and reproduction in anaerobic environments. *Nelumbo nucifera* is able to survive with a gas transport system which supplies oxygen to the roots and rhizomes submerged in the anaerobic sediment Schutz and Wolfgang (1988). It was possible to demonstrate that this gas transport system is based on a purely physical phenomenon. Thermo-osmotic oxygen transport was first demonstrated on fresh leaves with the help of an oxygen-sensitive electrode. A definite enhancement of oxygen flow was obtained through excised leaves when a temperature difference between the ambient and lacunar air was present in light. These leaves were then dried to brittleness and the enhanced oxygen flow was still detectable. This shows that not only photosynthetic oxygen, but also atmospheric oxygen can be transported to the buried organs. The absolute flow of oxygen through dry leaves was much lower than through fresh leaves, but the thermo-osmotic transport of oxygen still functioned. Furthermore, the process of thermo-osmosis need not rely on a difference in humidity between the two sides of a porous partition, but may be linked causally to the temperature difference and the pore size (Schutz and Wolfgang, 1988).

9.0 Cultivation

9.1.0 Planting

Tian *et al.* (2009b) conducted studies to find a suitable planting time and to determine best management practices that are of importance for container lotus production. Effects of planting time and disbudding on plant growth indices in southeast Alabama were evaluated in a container production system for the ornamental lotus, *N. nucifera* 'Embolene'. Results indicated that plant growth indices were little influenced by different planting dates in March, but were much influenced by planting dates with a difference over a month between February and May. Plants potted and placed outdoors in March and April performed best, and lotus planted in the greenhouse in February and planted outdoors in February and May performed worst. Flower number was not largely influenced by the planting time, but flowering characteristics, especially the flowering peaks, were different among treatments. Planting lotus outdoors between March and May produced the largest return. Influence of planting time on plant growth indices of lotus appeared to be explained by effects of growth-season climate conditions after planting. Disbudding had no impact on plant height but significantly increased underground fresh weight and the number of propagules. Therefore, disbudding should be considered a best management practice to maximize the yield of rhizomes or propagules. Positive linear, quadratic, or cubic relationships were detected among emerging leaf number, underground fresh biomass, and propagule number. Based on the regression models, the yield of lotus rhizomes or propagules can be predicted by the number of emerging leaves. This research provided a guide for nurseries, researchers, and collectors to select the best time to plant lotus outdoors (Tian *et al.*, 2009b).

The seedlings should be transplanted at 2-3 m apart. When seedlings are raised from seeds, the plants may take about 1-2 years to come into flowering while matured rhizomes regenerate strong and vigorous plants which generally bear flowers in the same year. Transplanting of rhizomes or seedlings is generally donein February-March. The requirement of lotus seeds for planting in 1 acre area of pond or lake is about 4-5 kg.

Apart from the natural pond or lake, the other artificial aquatic bodies should be filled with clay soil and farmyard manure (FYM) in 3:1 ratio to a depth of at least 45 cm mixed with *neem* cake, diammonium phosphate (DAP) 25 g, single super-phosphate 50 g and muriate of potash 25 g per sq m area should be mixed with layer of soil (Goel *et al.*, 2001). In China, the combined application of N, K and boron @ 270, 120 and 15 kg/ha, respectively was found to increase yield of rhizome (5.7 t/ha) (Bao *et al.*, 1998). In an experiment on *N. nucifera* cv. Huaon, Qu and Zhao (1991) observed that during initial rapid growth, dry matter (DM) was mainly accumulated in the leaves and rhizomes which lasted about 35 days during July-August and reported nutrient absorption ratio of N:P:K as 2.73:1:5.2 respectively.

9.2.0 Container Production

Lotus can be grown in soilless culture containing different nutrient solutions. Leaf and flower growth was better in soilless culture than in soil (Li and Qian, 1994).

Container ornamental lotus has a long history in Asian countries and has been recently becoming a nursery crop in the United States (Tian *et al.*, 2009a). Soil volume is an important consideration for container production of lotus. Generally, soil volume in a container ¼ to ¾ height was suitable for container lotus production depending on cultivar. Inadequate soil in containers did not anchor plants well, while too much soil inhibited plant growth and increased labour cost. Soil volume influenced plant growth not only through nutrient availability, but also indirectly by influencing water level, electrical conductivity (EC), and pH of water. EC was remarkably affected by soil volume to water relationship when the same rate of fertilizer was applied. Therefore, fertilization of lotus should be dependent on soil or water volume in containers for optimal growth of plants (Tian *et al.*, 2009a).

The effect of soil volume on containerized lotus (*Nelumbo*) production has been underreported. Tian *et al.* (2009a) studied American lotus (*Nelumbo lutea* Willd.) and three cultivars ('Embolene', '98 Seed' and 'Karizma') of Asian lotus (*N. nucifera* Gaertn.) were investigated for growth response to container soil volume in this study. Electrical conductivity, pH, plant growth indices, and plant nutritional content were influenced by container soil volume. Differences in some plant growth indices were significant between treatments with ¼ and higher (½ and ¾) container height soil (CHS) in 21 or 29 litre (#5 or #7) containers. However, plant growth indices were generally not different between treatments with ½ and ¾ CHS. Lotus planted in containers with ¼ CHS usually produced the greatest plant height and underground fresh weight, while the largest number of propagules often occurred in containers with ½ or ¾ CHS. The highest number of emerging leaves was observed in plants with ¼ or ½ CHS treatments, with no significant difference in emerging leaf number between lotus grown in containers with ½ and ¾ CHS. Flower number generally decreased as soil level increased. The ¼ and ½ CHS were more efficient than ¾ CHS for lotus production in containers (Tian *et al.*, 2009a).

9.3.0 Manuring and Fertilization

Seo *et al.* (2010) studied the uptake and release of nutrients from ponds used for lotus cultivation was measured in ponds under short-term (1 yr) cultivation with compost application (pond I) and under long-term (20 yr) cultivation without compost application (pond II). Total inflow loads of TN (irrigation water, rainfall and compost) during lotus cultivation period in ponds I and II were 72.3 and 34.3 kg/ha 182/day, respectively. TN removal rates in ponds I and II were 77.3 and 49.8 per cent of total inflow load, respectively. Major removal mechanisms of TN were attributed to microbial processes and uptake by lotus. The total outflow loads (infiltration and runoff) of TN during the lotus cultivation period were 13.9 kg/ha 182/day (19.2 per cent of total inflow TN load) for pond I, and 11.3 kg/ha 182/

day (32.9 per cent of total inflow TN load) for pond II. For TP the total inflow loads (irrigation water, rainfall and compost) during lotus cultivation in ponds I and II were 80.8 and 1.9 kg/ha 182/day, respectively. TP removal rates in ponds I and II were 84.9 and –274.1 per cent of total input, respectively. Phosphorus removal was attributed to lotus uptake and soil adsorption. The total outflow loads (infiltration and runoff) of TP during lotus cultivation period were 10.1 kg/ha 182/day (12.5 per cent of total inflow TP load) for pond I, and 6.6 kg/ha 182/day (355.6 per cent of total inflow TP load) for pond II. TN and TP in run off from pond I (with compost) was higher than that in pond II (without compost), showing that TN and TP in runoff were strongly influenced by compost addition. Therefore, in order to satisfy established water quality standards, the amount of compost used in lotus cultivation should be evaluated (Seo *et al.*, 2010).

Tian *et al.* (2009b) followed the following protocol for cultivation and disbudding in containers. All containers were filled to two-thirds container level (21 cm, 0.018 m^3) with natural sandy loam soil and immediately filled to three-fourths container level with municipal tap water (pH 7.0, EC 0.13 mS/cm) after planting. Fertilization began when several floating leaves developed. After containers were filled to full container level with water, each plant was fed 8 g of soluble fertilizer 20 N,4.4 P, 16.6 K (ProSol 20-10-20; ProSol Inc., Ozark, AL) at 20 days intervals. Fertilizerwas applied four times and each treatment received the same amount of fertilizer. During the flowering period, flower buds were discarded as found above water once every 3 days for one half of plants from each planting date, whereas the other half was not pruned allowing flower buds and fruits to mature.

Growth of lotus, yield and quality of rhizome, soil nutrient dynamics and irrigation were evaluated on the shallow soil cultivation system in Japan (Toma *et al.*, 2015). Chemical (C) and organic (O) fertilizer application plots were set up at the pool installed 10m depth of soil in 2011 and 2012. Nitrogen (N) and phosphorus application rate were same between two plots, while potassium (K) application in O plot was only 30 per cent of that in C plot due to low concentration of K in organic fertilizer. Yields of rhizome (3.15 to 3.93 kg/m^2) were larger than that in conventional cultivation, and there was no difference in the yield between the plots. In C plot, rhizome of 100-200 g was tended to larger than in O plot might be due to much K application in C plot. That is supported by the result more than 50 per cent of K in lotus was concentrated in the rhizome. Color of the rhizome was almost same between two plots. However, color of the rhizome in 2013 was significantly white compared with that in 2012, because soil iron oxide, which was causal material for brown color of rhizome, might be reduced during the winter. Cation exchange capacity, soil total carbon and N concentration, and exchangeable K concentration reduced 14-55 per cent after two cultivations. This showed that soil nutrient reduced by plant uptake should be supplied by organic matter or other materials. Irrigation amount of water during the cultivation was approximately 1,000 mm. However, water should be supplied when evapo-transpiration was greater than precipitation, especially in August and September. This study showed that high yield and quality of lotus rhizome was promising at the soil shallow cultivation system, though management for maintaining soil fertility would be required (Toma *et al.*, 2015).

9.4.0 Intercropping

Lotus (*N. nucifera*) may also be intercropped with wheat replacing rice usually followed after wheat cultivation in China (Gang, 1999). He suggested selection of early varieties of wheat and planting of lotus rhizomes at 10-15 cm deep 20 days earlier of wheat harvest followed by controlled flooding. A yield of rhizomes of 30-52t/ha may be obtained under such conditions which had higher economic value.

9.5.0 Water Management

Once the plants get established and start growing, the water level in the soil is to be increased gradually. If there is any casualty, replanting should be done immediately. Regular cleaning of the water bodies by removing dead and decomposed leaves is essential besides filling of fresh water as and when water level drops. In an experiment conducted in China in artificial environment of concrete ponds, it was observed that water depth and flooding had great influence on growth and yield of *N. nucifera*. Out of the seven water depths (0.2-3.0 m) studied, number of floating leaves, total number of leaves, leaf area index of emergent leaves were greatest in tanks having 0.5 m depth. A maximum depth where lotus can grow was found to be 2.4 m (Nohara and Kimura, 1997).

9.6.0 Weeding

Aquatic weeds are sometimes a major problem of lotus cultivation particularly during summer months as the growth of the weeds is faster. Commonly found weeds in lotus pond are water-hyacinth (*Eichhornia crassipes*) which spreads fast and covers entire water surface. Species of *Typha, Lemna* and *Wolfia* are the other common weeds found in the lotus ponds. Fresh water algae mainly *Spirogyra* also causes much damage to the crop. Mechanical removal of the weeds is a safe and easy method. However, in severe weed problem in large growing area herbicides like simazin (3-6 ppm) and monuron (4-12 ppm) may be applied to control the weeds.

9.7.0 Pollination and Pollinators

The flower can physiologically regulate the floral temperature during anthesis by self-heating via a large obconical receptacle, maintaining a relatively favourable temperature between 30 and 36.8°C when the surrounding temperature fluctuates between 10 and 45.8°C (Seymour *et al.*, 1998). This homoeothermic flower has been considered as a warm night shelter for beetle activity (Seymour *et al.*, 2003) and a type of beetle-pollinated chamber blossom (Vogel and Hadacek, 2004). The lotus flower closes during the night by petal movement, but it is accessible to diverse insects because it opens by day and provides a large pollen reward. Although the effective pollinators in the Asian lotus remain unknown (Vogel and Hadacek, 2004), a study on the American lotus *Nelumbo lutea* (syn. *N. pentapetala*) indicated that bees as well as beetles are important pollinators in some populations (Schneider and Buchanan, 1980).

Pollinators of *N. nucifera* were investigated in 11 wild populations and one cultivated population, and pollination experiments were conducted to examine the pollinating role of two major pollinators (bees and beetles) in three populations

(Li and Huang, 2009). Lotus flowers are protogynous, bowl shaped and without nectar. They found that the fragrant flowers can be self-heating during anthesis and produce around 1 million pollen grains per flower. It was found that bees and flies were the most frequent flower visitors in wild populations, contributing on average 87.9 and 49.4 per cent of seed set in Mishan and Lantian, respectively. Beetles were only found in one wild population and in the cultivated population, but the pollinator exclusion experiments showed that beetles were effective pollinators of Asian sacred lotus. They concluded that in their pollinating role, beetles, probable pollinators for this thermo-regulating plant, had been replaced by some generalist insects in the wild. This finding implies that contemporary pollinators may not reflect the pollination syndrome (Li and Huang, 2009).

10.0 Diseases and Pests

The most common disease is fungal leaf spot caused by *Cercospora* sp. and *Ovularia* sp. When grown for ornamental purpose, it is necessary to control the diseases by spraying fungicide like bavistin or bengard @ 2 g per litre of water. Root rot caused by *Phytopthora* sp. is also sometime problematic.

Aphids and beetles are common insects found to attack lotus which can be controlled by spraying 0.01 per cent rogor or monocrotophos. However, these insecticides should not be sprayed in natural lakes where fishes are grown simultaneously.

11.0 Harvesting and Yield

Lotus is generally harvested when flower buds is almost fully developed. The flowers should be cut with a long petiole as possible.

The harvesting of rhizomes is done in October-November when they get well matured. A well maintained and mature crop may yield on an average 30-35 quintals of rhizomes per hectare (Sharma and Goel, 2000). However, the yield may go up to 36-40 quintal per hectare (Anon., 1966).

Drying is the main process used to treat lotus seeds for storage. Tian *et al.* (2012) studied the response surface methodology was used to optimize processing methods for microwave vacuum (MWV) drying of lotus seeds and to create a desirable product. A central composite design with three important factors: microwave output power (2.0–4.0 kW), vacuum degree (–0.070 to –0.090 MPa) and on/off ratio (68/52s to 99/21s) was used to study the response variables of drying time, shrinkage ratio, rehydration ratio and whiteness index. The optimum conditions for MWV drying of lotus seeds were determined to obtain a minimal drying time, minimal shrinkage ratio, maximum rehydration ratio and maximum whiteness index. The optimum drying conditions were found to be: microwave output power 3.2 kW, vacuum degree –0.083 MPa and on/off ratio 94/26s. Under these optimal conditions, drying time, shrinkage ratio, rehydration ratio and whiteness index were found to be 10 min, 37.66 per cent, 157.1 per cent and 68.83, respectively (Tian *et al.*,2012).

Harvesting and Marketing of Lotus

Harvesting

Harvested Lotus

Cut Flowers

Loose Flowers

Harvested Rhizomes

Lotus in Local Markets

12.0 Biotechnology

12.1.0 Tissue Culture

12.1.1 Micropropagation

In vitro culture in vegetable lotus is rarely reported possibly because its recalcitrance to regeneration *in vitro* (Zhao, 1999). However, a protocol for flower lotus regeneration has been reported (Arunyanart, 1998,Arunyanart and Chaitrayagun 2005). Effect of plant growth regulators, explant size, season of explant collection, temperature (20, 25 and 30°C) and photoperiod on *in vitro* lotus (*Nelumbo nucifera* Gaertn.) shoot formation and growth were examined (Shou *et al.*, 2008). Shoots formation was greatly influenced by growth regulators, explant size and season of explant collection. The maximum number of shoots were induced from bud explants on Murashige and Skoog (MS) medium containing 4.44 µM benzyladenine (BA) + 0.54 µM α-naphthalene acetic acid (NAA). Explants formed by bud of one expanded and one unexpanded leaf, which was collected in spring gave encouraging results of shoot production. Higher temperature favoured shoot induction and subsequent growth was much better at 25 °C compared to that at 20 and 30 °C (Shou *et al.*, 2008).

In vitro direct regeneration of *Nelumbo nucifera* Gaertn. was successfully achieved from immature explants (yellow plumule) cultured on a solid MS media Harvesting and marketing of lotus supplemented with combinations of 0.5 mg/l BAP and 1.5mg/l NAA which resulted in 16.00 ±0.30 number of shoots per explant and exhibited a new characteristic of layered multiple shoots, while normal roots formed on the solid MS basal media (Mahmad *et al.*, 2014). The double-layered media gave the highest number of shoots per explant with a ratio of 2:1 (liquid to solid) with a mean number of 16.67 ± 0.23 shoots per explant with the formation of primary and secondary roots from immature explants. In the study involving light distance, the tallest shoot (16.67 ± 0.23 mm) obtained from the immature explants was at a light distance of 200mm from the source of inflorescent light (1000 lux). The plantlets were successfully acclimatized in clay loam soil after 8 months being maintained under *in vitro* conditions (Mahmad *et al.*, 2014).

An efficient *in vitro* method that ensures genetic stability was reported by Liu *et al.* (2019). An optimized method was developed to produce intact and sterile mature embryos by soaking fruits in sterile water containing 2 per cent (v/v) plant preservative mixture. Using mature embryos as explants, an improved micropropagation procedure was established in *N. nucifera* 'Weishan Hong.' The growth patterns of explants differed with the types and concentrations of cytokinins. Media containing 0.5 to 2.0 mg/l kinetin were suitable for elongation of unexpanded stems, while media with 1.0 – 2.0 mg/l 6-benzylaminopurine (BAP) the stems were shorter. Higher concentrations of about 3.0 mg/l BAP promoted shoot cluster formation. Liquid–solid, double-layer media promoted plantlet growth. The well-rooted plantlets could normally grow after being transplanted. Lotus is known to be recalcitrant to regeneration from callus. Apices of mature embryos, aseptic plantlets, and young embryos formed callus successfully when the medium was supplemented with 2,4-dichlorophenoxyacetic acid and thidiazuron. Callus morphology differed with respect to the source. However, only a few of the calluses developed axillary shoots (Liu *et al.*, 2019).

12.1.2 Somatic Embryogenesis

Callus induction and somatic embryogenesis of lotus (*Nelumbo nucifera* Gaertn.) *cv.* Satabankacha were studied (Arunyanart and Chaitrayagun, 2005). Callus was initiated by culturing bud, cotyledon, and young leaf explants on Murashige and Skoog (MS) medium containing a combination of 0, 4, 8 and 10 µM 2,4-dichlorophenoxy acetic acid (2,4-D) and 0, 1, 2 and 3 µM 6-furfuryl amino purine (kinetin) or substituting 0, 0.5 and 1 µM benzyladenine (BA) for kinetin. Bud explants cultured on medium containing 4 µM 2,4-D and 1 µM BA gave the best callus growth. For somatic embryogenesis, the calli initiated on MS medium containing a combination of 4, 6, 8 and 10 µM 2,4-D and 1 µM BA and subsequently transferred to media containing 2–4 µM 2,4-D and 0 or 0.5 µM BA produced the most somatic embryos. When cultures were 12-week-old, callus produced on medium with 6 µM 2,4-D and 1 µM BA showed the best growth for somatic embryo regeneration. When transferred to a medium with 2 µM 2,4-D and 0.5 µM BA somatic embryos were produced from 33 per cent of the calli. Embryos developed to the stage proembryo within 4 weeks and formed globular, heart, torpedo and mature embryos within 16 weeks.

12.2.0 Genetic Transformation

Only a couple of reports are available on genetic transformation in Lotus. The effects of explant and plant growth regulators on callus induction and plant regeneration and particle bombardment-induced transformation were examined in lotus (*Nelumbo nucifera* Gaertn.) cv. Buntharik (Buathong *et al.*, 2013). The apical buds from embryos (<"3 mm) formed an embryogenic callus when cultured on MS medium supplemented with 40 µM 1-napthaleneactic acid (NAA) and 0.5 µM 1-phenyl-3-(1, 2, 3-thiadiazol-5-yl)-urea (TDZ) for 8 weeks. All of the embryogenic calluses were transferred to MS medium supplemented with 0, 40, 50 and 60 µM 6-benzyladenine (BA) and the highest number of shoots was achieved in the medium supplemented with 50 µM BA after 8 weeks of culture. The transformation of lotus using a particle bombardment device was examined with the shoot clusters from embryogenic apices. The pCAMBIA2301anti-DFR plasmid contained β-*glucuronidase* (*GUS*) as a reporter gene, *neomycin phosphotransferase* (*NPTII*) as a selectable marker, and antisense dihydroflavonol 4-reductase (anti-*DFR*) as the gene of interest. The particle bombardment procedure with a helium gas pressure of 1100 psi and a target distance of 6 cm yielded the highest number of blue spots (4.8 spots/shoot cluster). However, stable transformations, as confirmed by polymerase chain reaction and reverse transcription polymerase chain reaction analysis, were only achieved using a helium gas pressure of 1100 psi and a target distance of 9 cm. Our work illustrates that the stable transformation of lotus is possible, and we suggest that this methodology can be used to target and modify specific characters of this plant (Buathong *et al.*, 2013).

Chalcone synthase (*CHS*) is a key enzyme in the flavonoid biosynthesis pathway. *CHS* genes were cloned from genomic DNA and cDNA from the petals of *'Buntharik'* white lotus and *'Sattabangkacha'* pink lotus by the PCR technique using a specific primer of the *CHS* gene designed from the GenBank database (Kanjana *et al.*, 2017). Semi-quantitative RT-PCR analysis revealed that the highest *CHS* gene expression was found in the early budding stage of the pink lotus and was reduced in later stages. Shoot tips from embryos of Buntharik and Rachinee lotus were used to induce shoot clusters by cultivation on a MS medium supplemented with 40 μM NAA and 0.5 μM TDZ for 8 weeks and a MS medium supplemented with 50 μM BA for 8 weeks. An antisense *CHS* gene (450 bp) from the cDNA of Buntharik lotus was used to construct a plant transformation vector; pCAMBIA1302CHSA. The vector construct was transformed into Buntharik and Rachinee shoot clusters by particle bombardment. After transformant selection and regeneration, two transformants of Buntharik shoot clusters showed *GFP* green spots and existence of the *GFP* gene and *hpt*II gene in the genomic DNA amplified by the PCR technique. In the Rachinee transformants, 3 of 5 showed the *GFP* green spots and the *GFP* and *hpt*II genes were identified in amplification by PCR. After *CHS* gene expression analyses by semi-quantitative RT-PCR, two transformed Rachinee shoot clusters had a reduction in *CHS* gene expression (Kanjana *et al.*, 2017).

12.3.0 Molecular Markers

12.3.1 Genetic Diversity

Genetic and genotypic variabilities of a rare relict aquatic plant *Nelumbo komarovii* Grossh. from six natural populations of Primorsky krai (Russia) were assessed using isozyme analysis (Nakonechnaya and Yatsunskaya, 2018). The highest genetic and genotypic diversities were observed in the Razdol'noe and Okeanskaya populations, which can be explained by cross-pollination participation in the seed formation for maintenance of the Razdol'noe population size and by sowing seeds from different habitats for the Okeanskaya population. Low polymorphism at the population level ($P = 8.97$, $H_O = 0.055$, $H_E = 0.039$) is determined by the history of existence of the species from the Tertiary Period as well as by the influence of gene drift. Clonal renewal plays a decisive role in the propagation of *N. komarovii* in most natural populations (Nakonechnaya and Yatsunskaya, 2018).

Genetic diversity and genetic relationships of lotus (*Nelumbo* Adanson) cultivars were evaluated using allozyme and ISSR markers (Tian *et al.*, 2008). The samples used covered 11 accessions of possible hybrids between *Nelumbo nucifera* and *Nelumbo lutea* and 92 accessions of *N. nucifera* including 69 flower lotus, 13 rhizome lotus, 5 seed lotus and 5 wild lotus. For allozyme studies, a total of 31 alleles at 23 loci of 18 enzyme systems were detected of which 5 (21.7 per cent) loci *Aat, Idh, Mdh-2, Pgd, Sod* were polymorphic. The loci of *Aat* and *Idh* included two alleles, *Mdh-2, Pgd* and *Sod* included three alleles. Eighteen genotypes were detected with the 13 alleles of the 5 polymorphic loci. The parameters of average allele number, observed heterozygosity, expected heterozygosity and Shannon information index of 92 *N. nucifera* samples were 1.35 ± 0.71, 0.06 ± 0.21, 0.05 ± 0.14, 0.10 ± 023, respectively. Thirteen ISSR primers generated 93 loci, of which 37.63 per cent were polymorphic across all samples. The percentage of polymorphic loci, average allele number, expected heterozygosity and Shannon information index of 92 *N. nucifera* samples were 26.67 per cent, 1.30 ± 0.46, 0.10 ± 0.18 and 0.15 ± 0.25, respectively for the ISSR data. The 'Bottleneck effect' and rapid propagation of clones after the ice ages may explain the low genetic diversity of lotus. The dendrograms based on ISSR and allozymes were not congruent. Based on the ISSR data, the 103 samples were divided into the *N. nucifera* group (Group I), and the group containing inter-specific hybrids between *N. nucifera* and N. *lutea* (Group II). The flower lotus, rhizome lotus, and seed lotus each has multiple sources of origin. Plant size, a criterion commonly used in the classification of cultivars of lotus, is not correlated with genetic variation. Flower color is correlated with the cultivar classification to some degree, but its variation is complex in the hybrids (Tian *et al.*, 2008).

Han *et al.* (2007b) studied 38 accessions were sampled including 37 accessions of *N. nucifera* or hybrids between *N. nucifera* and *Nelumbo lutea* and a single accession of *N. lutea*. In the ITS analyses, Chinese and Japanese lotus comprise a single cluster with a moderate bootstrap support 68 per cent indicating there is very high similarity between them. Moreover, these ISSR and RAPD results also indicate that there is very close genetic relationship between Chinese and Japanese lotus. In the ISSR and RAPD analyses, although 38 accessions all are distinctly separately into two groups, *viz. N. nucifera* and *N. lutea*, there is a high Jaccard similarity coefficient (0.785 and 0.656) between the two species. In *N. nucifera* the two different groups of the species, *viz.* flower lotus and rhizome lotus accessions show clear genetic variations. Seed lotus accessions do not form a distinct cluster but are interspersed among the flower accessions indicating that seed lotus is phylogenetically close to flower lotus and they might originate from close wild lotus in genetic relationship. In flower lotus, big-flower type accessions and medium-small type accessions have obvious genetic variation, indicating height is an important criterion in the classification system of flower lotus (Han *et al.*, 2007b).

Hu *et al.* (2011) developed twenty novel sacred lotus SSR markers, and used AFLP and SSR markers to investigate the genetic diversity and genetic relationships among 58 accessions of *N. nucifera* including 15 seed lotus, 12 rhizome lotus, 24 flower lotus and 7 wild lotus. The results showed that sacred lotus exhibited a low level of genetic diversity, which may attribute to asexual reproduction and long-term artificial selection. A dendrogram based on both AFLP and SSR clustering data showed that: (1) the seed lotus accessions and rhizome lotus accessions were distinctly clustered into different groups, which indicated the significant genetic differentiation between them. This may be attributed to the two modes of reproduction and lack of genetic exchange; (2) the accessions of Thailand wild lotus were separated from other wild lotus accessions. This implied that the Thailand lotus might be genetically differentiated from other wild lotuses. In addition, Mantel test conducted gave highly significant correlation between AFLP-SSR data and each of the AFLP and SSR ones, with the values of $r = 0.941$ and $r = 0.879$,respectively, indicating the higher efficiency of the combination of these techniques (AFLP and SSR) in estimation and validation of the genetic diversity among the accession of sacred lotus. This knowledge of the genetic diversity and genetic relatedness of *N. nucifera* is potentially useful to improve the current strategies in breeding and germplasm conservation to enhance the ornamental and economic value of sacred lotus (Hu *et al.*, 2011).

Based on 47 accessions from Ussuri River Valley, Songhua River Valley and Heilong River Valley together with 2 accessions from Russia and 27 accessions of cultivated lotus from other provinces in China, genetic diversity of wild lotus (*Nelumbo nucifera*) were revealed using RAPD and ISSR markers (Xue *et al.*, 2006). Twenty RAPD primers generated 113 loci, of which 71.68 per cent were polymorphic across all samples. The expected heterozygosity was 0.1583. The percentage of polymorphic loci and expected heterozygosity in the wild lotus were 50.44 per cent and 0.1241, respectively. The parameters of the cultivated lotus were slightly higher, 53.98 per cent and 0.1651 correspondingly. Sixteen ISSR primers produced 90 loci. The percentages of polymorphic loci and expected heterozygosity were 41.11 per cent and 0.0851 at species level, 28.89 per cent and 0.0661 for the wild lotus, and 32.22 per cent and 0.0963 for the cultivated lotus. AMOVA analysis of the wild lotus showed that a small number of variances exist among the 3 river valleys (21.68 per cent for RAPD with Gst=0.1312 and 15.11 per cent for ISSR with Gst=0.1352). The molecular variances of both the wild and the cultivated lotuses came predominantly from within the 3 river valleys and the cultivated samples (73.25 per cent for RAPD and 81.11 per cent for ISSR). Variance components from the wild and the cultivated lotus accounted for 19.17 per cent for RAPD and 13.17 per cent for ISSR, and variations among the valleys and the culta seemed the least important (7.585 for RAPD and 5.725 for ISSR). Neighbour-joining analysis demonstrated that considerable differentiation happened between the wild and the cultivated lotus. The wild lotus at middle reaches of the Songhua River Valley seemed to be the centre of remnants, from which it spread to the Ussuri River Valley and the Heilong River Valley. The very limited genetic diversity suggests that the wild lotus has experienced severe bottleneck effect, founder effect and rebirth effect. For conservation purpose, special attention should be paid to middle reaches of the Songhua River Valley (Xue *et al.*, 2006).

In Thailand, two major cultivars, Sattabongkot and Saddhabutra, are used and exported to foreign countries (Salaemae *et al.*, 2018). To investigate the difference between these two cultivars, their molecular phylogeny and postharvest morphology were examined. Using 25 simple sequence repeat (SSR) markers, the cultivar Saddhabutra was very close to the cultivar Satabuto. These results showed a high bootstrap (BS) value of 96 per cent. In contrast, lotus cultivars in Thailand including Sattabongkot and Saddhabutra, showed a BS value of 90 per cent. Saddhabutra and Satabuto are classified in the same group with high similarity, whereas Sattabongkot shows relatively lower similarity. Petal blackening started earlier in Saddhabutra than Sattabongkot, indicating a vase-life difference among cultivars, with Sattabongkot and Saddhabutra having a vase life of 72 and 54 h, respectively. The epidermal cells of normal petals of Sattabongkot and Saddhabutra showed freshness and a round shape with turgid cells. The petal blackening was accompanied with a reduction of the area and perimeter in both cultivars. Sattabongkot had fewer stomata than Saddhabutra. The study provided a molecular classification of the Thailand lotus cultivars and provides a useful technique for the quantification of the postharvest quality of lotus cultivars (Salaemae *et al.*, 2018).

Mekbib *et al.* (2020) sampled the population for genetic diversity and structure of 15 tropical lotus the population genetic diversity and structure of 15 tropical lotus populations from the previous understudied natural distribution ranges, including India, Thailand, and Australia, and were assessed using nine polymorphic SSR markers. They found the SSR markers used to genotype the 216 individuals yielded 65 alleles. The highest and lowest genetic diversity estimates were found in Thailand and Indian populations, respectively. STRUCTURE analysis revealed three distinct genetic clusters, with relatively low admixtures, supported by PCoA cluster analysis. Low levels of gene flow (mean Nm = 0.346) among the three genetic clusters signified the Mantel test for isolation by distance, revealing the existence of a positive correlation between the genetic and geographic distances (r = 0.448, *P* = 0.004). Besides, AMOVA analysis revealed a higher variation among populations (59.98 per cent) of the three groups. Overall, the populations used in this study exposed a high level of genetic differentiation (FST = 0.596). They concluded that the nine polymorphic microsatellite markers used in their study sufficiently differentiated the fifteen tropical *N. nucifera*populations based on geography. These populations presented different genetic variability,thereby confirming that populations found in each country are unique. The low genetic diversity (HE = 0.245) could be explained by limited gene flow and clonal propagation. Conserving the available diversity using various conservation approaches is essential to enable the continued utilization of this economically important crop species. Mekbib *et al.* (2020),therefore, proposed that complementary conservation approaches ought to be introduced to conserve tropical lotus,depending on the genetic variations and threat levels in populations.

Xue *et al.* (2012) studed population genetics, phylogeography, and hybridization of *Nelumbo*(Nelumbonaceae), by developing chloroplast microsatellite markers. Seventeen microsatellite loci were identified from the chloroplast genomes of *N. nucifera* and *N. lutea*. Polymorphisms were assessed in three populations of *N. nucifera* and one population of *N. lutea*. Nine loci were found to be polymorphic in *N. nucifera*, and all 17 loci were found to be polymorphic in *N. lutea*. In *N. nucifera*, the number of alleles per locus ranged from two to six, and the unbiased haploid diversity per locus ranged from 0.198 to 0.790. In *N. lutea*, the number of alleles ranged from two to four, and the unbiased haploid diversity per locus ranged from 0.245 to 0.694. They concluded that the identified chloroplast simple sequence repeat markers will be useful for the study of genetic diversity, phylogeography, and identifi cation of *Nelumbo* cultivars (Xue *et al.*, 2012).

12.3.2 Transcriptomics

Rhizome is the storage organ of lotus derived from modified stems. The development of rhizome is a complex process and depends on the balanced expression of the genes that is controlled by environmental and endogenous factors (Yang *et al.*, 2015). However, little is known about the mechanism that regulates rhizome girth enlargement. In this study, using RNA-seq, transcriptomic analyses were performed at three rhizome developmental stages—the stolon, middle swelling and later swelling stage —in the cultivars 'ZO' (temperate lotus with enlarged rhizome) and 'RL' (tropical lotus with stolon). About 348 million high-quality reads were generated and 88.5 per cent of the data were mapped to the reference genome. Of 26783 genes identified, 24069 genes were previously predicted in the reference and 2714 genes were novel transcripts. Moreover, 8821 genes were differentially expressed between the cultivars at the three stages. Functional analysis identified that these genes were significantly enriched in pathways carbohydrate metabolism and plant hormone signal transduction. Twenty-two genes involved in photoperiod pathway, starch metabolism and hormone signal transduction were candidate genes inducing rhizome girth enlargement. Comparative transcriptomic analysis detected several differentially expressed genes and potential candidate genes required for rhizome girth enlargement, which lay a foundation for future studies on molecular mechanisms underlying rhizome formation (Yang *et al.*, 2015).

High-throughput transcriptome sequencing provides large amounts of transcriptome data for promoting gene discovery and development of molecular markers. *Arabidopsis* transcription factor APETALA2 (AP2) controls multiple aspects of plant growth and development, including seed development, stem cell maintenance, and specification of floral organ identity. Based on sequence similar of *Arabidopsis AP2* and its homologues genes from other plant species, degenerate RT-PCR and rapid amplification of cDNA ends assay were used to clone *AP2* genes from lotus (*Nelumbo nucifera*). A 2,048-bp cDNA fragment was obtained, which contains a 1,536-bp open reading frame encoding a protein of 511 amino acids (Liu *et al.*, 2012). The protein contains two AP2 domains that are conserved in AP2 proteins from other plant species, thus was named as *N. nucifera* APETALA2 (*Nn*AP2). Quantitative RT-PCR revealed that *NnAP2* gene was expressed in flowers, roots, leaves, and stems of *N. nucifera*, with flowers which have the highest transcript levels. Further analysis showed that in all five lotus cultivars examined, including "Zhongguogudailian," "Yaoniangyujiao," "Jinxia," "Hongtailian," and "Yiliangqianban," petals always have the highest expression levels when compared with the other four flower organs, though the number of petals in these cultivars ranged from simple to thousands. However, *NnAP2* expression level in four non-simple petal flower cultivars was higher than that in the simple petal flower cultivar Zhongguogudailian, indicating that *NnAP2* may play a role in specification of petal identity during the evolutionary process of the ancient species *N. nucifera* (Liu *et al.*, 2012).

Zhang *et al.* (2014) were assembled 68,593 unigenes from 1.34 million 454 GS-FLX sequence reads of a mixed flower-bud cDNA pool derived from three accessions of *N. nucifera*. A total of 5,226 SSR loci were identified, and 3,059 primer pairs were designed for marker development. Di-nucleotide repeat motifs were the most abundant type identified with a frequency of 65.2 per cent, followed by tri- (31.7 per cent), tetra- (2.1 per cent), penta- (0.5 per cent) and hexa-nucleotide repeats (0.5 per cent). A total of 575 primer pairs were synthesized, of which 514 (89.4 per cent) yielded PCR amplification products. In eight Nelumbo accessions, 109 markers were polymorphic. They were used to genotype a sample of 44 accessions representing diverse wild and cultivated genotypes of Nelumbo. The number of alleles per locus varied from 2 to 9 alleles and the polymorphism information content values ranged from 0.6 to 0.9. They performed genetic diversity analysis using 109 polymorphic markers. A UPGMA dendrogram was constructed based on Jaccard's similarity coefficients revealing distinct clusters among the 44 accessions. Conclusions: Deep transcriptome sequencing of lotus flower buds developed 3,059 genic SSRs, making a significant addition to the existing SSR markers in lotus. Among them, 109 polymorphic markers were successfully validated in 44 accessions of Nelumbo. This comprehensive set of genic SSR markers developed in our study will facilitate analyses of genetic diversity, construction of linkage maps, gene mapping, and marker-assisted selection breeding for lotus.

12.3.3 Proteomics

The molecular mechanism of pigmentation in lotus flower has been barely reported. Previously, Chen *et al.* (2013) proposed a flavonoid biosynthetic pathway in lotus based on the identification of different compositions of anthocyanins, flavones and flavonols in various tissues. However, the regulatory mechanisms underlying anthocyanin biosynthesis in lotus are still elusive. Moreover, there are hardly any reports focusing on this as yet (Deng *et al.*, 2015).To explore the flower coloration mechanism in lotus, the constituents and contents of pigments in two lotus cultivars with red and white flowers were analyzed. Although flavones and flavonols were detected in both cultivars, anthocyanins could only be detected in the red cultivar (Deng *et al.*, 2015). A comparative proteomics analysis on the flower petals between these two cultivars was conducted. A total of 88 differentially expressed proteins were identified with 36 more abundant and 52 less abundant in the red than in the white cultivar. Among them, four enzymes involved in the anthocyanin pathway were identified, *i.e.* flavanone 3-hydroxylase, anthocyanidin synthase, anthocyanidin 3-Oglucosyltransferase and glutathione S-transferase. Analysis of the expression patterns of anthocyanin biosynthetic genes indicated that the anthocyanindin synthase (ANS) gene might be the critical gene determining anthocyanins biosynthesis and accumulation in lotus flower. Further analysis showed that different methylation intensities on the promoter sequence of the ANS gene might result in the different flower coloration in the red and white cultivar. This study provides new insights into the mechanism of flower coloration in lotus, and may be helpful in its breeding and germplasm enhancement (Deng *et al.*, 2015).

12.3.4 Genome Sequencing

A team of 70 scientists from the U.S., China, Australia and Japan today reports having sequenced and annotated the genome of the "sacred lotus," which is believed to have a powerful genetic system that repairs genetic defects, and may hold secrets about aging successfully. The scientists sequenced more than 86 percent of the nearly 27,000 genes of the plant, *Nelumbo nucifera*, which is revered in China and elsewhere as a symbol of spiritual purity and longevity (Wolpert, 2013). According to Shen-Miller, one of the corresponding author of the work on Lotus genome paper, informed Wolpert (2013) that the lotus' genetic repair mechanisms could be very useful if they could be transferred to humans or to crops such as rice, corn and wheat whose seeds have life spans of only a few years. "If our genes could repair disease as well as the lotus' genes, we would have healthier aging. We need to learn about its repair mechanisms, and about its biochemical, physiological and molecular properties, but the lotus genome is now open to everybody."

Sacred lotus is a basal eudicot with agricultural, medicinal, cultural and religious importance. It wasdomesticated in Asia about 7,000 years ago, and cultivated for its rhizomes and seeds as a food crop. It is particularly noted for its 1,300-year seed longevity and exceptional water repellency, known as the lotus effect. The latter property is due to the nanoscopic closely packed protuberances of its self-cleaning leaf surface, which have been adapted for the manufacture of a self-cleaning industrial paint, Lotusan.

Ming and 70 scientists (Ming *et al.*, 2013) together sequenced the genome of the China Antique variety of the sacred lotus with Illumina and 454 technologies, at respective depths of 101× and 5.2×. The final assembly has a contig N50 of 38.8 kbp and a scaffold N50 of 3.4 Mbp, and covers 86.5 per cent of the estimated 929 Mbp total genome size. The genome notably lacks the paleo-triplication observed in other eudicots, but reveals a lineage-specific duplication. The genome has evidence of slow evolution, with a 30 per cent slower nucleotide mutation rate than observed in grape. Comparisons of the available sequenced genomes suggest a minimum gene set for vascular plants of 4,223 genes. Strikingly, the sacred lotus has 16 COG2132 multi-copper oxidase family proteins with root-specific expression; these are involved in root meristem phosphate starvation, reflecting adaptation to limited nutrient availability in an aquatic environment.Ming *et al.* (2013) concluded that the slow nucleotide substitution rate makes the sacred lotus a better resource than the current standard, grape, for reconstructing the pan-eudicot genome, and should therefore accelerate comparative analysis between eudicots and monocots.

12.3.5 Genetic Maps

Genetic and physical maps are powerful tools to anchor fragmented draft genome assemblies generated from next-generation sequencing. Currently, two draft assemblies of *Nelumbo nucifera*, the genomes of 'China Antique' and 'Chinese Tai-zi', have been released. According to Gui *et al.* (2018), the first genetic linkage maps of Nelumbo were constructed based on the genotype of 171 Simple Sequence Repeat (SSR) markers and 53 sequence-related amplified polymorphism (SRAP) markers in *N. nucifera* 'Chinese Antique', *N. lutea* 'AL1' and their 51 F_1 populations (Yang *et al.*, 2012). Using the same map population, Zhang *et al.* (2014) refined both the linkage maps of *N. nucifera* and *N. lutea* based on SNPs that were identified using restriction site associated DNA sequencing (RAD-Seq) data. The resulting linkage map of *N. nucifera* was 656.9 cM with 23 SSR markers and 73 SNPs distributed in eight linkage groups (LGs), and the linkage map of *N. lutea* was 494.3 cM with 562 bins (3894 SNPs) and 136 SSRs distributed in nine LGs. Using the 96 F_2 individuals of two *N. nucifera* cultivars, Liu *et al.* (2016) constructed a linkage map of 581.3 cM, with 791 bin markers (8971 SNPs) sorted into 8 LGs (Gui *et al.*, 2018).

However, there is presently no information on how the sequences are assembled into chromosomes in *N. nucifera*. The lack of physical maps and inadequate resolution of available genetic maps hindered the assembly of *N. nucifera* chromosomes. Hence, linkage map of *N. nucifera* containing 2371 bin markers [217 577 single nucleotide polymorphisms (SNPs)] was constructed using restriction-site associated DNA sequencing data of 181 F2 individuals and validated by adding 197 simple sequence repeat (SSR) markers (Gui *et al.*, 2018). Additionally, a BioNano optical map covering 86.20 per cent of the 'Chinese Tai-zi' genome was constructed. The draft assembly of 'Chinese Tai-zi' was improved based on the Bio-Nano optical map, showing an increase of the scaffold N50 from 0.989 to 1.48 Mb. Using a combination of multiple maps, 97.9 per cent of the scaffolds in the 'Chinese Tai-zi' draft assembly and 97.6 per cent of the scaffolds in the 'China Antique' draft assembly were anchored into pseudo-chromosomes, and the centromere regions along the pseudo chromosomes were identified. An evolutionary scenario was proposed to reach the modern *N. nucifera* karyotype from the seven ancestral eudicot chromosomes. The study provides the highest-resolution linkage map, the optical map and chromosome level genome assemblies for *N. nucifera*, which are valuable for the breeding and cultivation of *N. nucifera* and future studies of comparative and evolutionary genomics in angiosperms (Gui *et al.*, 2018).

12.3.6 Expressed Sequence Tags (ESTs)

Expressed sequence tags (ESTs) are particularly useful for gene identification, genome annotation and alternative splicing (AS) study. The genome of sacred lotus variety 'China Antique' was recently sequenced (Ming *et al.*, 2013). It lacks the lambda triploidization event seen in all core eudicots, making sacred lotus an excellent model for reconstructing the pan-eudicot genome (Ming *et al.*, 2013). It is also expected to serve as a reference for comparative genomics due to its remarkably low mutation rate and a higher retention of duplicated genes compared with most other angiosperm genomes (Ming *et al.*, 2013). EST annotation and mapping ESTs to the genome will help to refine gene models predicted from genome sequence. According to VanBuren (2013) the identification of AS genes and AS events is the first step for our understanding of how these genes are regulated post-transcriptionally, and may provide insights into the exceptional seed longevity, thermogenesis, water repellency, and other biological phenomenon seen in sacred lotus.VanBuren *et al.* (2013) generated 626,137 reads by 454 pyrosequencing of a cDNA library, which was constructed from mRNA collected from 7 different tissues of sacred lotus. The reads were assembled into 16,349 non-redundant EST contigs with an average length of 715 bp. Aligning ESTs to sacred lotus genome revealed the internal exons and introns had an average size of 140 and 2,005 bp, respectively. The average intron length in the sacred lotus is much larger than most sequenced plant genomes. We also identified 174 AS events involving 161 genes. Classification of the AS events showed that 62.6 per cent of the AS events were retained introns, 8.0 per cent were alternative acceptor sites, 7.5 per cent were alternative donor sites, 4.0 per cent were skipped exons, and 17.8 per cent had "complex events" which had more than one basic event. Comparison of AS genes in sacred lotus, rice and Arabidopsis identified 7 commonly conserved genes between lotus and rice and 9 between sacred lotus and Arabidopsis. A gene encoding a 40S ribosomal protein was commonly conserved in all three species. These AS events and annotated ESTs are available through the Plant Alternative Splicing Database website (http://proteomics.ysu.edu/altsplice/plant/).

13.0 References

Anonymous (1966) *The Wealth of India:* Raw Materials, 7: 7-9, Publication and Information Directorate, CSIR, New Delhi.

Anonymous (1989) Collins Guide to Water Lilies and Other Aquatic Plants, Nelumbo -Lotus, pp.98-105. William Collins Sons and Co Ltd., London.

Arunyanart, S. (1998). *J. Jap. Soc. Hort. Sci.*, **67** (Suppl.): 257.

Arunyanart, S. and Chaitrayagun, M. (2005) *Sci. Hort.*, **105:** 411-420.

Bailey, L.H. and Bailey, E.Z. (1976) In: *Hortus Third - A Concise Dictionary of Plants Cultivated in the United States and Canada*, MacMillian Pub. Co. Inc., New York, USA.

Bao, G.L., Gang, S.K. and You, J. (1998) *China Vegetables*, **1**: 12-14.

Battaglia, S. (2019) Lotus - The Sacred flower. www.perfectpotion.com.au

Borsch, K. and Barthlott, W. (1994) *Beitr. Biol. Pflanzen.*, **68**: 421-450.

Borsch, T. and Wilde, V. (2000) *Pollen and Pores: Morphology and Biology* (eds. Harley, M.M., Morton, C.M. and Blackmore, S.), R, BG, Kew, pp. 285-299.

Buathong, R., Saetiew, K., Phansiri, S., Parinthawong, N. and Arunyanart, S. (2013) *Sci. Hortic.*, **161 :** 216-222.

Buddhadev, S.G. and Buddhadev, S.S. (2014) *Pharma Sci. Monit.*, **5**(3): 1-12.

Chen, S., Fang, L., Xi, H., Guan, L., Fang, J., Liu, Y., Wu, B. and Li, S. (2012). *Anal. Chim. Acta*, **724**: 127-135.

Chen, S., Xiang, Y., Deng, J., Liu, Y.L. and Li, S.H. (2013) *PLoS One*, **8:** 13.

Chen, G., Zhu, M. and Guo, M. (2019). *Crit. Rev. Food Sci. Nutr.*, **59**(sup1) S189-S209, doi: 10.1080/10408398.2018.1553846.

Chopra, R.N., Nayar, S.L. and Chopra, I.C. (1986) *Glossary of Indian Medicinal Plants*, CSIR, New Delhi.

Deng, J., Fu, Z., Chen, S., Damaris, R.N., Wang, K., Li, T. and Yang, P. (2015) *Plant Cell Physiol.*, **56**(8): 1546–1555.

Diao, Y., Chen, L., Yang, G., Zhou, M. and Song, Y. (2006) *Caryologia*, **59:** 25–30.

Everett, Thomas H. (1980) *The New York Botanical Garden Illustrated Encyclopedia of Horticulture*, Garland Pub. Inc., New York, USA.

Fritz, D., Ortmeier, A. and Habegger, R. (1992) *Acta Hortic.*, **318**: 179-186.

Gang, C. (1999) *J. Henan Agric.Sci.*, **5**: 40.

Goel, A. K. (2001) Lotus : God's favoriete bloem (Dutch), Hindorama (Netherlands), **2**: 26-29.

Graf, A. B. (1973) Exotica - Colour Cyclopedia of Exotic Plants and Trees, 6th Edn., Roehrs Co. Inc., USA.

Green, C. (2002) Jewels of the pond, *Hortic.*, **99**: 50-55.

Gui, S., Peng, J., Wang, X., Wu, Z., *et al.* (2018)*Plant J.*, **94**: 721–734.

Guo, H. B. (2009) *Genet. Resour. Crop Evol.*, **56**: 323–330.

Gupta, S.C., Paliwal, G.S. and Ahuja, R. (1968) *Amer. J. Bot.*, **55**: 295-301.

Han, Y.C., Teng C.Z., Chang, F.H., Robert G.W., Zhou, M.Q., Hu, Z.L. and Song, Y.C. (2007b) *Aquat. Bot.*, **87**: 141-146.

Hamana, K., Matsuzakii, S., Nitsu, M. and Samejuma, K. (1994) *Canadian J. Bot.*, **72**: 1114-1120.

Hayes, V., Schneider, E.L. and Carlquist, S. (2000) *Int. J. Plant Sci.*, **161**(6 Suppl.): S183-S191.

Hongpakdee, P. and Ruamrungsri, S. (2017) *Acta Hortic.*,**1171**: 47-52.

Hongpakdee, P., Samranyat, C. and Ruamrungsri, S. (2019) *Acta Hortic.*,**1263**: 233-240.

Hu, J., Pan, L., Liu, H., Wang, S., Wu, Z., Ke, W. and Ding Y. (2011) *Mol. Biol. Rep.* doi: 10.1007/s11033-011-1138-y.

Im, M.H., Kim, B.W., Park, Y.S., Yang S.R., Song, C.E. and Heo, B.G. (2012) *Korean J. Plant Res.*, **25**(1): 7-13. https://doi.org/10.7732/KJPR.2012.25.1.007.

Ji, H.W., Tang, C.Q., Li, L.B. and Ting-Yun, K. (2001) *Acta Bot. Sin.*, **43**: 1129-1133.

Kanjana, S., Prissadang, A., Nanglak, P. and Sumay A. (2017). *The Open Biotechnol. J.*, **11**: 1-8.

Kasumi, M. and Sakuma, F. (1998) *J. Japanese Soc. Hort. Sci.*, **67**: 595-599.

Lakhanpal, R N. (1955) *J. Indian Bot. Soc.*, **34**: 222-224.

Li, Y., Smith, T., Svetlana, P., Yang, J., Jin, J.H. and Li, C.S. (2014). *Palaeogeogr. Palaeoclimatol. Palaeoecol.*, **399**: 284–29.

Li, J.K. and Huang, S.Q. (2009) *Ann. Bot.*, **104**: 845–851.

Li, L.J., Cao, P.S., Cheng, J.L., Zhang, X.D., Xie, K., Sun, L., and Liu, R.H. (2005) *Mgt. Agri. Sci. Technol.*, **24**: 33–36.

Li, Z. Y. and Quian, P. (1994) *Acta Agriculturae Zhejiangensis*, **6**: 127-130.

LiangJun Li and YouWen Zhao (1998) *J. Nanjing Agril. Univ.*, **21**: 113-115.

Lin, Z., Zhang, C., Cao, D., Damaris, R.N. and Yang, P. (2019) *Intl. J. Mol. Sci.*, **20**: 3680-3692.

Liu, Q., Zhang, D., Liu, F. *et al.* (2019). *In Vitro Cell. Dev. Biol. –Plant*, **55L**: 305–312.

Liu, Z., Gu, C., Chen, F. *et al.* (2012) *Appl. Biochem. Biotechnol.*, **168**: 383–391.

Liu, Z., Zhu, H., Liu, Y. *et al.* (2016) *BMC Genom.*,**17**: 466.

Mahmad, N., Taha, R.M., Othman, R., Saleh, A., Hasbullah, N.A. and Elias, H. (2014) *Sci. World J.*, http://dx.doi.org/10.1155/2014/745148.

Mandal, R.N. and Bar, R. (2013) *Resonance.* pp.732-737. https://www.ias.ac.in/article/fulltext/reso/018/08/0732-0737.

Mandal, R.N., Saha, G.S. and Mukhopadhaya, P.K. (2007) *Aquac. Asia*, **12**: 11–13.

Mao, Wang., Sun, Ke-lian., Wang, Li and Shi-jie (2002) *Xibei Zhiwu Xuebao*, **22**: 146-149.

Masuda, J., Ozaki, Y. and Okubo, H. (2007) *Planta*, **226**: 909–915. https://doi.org/10.1007/s00425-007-0536-9.

Masuda, J.I., Urakawa, T., Ozaki, Y. and Okubo, H. (2006) *Ann. Bot.*, **97**: 39–45.

McDonald, A. (2004) *Econ Bot.*, **58**: S147–S173. https://doi.org/10.1663/0013-0001(2004)58 [S147:ABPOTI] 2.0. CO;2.

Mekbib, Y., Huang, S.X., Ngarega, B.K., *et al.* (2020) *Bot. Stud.*, **61**: 15. https://doi.org/10.1186/s40529-020-00293-3.

Meyer, W. C. (1930) *Plant Physiol.*, **5**: 225-234.

Ming R., Vanburen R., Liu Y., Yang M., Han Y., Li L.T., Zhang, *et al.* (2013) *Genome Biol.*, **14**: R41.

Mitchel, G. (2010). *The Water Garden J.*, **25**: 23-24.

Morozova, A. M. and Vasileva, V. E. (1993) *Botanicheskii Zhurnal*, **78** : 47-50.

Nadkarni, A.K. (1954) In : K.M. Nadkarni's Indian Materia Medica, Popular Book Depot, Bombay, **1** : 844-845.

Nakonechnaya, O.V. and Yatsunskaya, M.S. (2018) *Russ. J. Genet.*,**54**: 816–824.

Niklitschek, A. (1991) *Water Garden J.*, **7**: 28-32.

Nix, M. (1987) Chinese Lotus, Wuhan, Scientific Press, Petin, Weehan Research Inst. of Botany, Academia Sinica.

Nix. M. (1989) *Water Garden J.*, **5**: 34-43.

Nohara, S. and Kimura, M. (1997) *Ecol. Res.*,**12**: 11-20.

Nohara, S. and Tsuchiya, K. (1990 *Japan Ecol. Res.*,**5**: 237-252.

Ohga, I. (1926) *Amer. J. Bot.*, **64**: 154-157.

Ohno N. (1910) *Z. Bot.***2**: 650–651.

Pal, I and P. Dey. (2013) *Int. J. Sci. Res.*, **4**:1659-1667.

Poterfield, W.M. (1941) *J. New York Bot. Garden*, **42**: 280-286.

Qu, X. J. and Zhao, Y.W. (1991) *Acta Hortic. Sincia*, **18**: 335-339.

Raffeneau-Delile A. (1841) *Ann. Sci. Nat. Bot. 2. se´ r.* **16**: 328–335.

Riya, I.S., Sun, H., Furuhata, H., Shimamura, M., Nasu, H., Imano, R., Zhou, S. and Hosomi, M. (2020) *Aquat. Bot.*, **163**: 103203. https://doi.org/10.1016/j.aquabot.2020.103203.

Salaemae, N., Takeda, S., Kubo, N. and Kaewsuksaeng,S. (2018) *Agric. Nat. Res.*, **52**: 45-50.

Sayre, J. (2004) *Native Plants*, 14-17.

Schneider, E.L. and Buchanan, J. D. (1980) *Amer. J. Bot.*, **67**: 182-193.

Schutz, O.M. and Wolfgang, G. (1988) *J. Exp. Bot.*, **39**:1231–1236, https://doi.org/10.1093/jxb/39.9.1231.

Seo, D.C., DeLaune R.D., Han, M.J., Lee, Y.C., Bang, S.B., Oh, E.J., Chae, J.H., Kim, K.S., Park, J.H. and Cho, J.S. (2010). *Ecol. Eng.*, **36**: 1373-1382.

Seymour, R.S. and Schultz-Motel, P. (1996) *Nature*, **383**: 305.

Seymour, R.S. and Schultz-Motel, P. (1998) *Series B. Bio. Sci.*, **353**: 935-943.

Seymour, R.S., White, C.R. and Gibernau, M. (2003) *Nature*, **426**: 243–244.

Sharma, S.C., Goel, A.K. and Roy, R.K. (1998) *NBRI Newsltr.*, **25**: 34-35.

Sharma, S.C. and Goel, A.K. (1997) FTR on the Project - Introduction, Multiplication, Conservation and Documentation of *Nelumbo* (Lotus) and Euryale spp./cvs., *Min. of Environ. and For.*, New Delhi.

Sharma, S.C., Goel, A.K. (2000) *International Sociey of Environmenal Botanists. Environment News*, **6**(1)

Sheikh, S.A. (2014) *J. Med. Plants Stud.*, **2**(6): 42-46.

Shen Miller, J. (2002) *Seed Sci. Res.*, **12**(3):131-143.

Shen Miller, J., Aung, L.H., Turek, J., Schopf, J.W., Tholandi, M., Yang, M. and Czaja, A. (2013) *Trop. Plant Biol.*, **6**(2/3): 58-68.

Shen-Miller, J., Mudgett, M.B., Schopf, J.W., Clarke, S. and Berger, R. (1995) *Am. J. Bot.*, **82**: 1367–1380.

Shou, S.Y., Miao, L.X., Zai, W.S., Huang, X.Z. and Guo, D.P. (2008) *Biol. Plant.*, **52**(3): 529-532.

Slocum, P.D. and Robinson, P. (1996) Water gardening, water lilies and lotuses, Timber Press, USA, pp. 434.

Slocum, P.D. (1985) In : Propagating Water Lilies and aquatics, Brooklyn Botanic Garden Record, **41**: 25-28.

Stefferud, A. (1959) The Wonders of Seeds, MacMillan and Co., Ltd, London, pp.112.

Tian, D., Tilt, K.M., Sibley, J.L., Woods, F.M. and Dane, F. (2009b) *HortScience.*, **44**(3): 656–659.

Tian, D., Tilt, K.M., Sibley, J.L., Dane, F. and Woods, F.M. (2009a) *J. Environ. Hort.*, **27**(2): 79–79.

Tian, H.L., Xue, J.H., Wen, J., Mitchell, G. and Zhou, S.L. (2008) *Scientia Hortic.*,**116**: 421–429.

Tian,Y., Zhang, Y., Zeng, S., Zheng, Y., Chen, F., Guo, Z., Lin, Y. and Zheng, B. (2012) *Food Sci. Technol. Intl.*, **18** : 477-488.

Toma,Y., Iwamoto, R., Inayoshi, K., Nagasaki, N. and Ueno, H. (2015) *Jpn. J. Soil Sci. Plant Nutr.*, **86**: 89–97.

Tóth, G., Gáspári, Z., and Jurka, J. (2000) *Genome Res.*, **10**: 967-981.

Tsuchiya, T. and Nohara, S. (1989) *Aquat. Bot.*, **36**:87-95.

Van Bergen, P.F., Hatcher, P.G., Boon, J.J., Collinson, M.E. and De Leeuw, J.W. (1997) *Phytochemistry*, **45**: 601-610.

VanBuren, R., Walters, B., Ming, R. and Min, X.J. (2013) *Plant Omics J.*, **6**(4) : 311-317.

Vogel S. and Hadacek, F. (2004) *Plant Syst. Evol.*, **249**: 173–189.

Vogel, S. (2004) *Plant Syst. Evol.* **249:** 9–25.

Wahdah, R. and Ahmad, Z. (2004) *Agrivita* (Indonesia): *Jurnal Ilmu Pertanian.,* **22**: 134-137.

Wang, J. and Zhang, G. (2010) *Hubei JTCM.,* **32**: 75–76.

Wang, Q. and Zhang, X. (2010) *The Water Garden J.,* **25**(2): 9-14.

Watling, J.R., Robinson, S.A., Grant, N.M. and Miller, R.A. (2010) *Funct. Plant Biol.: An Int. J. Plant Fun.,* **37**(11): 1085-1095.

Willams, P.S. and Schneider, E.L. (1993) In : *Families and Genera of Flowering Plants,* **2**: 470-491.

Wolpert, S. (2013) https://newsroom.ucla.edu/releases/scientists-sequence-genome-of-245577.

Wong, C.F. (1987) Origin and historical record of *Nelumbo nucifera,* Eds. China Institute of Botany, China lotus, Beijing, *Academia Sinica*, Science Publishers, pp.1–8.

Xue, J., Zhuo, L. and Shiliang, S.Z. (2006) *Chinese Sci. Bull.,* **51**: 1-12.

Yang, M., Han, Y., VanBuren, R. *et al.* (2012) *BMC Genom.,* **13**: 653.

Yang, M., Zhu, L. and Pan, C. (2015) *Sci. Rep.,* **5** :13059, https://doi.org/10.1038/srep13059.

Zhao, Y.W. (199). Aquatic Vegetables in China. - China Agriculture Press, Beijing 1999.

Zhang, D., Chen, Q., Liu, Q., Liu, F., Cui, L., Shao, W., Wu, S., Xu, J. and Tian, D. (2019) *Int. J. Mol. Sci.,* **20**: 1015. doi:10.3390/ijms20051015.

Zhang, W., Tian, D., Huang, X., Xu, Y., Mo, H, *et al.* (2014) *PLoS ONE* **9**(11) : e112223. doi:10.1371/journal.pone.0112223.

Zhang, Q., Li, L., VanBuren, R. *et al.* (2014) *BMC Genom,* **15**: 372.

Zhang, Y., Lu, X., Zeng, S. *et al.* (2015) *Phytochem Rev.,* **14**: 321–334.

Zheng, Z. and Tamura, Y. (2006) *Hortic. Res.* (*Japan*), **4** : 385-390.

3
TULIP

Safeena Sandeep, Rajiv Kumar and Debmala Mukherjee

1.0 Introduction

Tulip (*Tulipa* spp.) is a commercially important bulbous ornamental crop which occupies a prominent position among the top ten cut flowers in global floriculture trade. In the international flower market, tulips have good demand owing to its unsurpassed beauty and economic value coupled with attractively coloured elegant and exquisite flowers of different hues and shapes. Tulip is the major flower bulb produced all over the world. The Netherlands is considered as the globally famous tulip-growing region. The Netherlands share world tulip production by 60 per cent. Ninety per cent of the global area planted with flowering bulbs is designed for reproduction of six bulbous flower crop species *viz.*, Tulip, Lily, Narcissus, Gladiolus, Hyacinth and Iris (Le Nard and De Hertogh, 2002). Tulip occupies the largest area under a true bulb crop in the world followed by Narcissus, Iris, Hyacinth and Lily. Tulip production is concentrated in a dozen countries around the world. The largest growing area is in the Netherlands, about 87 per cent of the global area. Other countries with major area of tulips cultivation are Japan (300 ha), France (293 ha), USA (280 ha) and Poland (200 ha) (Buschmann, 2005). In India, the cultivation of tulips is mostly limited to temperate regions of Jammu and Kashmir. The hilly temperate regions of Himachal Pradesh and Uttarakhand also have a huge potential for growing this incredibly important commercial flower crop. Areas of Himachal Pradesh between 5,000 to 8,000 metre above sea level, including Jhanjheli and Seraj regions in Mandi, Bharmour, Tissa and Pangi in Chamba, Chail in Solan, Shimla, Kullu, Lahaul and Spiti, Kangra, Sirmour and Kinnaur are reported to be ideal for tulip cultivation.

2.0 Importance and Uses

Several attractive and bewitching cultivars in a large assortment of delicate and brilliant shades are available in Tulips. Tulips are grown widely either for bulb production, forcing as cut flower or potted plant or for landscaping. They are suitable for pot culture, in landscaping of beds or border both in formal and informal gardens. Tulips can also be grown in the basins in apple, orchards, lawns, rockeries, and wild gardens. The flowers of tulip play an important role in decorating homes, offices, hotels and even shops in this modern era. The farms producing flower bulbs of tulips use them for two purposes – for further reproduction and for protected cultivation of cut flowers (Wroblewska, 2009). Tulips are grown in open and as well as under protected conditions. The cultivation of tulips will not only improve the financial condition of growers substantially, but also will promote tourism industry in the area.

3.0 Origin and History

The primary centre of origin of the genus *Tulipa* L. is located in the Pamir Alai and Tien Shan mountain ranges in Central Asia (Hoog, 1973), Further, a diversification occurred from this region, resulting in a distribution from Morocco to Western Europe and eastward to western China, to the north and northeast (Siberia, Mongolia, and China), south to Kashmir in India, and west to Afghanistan, Iran

Uses of Tulip

In Border

In Garden

In pots

Pathway of Residential Garden

In Floral Arrangement

In Flower Vase

In Indoor Decoration

Near Water Body

In Beds

and Turkey. A secondary gene centre of the genus *Tulipa* L. has been found in the Caucasus. But now Holland is regarded as home of the tulip (Debrowski, 1964). Tulip flower resembles the turbans which are worn in the Middle East and hence the name "turban," in Latin, becomes "tulipa."

Tulips are associated with the Dutch because a famous Dutch gardener named Carolus Clusius, who was head gardener at the University of Leiden in Holland in 16th century, first planted tulips which has today become a land well-known for its fields of tulips and daffodils. The largest producer of tulip bulbs and flowers is Holland. Tulips were introduced from Turkey into Europe. Around 1630, tulips were extremely popular in The Netherlands. The highest price recorded for one tulip bulb in 1637 was fl 5200 - (€ 2360). This turned out to be about four times the yearly salary of a middle-sized businessman (Dash, 1999). The time period between 1634 and 1637, commonly known as "Tulipmania," is often compared to the Stock Market surge of the 1920s. Tulips, whose original species were not determined, are grouped together and are called *T. gesneriana.* The current commercial assortment mainly consists of cultivars from *T. gesneriana*. Darwin hybrids, the second group of tulip cultivars is obtained from the interspecific crosses between cultivars of *T. gesneriana* and genotypes of *T. fosteriana* Hoog ex W. Irving.

About 85 per cent of tulip bulb production in the world is grown in Netherlands (Le Nard and De Hertogh, 1993). Another major area of tulip production is in France. Annual turnover for bulb production and cut flower production was about 600 million Dutch guilders and 274 million Dutch guilders respectively in 1995 (Anon., 1996). More than 70 per cent of the flower bulbs and cut flowers produced in The Netherlands are exported.

The planted acreage of tulips for the season 2001/2002 was about 10.700 ha in The Netherlands (Anon., 2002). Van Scheepen, 1996 reported that *Tulipa gesneriana* and Darwin hybrids consist of more than 1100 cultivars. The 10 most popular cultivars, however, occupy more than 35 per cent of the planted acreage. Only 643 ha (7 per cent) of the total tulip area consist of other species, of which *T. fosteriana*, *T. greigii* Regel and *T. kaufmanniana* Regel are the primary species.

4.0 Morphology

The tulip is a monocotyledonous plant belonging to the family *Liliaceae*. The species are perennials from bulbs. Tunicate bulbs are often produced on the ends of stolons and are covered with hairless to variously hairy papery coverings. The species comprises of short low-growing plants to tall upright plants (10-70 cm tall). Plants in general have 2 – 6 leaves, with some species having up to 12 leaves. Foliage is strap-shaped, waxy-coated, alternately arranged and usually light to medium green in colour. Leaf blades to some extent are fleshy and are linear to oblong in shape. The large flowers are produced on scapes or sub scapose stems normally lacking bracts. In general, tulip species have one flower per stem, but a few species have up to four flowers. The flower of tulips is shaped like a cup or egg. Tulip flowers usually are self coloured or a combination of two or more colours and some are striped and marked with contrasting colours. Flower colour in tulips range from white to black including various other hues like red, scarlet, crimson, terracotta, orange, pink, purple, violet, chocolate, brown, cherry, magenta, salmon, carmine, rose, cream, yellow, apricot, lilac, mauve, blue *etc*. The flowers are colourful and attractive and have three petals and three sepals, which are most often, termed tepals because they are nearly identical. Six petaloid tepals are marked near the bases with darker markings. The flowers have six basifixed, distinct stamens with filaments shorter than the tepals and the stigmas are distinctly 3-lobed. The ovaries are superior with three chambers. The 3 angled fruits are ellipsoid to sub-globose in shape, leathery textured capsules which contains numerous flat disc-shaped seeds in two rows per locule. Seeds are flat, light to dark brown in colour and are arranged in two rows per chamber and have very thin seed coats and endosperm that does not normally fill the entire seed coat.

5.0 Species, Classification and Cultivars

The number of species ranges from about 45 (Stork, 1984) to more than 100-150 (Hall, 1940, Botschantzeva, 1962) in *Tulipa*. A few species like *Tulipa stellata* and *T. aitchisonii* are natives of Himalayan region. *T. stellata* is commonly known as Star Tulip. *Tulipa species* are distributed in two sub-genus, Eriostemones and Leiostemones, which up to recently could not be inter-crossed. The genus probably separated at an early stage of the evolution. The Eriostemones, which includes about 30 species, generally are low growing, having narrow leaves and possess hairs at the inner base of the perianth parts. Garden tulips of the present day belong to the Leiostemones and are originated from types introduced into Europe from Turkey after the middle of the 16th century. These tulips show great diversity in flowering, growth, and vigor and flower shape. These tulips, whose original species have been not been determined, have been placed in *Tulipa gesneriana*.

The genus *Tulipa* is divided into two subgenera: *Tulipa* and *Eriostemones* (Boissier) according to the taxonomic classification by Van Raamsdonk and De Vries (1992, 1995). These sub-genera are classified into eight sections. The sub-genus *Tulipa* and Eriostemones consist respectively of five and three sections.

5.1.0 Taxonomic Classification of Species of the Genus *Tulipa* in the Sections of Two Sub Genera, *Tulipa* and *Eriostemones*, According to Van Raamsdonk and De Vries (1992, 1995)

A. Subgenus *Tulipa*

Tulipa

T. gesneriana L.

T. hungarica Borbas

T. didieri Jord.

T. armena Boiss.

T. suaveolens Roth

Eichleres (Hall) Van Raamsdonk

T. ingens Hoog

T. tubergeniana Hoog

T. fosteriana Hoog ex W. Irving

T. lanata Regel

T. eichleri Regel

T. greigii Regel

T. albertii Regel
T. praestans Hoog
T. tschimganica Bochantzeva
T. subpraestans Vvedensky
T. sosnovskyi Akhverdov et Mirzojeva
T. kaufmanniana Regel
T. dubia Vvedensky

Tulipanum de Reboul

T. agenensis DC.
T. kuschkensis B. Fedtschenko
T. aleppensis Boiss. ex Regel
T. systola Stapf
T. julia C. Koch
T. praecox Tenore

Kolpakowskianae (Hall) Van Raamsdonk

T. altaica Pall. Ex Sprengel
T. tetraphylla Regel
T. lehmanniana Mercklin

Clusianae Baker

T. clusiana DC.
T. linifolia Regel
T. montana Lindley

B. Subgenus *Eriostemones* (Boissier) Van Raamsdonk

Australes sensu Hall

T. australis Link
T. biebersteiniana Schultes
T. whittalii (Dykes) A.D. Hall
T. hageri Heldr.
T. primulina Baker
T. sylvestris L.
T. ophanidea Boiss. Ex Heldr.

Saxatiles sensu Hall

T. humilis Herb.
T. saxatilis Sieb. ex Sprengel
T. aucheriana Baker
T. pulchella Fenzl.
T. bakeri A.D. Hall

Biflores sensu Hall

T. turkestanica Regel
T. biflora Pallas
T. neustrueva Pob.
T. dasystemon Regel
T. polychroma Stapf
T. sogdiana Bunge
T. tarda Stapf

5.1.1 Classification Groups of Tulips

In accordance with the international nomenclature, Tulips are divided into 15 classification groups (Hoog, 1973). The characteristics and decorative capacities of 12 classification groups were reported by Juodkaite *et al*. (2005) and are described below.

Group 1 Single Early Tulips (Early flowering)

This is genetically an old group of cultivars. At the end of the 17th century it was formed of Djuk van Tol's group of taller seedlings. Most cultivars of this group are average in height (30-50 cm), flower in the end of April-first half of May.

Group 2 Double Early Tulips (Early flowering)

The first complete double tulips did not develop until 1665 (Tijmen, 2019). *Tulipa lutea centifolia* – "the double yellow monster" was the first double early tulip and it refers to flowers with more than 200 petals. The cultivars of this group burst into flower in the first half of May, are not high (20–30 cm), with firm stems bearing a well upraised double flower till the end of blossom. 'Monsella' and 'Peach Blossom' belongs to this group. Blossoms of the Double tulip are nicknamed as Peony Tulips since they have so many petals. They don't have a large selection of colours like other types but are very long lasting. When completely open, the blossoms can be as wide as 4 inches in diameter.

Group 3 Triumph Tulips (Mid-season flowering)

The group was formed in 1915 of seedlings obtained by cross breeding the Single Early and Darwin cultivars. Cultivars of this group are tall (50–60 cm) with a wide array of colours and the majority have cup-shaped flowers, petals resist to unfold even in the sun. The Triumph tulips/mid-season tulips are slightly smaller with their single cup-shaped flowers up to 2.5 inches (6 cm) wide. They make good cut flowers and work well for forcing to bloom indoors.

Group 4 Darwin Hybrid Tulips (Mid-season flowering)

The first cultivars in this group were obtained by crossing *T. fosteriana* 'Madame Lefeber' with a Darwin cultivar by D.W. Lefeber in 1936. Darwin hybrid tulips burst into flower somewhat earlier than Triumph tulips. They are high (60–80 cm), with firm and strong stems, big leaves, cup-shaped flowers. Blossoms have perfect pyramid shape when closed, and measure 6 inches (15 cm) in diameter when fully open. Darwin Hybrids are often considered the very best type of tulip to raise for cut flowers because of their long stems, lovely pyramid shape, and brilliant colours.

Group 5 Single Late Tulips (Late-flowering)

'Single' classification comes from the unique cup-shaped six-petal flower. Obtained at the end of the 19th century, the Cottage Tulip group is called Single late now. It was often called "cottage tulips" because of their perfect height to plant along traditional white cottage fencing throughout the U.S. and Europe. Cultivars of this group are high tulips (60–80 cm), various in colour (from white to purple blackish); they flower rather late (in the second half of May). 'Rosy Wings' is the first in the group to unfold. One of the darker and bursting into flower later than others and flowering rather long is 'Black Swan'.

Group 6 Lily-Flowering Tulips (Late-flowering)

The parentage of the first Lily-flowering tulips was *T. retroflexa* crossed with the Cottage tulip. Flower petals are sharp and inflected. Late flowering Lily tulips bears slender flowers often sporting pointed, recurving petals which are reflexed and resemble lily flower or urn. The open flower looks like a six-pointed star from above. The colour array is wide from white to purple. Their 40–70 cm high stems are rather slender. Lily-flowered tulips flower in the second half of May. This is the tulip of choice in Turkey.

Group 7 Fringed Tulips (Late-flowering)

In 1981, when amendments to the international classification of tulips were made, fringed tulips which earlier belonged to the Parrot, then to Darwin and Single late groups, were singled out into a separate group. The late flowering fringed tulips have petals topped with fringes which resembles the frayed edge of a piece of satin. Although they tend to be shorter tulips, the blooms are quite long-lasting. Flowers of these tulips are most often of regular cup-shape or oval. Colours of most of them are pastel, rich in undertones. Tulips grow up to 50–70 cm high and flower in the second half of May. 'Sundew' tulips have longest lasting flowers.

Group 8 Viridiflora Tulips (Late-flowering)

This is a new group singled out into a separate one only in 1981. It was obtained by cross breeding Viridiflora (*Tulipa viridiflora*) with other cultivars. Viridiflora is derived from two Latin words, 'viridis' meaning green and 'flos' meaning flower. Cultivars of this group are very impressive tulips, the back of their petals is green, and the rimming varies in undertones (white, yellowish, red). This contrasts with the basic flower colour (white, pink, gold, *etc*.). These tulips have a green streak on each petal. While flowering, the green colour disgorges into the main colour creating various tones. Plants are 30–70 cm tall; stems are strong and perfectly survive nasty weather. Theses tulips flower in the middle of May and are known for their exceptionally long-flowering capability.

Group 9 Rembrandt Tulips (Late-flowering)

Tulips of this group originate from Darwin or Cottage groups of cultivars. Flowers of Rembrandt tulips are stripped and late flowering. Multi-flowered tulips of this group are on the verge of extinction in the world.

Group 10 Parrot Tulips (Late-flowering)

Parrot tulips were started cultivating as back as the 17th century. Parrot tulips are flamboyant tulips which resemble the colourful feathers of the tropical bird of the same name and have curly, chiselled twisted, and fringed petals. Nearly all varieties of parrot tulips are vibrantly mutli-coloured. Flowers are very big, hence their supple stems cannot bear them and after wind and rain they incline. They grow up to 40–70 cm high. Mostly they flower late – in the second half of May.

Group 11 Double Late Tulips (Late-flowering)

The first Double late tulips were described at the end of the 17th century. Their flowers are big, double, similar to those of peonies. The big flowers are sensitive to rain, after it their stems cannot bear flowers, stems of higher cultivars break or become sinuous. Double late tulips grow to 60 cm in height and flower in the second half of May.

Group 12 Greigii Tulips

Greigii tulips are noted for their brightly-coloured flowers and purple striped or mottled foliage. Usually it is having marbled or striped foliage, flowering later than those in the Kaufmanniana group. Leaves usually spread out and bend down toward the ground. These have large brightly coloured flowers. Plant height varies from 8 to 12 inches. Because of their short stature, Greigii tulips are excellent choices for borders or rock gardens. They also perennialize well.

Group 13 Kaufmanniana Tulips

This group of tulips are long-lived perennial tulips. In sunlight, the flowers open fully. The open flowers resemble a star or waterlily. Flower colors include white, yellow, pink, and intermediary colors. The foliage is bluish green or chocolate brown striped. Kaufmanniana tulips are small plants. Average plant height is 4 to 8 inches. Their compact size makes them good choices for border edges and rock gardens.

Group 14 Fosteriana Tulips

This tulips produce some of the largest flowers of the genus. They also perennialize well. 'Red Emperor' is a widely grown variety in this class. Fosteriana tulips are sometimes referred to as emperor tulips.

Group 15 Multi Flowering Tulips

The tulips belonging to this group are bouquet-flowering and produce at least four full-sized flowers per stem. Technically, it is not a class of tulips, but part of the single late group, they make for a full garden display as well as lush cut flower arrangements.

5.1.2 Cultivars

The tulip is an important bulbous flower crop having more than 3000 varieties. There are many improved varieties under various classes of garden tulips that are in cultivation that differ in their flowering time, flower shape, size and colour. Garden tulips are again classified as early flowering tulips, mid-season tulips and late-flowering tulips.

Some of the cultivars under early flowering tulips are Duc van Tol, Single Early and Double Early. The mid season types include cultivars such as Mendal and Triumph. The late flowering tulips include cultivars like Darwin, Darwin hybrids, Breeders, Lily flowered, Cottage, Rembrandt, Bijbloemen, Parrots, Double Late and Species tulips and their hybrids. Some of the notably introduced Tulip varieties in India are Avignon, Adrem, Apeldoorn elite, Bright Parrot, Laptop, La Courtine, Golden Parade, Purissima, Parade, Parade(sr), Van Der Hoef, Niigata, Red Impression, Carlton, Candela, Spring-Green, Strong gold, Ile de France, Exotica sur, Fabio *etc*. Kashmiri growers mostly procure tulip bulbs from Holland. Nearly 1.2 million tulip bulbs in 68 varieties was imported to Kashmir since 2008.

A classification of Tulip varieties for Field Production grouped by flowering period and tulip varieties for Greenhouse Production grouped by Flower Type was reported by Stanton *et al*. (2006), and is given below.

5.1.3 Tulip Varieties for Field Production

Tulip varieties for field production grouped by Flowering Period are mentioned below along with the flower colour and natural flowering season.

Early Season Varieties

Pink Impression: Pink, early

Red Impression: Red, early

Van Eijk: Pink, early

Othello: Purple, early

Bastogne: Red, early/mid

Early/Mid-Season Varieties

Bastogne Parrot: Red parrot, early/mid

Oxford: Red, early/mid

Oxford Elite: Orange/yellow, early/mid

Mid-Season Varieties

Apricot Parrot: Apricot parrot, mid

Ben van Zanten: Red, mid

Blushing Beauty: Yellow, red blush, mid

Don Quichotte: Pink, mid

Golden Oxford: Yellow, mid

Gwen: White, mid

Jan van Nes: Yellow, mid

Laura Fyji: Red/yellow, mid

Leen van der Mark: Red/white, mid

Lucky Parrot: Red/white parrot, mid

Negrita Favourite: Purple, mid

Purple Lady: Purple, mid

Mid/Late Season Varieties

Barcelona: Pink, mid/late

Strong Gold: Yellow, mid/late

Synaeda Blue: Lavender/white, mid/late

Late Season Varieties

Avignon: Orange, late

Big Smile: Yellow, late

Flaming Parrot: Yellow/red, late

Grand Style: Red, late

Ile de France: Red, late

Lady Margot: Yellow, late

Maureen: White, late

Menton: Salmon/pink, late

Mrs. JT Scheepers: Yellow, late

Renown: Pink, late

Texas Flame: Yellow/red, late

Texas Gold: Yellow, late

Queen of Night: Black, late

Clearwater: Pure white, very late

5.1.4 Tulip Varieties for Greenhouse Production

Tulip varieties for Greenhouse production grouped by Flower Type are mentioned below along with the flower colour.

Double Early Tulips

Abba: red

Holland Ballet: white

Largo: red

Margarita: lavender

Monte Carlo: yellow

Mondial: white

Double Late Tulips

Abigail: dbl Attila

Angelique: soft pink

Don Camillo: dbl Don Quichotte

Mount Tacoma: white

Upstar: pink

Fringed Tulips

Arma: red

Curly Sue: purple

Louvre: pink

Madison Garden: deep pink

Multi-Flowering Tulips

Cloud Nine: pink

H.D. Genscher: crème

Silhouette Bouquet: red

Parrot Tulips

Apricot Parrot: apricot

Erna Lindgreen: red

Flaming Parrot: red/yellow flame

Prof. Rontgen: orange

Rococo: red

Salmon Parrot: salmon

Texas Flame: yellow/red flame

Texas Gold: yellow

Webers Parrot: pink/white base

Regular Cut Tulips

Ad Rem: orange/yellow edge

Attila's Record: lavender

Barcelona: pink

Baronesse: pink

Bastogne: red

Ben van Zanten: red

Bright Pink Lady: soft pink

Café Noir: dark purple

Carola: pink

Cassini: red

Charmeur: red/white

Cheers: crème

Cherida: crème yellow

Clearwater: white

Deshima: red

Don Quichotte: pink

Dynasty: pink/white base

Dreaming Maid: lavender/white

Early Glory: pink

Friso: red

Gabriella: soft pink

Gander: pink

Gander's Overture: crème/pink flame

Gander's Rhapsody: pink/white base

Golden Brigitta: yellow

Happy Generation: white/red flame

Holland Beauty: pink

Ile de France: red

Jack Pot: dark purple/white

Jan Reus: burgundy red

Jan van Nes: yellow

Kais. Maria Theresia: soft pink

Kees Nelis: red/yellow

La Nouba: white

Leen vd Mark: red/white

Leo Visser: red/white

Lucky Strike: red/white

Negrita: purple

Negrita's Favourite: purple

Niigata: pink

Orange Cassini: orange

Othello: purple

Peerless Pink: pink

Pink Impression: pink

Prima Vista: red

Primavera: pinkish-red

Purple Flag: purple

Purple Prince: purple

Red Impression: red

Rosalie: pink

Royal Virgin: white

Sevilla: red

Shirley: white/purple edge

Species of Tulip

Tulipa agenensis

Tulipa armena

Tulipa clusiana

Tulipa didieri

Tulipa dubia

Tulipa eichleri

Tulipa gesneriana

Tulipa greggii

Tulipa humilis

Tulipa hungarica

Tulipa ingens

Tulipa kaufmanniana

Species of Tulip

Tulipa kolpakowskiana

Tulipa lanata

Tulipa linifolia

Tulipa pulchella

Tulipa stellata

Tulipa suaveolens

Cultivars of Tulip

Apricot Parrot

Avignon

Barcelona

Bastogne

Ben van Zanten

Big Smile

Cultivars of Tulip

Blushing Beauty

Clear Water

Don Quichotte

Lady Margot

Laura Fyji

Maureen

Negrita Favourite

Oxford Elite

Queen of Night

Synaeda Blue

Texas Flame

Pink Impression

Cultivars of Tulip

Renown

Texas Gold

Van Eric

Groups of Tulip

Double Early Tulip

Abba

Holland Ballet

Largo

Margarita

Monte Carlo

Mondial

Groups of Tulip

Fringed Tulips

Arma

Curly Soe

Louvre

Parrot Tulips

Apricot Parrot

Erna Lindgreen

Flaming Parrot

Double Late Tulips

Abigail

Angelique

Mount Tacoma

Groups of Tulip

Multi-flowering Tulips

Cloud Nine

H.D. Genscher

Silhouette Bouquet

Regular Cut Tulips

Carola

Dreaming Maid

Kees Nelis

Leo Visser

Royal Virgin

World's Favourite

Silver Dollar: white

Strong Gold: yellow

Swinging World: pink

Synaeda Show: white

The Mounties watermelon: red

Washington: yellow/red flame

White Dream: white

White Flight: white

World's Favourite: orange

6.0 Genetics and Breeding

6.1.0 Genetics

The number of chromosomes of many *Tulipa species* and cultivars have been studied (Zeilinga and Schouten, 1968). Tulip assortment mainly comprises of diploids (2 sets of 12 chromosomes; 2n=2x=24), some triploids (mainly Darwin hybrids) and a rare tetraploid (2n=4x=48). Efforts have been made by the breeders to improve the colour, shape, and size of the flower apart from general aspect of the plant like leaves, height, growth type *etc*. With the development of improved techniques for cut flowers and pot plants forcing the physiological characteristics of the cultivar have become very important. Thus, breeding for forcing ability and suitable post greenhouse characteristics is one of the major goal for tulips.

6.1.1 Pre-selection

Because of the long juvenile period and the slow propagation rate of tulips much attention was initially paid to pre-selection at Centre for Plant Breeding and Reproduction Research, Netherlands. Pre-selection is a method of selection at the juvenile stage from small non flowering seedlings. Main goals of pre-selection include bulb production, forcing for cut flower production and disease resistance. Breeders can save money with an appropriate pre-selection system, because of less labour input, and reduction of field and greenhouse space.

6.1.1.1 Pre-selection Based on Bulb Production

Bulb production can be assessed as the yearly increase of the main bulb diameter and the number of daughter bulbs. Research on pre-selection of seedlings with a high bulb production resulted in the following selection procedure at Centre for Plant Breeding and Reproduction Research, Netherlands. Only the main bulbs from each population are cultivated in the first three years. By separating the main bulbs during those years into bulb diameter classes (or weight classes) and by replanting only the larger bulbs, the best growing bulbs within a population are selected. Only plants which produced two or more daughter bulbs are selected in the fourth year. Thus, only seedlings which grow very well and give enough daughter bulbs are selected. This selection procedure is useful for populations with a minimum of 1500 seedlings since many seedlings can be discarded at an early stage. Another selection procedure which gives comparable results involves looking at the number of years from seed to flowering plant. It is observed that the first seedlings to flower in a population grow better and give more daughter bulbs than seedlings which flower for the first time one or two years later. This procedure of selection is useful only for small populations as all seedlings have to be grown until the first year of flowering. Pre-selection based on bulb production does not interfere with selection for early forcing.

6.1.1.2 Pre-selection Based on Forcing

A cultivar possessing a good forcing ability makes it possible to produce, usually under greenhouse conditions, a maximum number of high quality flowers per square unit and in the shortest period of time. Forcing for cut flower production is considered as the most important quality for tulip breeders.

Forcing ability has at least three major components (Nayeem and Qayoom, 2015) and are described below:

1. The ability of the cultivar to be forced either very early or very late in the market season.
2. The quality of the flower, which must have ornamental value and meet commercial requirements vary with the country and/or the market. For example, in Europe generally requires very long stemmed tulips in contrast to the Northern European countries.
3. The number of tulips flowers per square unit. It depends on planting density and flowering percentage, which should be as close as possible to 100. The planting density is partly determined by the leaf habit. Thus, planting density can be higher with erect leaves. After flowers are produced, it should have a good vase-life (>5 days) after having been subjected to cold storage, transport, and delivering conditions. Potted tulips must also have a long life.

Nearly 50 per cent of the bulb growing acreage in the Netherlands is planted with 20 cultivars which are all suitable for forcing. Furthermore, ten most important cut flower cultivars account for approximately 50 per cent of the tulip cut flower production in the Netherlands. Research on pre-selection at Centre for Plant Breeding and Reproduction Research, Netherlands showed that the juvenile seedlings which sprout and die early in comparison with the other juvenile seedlings can be forced as the earliest flowering ones at the adult bulb stage. This information is hardly used by the hybridizers in breeding programmes. Besides earliness, other characteristics like stem length, firmness of leaves and stem, leaf position and leaf number, flower size, bud presentation and the proportion between flower and leaves are also important. Consequently, in practical breeding, seedling populations consisting of bulbs large enough to initiate a flower are all forced and judged for the characteristics mentioned above. 'Debutante' and 'Silver Dollar' are the best known results at present which also have a long vase life.

6.1.1.3 Shortening of the Juvenile Period

Besides pre-selection, another approach to speed up tulip breeding is shortening of the juvenile period (Straathof and Eikelboom., 1997). The growth cycle could be reduced from one year to eight months by cultivation in greenhouses and storage in climate rooms with standardised conditions. Three growth cycles could be obtained in two years. In spite of additional lightning, seedling bulbs from the greenhouse with three cycles of cultivation were not larger than the bulbs from outside with only two cycles of cultivation. By applying optimal field and storage conditions, the juvenile period can be shortened from 4-5 years to 3-4 years.

6.1.1.4 Inheritance of Flower Colour

Flower colour in Tulips is determined by the composition and proportion of different flower colour pigments (Straathof and Eikelboom). Yellow flowers contain carotenoids, while orange flowers contain both carotenoids and cyanidin. Red flowers contain cyanidin and pelargonidin, whereas, purple flowers contain a combination of cyanidin and delphinidin. Pink flowers contain the greatest variation in flower colour pigments but the level is usually smaller compared to the other colours. White flowers contain almost no colour pigments. Apart from single-coloured tulips, there are also multi-coloured tulips *e.g.* pink-red and purple-red. Flowers with nearly the same colour differed in the flower pigment composition and flowers with identical flower pigment composition differed in flower colour. Furthermore, in several seedlings flower pigments were found which were not present in either of the parents. It is not possible to predict the flower colour of the progeny on the basis of the flower pigment composition of the parents. For example, a yellow x yellow cross can result in some red seedlings.

6.1.1.5 Inheritance of Flower Shape

Crosses were made between lily-flowered, double-flowered, edge-coloured, single-coloured and base-coloured cultivars to study the inheritance of flower shape and other flower characteristics (Straathof and Eikelboom). As with the inheritance of flower colours, prediction of the inheritance of flower shapes in relation to that of their parents is difficult. Lily-flowered cultivar 'Dyanito' combined with a non lily-flowered cultivar gave almost no pure lily-flowered seedlings. Populations with *Tulipa accuminata* as parent gave many beautiful lily-flowered seedlings. 'Novired' and 'Talbion' were released from this programme by PRI. Double-flowered tulip 'Abodement' gave hardly any pure double-flowered seedlings. Crosses between edged-coloured cultivars gave high percentages of edged-coloured seedlings and when combined with non-edged cultivars the offspring resulted in approximately 50 per cent of edged-coloured seedlings.

Parrot tulips are popular for their bizarre flower shape. Previously, this shape was obtained only by spontaneous mutation of cultivars. From an inbreeding programme at PRI a "genetic" parrot was found in a selfing of the cultivar 'Cordell Hull'. This cultivar, named 'Amethyst', and other "genetically" based parrots were released to the breeders in 1978.

6.2.0 Breeding

Morphological variations among and within species of Wild Tulip (*Tulipa* L.) from Iran, one of the most diverse centers of wild tulip were studied (Khaleghi *et al.*, 2018). Two hundred and eight accessions were gathered from nine natural regions of Markazi province and two areas of Isfahan province. They were categorized into six species including *Tulipa systola* Stapf, *T. stapfii* Turill, *T. humilis* Herbert, *T. polychroma* Stapf, *T. biebersteiniana* Schultes et Schultes f., *T. montana* var. *montana* Lindley, and *T. montana* var. *chrysantha* (Boiss.) Wendelbo according to the morphological traits and diagnostic description of the species. The species were evaluated using 44 morphological characters. Most of the traits such as stem length, stem diameter, leaf dimensions, flower length, tepal length and width, diameter and length of the bulb showed a high degree of intra-species variabilities. In addition, considerable inter-species variability was found for blotch edge and shape of the inner tepal, leaf margin color, hair present in the middle part of the tunic, anther color, and stigma color. Seven significant components were obtained that explained 78.35 per cent of the total variation among the species based on the principal component analysis. Cluster analysis showed great variability among the accessions of each species and indicated that the studied plants can be a rich genetic source for breeding programs. Few accessions had unique floral scent, flower colour and size, stem length and bulb size which can be used in breeding programs.

The genetic variability, heritability, genetic advance and correlation in twenty one genotypes of tulip (*Tulipa gesneriana* L.) were estimated (Bhatia *et al.*, 2013). The study revealed wide variation for characters indicating sufficient genetic variability to be exploited in breeding programme. Blushing Apeldoorn was earliest in flowering and Golden melody took maximum days for flowering. The highest scape length and flower size was recorded in Ganders Rhapsody and Monte Carlo, respectively. Maximum number of bulbs and bulblets was produced by Golden Melody. Promising genotypes for cut flower production were Ali Bi, Blushing Apeldoorn, Ganders Rhapsody, Pretty Women, Strong Gold and Tulip Hb. The maximum value of PCV and GCV was recorded for number of bulbs per plant. High heritability coupled with high genetic advance was noticed for spike length, plant height, and leaf area. Selection based on these characters would be more effective for tulip improvement. Scape length revealed a highly significant and positive correlation with plant height and bulb size. Highly significant and positive correlation of number of bulbs was noticed with number of leaves per plant and leaf area. Therefore, direct selection from germplasm lines is effective for improvement in closely associated traits.

Van Der Meulen-Muisers (1997) studied the segregation of postharvest quality characters in tulip. A large variation in longevity of tulip flowers was found within populations tested at individual plant level. Number of flowers per stem, male sterility and stem elongation also varied within the populations tested. Due to the high correlations between the offspring and their corresponding parental genotypes for numerous characters, breeding for a considerable improvement of postharvest quality in future cultivars seems promising.

6.2.1 Developing F_1 Hybrids

Custers *et al.* (1997) stated advances in embryogenesis from tulip microspores in culture. High temperature pre-treatment (32°C) of bulbs had a positive effect on the production of microspore embryos. Pre-treatment reduced the time for embryo formation and stimulated the progression from the multicellular microspore stage into the embryo proper. Early unicellular microspores were observed to be better starting material for culture than mid unicellular microspores. Microspore embryogenesis technique was applicable to various tulip cultivars. Improved protocol now allows the production of large numbers of microspore-derived plants, and further evaluation of doubled haploid plant technology for tulip.

Barba-Gonzalez *et al.* (2006) reported that for overcoming F_1 sterility in inter-specific hybrids, mitotic and meiotic polyploidisation is applied and can result in fertile allopolyploids. Mitotic polyploids are obtained by artificial chromosome doubling and results in normal pairing of homologous chromosome set, which enables the formation of 2x gametes during the meiosis. Meiotic polyploidisation leads in rare cases of restitutional chromosome division, to the formation of unreduced gametes (2n gametes). The use of 2n gametes is therefore the most promising approach for the introgression of desirable characters in using interspecific hybrids in breeding. In both cases, the frequency of viable gametes appeared to be low and limited to a few F_1 hybrids. A new and promising method is developed in which 2n gametes can be induced by the application of laughing gas (N_2O). In tulip where the production of mitotic polyploids requires at least five years, it will speed up breeding with interspecific hybrids enormously.

6.2.2 Inter-specific Hybridization

A great deal of attention has always been paid to hybridization between different *Tulipa species* (interspecific hybridization) in combination with the taxonomy of the genus Tulipa. Several characteristics like extreme earliness and resistance against tulip breaking virus are almost not available in the *T. gesneriana* assortment. Hence, *Tulipa species* can be used for introducing these characteristics. In addition, many *Tulipa species* or interspecific hybrids can be used to obtain a larger garden tulip assortment. PRI has the largest collection of *Tulipa species* in the world. PRI released several interspecific hybrids exhibiting characteristics like earliness and resistance against tulip breaking virus and *Fusarium*. 'Beau Monde', 'Come Back', 'Explosion', 'Lefeber's Memory', 'Pink Impression, 'Purple World' and 'Spryng' are some of the well known cultivars.

Many cultivars of *T. gesneriana* were pollinated with most of the species in PRI collection. In this way a thorough idea was obtained of which species can be used for interspecific hybridization. Crosses between *T. gesneriana* and species from the Eriostemones, like *T. tarda*, *T. pulchella* and *T. turkestanica* were not successful at all. Crosses are possible within the sub-genus Tulipa, like the hybrids obtained between *T. gesneriana* and *T. fosteriana* (Darwin hybrids), *T. kaufmanniana*, *T. greigii*, *T. eichleri*, *T. ingens*, *T. albertii* (formerly *T. vvedenskyi*) and *T. didieri*. New *in vitro* techniques have been developed recently, by which interspecific hybrids can be obtained which were not possible by *in vivo* hybridization.

Xing *et al.* (2020) performed inter-specific hybridization to explore the cross-compatibility between tulip cultivars and wild species. It was done between five tulip cultivars *viz.*, 'Judith Leyster,' 'World's Favorite,' 'Heart of Poland,' 'Crispa Fabio' and 'Jumbo Pink,' and five wild species *viz.*, *Tulipa thianschanica* Regel, *T. sinkiangensis* Z. M. Mao, *T. altaica* Pall. ex Spreng, *T. buhseana* Boiss. and *T. schrenkii* Regel. Pollen germinability, size and storage life, stigma receptivity, fruit-setting, hybrid seed germination, bulblet formation, chromosome count and karyotype of hybrids were examined. *T. altaica* and five cultivars had closer genetic relationship, with the largest pollen and the highest germination rate (65.23 per cent) compared with other four wild species. Pollen viability of all five pollen parents decreased slowly across all stages at -20°C and decreased rapidly at 25°C and 4°C. Almost 1066 hybrids were obtained from interspecific hybridization, including 'Heart of Poland' × *T. thianschanica*, 'Heart of Poland' × *T. altaica*, 'Crispa Fabio' × *T. altaica*, 'Jumbo Pink' × *T. altaica*, 'Crispa Fabio' × *T. schrenkii* and 'Jumbo Pink' × *T. schrenkii*. 20 per cent and 12 per cent aneuploidy hybrids were produced from 'Heart of Poland' (2n=3x=36) × *T. thianschanica* (2n=2x=24) and 'Heart of Poland' (2n=3x=36) × *T. altaica* (2n=2x=24), respectively, and the any other hybrids were diploid (2n=24) in these two crosses. All hybrids from crosses of diploid parents were diploid (2n=24). The study also elaborated karyotype characteristics of six parents and their six progenies, which had similar 3A and 3B karyotypes. These results may facilitate tulip interspecific hybridization in the future.

Marasek-Ciolakowska *et al.* (2009) reported that inter-specific hybridization is an important tool in tulip breeding to transfer some important horticultural traits into new cultivars. Diploid (2n=2x=24) F_1 interspecific hybrids between *T. gesneriana* × *T. fosteriana* (GF) were backcrossed to diploid *T. gesneriana* cultivar 'Yellow Flight' (G). BC_1 plants were analysed by genomic *in situ* hybridization (GISH) to estimate their genome constitution and extent and position of intergenomic recombination. All the BC_1 progenies analysed were diploid (2n=2x=24). GISH enabled clearly the identification of parental chromosomes as well as the recombinant chromosomes. Since *T. gesneriana* parent was used for back crossing, number of G genome chromosomes (chromosomes with centormere of *T. gesneriana* genome) predominated in the BC_1 progenies. Total number of *T. fosteriana* chromosomes in hybrids ranged from four to ten. There were two types of recombinant chromosomes. Those with a centromere of *T. fosteriana* chromosome with recombinant segment of *T. gesneriana* (F/G) and vice versa (G/F). Most recombinant chromosomes contained a single crossover. The percentages of *T. fosteriana* chromatin in BC_1 hybrids ranged from 20 to 24 per cent.

Marasek-Ciolakowska (2011) reported that hybridization between *Tulipa gesneriana* cultivars (G, 2n=2x=24) and Darwin hybrids (*T. gesneriana* × *T. fosteriana*; 2n=2x=24) was made with the aim of transferring desirable horticultural traits from *Tulipa fosteriana* (F) to *T. gesneriana*. Genome composition and extent of intergenomic recombination in a number of diploid progenies (2n=2x=24) was evaluated by genomic in situ hybridization technique (GISH) using differently labelled total genomic DNAs of 'Princeps' (*T. fosteriana*) and 'Ile de France' (*T. gesneriana*). In all the progenies analysed, it was possible to distinguish chromosomes from both parental genomes as well as the recombinant chromosomes. Number of *T. fosteriana* chromosomes as well as the number and type of recombinant chromosomes differed among progenies. Total number of intact *T. fosteriana* chromosomes in hybrids ranged from zero to three whereas maximum number of recombinant *T. fosteriana* chromosomes was five. Among recombinant chromosomes, most contained a combination of a single *T. gesneriana* and a single *T. fosteriana* fragment. The maximum number of recombinant segments per chromosome was two and their position ranged from distal to highly interstitial. Darwin hybrids were very useful intermediate parents for introgressing the *T. fosteriana* germplasm into the *T. gesneriana* assortment.

6.2.3 Mutation Breeding

New cultivars of Tulips can be obtained by mutation breeding as well as by hybridization. Van Scheepen (1996) reported that Tulip being a vegetatively propagated crop, during cultivation hundreds of natural mutants (sports) was selected. Spontaneous mutations are found regularly in tulips. Mutations can exhibit variation in flower colour or flower shape (parrots, fringed and double). Differences exist in the mutation sensitivity of cultivars. Many mutants are known from specific cultivars, *e.g.* 'Bartigon', 'William Copland', 'Murillo' and 'Apeldoorn'. In order to produce mutations artificially, the possibilities of Röntgen (X-rays) for mutation breeding in tulips have been investigated (Broertjes and Alkema, 1970, Van Harten and Broertjes, 1989). Both mother bulbs and daughter bulbs can be used for the induction of mutation. The dosage requirement varies from 350 to 550 rad. Triploids and tetraploids need a higher dose than diploids. Radiation can be applied early (August) or late (November) in the planting season. Plants should be cultivated three to four years, before selection for mutations can begin. Differences between cultivars can be found in the number of mutations obtained.

In addition to mutations in flower colour and flower shape, mutations also occur in the colour of the leaf edge, in plant height and in bulb production. PRI released several radiation mutants of 'Preludium' and 'Lustige Witwe' during 1970's. One of the mutants was 'Santina', a leaf edge mutant of 'Lustige Witwe'. Many private breeders are still using the mutation breeding techniques (Straathof and Eikelboom, 1997).

6.2.4 Ploidization Techniques

6.2.4.1 Tetraploids

Tetraploid cultivars are more robust and firm than diploids. Hence, research has been carried out to obtain more tetraploids. By making crosses between tetraploids new tetraploids were obtained. Several tetraploids were released by PRI in 1974 of which 'Judith Leyster' is well known.

A problem with the existing tetraploids was that their progeny flowered late in the season, which made them unsuitable for early forcing. Consequently, other techniques for obtaining tetraploids were used. Zeilinga and Schouten (1968) described the production of tetraploids using laughing gas. Tetraploid seedlings were obtained by placing young seed buds, which were pollinated a week before, for one day in a cylinder with laughing gas (N_2O) with five to six atmospheric pressure. They were crossed mutually to improve the fertility of these tetraploids. Highly fertile tetraploid cultivars of *T. gesneriana* and tetraploids of *T. fosteriana* and *T. kaufmanniana* were released to Dutch tulip companies in 1989 and 1991 (Straathof and Eikelboom, 1997). In 1991, tetraploid lily-flowered *T. gesneriana* tulips were released by PRI. Crossing of tetraploids with diploids can result in strong growing triploids. But, these triploids are mostly sterile and cannot be used in further breeding. Tetraploid *T. gesneriana* cultivars can be used in combination with diploid *T. fosteriana* to obtain Darwin hybrids with a better vase life and suitable for early forcing. In 1988, the cultivar 'World's Favourite' was released, bred from a tetraploid *T. gesneriana* seedling ('Denbola' x 'Lustige Witwe') crossed with a diploid *T. fosteriana* seedling. Of late, using in vitro polyploidization techniques, tetraploid tulip cultivars have been produced (Eikelboom *et al.*, 2001, Van Tuyl *et al.*, 2002). For future tulip breeding, polyploidization technique will definitely have a significant role in overcoming interspecific crossing barriers.

Marasek-Ciolakowska *et al.* (2012) selected 2n gamete producing F_1 hybrids of Darwin hybrids (*T. gesneriana* × *T. fosteriana*) and used as pollen donor with the aim of producing of polyploid tulips. 308 BC_1 seedlings (one-year-old) resulting from crosses between diploid and triploid *T. gesneriana* cultivars and diploid 2n gametes producers were tested by flow cytometry analysis. Chromosome number and genome composition was evaluated for selected genotypes through genomic in situ hybridization (GISH) analysis. Progenies from crosses at diploid level (2x × 2x) were mostly diploids, while few seedlings were triploids. In crosses 3x×2x, 81 genotypes were tetraploids and 25 seedlings were penta ploids. Cytological analyses using GISH of sexual polyploid progenies have shown substantial amount of intergenomic recombination in BC_1 generations which is desirable for introgression breeding.

To improve bulb productivity and ornamental characteristics in tulips, it is required to increase genetic variation by using breeding techniques such as overcoming of interspecific barrier, embryo rescue of abortive embryos and induction of polyploids. To identify interspecific barrier between species, crossing experiments have been studied about fifty years ago and recently embryo rescue technique was attempted in Holland. Colchicine treatment is a common technique to produce polyploids, whereas nitrous oxide has been applied to zygotes as a polyploidizing agent in various crops. Usefulness of 2n pollen induced with nitrous gas in tulip breeding was also reported (Okazaki *et al.*, 2005).

6.2.5 Haploidization

Production of in vitro haploid plants using micropore culture is a technique to produce homozygous plants and is used in many plant species. Research showed that tulip was more promising to develop via this technique these plants than lily (Van den Bulk *et al.*, 1992, Van den Bulk and Van Tuyl, 1997). Embryo-like structures from young pollen (microspores) were obtained in tissue culture. But, due to the difficulties with the *in vitro* propagation of tulips, this research was completed without obtaining homozygous doubled haploids.

6.2.6 Disease Resistance

The use of resistant tulip cultivars gives better bulb production, less rejection of infected bulb lots, lower costs for chemical disinfection and needs less labour. Resistance is important both for production and also for forcing, since bulbs infected with tulip breaking virus will give visible symptoms resulting in a low price at auction. Resistant cultivars are also important for bulb exports, since bulbs infected with *Fusarium* cause extra efforts to clean the stock and give rise to complaints from buyers. Resistant cultivars can reduce the amount of chemical disinfection of bulbs.

6.2.7 *Fusarium* Resistance

The *Fusarium* research, carried out in the eighties, has resulted in reliable screening tests for clones as well as for juvenile seedlings at the pre-selection stage. Tulip bulbs were planted in *Fusarium*-infested soil and grown under standardized conditions. After harvest, bulbs are examined for *Fusarium* infection. Selection of one-year old seedlings can reduce plant population considerably at an early stage of the selection programme. Since seedling selection gives some susceptible plants, selected plants must be re-tested at the clonal level. It was observed that the use of one resistant parent resulted in resistant descendants. In 1988, *Fusarium* resistant *T. gesneriana* material was released. Several of these genotypes like 'Furand', 'Furanel', 'Fusarino' and 'Fusor' are becoming successful as new cultivars.

Miller *et al.* (2005) reported that ethylene evolution is a well-known consequence of *Fusarium* infection of tulip (*Tulipa gesneriana*) bulbs. But only little is known about bulb-pathogen interactions involved in the induction or the time course of ethylene production in the infected bulb. Bulbs of 36 tulip cultivars were wounded, inoculated with a liquid *Fusarium* suspension, and held at 21°C. Control bulbs were wounded, but not inoculated. Production of ethylene was monitored by headspace accumulation and gas chromatography. The results indicated that ethylene could increase rapidly after a lag phase of at least 8 days after inoculation, but there was a great variability in ethylene production among cultivars. Of the cultivars tested, highest ethylene producer was 'Furand', which evolved more than 340 µl/kg fwt wt/h (ca. 250 µl/bulb/day) on the 12th day after infection. Inoculated 'Furand' and 'Nashville' produced ethylene at ca. 800 µl/bulb/day by 19 days. 40 per cent of cultivars produced ethylene at rates greater than 150 µl/bulb/day, only 11 per cent produced less than 5 µl/bulb/day. Results suggest that high ethylene producing tulips need to be stored separately from other cultivars and increased ventilation should be maintained during storage or transportation.

6.2.8 Tulip Breaking Virus Resistance

The tulip breaking virus (TBV) research also resulted in reliable screening tests at clonal and seedling level. Absolute TBV resistance was found in several *T. fosteriana* cultivars such as 'Cantata' and 'Princeps'. In 1989 and 1991 TBV resistant *T. gesneriana* x *T. fosteriana* hybrids were released. The restoration of the F_1 sterility of these hybrids makes further research necessary.

Deligoz (2019) investigated the occurrence of Tulip breaking virus (TBV) in the tulip cultivars Strong Gold, Pretty Woman and Purple Prince in Samsun province of

Turkey. TBV causes striping of the leaves, flames of different colours on the petals and mosaic patterns on the leaves. A total of 212 samples were collected from four locations and checked by biological, serological, and molecular methods for the presence of TBV. TBV was detected in leaves and flowers by DAS-ELISA (Double-antibody sandwich enzyme-linked immune sorbent assays) in the tulip cultivars (15.5 per cent) tested from Samsun province. TBV infection was highest in Strong Gold (19.7 per cent), followed by Pretty Woman (14.1 per cent) and Purple Prince (12.8 per cent). Presence of TBV in samples was further confirmed by RT-PCR assays.

Tulip collection of the Department of Plant Systematics and Geography of the Botanical Garden of Vilnius University comprising 206 tulip cultivars, classified into 1–11 groups, were analysed estimating their decorative capacities and resistance to Tulip breaking potyvirus (TBV) during 1997–2005. TBV was identified by test-plant, electron microscopy and serological methods. 117 tulip cultivars scored highest point (4–5) by their decorative capacities, 25 per cent of the cultivars were selected as resistant to TBV, 35 per cent as average resistant, and 40 per cent as not virus resistant. Darwin hybrid tulips were most resistant to TBV (Juodkaite *et al.*, 2005).

6.3.0 Molecular Breeding Methods

Tulip is a bulbous ornamental crop with a long juvenile period. It takes approximately 25 years to develop a new cultivar after carrying out a cross. This will take even longer when disease resistance tests are performed. To reduce the use of pesticides, restricted by many governments, disease resistance will be required in the future. Currently, *in vitro* propagation is not an economical method to speed up this process. Application of molecular assisted breeding techniques can be an important tool for reaching the goal of developing a tulip assortment, which can be grown without large disease problems.

6.3.1 Transformation

The genetic modification of tulips was investigated by Wilmink *et al.* (1992 and 1995). Transient expression of the reporter gene for beta-glucuronidase was demonstrated using a Particle Delivery System. It was shown that the CAMV 35S as well as the TR2′ promoter were active in flower stem explants. Regeneration and bulb formation of tulip *in vitro* is, however, extremely difficult and it takes many years for a flowering tulip to be obtained. Seven years after selecting Gus-positive and PPT-resistant explants only one complete transgenic tulip plant flowered. It proved, however, that stable integration was established.

6.3.2 Marker-Assisted Breeding

Tulip and the long juvenile period is highly suitable for using molecular marker techniques and this makes pre-selection possible. Marker-assisted breeding has been investigated using AFLP-markers in lily and tulip (Van Heusden *at al.*, 2002). AFLP-markers were found over the entire tulip genome (12 linkage groups). Several AFLP-markers were linked with TBV-resistance. In the future, the application of PCR-derived markers for various characters will speed up the breeding process of tulips.

7.0 Growth and Flowering

Four factors *viz.*, temperature, light, humidity and air composition plays a vital role for the production of tulips. Tulip seeds are harvested from June to August and sown in August. Germination starts during winter and active growth of leaf occurs in spring. Bulb formed in the extremity of a dropper enlarges in spring and growth ends in May or June.

Tulips require hot dry summers followed by cold and wet winters for their optimum growth and development. Tulip bulb has a cold requirement of many weeks which must be satisfied before it can grow above ground. This cold requirement prevents the emergence of tulip in autumn although temperature and soil moisture conditions are favourable for its growth. After the cold period of winter, plant grows with the rise in temperature and flowers in warm conditions of spring. During hot periods of summer, it survives as an underground organ in soil or is lifted for storage.

7.1.0 Environmental Factors

7.1.1 Temperature

Among various environmental factors, temperature is the most important factor affecting growth physiology of tulip. There is a mandatory temperature requirement for tulip growth. It requires warm (17-20°C), cold (2-9°C) and warm (17-20°C) temperature in sequence.

The tulip bulb has an annual replacement cycle. The annual cycle can be divided into the following main phases (Nayeem and Qayoom, 2015).

1. Root growth occurs rapidly after planting in autumn, when the soil temperature decreases, the apical bud, which has already differentiated the aerial organs also starts to elongate, but this growth is very slow during winter.
2. During early spring, when the temperature rises, plant growth become active, rapid scape and floral bud elongation leads to flowering. In the case of small non-flowering bulbs, the apical bud only produces one leaf. Flowering tulips form two or more leaves. The transformation of the vegetative buds has been initiated and is epically rapid after flowering. Simultaneously, the mother-bulb scales shrivel and progressively disappear.
3. At the end of spring, the aerial organs senesce, and daughter bulb is in an apparent state of dormancy. However, periodical dissection of bulbs shows that an active bud differentiation takes place during this period. Also, the root in the basal plate becomes very prominent.

This demonstrates that continuous changes are taking place nearly all year round and that there are always organs being initiated, growing, senescing, or developing.

7.1.2 Light

Light plays a significant role in the plant growth and flowering in the flowering plants. A study was conducted to investigate the effectiveness of using various light colors on the sprouting, morphological, biochemical and flower quality of tulip *cv.* 'Princess Catharina-Amalia'. The light treatments included white, blue, red light and dark conditions. The results indicated that days to sprouting and flowering reduced by all light treatments especially at blue light (9.9 and 10.7 per cent, respectively) compared to dark conditions. Besides, in light treatments, sugar compounds and biomass of tulip plants was better in comparison to dark conditions. Tulip plants possessed a higher plant yield under blue, red, and white light treatments than dark conditions, especially under blue light. Since light can alter morphological, biochemical and flowering traits of plant, it is recommended to use the changing of light qualities rather than using hormone for various purposes such as cut flowers or pot plants (Kafi *et al.*, 2018).

In contrast Sumitomo *et al.* (2012) reported that the number of days to flowering is not affected by light treatments. They noticed that the trend was the same for all light conditions in tulip cultivars including 'Leen van der mark', 'Murasaki suisho' and 'Kikomachi'.

Okubo and Uemoto (1985) reported that dark conditions lead to an increase in length of the first internode, whereas the last internode was the longest one when grown under natural light conditions.

It was reported that under Kashmir conditions, partial reduction in sunlight results in healthier plants of tulips with longer flower stalks (Anon., 2012).

7.1.3 Growth Regulators

7.1.3.1 Effect on Vegetative Growth

Kumar *et al.* (2013) evaluated the effect of plant growth regulators *viz.*, gibberellic acid, 2-chloroethyl trimethyl ammonium chloride and maleic hydrazide as dip treatment and foliar spray on growth of tulip under Karewa conditions of Kashmir Himalaya. Maximum plant height was recorded with 400 ppm GA_3 followed by 200 ppm GA_3.

Ahmed *et al.* (2013) investigated the effect of plant growth regulators *viz.*, indole acetic acid, 2-choloroethyl trimethyl ammonium chloride, 3,5 tri-iodobenzoic acid on growth of two tulip cultivars (Apeldoorn and Golden Oxford). Bulbs were given dip treatment in three different growth regulator solutions. Indole acetic acid at 500 ppm recorded early bulb sprouting, maximum plant height, number of leaves/plant, leaf area, leaf area index and scape length in both the cultivars.

7.1.3.2 Effect on Flowering

A study was carried out to evaluate the effect of plant growth regulators on flowering of tulip under Karewa conditions of Kashmir Himalaya (Kumar *et al.*, 2013). GA_3 at 400 ppm recorded earliest flowering followed by 200 ppm GA_3 while, delayed flowering were observed by 500 ppm MH followed by 200 ppm MH. Blooming period was longest in 200 ppm GA_3 (28.46 days) followed by 400 ppm

GA_3 (27.76 days) in comparison to control (21.59 days). Maximum vase life was obtained with 400 ppm GA_3 (11.26 days) followed by 200 ppm GA_3 (10.43 days) over the control (7.30 days).

A study was carried out to investigate the effect of plant growth regulators on flowering of tulip using two cultivars *viz.*, Apeldoorn and Golden Oxford. Bulbs were given dip treatment in three different growth regulator solutions *viz.*, indole acetic acid, 2-choloroethyl trimethyl ammonium chloride, 2,3,5 triiodobenzoic acid along with control for two hours. Results revealed that indole acetic acid at 500 ppm recorded early floral bud appearance, colour break and flower opening, while 2-choloroethyl trimethyl ammonium chloride at 500 ppm recorded maximum flower diameter, scape thickness, scape weight, flower duration and vase life in both cultivars. Indole acetic acid was most effective for enhancing the early flowering and 2-choloroethyl trimethyl ammonium chloride improved quality parameters of tulip. 2-choloroethy trimethyl ammonium chloride and 2,3,5 triiodobenzoic acid can be used for producing dwarf tulips suitable for bedding and pot plant production. Apeldoorn was observed to perform better than Golden Oxford for most of the characters (Ahmed *et al.*, 2013).

Khan *et al.* (2007) found that gibberellic acid (GA_3) as pre-soaking plus foliar spray resulted in improved growth and flowering attributes while as paclobutrzol (*PP333*) as bulb dip plus foliar spray supress plant growth but improved scape thickness and flower diameter. GA_3 caused hastening of flowering while *PP333* resulted in delayed flowering in tulip. The field life of tulip flower was extended by GA_3 treatments. Bulb dip was found more superior over foliar sprays.

7.1.3.3 Effect on Bulb Formation

Production of larger commercial sized (10-12 cm or more) bulbs capable of producing marketable flowers is a major problem as observed in many parts of the world. Deflowering at bud initiation is partially beneficial in increasing the bulb size. Foliar applications of CCC 1000 ppm (cycocel) with Tracel 4 g/l immediately after removal of the flower buds at green bud stage shows marked increase in number, weight and circumference of bulbs and bulblets. Two to three foliar applications of CCC and Tracel at an interval of 3-4 weeks result in larger bulbs (Anon., 2012).

A study was carried out to evaluate the effect of plant growth regulators on bulb production of tulip under Karewa conditions of Kashmir Himalaya (Kumar *et al.*, 2013). Maximum number of bulbs and daughter bulbs per plant were recorded with 400 ppm GA_3 (1.43 and 3.03) followed by 500 ppm CCC (1.41 and 2.65) over the control (1.07 and 1.72), respectively and thereby enhanced propagation coefficient was obtained in 400 ppm GA_3 (258.66 per cent) followed by 500 ppm CCC (237.73 per cent) as against the control (170 per cent).

Mukherjee *et al.* (1999) studied the effect of foliar sprays of 2-chloroethyl trimethyl ammonium chloride, after carrying out deflowering in the early stage of flower emergence on the production and development of bulbs and bulblets of 4 cultivars of tulip (*Tulipa gesnerana* L.). Treatment with 2-chloroethyl trimethyl ammonium chloride was found to be effective in increasing the production and size of bulbs and bulblets with 2-chloroethyl trimethyl ammonium chloride 1000 ppm being the most effective. This treatment has increased the bulb yield by 48.2 per cent, bulb weight by 116.2 per cent, bulb circumference by 193 per cent, bulblet production by 29.1 per cent and bulblet weight by 56.8 per cent compared with the untreated plants.

8.0 Propagation

Tulips grown for bulb production, commercial forcing and landscaping are vegetatively propagated by bulblets and bulbs. Daughter bulb offsets develop from vegetative axillary buds which are located in the axils of the tunicated bulb scales are used for vegetative propagation. Average propagation rate of most tulip cultivars is between two and three new bulblets per year (Le Nard and De Hertogh, 1993). It takes 2 to 3 years to produce a commercial size bulb capable of flowering.

8.1.0 Seed Propagation

Seed is also used to propagate but it produces different shades of flowers. It is only for breeding purposes that tulips are crossed, and seeds are harvested. After sowing, seeds require a period of low temperature to induce germination and to initiate a bulb primordium (Niimi, 1978). The embryo produces one cotyledonary leaf, a primary root, and a hollow diverticulum called a 'dropper'. The bulb primordium is positioned at the tip of the dropper. As the dropper grows further into the soil a bulblet is produced at the end of the dropper (Taillandier and Riviere, 1981). This small tulip bulb requires 4 –5 years of growth before it reaches the critical minimal size for flowering. The minimal size depends on the genotype, but normally *T. gesneriana* bulbs reach a circumference between 6 and 8 cm.

8.2.0 Vegetative Propagation

A mature tulip bulb consists of five to seven tunicated scales attached to the basal plate. The outer bulb scales are generally fleshy and contain reserve food material. Axillary bud is formed on the basal plate in the interior of each tunicated scale. Usually most of the axillary buds abort and only 2 or 3 bulblets develop annually along with the daughter bulb. In late spring or early summer, growth of the new daughter bulb is complete. After the flowering, the old mother bulbs disintegrate and are replaced by a cluster of new bulb and bulblets, which were initiated during the previous flowering season. Largest of these bulbs, the daughter bulb may attain the flowering size within one year, but smaller bulblets require 3-4 years, depending on the size and cultivar (Anon, 2012).

8.3.0 Adventitious Shoots and Bulblets Formation in Tulip

Wilmink *et al.* (1995) reported that adventitious shoot formation was induced on explants taken from young floral stems of tulip (*Tulipa gesneriana* L.). The shoots were formed directly from the explants without an intermediate callus phase. More shoots are formed on explants taken from dry-stored bulbs than from bulbs stored in ice. At the basis of the shoots a meristem was formed that developed into a new bulb. Ability to form this bulb meristem was compared in 8 cultivars, and was highest in Lucky Strike and Monte Carlo. Adventitious shoot formation was initiated in the first two subepidermal cell layers and a number of cells of the original explant

Propagation of Tulip

Seed Pods and Seeds

Bulbs

Planting of Bulbs

Planting of Bulbs in Pots

Sprouting of Bulbs

Micropropagation

Large Scale Multiplication

Source: Popescu (2012).

contributed to the shoot formation. To avoid the formation of chimeric transformed shoots in future transformation, an optimal selection system was established. For selection purposes, the aminoglycoside antibiotics were not effective in inhibiting shoot formation, since tulip tissue showed a high tolerance for kanamycin, while G418 and hygromycin induced severe necrosis of the explants at low concentrations. The herbicides phosphinothricin and glyphosate were highly effective and offer good perspectives to be used for selection of transformed shoots.

Nabieva *et al.* (2019) tried to clarify the effect of cytokinins (TDZ, BAP) and auxin (NAA) on the regeneration capacity of isolated *Tulipa kaufmanniana* (waterlily tulip) embryos for adventitious shoots and bulblets formation. Direct shoot organogenesis was induced by the combined action of chilling and BAP or TDZ treatment. Among the variants tested, TDZ promoted the formation of adventitious bulbs in 48 per cent of the chilled embryos, compared to non-chilled embryos (0.0 per cent) and embryos exposed to the same concentration of BAP (39 per cent). The best culture condition for *T. kaufmanniana* embryos growth and bulblets formation consisted of chilling for 10 weeks in the dark at 4°C, 4 weeks at 20°C under 16 h light photoperiod and 10 weeks at 4°C in the dark. Sufficiently effective protocol for clonal micropropagation was developed for the first time for *T. kaufmanniana* using immature zygotic embryos as primary explants.

9.0 Forcing

For uniform and proper flowering, the tulip bulbs should be subjected to low temperature treatment, which is known as forcing. The bulbs should reach the 'G' stage (gynoecium clearly visible on dissection) before forcing the bulbs, otherwise flower may abort. Depending on the temperature (10°C/32°C) and duration (10-15 weeks) of the forcing treatment, the plants will flower either in March or in early April. Nutrient Film Technique also could be a good choice for tulip forcing by hydroponics.

10.0 Soil and Climate

A well drained, light sandy loam soil is most suitable for growing tulips. Well-decomposed manure should be thoroughly mixed in heavy soils. Tulips prefer light soil of pH of 6-7 with a low salt content. Tulips are indigenous to mountainous areas with temperate climates and need a period of cool dormancy. They perform well in climates with long cool springs and early summers but are often grown as spring blooming annual plantings in warmer regions of the world.

Tulips require several weeks of cold weather to gear up for bloom time in the spring. On an average, soil temperature should be below 10°C for about eight weeks. Tulip is planted in hills where night and day temperatures range between 5-10°C and 20-25°C, respectively during growing season. Tulips do not grow well in plains owing to its high chilling requirements. They grow in the cold and snowy winter. Direct sun during morning and evening is beneficial for improving its flower quality, whereas partial shade is required during mid-day (12 pm – 4 pm). Frost is harmful mainly during bulb emergence. Since tulips require a cold winter season to grow successfully, they do not thrive in the open in tropical climates. However, pre-cooled bulbs can be made to flower during winter in plains.

11.0 Cultivation

11.1.0 Soil Preparation

Planting of tulips by digging a trench is easier than to plant the individual bulbs. To form a trench, cultivate the planting area to a depth of 30 cm and then shovel the soil to one side of the bed. Tulip bulbs can be planted up to 20 cm deep in field production. The deeper the bulb, the longer the stem will be when pulled at harvesting. If a deeper trench is needed, cultivate the area again and remove more soil. Place the bulbs in the bottom of the trench 15–20 cm deep, leaving about as much space between each bulb as the size of the bulb itself. Cover with soil, being careful not to overturn the bulbs in the process. Loosened soil can be used to create a raised bed over the tulip bulbs. Pre-emergent herbicide can also be applied at this time to prevent winter and early spring weed germination.

11.1.1 Planting

In mid-hills (1,000-1,800 m above mean sea level) October-December is the time of planting, whereas in high hills (1,800 m above mean sea level) November-December and February is the time of planting. Staggered planting at 15 days intervals ensures regular supply of cut flowers. To get a good crop of flower, bulbs of 10-12 cm size or more should be planted 5-8 cm deep at 15cm x 30cm spacing in beds. The tunics should be removed before planting the bulbs. The flowering plants will produce 4-6 leaves. Smaller (less than 6-8 cm) bulbs will produce plants with single leaf and will not produce any flower until the bulbs reach proper size.

In containers, the bulbs should be planted close together, but they should never touch. In a 15 cm pot, 3-5 bulbs should be planted. Partial shading during mid-day is beneficial for improving scape length and flower longevity.

11.1.2 Depth of Planting

The depth of the planting hole will depend upon the size of the bulb. Generally, the planting hole should be made two to three times the height of the bulb. Bulbs are typically planted in late summer and fall, normally from 10–20 cm deep, depending on the type planted, in well drained soils. In regions which do not have long cool springs and early summers, bulbs are often planted up to 30 cm deep. This provides protection from the heat of summer and tends to force the plants to regenerate one large bulb each year instead of many smaller non-blooming ones. This can further extend the usefulness of the plants in warmer areas a few years but not stave off the degradation in bulb size and eventual death of the plants.

11.1.3 Watering

Watering in glass house/polyhouse every alternate day is beneficial, whereas, in open areas irrigation for the crop should be done at weekly interval.

11.1.4 Manuring and Fertilization

Proper fertilization for tulips used as cut flowers is important. Tulips are not considered high feeders, and the bulbs themselves store many nutrients for the plants' initial growth. Excessive fertilization can lead to reduced plant height, which may affect marketability. The use of slow or controlled release fertilizers is not recommended as the plant will be harvested before most of the nutrients are released. No additional manures are required if soil is sufficiently rich. However, well-rotten farmyard manure or compost at 5-10 kg/m^2 should be mixed thoroughly depending on the soil condition, to enrich the soil. The best inorganic fertilizer for tulips is a general purpose one such as NPK 5-10-5, with 5 per cent nitrogen, 10 per cent phosphorus and 5 per cent potassium. The higher percentage of phosphorus puts the emphasis on the development of the flower and the root system. Plants need nitrogen for healthy foliage, phosphorus for a strong flower and root system and potassium for firm stems and resistance to disease and cold. Other NPK ratios include 7-14- 28 and 12-10-18 both putting more emphasis on firm growth (Anon., 2012). Spraying of micronutrient rich solution Multiplex at 50 ppm (once or twice) before colour-breaking is beneficial.

11.1.5 Mulching and Weed Control

Inter-culture is necessary to keep the field of tulips weeds free and make the soil porous. An experiment was conducted to study the effect of various mulching terms of 'Foxtrot' tulips with wheat straw in combination with spraying of plantations with herbicides using various doses and different amounts of working liquid (Barbara and Wojciech, 2019). The mulch was laid out just after planting and after the soil had frozen. The experiment included combinations, in which the mulch was left until the end of the growing season or removed in spring. Spraying with single (linuron 675 g/ha + lenacyl 1000 g/ha) or double herbicide dose (linuron 1350 g/ha + lenacyl 2000 g/ha) was made before covering the soil with mulch and also on mulch-two weeks after planting the bulbs, or in the spring after thawing the soil. Volumes of 300 dm^3 and 600 dm^3 of working liquid were used for spraying. Tulips covered with mulch bloomed 2–3 days later and formed longer shoots as compared to non-mulched ones. Highest yield of marketable bulbs and first-selection ones was obtained from tulips mulched immediately after planting and sprayed 2 weeks later with herbicides used in a single dose of 600 dm^3 of a working liquid. Mulching of tulips after soil freezing and removal of mulch in spring adversely affects the number and weight of commercial bulbs. Lowest yield was obtained in combinations, in which the mulch was not used.

11.2.0 Greenhouse Cultivation

Tulips are a refreshing reminder to customers that winter is almost over. Customers' desire for color at the end of winter creates a profitable crop for cut flower growers using minimally heated greenhouses and high tunnels. Tulip bulbs can be forced to bloom out of season (from December through May) for early spring sales at Valentine's Day, Easter, and Mother's Day.

11.2.1 Selection and Treatment of Bulbs

The selection of bulb is important in green house tulip production. The ideal bulb size for tulips is top size 12/+ cm, although some growers use a smaller 11/12 cm size. However, for optimum flower production the 12/+ cm top size is highly recommended because smaller bulbs result in a poorer quality crop with smaller

Cultivation of Tulip

Seedlings in Open

Cultivation of Tulip in Open

Bud Stage in Greenhouse

Flowering Stage in Greenhouse

Soilless Cultivation

Forcing of Tulip

Nutrient Film Technique

flowers and shorter stems. There are two main types of bulbs, precooled and non-precooled. The *Holland Bulb Forcer's Guide* defines precooling as 'the dry storage of spring flowering bulbs at temperatures between 1.7 and 9°C after floral initiation and development is completed, but prior to planting'. Precooled bulbs need to be planted fairly quickly upon receipt. If precooled tulip bulbs must be stored for a short period, keep them at the temperature of 4.4-7.2°C. The second type, non-precooled bulbs require the correct amount of chilling by either allowing them to sit in an unheated greenhouse, or by placing them in a cooler (Nayeem and Qayoom, 2015). They must not be exposed to light during the chilling time, or they may sprout. The length of chilling required for most tulip varieties is 14 to 16 weeks, depending on the variety. The temperature must stay between 1.2–7.5 °C during this time. The bulbs should not be allowed to freeze during the chilling period. If chilling is done in a cooler, inspect the bulbs every few days and move them around. This prevents the attack of mold and mildew from becoming established between the bulbs. No fruits and vegetables in the cooler should be stored with the tulip bulbs. Fruits and vegetables give off ethylene gas as they ripen that kills tulip bulbs. The amount of time from chilling to bloom time will take about 21 weeks, depending on the tulip variety.

11.2.2 Planting and Management of Temperature and Light

Plant the bulbs in the ground inside the greenhouse, or planting trays filled with potting soil inside the greenhouse, after the chilling period is finished. An indoor, raised bed can be constructed using pressure treated wood to create six-inch high sides. When constructing raised beds in a high tunnel or greenhouse, consider the space between the beds for manoeuvring equipment. Raised beds are useful for tulip production because they help to improve drainage and prevent disease problems caused by root rots. Do not replant in beds that have had tulips in the past unless steam sterilize the substrate at 70–82°C for 30 minutes. During the heat of the summer when a greenhouse is not in use, plastic covers can be used to solarize (using the sun to heat the soil) the beds for 4-6 weeks.

It takes four to six weeks for the tulips to bloom after planting in the ground or trays inside the greenhouse once they are exposed to the correct temperatures and light levels. When planting the bulbs directly in the ground or in planting trays, plant as close as 5 cm apart and 2.5 cm below the surface of the soil ensuring longer and cleaner stems.

Expose the planting area in the green house to a steady temperature of 18 to 21°C if tulip is grown in the early spring when outdoor temperatures are still cool. If the temperatures are too warm, provide ventilation. If tulip is grown in hot summer, air conditioner, exhaust fans or other cooling devices are needed to keep the temperatures down to the desired level. Temperatures that are below the required level will slow down tulip growth and cause the stems to be short. Too warm temperatures will cause the bulbs to bloom quickly on long, rangy stems, so maintaining the temperature at 18 – 21°C is required for optimal flower production.

It is required to allow unobstructed sunlight through the top and sides of the greenhouse to reach the area where the tulip bulbs are planted. No artificial light is necessary once the chilled bulbs are planted and exposed to the right temperatures and normal green house light levels.

Proper fertilization for tulips used as cut flowers is important. Tulips are not considered high feeders, and the bulbs themselves store many nutrients for the plants' initial growth. Excessive fertilization can lead to reduced plant height, which may affect marketability. The use of slow or controlled release fertilizers is not recommended as the plant will be harvested before most of the nutrients are released. In greenhouse and high tunnel production fertilization should begin after shoot emergence. When the shoots are 2 – 2.5 inches tall use a fertilizer with a 2:1 ratio of calcium nitrate to potassium nitrate after amending the substrate for phosphorus and potassium. Weekly application of this ratio can be used, or a fertilizer injector can be utilized to supply 200 – 250 ppm of N on a constant basis using a well-balanced fertilizer (either 2:1:1 or 3:1:1 nitrogen to phosphorus to potassium ratio) that includes a small amount of phosphorus and potassium. With a soluble salt meter, use the 1:2 dilution methods to monitor the electrical conductivity. Acceptable readings should be between 1 and 1.5 mS/cm (mmhos/cm).

12.0 Diseases, Pests and Disorders

12.1.0 Diseases

Fungal Diseases

12.1.1 *Botrytis cinerea*

The disease is caused by fungus *Botrytis cinerea,* also known as grey rot. It is distributed by spores and moist conditions promote the infection. It is most found in long-stored bulbs and where bulbs were planted late. Infection is encouraged by using steamed soil or fresh peat-based substrate, as this does not contain any of its natural competitors. In soil cultivation, presence of loose or lumpy soil and addition of organic material further add to the problem. This fungus occurs in the damaged or weakened plant parts and beneath the tunic between the roots. Both bulbs and roots beneath the tunic may be infected.

Symptoms of infected bulbs appear as one or more bulb scales becomes completely or partly soft and dark brown. Large (2- 3 mm), mat black flat sclerotia can be seen on the diseased tissue. In parts above soil the infected plants become brittle and break readily. Flowers will have an unusual bright shade and leaves will be shiny because the wax layer is missing. Heavily infected plants remain short or don't emerge at all. The fungus thrives in thick layers of roots at the bottom or on parts of the root that lie beneath the tunic of the bulb. A white fungal weft develops on these roots, which later turns grey and often covered with black sclerotia (2-3 mm in diameter). Infected roots touch the shoots emerging below and cause marks there, leading to 'Ghost spotting'. Following points should be considered for managing the disease.

- ✰ Do not plant tulips in a purely peat-based substrate.
- ✰ Mix coarse-grain sand or disease-free garden soil in the peat product (50 per cent).

- ✰ Treat the bulbs with a fungicide before planting.
- ✰ Prevent the bulbs from being damaged and drying out.
- ✰ After planting, cover the bulbs with a layer of course sand.

12.1.2 *Botrytis tulipae*

The disease is caused by the fungus *Botrytis tulipae*. Sclerotia and spores only germinate in moist conditions, at temperatures starting from 1°C. The spores spread quickly by watering and results in feared spots. The fungus can survive up to two years in the soil by means of its sclerotia. Heavily infected plants do not open or are retarded in growth ("Primaries"). The lowermost, often twisted, leaves at the bottom of the stem develop greyish brown fungal hyphae with spores. Plant parts beneath the soil develop 1-2 mm large, mat black sclerotia. Germinating spores cause small, watery marks on leaves and flowers. These marks are green at first which turn later to large white and brown spots ("Spotting"). The disease can be prevented by following control measures.

- ✰ Treat greenhouse and standing ground soil with steam, flooding or apply a general soil disinfecting treatment.
- ✰ Disinfect the bulbs before planting and avoid dense planting.
- ✰ Ensure that the crop remains dry during cultivation, especially at night.
- ✰ Apply crop heating 40 cm above the crop or use horizontally aimed fans (either of these measures dramatically reduces the risk of Botrytis infection).

12.1.3 *Fusarium*

The disease is caused by fungus *Fusarium oxysporum* f-sp. *tulipae*. Disease develops from *Fusarium*-diseased bulbs, healthy-looking bulbs but ones that are already infected with the fungus, and severely infected greenhouse soil. Fungus thrives in high planting temperatures (above 13°C). In the cultivation of 5°C tulips there is a high infection risk especially in early crops, as the bulbs are subjected to temperature treatments that encourage the development of the fungus. Infected bulbs develop greyish brown spots during storage, occasionally with concentric rings and a clear yellow rim. They give off a distinctive, acrid smell and release ethylene. Bulbs shrivel up and become "loose" in their tunics. In less serious cases growth will be retarded, flower tips will turn yellow and flower buds will desiccate. Longitudinal bulb dissection clearly shows that the stem turns brown from the base upwards. Bulbs infected with *Fusarium* release ethylene into the soil, which may lead to retarded growth and even bud blasting in neighbouring plants. The disease can be prevented by following control measures.

- ✰ Infected batches should be given proper ventilation treatment
- ✰ Always remove diseased bulbs at planting.
- ✰ Never plant in temperatures above 13°C. If necessary, plant later.
- ✰ Only use potting soil that has reached a low enough temperature.

☆ Make sure that the tulips have been given the appropriate weeks of cold treatment so that the greenhouse period will not have to be extended.

12.1.4 *Penicillium*

The disease is caused by fungus, *Penicillium hirsutum*, which is commonly found in the environment. Infection develops particularly on bulbs that have suffered mechanical damage that has occurred late (after August). It can also affect bulbs that have been lifted early and which are then stored at low temperatures and under sufficient dry. *Penicillium* infection is recognised by brown discoloration on the inside of the outermost bulb scales. Spots, which have developed as a result of mechanical damage during warm dry storage, become covered with a blue-green mass of weft and spores, and are visible through the translucent bulb scale as grey, violet-brown spots. The infection slowly continues to spread within the bulb after planting. Slightly damaged bulbs will produce a normal healthy plant. Moderately to heavily damaged bulbs will produce no plant or a stunted, weak, spindly one with pale green leaves. If the infection starts in a bulblet located under the tunic, fungus will penetrate the basal plate and pass from there into healthy undamaged bulb scales, thus leading to the loss of the entire bulb. The disease can be managed by following points.

☆ Store bulbs in a well ventilated storage room with a sufficiently low RH.

☆ Avoid damaging the bulb and shoot.

☆ Disinfect the bulbs immediately before planting.

12.1.5 *Pythium* - Root Rot

The disease is caused by soil borne fungi of *Pythium* genus, especially *P. ultimum*. The fungi are active in soil temperatures above freezing. Risk increases as temperature and moisture content of soil increases and also when the soil contains more water, has a looser structure or has a higher EC. Serious cases may lead to flower blasting and to shorter plants that occur in patches in the greenhouse. The infected roots in the soil are glassy or watery, often marked with a narrow brown stripe, and break off easily. When the disease develops the entire root system turns brown. The disease can be prevented by following control measures.

☆ Use fresh and sufficiently cool soil for the standing ground and greenhouse.

☆ Soil infected with *Pythium* requires an additional soil treatment.

☆ Soil of the greenhouse must have good structure and be free draining.

☆ Use only peeled bulbs.

12.1.6 *Pythium* - Soft Rot

The disease is caused by certain strains of the soil borne fungus *Pythium ultimum*. The fungus attacks bulbs in the first few weeks after planting, in soil temperatures of 12°C or higher. Bulbs that are infected at an early stage often develop short shoots. The bulb tissue is weak and frequently coloured pink. The bulbs give off a typical, unpleasant odour and shoot and root will rot eventually. If the attack occurs later during greenhouse cultivation, the bulbs are retarded in growth, leaf tips turn yellow; the plants fall over and flowers may desiccate. The following points are effective in managing the disease.

- For greenhouse production, soil temperature during first two weeks after planting should be kept below 12°C and preferably lower than 10°C.
- Remove the tunic from the root wall and dip the bulbs before planting.
- Soil of the greenhouse must have good structure and be free draining.
- Ensure the temperature in the greenhouse is below 12°C, preferably even below 10°C, in the first two weeks after planting the tulips.
- Use only peeled bulbs.

12.1.7 *Rhizoctonia solani*

The disease is caused by the fungus *Rhizoctonia solani*, which attacks the tulip shoot from the soil. Unlike *Rhizoctonia tuliparum*, this fungus thrives at higher temperatures (15-18°C). Orange-brown spots and stripes develop on the outside of the shoot. The tissue splits later as if it has been gnawed. Plants flower normally, but tips of bottom leaves are curved outward. In severe cases, the leaf is damaged and the bottom part of the stem develops oval, deeply set patches. Plants will be retarded in growth and break easily when handled. It can be managed by following measures.

- Treat the soil of the greenhouse with an efficient fungicide.
- Incorporate the fungicide carefully through the top layer (10 cm) of the soil.

12.1.8 *Rhizoctonia tuliparum*

The disease is caused by the fungus *Rhizoctonia tuliparum*. This fungus, which does not produce spores, usually attacks bulbous plants in winter, especially at soil temperatures of 13°C or less. The bulbs and shoot are occasionally covered with mycelium-coated soil often containing whitish grey to blackish brown sclerotia. The shape and size of these sclerotia vary greatly (from 1 mm to 1 cm). Bulb scales develop large, greyish-brown marks with grey fungal tissue. Characteristic, brown rings are visible upon cutting the bulb in half. The entire bulb usually dies through rot. It can be control by

- Remove, flood or steam infected soil or treat it with a suitable fungicide.
- Do not store infected soil in the vicinity of the standing ground.
- Treat the bulbs with a suitable fungicide before planting.
- Planting the bulbs late will reduce the risk of disease attack.

12.1.9 Root - *Fusarium*

The causes of these diseases are two fungi: *Fusarium culmorum* and *Fusarium avenaceum*. These fungi are commonly found in soil and peat substrates. The fungi produce a poisonous substance that is absorbed through the roots and which then causes the symptoms observed in the upper parts of the plant. Damage caused by *Fusarium culmorum* produces shorter tulips with brown vascular bundles at the base of the bulb and the foot of the stems and sometimes white dried-out tissue and cavities. An infection by *Fusarium avenaceum* causes the tulips to grow faster and produce light green, pinched and twisted leaves with partially green and sometimes strangely dented flowers. It can be managed by following measures.

- ✰ Never plant tulips in purely peat-based substrate and do not use previously used infected potting soil.
- ✰ When using peat products, mix these with an equal quantity of course-grain sand or disease-free garden soil (50 per cent).
- ✰ Disinfect the bulbs before planting.

12.1.10 *Trichoderma*

The disease is caused by the fungus *Trichoderma viride*. It releases a toxin that is transported through plants and causes the symptoms described in the tip of the leaves. The fungus is naturally present in all soil types and in peat-based substrate. The roots develop a glassy look and are covered in fungal hyphae and later start to rot and turn light to dark brown in colour. The tips of the leaves of the diseased plants turn bright grey. At a later stage the tissue turns white and rapidly dries out. Bulbs can also become diseased through the root crown. It is most common during the late cultivation of tulips planted in peat substrate. Susceptible cultivars are 'Ad Rem', 'Angélique', 'Capri', 'Prominence', 'Rosario' *etc*. It can be managed by following measures.

- ✰ Never plant tulips in purely peat-based substrate and do not use previously used infected potting soil
- ✰ When using peat products, mix these with an equal quantity of course-grain sand or disease-free garden soil (50 per cent).

Viral Diseases

12.1.11 Augusta Disease

The Augusta disease is caused by the tobacco necrosis virus, which is transferred to the roots by the swarm cells of the fungus *Olpidium brassicae*. The virus transfer from soil to the root depends on the moisture level and the temperature (above 9°C) of the soil. Augusta disease occurs mainly in tulips planted too early in the standing grounds when temperatures are too high. Diseased plants grow crookedly and remain short while necrotic brown, long patches and stripes appear on the leaves. The flowers of red-flowering species develop thin, dark veins. Brown spots can be observed on the new young developing bulbs. Roots display brown rotting tissue. Susceptible cultivars include 'Angélique', 'Apricot Beauty', 'Blenda', 'Inzell' and 'Prominence'. It can be managed by following measures.

- ✰ Give the soil of the standing ground or soil of a unheated greenhouse a steam treatment (100°C for 30 minutes duration) or a chemical treatment.

- ☆ When using soil suspected of being infected, do not plant the bulbs until the ambient temperature has dropped to below 9°C.

12.1.12 Veinal Streak

This disease is often thought to be connected to tobacco ring spot virus. Infection will be severe in susceptible cultivars under wet and too cold conditions (< 15°C). Veinal streak emerges as twisted leaves, retarded growth and discoloured leaf veins. At an advanced stage, the leaf develops glassy, withered patches. The flower buds of the diseased plants develop watery, sunken stripes and often green marks. Susceptible cultivars include 'Montreux', 'Lustige Witwe' and 'Monte Carlo' and sports. It can be managed by following measures.

- ☆ Do not water as long as the soil below the bulbs is still moist.
- ☆ Only water when really necessary and never too much in a single irrigation
- ☆ Keep the greenhouse temperature constant and never allow it to drop below 16°C.

12.2.0 Pests

12.2.1 Aphids

An infestation is caused by various species of aphids *viz.*, tulip aphid (*Dysaphis tulipae*), the lily aphid (*Aulacortum circumflexum*), the green peach aphid (*Myzus persicae*) and the black bean aphid (*Aphis fabae*). Tulip aphid is most commonly found during storage; the lily aphid is encountered during the greenhouse period; green peach aphid and black bean aphid are primarily found on plants in the field. Aphids reproduce particularly well on bulbs stored for long periods at temperatures higher than 15°C. The presence of aphids can be detected by the development of red or brownish spots on the bulbs that are sticky with honeydew. These bulbs will later turn black due to the development of honeydew fungi (sooty mould). Aphids can be found primarily around the top of the bulb and at places where the brown tunic is loose or missing. If the bulbs have sprouted during storage, large numbers of aphids can be found on the shoots. This will result in the upper part of the lower leaf being damaged, misshapen, and sometimes thickened after emergence. Round light green to yellow spots will be found on the leaves. It can be managed by following measures.

- ☆ Regularly check the bulbs during storage and the plants in the greenhouse or in the field for the presence of aphids and apply aphid control measures as needed.

12.2.2 Bulb Mites

Bulb mites (*Rhizoglyphus echinopus, Rhizoglyphus robini* and *Thyrophagus species*) infest bulb tissue damaged by fungi, bacteria, nematodes, or the surrounding environment and can also infest healthy bulb tissue. These mites feed primarily on diseased bulb tissue that has already been damaged by fungi, bacteria and nematodes and remove diseased bulb regularly. In combination with ethylene, they can reach young flower parts in the shoot and cause bud necrosis. Damages in young shoot results in shallow pits, scratches on the outer sides of the shoot near the edge of the first leaf. The damaged tissue then turns a brownish colour. It can be managed by following measures.

- ✰ Store bulbs under dry, properly ventilated conditions until the end of the storage period. Provide a properly ventilated storage room so that no ethylene can accumulate from *Fusarium*-diseased bulbs.
- ✰ In serious cases, apply a miticide.

12.2.3 Wheat Curl Mite (Tulip gall mite)

This problem is caused by *Eriophyes tulipae*, a long slender mite up to 0.2 mm. in length. Infestation by wheat curl mites can spread very quickly when the bulbs are stored for a long time under warm conditions. In a storage room, it is easy for tulip bulbs that are not infested to become infested by the presence of infested batches. Susceptible cultivars include 'Leen van der Mark', 'Esther', 'Oxford', 'Yokohama' and 'Rosario'. Outer bulb scale will become dull and creamy to purplish-red in colour late in the season during warm storage. Badly affected bulbs become limp, fail to produce roots and fail to emerge. Bulbs not so badly affected will root slowly and produce spindly plants. Leaves remain curled up and stuck inside the nose of the bulb. Dark-flowering cultivars will display yellow or white stripe-like spots on their petals. It can be managed by following measures.

- ✰ Treat the storage room with an appropriate agent immediately after delivery of infested batches of bulbs.
- ✰ Cool storage and early planting will prevent damage to the bulbs.
- ✰ Keep the storage room clean.

12.3.0 Physiological Disorders

12.3.1 Flower Blasting

Flower blasting in the bulb can develop as a result of the bulbs having been temporarily exposed to an excessively high storage temperature after flower initiation or during transport (heating in transit), or have been stored too long during the autumn (after 15 November) at an excessively warm temperature (20°C). There is greater risk for flower blasting in cooled bulbs and susceptible cultivars. The elongation of the neck stays entirely or part back, so that the blasted flower sits between the leaves. Flower blasting is caused by a variety of factors, such as cultivar-dependent characteristics, small bulb size, too brief intermediate temperature, too brief cold period, ethylene damage during storage or in the greenhouse released by *Fusarium* bulbs or heaters, high relative humidity and/or temperature in the greenhouse, water shortage, root suffocation and diseases. Flower blasting is a phenomenon in which some or all parts of the flower bud become desiccated. Symptoms can first be observed at the tips of the stamens and sepals and then until the base of the flower. Common symptoms are bud blasting, green petals, whitening of leaf tips, desiccation of pistil and stamen and inadequate flower opening. It can be managed by following measures.

- ✰ Avoid ethylene concentrations of more than 0.1 ppm by removing *Fusarium* bulbs from the batches, provide sufficient aeration, keep bulbs apart from cut flowers, vegetables and fruit, and avoid exhaust fumes.
- ✰ Avoid ethylene gas in the greenhouse by ensuring proper annual maintenance checks on the heating unit checks and by using heating equipment that uses outside air for combustion.

12.3.2 Chlorosis

Chlorosis (yellow leaves) is attributed to iron deficiency and occurs in all types of soil, but more commonly in sandy soils with a pH higher than 6.5. Cold wet growing conditions encourage the occurrence of chlorosis and it arises primarily in tulips produced in cold greenhouses. Symptoms disappear when the greenhouse temperature rises. Leaves are light green to yellow-green in colour except for the leaf veins that retain their normal leaf colour. In some cases, the plants can be lighter in weight and flowering can be delayed. It can be managed by maintaining the greenhouse temperature.

12.3.3 Ethylene Damage

Bulbs infected with *Fusarium* fungus produce ethylene gas which is a hormone that affects processes such as respiration and organ initiation. Ethylene damage will occur when concentrations are too high (> 0.1ppm). Damage depends on stage in plant's development, concentration of the gas, how long the plant is exposed to the gas, and temperature. Damage will be negligible at a temperature lower than 13°C. Ethylene is also produced by ripening fruit, vegetables, and flowers as well as the incomplete combustion of fuels. Flower blasting, gum formation, bud necrosis, short plants, poor rooting, and thin plants are symptoms of ethylene damage. It can be managed by following measures.

- ✰ Prevent ethylene from entering storage rooms and the greenhouse.
- ✰ Reducing the temperature will limit any possible damage.

12.3.4 Gum Formation

Gum formation is caused by the presence of ethylene. Blisters in various sizes and shapes filled with a clear liquid that rapidly turns into a brown gummy substance form on the bulb scales. When a blister breaks it exudes the gummy mass that will then harden into a resinous substance. In some cultivars, internal gum formation can occur around the young shoot. Most symptoms of gum formation are observed in the outer bulb scale. In some cultivars, gum forms inside the cells of the bulb scale but does not escape; this causes the colour of the bulb to change from white to beige. It can be managed by following measures.

- ✰ Store uncooled bulbs in a room without any ethylene or at least where concentration of ethylene is less than 0.1ppm and also provide sufficient air circulation between the bulbs. Batches displaying excessive *Fusarium* symptoms should receive extra ventilation or even be stored separately.
- ✰ Discard *Fusarium*-diseased bulbs as quickly as possible.

12.3.5 Heart Rot (Bud necrosis)

Bud necrosis develops due to the presence of ethylene results in space developing between the bulb scales earlier than normal and the reduction in the rate at which the young leaves surrounding the shoot elongate. Young flower bud is not completely surrounded by leaves after reaching Stage G (a phenomenon also known as "open shoots"), thus making the stamens accessible to bulb mites. The bulb mites feed on the stamens causing rotting symptoms that lead to the destruction of the flower and/or shoot. Inside the main shoot of a bulb with bud necrosis a black rotten flower remnant will be present (flower necrosis) at the end of a stem that looks like a black stump. In shoot necrosis, the entire main shoot is black and rotten; wispy leaves or shoots that produce flowers ("wispy plants") develop from the daughter bulbs. It can be managed by following measures.

- Store uncooled bulbs in an ethylene-free store or at least where any concentration of ethylene is less than 0.1ppm and also provide sufficient air circulation between the bulbs. Batches showing excessive *Fusarium* symptoms should receive extra ventilation or even be stored separately.
- Remove *Fusarium*-diseased bulbs from the batch as quickly as possible.
- Treat the store in which the bulbs are to be stored with a miticide.

12.3.6 Leaf Topple

Leaf topple develops due to an insufficient supply of calcium to the rapidly developing zones halfway down the leaves. When using hydroponic forcing, using large bulb sizes of certain cultivars will result in more problems with leaf topple. When forcing in soil, the problem develops particularly among cultivars with more or less rotten bulbs, bulbs with poor roots, or roots that have not properly emerged through the tunic. One cultivar particularly susceptible to leaf topple when forced in soil is 'Monte Carlo'. Leaf topple occurs primarily during hydroponic forcing and to a lesser degree when forcing in soil. It starts when the cell walls become weakened. The first symptoms watery-looking areas, later, the surfaces on the leaves of some cultivars such as 'Purple Prince' can easily split open or dry out and then, if extremely affected, can constrict and bend over. The rapidly growing cultivars are often the most susceptible. It can be managed by following measures.

- Do not use any large-sized bulbs when using hydroponic forcing.
- Ensure proper transpiration by keeping the RH low enough.
- Provide very effective air circulation through the greenhouse.
- Keep planting density low enough.
- When planting, use water with an EC of 1.5 (or slightly higher if using fertilisers contain calcium such as calcium nitrate).
- When forcing in soil, peeling the bulbs before planting if you are using bulbs that have undergone prolonged storage at low temperatures (*e.g.* 5°C production).

12.3.7 Topple

Topple is caused by high relative humidity in the greenhouse and/or poor rooting. This means little water is transported through the plants, leading to a calcium deficiency in the fast-growing parts of the plant. This increases the permeability of the cell membranes so that cell moisture leaks out of the cells. Leaf topple develops especially with very big, poorly rooted or tunic-less partly moulded bulbs. Highly susceptible cultivars include Kees Nelis, Leen van der Mark, Prinses Irene and White Dream. Calcium deficiency may produce glassy stems during the growth phase, which later fall over. The top part of the stem becomes dark green and watery. The tissue shrinks, and the stem above the shrunken tissue, with the flower, topple. Topple in leaves shows as watery, dark marks in the middle of the second or third leaf. Another symptom is leaves displaying grey in middle parts. It can be managed by following measures.

- Avoid high RH levels (over 80 per cent) in the greenhouse at all temperatures.
- Avoid excessive cold phases in the greenhouse.
- Provide optimum evaporation between the plants by applying crop heating 40 cm above the crop or by using horizontally-directed fans.
- Apply nitrogen in the greenhouse.
- Place the batches with topple in a solution containing 1 per cent calcium nitrate after cutting.

12.3.8 Poor Rooting and Hard Base

Poor rooting and hard base can be due to the presence of ethylene in the soil as the result of *Fusarium*-diseased bulbs planted along with the rest of the bulbs; starting the cooling process too early; starting the cooling process at the right time, the presence of wheat curl mites; and/or the after-effects of herbicide applications. After planting, the roots fail to emerge from the bulbs; the root wall exhibits excessive swelling. A few roots sometimes develop but often grow upward between the first and second bulb scales and emerge at the top of the bulb. It can be managed by preventing the causes mentioned above.

12.3.9 Salt Damage

Salt damage occurs in soils with an excessively high salt content (EC > 2) or in excessively acidic soils (pH < 4). It can also occur following the incorrect application of fertilisers and/or incorrect use of fungicides. The roots remain short, grow crookedly and usually turn a light brown. Root tips are dark brown, sometimes thickened, and break-off easily. It can be managed by following measures.

- Use a potting soil of good quality that does not have an excessively high salt content and/or is not excessively acidic.
- Add lime to substrates with an excessively low pH.
- Do not apply too much fertiliser (*e.g.* calcium nitrate to prevent "oedema" and hollow stems). Always uses the recommended rate of fungicides.

12.3.10 Oedema and Hollow Stems

This disorder occurs mainly in batches planted early (September/October). Plants that grow from the heaviest bulbs are particularly susceptible. The cause of the disorder is excessive water absorption combined with limited evaporation. Climatic conditions that hinder evaporation of the plants or bulbs stimulate the development of 'oedema' and hollow stems. Potting soil is not normally used as this entails a higher risk of disease. Immediately after housing the plants in the greenhouse, they develop dark-green, watery patches that resemble frost damage. Thick drops of moisture are released in these damaged places. In forcing, this phenomenon is known as 'oedema'. Susceptible cultivars include Monte Carlo and sports, Cassini and Snowstar. During the rapid development of the plants in the greenhouse, a tear occurs lengthways in the stem which then develops into a hollow stem. Outside of hollow stem contains small tears. Although the plant will flower, it flowers between the leaves, the bottom part of the stem is unusually thick and remains short. Although the hollow stems may occur in all tulip cultivars following frost damage, the two diseases only occur together in Monte Carlo and sports. It can be managed by following measures.

- Never use peat products as planting substrate without mixing with 50 per cent coarse-grain sand or potting soil.
- If glassy plants emerge, add 0.5 to 1.0 kg calcium nitrate per cubic metre substrate mixture.
- Stimulate the evaporation of the plants by applying crop heating or horizontally directed fans and maintain a relative humidity of no more than 75 per cent.

13.0 Harvesting and Postharvest Management

13.1.0 Harvesting

Cut tulips are one of the most popular of the spring bulbs. Tulips are harvested when petals show color, but are not completely coloured or open. When harvesting, tulips can be cut or pulled. If the soil is soft, the bulb usually comes up when the stem is pulled. When pulling tulips, grasp the stem at the soil line and pull straight up. The stem should snap off at the bulb, producing a longer stem. The stems of pulled tulips should be cleaned after picking. If the tulips experience wilting after harvest, any bent stems must be wrapped or somehow supported in the bucket while rehydrating. Otherwise, they will remain bent even after being properly hydrated.

Harvest the flowers when the entire bud is colored but still closed. If tulip buds are harvested at an earlier stage than recommended, flowers will not fully develop in the consumer home. Harvesting the buds too late, on the other hand, reduces the vase life of the flowers. Keep the crop dry to reduce the occurrence of Botrytis that often attacks cut tulips.

Cut tulips are one of the most popular of the spring bulbs. The short postharvest life of this flower, however, requires very careful handling at all steps in the marketing chain.

Harvesting of Tulip

Harvesting Stage

Harvesting in Open

Harvesting in Greenhouse

Harvesting of Bulbs

Grading of Bulbs

Packaging and Marketing of Tulip

Packing of Cut Flowers

Cardboard Box Packing

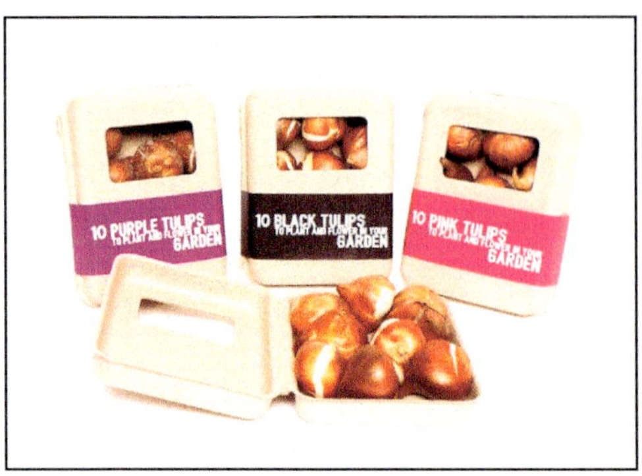

Packing of Bulbs

Cut Flowers in Local Market

Packaging and Marketing of Tulip

Cut Flowers in International Markets

Exhibition of Tulip

Exhibition of Tulip in Kashmir, India

Indoor Exhibition

Diseases, Pests and Disorders of Tulip

Fusarium Basal Rot

Crown Rot

***Botrytis tulipae* Infection**

Gray Mold

Gray Bulb Rot

***Penicillium hirsutum* Infection**

Tulip Breaking Virus

Tulip Fire

Topple of Tulip

Tulip Aphids

Bulb Mites

13.2.0 Storage

The stems of pulled tulips should be cleaned after picking. Swish a handful of stems in a large bucket of water to clean the soil off the stems before placing them in a bucket of water with preservative. Tulips can be stored in a cooler in water or dry at 1.6°C for up to one week. If stored dry, they must be supported when rehydrating. Tulips continue to grow after cutting, up to six inches after placing them in a vase. Place the flowers in a 0°-1.6° C cooler as soon as possible with a relative humidity of more than 90 per cent. This process causes the petals to quickly close up, thereby reducing the mechanical bruising of petals that occurs during bunching. In addition, storage at this temperature reduces the consumption of the limited amount of stored carbohydrates as well as reduces water loss from the flowers.

Bunch the flowers by lining up the buds to the same height. Trim 1.25 – 2.2 cm off the end of the stems. Place the flowers upright in clean, cold (0° -1.6° C) water for at least 30 minutes. It is critical that the stems are kept upright during this time (especially of those that have been left dry for some time) as tulip stems will assume whatever shape the stems are in during the rehydration process. Make sure that the buckets are clean by washing them with a dilute bleach solution before use.

Store the flowers in an upright position for, preferably, no more than 2 days. If flowers are to be stored for more than 2 days, extra handling steps should be taken to assure good postharvest quality. Steps include leaving the bulb attached to the flowers when harvested (this will allow the continued transport of the stored carbohydrates from the bulbs to the flowers), wrapping the bunched flowers tightly before storage, and storing the flowers dry (not in water) and in a horizontal position. Make sure that the temperature in the cooler is around 0°C and the relative humidity is high in order to prevent desiccation of the petals. With this method, the flowers can be stored for several days in a cooler. Before selling the flowers, cut off the ends but leave the wrapping on the bunch. Bunches should remain in an upright position and tightly wrap during rehydration (15-20 cm of 37°C lukewarm water) to prevent stems from becoming wavy.

13.3.0 Storage of Bulbs

Bulbs are harvested when leaves start turning yellow or 40-45 days after flowering. Bulb harvesting should not be followed for the next season cut flower production from the same bulb. Old bulb scales and roots should be removed. They should be air-dried in partial shade. After putting them in wooden trays in single or double layers they are marketed.

For quality flower production by succeeding crop, bulbs should be stored at 7-9°C for 6-8 weeks during September-October for proper development of flower primodia, since tulips are extremely sensitive to fluctuating temperature which otherwise leads to flower abortion. Do not store the bulbs in damp or moist conditions. Dampness leads to rotting and tulip bulbs should be dried out properly before storage. Store tulip bulbs in a location that provides ventilation to remove excess humidity since extreme humidity will also cause the bulbs to rot or decay. Dry sand, vermiculite, or peat moss is ideal storage mediums for tulip bulbs. Do not store tulip bulbs in an area that is susceptible to rodents or pests.

13.4.0 Vase Life

Flower longevity or vase life of the cut flower is one of the most important characteristics for the consumer. Research has shown that a large genetic variation is available in the cultivar assortment of tulips (Van Eijk and Eikelboom, 1976; Van der Meulen and Van Oeveren, 1993, Van der Meulen *et al.*, 1997). Screening tests were developed in which the vase life of a genotype could be estimated. A good correlation was found between the longevity of the flower still attached to the bulb and the vase life of the flower (Van Eijk and Eikelboom, 1976). The variation in flower longevity among cultivars varied from 8 to 16 days, when evaluated at 14°C. Studying segregating populations derived from these cultivars, the flower longevity varied from 6 – 22 days (Van der Meulen *et al.*, 1997). This means that breeders can select for vase life during the forcing period of the best genotypes. When a clone of a genotype has been obtained, the vase life can exactly be determined using a cut-flower test in water. The time between colouring of the flower bud and the moment that the ornamental value of the flower drops below a certain level, can be taken as a selection criterion. The vase life has been determined for many cultivars and the results made available to breeders and forcers.

The influence of different floral preservatives was assessed to determine their effect on the water relations and vase life of cut tulip cv. Yellow Purissima. Uniform size scapes of tulip at bud colour break stage were kept in ten different treatments of floral preservatives comprised of sucrose, aluminium sulphate and 8-HQS along with control (distilled water). The greatest cumulative water balance and maximum vase life were recorded in 8-HQS 300 ppm (10.5 g/scape and 10.1 days) followed by aluminium sulphate 300 ppm (9.67 g/scape and 8.9 days) over control (2.53 g/scape and 5.4 days), respectively. Maximum fresh weight change (10th day) was recorded in 8-HQS 300 ppm (105.13 per cent) followed by aluminium sulphate 300 ppm (103.75 per cent) in comparison to control (89.91 per cent). Floral preservatives delayed the senescence of cut tulip by improving water uptake and postharvest physiology, thereby maintained better water balance leading to improved fresh weight and vase life (Kumar *et al.*, 2012).

The tepals of cut tulips (*Tulipa gesneriana* cv. Ile de France) kept at 20°C had severely wilted 7 days after flower opening. Suppression of abscission and undesirable growth of tepals is required to extend vase life. Treatment with 50 mM trehalose in combination with 50 μM chloramphenicol (CAP) delayed abscission by 4 days compared with stems placed in distilled water or CAP without trehalose. Only 4 per cent of trehalose+CAP-treated flowers exhibited tepal abscission 7 days after harvest (Iwaya-Inoue and Takata, 2001).

It is well documented that microorganisms develop at the stem base of cut flowers or in the vase solution, shortening their longevity (Jiang and Reid, 2012). Silver nitrate and silver thiosulphate as antibacterial agents are very effective in limiting development of aqueous microflora but they are not used in commercial vase solutions because of highly toxic for the environment. An alternative might be silver in the form of nanoparticles with broad spectrum activity (Carrillo-Lopez *et al.*, 2016). Recent studies provided increasing evidences that silver nanoparticles treatments prolong the vase life of cut flowers and foliage due to inhibit bacteria growth in the vase solution and at the cut stem-ends in cut flowers (Liu *et al.*, 2009; Lu *et al.*, 2010; Nazemi Rafi and Ramezanian, 2013). The efficacy of nano silver (NS) in extending the vase life of cut tulip 'White Parrot' was evaluated by Byczyñska (2017). Cut flowers were kept in solutions containing 0, 10, 20 and 40 mg/l NS. The NS treatments at all concentrations extended the vase life of cut flowers. The 10 mg/l NS treatment produced the longest vase life, as compared to the control. On the last day of decorative value, the greater fresh weight was observed in the flowers exposed to the solution of NS in 10 or 20 mg/l. NS at 20 or 40 mg/l prevents the leaves from yellowing.

Humic acid and NPK are excellent preservative solutions which have the significant postharvest effect for cut flowers. The study of Ali *et al.* (2014) reported that humic substances played a major role in plant physiological process and possesses auxin-like hormonal activity being an integral part of reproductive growth. Application of HA in combination with NPK nutrients improved the postharvest life of cut tulip flowers. Pulsing treatment with a high concentration of HA (75 ml/l) with 10 gram macro nutrients enhanced the scape length, fresh weight, bud diameter, flower longevity and vase life attributes as compared to control. Furthermore, flower senescence and premature wilting have also been reduced significantly.

Tulip is one of the most important cut flowers in the world market and abscission of the tepals is the most common symptom during vase life. To prevent these symptoms and extend postharvest life, passive and active modified atmosphere packaging (MAP) was applied at 0°C (Aros *et al.*, 2017). Respiratory rate (CO_2 kg/ha), ethylene production (µL kg/ha) and concentration of CO_2 and O_2 (per cent) inside the packaging were evaluated during storage, while fresh weight loss (FWL) (per cent) and vase life (days) were assessed. Flowers stored using MAP performed significantly better than conventional packaging. A lower FWL (only up to 0.3 per cent) was observed on MAP while FWL was as high as 18 to 21 per cent for conventional packaging after 20 and 31 days, respectively. Vase life was 5.7 and 6.0 days for active and passive MAP, respectively compared to 3.3 days for conventional packing.

14.0 Biotechnology

The process of developing a new tulip variety takes about 25-30 years (Podwyszynska and Marasek, 2003), not only because the time interval from planting to the flowering of the hybrids is long, but also because the period for bulbs propagation until the level of commercial scale is very long (Juodkaite *et al.*, 2008). This is the main explanation for the low rate of replacing varieties within the tulip assortment. In the last decades, biotechnology methods contributed significantly to the shortening of breeding cycles. Similarly to the other flower species, the first application of the biotechnology in tulips was the *in vitro* clonal propagation.

14.1.0 Micropropagation

Micropropagation may produce very fast large numbers of vigorous plants with high quality and free of endogenous pathogens. It can be achieved by inducing formation of adventitious shoots (organogenesis) from the bulb scales or by somatic embryogenesis. During the last three decades, optimal conditions for the *in vitro* culture of tulip genotypes and factors influencing the efficiency of plant micropropagation and shoot regeneration have been studied extensively (Riviere and Müller, 1976; Gabryszewska and Saniewski,1983; Taeb and Alderson, 1990; Wilmink *et al.*, 1995; Podwyszynska and Rojek, 2000; Podwyszynska and Ross, 2003; Kalyoncu *et al.*, 2006; Minas, 2007; Podwyszynska and Sochacki, 2010). Not surprisingly, the first attempts of *in vitro* cultures of tulips have been carried out not in The Netherlands, but in Japan, this country having also a long and impressive tradition in growing and breeding of *Tulipa gesneriana* varieties. The studies carried out by Nishiuchi (1986, 1990), Nishiuchi and Myodo (1976), and Nishiuchi and Koster (1988) contributed by their results to the understanding of the physiological factors which influence the *in vitro* culture of tulips. They allowed establishment of the influence of some growth regulators on either *in vitro* plant development or multiplication from bulb scales, or formation of adventitious shoots (by organogenesis) from such type of explants. Alderson *et al.* (1983) developed an efficient method for regeneration of adventitious shoots from immature flower stems formed from dry bulbs. It was reported that by using this method, the formation of bulblets from such shoots can be induced by the cultivation at 20°C for 14 to 18 weeks, followed by the transfer to 4°C for 8 weeks and subsequent incubation at the temperature of 20°C.

Hulscher *et al.* (1992) developed a method for the micropropagation of tulips from stem and axillary bud explants cultivated on media containing α-naphthelene acid (NAA), N6-2-isopentenyl-adenine (2iP) and 6-benzylaminopurine (BAP) or zeatine. The regenerated shoots are able to form bulbs after cold treatment on a medium supplemented with 70 g/l sucrose and lacking growth regulators. Kuijpers and Langens-Gerrits (1997) have reported the regeneration of tulip shoots on a culture medium containing 5μM zeatine and 5μM α-naphthelene acetic acid (NAA), pointing out the significant increase of the rate of meristems formation on stem explants when silver thyosulfate and, especially, paclobutrazol and methyl jasmonate were used. Recently, Podwyszynska and Sochacki (2010) have described a new method of tulip micropropagation, based on cyclic multiplication of shoots at intervals of 2-3 years, using thidiazuron (TDZ) instead of the currently used cytokinins (N6-2- izopentenyl-adenine (2iP) and 6-benzylaminopurine). Fragments of flower stems are used as initial explants, and the adventitious shoots regenerated on medium containing TDZ are sub-cultivated at every 8 weeks. Formation of bulblets by the shoots resulted by *in vitro* multiplication is induced by the cold treatment, followed by cultivation on a sucrose rich medium, at the temperature of 20°C. In order to avoid the risk of mutations (somaclonal variation), it was recommended that the number of multiplication subcultures should not exceed 5-10. This method makes possible the production of 500-2000 micro bulblets from a single healthy plant.

The regeneration ability is dependent on cultivar (genotype) and growth regulators. For instance, a study carried out by Podwyszynska and Marasek (2003) clearly showed different regeneration potential of different tulip cultivars. Thus, the percentage of explants forming leaf-like structures ranged from 80 per cent in 'Blue Parrot' and 'Prominence' to below 30 per cent in 'Apeldoorn' and 'Mirjoran'.

14.1.1 Plant Regeneration by Organogenesis and Somatic Embryogenesis

Several types of explants (flower stalk, flower stem, ovary, isolated microspores) have been investigated for their potential to regenerate plants *in vitro* by either organogenesis or somatic embryogenesis. The results have shown that the regeneration capacity is influenced by many factors, among the most important being the genotype (Podwyszynska and Marasek, 2003), the basal culture medium (Podwyszynska and Marasek, 2003), the growth regulators (Ptak and Bach, 2007), the type of explants, the physical culture conditions (Bach and Ptak, 2005), the developmental stage and physiological state of the initial explants (van den Bulk *et al.*, 1994). Both cytokinins and auxins are necessary to induce organogenesis or somatic embryogenesis in tulip somatic tissue explants. For instance, the formation of somatic embryos in ovary tissue cultures was induced in the presence of 5 µmol BA and 0.5 µmol NAA (Bach and Ptak, 2001).

In experiments carried out with tulip flower stalk explants of six cultivars, excised from cooled and subsequently forced bulbs, Podwyszynska and Marasek (2003) were successful in inducing plant regeneration by somatic embryogenesis with a frequency ranging (depending on the cultivar and medium composition) from 14.3 per cent to 100 per cent, and with a rate of leaf-like structures formed per explant ranging from 13.7 to 22.8. Subsequently, adventitious shoots developed from this primary regenerants, and formed clusters. An important advance resulted from the studies carried out by Bach and Ptak (2005), who reported that a high frequency of adventitious bulb formation took place on stolons developing from embryos converted into plantlets in the liquid medium lacking growth regulators and containing 6 per cent of sucrose. Both the number of bulblets per explants and individual fresh weight significantly increased in liquid medium as compared to control.

14.1.2 Embryo Rescue

In vitro culture for the rescue of embryos resulting from interspecific crosses is another important application of biotechnology in tulips. This is actually the only way to overcome the post fertilization barriers, and to avoid the abortion of embryos formed as result of such crosses. Therefore, considerable efforts have been made towards the establishment of optimum culture conditions and most favorable composition of culture media for inducing high rates of embryo rescue (Van Creij *et al.*, 1999; Van Creij *et al.*, 2000). Embryo-rescue was found to be successful in the interspecific cross *T. gesneriana* × *T. kaufmanniana*, which is hampered by embryo breakdown. Custers *et al.* (1995) reported that optimal embryo-rescue was achieved in cultures initiated seven weeks after pollination. Compared to embryo culture, ovule culture allowed a higher efficiency of embryo rescue at earlier developmental stage (four weeks after pollination). Also, more embryos could be rescued at each culture date. A very important finding was that ovary-slice culture, started at various dates after pollination, allows obtention of comparable or even better results than that of direct ovule culture.

14.1.3 Polyploidization

Due to their vigorous growth, very attractive flower colour and large flower size, triploid tulips are gaining more and more popularity and, therefore, are in great demand (Okazaki, 2005). Consequently, there is a great demand for tulip polyploids as initial breeding material. Since the 90′, methods for *in vivo* polyploidization were gradually replaced by the techniques of *in vitro* polyploidization, which were reported to contribute significantly to the efficiency of creating tetraploid varieties of tulips (Van Tuyl *et al.*, 2002). Although the *in vivo* chromosome doubling induced by treatment with polyploidization agents is still largely used within the breeding programs (especially for obtaining triploid varieties), doubling of chromosomes by the *in vitro* culture offer some important advantages, including that the process is more easily to be programed and measured. Moreover, in the *in vitro* tissue and organ culture, can be induced the spontaneous doubling of chromosomes by explants wounding and forcing the production of abnormal mitotic divisions, generating of endopolyploidy. Regeneration of shoots from the cells situated in the areas of explant's wounding and selection of the tetraploid shoots by using flowcytometry, allows formation of the stock of parental plants necessary for crosses with the different cultivars of diploid tulips, as a stage in obtention and subsequent selection of triploid hybrids with valuable traits. Van Tuyl *et al.* (2002) reported that tetraploid tulip cultivars have been produced after treating *in vitro* grown tulip bulbs with oryzalin or colchicine. A major achievement resulted from the studies of Chauvin *et al.* (2005), who developed a method for obtaining tetraploid clones from diploid varieties of tulip by using a system for plant regeneration from stem discs cultivated *in vitro* on medium containing oryzalin. It was reported that this method allowed the production of tetraploid clones in all the tulip varieties tested. Although the tetraploidy induction rate was low in some cultivars, this was sufficient for initiating crossing experiments between diploid and tetraploid for obtaining triploid offspring. An important finding was that the regeneration efficiency is dependent on genotype, this explaining the significantly higher regeneration ability of 'Lucky Strike' and 'Gander' varieties, compared to that of 'Lustige Witwe' and 'Don Quichotte' varieties.

14.1.4 Somaclonal Variation

Somaclones regenerated from tissue culture can be an interesting material for tulip breeding. Podwyszynska (2005) and Podwyszynska *et al.* (2006, 2010) reported that all the plants from 'Blue Parrot' cultivar regenerated *in vitro* from four-year-old cultures had modified flower color (red-purple instead of purple-violet), when they used a new protocol for tulip micropropagation, based on cyclic multiplication of adventitious shoots in the presence of thidiazuron. Also, among the juvenile plants derived from shoot cultures which had been propagated *in vitro* for four years or longer occurred variants with variegated leaves or altered morphology. RAPD and ISSR analyses revealed that the phenotypic variation shown by regenerants is the consequence of some genetic changes. Thus was demonstrated that *in vitro* plant regeneration in long term cultures gives chances of selecting somaclonal variants with modified characteristics.

14.2.0 Molecular Markers

As in many other horticultural species, DNA markers became extremely useful tools, applied for either controlling the genetic uniformity of *in vitro* micropropagated tulip plants, or detecting somaclonal variation in micropropagated tulips or plantlets regenerated from somatic tissues. RAPD and ISSR markers have been proven efficient in detecting somaclonal variation among the micropropagated plants of the 'Blue Parrot' cultivar (Podwyszynska *et al.*, 2006). While no phenotypic off-types were noted in the progeny-line derived from a two-year-old culture and polymorphism on DNA level was not detected, genetic changes resulting in phenotypic variation (a changed color of flowers, from purple-violet to red purple) were revealed by the RAPD and ISSR analyses within the juvenile plants derived from 4 – 7-year-old cultures. A proof of the reliability of this molecular markers is the fact that they easily distinguished distantly related cultivars of tulips (such as 'Blue Parrot', 'Prominence', and 'Giewont'), but within the family of sports both RAPD and ISSR did not reveal polymorphism. Podwyszynska *et al.* (2010) reported that ISSR technique confirmed that the variants with major phenotypic changes (highly malformed leaves and/or flowers), which occurred among the tissue culture-derived plants of several tulip cultivars, resulted from genetic changes. DNA markers were recently recommended as efficient molecular tools for the identification of tulip cultivars. Bondrea *et al.* (2007) have been successful in using AFLP (amplified fragments length polymorphism) as molecular tool for distinguishing the various genotypes of *Tulipa*. It was emphasized that AFLP can be tailored by varying the number of selective nucleotides added to core primers and can allow accurate amplification even in complex template mixtures generated from plant species with very large genomes, such as tulip. As shown by the results obtained with several cultivars and hybrid progenies, reproducible AFLP patterns can be obtained with only seven selective nucleotides.

14.3.0 Genetic Transformation

Considerable enrichments of the cultivated tulip with foreign genetic material has been accomplished using interspecific hybridization (van Eijk *et al.*, 1991; Wilmink *et al.*, 1995), but further transfer of germplasm from interesting related species into the cultivated varieties is hampered by strong crossing barriers.

Gene transfer to the tulip was obtained both by particle bombardment and *Agrobacterium* mediated transformation. Using a Particle Delivery System, transient expression of the reporter gene for β-glucuronidase was demonstrated in flowers stem explants of 14 cultivars and *Tulipa* species (Wilmink *et al.*, 1992). However, successful tulip transformation requires efficient regeneration and selection systems. Chauvin *et al.* (1997) have analyzed *in vitro* development of tulip floral scape segments in the presence of kanamycin, hygromycin and phosphinithricin. Unlike kanamycin, to which the tulip proved to be insensitive even at high concentrations, hygromycin was effective in controlling the regeneration of plantlets. The most effective selective agent was phosphinithricin, which totally inhibited regeneration for tulip even at low medium concentration.

15.0 References

Ahmed, Z., Sheikh, M.Q., Siddique, M.A.A., Jeelani, M.I., Singh, A., Nazir, G., Laishram, N. and Rehman, S.I. (2013) *Afr. J. Agric. Res.*, **8**(38): 4780-4786.

Alderson P.G., Rice R.D. and Wright N.A. (1983) *Acta Hortic.*, **131**: 39-48.

Ali, Ahmad, Shoaib-ur-Rehman, Hussain, Rashid, Raza, Sajjad, Sarwar, Mubeen, Bashir, Ajmal and Khan, Muhammad Abdulsalam (2014) *IJPAES*, **4**: 193-200.

Anonymous (2012) *Tulip Cultivation*. Department of Floriculture- Kashmir. *www. dirflorikmr.in*

Anonymous (1996) *Tuinbouw statistiek 1995: handel-teelt-industrie. Afdeling Documentatie/Statistiek van PGF* (Produktschap voor Groenten en Fruit) en PVS (Productschap voor Siergewassen), Den Haag, The Netherlands.

Anonymous (2002) *Productschap Tuinbouw/BKD Beplante oppervlakten Bloembollen 2002, voorjaarsbloeiers.*

Aros, Danilo, Orellana, Karina and Escalon,Víctor (2017) *N.Z. J. Crop Hortic. Sci.*, **45**: 202-215.

Barba-Gonzalez, R., Miller, C.T., Ramanna, M.S. and van Tuyl, J.M. (2006) *Acta Hortic.*,**714** : 99-106.

Bach,A. and Ptak, A. (2001) *Acta Hortic.*, **560**: 391-394.

Bach, A. and Ptak, A. (2005). *Induction and growth of tulip 'Apeldoorn' bulblets from embryo cultures in liquid media. In: Liquid Culture Systems for In Vitro Plant Propagation*, pp. 359-364.

Barbara, Marcinek and Wojciech, Durlak (2019) *Acta Sci. Pol.*, Hortorumcultus Ogrodnictwo, **18**(4): 71-82.

Bhatia, R., Dhiman, M.R. and Parkash, C. and Dey, S.S. (2013) *Indian J. Agric. Sci.*, **83**(7): 773–80.

Botschantzeva, Z.P. (1962) *Tulips. Taxonomy, morphology, cytology, phytogeography and physiology*, (Russian edn), English translation.

Bondrea, I.O., Pamfil, D., Van Heusden, S., Van Tuyl, J., Meijer-Dekens, F., Bondrea, M., Sestras, R., Rusu, A.R., Lucaci, M., Patrascu, B.I. and Balteanu V.A. (2007) *Bull. USAMV-CN* 63-64.

Broertjes, C. and Alkema, H.Y. (1970) *Acta Hortic.*, **23**: 407-412.

Buschmann, J.C.M. (2005) *Globalisation –Flower- Flower Bulbs- Bulb Flowers. Acta Hortic.*, **673**: 27- 33.

Byczyñska, A. (2017) *World Sci. News*, **83**: 224-228.

Carrillo-Lopez, L.M., Morgado-González, A. and Morgado-González, A. (2016) *J. Nanomater.*, Article ID 1769250, 1-10 http://dx.doi.org/10.1155/2016/1769250.

Custers, J.B.M., Ennik, E., Eikelboom, W., Dons, J.J.M. and van Lookeren Campagne, M.M. (1997) *Acta Hortic.*, **430**: 259-266.

Chauvin,J.E., Label A. and Kermarrec M.P. (2005) *J. Hort. Sci. Biotechnol.*, **80**: 693-698.

Custers, J.B.M., Eikelboom, W., Bergervoet, J.H.W. and Van Eijk, J.P. (1995) *Euphytica*, **82**: 253-261.

Dash, M. (1999) *Tulpengekte. Spectrum B.V.*, Utrecht, The Netherlands.

Debrowski, J. (1964) *Acta Agrobot.*, **24** : 171-204.

Deligoz, I. and Sevik, M.A. (2019) *Folia Hortic.*, **31**(2): 263-268.

Eikelboom, W., Straathof, Th.P., and Van Tuyl, J.M. (2001) *Tetraploide "Christmas marvel" methoden om tetraploide tulpen te verkrijgen, Bloembollencultuur,* **112**(12) : 22-23.

Gabryszewska, E. and Saniewski, M. (1983) *Sci. Hortic.*, **19**: 153- 159.

Hall, A.D. (1940) The genus *Tulipa*. The Royal Horticulture Society, London, England.

Hoog, M.H. (1973) *On the origin of Tulipa. Lilies and other Liliaceae*, Royal Horticulture Society, London, England, pp. 47-64.

Hulscher M., Krijgsheld H.T. and Van der Linde P.G. (1992) *Acta Hortic.*, **325**: 441–446.

Iwaya-Inoue, M. and Takata, M. (2001) *HortScience*, **36**(5) : 946-950.

Jauron, R. (1998) *Tulip classes. Horticulture and Home pest news,* Lowa State University Extension and Outreach., pp. 118-119.

Jiang, C. Z. and Reid, M. S. (2012) In: Horticultural Reviews, Vol. 40. (Ed. Jules Janick), Wiley-Blackwell. John Wiley and Sons, Inc. Hoboken, New Jersey, USA.

Juodkaite, R., Naujalis, J. R., Navalinskiene, M. and Samuitiene, M. (2005) *Biologija,* **4** : 64-70.

Juodkaite, R., Baliuniene, A., Naujalis, J.R., Navalinskiene, M. and Samuitiene, M. (2008) *Biologija.,* **54**: 139-146.

Kafi, A.M., Kalate-Jari, S. and Matinizadeh, M. (2018) *J. Anim. Plant Sci.,* **28**(2): 539-545.

Khaleghi, A., Khadivi, A. and Zonneveld, B.J.M. (2018) *Genet. Resour. Crop Evol.,* **65**(8) : 2241-2266.

Khan, F.U., Maeik, F.A., Khan, F.A. and Siddique, M.A.A. (2007) *Indian J. Agric. Sci.,* **77** (10) : 695-697.

Kalyoncu, D.D., Ipek, A., Parmaksiz, I., Arslan, N., Sancak, C. and Ozcan, S. (2006) *Proceedings of the International Symposium*, pp. 289-292.

Kuijpers, A.M. and Langens-Gerrits, M. (1997) *Acta Hortic.*, **430**: 321-324.

Kumar, R., Ahmed, N., Singh, D.B. and Sharma, O.C. (2012) *J. Appl. Hortic.*, **14** : 152-156.

Kumar, R., Ahmed, N., Singh, D.B., Sharma, O.C. and Salmani, M.M. (2013) *Afr. J. Biotechnol.,* **12**(2): 168-174.

Le Nard, M. and De Hertogh, A.A. (1993) *Tulipa*, In: A.A. De Hertogh and M. Le Nard (eds.), *The physiology of flower bulbs*, Elsevier, Amsterdam, The Netherlands, pp. 617-682.

Le Nard, M. and De Hertogh, A.A. (2002) *Acta Hortic.*, **570** : 121-127.

Liu, J., He, S., Zhang, Z., Cao, J., Lv, P., Cheng, G. and Joyce, D. C. (2009) *Postharvest Biol. Technol.*, **54** : 59-62.

Lu, P., Cao, J., He, S., Liu, J., Li, H., Cheng, G., Ding, Y., Joyce, D. C. (2010) *Postharvest Biol. Technol.*, **57** : 196-202.

Marasek-Ciolakowska, A., Ramanna, M.S. and van Tuyl, J.M. (2009) *Acta Hortic.*, **836** : 105-110.

Marasek-Ciolakowska, A., Ramanna, M.S. and van Tuyl, J.M. (2011) *Acta Hortic.*, **886** : 175-182.

Marasek-Ciolakowska, A., Xie, S., Ramanna, M.S., Arens, P. and van Tuyl, J.M. (2012) *Acta Hortic.*, **953** : 187-192.

Miller, W.B., Verlouw, M., Liou, S.S., Cirri, H.O., Watkins, C.B. and Snover-Clift, K. (2005) *Acta Hortic.*, **67** : 229-235.

Minas G.J. (2007) *Acta Hortic.*, **755**:313-316.

Mukherjee, D., Verma, R.R., Singh, D. and Manuj, S. (1999) *Indian J. Agric. Sci.*, **69**(3) : 189-192.

Nabieva, Alexandra Yurievna, and Gerasimovich, Lyudmila Vladimirovna (2019) *Ornamental Horticulture*. ISSN 2447-536X. *https://ornamentalhorticulture.emnuvens.com.br/rbho*

Nayeem, M. and Qayoom, A. (2015) *Int. J. Adv. Prod. Mech. Eng.*, **1**(2) : 2394-6202.

Nazemi Rafi, Z. and Ramezanian, A. (2013) *S. Afr. J. Bot.*, **86** : 68-72.

Nishiuchi Y. and Myodo H. (1976) *J. Japan Soc. Hort. Sci.*, **45**: 59-64.

Nishiuchi Y. (1990) *J. Hokkaido Univ. of Education.*, **40**: 47-51.

Nishiuchi Y., Koster J. (1988) *J. Hokkaido Univ. of Education*, **39**: 51-56.

Niimi, Y. (1978) *Sci. Hortic.*, **9** : 61-69.

Okazaki, K. (2005) *Acta Hortic.*, **673** : 127-140.

Okubo, H. and S. Uemoto (1985) *Plant Cell Physiol.*, **26** : 709-719.

Podwyszyñska M. and Marasek A. (2003) *Acta Soc. Bot. Pol.*, 72: 181-190.

Podwyszynska M. and Rojek. (2000) *Instytutu Sadownictwa i Kwiaciarstwa w Skierniewicach*, **7**: 130-137.

Podwyszynska M. (2005) *J. Fruit Ornam. Plant Res.* **13**: 109-122.

Podwyszynska M., Niedoba K., Korbin M. and Marasek A. (2006) *Acta Hortic.*, **714**: 211-220.

Podwyszynska M. and Ross H. (2003) *Acta Hortic.*, **616**: 413-419

Podwyszynska M. and Sochacki D. (2010) *The Humana Press Inc.* USA, pp. 243-256.

Popescu, A. (2012) *Current Trends in Nat. Sci.*, **1**: 147-160

Rivière S. and Muller J.F. (1976) *CR Séances Acad. Sci Paris.*, **282**: 533-536.

Stanton, G., Schuster, C., Anacker, R. and Klick, S.W.S. (2006) *https://extension.umd.edu/ipm/cut-flower-production/cut-flower-publications.*

Stork, A.L. (1984) *Tulipes sauvages et cultivees. Serie documentaire 13 des Conservatoire et Jardin botaniques*, Geneve.

Straathof and Eikelboom (1997) *Tulip breeding at PRI. DLO-Centre for Plant Breeding and Reproduction Research (PRI)* Department of Ornamental Crops, Wageningen, The Netherlands. Daffodil and Tulip Yearbook 1997-8, 27-33.

Sumitomo, K., Tsuji, T., Yamagata, A., Ishiwata, M., Yamada, M., Shima, K. and Hisamatsu, T. (2012) *Jpn. Agri. Res. Q.*, **46** : 95-103.

Taillandier, J. and Riviere, S. (1981) *Can. J. Bot.*, **59** : 1322-1330.

Taeb, A.G. and Alderson, P.G. (1990) *J. Hort. Sci.*, **65**: 71-74.

Tijmen, V. (2019) *Flower Bulbs. Create. Verver Export Hasselaarsweg 30 1704 DX Heerhugowaard* The Netherlands.

Van den Bulk, R.W. and Van Tuyl, J.M. (1997) In: *In vitro haploid production in higher plants* 5 (Kluwer Academic Publishers) pp. 73-88.

Van den Bulk, R.W., De Vries-Van Hulten, H.P.J., Custers, J.B.M. and Dons, J.J.M. (1992) *Plant Sci.*, **104** : 101-111.

van den Bulk R.W., de Vries-van Hulten H.P.J., Custers J.B.M. and Dons J.J.M. (1994) *Plant Science*, **104**: 101-111.

Van der Meulen, J. J.M., Van Oeveren, J.C. and Van Tuyl, J.M. (1997) *Acta Hortic.*, **430 :** 569-575.

Van der Meulen-Muisers, J.J.M. and Van Oeveren, J.C. (1993) *Proc. Eucarpia Symposium* ed. T. Schiva and A. Mercuri, pp. 191-198.

Van Der Meulen-Muisers, Van Oeveren, J.C. and Van Tuyl, J.M. (1997) *Acta Hortic.*, **430** : 569-576.

Van Eijk, J.P. and Eikelboom, W. (1976) *Euphytica*, **25** : 353-359.

van Creij, M.G.M., Kerckhoffs, D.M.F.J. and van Tuyl, J.M. (1999) *Euphytica*, **108**: 21-28.

van Creij, M.G.M., Kerckhoffs, D.M.F.J., De Brijn, S.M., Vreugdenhil, D. and van Tuyl, J.M. (2000) *Plant Cell, Tissue and Organ Cult.*, **60**: 61-67.

Van Harten, A.M. and Broertjes, C. (1989) *Plant Breed Rev.*, **6** : 55-91.

Van Heusden, A.W., Jongerius M.C., Van Tuyl, J.M., Straathof, T.P. and Mes, J.J. (2002) *Acta Hortic.*, **572** : 131-138.

Van Raamsdonk, L.W.D. and De Vries, T. (1992) *Pl. Syst. Evol.*, **179** : 27-41.

Van Raamsdonk, L.W.D. and De Vries, T. (1995) *Pl. Syst. Evol.*, **195** : 13-44.

Van Scheepen, J. (1996) *Classified list and International Register of tulip names*, Royal General Bulbgrowers Association KAVB, Hillegom, The Netherlands.

Van Tuyl, J.M., Lim, K.B. and Ramanna, M.S. (2002) *Interspecific hybridization and introgression. In: Breeding for ornamentals*: Classical and molecular approaches, pp. 85-103 (ed. A. Vainstein), Kluwer Academic Publishers Dordrecht/Boston/London.

Wilmink, A., Van de Ven, B.C.E., Dons, J.J.M. (1992) *Plant Cell Rep.,* **11**(2) : 76-80.

Wilmink, A., Van der Ven, B.C.E., Custers, J.B.M., Van Tuyl, J.M., Eikelboom, W. and Dons, J.J.M. (1995) *Plant Protoplasts and Genetic Engineering,* VI, II.13 : 289-298.

Wroblewska W. (2009) *Electron. J. Pol. Agric. Univ.* (*EJPAU*) **12**(2) : 6. *www.ejpau.media.pl.*

Xing, Guimei, qu Lianwei, Zhang, W., Zhang, Y., Yuan, X. and Lei, J. (2020) *Nature,* **216** (4). doi :10. 1007/s10681-020-02594-x.

Yamasaki, A., Uragami, A. and Yamada, M. (2002) *Acta Hortic.*, **570**: 423-427.

Zeilinga, A.E. and Schouten, H.P. (1968) *Euphytica,* **17** : 252-264.

4

LILIUM

C. Aswath, V.A. Parthasarathy, Rajiv Kumar, N. Roychoudhury and Solanki Bal

1.0 Introduction

The genus *Lilium* comprises of a large number of species, which can be found in temperature regions of the northern hemisphere. Lilium is being cultivated for centuries as an ornamental plant. The majority of the cultivated lilies are either hybrids or selections. A large number of species and varieties can be used in ornamental gardens even though the lily requires a little more attention than many other bulbs.

2.0 Importance and Uses

Lilies are wonderful ornamental plants with varied uses, grown in border, beds, pots and are excellent cut flowers of magnificent appearance and beautiful colours. Roman naturalist and writer, Plinius, recorded that salves and oils were prepared from the leaves and flowers of the Madonna lily, or *L. candidum* (candidum means 'dazzling white'). Dioscorides of ancient Greece indicated that lily leaves were placed on burns, injuries and snake bites. The white lily was used as a medicinal plant until the 17th century. Lily oil was an aid in childbirth. Distilled lily extract was used as a remedy for neck ailments and leprosy. Bruised bulbs cool infections. A diuretic, the lily was prescribed as a remedy for fluctuating fevers. As a cure for wounds, they were highly valued. The plant was really more of a miniature household pharmacy. The white lily was also famous as a cosmetic. A product obtained by mixing the petals with honey was used to remove wrinkles from the face. Lily oil from Persia sunsinion, was a costly beauty treatment in the ancient world and lily juice was considered by women to preserve youthful beauty. Lily oils were used as aphrodisiacs.

3.0 Origin, History and Distribution

Lilies are natives of the Northern Hemisphere up to South Canada and Siberia and their southern limit is Florida and the Nilgiri mountains of India. The east coast of Asia, the west coast of North America and the Mediterranean region are the three most richly garnished places. The Far East is the home of almost half of the world's lilies and the majority of our finest garden lilies originated there. A few species have been found in North-East India also. The diversity of flower color, shape, fragrance and other phenotypic and physiological characteristics are found in the wild species which are dispersed in the Northern Hemisphere (10° to 60°), mainly in Asia, North America and Europe. Especially, China, Nepal, Korea and Japan are the gene centers of this genus around the world (Lim *et al.*, 2008).

The first authentic record comes from Assyrian monuments dating back to 1000 BC with sculptured lily forms. By 143 BC, Palestinian coin-makers were stamping lily on shekels. In myth and poetry, the lily appeared much sooner, besung by Homer of Greece and Solomon of Jerusalem. To the lands, north of Egypt, lily became what the lotus had long been in the thought and art forms of the Pharaonic country. The mountains and meadows of the whole region were studded with lilies of various sorts; the white among them, in prehistoric times, acquired religious significance, and was included in art, literature, and gardens everywhere. For

Uses of Lilium

Lilium in Garden

In Park

In Pathway

In Resident

In Greenhouse

In Pot

Oil from lilium

In Flower Arrangement

Species of Lilium

Lilium bulbiferum

Lilium formosanum

Lilium concolor

centuries, artists stressed the white lily's significance, climaxed by Rossetti's *Ecce Ancilla Domine,* hung in 1850 in the National Gallery, London. Easter lily (*Lilium longiflorum*), producing long-trumpeted white flowers, became a florists; flower, yet for countless seasons it perfumed quiet Japanese gardens before trade opened the Orient to the West. Then, in 1819, the Royal Horticultural Society brought it to England. British ships might have taken it to Bermuda where it flourished. Perhaps the stately lily family stepped from Asia into America in that distant time when the two continents were still land-linked. Whatever route they reached, lilies have become a precious heritage. Poets have sung of the lily since the Biblical days, Shakespeare and Tennyson among them; and in a charming Chinese tale Lafcadio Hearn recounts how Kuan-yin, Goddess of Mercy, rested her foot upon a lily-flower (Embury, 1960).

4.0 Morphology, Species and Cultivars

4.1.0 Morphology

Lilium belong to the order Liliales. Plants in the Liliales grow from bulbs. They have broader leaves than most monocots, many with the networking veins usually associated with dicots. Liliales have some of the showiest flowers in the plant world. Flowers are terminal, solitary, racemose or umbellate and perfect. Their large colourful tepals come in whites, yellows, browns, oranges and reds, blues and violets. The flowers are often variegated, having spots or streaks of other colours mixed in, and often have nectar-producing glands on the tepals or stamens, making them popular with bees and other such flower visitors (Dahlgren and Clifford, 1982).

The lilies may be divided into two categories according to the mode of their rooting habit: (i) those developing roots from the base of the bulb, *e.g., Lilium candidum, L. camiolicum, L. giganteum, L. martagon, L. pardalinum, L. parryi, L. superbum* and *L. washingtonianum,* and (ii) those in which the roots at first develop from the base of the bulb, but later from the stem as well, *e.g., L. auratum, L. croceum, L. longiflorum, L. regale, L. rubellum, L. speciosum, L. tigrinum* and *L. wallacei* (Bailey, 1963).

The genus was divided into seven sub-genera depending on the shape and arrangements of flowers (Bailey, 1963); these are:

Archelirion: flowers open

Cardiocrinum: leaves heart-shaped

Eulirion: true lilies, flowers trumpet or funnel-shaped

Isolirion: flowers erect or upright

Martagon: flowers Turk's cap or turban-shaped

Notholirion: flowers fritillaria-like

Pseudomartagon: flowers bell-shaped

4.2.0 Species

About 300-400 species have been described but there are probably less than 100 entitled to rank as such (Bailey, 1963). A brief description of some important species is given below:

Lilium auratum

Bulbs spherical, diameter 8-12 cm, broad thick scales; stem 90-180 cm high, smooth; leaves 8-12 cm long, horizontal, 2.5 cm wide; flower diameter 15-25 cm, delicately fragrant, white spotted crimson with yellow band on midrib of each segment; anthers red. Flowers bloom from July to August.

L. candidum

Bulb globular, 10 cm in diameter, white tinged pale yellow; stem 60-120 cm high, smooth; leaves 20 cm long, 6 cm wide; flowers 10 cm long, delicately fragrant, pure waxy white; anthers yellow. Flowers bloom from June-July.

L. elegans

Bulb spherical, 5 cm in diameter, white or pale yellow, often tinged red or purple; stem 45 cm high, smooth, top slightly pubescent; leaves 10 cm long, 1.5 cm wide, horizontal or semi-erect; flowers 10-15 cm diameter, orange-red, slightly spotted purplish black; anthers red. Flowers bloom from June-July.

L. formosum

Bulb spherical, white with red tinge, 3 to 5 cm in diameter; stem 30 to 45 cm high, smooth and green; leaves 5 to 10 cm long, 0.3 to 0.8 cm wide, horizontal, upper leaves semi-erect and green; flowers 2 to 4 per plant, diameter 10 to 15 cm, orange-red, slightly spotted black; anthers red. Flowers bloom from June-July.

L. giganteum

Bulb oval, deep green, thick; stem 1.5 to 2 cm high, stout base, diameter 6 to 8 cm; leaves deep green, 30 to 45 cm long, 20 to 25 cm wide; flowers racemes, 4 or 5, pure white, 12 to 15 cm long; anthers yellow. Flowers bloom from July to early August.

L. hansonii

It is native of Korea and Siberia. Stem 120 to 150 cm high, slender, terete; leaves in whorls, 10-12.5 cm long; flowers fragrant, golden yellow with dark spots; petals reflexed about 1 m in length. Flowers bloom in the month of June.

This extremely winter-hardy species thrives in a lightly shaded location. Due to the formation of stem roots, the bulb should be planted deeply. *L. hansonii* has been used extensively in breeding in the production of Hansonni and Backhouse hybrids.

L. henry

This species is named after the famous plant collector Dr. A. Henry who gathered plants in China at the end of the nineteenth century. Plants about 2 m tall; stem arching, purple-brown, 1.5-2.4 m high; leaves closely set, lanceolate; flowers 4 to 20, orange-yellow, reddish-brown dotted, drooping, 7.5-9.0 cm long. Flowers bloom from July-August. The flowers fade in the sun hence it is best to plant them in light shade.

L. longiflorum (Easter lily, Trumpet lily)

It is the most important *Lilium* species, especially for pots and cut flower production particularly favoured for church decoration. The species is native to

Japan. Plant 50-90 cm high; scattered leaves, 10-12 cm long, linear, acute; flowers pure white, trumpet-shaped, delightfully scented. Flowers bloom from May-June.

L. martagon

Unlike other lilies from Asia, *L. martagon* is native to southern and middle Europe. Because of its unique flower shape and sharply reflexed petals it is also known as the Turk's-cap lily. Bulbs are ovoid, 2.5 to 3.0 cm thick, bright yellow scales; stem 60 to 120 cm long, erect, hairy; flowers wine red with many speckles and spots. Other such variantis *Album*.The flower stems about 1.0 to 1.5 m long sometimes about 2.0 m; flower white, pendulous, many. This species is winter-hardy with wide adaptability for soil. Flowers bloom from June-July.

L. nepalense

Bulb globular, diameter 8 to 10 cm, scales broad thick, purple; stem smooth, 60 to 80 cm high; leaves erect, 8 to 10 cm long, 3.5 to 4.0 cm wide; flowers greenish yellow, dark red inside, fragrant pendulous, 4 or 5, 10 to 12 cm long; anthers yellow. Flowers bloom from June-July. Profuse fruiting is observed.

L. neilgherrense

Bulb globular, scales broad, thick, white, tinged purple, diameter 5 to 8 cm; stem smooth, stout, deep green, 60 to 80 cm high; leaves erect, 5 to 8 cm long, 0.5 to 1.5 cmwide, green; flowers 2 to 3 long, segments waxy, light yellow colour inside, edge white, tinged purple outside, 15 to 20 cm long; anthers yellow. Flowers bloom from July-September.

L. pardalinum

Bulb isa scaly rhizome. Stem 90-180 cm high, smooth; leaves 10 cm long, 2.5 cm wide, in 3 or 4 whorls of 9-15 leaves; flowers 10 cm in diameter, segments reflexed, bright yellow, spotted brownish purple; anthers red. Flowers bloom from late June to early August.

L. philippinense

Bulb globular, 4 to 5 cm in diameter, stem smooth, slender, 60 to 80 cm high, green; leaves 8 to 10 cm long, 1.5-2.0 cm wide, upper ones slightly recurved; flowers 4 to 5, solitary, pure waxy white, base tinged green, fragrant, 15 to 20 cm long; anthers yellow. Flowers bloom from late June to July. Profuse fruiting is observed.

L. polyphyllum

Bulb oblong, 6 to 8 cm long, 3 to 5 cm wide, scales narrow white closely clasped together; stem 60 to 100 cm high, smooth; leaves scattered, erect, 10 to 12 cm long; flowers 4 to 5, pale creamy yellow, tinged and dotted purple, fragrant, diameter 3 to 5 cm; anthers red. Flowers bloom from June-July.

L. regale

Bulb globular, diameter 5 to 12 cm, purple; stem 60 to 80 cm high, purplish leaves thickly scattered, recurved, 10 to 15 cm long, 0.5 to 0.8 cm wide; flowers 3 to

4, small, white, base tinged yellow, midribs purple, throat sulphur-yellow, funnel-shaped, 10 to 15 cm long; anthers reddish brown. Flowers bloom from mid-June to mid-July. Profuse fruiting is observed.

L. speciosum

Bulbs are globular, 10 cm in diameter, reddish purple or brownish, rarely yellow or orange, scales broad and thick. Stem 120 cm high, smooth, green, often tinged and dotted purple; leaves 15 cm long, 3 cm wide; flower 15 cm diameter, delicately fragrant, segments reflexed, white or suffused rose-pink, spotted blood-red, green stripe at base; anthers red. Flowers bloom from late August to September.

L. thomsonianum

Bulb oblong, 6 to 8 cm long, 3 to 4 cm wide; stem 50 to 60 cm high, erect; leaves crowded at stem base, 45-50 cm long, 1.8-2.0 cm wide; flowers rose-pink, 10 to 12, raceme, 4 to 5 cm long, bell-shaped; anthers purple. Flowers bloom from July-August. Profuse fruiting is observed.

L. tigrinum

It is called as 'The Tiger-lily of Japan'. Bulb spherical, diameter 5 or 6 cm, white, sometimes pale yellow; stem stout, 80 to 100 cm high, deep purple-brown and hairy, producing deep purple bulbils in the upper leafaxils; leaves deep green, 8 to 10 cm long, 1.5 to 2.0 cm wide; flowers 10 to 12, bright salmon-red, spotted purple-black, diameter 8 to 10 cm; anthers red. Flowers bloom from July-August. Profuse fruiting is observed.

L. tigrinum var. *flore-pleno*

It is identical of *L. tigrinum* except the flowers are double. It is the only desirable double-flowered lilium in cultivation.

L. wallichianum

Bulb globular, diameter 5 to 8 cm, scales, edges slightly notched, broad, thick reddish; stem 1.5 to 2.0 m high, green and smooth; leaves erect, 10 to 12 cm long, 1.5 to 2.0 cm wide, ascending, linear, lower 20 to 25 cm long; flowers 2 to 3, 20 cm long, delicately fragrant, waxy, creamy white, greenish towards base; anthers yellow. Flowers bloom from July-August.

L. wilmottiae

Bulb globular, diameter 4 or 5 cm, purple; stem 1 to 1.5 m high, smooth, stout; leaves numerous, linear, spirally arranged, 8 to 12 cm long, 1.0 to 0.7 cm wide; flowers 8 to 12, pendulous, small, 8 to 10 cm diameter, recurved, orange-red and spotted reddish brown; anthers yellow. Flowers bloom from July-August. Profuse fruiting is observed.

From bulb morphogenesis, Baranova (1971) classified 35 introduced *Lilium* spp. into 4 groups according to the rate of development. Species within each group showed similarities in terminal bud formation and dates of initiation, life span of organs and geographical origin. The earliest flowering species were *L. kesselringianum*

Species of Lilium

Lilium columbianum

Lilium davidii

Lilium formocerum

Lilium henryi

Lilium martagon

Lilium speciosum

Lilium japonicum

Lilium longiflorum

Lilium pumilum

Lilium maculatum

Lilium auratum

Lilium amabile

and *L. monadelphum* and the latest ones are *L. speciosum* and *L. auratum.* Ohkawa (1982) in his study with 15 lilies which are native to Japan, classified the minto 3 distinctly separate groups according to flower differentiation.

4.3.0 Cultivars

Most lily cultivars originating from intra-sectional hybridizations of the genus *Lilium* are classified into four groups: Asiatic (A), Longiflorum (L), Oriental (O) and Trumpet (T) (Van Tuyl *et al.,* 2000). Hybridizations within each group are usually straightforward and their F_1 hybrids are fertile, however, those between different groups need cut-style pollination and embryo rescue, and such distant F_1 hybrids are highly sterile (Zhou *et al.,* 2008).

Ornamental lilies could be classified into five major groups as given below:

Asiatic lilies: They are the early in flowering, usually in June with as many as 12 fragrant flowers per stem. They are useful for cutting, forcing and borders, good for a late summer garden. Large flowers bloom a top tall stems. Asiatic lily cvs. are White Kiss, Latoya, Barcelona.

Oriental lilies: They bloom between July and September. They come in white, pink and red, and are usually spotted or striped with strong fragrance in the evenings. Oriental lily cvs. are Antonia, Casa: Blanca, Muscadet, Star Gazer.

Tiger lilies: They bloom from July through September, and grow to a height of about 1.4 m. The usual colours are orange, pink, yellow and red.

Trumpet lilies: They are also called Aurelian lilies. They have large, trumpet shaped blooms, and grow up to 1.6 m. Blooms appear from late June through August. Such cultivarsareAfrican Queens, Regale, Golden Splendour.

Turk-cap lilies: This species of lily is turban shaped and shorter in stature than the other lilies, reaching a height of only 2.5 m.

The Royal Horticultural Society of UK and the North American Lily Society have classified lilies as follows:

Division 1: Asiatic hybrids were once known as Mid-Century hybrids because they were developed around the mid-point of this century by hybridizer *Ian* de Graaff of Oregon. Enchantment, his first of the mid-century hybrids produced erect flowers. This characteristic revolutionized the use of lilies in the cut flower industry. By the 1980s, the category of mid-Century hybrids had grown so large that the name was changed to Asiatic. They should be planted in late spring at 10-12 cm deep. Roots develop along the stem. The following is a summary of the most important varieties.

Alaska - white (100-110 cm); Apeldoom - orange (100-110 cm); Compass - orange (80-90 cm); Connecticut Kind - yellow (90-100 cm); Cordelia - yellow (100-100 cm); Dreamland- yellow (90-100 cm); Elite - orange (130-140 cm); Enchantment - orange (90-100 cm); Gran Paradiso - red (110-120 cm); Hilde -yellow (90-100 cm); London-yellow (120-130 cm); Mona- yellow (100-110 cm); Monta Rosa - pink (120-130 cm); Montreux - pink (110-120 cm); Nove Cento - yellow (110-l20 cm); Pollyanna- yellow (120-130 cm); Roma-creamy white (130-140 cm); Sancerre - white (80-90 cm); Sunray

Cultivars of Lilium

African Queen

Amarone

Black Beauty

Bernini

Black Spider

Citronella

Destiny

Gelria

Gironde

Cultivars of lilium

Eyeliner

Harmony

Cassa

Vermeer

Fata Morgana

King Peter

Pink Pixie

Morning Star

Sorbone

- yellow (70-80 cm); Toscana - pink (100-110 cm); Vivaldi - pink (100-110 cm); Yellow Submarine - yellow (90-100 cm).

Division 2: Turk's cap hybrids of *L. martagon* and *L.hansonii* were raised in Holland by van Tubergen. Turk's cap is mainly woodland plants. They prefer shade.

Division 3: Includes hybrids of *L. candidum,* the madonna lily.

Division 4: Hybrids of native North American lilies, mostly progenies of tall-growing *L. parryi* and *L. parda/inum.* Flowers are reflexed, which means that the petals curved back.

Division 5: *L. longiflorum.* The white Easter lily commonly forced for spring bloom. In nature it blooms mid-summer.

Division 6: Trumpet lilies, derived from Asiatic species. The Aurelia and Olympic hybrids are part of this group. Some have pendant flowers.

Division 7: Oriental hybrids, mostly derived from *L. auratum* and *L. speciosum.* Flowers are usually very fragrant, large and shaped like bowls.

The following are the most commonly grown cultivars:

Acapulco - cyclamen pink (120-140 cm) upward flowers; Casa Blanca - white (120140 cm) horizontal flowers; Cascade - light pink/dark pink (130-140 cm) upward flowers; Dame Blanche - white upward flowers; Laura Lee - pink (100-120 cm) horizontal inflorescence; Le Reve (Joy) - 80-100 cm long, upward inflorescence; Marco Polo - white with light pink edge (100-110 cm) upward inflorescence; Mona Lisa - pink (50-60 cm) horizontal inflorescence; Olympic Star - pink (100-120 cm) upward inflorescence; Star Gazer - pink-red (80-100 cm) upward inflorescence.

Division 8: Includes all other hybrids.

Division 9: Lily species, some are still commonly available.

5.0 Genetics and Breeding

5.1.0 Genetics

The diploid chromosome number of *L. longiflorum* is 24. Creole Easter lily was also found to be diploid. All the diploid cultivars are cross-fertile and the seedlings obtained by crossing are highly variable, indicating considerable heterozygosity of the parents.

Noda (1978) noted that *L. lancifolium* (*tigrinum*) is normally a triploid (2n=36). Diploid (2n=24) and triploid forms were found growing in the Tsushima Islands in Japan. Their general morphology was very similar. A diploid yellow-flowered form, var. *flaviflorum,* had a similar karyotype to that of the normal reddish orange diploids.

5.2.0 Breeding

According to Shimizu (1987), as quoted by Lim *et al.* (2008), the breeding history of lily traces back to more than 200 years.However, the real breakthroughs on lily breeding are only recent 50 years when assortments of Asiatic hybrids were bred. According to International Lily register, more than 9,465 lily cultivars have been registered till 2007 (Matthews, 2007). Among them, number of division I (Asiatic hybrids) is the most superior as compared to other divisions. Recently the number of cultivars involving interspecific hybrid has increased rapidly. The reason is that interspecific hybrids that have more genetic variation and show distinct characteristics than the existing cultivars bred from the cultivars or species of each section alone (Lim *et al.*, 2008). Lily comprises more than 80 species belonging to 7 sections. Within the sections cultivars bred from Sinomartagon, Archelirion, and Leucolirion are the most important in the commercial market. At this moment, the most promising breakthrough in lily breeding is the raising of new cultivar through interspecific hybridization with introgression of useful genetic traits from species or breeding materials belonging to the wild species which are not commonly used so far for commercial breeding. Lim *et al.* (2008) reported that they have been crossing almost all different cross combinations and have succeeded in more than 28 cross combinations since 1980. The F_1 hybrids between the species have shown the intermediate phenotypic characteristics. Making interspecific or intergeneric hybrid is laborious but finding the clues, affecting to the most successful embryo formation, and growth is even more difficult and time consuming. Therefore, research on the successful interspecific hybridization between distantly related species was carried out at several laboratories in the world. For the successful interspecific hybridization breeding, not only production of F_1 inter-specific hybrids but also successful production of subsequent generations using interspecific hybrids to introgress valuable trait(s) is/are important. One of the promising crosses is of course between Orientals and several species such as *L. henryi*, Asiatics, and trumpet lilies. In this context, Lim *et al.* (2008) demonstrated the possible methods and some valuable instances of the interspecific hybridization in lilies.

Breeding research in lilies is being carried out in many countries. An ideal lily should be easy to propagate and induced to flower and have a short forcing time, long postharvest life, genetically controlled height, low nutritional requirement, pest resistance with wide array of flower colours and forms (Dole and Wilkins, 1996). The breeding objectives, though many are common, vary in different countries depending upon the local needs. According to Van Tuyl and Holsteijn (1996) the main areas of research in Netherlands are breeding for resistance to *Fusarium oxysporum*, *Pythium* and viral diseases, breeding for quality traits (flower longevity, forcing ability, bulb growth of *L. longiflorum*), interspecific hybridization and genetic modification. A molecular marker system using RAPD is being developed to link *Fusarium* resistance with molecular markers and to construct a genetic map of lily. By combining pollination techniques to overcome pre-fertilization barriers with *in vitro* methods to overcome post-fertilization barriers, wide interspecific crosses can be made more efficiently. Fiedler (1999) suggested that flowering ability, rapid propagation and disease resistance should be the main objectives.

5.2.1 Interspecific Hybridization

A large number of reports are available on interspecific hybridization in lilies to meet the breeding objectives. The importance of lily has increased due to the development of Asiatic hybrids, which are derived from about 8 species of a total of more than 80 species in the genus *Lilium*. Widening the genetic basis of the commercial cultivars by introducing desired characters (*e.g.*, virus and *Fusarium* resistance from *L. henry;* low temperature and low light tolerance from *L. candidum*, year-round forcing ability and vigorous growth from *L. longiflorum*)is one of the major objectives of interspecific hybridization (Van Tuyl, 1985; Van Tuyl *et al.*, 1986b).

Lilies are widely distributed in Japan. From analysis of karyotypes and numerical studies, *Lilium* species native to Japan are classified into 4 or 5 groups. Eight out of the 15 species grown in Japan are native to Japan, including *L. longiflorum* and *L. speciosum*, which are also distributed in Taiwan and China. Private plant breeders and research stations have developed many varieties of Asiatic Hybrid. Some breeders started crossing *L. formosanum* with *L. longiflorum* as early as 1928 to develop *L.* × *formolongi* which flowered within a year from sowing seeds and has broad leaves like *L. longiflorum*. As *L.* × *formolongi* is propagated by seeds, there is no problem of virus infection during forcing for cut flower production (Okazaki, 1996). Van Tuyl*et al.* (1997) used cut style and grafted style methods to produce wide interspecific crosses between species and cultivars from different sections of the genus *Lilium*. Later they (Van Tuyl *et al.*, 2000) used genomic *in situ* hybridization to produce wide interspecific lily crosses with species and cultivars from the different sections of the genus *Lilium* (*L. longiflorum, L. henryi, L. canadense, L. pardalinum, L. concplor, L. dauricum* [*L. pensylvanicum*], *L. candidum, L. rubellum, L. martagon*, Asiatic and Oriental hybrids). Hybrids originating from intersectional crosses have been produced (*e.g.*, *L. longiflorum* ×*L. concolor, L. longiflorum* × *L. dauricum, L. Longiflorum* × *L. henryi, L. longiflorum* × *L. martagon, L. longiflorum* ×*L. candidum, L. longiflorum* × Asiatic hybrids, *L. longiflorum* × Oriental hybrids, *L. longiflorum* × *L. rubellum, L. longiflorum* × *L. canadense*, Oriental ×*L. pardalinum*, Oriental × Asiatic hybrids and *L. henryi* × *L. candidum*). The production of Oriental × Asiatic hybrids is a breakthrough in lily breeding. In general, wide interspecific lily hybrids show F_1sterility. Using chromosome-doubling techniques, tetraploids with restored fertility are produced from these diploid hybrids. Kim *et al.* (1998a, 1998b) produced an intermediate type from an interspecific hybrid between *L. longiflorum* cv. Nellie White (NW) and *L. callosum* (LC). NW flowers are white; LC flowers are dark orange, whereas flowers of the F_1 hybrid are light orange with backward rolled petals. Chin *et al.* (1997) accomplished distant hybridization between 5 local *L. longiflorum* selections and 3 Asiatic or 1 Oriental hybrid lilies by *in vitro* pollination and ovary culture as well as by *in vivo* pollination combined with ovary slice culture. *L. longiflorum* was preferred as the maternal parent. More than 15 combinations were successfully rescued and multiplied by micropropagation or by scaling. Floral bud development and anthesis in 3 clones were heat tolerant. In the current cycle, distant hybrids produced bulbs in size ranges of 12.7 -16.4 cm and 10.813.9 cm under conditions of 20/15°C and 35/10°C, respectively. Additionally, the outfacing, funnel flower pattern of the longiflorum group was expressed as recessive or intermediate to the up-facing, split perianth type. The strong expression of rosette growth and heat tolerance indicated that breeding between distant types might facilitate the development of lines suitable for cut flower and bulb production in one cycle, as well as, adaptation to subtropical conditions. *L. longiflorum* produces white flowers and is widely grown as both potted plant and cut flower. *L. callosum* produces small red flowers and is difficult to grow in a pot. Roh *et al.* (1996a) successfully produced interspecific hybrids between *L. longiflorum* and *L. callosum* rescuing the hybrids through *in vitro* separated carpels and whole ovule culture. Roh *et al*. (1996b) produced interspecific hybrids between *L. longiflorum* cv. Nellie White and a selection of *L.* x *elegans* [*L. maculatum*](D), by cut style pollination ovary slice followed by ovule culture. The growth and flowering of the hybrids resembled that of *L. longiflorum*, although *L.* × *elegans* influenced flower colour, the arrangement of sepals and petals, and bulbil forming characteristics. In order to introduce pink flower colouration into normally white-flowered *L. longiflorum* and *L. jormolongi* (*L. longiflorum* × *L. jormosanum*) Okazaki and Jong (1990) made crosses with *L. rubellum* and *L. japonicum*. Cut style pollination with subsequent embryo culture was used. Hybrids had flowers from nearly white to lighter than in *L. rubellum*. Similar techniques were used to introduce small flower size from *L. concolor* and flowering ability from a small bulb from *L. jormolongi* into Asiatic hybrids. Dwarf hybrids with fragrant pink flowers following *in vitro* embryo culture were produced when *L. amoenum* used as pollen parent in crosses with *L. regale* and *L. longiflorum* (Huang *et al*., 1990).

Zhou *et al*. (2013) studied the female fertility of the odd-tetraploid cultivar Honesty of *Lilium* containing one set of *Longiflorum* chromosomes and three sets of Asiatic chromosomes (LAAA) to open a new approach to Lilium breeding. To assess its female fertility, 'Honesty' was hybridized with four autotetraploid Asiatic lily cultivars, showed that the fruit of all 'Honesty' x tetraploid (4x x 4x) combinations developed well, and obtained viable seedlings, suggesting that 'Honesty', despite being malesterile, has considerable female fertility. Genomic *in situ* hybridization showed that the progenies of the 4x × 4x hybridizations were aneuploid. Considering that lily is vegetatively propagated and aneuploids often demonstrate considerable phenotypic variation, odd-tetraploid lilies such as 'Honesty' may be useful maternal parents for breeding new lily cultivars.

5.2.2 Disease Resistance

Loffler *et al*. (1996) crossed a *Fusarium oxysporum* resistant accession of *L. dauricum* [*L. pensylvanicum*](section Sinomartagon), with the susceptible *L. longiflorum* cultivars Gelria and Flevo (section Leucolirion). The assessment of the progenies indicated that the level of resistance in the most resistant progeny equalled that of the resistant parent. *L. dauricum* can be used as a source of *Fusarium* resistance in interspecific crosses with *L. longiflorum*.

5.2.3 Breeding for Vase Life

Plant characteristics cause only small non-genetic variation in individual flower longevity when compared to inflorescence longevity. Therefore, individual flower longevity is the best criterion to discriminate among longevity levels for a lily breeding programme (Meulen-Muisers and Oeveren, 1996). Meulen-Muisers *et al*. (1997) observed a large variation in vase life longevity of lily within populations tested at the individual plant level. Other postharvest quality characters such as number of flowers per stem, male sterility (lily) also varied within the populations tested. Due to the high correlations between the offspring and their corresponding parental genotypes for several characters, breeding for a considerable improvement of postharvest quality in future cultivars seems promising. Meulen-Muisers *et al.* (1998) observed that considerable genotypic variation in individual flower longevity and estimates of the degree of genotypic determination were high. The rank order of the genotypes with respect to individual flower longevity was similar between years using standardized test conditions. Screening results for flowers forced in a growth chamber were similar to those obtained in a greenhouse. Individual flower longevity of a genotype can be used as an indication for its breeding value. High broad sense heritability (0.79) at the individual plant level indicated that selection for long individual flower longevity can be expected to be very effective in Asiatic hybrid lilies (Meulen-Muisers *et al.*, 1999). General combining ability (GCA) effects were highly significant ($P<0.001$), and the estimated narrow sense heritability was high (0.74). However, deviating results could be expected as specific combining ability (SCA) effects were also significant.

5.2.4 Incompatibility

Self-incompatibilities in *Lilium* have been reported by several workers (Sagawa, 1958, 1959). In general, the effect of genes for self-incompatibility is to retard the growth rate of pollen tubes in the style. This results in either failure of the pollen tubes to reach the ovary or degeneration of ovules and causes a serious obstacle in the improvement of this crop.

Ascher and Drewlow (1975) reported that the styles of *L. longiflorum* express two types of interspecific incompatibilities, short-growth interspecific incompatibility (pollen tube growth inhibited at the base of the stigma) and half-growth incompatibility (pollen tubes grown half way down the style). Short-growth interspecific incompatibility in *L. longiflorum* styles is often associated with species combinations exhibiting unilateral interspecific incompatibility where the reciprocal combination results in normal compatible pollen tube growth. Half-growth incompatibility in *L. longiflorum*styles usually occurs in species combinations resulting in reciprocal incompatibility. They also recorded that injection in the styles of *L. longiflorum* cvs. Ace and Nellie White with stigmatic exudate (collected from *L. longiflorum*flowers) before pollination by *L. regale, L. formosanum* and 14 hybrid lily cultivars significantly increased the length of pollen tubes of *L. regale, L. formosanum* and 7 of the hybrids, all of which expressed short-growth unilateral interspecific incompatibility in non-injected styles. Stigmatic exudate injection had no effect on, nor significantly retarded pollen tube growth of the 7 hybrids expressing reciprocal interspecific incompatibility with *L. longiflorum,* indicating that the inhibition of short-growth pollen tubes in unilateral incompatible combinations was achieved through a different process than the inhibition-limiting pollen tube growth in reciprocal interspecific incompatibility.

There is a possible relationship between the intensity of the self-incompatible reaction and the rate of deformed pollen tubes having bulbous or rugged tips (Li and Niimi, 1995). *Lilium longiflorum* shows gametophytic self-incompatibility and the self-incompatibility response of *L. longiflorum*does not involve a pollen tube growth inhibitor (Amaki and Yamamoto, 1988).

5.2.5 Overcoming Incompatibility Barrier

Techniques to overcome the barriers between species, cut style pollination, embryo culture, *etc.*, have been developed in Japan. In many of the interspecific crosses pollen tube growth ceases in the style and pollinating cut styles results in only a low percentage of fertilization (Bino *et al.*, 1988).

Taxonomically, the genus *Lilium* can be divided into 7 sections. *L. longiflorum* can be crossed successfully with *L. candidum,* the Asiatic hybrid Mont Blanc and *L. rubellum.* To restore the fertility of the F_1 hybrids, tetraploid forms are produced by colchicine treatment. The cross between Shikayama ×*L. henryi* and *L. speciosum* and the oriental hybrids Allegra and Journey's End resulted in 11 (sterile) triploid hybrids. Backcrosses with oriental hybrids and *L. auratum* ×*L. henry;* were successful in a few cases, even without embryo culture (Van Tuyl*et al.,* 1986b). According to Brandham (1982), as quoted by Zhou *et al.* (2014), it is well known that most Polygonum-type triploid plants, such as triploid watermelon and banana, are sterile and seedless, so are usually not the ideal sources for further introgression breeding. In contrast, the triploid lilies ($2n = 3x = 36$), both autotriploid (AAA) and allotriploid (AOA, LAA, LLO), can be used as maternal parents to hybridize with appropriate diploid ($2n = 2x = 24$) or tetraploid lilies ($2n = 4x = 48$), although they are also male sterile due to abnormal meiosis (Zhou *et al.*, 2011, 2012). The basis for this is the difference in the embryo sac formation between Polygonum-type and Fritillaria-type plants. From normal megasporogenesis of the Fritillaria-type embryo sac, Zhou (2007), quoted by Zhou *et al.* (2014), deduced that triploid lilies produce aneuploid eggs and hexaploid central cells (secondary nuclei). Based on the crossability of these $3x \times 2x/4x$ interploid hybridizations, the "Five same genomes of endosperm are essential for its development in *Lilium*" ("five-same-genomes") hypothesis was proposed,to explain the success or failure of $3x \times 2x/4x$ crosses in *Lilium* (Zhou *et al.*,2012). Based on this hypothesis, "Five same genomes of endosperm are essential for its development in *Lilium*", it is expected that allotriploid lily (OTO) can be hybridized with diploid Oriental lily (OO) for introgression breeding in *Lilium* L. To test the hypothesis, OTO lilies, 'Belladonna', 'Candy Club' and 'Travatore', were used as the maternal parents and crossed with two diploid OO cultivars, 'Siberia' and 'Sorbonne', and the species *L. regale* Wilson (TT). Results showed that capsules of all OTO × OO hybridizations developed well and 0.8~3.3 viable seedlings per ovary were obtained through normal pollination and embryo rescue; however, all OTO × TT crosses failed. Genomic *in situ* hybridization showed that the progenies of the OTO × OO hybridizations were aneuploid and a variable number of T-genome chromosomes were introduced into the progenies through the allotriploid lilies. These results not only demonstrate that allotriploid OTO lilies, although male sterile, can be used as maternal parents to produce aneuploid progenies, but also strongly support the new hypothesis in lily breeding (Zhou *et al.*,2014).

The presence of recombinant chromosomes in the progeny plants can be translated into genotypic and phenotypic variation, which contrasts with the lack of phenotypic variation in the progeny plants obtained by the use of induced allopolyploids. But more important is that the recombinant segments of chromosomes from different parental origin will be maintained in the progeny through generations, conserving the desired agronomical traits. As was mentioned earlier, the use of *2n* gametes was discarded by breeders due to their sporadic production, however, with N_2O treatments of flower buds in early meiotic stages it has been possible to increase their production and in some cases to induce them in completely sterile hybrids, bringing new possibilities for interspecific hybridization in lilies (Gonzalez *et al.*, 2008).

Emsweller *et al.* (1960) reported that naphthalene acetamide (NAAm) used at the time of pollination delayed senescence of the embryo sacs for 26 days while the ovules of the untreated capsules were completely shrivelled. Capsules from treated flowers increased in length and diameter during the 26-day period. The use ofNAAm also greatly increased seed set. When sap was collected by centrifugation from the stigma or style and applied to the stylar canals of bisected pistils, pollen tube growth was accelerated even following self-pollination (Amaki *et al.*, 1989a). The movement of sap from thestylar tissue to the stylar canal differs between self- and cross-combinations, even in the self-combination there is no substance inhibiting pollen tube growth, and certain proteoglycans are associated with self-growth (Amaki *et al.*, 1989b). Self-incompatibility of *L. longiforum* has been reported to be overcome by heating the styles and overcoming self-incompatibility by heating is temporary and the incompatibility mechanism is already induced at about 3 h after pollination (Hiratsuka *et al.*, 1989). Identification of interspecific hybrids and determination of relationships between species in the genus *Lilium* by isoelectric focusing was reported by Van Tuyl *et al.* (1986a, 1989).

6.0 Growth and Flowering

6.1.0 Environmental Factors

Pre- and post-planting environments are important for growth and flowering of lilies. Among various factors, light, temperature and their interactions play major roles.

6.1.1 Light

Both intensity and duration of light influence the growth and development. Low light intensity, though increased the plant height reduced the percentage of dry weight in *L. longiflorum* (Kohl and Nelson, 1963). Einert and Box (1967) observed that flowering in this species was adversely affected when the plants were grown under 50 per cent shade. Daily supplementary lighting of 3200 and 320 lux for 6 hours inhibited flower abscission and flower abortion (blasting) in cultivars Enchantment and Corina, respectively, grown between October and December (Sponga and Lercari, 1986). Bulblet formation was greater on basal tissues than on distal tissues in light intensity of 1000 lux and dark, those formed in dark conditions were large than those in the light (Schenk, 1975).

Long-day treatments promoted vegetative growth (Langhans and Smith, 1965). Heins *et al.* (1982) reported that height of the plants was reduced when they were forced under 8 h photoperiod compared with natural-day-forced plants. Plant height of cvs. Ace and Nellie White were reduced by 29 and 45 per cent, respectively, when forced under short day from emergence to flowering, by 19 and 42 per cent when forced from 30 days after emergence to flowering and by 20 and 20 per cent when forced from visible bud to flowering. Receptive light/dark cycles of 4, 6 or 12 h had no effect on flower bud development rate from the time buds were 6-12 cm long to anthesis.

Table 1: The Effect of Photoperiod and Duration of 17.2·C (63°F) Growing Periods Before 4.4°C (40°F) Storage on the Growth and Flowering of Japanese-grown Georgia Lilies*

Environmental Regimes	*Days to Flower*	*Avg. No. of Flowers*	*Plant Height at Flowering (cm)*	*No. of Leaves*
16-h day length				
3 weeks at 63° + 3 weeks at 40°	129	4.5	83	87
3 weeks at 63° + 6 weeks at 40°	139	6.0	92	86
6 weeks at 63° + 6 weeks at 40°	140	5.6	83	95
10 weeks at 63° + 6 weeks at 40°	156	6.0	103	118
Continuous 63°	100	5.4	126	150
Natural day length				
3 weeks at 63° + 3 weeks at 40°	148	5.6	68	100
3 weeks at 63° + 6 weeks at 40°	154	7.0	70	89
6 weeks at 63° + 6 weeks at 40°	168	7.8	62	115
10 weeks at 63° + 6 weeks at 40°	183	7.5	61	132
Continuous 63°	220	6.6	72	209
HSD5 per cent	17	1.8	13	25
HSDI per cent	20	2.1	16	19

* Carlson and Catpenter (1970).

According to Wilkins and Waters (1967), increasing photoperiod, when plants first emerged, caused earlier floral initiation and flowering of non-refrigerated Georgia Easter lilies. This caused non-precooled bulbs to flower a few days earlier than pre-cooled ones and 8 weeks earlier than non-precooled bulbs grown at natural winter night lengths. Wilkins *et al.* (1968) later suggested lighting for 2 weeks at 15 to 30 ft candle from 10 p.m. to 3 a.m. beginning at plant emergence. Carlson and Carpenter (1970) reported that artificial fluorescent light of 5 watts per ft^2 to create a 16 h day length was found adequate to reduce the 'controlled temperature forcing' (CTF) effects. The day length and duration of 11.2°C growing period prior to cold treatment affected the days from bulb planting to flowering, number of flowers, plantheight and number ofleaves (Table 1). These results were similar at both 10 and 16 h day lengths. The effect of 6-weeks 4.4°C cold period declined at both photoperiods as the duration of 17.2°C period before storage increased. Delaying by 4.4°C cold period until 10 weeks after planting at 16 h day length resulted in flowering only 4 days earlier than those receiving no cold treatment. The conventional forcing period of *L. speciosum* cultivars in glasshouse is from 150 to 160 days. This period was shortened by 24 days in the winter by long day treatment with incandescent light. Flower production in the cv. Enchantment in winter is risky owing to sufficient sunlight. Flower bud abscission and flower bud blast can be prevented by irradiating the plants intensively with high pressure mercury lamps (Boontjes *et al.*, 1975). Bulbs of *L. Zongiflorum* cv. Nellie White, forced under an 8 h duration of fluorescent light at 400 µmol/m^2/s, produced acceptable plants (in terms of flower number) for commercial purpose, but unacceptable plants under a 2 or 4 h duration treatment (Wilkins *et al.*, 1986).

Lercari *et al.* (1987) reported that cultivars Enchantment and Connecticut King were grown in glasshouse under mercury and sodium vapour lamps (3350 lux) which were lighted for 8 h/day prolonging the photoperiod from 16 to 24 h. The growth cycle was reduced by 2 days and the flower stalks were shortened and reduced the number of aborted flowers. It was also observed that Connecticut King was very sensitive to light and supplementary lighting reduced the number of blind shoots and blasted buds (Anon., 1979). According to Roh and Wilkins (1974a), total phytochrome decreased by 25 per cent in 4 h after the initial 10 min red light treatment.

6.1.2 Temperature

Vernalization stimulates early emergence. Growth acceleration, as reflected in earlier shoot emergence, internode elongation and rapid leaf unfolding were evident following 6-weeks storage at 4.4°C (40°F) but prolonged (18 weeks) treatment drastically reduced subsequent initiatory activity and rate of leaf unfolding. Although 6-weeks vernalization at 4.4°C (40°F) reduced the number of leaves, increased internode length and the vernalized plants were taller (74 cm) than the controls (50 cm). Extending vernalization to 18 weeks markedly reduced leaf number but did not further increase internode length, consequently these plants were shorter than the non-vernalized ones (Wang *et al.*, 1970). Pergola and Roh (1987) observed that bulbs offully vernalized Asiatic hybrid lily cvs. Inferno flowered in 126 days and Sunray in 143 days, while Red Carpet flowered in 138 days. The cultivars also showed variation in leaf number ranging between 48 and 86 (Table 2).

Table 2: Number of Days to Shoot Emergence and Flower and Number of Leaves of three Asiatic Hybrid Lilies (Number of leaves unfolded was calculated by dividing the number of leaves by the number of days from shoot emergence to flowering)*

Cultivar	*Weeks at 5°*	*Shoot Emergence (A)*	*Days to Flowering (B)*	*From A to B*	*No of Leaves*	*No. of Leaves Unfolded per Day*
Inferno	8	80	126	46	80	1.7
Red Carpet	8	82	138	56	86	1.5
Sunray	8	85	143	58	48	0.8

* Pergola and Roh (1987).

Precooled bulbs from the July and August harvests produced plants with longer leaves, and those from the September and October harvests were shorter than those from corresponding non-precooled bulbs. Plants from July harvested pre-cooled bulbs had the longest leaves and with successively later harvests leaf length decreased sequentially (Blaney and Roberts, 1966). Matsuo and Van Tuyl (1984) found that a higher storage temperature promoted leaf emergence from scale bulblets of *L. longiflorum* cv. White American, whereas a lower temperature delayed it. Higher storage temperature produced more epigeous type plants, especially from outer and middle scales. The variation in temperature also influenced the bulb formation. Wider temperature variation gave better results than constant high temperature which failed to induce bulb formation (Gries, 1970).

Shimizu and Asahira (1979), in *L. longiflorum* cvs. Georgia and Hinomoto found that rosette formation occurred when non-chilled bulbs received light at the apical part of the shoot and were kept at high temperatures (20° or 30°C) and in day lengths of 9 or 15 h. Treatment less than 20°C and photoperiod longer than 15 h induced bolting, the optimum temperature being 10°C.

Temperature also plays an important role in flowering. Commercially forced Easter lily bulbs are precooled to stimulate flowering in a particular length of time. As precooling time is increased, both flower number and time to flowering are decreased (Laurie *et al.*, 1958). Gill (1964) recorded that precooling of bulbs at 1.1-1.7°C (343°F) for 2 or 4 weeks before potting and plunging the pots outside resulted in earlier flowering when the plants were brought in for forcing in November and December, but did not flower so consistently when they were brought in during January. According to Lin and Wilkins (1975) bulbs treated at 4.5°C/30 days, followed by 21°C/30 days had higher GA activity and the shoots were slower to flower than from those first treated at 21°C/30 days then given 4.5°C/30 days. Earliest flowering of mature bulbs was obtained following bulb storage at 7.2-10°C (45-50°F) in moist peat for 5 to 6 weeks (Stuart, 1967). Roh and Wilkins (1975) reported that from floral differentiation to the stage of visible buds and from visible buds to anthesis day/night temperatures of 18.3/15.6°C and 21.1/18.3°C, respectively, were optimum in Easter lily cvs. Nellie White and Ace.

Healy and Wilkins (1984) found that flower buds of Nellie White developed from visible bud to open flower more rapidly at a constant temperature of 27°C than at 15°C. This increased growth rate was more pronounced when buds were less than 6 cm long. Once buds were longer than 10 cm, the difference in rate of flower bud growth at 15°C compared with that at 15°Cwas not significant. Bud development was biphasic, with a relatively slow growth rate up to 6 cm and then an accelerated rate from 6 cm to open flower.

Interaction between temperature and day length showed various responses in lily. Roh and Wilikns (1977a) recorded that under 16 h photoperiod a temperature of 7.2°C favoured secondary bud formation, while at 15.6°C night temperature favoured tertiary bud formation. The increase in flower number was due to increased number of secondary and tertiary flower buds. A 12 h photoperiod from the visible bud stage to anthesis, and 21.2/18.3°C day/night regime were optimal for early flowering with the fewest aborted tertiary buds. They (1977b) suggested that bulb vernalization and photoperiod treatments led to rapid flowering through two inducing mechanisms, the phytochrome and high energy reaction systems.

Durieux (1978) reported that forcing of Enchantment lilies in the winter was impeded by the occurrence of flower bud abscission and blasting. Higher temperature in the glasshouse promoted bud abscission, but the light factor was critical. The flower buds are most likely to undergo abscission when they are about 1-3 cm long. On the basis of a temperature programme, forcing period and an obligatory period of supplementary illumination, a schedule can be made for additional lighting for several batches of lilies during winter forcing.

6.2.0 Growth Regulators

Influence of growth regulator on growth and flowering of lilies has been studied. Rob (1982a) reported that emergence of second scale (breaking of dormancy) of *L. lancifolium* was promoted with IAA but GA_3 and kinetin were ineffective. Thirty days' cold treatment was needed to break dormancy and it also produced the highest bulb weight. Henny and Ascher (1975)found that pre-pollination injectionsof 50ppm kinetin produced variable results while injection of 1 or 10 ppm kinetin had no effect on pollen tube growth. Bulbs of *L. speciosum* cv. Uchida soaked for 48 h in solution of GA_{4+7} at 1000 mg/l alone or in combination with BA at 100 mg/l induced shoot emergence and flowering in dormant bulb (Ohkawa, 1979). Roh (1978) recorded that bulbs of *L. lancifolium* treated with abscisic acid at 100 or 200 ppm delayed sprouting and counteracted the effect of cold.

Treatment with IAA promoted bulblet formation (Hackett, 1969) while CCC @ 100 mg/l inhibited bulbil formation (Ivanova, 1983). A concentration of 500 ppm PBA applied 3 times beginning at 15cm shoot length produced the largest number of bulbils in *L. longiflorum* cv. Nellie White (Nightingale, 1979). The greatest number and size of bulbil were obtained with succinic acid at 100 mg/l. NAA at 25 mg/l, however, gave a similar number of bulbils but smaller in size.

Chemical height control is required for most pot-grown hybrid lily cultivars. Ancymidol is a commercially preferred growth retardant and is normally applied at 0.25 – 0.50 mg *a.i.* per pot as a soil drench after shoot emergence (Ball, 1985). In an experiment on the effect of ancymidol, CCC and ethrel, Simmonds and Cumming (1977) recorded greater retardation of shoot length by immersing the bulbs for 12 h in aqueous solution of the chemicals than by applying as soil drench. Ancymidol was the most effective retardant. Ethrel, though inhibited stem growth in the first season, increased elongation of the renewal shoots and induced early senescence, whereas ancymidol and CCC inhibited growth of the renewal shoots in the following season. Pre-plant bulb-dip in these retardants has been found more effective to regulate the height of Easter lily than either foliar spray or soil drench (Lewis and Lewis, 1982). Commercially produced pre-cooled bulbs of cv. Gold Nugget were treated with ancymidol at 0.25, 0.50 and 1.0 mg *a.i.* per bulb as a bulb-dip prior to planting, as a soil drench at planting and as a soil drench when shoots were 2-3 cm in length. Pre-plant dips were more effective than drenches in reducing the final plant height. Flower number and flowering date, however, were not significantly different among the rates or methods of application (Lewis and Gilbertz, 1987). Gianfagna and Wulster (1986) compared the effects of ancymidol and paclobutrazol on cv. Nellie White by using the chemicals as soil drench (1, 2.5 or 5 mg/plant).

Ancymidol significantly retarded plant height at all concentrations and for both soil drench and foliar methods of application. In addition, ancymidol at 5 mg/pot increased days to anthesis. Paclobutrazol did not retard plant height. Working with *L. regale*, Woodley and Thomas (1973) found that CCC and SADH applied for dwarfing were not satisfactory whereas phosfon was more successful and reduced the height of plants without affecting flowering or distorting foliage when used at low concentrations. Sytsema (1968) recorded that plants of cv. Enchantment sprayed with 50 ml 5 per cent CCC per pot were 15 cm shorter than control and flowered well, whereas *L. speciosum* cv. Rubrum was less responsive to the treatment. Marked improvement in flower production of *L. tigrinum* was also obtained with CCC at 2500 ppm, SADH at 2500 ppm or GA_3 at 10 ppm (Bhattacharjee, 1983). Phosfon treatment of bulbs, before or after storage at 7.2°C (45°F) or during forcing, reduced plant height without much effect on flowering time or flower quality. Bulbs potted in late August and held in a cold frame until late December grew to 52.2 cm when forced, but when phosfon was used (as a drench at 2.25 g per pot) stem height was reduced to 39.2 cm. Phosfon applied as a soil amendment at potting reduced stem height to 25.5 cm (Stuart *et al.*, 1961). In Easter lily cv. Harson, bulbs were precooled at 7.2-10°C (45-50°F) for 0, 2,4 and 8 weeks in damp peat (68 per cent moisture) or dry peat (8 per cent moisture) and then soaked for 2 hours in a 900 ppm solution of GA_3 or tap water prior to forcing. Bulbs precooled in damp peat for 4 to 6 weeks had the shortest stems and earliest flowering. Four weeks' precooling in damp peat hastened stem emergence and flowering but reduced flower number compared with those pre-cooled for 6 weeks in dry peat. Bulbs treated with GA_3 and pre-cooled for 2 weeks flowered about 4 weeks earlier without a decrease in flower number compared with bulbs pre-cooled for 0,2,4 and 6 weeks (Laiche and Box, 1970).

Roh and Wilkins (1977c) also noted significant interaction between photoperiod and treatment with ancymidolas evidenced by increased plant height in cv. Nellie White.

Meeteren and Slootweg (1986) in a greenhouse experiment reported that flower bud abscission was reduced from 94 per cent to 17 per cent by foliar spray with silver thiosulphate (STS). When plants with 3 cm long flower bud were placed in darkness for 5 days, a 10-fold increase in ethylene emanation from the bud was observed. This did not occur when the plants were sprayed with STS prior to dark period. Sponga *et al.* (1986) reported that abscission of flower buds of cvs. Enchantment and Corina, induced by low intensity light was inhibited by immersing the bulbs for 24 h in a 2 mM solution of STS before planting.

7.0 Propagation

Generally lilies can be propagated by seed and asexually through vegetative propagation (from bulblets obtained from bulb scales, by bulb division, from hypogeal bulblets, from bulblets obtained from leaves, and from bulbils) (Grassotti and Magnani, 1988).

7.1.0 Seed Propagation

Seed propagation is not a commercial practice for ornamental lilies. Light and temperature are known to affect seed germination in six species of *Lilium* (*L. callosum*,

L. leichtlinii var. *maximowiczii*, *L. concolor* var. *pulchellum*, *L. davidii*, *L. dauricum* and *L. distichum*). Treating the seeds with 24 h illumination gave early germination in all the species tested. The optimum temperature needed for seed germination of *L. callosum* was found to be 20°C. Application of light at 9000 lux was found to be a substitute for low temperature treatment in *L. dauricum* (Yang *et al.*, 2000). Seed priming with 200 mM $Ca(NO_3)_2$ improved the growth and earliness of seedlings (Lee *et al.*, 1999).

7.2.0 Vegetative Propagation

7.2.1 Bulb/Scales

Lilies are commonly multiplied by bulb/scales. Mature bulbs are needed for successful forcing of lilies. Mature bulbs force rapidly and predictably after low temperature treatment. Maturity of the bulbs could be determined by the weight, extent of scale filling, degree of summer dormancy remaining and responsiveness of flower inducing (coldlight) treatments (Roberts and Tomasovic, 1975). Lateral bud removal is a common practice to obtain large bulbs (Blaney and Roberts, 1965). In cultivars Pollyanna and Star Gazer, Dilta *et al.* (2000) reported highest bulblet formation (80 to 100 per cent) with 100 ppm NAA or IBA in FYM + Soilrite (Table 3). They could break the dormancy of regenerated bulblets by treating at 4°C for 6 weeks, but 8 weeks was the most effective, producing 100 per cent break in dormancy in both cultivars (Table 4). Kapoor *et al.* (2000) found vermiculite as the best medium of propagation for bulb scales in lilies. Wang and Breen (1986) found the removal of flower bud when the length of the largest bud was 1 cm to promote premature sprouting, affects plant growth and bulb weight than the bud was at 3.5 cm. Kumar *et al.* (1999) recorded maximum bulb size when flowers were harvested at 10 cm above ground level. Bulb production suffers from problems such as double nose bulbs (*i.e.*, bulbs having daughter meristems) and premature sprouting of the daughter bulbs due to low temperatures during the summer (Van Tyul, 1985). Soaking the bulbs of cv. Hinamoto in hot water (47-48°C) before dry storage promoted the sprouting of bulbs (Shimada *et al.*, 1999).

Grassotti (1997) treated scales at 22°C, followed by a cold treatment (0, 3 or 5°C), for a total of 14 or 11 weeks. Treatment at 22°C delayed emergence and promoted production of many small bulbs. When 22°C treatment was followed by 0°C, thenumber of bulbs decreased and their size increased. Epigeous type plants were found following high temperature treatments in certain varieties like Pollyanna but only at a very low level in some varieties like Snow Queen.

Marinangeli and Curvetto (1997) treated middle scales from bulbs of *L. longiflorum* cultivar Snow Queen, *L. lancifolium*, one Oriental hybrid (Stargazer) and 4 Asiatic hybrids (Enchantment, Connecticut King, Sunray, Cote d' Azur) by immersing in 1 µM aqueous solution of traumatic acid (TA) for 1 h and incubating in peat for 2 months in darkness. Genotypes showed a wide variability in the number (1.1-2.6 bulblets per scale) and biomass (83-295 mg per bulblet) of bulblets. TA produced an increase of 20-40 per cent in the number of bulblets per scale and also a significant increase in fresh mass (20-60 per cent). Using poor quality Cote

Table 3: Effect of Bio-regulators and Potting Mixture on Bulblet Scale Propagation of Lily

Potting Mixture	*Bio-regulators (ppm)*	*Cultivars*											
		Pollyanna						*Star Gazer*					
		No. of Bulblets/ Scale		*Per cent of Bulblet Vormation*		*Diameter of Bulblets (cm)*		*No. of Bulblets/ Scale*		*Per cent of Bulblet Formation*		*Diameter of Bulblets (cm)*	
		NAA	*IBA*	*NAA*	*IBA*	*NAA*	*IBA*	*NAA*	*IBA*	*NAA*	*IBA*	*NAA*	*IBA*
Sand + FYM + Soilrite	0	1.1	I.I	30	40	3.8	3.9	0.4	0.4	20	20	2.0	3.2
	50	1.2	1.9	40	60	4.2	4.4	1.0	1.1	40	50	3.4	4.0
	100	2.6	2.4	90	90	4.3	4.6	1.4	1.4	60	65	3.2	4.4
FYM + Soilrite	0	1.3	1.3	50	50	3.0	4.1	0.7	0.7	30	30	3.2	4.0
	50	2.5	2.2	70	70	4.6	5.0	1.2	1.4	50	60	3.4	4.7
	100	2.9	2.7	100	100	4.8	5.2	1.7	1.5	70	80	3.6	5.0
C.D. at 5 per cent		0.2	0.2	8.0	7.0	0.1	0.2	0.3	0.2	10	12	0.1	0.2

* Dilta *et al.* (2000).

d' Azur bulbs adversely affected the biomass and number of bulblets produced on the scales, and this effect was not overcome with TA treatment (Table 5).

Table 4: Effect of Cold Treatment and GA_3 on Dormancy Breaking of Bulblets in *Lilium* Hybrids*

Cultivars	*Per cent of Dormancy Breaking*				
	Cold Treatment (4°C)			*GA_3*	
	4 Weeks	*6 Weeks*	*8 Weeks*	*50 ppm*	*100 ppm*
Pollyanna	56a[z]	98a	100a	36a	62a
Star Gazer	45b	93b	100a	24b	53b

z: Mean separation in columns by Duncan's multiple range test at 5 per cent level.

* Dilta *et al.* (2000).

Table 5: Effect of Traumatic Acid (TA) on the Number of Bulblets per Scale and Fresh Weight per Bulblet for Several *Lilium* Genotypes*

Lilium Genotypes	*Bulblets/Scale*			*Fresh Mass/Bulblet (mg)*		
	–TA	*+TA*	*Significance*	*–TA*	*+TA*	*Significance*
L. longij1orum	1.5bz	2.0bc	**	127b	251b	**
L. lancifolium	2.2 cd	3.l d	**	136b	230b	**
L × Stargazer	1.3 ab	1.7 b	*	216 c	262 b	**
L × Enchantment	1.1 a	1.2 a	**	83a	144 a	**
L. × Connecticut King	2.5d	3.2d	**	172 c	206 b	NS
L. × Sunray	2.6 d	3.4 d	*	225 c	367 c	*
L × Coted' Azur	2.0 c	2.4 c	**	295d	346 c	**

z Genotypes mean separation within columns using Fisher's protected LSD at $P < 0.05$.

NS, *, ** Non-significant or significant at $P<0.05$ or *0.01*, respectively.

* Marinangeli and Curvetto (1997).

Propagation through scales depended largely on variety. Seradix B no. 1 stimulated rooting but increased bulblet yield in certain varieties (Krause, 1996a). Incubation at 21°C (17-18°C for Arai 5) for 13-14 weeks gave better results than a shorter incubation period. Cooling for short periods (around 6 weeks) was sufficient for rapid growth (Krause, 1996b). Darkness inhibited leaf emergence from bulblets *in vivo* but this was reversed by chilling treatment (4 ± 1°C) for 1 week. Day length *in vitro* is an important factor for the control of bulblet formation from scales, bulblet enlargement and leaf emergence from bulblets *in vitro* (UnNo and Wagatsuma, 1996).

The season of taking the scales for propagation also influences the propagation. Suh and Lee (1995) found that the number, weight and diameter of bulb lets were higher on scales taken on 1 June than on those taken on 1 March or 1 September and the numbers of bulb lets were promoted by IDA (500 or 1000 mg/l), IAA (500 mg/l) or NAA (up to 500 mg/l); bulblet weight was promoted by Rootone (IDA) as well as by BA (1000 mg/l) and kinetin (500 or 1000 mg/l). Low temperature treatment of the mother bulb (2-8 weeks at 4°C) before scaling did not increase bulblet formation, while the number and weight of bulb lets decreased with increasing stratification. Bulblet development was slightly enhanced at 20 and 25°C. Bulblet weight was highest on scales placed in vermiculite. Bulblet formation and root growth were promoted by NAA (10-500 mg/l) in leaf cuttings. However, they observed variation between varieties. Iapichino *et al.* (1994) detached scales from bulbs of 5 different *Lilium* species and hybrids and separated them into 3 groups according to their anatomical positions on the mother bulb. Irrespective of the species and hybrids tested, outer and middle scales performed better than inner scaies. In*L. longiflorum* cv. White American outer scales gave the highest bulblet number per scale (2.1), whereas hybrid *Lilium* cultivars Enchantment and Star Gazer produced 1.7 bulb lets from outer scales. *L. speciosum* cv. Uchida and *L. regale* showed the least regenerative capacity from the inner scales. Careful consideration to scale position on the mother bulb must be given to achieve satisfactory results. Bulblets harvested in the autumn could be immediately replanted to give small grade bulbs the following year.

7.2.2 Bulbils

Bulbils of tiger lily do not have a dormant period when cultivated at temperatures of >0°C, but under open cultivation at 0°C they become dormant. Plants grown from bulbils produced successive adventitious roots, each with a life not exceeding 4 months. It was not possible to establish any links between development and the number of bulbils. With plants grown from bulbils, it was found that vegetative propagation preceded reproductive (Ignat'eva, 1998).

7.2.3 Micropropagation

Due to the fact that lilies are monocotyledonous bulbous species, they are comparatively more difficult to establish and manipulate in tissue culture than dicotyledonous ornamental plants. Nevertheless, as a result of lilies having been the object of intense floricultural interest for centuries,techniques for micro-propagation of these bulbous plants are well developed and have been extensively studied (Bakhshaie, 2016). Among the various explants tried, bulb scales are the most preferred for tissue culture of lilies.

7.2.3.1 Bulb-scale

Most of the works used bulb-scales as the explant source for micropropagation studies. The response has been encouraging, except for initial contamination. The highest number of adventitious bulb lets was obtained on explants from scales placed horizontally with abaxial side touching the media (Leshem *et al.*, 1982). Young (1995) found that the middle portion of bulb-scale explant in *L. orientale* hybrid Casablanca and *L. leichtlinii* cv. Nigrium produced more bulblets than distinct basal part. Kawarabayashi (1993) reported that dissecting *in vitro* bulb into small segments increased the total number and hastened the propagation of bulblets compared with using scale segment cut from *in vitro* bulbs of a similar size.

The effects of the several constituents of the nutrient medium have been studied extensively. Number of buds on *L. auratum* scales was increased with

Propagation of Lilium

Lilium Seeds

Lilium Pods

Lilium Seedlings

Lilium Bulbs

Planting of Bulbs in Pots

Planting in Field

Micropropagation of Lilium

Micropropagation of ***Lilium polyphyllum***: a, Young callus showing shoot primordia. b, c, Callus showing shoot proliferation and well-developed multiple shoots. d, Well-developed shoots inoculated in the rooting medium. e, f, Well-developed roots. g, Hardening in vermiculite and soil (1 : 1 v/v) ratio. h, i, Hardening in compost-enriched soil (Source: Panwar *et al.* (2017).

increasing strength of MS medium (Takayama and Misawa, 1979). Van Aartrijk and BlomBarnhoom (1980) observed that organic growth factors like nicotinic acid, pyridoxine and glycine are not essential but thiamine and inositol were, however, beneficial. The type of media like liquid or solid seems to have marked effect on bulblet formation. Kawarabayashi (1996) cultured scales of *L. japonicum* on solid medium with horizontal rotation and liquid medium with aeration. He found that adventitious bulblet formation was greatest in the dark. The growth hormones are found to influence the regeneration to a greater extent. There are very few reports on the effect of either auxin or cytokinin alone. Auxin is mainly applied as NAA (0.5-2 mM) which increased the number of adventitious buds; some reports showed a stimulatory role for cytokinins when applied in combination with auxin (Hackett, 1969; Takayama and Misawa, 1982). The effect of growth hormones also depends on wounding (Van Aartrijk and Blom-Barhoom, 1984).

Yongmin and Takeshri (1999) worked on effect of cytokinins in *L. japonicum*. They found that addition of BA above 0.05 µM induced abnormal bulblets and addition of NAA showed no beneficial effects, Kin proliferated more of abnormal bulblets and high concentration produced callus-like tissue. Hence they recommended 5.0 µM 2iP for regeneration and subsequent growth. Chaudarov and Denkova (1994) obtained the optimum number of bulb-scales on LS basal media with 1.5-2 mg/l BAP + 0.5 mg/1 NAA. In addition, they found highest percentage of lily symptomless carlavirus-free regenerants on media containing BAP and NAA, while 2,4-D reduced virus concentration, all bulblets without growth hormone were infected by viruses.

Bhandari and Aswath (2018) worked on three different types of explants *viz.*, leaves, outer and inner bulb scales were exposed to two sterilants: sodium hypochlorite (NaClO) (2 per cent, 3 per cent, 4 per cent and 5 per cent) and hydrogen peroxide (H_2O_2) (0.05 per cent, 0.10 per cent, 0.15 per cent and 0.20 per cent). NaClO @ 3 per cent was highly effective over different regime of H_2O_2 for minimizing contamination (18.66 per cent), low necrosis (2.67 per cent) and maximum survivability (79.13 per cent). Highest bulblets induction (4.66) and maximum bulblets diameter (7.85 mm) was recorded in Murashige and Skoog (MS) medium supplemented with BAP (0.5 mg/l) and NAA (0.3 mg/l). The maximum length of shoots (38.33 mm) and fresh weight (239.51 mg) were recorded with BAP (1.0 mg/l) and NAA (0.1 mg/l). The maximum numbers of basal roots (5.83/bulblets) were obtained with 0.3 mg/l NAA. The maximum length (49.00 mm) of basal roots was recorded with BAP (1.0 mg/l) and NAA (0.3 mg/l).

7.2.3.2 Meristem Culture

In spite of early reports of meristem culture by Nishizawa and Nishi (1966) and Mori (1971), the production of virus-free lilium through *in vitro* meristem culture was published by Allen and Fernald (1972) and Asjes *et al.* (1974). They indexed the virus using electron microscope and found LSV concentration with meristem-cultured lilium plants in the culture tubes and those grown in soil, under the detection limit. There is considerable variation among lilium cultivars as to the ability to produce LSVfree plants by meristem culture. Allen (1972), working with nine Asiatic and one *Aurelean* hybrid, found the chance of obtaining healthy plants to a range between 14 per cent (*L. cinnaber*) and 89 per cent (*L.* × Enchantment). Asjes *et al.* (1974) concluded that 50-70 per cent of their plantlets were LSV -free. The size of the meristems influenced the chances of elimination of virus too. LSV -free plants of the *L. longiflorum* cultivars Nelle White and Ace could be obtained only by small shoot-tip meristems (Allen and Anderson, 1980).

7.2.3.3 Leaf Culture

Niimi (1985) cultured transverse section of 5 mm strips ofleaves numbered LS I, LS II and LS III in order of location in basal region of *L. maculatum, L. rubellum, L. auratum, L. speciosum, L. platyphyllium* and *L. formosum* in MS supplemented with glycine, myoinositol, pyridoxine-HCl, thiamine HCI, sucrose and agar. He found that the regenerated ability of the 3 different strips was independent on species with a bulblet production being more or less restricted to LS I in *L. rubellum, L. maculatum* and *L. formosum* in contrast to the other 3 species which produced bulblet easily from all strips.

Table 6: Effect of BAP and NAA on Direct Organogenesis from *In vitro* Leaf Petiole of Lilium Hybrid 'Pavia' in Semi Solid Media*

MS Medium + Hormones (mg/l)		*Regeneration (per cent)*	*Days to Shoot initiation*	*Adventitious Shoot per Explant*
BAP	*NAA*			
0.00	0.00	3.620 (10.968)[k]	26.590 (31.040)[a]	1.900 (7.923)[h]
0.00	0.10	26.460 (30.957)[j]	24.430 (29.619)[b]	2.500 (9.097)[fg]
0.00	0.20	42.760 (40.837)[h]	22.760 (28.492)[c]	2.400 (8.911)[g]
0.00	0.30	33.460 (35.341)[i]	22.340 (28.204)[c]	2.600 (9.729)[f]
0.50	0.00	31.950 (34.419)[i]	24.930 (29.954)[b]	4.300 (11.967)[d]
0.50	0.10	49.460 (44.691)[f]	18.750 (25.659)[ef]	4.400 (12.106)[d]
0.50	0.20	57.620 (49.383)[d]	17.740 (24.907)[gh]	4.400 (12.108)[d]
0.50	0.30	53.860 (47.214)[e]	19.530 (26.227)[de]	4.300 (13.968)[d]
1.00	0.00	45.490 (42.412)[g]	19.850 (26.457)[d]	3.900 (11.389)[e]
1.00	0.10	61.950 (51.914)[c]	18.190 (25.224)[fg]	4.900 (12.789)[c]
1.00	0.20	72.860 (58.623)[a]	16.430 (23.911)[i]	5.800 (13.936)[a]
1.00	0.30	65.430 (53.995)[b]	16.920 (24.289)[hi]	5.400 (13.436)[b]

* Values in parenthesis indicates Arc sin transformation

** Different letter within the columns indicates significant difference between treatments according to Duncan's multiple range test (P≤0.01).

* Bhandari *et al.* (2019).

Bhandari *et al.* (2019) worked on shoot regeneration from *in vitro* cultured leaves on the 11 MS media combination consisting of different concentrations of NAA and BAP. Shoot organogenesis occurred directly from the base of leaves without forming callus. The highest frequency of shoot regeneration (72.86 per cent) andthe largest number of shoots per leaf (5.80) were obtained on MS medium supplemented with 1.00 mg/l BAP and 0.20 mg/l NAA (Table 6). Highest rooting (99.33 per cent) along with maximum number of roots (11.40) per bulblets was recorded in

½ MS medium containing 1.00 mg/l IBA after 4 weeks of subculture. They strongly suggest that the leaf petioles have an organogenic potential for shoot induction and bulblets development. This protocol can be used in micropropagation and genetic transformation studies of other *Lilium* spp.

7.2.3.4 Stem Node Culture

Nhut (1998) cultured stem segments with 2 to 3 nodes of 1 cm which were taken from *in vitro* shoots, developed from shoot tips. Within 30 days, structures which resembled bulblets were formed. These are called pseudobulblets from axillary bud and they did not exhibit dormancy or an arrested phase in their development. In another 30 days, each bulblet developed into a rosette of shoots.

7.2.3.5 Flower Parts Culture

Kruczkowska (1989) carried out studies with *L. candidum, L. davidii, L. henryi* and Australian hybrid and the cultivars Bright Star, Cinnabar, Joan Evans and Star Dust using flower buds, peduncle, perianths, stems and leaves. He found flower bud as the best regenerative explants among all and also observed that 1 mg/l NAA and 5 mg/l BA gave more bulb lets and in majority of cases light was better than dark cultures. Kubalakova and Gavel (1991) found filaments and receptacle isolated from unopened flower buds of *L. lancifolium, L. pardalinum, L. davidii, L. henryi* and 10 other cultures produced plantlets.

7.2.3.6 Thin Cell Layer Culture System

In an attempt to obtain a higher frequency by mass propagation in a short time frame in *L. longiflorum,* thin cell layer system was developed to induce differentiation of buds *in vitro.* Thin cell layer system was found to be the most responsive explant source for callus induction and plantlet regeneration, which, in turn, derived from different parts ranging from floral organs to root of lilies. As such, the idea of using TCL system as a novel explant for initiation of culture was borne out of its various potential advantages in combination with the conventional technique of tissue culture, including the potential for rapid clonal propagation, production cost reduction and elimination of viruses and other pathogenic diseases in combination with antiviral compounds as well as being a candidate source of material for genetic transformation and crop improvement (Bakhshaie *et al.,* 2016).

Thin cell layers (TLCs) are excised longitudinally (lTCL) or transversely (tTCL) from different organ types (epidermal, subepidermal, stems, stem nodes, leaf veins, leaf blades, floral organs such as sepals, petals, anthers, filaments, pistils, young spikelets, meristems, petioles, root rhizomes and bulb-scale).

7.2.3.7 Receptacle Culture

Closed flower buds of *L. longiflorum,* 8 cm in length, were used. tTCL explants from receptacle and ovary of a flower stalk (2-3 mm) denuded flower, bracts and perianth were cultured on MS media containing 3 per cent sucrose, 2 μM BAP and 5 μM NAA (optimal condition) to induce bud proliferation. Buds appeared in receptacle tTCL explants within 3 weeks of culture but not in the ovary or flower stalks. These buds continued to develop into bulblets. The receptacle culture of *Lilium* flower stalks was shown to be a suitable method for the successful micropropagation of *Lilium*.

Culture of young *Lilium* leaf explants was efficient for shoot regeneration. Pseudobulblets formed in tTCL leaf explants and more bulbs were formed on the adaxial surface of leaf sections (Nhut, 1998). Stem node (2-3 mm) tTCL sections from plantlets derived from shoot-tips of dormant bulbs were cultured on half-strengthMS medium with 20 g/l sucrose, 0.8 per cent agar and1 µM BAP (Nhut, 1998). Bakhshaie *et al.* (2010) succeeded in plant regeneration via somatic embryogenesis (SE) from tTCLs of both roots and bulblet microscales (1.0 mm in thickness) derived from *in vitro* bulblets of *L. ledebourii* (Baker) Boiss, an endangered species. However, the higher rate of regeneration was observed from *in vitro* bulblet microscales.

7.2.3.8 *In vitro* Bulblet Formation

Several factors have been shown to affect bulblet growth, especially media constituents like carbon source, growth regulators, temperature and light. Conditions for promoting bulb enlargement using different carbon source were investigated by Haruki *et al.* (1996b). They found that sucrose was hydrolyzed into glucose and fructose which resulted in an increase in osmotic pressure and subsequent suppression of bulb growth. It was reported that addition of 0.05 mg/l NAA was required forthe formation and growth of bulb in *L. rubellum* (Niimi and Saito, 1991) and *L. japonicum* (Niimi, 1986). Similarly, it has been reported that the development of bulblets was stimulated at 0.03 mg/l NAA in *L. longiflorum* and *L. speciosum* (Rybezynski and Gomolinska, 1989; Panizza *et al.*, 1996). It was also reported that BA was not an important regulator to increase the number of bulblets in *L. rubellum* but 0.01 mg/l BA in combination with 0.05 mg/l NAA was best for *L. longiflorum* (Niimi, 1985). Lim *et al.* (1998) found that in general flower stems and bulb-scale segments were suitable explant sources for bulblet production. Bulblet formation and growth were induced by BA and inhibited by kinetin. Relatively low concentrations of NAA stimulated bulblet formation in Dame Blanche and Casablanca, and enhanced bulblet weight in Casablanca and Acapulco.

7.2.3.9 Stability of *Lilium* Cultures

Stability of cultures is one of the important aspects of *in vitro* culture, especially for commercial production. Caputo *et al.* (1990) observed stable plantlets with a chromosome number (2n=24) up to the 4th generation in lilium 'Casablanca' in spite of growing in culture with growth hormones. Takayama *et al.* (1982) obtained plants and flowers from *in vitro* cultured bulblets phenotypically similar to the parent plants. Qu *et al.* (1998) regenerated plants for callus induced on anthers and observed the presence of mixoploids, aneuploids and other abnormalities in the regenerants.

8.0 Soil and Climate

8.1.0 Soil

Lilies prefer a well-drained soil. Water logging particularly in the dormant period, can be lethal to the plant. Some species require an acid soil (*e.g.*, *L. auratum*). Saline soils are not suitable for growing lilies. As salinity increased, plant height, flower bud length and flower diameter decreased (Hwang *et al.*, 1998). The critical electrical conductivity (EC) for growth was 2.8 dS/m. Lilies cannot survive in soils with an EC of>10.8 dS/m. Application of NPK fertilizer to the soil would increase EC. At 55 days after transplanting, nutrient uptake was in the order K > N > Ca > Mg > P. Nutrient uptake decreased with increasing salinity, and significantly decreased when EC was >4.4 dS/m. A significant increase of mite populations was observed when organic amendments were added prior to planting. Close examination revealed the presence of a small number of mites in bark compost, dairy compost, organic fertilizer and soils, but none in peat, guano and fish meal (Liu, 1996).

8.2.0 Climate

Lilies love sunny climate. But the root system of lilies cannot tolerate direct sunlight. In original habitats *L. longiflorum* grows at mild temperatures of 16 to 12°C: Knowledge of climatic factors on growth and flowering of lilies is important for successful forcing. Temperature strongly affects the growth, stem elongation and bud development. Temperature above 30°C and photoperiod in the early stage prevent loss of flower buds. Lilies are susceptible to high temperatures (>30°C) but they tolerate and respond to high light intensity (2000 µmol/m^2/s Average daily temperature (ADT) affects flowering rate in *L. lanciflorom,* which profoundly decreases as ADT increases and flower abortion occurs when ADT is >24°C. The optimum day/night temperature combination that could shorten stem length without interfering with flowering was found to be 16/24°C day/night (Son and Han, 2000). High temperature being detrimental to the crop growth, temperature-dropping methods was tried in lily. Of the five different methods tried, water sprinkling over the white shading net over roof gave best results (Lee *et al.*, 2000).

Shading in the early vegetative growth phase helps better growth (Sorrentino *et al.*, 1997a). Effect of shading (82, 79 and 72 per cent) was tried on five different lily varieties. All the varieties recorded favourable height, fresh and dry weight, number of flowers, stem diameter and leaf area by increasing shading percentage. Oriental hybrid cv. Casablanca had shown earliest flowering (reduction by 7-9 days) by 82 per cent shading (Yih and Huang, 1998).

9.0 Cultivation

9.1.0 Planting

The need of planting clean, healthy and disease-free bulbs can hardly be overstressed. Time, depth and method of planting depend on the type of lily to be grown, although there are certain general rules for all members of the lily groups.

The planting depth affects the stem, leafand root growth of lilies in forced cultivation. With shallow planting, basal roots establish well, and flower stems become shorter and stem diameter more compared with deeper planting. When bulbs are planted deeply, the growth of stem roots is promoted (Choi *et al.*, 1998a). However, number of days to shoot emergence and flowering, and plant height are reduced by longer periods of cold treatment (Choi *et al.*, 1996; Choi *et al.*, 1998b). Percentage of winter sprouting decreased at later planting times and at deeper planting depths. Percentage of winter sprouting also decreased as bulb size increased. Generally, plants in the unheated plastic house produced more seed bulbs than plants in the field (Kim *et al.*, 1994).

9.2.0 Spacing

Roh (1979) found the need to plant bulbs with a minimum circumference of 15 cm and a weight of 13-14 g. Ten flowers per plant were produced even at the closest spacingof 10×10 cm. But, in India (Ludhiana) plant height and flower diameter were highest with a spacing 30 × 15 cm compared to spacing 30 × 25, or 20 (Singh *et al.*, 2000). Decreasing plant density increased soil temperature and improved cut flower quality (Aoki and Yashino, 1988b). Planting density affected the dry matter production from the appearance of the flower bud until flowering.

9.3.0 Irrigation

The number of irrigations needed for *L. longiflorum* cv. Nellie White ranged from 12-20 and the amount of applied water ranged from 5.3 to 6.8 litres for the uncovered media (Argo and Biembaum, 1994). When Easter lilies are irrigated only once weekly, flower buds were aborted on several of the plants (Kiplinger *et al.*, 1975). Subirrigation of *L. longiflorum* cultivars Ace and Nellie White, growing in soil-based, peatlite or peatwool (mixture of peat and rockwool) media, with 1/4-strength nutrient solution gave higher plant DW and FW than sub irrigation with half or full-strength nutrient solution or hand-watering with full-strength nutrient solution (Choi and Lee, 1995). When the root-medium of surface sub-irrigated, Nellie White plants were covered with an evaporation barrier, the average amount of applied water was reduced by 35 per cent compared with the uncovered media (Argo and Biembaum, 1994). Sprinkler irrigation of the hybrid lily cv. Enchantment (field grown for bulb production) with water containing 250 mg Cl/l increased total bulb weight (main bulb + stem bulbs) by 7-11 per cent depending on the size of bulb initially planted (Boontjes and Ploegman, 1981).

Quality plants of *L. longiflorum* cv. Nellie White were produced in all media *viz.*, soil (1 clay loam: 2 sphagnum peat: 1 humph peat: 1 perlite), peatlite or loose rockwool, and irrigated with warm (21-26.6°C) or cold (4.4-10°C) water regardless of water temperature, but plants irrigated with warm water reached flowering height faster. Soil-grown plants in both water temperature treatments flowered significantly later than those in rockwool or peatlite. There were no differences in bud count and the vase life (Goldsberry and Maffei, 1986). Root masses were reported to be denser in pots watered at 23.8 and 29.4°C and, except for plants at 18.3°C, the fresh plant weight rose with increasing water temperature (McMurry, 1978).

Shading reduces transpiration and soil evaporation hence shading net was used to reduce irrigation requirements of *Lilium* cv. Casablanca grown in full light (Sorrentino *et al.*, 1997b). Applying ancymidol 0.25 mg of ancymidol-charged hydrophilic polymer placed at different locations in relation to the bulb of Asiatic hybrid lily either by dipping or as a charged polymer was the most effective height control method. Marked differences in response between irrigation methods were observed for ancymidol drenched plants (White, 1972; Beattie *et al.*, 1993). Easter lily grown in clay or plastic pots with capillary irrigation (water supplied from 8 a.m. to 4.30 p.m.) or at mist propagation frequency by a solenoid system showed that plant height was similar for both watering systems, but when growth retardants (ancymidol or chlorphonium) were applied the reduction in height was slightly greater with capillary watering (Koths and Judd, 1976).

9.4.0 Manuring and Fertilization

9.4.1 Inorganic Fertilizers

The nutrient requirement depends on the soil type. Sandy soils need 125-200 kg N/ha followed by top dressing of same amount. However in sandy loam soils highest yield of bulbs was obtained with 25 kg N/ha followed by top dressing of 50 kg N/ha (Boon and Niers, 1987).

There was a significant interaction between growing medium and fertilizer application on the number of buds per stem for plants flowering in June and on stem length for flowering in November-December (Szlachetka *et al.*, 1997). Cut flowers with 34 buds constituted >90 per cent of the crop in both summer and winter seasons. Giustiniani *et al.* (1988) found that a high proportion of the mineral requirement was derived from the bulbs, and most mineral uptake from the growing medium occurred late in the growth cycle. A study in the autumn with 2 rates each of N and K_2O showed that the higher rates (16 and 42 g/m^2, respectively) gave better quality flowers. Kawagishi and Miura (1996) classified the growing period into 4 stages and indicated the stage for fertilizer application. In the first stage, from planting to sprouting, the foliage grew using the materials stored in the bulb. In the second stage, from leaf expansion to flower bud appearance, the foliage grew vigorously and the transfer of photosynthetic products from the above ground parts to the subterranean parts began. The third stage was from flower bud expansion to mid-August which would be flowering time if disbudding were not carried out. In this stage, leaf area and top dry weight (DW) continued to increase and the old bulb increased in DW more than other subterranean organs. In the fourth stage, from mid-August to harvesting, the growth of the foliage almost stopped and the main DW increased was in the new bulb. With the application of N and K top dressings after flower bud appearance, root growth became more active and nutrient absorption and DM production were prolonged until the later growth stages, resulting in increased yield. Aoki and Yoshino (1988a) found that fertilizer application had little effect on DM production, but bulb DW at flowering was increased by increasing rates of fertilizer application.

Leaching of fertilizers applied before planting induced low nutrient efficiencies in sandy soils. Dividing the total N-dressings into 4 monthly applications from March to June or April to July led to a higher N-efficiency, though fertilizers were easily leached with high rainfall. A total of 150 kg N/ha appeared to be adequate (Slangen *et al.*, 1989). The response to N fertilization was greater on sand than on sandy loam. On sand, a total amount of 125-200 kg N/ha followed by a top dressing of the same amount is recommended. On the sandy loam, the highest yield of bulbs was obtained with 25 kg N/ha before planting and ridging, followed by a top dressing of 50 kg N/ha (Boon and Niers, 1987). There was no yield response when

Cultivation of Lilium

Cultivation in Open

Vegetative and Bud Formation Stage in Greenhouse

Flowering Stages in Greenhouse

FYM was applied, in addition to N fertilizer, on the sandy loam. After lifting, the bulbs were forced in pots containing a 60:40 mixture of black and white peat. In a comparison of 0-6 g PG-Mix/1, a total application of between 0.75 and 1.5 g gave the best results (half being applied before planting and the remainder as 2 equal top dressings 1 month apart); large amounts of NPK, however, retarded bulb emergence, reduced longitudinal growth and delayed the time of cutting of the flower stems. Fertilization of the pot soil markedly improved flower quality and vase life. The optimum leaf N content in late August was found to be 1.9-2.2 per cent of DM. For optimum yield and forcing quality, the bulb N content at harvest should be 1.1-1.4 per cent. Nitrogen fertilization in the field, prior to forcing, had little or no effect on vase life. Boon and Niers (1986) also reported that the N requirement of cv. Enchantment (for bulb production) growing on sand varied with the season. Yields of bulbs were highest after application of 150 kg N/ha, either as basal or top dressing, in two dry seasons (with less rainfall than average between planting and end of June). In a wet season, N when applied as basal dressing was much less effective than when applied as a top dressing. In sandy clay soils, a basal dressing of 50 kg N/ha was best where the moisture content of the soil was good, otherwise a dressing of 100 kg N/ha was needed. A top dressing of 50 kg/ha was most effective when applied in 3 stages, *i.e.*, mid-May, early June and early July.

Phosphorus nutrition did not affect flowering, bud count or plant height but significantly affected dry weight and phosphorus content of roots and shoots (Tsujita *et al.*, 1979). The ornamental value of the *L. davidii* plants was enhanced most in the third year by applying 45 kg N + 60 kg P_2O_5per ha at the start of growth, 45 kg N + 40 kg K_2O/ha during most vigorous growth and 45 kg N + 60 kg P_2O_5 + 40 kg K_2O per ha at bud development (Gindina, 1976). Triple superphosphate contains about 1 per cent fluoride as an impurity, and the lilies showed leaf scorch at the time of flowering when triple superphosphate was added, especially at the higher rate (Roorda van Eijsinga, 1980).

Among different potting composts supplied with different fertilizers growth and flowering of *L. longiflorum* cv. Nellie White was satisfactory in mixtures of 2 : 1 soil : moss peat and 1: 1: 1 soil : moss peat : perlite. Slow-release fertilizer incorporated at potting and regular applications of soluble fertilizers gave good results. Basal application of MagAmp (7-40-6) alone or plus a weekly application of liquid 25-0-25 fertilizer proved best with 2:1 soi/moss peat. Osmocote (19-0-20.5) and 25-0-25 fertilizer in 1: 1: 1 was most suitable (Powell *et al.*, 1975). Kiplinger *et al.* (1972) also found that potted bulbs of Ace, Nellie White and No. 44 lily cultivars grew well with Osmocote (18-9-9) or MagAmp (7-40-6) slow-release fertilizers incorporated in the soil mixture before potting.

Lilium plants are sensitive to manganese, bromine, fluoride and boron nutrients. High levels of Mn (>50 ppm) produces short plants, leaf chlorosis and low dry weight (Holmes and Coorts, 1980).

9.4.2 Organic Manures

Organic manures like farmyard manure and arbuscular mycorrhizal fungus have been used for improving the yield and quality of lilies. In an alkaline growth medium, inoculation with arbuscular mycorrhizal fungus *Glomus etunicatum* did not significantly enhance bulblet growth of Asian hybrid lily, however, amendment with rice bran increased the growth of bulblets (Lin *et al.*, 1999). With the same amendment, dual inoculation of VA mycorrhizal fungus and phosphate-solubilizing bacteria showed the best promotion of growth. Wu *et al.* (1999a) inoculated VA mycorrhizal fungi on Oriental lily (*cv.* Casablanca) scales in vermiculite, and evaluated interactions with selected fungicides. Lily scales were stored at a temperature of 25-17-5°C to induce bulblet production. As bulblets and new roots formed, mycorrhizas were established and this innovation is helpful in increasing bulblet size, especially under the treatment of biocide and no fertilizers. Fertilizer, if incorporated in medium, may diminish the effect of VA mycorrhizal inoculation on the growth of bulblets.

Hao *et al.* (1991) showed that infections of indigenous VAM fungi in most soils were low and there was usually a lag phase. Lag phases of infections were decreased or even disappeared byintroducing mycorrhizal fungi into the soil, and plant growth was greatly stimulated. Phosphorus encouraged mycorrhizal infection in most soils deficient in available P, but the optimum amount of P fertilization was different between alluvial soil and red soils. For alluvial soils, the optimum level is lower than red soils. Mycorrhizal dependencies of various plants were quite different. According to the calculation of Menge, mycorrhizal dependencies of lily was <150 per cent. Logan *et al.* (1984) recommended a commercial compost utilizing sludge, a formulation of 4: 4: 2 of sludge: sphagnum moss: peat or pine bark: perlite. The peat or bark would keep the pH down. Kruijer (1982) reported that FYM improved soil porosity and moisture holding capacity and the plants of *L. speciosum* cv. Uchida were more vigorous and darker green as the rate increased. However, earlier dying-back due to increased infection by *Nectria radicicola* was associated with high rates of FYM or N and thus bulb yield and quality were reduced.

Physical and chemical analyses, including bioassay tests of selected cultivation media, showed that a medium for cultivating *Lilium* cv. Casablanca bulbs comprising filter cakes, rice hull cake, with some pig compost (farmyard manure), supplied sufficient nutrients. Such a growing medium had pH 5, EC 1.74 dS/m,porosityof56 per cent, specific gravity of 1 mg/m^3, a moisture holding capacity of 60 per cent, and exhibited small changes in bulb density between planting and harvest (only 5 per cent increase). Plants grew well in the medium (height, leaf area and leaf numbers per stem), and produced a good yield of quality bulbs compared with plants grown in a vermiculite-perlite-peat moss mix (Wu *et al.*, 1999b).

To plan a commercial crop, it is necessary to understand tools that relate environmental variables to crop growth and development (Lorenzo *et al.*, 2018). One of these tools is leaf expansion, which can be determined from the evolution of the leaf area index (LAI). The technique known as sand mulching substantially modifies the soil temperature but can also alter other environmental variables such as light quality. They determined whether the application of sand mulching affects the growth rate and leaf expansion of a lilium. The experiment was carried out in a greenhouse, where 5 cm layer of fine sand spread over the entire soil surface to reduce soil temperature. Although there were no significant differences in the cycle length or leaf appearance, sand mulching led to differences in the stem growth rate, LAI and final plant height, resulting in a better quality of the product (Lorenzo *et al.*,2018).

9.5.0 Weed Control

Weeds pose hindrance to growth and flowering of lilies. Common weeds in lily fields are *Amaranthus, Portuluea, Phalaris, Diplotaxis, Erigeron, Malva* and *Convolvulus* spp. Use of herbicides although controls weeds and reduce competition, they also cause phytotoxicity. Simazine @ 0.5 kg/ha and Diuron@ 0.96 kg/ha caused slight scorching and stunting after scale germination (Valis *et al.,* 1976). Application of herbicide Goltrix (Metamitron) @ 3 kg/ha was less phytotoxic and effective on cv. Enchantment grown on sandy soils (Koster and Kruyer, 1983). CIPC (chlorpropham) @ 3.5 l/ha gave better weed control than DCPA (chlorthal-dimethyl) @ 11.2 kg/ha. Use of colourless polythene mulch with herbicide gave less phytotoxicity and high yield of flowers with long stems (Grassoti *et al.,* 1982).

10.0 Diseases, Pests and Disorders

Many diseases and pests affect *Lilium.* Disease incidence was closely related to bulb quality, growing environment and culture measures. As high as 22 diseases caused by bacteria, fungi, viruses and physiological factors affect the lilies. Among the diseases, *Colletotrichum liliacearum, Fusarium oxysporum* and *Botrytis liliorum* are important (Chen, 2000). Mites and nematodes are major pests of lilium.

10.1.0 Diseases

Fungal Diseases

Yellowing, leaf blight and stem rot are the major fungal diseases afflicting the lily crop. Use of suitable fungicide or resistance cultivar is suggested to combat these problems.

10.1.1 *Phytophthora* Blight

This disease is caused by *Phytophthora parasitica.* Easter lily and half of the Asiatic lily varieties are resistant to this pathogen. About 26 isolates of this pathogen were isolated in diseased tissues oflily grown in Taiwan (Fang *et al.,* 1999). The stem rot oflily in Korea is caused by *Phytophthora cactorum* (Ryu *et al.,* 1998).

10.1.2 Leaf Blight

Leaf blight in lilies is caused by the fungal pathogen *Botrytis elliptica.* Film forming polymers such as polyelectrolytes were evaluated for control of the disease using laboratory assays. The polymers reduced the esterase secretion by the pathogen (Hsieh and Huang, 1999). Probenazole @ 40 mg/kg soil protected the plants from infection without reducing the growth (Lu and Chen, 1998). A weather based warning system was developed for preventing this disease by Ende van den *et al.* (2000). It mainly uses the data of leaf wetness duration and leaf temperature.

10.1.3 Yellowing

Control of this disease caused by *Fusarium oxysporum* f. sp. *lilii* or *Rhizoctonia solani* was attempted using soil steam at 60°C for 30 min or 80°C for 20 min and with methyl bromide. Treatment at 80°C increased the cut flower quality better than other two treatments (Lee and Leu, 1998). Resistance to this disease was studied and was found to be polygenic in nature (Straathof *et al.*, 1997).

10.1.4 Phytoplasma

A new disease causing stunted growth, leaf malformation and flowering bud deficiency was suspected to have been caused by phytoplasma. Kaminska *et al.* (1998) were able to detect phytoplasma in roots, leaves and bulbs with symptoms.

Bacterial Diseases

10.1.5 Bacterial Soft Rot

This bacterial disease is caused by *Pectobacterium carotovorum.* The organism causes a soft, wet decay of bulbs. *Bacillus lilii,* responsible for heavy losses in many lily-growing areasof Japan was earlier reported by Bokura (1919). To prevent the disease all affected bulbs should be discarded and healthy ones planted in a well-drained soil. Care should also be taken to avoid bulb injury during lifting from the soil.

Viral Diseases

Virus diseases cause a major loss in flower yield. Yield loss was as high as 53.7 per cent in cv. Casablanca was reported in Ichon area of Korea Republic. Major symptoms of the disease were reduction in height, mosaic, stunting, yellowing and reduction in flower yield. Lily symptomless carla virus (LSV) and lily mottle potyvirus were detected from these plants (Kim *et al.*, 1998c). In Taiwan, apart from these, CMV (cucumber mosaic cucumovirus), tulip breaking potyvirus (TBV) and lily X potyexvirus (LXV) were also found to affect the lilies (Chang *et al.*, 1998). Xu and Niimi (1999) found the use of chemotherapy (5 µM ribavirin) with thermotherapy (35°C for 4 weeks) to be an effective strategy to control virus problems in lily bulbs.

10.2.0 Pests

10.2.1 Aphids

Green peach, foxglove, melon, crescent-marked and purple-spotted aphids are found to attack lily plants. Spraying with Malathion or lindane checks aphids. Shoemaker and Teskey (1969) reported that purple-spotted lily aphids cause seedling injury and are controlled by spraying with one and half teaspoonful of nicotine sulphate plus 2 teaspoonful of soap powder per gallon of water.

10.2.2 Lily Beetle

Lily beetle (*Lilioceris lilii*)seems to be mainly confined to some parts of UK and Canada. Both larvae and adults feed on the leaves, stems and seed pods of lily, causing severe damage. Beetles are easily controlled by spraying with malathion (Smith, 1967).

10.2.3 Thrips

The lily thrips, *Liothrips vaneeckei* and *Taeniothrips simplex* are serious pests of lily bulbs, producing rust coloured sunken areas at the base of the outer scales, which become soft and break off easily. Plants grown from infected bulbs are stunted. To destroy thrips, the bulbs should be treated with monocrotophos and stored at a temperature below 19°C. Fumigation of infested bulbs with methyl bromide or dipping them for an hour in hot water at 43.9°C (111°F) to control the pest was suggested by Pirone *et al.* (1960).

10.2.4 Mites

Mites cause severe yield loss in lily bulbs. Addition of organic manures such as bark compost, fish meal and guano did not decrease the mite infestation. Growing of lily or gladioli in peat was recommended to reduce the mite infestation (Liu, 1996). Steam sterilization with 60°C for 30 min or 80°C for 20 min reduced the mite. Growing lilyin peat soil could reduce mite damage (Liu, 1996; Liu *et al.*, 1998). An integrated pest management for mites in lilies (IPM) through alternate cropping with rice, solar drying of soils, planting and ploughing in of *Sesbania,* long term submergence of water, chemical or warm water treatment of bulbs was also suggested by Liu and Chen (1998).

10.2.5 Stem Borer

This pest (*Papaipema nebris*) bores in lily stems and feeds on the pith, moving from one plant to another (Smith, 1967). Destruction oflarvae by splitting the stalks open saves the plant.

10.2.6 Weevil

The pest (*Agaspherops nigra*) damages the grub stage by feeding on the stem below the ground. Spraying of foliage with lindane or Malathion controls the weevil.

10.2.7 Nematodes

Being a bulbous plant, lily is highly prone to nematode infestations. Control of nematode *Pratylenchus penetrans* is effective with oxamyl applied as foliar application (Westerdahl *et al.*, 1994). Hot water treatment (for 2 h at 39°C) of seed and planting and marketable stock of different types of lilies, to control pathogenic fungi and nematodes had been found to be effective (Boontjes, 1973; Boontjes and Mey, 1975). Hot water treatment (HWT) also improved the effect of most of the subsequently applied fungicidal dips, the best of which were a phenol derivative, captafol and Zn-Mn carbamate. Boontjes (1973) suggested that for early treatment HWT should be followed by a 2 per cent Zn-Mg carbamate or 2 per cent captafol dip in the first and third years and by a 0.75-1 per cent phenol derivative dip in the second year; for late treatment the HWT should be omitted and the bulbs should be dipped each year in 2 per cent captafol.

10.2.8 Rats

Rats are very fond of most lily bulbs and in gardens where rats are apt to be troublesome it is useless to attempt to grow lilies unless adequate protection is

Diseases, Pests and Disorder of Lilium

Bacterial Soft Rot

Botrytis Blight

Phytophthora Blight

Lilium Aphids

Beetle

Thrips Infestation

Bud Blast

Leaf Scorch

provided (Bush-Brown and Bush-Brown, 1958). If the lilies are to be grown in beds or borders, the sides of the entire beds should be enclosed with galvanized wire. Celphos may be used in each hole to control the mice in the field.

10.3.0 Disorders

10.3.1 Leaf Scorch

Also known as leaf bum and tip bum (Miles, 1952; Seeley and Valasquez, 1952), it is caused by the presence of sufficient quantities of manganese and aluminium salts at low nitrate levels. A similar scorch due to excess lithiumin soil was reported by Furuata *et al.* (1963). The addition oflime to the soil reduces leaf scorch.

10.3.2 Bud Blast

The buds of lilies in greenhouse may blast or dry up and fail to open. Factors contributing to this disorder include shortage of water at the top of the plant (Smith and Langhans, 1961), competition for limited nutrient and high levels of soil nitrate, *i.e.*, above 50 ppm (Eastwood, 1952). Mastalerz (1965) demonstrated increased abortion by changing the carbohydrate supply to the developing flower buds, *e.g.*, by leaf removal, and showed a direct relationship between length of darkness and flower bud number when bud length was 6-12 mm. In the Netherlands, Schenk and Boontjes (1970) showed that time of subjecting Enchantment lily plants to a week of darkness during forcing at 21°C was related to bud blasting. About 27 per cent bud blast and 72 per cent bud abscission were observed when this was done 6 to 8 weeks after planting. Abscission can be eliminated by supplementary lighting during the month before flowering and about 450 watt/m^2 are required.

Frost injury resulting in puffy leaf, stunted plants, and non-infectious chlorosis are other disorders of the lilies. Mulching protects from frost injury in spring. Application of iron sulphate or iron chelate cures the chlorosis. Shoemaker and Teskey (1969) also recorded abnormal flat stems resulting in fasciation.

11.0 Harvesting and Postharvest Management

11.1.0 Harvesting

The flower stems are usually cut full length just when they are opening (Chittenden and Synge, 1956). According to Laurie *et al.* (1958), the stems are cut when the lower flowers are at least half open, but at this stage not all the remaining flowers may open, particularly if there are more than five buds present. Sometimes individual flowers or heads are cut when the bud count is high and later the stem is cut when fewer flowers are present. If the flowers are sufficiently open, the anthers should be removed to avoid the messy appearance, created by the sticky pollen.

Harvesting the lilies at 25 per cent bud colour stage gives the longest vase life in all cultivars (Kumar *et al.*, 1999). Buds which are longer than 4.5 cm develop normally and open in the preservative solutions (Song *et al.*, 1996).

Vase life has been found to be best in winter when the stems were picked with more than 3 buds showing colour and in summer when 3 buds were coloured (Swart, 1980). Song *et al.* (1996) found that the best harvesting stage of Asiatic hybrid lily (*Lilium* ×*elegans* [*L. maculatum*]) cultivars was when the first flower was fully open.

11.2.0 Postharvest Management

11.2.1 Postharvest Handling

The postharvest life of cut flowers is closely related to the temperature at which they are held after removal from the plant. As temperature decreases, the postharvestlife of cut flowers is extended. An attempt was made by Harkema and Doom (1986) to quantify the effects of postharvest handling and postharvest climate on water uptake and transpiration by the lily cv. Enchantment. The flowers were kept dry for 2-3 h after harvest, when placed in water the initial rate of uptake was high. After this initial high uptake rate which lasted 1-2 h, the uptake rate was constant in lilies. A glasshouse condition (20°C, 60 per cent RH and light) resulted in a much higher uptake and transpiration than a cold store environment (5°C, 95 per cent RH and darkness or 2°C, 90 per cent RH and darkness).

Yang *et al.* (1996) analyzed the hormone levels in cut lilies in cold storage. The results of these studies indicated that there was no striking peak of endogenous ethylene release during storage of cut lilies at low temperature (4 ± 1°C), and it was lower than that of flowers stored at room temperature (28 ± 2°C). The contents of IAA and GA_3 were higher at low temperature than at room temperature. The peak level of endogenous abscisic acid (ABA) appeared 6 days later in cold-stored flowers than in those stored at room temperature. ABA seems to be the primary factor affecting senescence of the cut flowers.

Vase life of cut *L. longiflorum* flowers at room temperature of 20°C was greatly affected by the temperature of previous storage for 1 week; 0°C was best in the cool season and 0°C or 5°C in the warm season and 10°C was too high. Use of the preservative Kagawa solution during cold storage prolonged subsequent vase life compared with storage in deionized water. Dry storage for 1 week after harvest resulted in poorer quality flowers than storage in Kagawa solution or deionized water (Hasegawa *et al.*, 1980).

11.2.2 Biochemical Changes

Carbohydrate redistribution is suggested to play a major role in the postharvest performance of Asiatic lily inflorescences. In inflorescences of lily postharvest translocation of substrate from the basal to the upper buds takes place. At the bud length of 70-75 mm the total carbohydrate content (fructose, glucose, glycerol glucoside, starch, sucrose) amounted to 75 per cent of the total tepal carbohydrate content just prior to anthesis. Postharvest flower longevity after anthesis of detached buds was well correlated with total carbohydrate content of the tepals at harvest. Longevity of inflorescence attached flowers remained constant within the inflorescence, likely due to postharvest redistribution of tepal carbohydrate. These findings indicated an important role for tepal carbohydrate content in postharvest bud development and flower longevity of Asiatic hybrid lilies (Meulen-Muisers *et al.*, 2001).

Ethylene production by florets increased steeply from the 5th day after flowering, which corresponded with loss in vase life. Ethylene production from stamens and pistils was higher than that from tepals (Son and Sub, 1999). In general, ethylene

Harvesting, Packing and Marketing of Lilium

Full Bloom Stage

Bunch of lilium for Sale

Packing in Polythene

Lilium Flowers in Market

Lilium Cut Flowers

Flower Vase for Sale

Lilium Pots for Sale

induces a forceto unfold flower buds, which could result in a shorter vase life (Woltering and Harkema, 1981). The vase life of most Oriental hybrids and *L. longiflorum*cultivars was unaffected by ethylene exposure. However, some Asiatic hybrids showed minor responses. Endogenous ethylene production by all lily cultivars was close to or below the limits of detection. CO_2 production rates during vase life of 7 days at 20°C varied from 0.5 to 1.5 µmol kg-^{1}S-^{1}, with Asiatic hybrids having rates in the upper end of the range. No climacteric increase in ethylene or CO_2 production was detected during 7 days at 20°C. Asiatic lily cultivars responding to ethylene exposure showed a reduced degree of flower opening and percentage of buds, which opened. *In vitro* l-aminocyclopropane-1-carboxylic acid (ACC) oxidase activity of some cultivars was high, indicating that the enzymatic machinery for ethylene production was present although endogenous levels were difficult to detect (Elgar *et al.*, 1999).

11.2.3 Other Factors

The number of bacteria was often high enough to reduce flower longevity. The number of bacteria found were low when chlorine was used as an antibacterial chemical. Use of floral preservatives (containing antibacterial chemicals and sugars) did not decrease the numbers of bacteria (Hoogerwerf and Doom, 1992). Certain diseases like lily symptomless carla virus (LSV) were considered to affect vase life. The characters considered to be of greatest economic importance to cut flower producers (the number of inflorescences/*st*em, flower quality and vase life) were only mildly affected by LSV infection in any cultivar and unlikely to be of major commercial concern (Blake and Wilson, 1996).

11.2.4 Vase Life

In greenhouse flower bud abscission during winter has been reported to be reduced to about 20 per cent of the control by foliar sprays with silver thiosulphate (STS). Also, foliar sprays with STS greatly improved flower bud opening when cut flowers were placed in water (Swart, 1979). This treatment was as effective as pulsing of cut flowers with STS. When the plants were grown without supplementary illumination but sprayed with STS, all flower buds opened during vase life, even after a dry storage in air containing 0.5 ppm ethylene at 20°C for 24 h (Meeteren and Slootweg, 1986). A continuous supply of 30 g/l sucrose together with 8-hydroxyquinoline citrate (8-HQC) to inflorescences of cv. Prima resulted in satisfactory bud opening and increased longevity. Cut inflorescences could be stored at 1°C for 4 weeks, without a great loss in potential vase life and decorative value, when they were pre-treated with silver thiosulphate + 100 g/l sucrose for 24 h, kept in a cold room in a solution containing 50 mg/l silver nitrate, and then kept in a solution containing 30 g/l sucrose and 200 mg/l 8hydroxyquinoline citrate. Such treatment greatly improved bud opening, increased the diameter of individual flowers and prolonged their vase life (Nowak and Mynett, 1985). Swart (1981) found that keeping cut flowers cv. Enchantment in water containing the flower preservative Bendien Anjer VB (used commonly for carnation) for 4-36 h immediately after cutting and before dry storage (prior to transport) increased the percentage of normal flowers opening, prolonged vase life and prevented damage due to ethylene emission by the smaller buds (a problem with this cultivar).

Brosnan and Sun (2001) found that by spraying adequate water on the cut flowers prior to vacuum cooling, weight loss could be eliminated besides extending the vase life. Spraying water also improved the cooling rate leading to lower temperatures Combined fumigation with pre-treatment at the pre-shipping stage provided an alternative method to minimize the loss of quality and longevity of cut flowers. Regardless of fumigation method, pre-treatment extended the vase life and the longevity of fumigated flowers was slightly shortened compared with the control, but there were no visible phytotoxicity symptoms (Son *et al.*, 2000). However, fumigation might damage flowers if traditional fumigation is applied 3 days after harvest (*i.e.*, during the quarantine period). Pre-treatment of cut flowers with ACC and STS induced changes in floret fresh weight. STS pre-treatment significantly prolonged vase life (7.4 days compared with 6 days in the control) (Sytsema *et al.*, 1984; Son and Suh, 1999). Pre-treatment with 1-methylcyclopropene (l-MCP) did not generally extend the vase life or improve flower quality of the Asiatic or *L. longiflorum* cultivars tested, although exposure to ethylene did reduce vase life of Elite and Cordelia by 1.5 days compared with flowers pre-treated with 1-MCP. Lee and Sub (1996) harvested flowers of cvs. Enchantment, Stargazer and *L. longiflorum* cv. Jinsil 3 or 4 days before anthesis and found that for Enchantment, the longevity of the first and second flowers was greatest when stems were harvested 4 days before anthesis, pre-treated with 4 mM STS, stored for 24 h and then held in 4 mM $AgNO_3$ + 10 per cent sucrose for 10 min. Controls (water treated) and flowers stored for 5 days failed to open or showed rolling of petal and sepal edges. For Stargazer, STS pre-treatment did not affect longevity, but flower colour became deep red with STS pre-treatment and holding solution treatment. For Jinsil, harvesting at the mature bud stage (3 days before anthesis) extended vase life by about 1.5 times compared with harvesting one day earlier, and holding in $AgNO_3$ + sucrose almost doubled the vase life compared with holding in water. The number of bacteria in the stem tissue after holding in $AgNO_3$ + sucrose was considerably lower than after holding in water. Pre-treatment with 1 mM STS + 10 per cent sucrose for 30 min or 0.2 mM STS + 10 per cent sucrose + 100 ppm GA_3 + $MnCl_2$ for 3 h effectively prolonged vase life and increased flowering percentage. Preservative solution with 150 ppm 8-HQS + 2 per cent sucrose + 50 ppm $AgNO_3$ or 200 ppm 8-HQC + 3 per cent sucrose + 50 ppm GA_3 extended vase life 1.8-fold compared with the control and improved quality, as judged by percentage of flowering, fresh weight and solution uptake. Buds which are longer than 4.5 cm developed normally and opened in the above preservative solutions. Soda water increased vase life, but yellowing of the foliage started after 6-7 days. Use of preservative solution containing 8-HQC + sucrose + GA_3 after pre-treatment with STS + sucrose + GA_3 + $MnCl_2$ kept the foliage green until the end of vase life (Song *et al.*, 1996). Flowers of lily cv. Enchantment picked before the correct stage for harvesting lasted better if they were stood for 24-48 h in a solution of the preservative compound Aadural AK at 30 g/l at a temperature above 10°C than if they were kept in water alone before marketing (Barendse *et al.*, 1978). Barendse (1975) also reported cutting lilies too early would result in poorer colour, fewer opened flowers and often poorer quality flowers. After storage the flower colour was often lighter and flowers did not open so well. The use of a vase life prolonging chemical Aadural AK at half the normal strength gave better results than full strength.

Lu *et al.* (2020) studied the role of sodium selenite (Na_2SeO_3) in improving cut flower's vase life of *Lilium longiflorum*. Results revealed that that Na_2SeO_3 remarkably enhanced the activities of superoxide dismutase (SOD), peroxidase (POD), catalase (CAT), ascorbate peroxidase (APX), glutathione reductase (GR), dehydroascorbate reductase (DHAR) and monodehydroascorbate reductase (MDHAR), improved relative water content (RWC) and the levels of soluble sugar, proline and soluble protein in cut flower's petals of *Lilium longiflorum*, compared with control. Meanwhile, Na_2SeO_3 remarkably decreased the production of malondialdehyde (MDA) and hydrogen peroxide (H_2O_2), compared with control. Furthermore, Na_2SeO_3 remarkably improved the vase life of *L. longiflorum* cut flower, compared with control. These findings suggested that Na_2SeO_3 improved the vase life by regulating the antioxidant system and osmotic adjustment ability of *L. longiflorum* cut flower.

Lilium pumilum featured long stalks containing orange inflorescences with several flower buds and numerous smooth as well as linear leaves. Due to its high perishability, appropriate techniques that extend flower longevity should be employed (Santos *et al.*,2020). Stalks were kept in maintenance solution with different concentrations of SA (0, 1, 5 and 10 mM) for 0, 6, 12, 18 and 24 h. The *L. pumilum* flowers had longevity decreased by 57.8 and 63.1 per cent when kept in 5 and 10 mM SA, respectively, compared with control. This effect may be associated with the observed decrease in the percentages of flower opening and chlorophyll content as well as the damping-off of stalks, indicating phytotoxicity. Therefore, the use of 5 and 10 mM SA is inefficient to extend the longevity of *L. pumilum* (Santos *et al.*,2020).

12.0 Biotechnology

Application of biotechnology in lilium has resulted in marked improvement in micropropagation particularly through thin cell layer culture. Incompatibility and failure in seed formation in hybrids of several *Lilium* sp. has been overcome through embryo and ovary culture. Molecular markers for establishing the genetic linkage and phylogenetic relationship among *Lilium* sp. and gene transfer has been reviewed by Aswad (2001).

12.1.0 Tissue Culture

Micropropagation and embryo culture in *Lilium* have been summarized in propagation and breeding respectively.

There are reports concerning plant regeneration in tissue culture from different organs of lily species using: (i) bulblet and scale, (ii) leaf, (iii) shoot apex, or (iv) stem node, anther or filament, style and pedicel. Plant regeneration can be achieved by the formation of bulb lets from cultured bulb-scales and callus cells are observed in some explants. Plants have also been obtained from bulbs or through somatic embryogenesis.

Plant regeneration with simultaneous callus formation can be achieved by the formation of bulblets from bulb-scale, leaf or anther culture (Kato and Yasutake, 1977; Arzate-Fernandes *et al.*, 1997a). Production of haploid plants can be obtained by anther culture (Han *et al.*, 1997), but results in a limited recovery of plant. Plants were also obtained with the bulbs through somatic embryogenesis (Tribulato *et al.*, 1997). Leaf explant cultures led to the production of bulblet plants but at a low frequency.

12.2.0 Anther Culture

In vitro haploid production from gametophytic cells enables the establishment of complete homozygous lines in a shortened time frame compared to conventional methods and has many parental applications in plant breeding to establish inbred lines rapidly for F_1 hybrid production and to observe recessive traits. Haploid plants have been produced from male or female gametophytes for over 200 species of higher plants and they have been utilized effectively for breeding programmes in certain crops (Morrison and Evana, 1988). Haploid, diploid and aneuploid plantlets were regenerated from anther cultures (Gu and Cheng, 1982).

Yarnagishi (1995) obtained calli from anthers of *L. longiflorum* and found that bulblets were diploids and derived from somatic origin. Sheng Han *et al.* (1997) opined that even non-haploids may be derived from somatic cells of anther wall of unreduced microspores as well as from nuclear fusion or endopolyploidys within microspores.

12.3.0 Protoplast Culture

Protoplast culture from microsporocytes in *L. longiflorum* was first reported by Ito (1973) who later observed fusion of these protoplasts.

Simmonds *et al.* (1979) obtained stable protoplasts from callus of the interspecific hybrid 'Black beauty' precultured on sucrose-free medium. The cell walls were removed by enzymatic degradation in a KCl solution as osmotic stabilizer. After the protoplasts were transferred from the high salt medium into a medium supplied with zeatin (1µM), NAA (10 µM) and 2,4-0 (0.1 µM), cell regeneration and cell division occurred, although at low frequency. But formation of plantlets from the resulting calli was not reported. One of the first successful results of regenerating fertile plants from protoplasts in lily was reported by Mii *et al.* (1994) when they employed 4.1 lM picloram, a chlorinated auxin, to induce creamy-white nodular calli from the seeds of L. x *formolongi* cv. Azusa; then the highest frequency of shoot regeneration was achieved from protoplast derived calli on MS media supplemented with 2.2 lM BAP after 2 months of transfer. Godo *et al.* (1996) had used different concentrations of various carbohydrates for cell suspension culture and plant regeneration from protoplast of *L.* x *formolongi*. They found that glucose produced the highest FW of cell clumps and protoplast yield, followed by sucrose, fructose andsorbitol. MS medium + 0.5 M glucose resulted in the highest plating efficiency. After 2 months of culture, shoots were regenerated from the protoplast-derived calli on half-strength MS medium with sucrose (2.5–10 g/l) or glucose (5–10 g/l).

12.4.0 Somatic Embryogenesis

Somatic embryogenesis was shown to occur from bulb-scales in some lily hybrids. Bulb-scale segments of *L. davidii* var. 'Willmottiae' and *L. tigrinum* on MS medium with 1.5 mg NAA or 0.5 mg BA/l and 0.5 per cent agar and 3 per cent sugar produced embryoids. on the surface or within the callus and formed complete plantlets (Wu and Wang, 1989). Embryonic cells originated from the centre of the multicellular meristematic mass derived from the callus. Meristem which could produce complete plantlets or only shoots or roots was derived from single or multiple cells in the callus. Haensch (1996) observed somatic embryos from bulb-scales of 23 genotypes of various lily hybrids. Somatic embryos formed better on media containing 4μM picloram. Yarnagishi (1995) obtained a creamy white nodular calli with high regeneration ability from somatic tissue of anthers of *L. longiflorum.*

12.5.0 Biochemical and Molecular Markers

DNA molecular markers have been developed as proper tools for rapid indirect selection. There are two important diseases in Lilium which cause serious damages including *F. oxysporum* and lily mottle virus (LMoV). Randomly amplified polymorphic DNA (RAPD) markers linked to partial resistance to *Fusarium* and corresponding genomic regions have been investigated in 150 descendants of a backcross population of Asiatic hybrids of lily (Straathof *et al.*, 1996).

Persson *et al.* (1998) used RAPD markers to estimate genetic variation within populations and relatedness among populations of *L. martagon.* Twelve domesticated populations from Sweden, Denmark, Norway and Lithuania, four native populations from Lithuania and two populations from Switzerland and Italy were analyzed with six decamer primers. The primers yielded a total of 55 polymorphic bands *i.e.*, an average of nine polymorphic bands per primer. The investigated populations were surprisingly heterogeneous in spite of a restricted population size. In an investigation on genetic analysis of major *Lilium* cultivars and hybrids, Choi *et al.* (1999) concluded that RAPD markers produced with random primers OPA02, OPA03, OPA04, OPA14, OPA16 and OPA17 could be used to classify *Lilium* cultivars and hybrids.

Bakhshaie *et al.* (2016), quoting the work of Van Heusden *et al.* (2002), detailed the development of a genetic map of 251 AFLP markers on 100 descendants of a lily backcross population was constructed and four Fusarium resistance QTLs mapped on linkage groups of 1, 5, 13 and 16. The resistance gene for tulip breaking virus also was detected on linkage group nine. This map was improved later in 2011 by application of diversity arrays technology (DArT) and nucleotide binding site (NBS) profiling and AFLP markers. Markers have been developed to map important ornamental traits including flower colour, flower spots, stem colour, antherless phenotype and flower direction (Shahin *et al.*, 2011). Although lilies are recalcitrant to genetic manipulation, superior genotypes are developed with improved flower colour and form, disease resistance and year round forcing ability. Different DNA molecular markers have been developed for rapid indirect selection, genetic diversity evaluation, mutation detection and construction of Lilium linkage map. Some disease resistance QTLs are already mapped on the Lilium linkage map (Bakhshaie *et al.*, 2016).

12.6.0 Genetic Transformation

In the genus *Lilium,* which contains a number of important ornamental species and hybrids, only limited work has been done on gene transfer. Lilies are monocotyledonous plants and generally recalcitrant to molecular genetic manipulation because of limitations that restrict utilization of transformation technologies (*e.g., A. tumefaciens*)that are routinely applied to dicotyledonous plants. Watad *et al.* (1998) reported the recovery of transgenic lily plants after microprojectile bombardment of morphogenic primary callus with plasmids containing *uida* reporter and *pat* selectable marker genes in *L. longiflorum.*

Agrobacterium transformation has more advantages over microprojectile bombardment. Cohen and Meredith (1992) reported on the transformation of *L. longiflorum* cv Nellie White cells by inoculation of bulb-scale sections with *A. tumefaciens* (strain C58). Langeveld *et al.* (1995) using stem internode explants of cv. Harmony inoculated with several *Agrobacterium* strains (C58-C9, C58-Cl, 1885 and ATCC, 15834) confirmed that strain C58 was effective in *Lilium* transformation, but transgenic plants were not obtained.

The advantage of TCLs for the study of transformation in tobacco was outlined by Tran Thanh Van *et al.* (1990). In lily species transformation by *Agrobacterium* was reported (Cohen and Meredith, 1992; Langeveld *et al.,* 1995). To increase the efficiency of transformation of lily species, it is important to target cells that are actively dividing because they have a greater hypersensitive response to *Agrobacterium* and 1 or a higher capacity for gene integration. This problem was overcome by determining the optimal conditions to enhance tissue survival, thus substantially facilitating transformation (Tran Thanh Van *et al.,* 1990). Watad *et al.* (1998) obtained transgenic lily (*Lilium longiflorum*) plants after bombardment with plasmids containing either *uidA* (gus)reporter gene or *pat* selectable marker under the control of maize Ubi promoter. PCR analysis of 19 plantlets out of 55 plantlets that survived (4 per cent) from 1800 morphogenic calli bombarded after having been exposed to 2 mg/l bialaphos showed the presence of the *pat* gene. Transgenic *Lilium* plants were obtained. However, the authors also indicated that the efficiency of transformation by biolistics was low. Thus, the tTCL method with a high frequency of shoot regeneration in *Lilium* with their efficient transformation protocol was reported in many dicot and monocot species.

The lack of an efficient genetic transformation system for *Lilium* has been an international obstacle (Yan *et al.,* 2019). Because existing model plants lack bulbs, bulb-related gene function verification studies cannot be carried out in model plants. Here, two stable and efficient genetic transformation systems based on somatic embryogenesis and adventitious bud regeneration were established in two *Lilium* species. Transgenic plants and T-DNA insertion lines were confirmed by *β-glucuronidase* (*GUS*) assay, polymerase chain reaction (PCR) and Southern blot. After condition optimization, transformation efficiencies were increased to 29.17 per cent and 4 per cent in *Lilium pumilum* DC. Fisch. and the *Lilium longiflorum*

'White Heaven', respectively. To further verify the validity of these transformation systems and apply the CRISPR/Cas9 (Clustered Regularly Interspaced Short Palindromic Repeats (CRISPR)-associated protein 9) technology in *Lilium*, the *LpPDS* gene in the two *Lilium* species was knocked out. Completely albino, pale yellow and albino–green chimeric mutants were observed. Sequence analysis in the transgenic lines revealed various mutation patterns, including base insertion, deletion and substitution. These results verified the feasibility and high efficiency of both transformation systems and the successful application of the CRISPR/Cas9 system to gene editing in *Lilium* for the first time. Overall, the study lays an important foundation for gene function research and germplasm improvement in *Lilium* spp. (Yan *et al.*, 2019).

12.7.0 *In vitro* Conservation

12.7.1 Encapsulation/Dehydration

Pre-culturing with high concentration of sugar or sorbitol for 1 or 2 days was reported to be very important in improving survival of cryopreserved cells, meristems and somatic embryos (Matsumoto *et al.*, 1994). Matsumoto and Sakai (1995) confirmed that a mixture of 1 M glycerol in combination with 0.8 M sucrose produced much higher levels and shoot formation in the encapsulated dried meristems of *L. japonicum* than those treated with 0.8 M sucrose alone. They found that encapsulated dried meristems following treatment with mixture of 1 M glycerol and 0.8 M sucrose produced the highest shoot formation after cooling to -196°C. These cryopreserved meristems resumed growth about 3 days after planting and showed normal recovery growth.

13.7.2 Vitrification

Matsumoto *et al.* (1995) successfully cryopreserved Japanese pink lily (*L. japonicum*) by vitrification. The excised apical meristems were precultured on a solidified MS medium containing 0.3 M sucrose for 1 day at 25°C and then loaded in a mixture of 2 M glycerol plus 0.4 M sucrose for 20 min at 25°C. Cryoprotected meristems were then sufficiently dehydrated with highly concentrated vitrification solution Pvs2 (30 per cent dimethyl sulphoxide in 0.4 mM sucrose solution) at 25°C for 20 min or at 0°C for 10 min prior to plunging into liquid nitrogen. They also successfully cryopreserved alginate-coated apical meristems derived from bulb scales of Japanese lily by dehydration and vitrification.

13.0 References

Allen, L.O. (1972) *CMI/IMB Plant Viruses*, **96**.

Allen, T.C. and Anderson, WC. (1980) *Acta Hortic.*, **10**: 245-249.

Allen, T.C. and Femald, K. (1972) *Lily Yb. N. Am. Lily Soc.*, **25**: 53-55.

Amaki, W. and Yamamoto, Y. (1988) *J. Japanese Soc. Hort. Sci.*, **57**: 269-272.

Amaki, W., Tezuka, T. and Yamamoto, Y. (1989a) *J. Japanese Soc. Hort Sci.*, **57**: 648-654.

Amaki, W., Tezuka, T. and Yamamoto, Y. (1989b) *Plant Cell Incompatibility Newslett.*, **21**: 1-6.

Ames, R.N. and Linderman, R.G. (1977) *Canadian J. Microbiol.*, **23**: 1663-1668.

Ames, R.N. and Linderman, R.G. (1978) *Canadian J. Bot.*, **56**: 2773-2780.

Anonymous (1979) *Vakblad. Voor de Bloemisterij*, **34**: 48-49.

Aoki, N. and Yoshino, S. (1988a) *Bull. Fac. Agril., Shimane Univ.*, **22**: 37-42.

Aoki, N. and Yoshino, S. (1988b) *Shimane Daigaku Nogakuba Kenkyu Hokiku Bull Fac. Agril.*, Shimane University, **22**: 37-42.

Argo, W.R. and Biembaum, J.A. (1994) *J. Amer. Soc. Hort. Sci.*, **119**: 1151-1156.

Arzate-Fernandez, A.M., Nakazaki, T., Okumoto, Y. and Tanisaka, T. (1997a) *Plant Cell Rep.*, **16**: 836-840.

Asano, Y. (1980) *J. Japanese Soc. Hort. Sci.*, **49**: 241-250.

Ascher, P.D. and Drewlow, L.W. (1975) *Plant Sci. Letter*,**4**: 401-405.

Asjes, C.J., Bunt, M.H. and Van Slogteren, D.H.M. (1974) *Acta Hortic.*, **36**: 223-228.

Azarte-Fernandez, A.M., Nakazaki, T. and Tanisaka, T. (1998) *Plant Breeding*, **117**: 479-484.

Bailey, L.H. (1963) *The Standard Cyclopedia of Horticulture*, The MacMillan Company, New York.

Bakhshaie, M., Babalar, M., Mirmasoumi, M. and Khalighi, A. (2010) *Plant Cell Tissue Organ Cult.*, **102**: 229–235.

Bakhshaie, M., Khosravi, S., Azadi, P., Bagheri, H. and van Tuyl, J.M. (2016) *Plant Cell Rep.*, **35**: 1799–1826.

Ball, V. (1985) *Ball Book*, West Chicago, pp. 569-574.

Baranova, M.V (1971) *Botanicheskii Zhurnal*, **56**: 1593-1604.

Barendse, L. (1975) *Vakblad voor de Bloemisterij*, **30**: 12-13.

Barendse, L., Slootweg, C. and Broek, W. van den. (1978) *Vakblad voor de Bloemisterij*, **33**: 21.

Beattie, D., Blodgett, A.M. and Holcomb, E.J. (1993) *Plant Growth Reg. Soc. Amer. Quart.*, **21**: 64-72.

Bhandari, N.S. and Aswath, C. (2018) *Int. J. Curr. Microbiol. App. Sci.*, **7**(4): 1183-1190.

Bhandari, N.S., Aswath, C., Manish, B. and Versha G. (2019) *Int. J. Agricult. Stat. Sci.*, **15**(2): 801-806.

Bhattacharjee, S.K. (1983) *Prog. Hort.*, **15**: 204-209.

Bino, R.J., Van Tuyl, J.M. van and Vries, J.N. de. (1988) *Prophyta*, **42**: 147-148.

Blake, M.R. and Wilson, C.R. (1996) *Annals Appl. Biol.*, **129**: 39-45.

Blaney, L. T. and Roberts, A.N. (1965) *Oregon Ornament Nursery Dig.*, **9**: 1-2.

Blaney, L.T. and Roberts, A.N. (1966) *Proc. Amer. Soc. Hort. Sci.*, **89**: 651-656.

Bokura, V. (1919) *The Standard Cyclopedia of Horticulture*, The MacMillan Company, New York.

Boon, J. Van Der. and Niers, H. (1986) *Acta Hortic.*, **177**: 249-254.

Boon, J. van der. and Niers, H. (1987) *Rep. Inst Voor Bodemvruchtbaarheid.* Netherlands, **2**: 75.

Boontjes, J. (1973) *Bloembollencultuur,* **83**: 662-664.

Boontjes,J. and Plocgman, C. (1981) *Bloembollencultuur,* **91**: 49-1307.

Boontjes, J. and Mey, H. van der. (1975) *Bloembollencultuur,* **86**: 423.

Boontjes, J., Durieux, A.J.B. and Kamerbeek, G.A. (1975) *Acta Hortic.*, **51**: 261-262.

Brosnan, T. and Sun, D.W. (2001) *J. Agril. Engineering Res.*, **79**: 299-305.

Bui, V.L., Nhut, O.T. and Tran Than Van, K. (1999) *C.R. Acad. Sci. Paris,* **322**: 303-310.

Bush-Brown, J. and Bush-Brown, L. (1958) *Americas Garden Book,* Charles Scribner's Sons, New York and London.

Caputo, G., Mercuri, A, Benedetti, L. and Giovanni, A (1990) *Plant Tissue Cult. Biotech.*, **2**: 94-104.

Carlson, W.H. and Carpenter, W.J. (1970) *HortScience,* **5**: 397-399.

Chang, C.A., Chen, C.C. and Deng, T.C. (1998) *Special Publication Taichung District Agril. Improv. Stn.*, **40**: 151-158.

Chase, S.S. (1963) *J. Hered.*,**54**: 152–158.

Chaudorov, I. and Denkova, S. (1994) In : *Inter. Conf Plant Virology Apilts.* Trogan, Baigenia,

Chen, Q.P. (2000) *J. Fujian College Forestry,* **20**: 162-164.

Chi, H.S. (2000) *Botanical Bull. Academia Sinica,* **41**: 143-149.

Chin, SW., Kuo, C.E., Liou, J. J. and Shii, C.T. (1997) *Acta Hortic.*, **430**: 495-502.

Chittenden, E.J. and Synge, P.M. (1956) *Dictionary of Gardening,* The Royal Horticultural Society, Oxford University Press, London.

Choi, H.S., Kim, K.S., Choi, J.K., Lee, K.K., Hong, D.K., Won, H. and Lee, Y.S. (1999) *Korean J. Hort. Technol.*, **17**: 144-147.

Choi, S.T. (1983) *J. Korean Soc. Hort. Sci.*, **24**: 42-48.

Choi, S.T., Cheong, S.T. and Ahn, H.G. (1996) *Acta Hortic.*, **414**: 235-242.

Choi, S.T., Jung, W.Y., Ahn, H.G. and Chang, Y.D. (1998a) *J. Korean Soc. Hort. Sci.*, **39**: 771-775.

Choi, S.T., Jung, WY., Ahn, H.G. and Chang, Y.D. (1998b) *J. Korean Soc. Hort. Sci.*, **39**: 765-770.

Ciolakowska, A.M., T. Nishikawa, D.J. Shea and K. Okazaki (2008) *Breed. Sci.*, **68**: 35–52.

Cohen, A and Meredith, C.P. (1992) *Acta Hortic.*, **325**: 611-314.

Creij, M.G.M. van., Kerckhoffs, D.M.F.J., Van Tuyl, J.M. van. and Cadic, A. (2000) *Acta Hortic.*, **508**: 267-274.

Dahlgren, R.M.T. and Clifford, H.T. (1982) *The Monocotyledons: A comparative study*, Academic Press, London and New York, p. 378.

Dilta, B.S., Pathania, N.S. and Sehgal, O.P. (2000) *J. Omam. Hort.*, **3**: 62-63.

Dole, J.M. and Wilkins, H.F. (1996) *Acta Hortic.*, **414**: 295-300.

Durieux, A. J. B. (1978) *Vakblad voor de Bloemisterij*, **33**: 24-25.

Eastwood, T. (1952) *Proc. Amer. Soc. Hort. Sci.*, **59**: 531-541.

Einert, A. E. and Box, C.O. (1967) *Proc. Amer. Soc. Hart.* Sci., **90**: 427-432.

Elgar, H. J., Woolf, AB. and Bieleski, R.L. (1999) *Postharvest Biol. Technol.*, **16**: 257-267.

Embury, L. (1960) *Hortic.*, **38**: 520-521.

Emsweller, S.L., Uhring, J. and Stuart, N.W. (1960) *Proc. Amer. Soc. Hart. Sci.*, **75**: 720-725.

Ende, van dan, J. E., Pennock, vos, M. G., Bastiaansen, C. and Koster, A T.H.J. (2000) *Acta Hortic.*, **519** : 215-220.

Fang, J. G., Yang, L., Hsieh, P. T. and Chen, W. H. (1999) *Rep. Taiwan Sugar Res. Inst.*, **65**: 29-40.

Fernandez, A.M., Nakazaki, T. and Tanisaka, T. (1996) *Plant Breed.*, **115**: 167-171.

Fiedler, H. (1999) *TASPO Gartenbaumagazin*, **9**: 30-31.

Furuata, T., Martin, WC. and Perry, F. (1963) *Proc. Amer. Soc. Hort. Sci.*, **83**: 803-810.

Gianfagna, T.J. and Wulster, G.J. (1986) *HortSci.*, **21**: 463-464.

Gill, D. L. (1964) *The Exchange*, **142**: 32,64-68.

Gindina, S.R (1976) *Okhrana Sredy I Ratsional'n Ispol'z. Rastitel'n. Resursov.*, pp. 232-233.

Giustiniani, L., Moschini, E. and Graifenberg, A. (1988) *Colture Protette*, **17**: 81-87.

Godo T., Matsui, K., Kida, T., Mii, M. (1996) *Plant Cell Rep.*,**15**: 401–404.

Goldsberry, K.L. and Maffei, H.M. (1986) *Res. Bull. Colorado Greenhouse Growers Assn.*, **436**: 3-5.

Gonzalez, R.B, Lim, K.B. and van Tuyl, J.M. (2014) *Acta Hortic.*, **1027**:129-142.

Gonzalez, R.B., Lim, K.B., Zhou, S., Ramanna, M.S. and Van Tuyl, J.M. (2008) In: Floriculture, Ornamental and Plant Biotechnology Volume V ©2008 Global Science Books, UK. Pp.138-145.

Grassoti, A, Magnani, G. and Ranieri, M. De. (1982) *Notiziario di ortoflorofrutticoltura*, **8** : 61-64.

Grassotti, A and Magnani, G. (1988) *Colture Protette*, **17**: 33-42.

Grassotti, A. (1997) *Acta Hortic.*, **430**: 221-226.

Gries, H. (1970) *Iris und Lilien*, No. 4, pp. 193-195.

Gu, Z.P. and Cheng, K.C. (1982) *Acta. Bot. Sinica.*, **24**: 28-32.

Hackett, W.P. (1969) *Comb. Proc. Intl. Plant Prop. Soc.*, **19**: 105-108.

Haensch, K. J. (1996) *Gartenbauwissenschafl,* **61**: 214-218.

Han, D.S., Niimi, Y, and Nakano. (1997) *Plant Cell Tissue Organ Cult.,* **47**: 153-158.

Hao, W.Y., Lin, X.G., Gu, XX. and Niu, J.Q. (1991) *Acta Pedologica Sinica,* **28**: 129-131.

Harkema, H. and Doom, WG. van. (1986) *Acta Hortic.,* **181**: 383-387.

Haruki, K., Yamala, K., Hosoki, T. and Ohika, K. (1996b) *J. Japanese Soc. Hart. Sci.,* **65**: 363-371.

Hasegawa, A., Goi, M. and Ogura, K. (1980) *Kagawa Daigaku Nogakubu Gakuzyutu Hokuku,* **31**: 113- 117.

Healy, W.E. and Wilkins, H.F. (1984) *HortScience,* **19**: 843-844.

Heins, R.D., Wilkins, H.F. and Healy, W.E. (1982) *J. Amer. Soc. Hort. Sci.,* **107**: 335-338.

Henny, R.J. and Ascher, P.D. (1975) *HortScience,* **10**: 25-26.

Hiratsuka, S., Tezuka, T. and Yamamoto, Y, (1989) *J. Amer. Soc. Hort. Sci.,* **114**: 505-508.

Holmes, R.L. and Coorts, GD. (1980) *HortScience,* **15**: 497-498.

Hoogerwerf, A. and Doom, W.G. van. (1992) *Postharvest Biol. Technol.,* **1**: 295-304.

Hsieh, T.F. and Huang, JW. (1999) *European J. Plant Pathol.,* **105**: 501-508.

Huang, J.M., Zhao, X.Y., Zhang, G.M. and Ni, Y.Y. (1990) *Acta Hortic. Sinica,* **17**: 153-157.

Hwang, K.S., Park, J.M., Yiem, M.S. and Ho, Q.S. (1998) *RDA J. Agro. Env. Sci.,* **40**: 1-5.

Iapichino, G., Amico Roxas, U. and Sciortino, A. (1994) *Advances Hort. Sci.,* **8**: 195-199.

Ignat'eva, I.P. (1998) *Izvestiya Timiryazevskoi Sel Skokhozyaistvennoi Akademii,* **2**: 66-89.

Ito, M. (1973) *Bot. Mag. Tokyo,* **86**: 133-141.

Ivanova, N.V. (1983a) *Byulleten Glavnogo Botanicheskogo Sada,* No. 127, pp. 62-64.

Jacobsen E., Ramanna, M.S., Huigen, D.J. and Sawor, Z. (1991) *Euphytica,* **53**: 247–253. doi:10.1007/BF00023276.

Kaminska, M., Korbin, M., Komorowska, B. and Fulawska, J. (1998) *Acta Physiol. Planta.,* **20**: 49-53.

Kapoor, M., Grewal, J.S. and Arora, J.S. (2000) *J. Ornam. Hort.,* **3**: 58-59.

Kato, Y. and Yasutake, Y. (1977) *Phytomorphology,* **27**: 290-396.

Kawarabayashi, W. (1993) *J. Japanese Soc. Hart. Sci.,* **62**: 611-618.

Kawarabayashi, W. (1996) *Lily Yb. N. Am. Lily Sco.,* **45**: 15-18.

Khan, N.S. Zhou, M.S. Ramanna, Arens, P., Herrera, J., Visser, R.G.F. and Van Tuyl, J.M. (2009). *Euphytica,* **166** : 399–409.

Kim, S.H, Lee, J.S., Hong, J. and Moom, D.K. (1998b) *J. Korean Soc. Hort. Sci.,* **39**: 819-823.

Kim, S.H., Moon, D.K., Lee, J.S. and Hong, J. (1998a) *J. Korean Soc. Hort Sci.,* **39**: 827-832.

Kim, S.R., Finkel, D., Chung, Y.Y. and An, G. (1994) *Sexual Plant Reproduction*, **7**: 76-86.

Kim, T.J. (1996) *Acta Hortic.*, **414**: 69-79.

Kim, Y.H., Han, Y.H., Soh, H.S., Lee, S.J., Kim, J.S. and Ra, J.Y. (1998c) *RDA J. Crop Prot.*, **40**: 58-65.

Kiplinger, D.C., Tayama, H.K. and Staby, G. (1972) *Ohio Florists' Asson. Bull.*, **509**: 9.

Kiplinger, D.C., Tayama, H.K. and Staby, G. (1975) *Ohio Florists' Asson. Bull.*, **543**: 6.

Kohl, H.C. Jr. and Nelson, R.L. (1963) *Proc. Amer. Soc. Hort. Sci.*, **83**: 808-810.

Koster, A.T.J. and Kruyer, C.J. (1983) *Bloembollencultuur*, **93**: 945-947.

Koths, J.S. and Judd, R.W. Jr. (1976) *Connecticut Greenhouse Newsl.*, **69**: 1-3.

Krause, J. (1996a) *Prace z Zakresu Nauk Rolniczych.*, **80**: 275-281.

Krause, J. (1996b) *Prace z Zakresu Nauk Rolniczych.*, **80**: 267-273.

Kruczkowska, H. (1989) *Prace instytutosodownictuer I Kwlockrstwa w. skierrewicach*, Seria B, Koshny, Ozdobne, **11**: 137-147.

Kruijer, C.J. (1982) *Bloembollencultuur*, **92**: 876-877.

Kubalakova, M. and Gavel, L. (1991) In: *Biological Kui Tiuriovan;ya Kletok I Biotechnology Vastenil*, pp. 206-209.

Kumar, R., Arora, J.S. and Singh, S. (1999) *J. Omam. Hort.*, **2**: 133-134.

Laiche, A.J. and Box, C.O. (1970) *HortScience*, **5**: 396-397.

Langeveld, S.A., Gerrits, M.M., Derk, AF.L.M., Boonekamp, P.M. and Bol, J.E. (1995) *Euphytica*, **85**: 97-100.

Langhans, R.W. and Smith, D.R. (1965) *Bull. N.Y.* St *Flower Grs.*, No. 237, pp. 1-2.

Laurie, A, Kiplinger, D.C. and Nelson, K.S. (1958) *Commercial Flower Forcing*, McGraw-Hill Book Co., New York.

Lee, A.K. and Suh, J.K. (1996) *Acta Hortic.*, **414**: 287-293.

Lee, J.S., Kim, T.H., Ryu, By. and Noh, H.S. (2000) *J. Korean Soc. Hort. Sci.*, **41**: 301-304.

Lee, J.W, Kim, T.J., Kim, J.H., Kim, H.H., Lee, C.H, Choi, K.S. and Paek, K.Y (1999) *Korean J. Hort Sci. Technol.*, **17**: 136.

Lee, M.L. and Leu, L.S. (1998) *Plant Protection Bull.*, Taipei, **40**: 251-264.

Lercari, B., Moschini, E. and Malorgio, F. (1987) *Informatore Agrario*, **43**: 87-90.

Leshem, B., Lilien-Kipnis, H. and Steinitz, B. (1982) *HortScience*, **17**: 129-136.

Lewis, A.J. and Gilbertz, D.A. (1987) *Acta Hortic.*,**205**: 237-240.

Lewis, AJ. and Lewis, J.S. (1982) *HortScience*, **17**: 336-337.

Li, T.J. and Niimi, Y (1995) *J. Japanese Soc. Hort. Sci.*, **64:** 149-159.

Lim, K.B. (2000) *Ph.D. Thesis*, Landbouwuniversiteit Wageningen (Wageningen Agricultural Univeristy), Wageningen, Netherlands.

Lim, K.B., Gonzalez, R.B., Zhou, S., Ramanna, M.S. and van Tuyl, J.M. (2008) In: Floriculture, Ornamental and Plant Biotechnology Volume V ©2008 Global Science Books, UK. pp.146-151.

Lim, S., Seon, J.H., Son, S.H., Han, R.H. and Paek, K.Y. (1998) *J. Korean Soc. Hort. Sci.,* **39**: 111-114.

Lin, S.C., Wu, C.G., Huang, S.N. and Houng, K.H. (1999) *J. Agril. Res. China,* **48**: 135-142.

Lin, W.C. and Wilkins, H.F (1975) *J. Amer. Soc. Hort. Sci.,* **100**: 6-9.

Liu, T.S. (1996) *Bull. Taichung Dist. Agril. Imp. Stn.,* **51**: 49-57.

Liu, T.S. and Chen, Y.H. (1998) *Special Publ. Taichung Dist. Agril. Improv. Stn.,* **40**: 139-150.

Liu, T.S., Huang, Y.C., Twu, J.S. and Hsieh, J.R. (1998) *Plant Protection Bull.,* Taipei, **40**: 241-249.

Loffler, H.J.M., Meijer, H., Straathof, Th, P., Van Tuyl, J.M. van. and Lee, J.S. (1996) *Acta Hortic.,* **414** : 203-208.

Logan, T.J., Faber, WR. and Smith, E.M. (1984) *Ohio Rep. Res. Dev. Agril. Home Eco. Natural Res.,* **69**: 37-40.

Lorenzo, G.A., Mascarini, L., Gonzalez, M.N. and Lalor, E. (2018). *Scientia Hortic.,* **240**: 453-459.

Lu, N., Wu, L. and Shi, M. (2020) *Sci. Hortic.,* **264**, https://doi.org/10.1016/j.scienta.2019.109172.

Lu, Y.Y. and Chen, C.Y. (1998) *Plant Pathol. Bull.,* **7**: 134-140.

Marinangeli, P. and Curvetto, N. (1997) *HortScience,* **32**: 739-741.

Mastalerz, J.W. (1965) *Proc. Amer. Soc. Hart. Sci.,* **87**: 502-509.

Matsumoto, T. and Sakai, A. (1995) *Cryo-letters,* **16**: 237-245.

Matsumoto, T., Sakai, A. and Yamode, K. (1994) *Plant Cell Rep.,* **13**: 442-446.

Matsuo, E. and Van Tuyl, J.M. Van. (1984) *Phytochemistry,* **26**: 1844-1845.

Matthews, V. (2007) The International lily register and checklist 2007, Fourth edition. https://www.rhs.org.uk/plants/pdfs/plant-register-supplements/Lilies/lily-register-and-checklist.pdf

McMurry, A. (1978) *North Carolina Flower Growers' Bull.,* **22**: 1-5.

Meeteren, U. van. and Slootweg, G. (1986) *Acta Hortic.,* **181**: 473-476.

Meulen Muisers, J.J.M. van der., Oeveren, J.C. van., Plas, L.H.W. van der. and Van Tuyl, J.M. van. (2001) *Postharvest Biol. Technol.,* **21**: 201-211.

Meulen-Muisers, J.J.M. van der., Oeveren, J.C. van., Jansen, J. and Van Tuyl, J.M. van. (1999) *Euphytica,* **107**: 149-157.

Meulen-Muisers, J.J.M. van. der. and Oeveren, J.C. van. (1996) *J. Amer. Soc. Hort. Sci.,* **121**: 33-36.

Meulen-Muisers, J.J.M. van. der. and Oeveren, J.C. van. (1997) *J. Amer. Soc. Hort. Sci.,* **122**: 368-372.

Meulen-Muisers, J.J.M. van. der., Oeveren, J.C. van. and Van Tuyl, J.M. van. (1998) *J. Amer. Soc. Hort. Sci.,* **123**: 283-287.

Mii, M., Yuzawa, Y., Suetomi, H., Motegi, T. and Godo, T. (1994). *Plant Sci.,* **100**: 221–226.

Miles, P. (1952) *Lily Year Book,* p. 16.

Mori, K. (1971) *Japanese Agric. Res. Q.,* **6**: 1-7.

Morrison, R.A. and Evans, D.A. (1988) *BioTechnol.,* **6**: 684-690.

Nhut, D.T. (1998) *Plant Cell Rep.,* **17**: 913-916.

Nightangle, A.E. (1979) *HartScience,* **14**: 67-68.

Niimi, Y. (1985) *Japanese Soc. Hart. Sci.,* **54**: 635-640.

Niimi, Y. (1986) *Acta Bot.,* **35**: 1889-1894.

Niimi, Y. and Saito, T. (1991) *J. Japanese Soc. Hart. Sci.,* **59**: 635-640.

Nishizawa, T. and Nishi, Y. (1966) *Exp. Sta.,* **12**: 139·157.

Noda, S. (1978) *Bot. Mag.,* Japan, **91**: 279-283.

Nowak, J. and Mynett, K. (1985) *Scientia Hort.,* **25**: 299-302.

Ohkawa, K. (1979) *Scientia Hort.,* **10**: 255-260.

Ohkawa, K. (1982) *Proc. 21st. Int. Hart. Congr.,* Hamburg, **2**: 1726.

Okazaki, K. (1996) *Acta Hortic.,* **414**: 81-92.

Okazaki, K. and Jong, J. de. (1990) *Proc. Symp. Integration of in vitro techniques in ornamental plant breeding, EUCARPIA,* Wageningen, Netherlands 10-14 November, pp 98-103.

Panizza, M., Sudi, A.M. and Frgnoni, F. (1996) *HortScience,* **4**: 103-106.

Panwar, G.S., Srivastava, S.K. and Uniyal, P.L. (2017) *Current Sci.,* 113 (5): 946-951.

Pergola, G. and Roh, M.S. (1987) *Acta Hortic.,* **205**: 241-247.

Persson, H.A., Lundquist, K. and Nybom, H. (1998) *Hereditas,* **128**: 213-220.

Pirone, P.P., Dodge, B.O. and Rickett, H.W. (1960) *Diseases and Pests of Ornamental Plants,* Constable and Co. Ltd., London.

Powell, M.V., Widmer, R.E. and Wilkins, H.F (1975) *Minnesota State Florists Bull.,* pp. 4-13.

Qu, Y. Mok, Mok, D.W.S. and Stang, J. (1998) *In Vitro Cell Dev. Biol.,* **24**: 471-476.

Roberts, A.N. and Tomasovic, B.J. (1975) *Acta Hortic.,* **47**: 339-345.

Roh, M.S., Griebach, R.J., Gross, K.C. and Line, M. (1996a) *Acta Hortic.,* **414**: 111-124.

Roh, M.S., Griesbach, R.J. and Lawson, R.H. (l996b) *Acta Hortic.,* **414**: 101-110.

Roh, S.M. (1978) *J. Korean Soc. Hart. Sci.,* **19**: 82-88.

Roh, S.M. (1979) *J. Korean Soc. Hart. Sci.,* **20**: 72-83.

Roh, S.M. (1982a) *Korean Soc. Hart. Sci.,* **23**: 1.

Roh, S.M. and Wilkins, H.E (1974) *HortScience,* **9**: 37-38.

Roh, S.M. and Wilkins, H.E (1975) *Acta Hortic.,***47**: 215-224.

Roh, S.M. and Wilkins, H.E. (1977a) *J.Korean Soc. Hort. Sci.,* **18**: 187-193.

Roh, S.M. and Wilkins, H.E. (1977b) *J.Amer. Soc. Hort. Sci.,* **102**: 229-235.

Roh, S.M. and Wilkins, H.E. (1977c) *J.Amer. Soc. Hort. Sci.,* **102**: 255-257.

Ronald, W.G. (1981) *Quarterly Bull. North Amer. Lily-Soc.,* **35**: 14-15.

Roorda van Eijsinga, J.P.N.L. (1980) *Vakblad voor de Bloemisterij,* **35**: 31.

Rybezynski, J.J. and Gomolinska, H. (1989) *Acta Hortic.,* **251**: 183-189.

Ryu, K.Y., Kim, R.H., Lee, Y.H. and Heo, N.Y. (1998) *Korean J.Plant Pathol.,* **14**: 459-462.

Sagawa, Y. (1958) *J. Hered.,* **49**: 195-198.

Sagawa, Y. (1959) *Proc. 9th Int Bot. Congr. II,* p. 339.

Santos, M.N.D., Martins, T.M., and Maria, M.A. (2018) *Ornamental Hortic.,* **24**(1): 44-49.

Schenk, P.C. (1975) *Bloembollencultuur,* **85**: 1131.

Schenk, P.K and Boontjes, J. (1970) *Lily Year Book,* pp. 47-57.

Seeley, J.G. and Velasquez, D. De C. (1952) *Proc. Amer. Soc. Hort. Sci.,* **60**: 459-472.

Shahin A., Arens P., Van Heusden A.W., Van Der Linden G., Van Kaauwen M., Khan N., Schouten, H.J., Van De Weg W.E., Visser R.G.F., Van Tuyl J.M. (2011) *Plant Breed.,* **130**: 372–382.

Sheng Han, D., Niimi, Y. and Nakano, M. (1997) *Plant Tissue Organ Cult.,* **47**: 153-158.

Shimada, Y., Imanishi, H. and Mori, G. (1999) *J. Japanese Soc. Hart. Sci.,***68**: 1027-1032.

Shimizu, T. and Asahira, T. (1979) *Stu.Inst. Hort,* Kyoto University, Japan, **9**: 128-134.

Shoemaker, J.S. and Teskey, B.J.E. (1969) *Practical Horticulture,* Wiley Eastern Private Ltd., New Delhi, pp. 82-92.

Simmonds, J.A. and Cumming, B.G. (1977) *Sci. Hort.,* **6**: 71-81.

Simmonds, J.A., Simmonds, D.H. and Cumming, B.J. (1979) *Canadian J. Bot.,* **57**: 512-516.

Singh, S., Kumar, R and Arora, J.S. (2000) *J. Ornam. Hort.,* **3**: 100-102.

Slangen, J.H.G., Krook, G.J, Hendriks, C.H.M. and Hof, NAA (1989) *Netherlands J. Agric. Sci.,* **37**: 269-272.

Smith, D.R. and Langhans, R.W. (1961) *Bull. N.Y. St. Flowers Grs.,* No. 192, pp. 1-4.

Smith, F.F. (1967) *Controlling Insects of Flowers,* Agriculture Information Bulletin No. 237, US, Dept. of Agriculture, p. 30.

Son, K.C. and Han, M.S. (2000) *J. Korean Soc. Hort. Sci.,* **41**: 207-211.

Son, K.C. and Suh, J.N. (1999) *J. Korean Soc. Hart Sci.*, **40**: 467-469.

Son, K.C., Kim, M.K., Yoo, M.H. and Pak, H.Y. (2000) *J. Korean Soc. Hort Sci.*, **41**: 83-86.

Song, C.Y., Bang, C.S., Chung, S.K., Kim, Y.J., Lee, D.C. and Lee, J.S. (1996) *Acta Hortic.*, **414**: 277-285.

Sorrentino, G., Ceiro, L. and alvino, A. (1997b) *Scientia Hort.*, **69**: 359-373.

Sorrentino, G., Cerio, L. and Alvino, A. (1997a) *Scientia Hort.*, **69**: 259-273.

Sponga, E, Ferrari, L. and Lereari, B. (1986) *Colture Protelte*, **15**: 63-68.

Sponga, F. and Lercart, B. (1986) *Culture Protette*, **15**: 45-51.

Straathof, T.P., Loffer, H.J.M., Linfield, CA and Roebroeck, E.J.A (1997) *Acta Hortic.*, **430**: 477-486.

Straathof, T.P., Van Tuyl, J.M., Dekker, B., Van Winden, M.J.M. and Sandbrink, J.M. (1996) *Acta Hortic.*, **414**: 209–218.

Stuart, N.W, Gill, D.L. and Hickman, M.G. (1961) *The Exchange*, **136**: 22-23.

Stuart, N.W. (1967) *The New York and Ohio Lily Schools*, pp. 47-58.

Suh, J.K. and Lee, J.S. (1995) *J. Korean Soc. Hort. Sci.*, **36**: 248-254.

Swart, A. (1979) *Bloembollencultuur*, **89**: 902-903.

Swart, A. (1980) *Vakblad voor de B1oemisterij*, **35**: 37.

Swart, A. (1981) *Bloembollencultuur*, **92**: 398-399.

Sytsema, W. (1968) *Vakblad voor de Bloemisterij*, **23**: 1703.

Sytsema, W., Elfering, K.G. and Swart, A. (1984) *Vakblad voor de B1oemisterij*, **39**: 28-29.

Szlachetka, W.I., Wojcik, B. and Prabucki, A. (1997) *Conf Recent trends growing ornamental plants*, Lublin, Poland, 17-18 June 1997, **449**: 119-212.

Takayama, S. and Misawa, M. (1979) *Physiol. Plant.*, **46**: 184-190.

Takayama, S. and Misawa, M. (1982) *Plant Cell Physiol.*, **23**: 67-74.

Tran Thanh Van., K., Richard, L. and Gendy, A. (1990) In: *NATO, Biotechnology Series* (Durzan, D. and Rodriguez, R. eds.), Plenum Press, New York, pp. 215-224.

Tribulato, A., Remotti, PC., Lomer, H.J.M. and Vtuys, J.M. (1997) *Plant Cell Rep.*, **17**: 113-118.

Tsujita, M.J., Murr, D.P. and Johnson, G. (1979) *Canadian J. Plant Sci.*, **59**: 757-761.

UnNo, Y and Wagatsuma, T. (1996) *J. Rakuno Gakuen Univ., Natural Sci.*, **21**: 73-77.

Valis, G., Herman, M., Brosch, S. and A1on, Y (1976) *Phytoparasitica*, **4**: 159.

Van Aartrijk, J. and Blom-Bamhoom, G.J. (1984) *J.Plant Physiol.*,**116**: 409-416.

Van Aartrijk, J. and Bloom-Barnhoom, G.J. (1980) *Acta Hortic.*, **109**: 297-301.

Van Heusden, A.W., Jongerius, M.C., Van Tuyl, J.M., Straathof, T.P., Mes, J.J. (2002) *Acta Hortic.*, **572** : 131–138.

Van Tuyl, J.M. (1985) *Sci. Hortic.*, **25**: 177-187.

Van Tuyl, J.M., Marcucci, M.C. and Visser, T. (1982) *Euphytica,* **31**: 613-619.

Van Tuyl, J.M. and Arens, P. (2011). *Acta Hortic.,* **900** : 223-230.

Van Tuyl, J.M., and Hoisteijn, H.C.M. Van. (1996) *Acta Hortic.,* **414**: 35-45.

Van Tuyl, J.M., Chi, H.S., Kronenburg, B.C.M. van. and Meijer, B. (1997) *Acta Hortic.,* **430**: 539-544.

Van Tuyl, J.M., Dijkenn, A. van., Chi, H.S., Lim, K.B., Villemoes, S., Kronenburg, B.C.E. van. and Cadie, A. (2000) *Acta Hortic.,* **508** : 83-88.

Van Tuyl, J.M., Franken, J., Jongerius, R.C., Lock, C.A.M. and Kwakkenbos, T.A.M. (1986b) *Acta Hortic.,* **177**: 591-595.

Van Tuyl, J.M., Keijzer, C.J., Wilms, H.J. and Kwakkenbos, A.A.M. (1989) *Lily yearb. North Amer. Lily Soc.,* pp. 103-111.

Van Tuyl, J.M., Dijk, A.J. van. and Raamsdonk, L.W.D. van. (1986a) *Acta Hortic.,* **177**: 601-605.

Wang, S.Y., Roberts, A.N. and Blaney, L.T. (1970) *HortScience,* **5**: 113-114.

Wang, Y.T. and Breen, P.J. (1986) *J. Amer. Soc. Hort. Sci.,***11**: 442-446.

Watad, A.A., Yun, D.J., Matsumoto, T., Wiv, X., Wu, Y., Kononowlcz, A.K., Bressan, R.A. and Hasegaua (1998) *Plant Cell Rep.,* **17**: 262-267.

Westerdahl, B.B., Giraud, D., Radewald, J.D., Anderson, C.A. and Darso, J. (1994) *J. Nematol.,***25** : 758-767.

White, J.W. (1972) *Bull. Penn. Flower Growers,* **247**: 12.

Wilkins, H.F. and Waters, W. (1967) *Proc. Amer. Soc. Hort. Sci.,* **90**: 433-439.

Wilkins, H.F., Grueber, K., Healy, Wand Pemberton, H.B. (1986) *J. Amer. Soc. Hort. Sci.,***111**: 384-387.

Wilkins, H.F., Widner, R.E. and Waters, W.E. (1968) *J. Series,Agri. Expt. Sta. Univ. Minn.* USA.

Wolf, S.T. and Van Tuyl, J.M. van. (1984) *HortScience.,* **19**: 696-697.

Woltering, E.J. and Harkema, H (1981) *Bedrijfsontwikkeling,* **12**: 193-196.

Woodley, J. and Thomas, M.B. (1973) *Ann. J. Royal New Zealand Institute of the Hortic.,* **1**: 27-29.

Wu, C.G., Lin, S.C., Lin, M.L., Wu, G.D. and Chen, W.H. (1999a) *J. Agric. Res.,* China, **48**: 128-134.

Wu, G.D., Liu, W.C., Chang, J.W., Tao, T.P., Chen, J.B. and Chen, W.H. (1999b) *Rep. Taiwan Sugar Res. Inst.,* **163**: 33-44.

Wu, H.M. and Wang, YF. (1989) *J. Agri. Sci.,* **5**: 31-37.

Xu, P.S. and Niimi, Y. (1999) *J. Japanese Soc. Hort. Sci.,***68**: 640-647.

Yan, R., Wang, Z., Ren, Y., Li, H., Liu, N. and Sun, H. (2019) Int. J. Mol. Sci., **20**: 2920.

Yang, L.P, Song, M. and Zhang, J. (2000) *J. Plant Res. Env.,* **9**: 14-18.

Yang, Q.S., Huang, X.S., Ji, Y, He, S.L. and Song, H.Y. (1996) *Acta Agriculturae Universitatis Henanensis*, **30**: 203-206.

Yarnagishi, M. (1995) *Bull. RIAR, Ishikawa Agri. Coll.*, **4**: 52-59.

Yih, M.S. and Huang, S.C. (1998) *Special Publ. Taichung Dist. Agril. Improv.* Stn., **40**: 105-112.

Yongmin, W. and Takeshi, Y. (1999) *J. Japanese Soc. Hort. Sci.*

Young, J.H. (1995) *RDA J. Agri. Sci. Hort.*, **37**: 378-383.

Zhou, S., Li, K. and Zhou, G. (2012) *Euphytica.*,**184**: 401–412.

Zhou, S., Ramanna, M.S., Visser, R.G.F. and Van Tuyl, J.M. (2008) *Theor. Appl. Genet.*,**57**: 5–9.

Zhou, S., Tan, X., Fang, L., Jian, J., Xu, P., and Yuan, G. (2013) *J. Amer. Soc. Hort. Sci.*, **138**(2): 114–119.

Zhou, S., Yuan, G., Xu, P. and Gong, H. (2014). *Breed. Sci.*, **64**: 97–102.

Zhou, S., Zhou, G. and Li, K. (2011) *HortScience.*, **46**: 558–562.

5

GLADIOLUS

Rajiv Kumar, R. L. Misra, B. Singh and S.K. Palai

1.0 Introduction

The name gladiolus was originally coined by Pliny the Elder (A.D. 23-79), from the Latin word *gladius*, meaning a sword, on account of the sword-like shape of the foliage. It was also the name of 'iris' which ancient Greeks used to call 'Xiphos'. As a genus it was christened by Tournefort and accepted in botanical literature by Linnaeus (Lewis *et al.*, 1972). The common name of gladiolus is 'sword lily' because of its sword-shaped foliage. 'Corn flag' is another common name in Europe because *Gladiolus illyricus* is found wild as weed in the corn-fields. *Gladiolus primulinus* is also known as 'waterfall gladiolus' as it was found growing near the Victoria Falls in the tropical forests of Africa. Gladiolus (*Gladiolus* × *grandiflorus* Hort.), also called as Palma-de-Santa-Rita, is culturally known in Brazil as the All Soul's' Day flower (Tomiozzo *et al.*, 2019). It was introduced into cultivation towards the end of the 16th century (Innes, 1985). In India, its cultivation dates back to nineteenth century, as Firminger's Manual of Gardening in India published in 1863 mentioned that Charles Gray of Coonoor grew some gladioli from corms and seeds in his garden (Apte, 1958). In India during 2016-17, gladiolus was grown in 10.44 thousand hectare with cut flower production of 112.69 lakh numbers. During 2013 to 2016, 74090 number of cut spikes of gladiolus were exported from India to UAE, Bahrain and Saudi Arabia (Anonymous, 2020a).

2.0 Importance and Uses

2.1.0 Aesthetic Uses

Gladiolus is a very popular flowering plant in international cut flower trade. Its magnificent inflorescence with a variety of colours has made it attractive for use in herbaceous borders, beddings, pots and for cut flowers. For cut flowers, primulinus types are better as more spikes often come out from a corm and they may be planted in isolated borders. While planting in mixed flower beds or herbaceous borders, care should be taken to plant only one cultivar in groups or drifts of a dozen or so to have an intriguing effect. Grandiflorus and primulinus types look very attractive in mixed flower borders, but primulinus types are preferred as these do not need staking, and so, are also good for bedding. The pixiolas (miniatures like nanus, colvillei, byzantinus, grandis, tristis, *etc.*) are the daintiest and preferred for forcing under glass or for growing in pots or bowls. These also make good flowers as they bear smaller spikes to look elegant on the tables but will not be attractive as bedding plants. They make very good cut flowers with long vase life.

2.2.0 Therapeutic and other Uses

Uphof (1955) stated that the natives consumed the corms of *Gladiolus edulis* by roasting them, and in Bassa (Africa), *G. quartinianus* corms were used as food when pounded in water with guinea-cornflour and into a cooking beverage; *G. spicatus* is used as food by Lakoja tribe Uses of gladiolus (Africa) and *G. zambesiacus* by Njelekwas (East Africa). Guillarmod (1977) described various uses of gladiolus in Lesotho (South Africa) the people residing at the high altitude mountains relished the flowers of certain species. The flowers of *Gladiolus saundersii, G. ecklonii,*

Uses of Gladiolus

In Garden

In Border

In Pot

As Cut Flowers

In Flower Vase

Interior Decoration

In Floral Arrangements

G. papilio, G. cruentus and *G. natalensis* are used as salad after nipping off the anthers. De Meyer, Jr. (1982) writes the culinary virtues of many flowers including gladiolus. It was found that the corms of Psittacinus hybrids contained high amount of carbohydrates mostly as starch (65.4 to 78.61 per cent) and protein (12.6 to 18.5 per cent). The corms contained pentosan (2.91 per cent), fat (0.58 per cent), saponin (present only before alcoholic extraction) and ash (sulphated, 3.48 per cent); after thorough extraction glucose, xylose and arabinose were found. Amino acid analysis revealed the presense of lysine, glycine, threonine, glutamic acid, alanine, proline, tyrosine, valine isoluecine and some unidentified compounds (Khan *et al.*, 1980). Ancient Greeks used the roasted corms of *Gladiolus italicus* as food.

Gladiolus crassifolius is used for headache and lumbago (the whole plant is crushed, heated and applied to the affected part). *Gladiolus saundersii* (cooked corms) mixed with food is effective against diarrhoea. The crushed and ground corms of *Gladiolus ecklonii* are decocted and drunk to get rid of rheumatism and allied pains.

Gladiolus imbricatus, a rare species spread in the hill region of Europe, contains high vitamin C and minerals in the leaves and the flowers are considered as edible (Rakosy-tican *et al.*, 2012). Physico-chemical analysis of *Gladiolus imbricatus* established the presence of polysaccharides, flavonoids, terpenoids and vitamin C mainly and fewer aromatic and quinones compounds. It can be a potential source of useful drugs exploiting the antimicrobial, antioxidant and antitumor activities (Krvavych *et al.*, 2014). Aqueous extract from the corm of *Gladiolus dalenii* has antidepressant properties which has used for the treatment of depression and psychotic disorders in African traditional medicine (Ngoupaye *et al.*, 2014). Corms from organic extracts of *Gladiolus watsonoides* showed moderate to high antifungal activity against the three fungi *Candida albicans, Aspergillus niger* and *Cryptococcus neoformans* (Judith *et al.*, 2014). Matraszek-Gawron *et al.* (2019) systemically reviewed the antidepressant activity of Gladiolus.

Gladilous psittascinus bulb aqueous extract has potential in enhancing the production of antioxidant enzymes required to scavenge reactive oxygen species. It is used as diet therapy in the management and control of hepatotoxicity and oxidative stress related disorders (Oyetayo *et al.*, 2020).

3.0 Origin and History

Gladiolus L., the largest genus of the Iridaceae in the flora, now has 39 species in the Flora Zambesiaca region alone, and a total of 254 species worldwide (Goldblatt, (2008). The species are scattered throughout tropical and South Africa; Zambesi river near Victoria Falls, Zimbabwe; Mt. Kilimanjaro, Tanganyika; Natal and Cape of Good Hope, Republic of South Africa; Madagascar and Abyssinia (Ethiopia); central Europe and Mediterranean regions (Italy, the Balkans and England); Mascarene Islands and western Asia or Asia Minor (Afghanistan, Armenia and Caucasus, Syria, Persia, *etc.*) but a large number of them belong to South Africa in the Cape region, and about 15 species grow only in north of Sahara (Misra, 1977). No species is known to be native to India.

Since the days of ancient Greece, the gladioli are said to be cultivated. History reveals that it is known since 1578, as evidenced by a record in *Lyte's Nievve Herball,* first introduced into France and soon after spread to England, Germany, Holland and North America. These were the only wild species of garden gladioli three hundred years ago. In 1596, John Gerard, gardener to Lord Burleigh, who also owned his own 'Physic Garden' in Holborn, had two European species, *Gladiolus communis* from the Mediterranean region and *G. segetum* from South Europe, the Canaries and Mediterranean region; as is evidenced by his catalogue of the plants he grew some 1030 kinds. In 1597, Gerard published his famous book, *General Historie of Plantes* describing these two species. In 1904, William Coys cultivated *Gladiolus imbricatum* in his garden at North Okington (Ockenden) in Essex and by 1629, John Parkinson introduced *G. byzantinus* into Europe from Mediterranean regions and published his *Paradisi in Sole Paradisus Terrestris* in which are noted many other new plants since *Gerards Herball*. This *Gladiolus byzantinus* in 1893 was crossed with *G. cardinalis* which resulted in a slightly fragrant hybrid known as *G. victorialis*.

No evidence was available on cultivation of any species of gladiolus of South African origin grown in Europe prior to 1700 (Lewis *et al.*, 1972). *Gladiolus aethiopian*, described first by Cornutus, is from Cape region bearing light red-scarlet flowers. Around the middle of the eighteenth century the first South African species were introduced. *Gladiolus tristis*, a sweet-scented species closely related to *G. grandis*, was first introduced in Chelsea Physic Garden, UK in 1745 from Cape region and Natal. In 1772, Kew Gardens sent overseas the first of its numerous plant collectors, the Aberdonian, Francis Masson, who travelled principally in Cape region and collected many wild growing species.

Gladiolus illyricus native to Mediterranean region was later on found growing in England. *Gladiolus grandis* from South Africa was naturalized in Spain in 1749. In 1784, C. P. Thunberg published *Dissertatio de Gladiolo* which might be taken as the starting point of post-Linnaean era. In 1789, W. Aiton published his *Hortus Kewensis* describing two *Gladiolus* species brought by Masson during 1773-75.

By 1816, France, Holland and Belgium started raising hybrids, the first being Ghent gladiolus (Gand in Belgium), *Gladiolus gandavensis*, in 1841, by crossing *G. cardinalis* and *G. psittacinus* or *vice-versa*. However, William Dean Herbert was of the opinion that it was evolved by crossing *G. oppositiflorus* with *G. psittacinus*. This was a beautiful gladiolus having bright rich crimson flowers with markings. Messrs Froebel of Zurich produced a race *G. turicensis* by crossing *G. Gandavensis* and *G. saundersii superbus*, with purplish crimson flowers and white marks on the lower segments. In 1823, *Gladiolus colvillei* race was evolved by Colville at his nursery in Chelsea, England from *G. tristis concolor* and *G. cardinalis*. Around 1900, *Gladiolus nanus* was created and in 1951, the Dutch firm K and M introduced the butterfly or mignon type.

Gladiolus primulinus is known as Maid of the Mist because its native haunts are the banks of the Zambesi river, where it grows on the spray of the Victoria Falls, so to protect the sex organs, especially pollen, the upper central petals became hooded. It was introduced into England in 1879 from Tanganyika and the name *Gladiolus nubicola* was proposed to it.

Some other races were also produced by crossing *Gladiolus alatus* with *G. vittatus* (syn. *G. vinulus*); *G. cardinalis* with *G. brachyandrus*, *G. cuspidatus*, *G. fioribundus*, *G. hirsutus*, *G. oppositiflorus* and *G. tristis*; *G. gandavensis* with *G. floribundus* resultant race *G. massiliensis; G. psittacinus* with *G. floribundus, G. pseudo-auratus* and *G. saundersii* var. *cooperi; G. recurvus* with *G. blandus* and *G. tristis; G. tristis* with *G. blandus* and *G. primulinus, etc*. Thousands of present-day cultivars of *G. hortulanus* or/and *G. hybridus* are the contributions of 8-12 species of the genus (Laurie *et al.*, 1958).

Homoglossum watsonianum and *Acidanthera bicolour murielae* (first introduced from Abyssinia (Ethiopia) in 1893 and named by Perry, an English nurseryman in 1930, a scented species) are the last used in hybridization programme. Collingwood Ingram, in 1931, used the former genus with *Gladiolus tristis* and produced Homoglad hybrids (Ingram, 1967), whereas Mrs. Joan Wright of New Zealand, in 1967, evolved *Gladanthera* 'Lucky Star' by crossing gladiolus cv. Filigree with latter genus, which is fragrant (Wright, 1971).

A.E. Kunderd of Indiana, USA reared the first ruffled gladiolus in 1907, together with a species *Gladiolus quartinianus*, however, 1920-40 is known as 'Kundred era' as during this period American Gladiolus Society helped much in popularization of gladiolus.

Goldblatt (2008) described *Gladiolus crassifolius* Baker (earlier name *G. mosambicensis* Baker) as spring-flowering, distinguished by the slightly smaller flowers, dry, obtuse floral bracts, and round, broadly winged seeds; bracts green, acute; wedge to pear-shaped seeds with the wing fully developed at one or both ends but not on the sides. Plants from northern Zimbabwe provisionally referred to *G. permeabilis* subsp. edulis (Burchell ex Ker Gawler) Obermeyer in the Flora Zambesiaca account of the genus do not belong in this species and are described as *G. metallicola* Goldblatt. It is a member of section Hebea (Persoon) Bentham and Hooker f., remains known from one collection from the Great Dyke in Lomagundi district.

Manning *et al.* (2011) reported new species *Gladiolus halophila* Goldblatt and J.C.Manning from near Keetmanshoop in southern Namibia. It is closely allied to *G. longicollis* and *G. tristis* from south western and eastern South Africa but is distinguished by its lilac flowers with shorter perianth tube, 35-40 mm long. The flowers last just one day and are self-compatible. The species is restricted to saline seepages and classified as Endangered.

Valente *et al.* (2011) examined spatial asymmetries of diversity in *Gladiolus* (Iridaceae), a large genus (>260 species) that is present in two mediterranean climate biomes: the Cape of southern Africa (106 species) and the Mediterranean Basin (7 species). Despite convergence of climatic conditions between the two regions, the species density of *Gladiolus* is over one order of magnitude higher in the Cape than in the Mediterranean Basin.

Rakosy-tican *et al.* (2012) expressed that *Gladiolus imbricatus*, a rare species spread in the hill region of Europe, is resistant to abiotic and biotic stress being one of the most cold-tolerant in the genus. Sagiroglu and Akgül (2014) reported a new species, *Gladiolus osmaniyensis* Sagiroglu, is from South Anatolia, Turkey. It is morphologically close to *G. attilae* and *G. atroviolaceus*.

Van Der Burgt (2019) reported new species from Guinea, West Africa is *Gladiolus mariae* Burgt. It is placed in Gladiolus sect. Decorati Goldblatt. Flowering plants stand

28–160 cm high with 1–6 bright orange flowers, opening one at a time. *Gladiolus mariae* is at present known only from two uninhabited sandstone table mountains in the Kounounkan Massif, Forécariah Prefecture on rocky soils at altitudes of 650–1100 m. The species is assessed to the IUCN category Endangered.

Goldblatt *et al.* (2019) described new species *Gladiolus virgineus* Goldblatt and J.C. Manning from the collection of made in northern Zambia. It is evidently allied to *G. atropurpureus* Baker but with the stem bearing four well developed foliage leaves, and flowering in late summer (February). The white flowers are ±28mm long, with a short style dividing opposite the base of the anthers.

Takatsu *et al.* (2002) observed wide variations in plant height, leaf length, leaf width, number of leaves, tepal color, diameter of floret and number of florets in 21 wild gladiolus species. Dwarf species could be suitable for pot culture, and blue gladiolus cultivars might be developed from wild species having blue tepals. More than half of these species (52.3 per cent) produced a good scent. Based on flow cytometric analysis, wild gladiolus species showed a fluorescence intensity 0.9-fold to 3.5-fold higher than that of rice plant, indicating that the genome size of these species showed a wide distribution, and this information might be used for the estimation of hybridity in some cross combinations, though it was difficult to estimate the ploidy of wild species using flow cytometry

4.0 Taxonomy

Gladiolus L. belongs to the subfamily Ixioideae with chromosome number n=15. Pax, in 1888, classified Iridaceae (Engler and Prantl) into Crocoideae, Iridoideae and Ixioideae, further dividing the latter one into lxieae, Watsonieae and Gladioleae based on the forms of the flowers, *i.e.*, zygomorphic or actinomorphic and styles, *i.e.*, complete or divided into 2; and mono- or bi-carpillary ovary. This latter subdivision comprises eight genera with *Gladiolus* L. as one of them. The genus has further been divided into four sections :

1. Section I: *Eugladiolus* which includes 100 species of Europe, Western Asia, Tropical and South Africa.
2. Section II: *Habea* which includes 12 species of South Africa (Cape Peninsula) and the Madagascar island.
3. Section III: *Schweiggeria* which includes only two species, *i.e.*, *Gladiolus montana* and *G. arenarius* from Cape.
4. Section IV: *Homoglossum* which includes five or six species of South Africa.

Baker, in 1892, through his treatise *Handbook of Irideae*, also adopted the same placing : 109 species under *Eugladiolus*, 15 under *Habea*, 2 under *Schweiggeria* and 6 under *Homoglossum*.

Lewis *et a1.* (1972) divided the genus *Gladiolus* into four groups. The species belonging to different groups along with their places of origin (in parenthesis) and flower colour are given below :

I. Plurifoliati

Twenty two species, having 5-8 or more well developed leaves, distichously arranged, forming a fan.

Gladiolus cardinalis (Cape), *G. sempervirens* (Cape), *G. cruentus* (Natal and Lesotho), *G. saundersii* (Natal and Lesotho) all of red or scarlet colour; *G. oppositiflorus* (Cape) flowers white or pink; *G. elliotii* (Natal, Botswana and Rhodesia) - white, speckled with purple; *G. sericeo-villosus* (Cape, Natal, Swaziland and Transvaal) - pink and/or yellow flowers, arranged to opposite directions; *G. varius* (Transvaal) - pink; *G. ochroleucus* (Cape) - cream to yellow; *G. ecklonii* (Cape) - white, red and mauve; *G. buckerveldii* syn. *Antholyza buckerveldii, Petamenes buckerveldii* (Cape) - yellow; *G. natalensis* (Cape and Natal) - red, orange, yellow and/or greenish; *G. papilio* syn. *G. purpureoauratus* (Transkei to Transvaal) - yellow, all have one side facing spike; *G. hollandii* (Transvaal) - white, speckled with pink; *G. exiguus* (Transvaal) - pink; *G. crassifolius* (Cape, Natal and Lesotho) - white, pink and mauve; *G. densiflorus* (Transvaal) white, with red spots; *G. invenustus* (Swaziland and Natal) - white to mauve, with blotches; *G. calcaratus* (Transvaal) - white; *G. appendiculatus* (Transvaal, Natal and Swaziland) - white, with mauve markings; and *G. pole-evansii* (Transvaal) - pinkish lilac, hairy on all vegetative parts.

II. Paucifoliati

Thirty four species, having 2-5 well developed leaves, distichously or spirally arranged.

Gladiolus stellatus (Cape) - fragrant, whitish to pale mauve; *G. gueinzii* (Cape and Natal) - pink, marked with purple; *G. acuminatus* (Cape) - fragrant, dull greenish yellow, *G. lapeirousiodes* (Cape) - white, with the flower lobes each bearing 2 red blotches; *G. leptosiphon* (Cape) - cream to cream-brown; *G. vigilans* (Cape) - pale rose-pink; *G. carneus* (Cape) - white, cream, pink or mauve; *G. macneilii* (Transvaal) – pale salmon-pink; *G. microcarpus* (Natal) - whitish, pink or mauve; *G. buckerveldii* (also described under Plurifoliati); *G. angustus* (Cape) - white to cream yellow; *G. fioribundus* (Cape) - fragrant, with white, cream, pink or mauve, closing at night; *G. undulatus* (Cape) - greenish to cream-white or pink; *G. caryophyllaceus* (Cape) - strongly scented, flowers pale to mauve; *G. lewisiae* (Cape) - fragrant, creamy white, flushed with pink; *G. involutus* (Cape) - pale pink or white; *G. scullyi* (Cape) - fragrant, cream, yellow, lime-green, pink, mauve or maroon; *G. arcuatus* (Cape) - fragrant, mauve or purple; *G. salteri* (Cape) - pink; *G. kamiesbergensis* (Cape) - mauve; *G. permeabillis* (Cape) - fragrant, white to cream, mauve, pink or brown; *G. vernus* (Transvaal) - pale magenta, pink; *G. pretoriensis* (Transvaal) - dull purple to pink; *G. marlothii* (Cape) - fragrant, mauve with lower lobes minutely speckled; *G. mostertiae* (Cape) - pale pink; *G. rufomarginatus* (Transvaal) - white or cream-yellow, densely flecked with red or purple; *G. orchidiflorus* (Cape) - fragrant, grey-green with purple sheen; *G. watermeyeri* (Cape) - fragrant, cream or greenish cream, flushed with dull purple; *G. virescens* (Cape) - fragrant, greenish yellow; *G. seresianus* (Cape) - fragrant, dull or greenish yellow; *G. uysiae* (Cape) - fragrant, cream or greenish; *G. alatus* (Cape) - scented, red, orange or salmon but lower leaves greenish; and *G. equitans* (Cape) - fragrant, orange, red or vermilion, with yellow-green patches on laterals and falls.

III. Unifoliati

Twenty nine species, having one well developed basal leaf, the following reduced, bract-like with short blades, spirally arranged on the scapes.

Gladiolus brevitubus (Cape) - pale pink or red; *G. quadrangulus* (Cape) - fragrant, white to pale blue or pinkish mauve; *G. citrinus* (Cape) - bright yellow; *G. tenellus* (Cape) - fragrant, yellow, cream or white tinged purple; *G. oreocharis* (Cape) - white to mauve; *G. inflatus* (Cape) - pale, pink or mauve; *G. robertsoniae* (Transvaal) - fragrant, mauve or white tinged pale mauve; *G. cylindraceus* (Cape) - pale pink or creamy pink, flushed with salmon-pink; *G. debilis* (Cape) - white to pale pink with red dots in throat; *G. longicollis* (Cape, Natal, Lesotho, Orange Free State, Swaziland and Transvaal) - evening-fragrant, white, cream or yellow; *G.tristis* (Cape) - night-fragrant, white to buff-cream or pale yellow tinged green; *G. liliaceus* (Cape) - night-fragrant, yellowish to dull yellow, flecked with brown, pink, red or purple; *G. hyalinus* (Cape) - slightly fragrant towards evening, yellow-brown, yellowish cream or greenish brown, spotted with brown and purple; *G. recurvus* (Cape) - fragrant, pale greenish grey or greenish mauve to pale yellow or cream; *G. symonsii* (Natal) - pink; *G. punctulatus* (Cape) - pink to mauve; *G. carinatus* (Cape) - fragrant, pale greyish blue to pale pink; *G. viridiflorus* (Cape) - fragrant, pale green; *G. blommesteinii* (Cape) - pale pink or mauve; *G. ornatus* (Cape) - pale to deep pink; *G. gracilis* (Cape) - fragrant, pale mauve or pink, sometimes white or yellow; *G. exilis* (Cape) - slightly fragrant, pale blue, mauve or white tinged pale mauve; *G. mutabilis* (Cape) - fragrant, greyish mauve to cream, brown or yellow; *G. violaceo-lineatus* (Cape) - mauve; *G. comptonii* (Cape) - bright yellow with reddish streaks; *G. rogersii*(Cape) - purple; *G. bullatus* (Cape) - deep mauve; and *G. pritzelli* (Cape) - fragrant, yellow, sometimes tinged red or grey outside.

IV. Exfoliati

Twentyfour species, having well developed leaves (or leaf) absent at the time of flowering, the scape bearing 1-3 (or 4) sheathing bracts without or with small free blades.

Gladiolus stefaniae (Cape) - red; *G. stokoei* (Cape) - scarlet; *G. nerineoides* (Cape) - pale salmon-pink to deep golden red; *G. guthriei* (Cape) - fragrant, deep pink or reddish; *G. carmineus* (Cape) - deep rose-pink or carmine, with a white or cream median stripe; *G. monticola* (Cape) - cream colour tinged pink; *G. maculatus* (Cape) - fragrant, yellow, pink or white, streaked and spotted with brown, purple or red; *G. engysiphon* (Cape) - strongly and sweetly scented, white with a crimson median line on 3 flower and upper lateral lobes; *G. bilineatus* (Cape) - white or cream, faintly flushed with pink; *G. emiliae* (Cape) - fragrant, yellow, densely speckled with red, brown or purple; *G. odoratus* (Cape) - fragrant, dull or brownish yellow, speckled and striped with purple, red or maroon; *G. vaginatus* (Cape) - fragrant, pale mauve to grey-mauve, rarely white; *G. brevifolius* (Cape) - scented, pale to pink, rarely white, mauve, brownish green or brown; *G. pillansii* (Cape) - fragrant, mauve or pink;

G. martleyi (Cape) - white, suffused with rose-pink ; *G. subcaeruleus* (Cape) - mauve or pinkish; *G. jonquilliodorus* (Cape) - fragrant, cream-coloured or pale yellow, flushed with pink or mauve; *G. aurantiacus* (Natal, Swaziland and Transvaal) - golden yellow or orange and yellow, streaked with red; *G. brachyphyllus* (Transvaal, Swaziland and Mozambique) - pink to mauve, with red spots; *G. unguiculatus* (Sierra Leone and Transvaal) - mauve, with purplish blotches on lower lobes; *G. woodii* (Natal, Transvaal and Swaziland) - dark brown; and *G. parvulus* (Cape, Natal and Transvaal) - mauve, pink or white.

Bentham and Hooker (1883) placed the family in the series Epigynae whereas Engler (1908) placed it under Liliflorae, and Hutchinson (1932) put it in a separate order Iridales through his treatise *The Families of Flowering Plants II: Monocotyledons*. In 1959 when this book was revised, the genus *Gladiolus* was placed in tribe 10-Gladioleae. Goldblatt (1971) placed the genus *Gladiolus* under subtribe Gladiolineae of the tribe 3-Ixieae.

The family resembles Amaryllidaceae in many aspects, hence has been placed under Liliflorae.

Singh *et al.* (2017) studied the nucleotide diversity and phylogenetic relationships among *Gladiolus* cultivars by sequencing the plastid genome regions of two intergenic spacers, psbA–trnH and trnL–trnF. Results revealed that sub-family Iridoideae have sister lineage with subfamily Ixioideae and Crocoideae. The psbA–trnH region are useful in addressing the phylogenetic relationships among the *Gladiolus* cultivars.

Gladiolus is a tender herbaceous perennial, growing from seeds and corms. The corms are bulb-like, globose or ovoid having series of nodes, wholly covered with tunic or husks and orient strictly in vertical fashion within the soil. The corms produce a number of cormels or offsets, sometimes called as spawn, which have very hard tunic. The leaves are sword-shaped phyllode, clustering at the swollen stem base, equitant in 2-ranks, prominently ribbed, radical and cauline. Flowers 2-whorled with six perianth segments, attached to a funnel-shaped cup shaping down wards into a bent floral tube, hermaphrodite, trimerous, irregular, medianly zygomorphic, sessile and often protandrous and epigynous, borne to one side on a simple or branched inflorescence subtended by two lanceolate, leaf-like valves. Stamens 3, placed opposite to the outer whorl of the petals, extrose, epiphyllous, inserted below throat, bithecous, dehiscing by vertical slits, filaments distinct and basifixed. Gynoecium is tricarpellary and syncarpous, having inferior ovary with axile placentation; style long, slender and simple with typically hinge-like creased trifid stigma and petaloid. Fruits oblong-ellipsoid to obtusely triangular, loculicidal capsule with 3 chambers each having 2 rows mostly with winged seeds.

4.1.0 Floral Biology

Messar *et al.* (2016) studied the floral biology of six gladiolus varieties. Anthesis was completed on the same day. Anthesis was early and started from 8.02 a.m. in American Beauty, while, it was late in Jester (10.28 a.m.). Anther dehiscence started simultaneously with anthesis in Nathan White, Careless and Red Beauty, whereas, in other varieties, anther dehiscence took place 1-2 hrs after anthesis. No variation was observed in stigma receptivity among different varieties. Pollen viability was observed maximum in Nathan White (64.95 per cent) followed by Nathan Red (64.45 per cent), while, the minimum viability of the pollen was observed in variety Red Beauty (56.69 per cent).

Poon *et al.* (2010) used pollen germination media consisted of 15 per cent sucrose supplemented with 300 ppm calcium nitrate, 200 ppm magnesium sulphate, 100 ppm potassium nitrate and 100 ppm boric acid for testing pollen viability of 16 gladiolus genotypes. The highest pollen germination (76.41 per cent) was recorded in 'Hybrid selection 88-10-22 which was on par with *Gladiolus callianthus* (75.41 per cent), Sapna (75.10 per cent), Hybrid selection 86-32-11, (73.28 per cent), Kum Kum (69.41 per cent), Poonam (69.22 per cent), Hybrid selection 87-22-1 (67.87 per cent), Hybrid selection 87-1-1 (67.61 per cent), *Psittacinus* hybrid (64.64 per cent) and Darshan (63.97 per cent).

Takatsu *et al.* (2001) suggested as long-term storage of pollen to facilitate crossing between *Gladiolus* ×*grandiflora* hort. ($2n = 60$), summer-flowering species, and wild species *G. tristis* L. ($2n = 30$), winter flowering species. The pollen of *G. tristis* could be stored at –20 °C for ≈1 year. Kaur and Dhatt (2019) determined the pollen viability response to storage method and period in *cv.* Yellow Stone and found that maximum pollen viability was recorded under freezer (82.87 per cent), however, pollen stored under freezer retained pollen viability for longer duration (85.09 per cent) after 11 weeks and pollen viability decreases (49.58 per cent) after 16 weeks of storage. The pollen of gladiolus can be stored in freezer (-4 °C) up to 2½ month for obtaining good pollen viability.

Zhygalova *et al.* (2014) investigated micro-morphological characteristics of the four *Gladiolus* L. species (*G. imbricatus* L., *G. italicus* Mill., *G. palustris* Gaudin and *G. tenuis* M. Bieb.) with regards leaves, seeds and pollens. The type of surface ultrastructure of the seed coat is a diagnostic feature as at genus level so for species. The shape and position of the cicatricle, the type of cuticle, the shape and boundaries of cells of testa, and the anticlinal cell walls as diagnostic features at genera level. The shape of seeds, the presence and disposition of wing, the level of the periclinal cell walls of the seed coat and types of relief are additional diagnostic features for distinguishing of *Gladiolus* species.

Heidmann *et al.* (2016) reported that impedance flow cytometry (IFC) is an efficient, label-free, reliable and non-destructive technique to analyse pollen quality in a species-independent manner. Different stages of developing pollen, dead, viable and germinating pollen populations could be detected and quantified by IFC. Pollen viability analysis by classical FDA staining showed a high correlation with IFC data.

Zhang *et al.* (2007) investigated the effects of 3MeVe electron beam with different doses on pollen traits in M_1 generation. The sterile pollen rate of M_1 generation is higher than that of control. With the increase of irradiation dose, the viability and germination rate of pollen was decreased. Mutagenic effects on pollen increased with the dose increasing in the angle of polynology.

5.0 Classification

It lacks universal synchrony for its classification because the Europeans, New Zealanders (also follow Australian system) and Asians used to classify them on the size of the flowers, *i.e.* (i) grandiflorus (large or exhibition type), which grows to a height of 90 – 150 cm with strong and erect stalks, bearing 14 – 20 cm closely arranged, triangular and symmetrical flowers, coming up late in the season; (ii) primulinus, which is free flowering and daintier than the grandiflorus, growing to a height of one metre, bearing flowers 5 –10 cm across with mostly hooded florets of soft colours, blooming generally in the mid-season; (iii) butterflys, which are upto 1.25 metre, bearing medium-sized flowers spaced closely and marked with attractive blotches and throat markings; and (iv) miniature, growing to a height of 75 – 90 cm with flowers 5 – 7.5 cm across, similar in arrangements to the 'prims' but frequently ruffled, and flowering early. The Australians classify these into formals, informals, intermediates and miniatures.

American system of classification (Table 1) is universal in application which is now being used everywhere as it is more scientific and comprehensive. This is based on 3-digit system which originated in 1940's. First digit (from 1 – 5) represents floral size, second (0 – 9) the basic hue or colour and third (0 – 9) the depth of colour.

Table 1: Size and Colour Classifications of Gladiolus as Designated by Gladiolus Council*

Class[a]	*Designation*		*Floret Size (cm)*		
100	Miniature		< 6.4		
200	Small or miniature		≥ 6.4 to < 8.9		
300	Decorative		≥ 8.9 to < 11.4		
400	Standard or large		≥ 11.4 to 14.0		
500	Giant		> 14.0		
Colour[b]	*Pale*	*Light*	*Medium*	*Deep*	*Other*
White	00				
Green		02	04		
Yellow	10	12	14	16	
Orange	20	22	24	26	
Salmon	30	32	34	36	
Pink	40	42	44	46	
Red	50	52	54	56	58 Black-red
Rose	60	62	64	66	68 Black-rose
Lavender	70	72	74	76	78 Purple
Violet	80	82	84	86	
Smokies		92	94	96	
Tan	90				98 Brown

a The first digit indicates floret size in the five classes; b: The last two digits in the classification indicate floret colour and shade. An odd last digit indicates a conspicuous mark or blotch. c: Includes cream.

* Wilfret (1980).

Species of Gladiolus

Gladiolus aureus

Gladiolus alatus

Gladiolus brevifolius

Gladiolus cardinalis

Gladiolus caryophyllaceus

Gladiolus dalenii

Gladiolus grandiflorus

Gladiolus hirsutus

Gladiolus imbricatus

Gladiolus orchidiflorus

Gladiolus parvulus

Gladiolus viridiflorus

First digit represents floret size from 100 (miniature, below 6.4 cm) to 500 (giant, above 14.0 cm) via 200 (small, 6.4 – 8.9 cm), 300 (decorative, 8.9 – 11.4 cm) and 400 (large, 11.4 – 14.0 cm).

Regarding the second and third digits, as white does not fit in the series, hence it is placed first, 00 means pure white whereas 01 white with markings. There is no pure green and blues, hence no number has been assigned to them, however, the depth of greenness is defined on the basic colour of 0, *i.e.*, 02 (light green), 04 (medium green), *etc*. Other numbers are 1 for yellow, 2 for orange, 3 for salmon, 4 for pink, 5 for red, 6 for rose, 7 for lavender, 8 for violet (no blue) and 9 for tan and 0 to 9, odd numbers standing for conspicuous markings.

For example, '498 Brown Orchid' Baxter, 64 means the cultivar Brown Orchid was bred by Baxter, released in 1964 and is of large size (11.4 – 14.0 cm floret size) which has inconspicuous deep smoky colouring of the flowers, mostly towards the brown type.

Europeans use prefatory letters P (Primulinus), B (Butterfly) or FU (Face-ups), and suffix letters E (early), M (mid-season), L (late) and V (very for extremes).

In order to distinguish the difference of gladiolus cultivars (White, Light Pink, Salmon Red, Red, Light Blue White) of different colors based on spectral reflectance of their petals, a spectrometer connected to a portable computer and configured for reflectance data acquisition in the 375 to 750 nm range was used to determine flower color. Thus, mainly in the region of visible spectrum, the reflectance can be used as a criterion for development of automatic systems for classification of gladiolus in a reliable and precise way (Neto *et al.*, 2009).

6.0 Cultivars

In different literatures, the parentages of many of the 30000 cultivars may be traced out, although the parentages of many of the earliest cultivars may not be known because the hybridizers did not maintain the records properly, or considered it secret. The North American Gladiolus Council first published its account in 1950 in *Gladiolus Variety Parentages* which was revised in 1957, 1964 and 1975, and every year the addenda are added for making the information up-to-date. Dr. Godon R. McKenny compiled the first one. Before 1931, such information was very scanty as only Lemoine, Childs, Kunderd and Salbach used to maintain the parentages.

Some of the earlier exotics (Buell, 1972; Misra, 1975) may well have arisen as mutants which in many cases are freakish. These include Face-ups, Dragons, Doubles, Lacinated, Orchids, Stars and Reflexes. Many of such forms are sterile. Dragon and Lacinated have originated in the USA by Doerr, Koerner, Gosling, Lins, *etc*. The double flowering gladiolus was first introduced by Arthur C. Koerner in 1947 with the name Multipetal No. 1, which later on was put on sound footing by E. H. Doerr (Minnesota) during 1950s. Bird of Paradise was produced by Carl H. Sommer in 1954 from a cross of Koerner seedling and Miss New Zealand. This has large, flat and saucer-shaped florets. The exotics like Doubles, Face-ups, Dragons, *etc.*, are more of curiosity than being beautiful, they are not stable for their genetic make-up and are even poor multiplier. The star-type was originated by Frank

Urwin of England, and the orchid-type by Ron Park in 1960s. Reflexes have three lower petals curving back in a curl or spiral until they press against the stem, first originated by Frank Urwin.

Fragrants first originated in the early nineteenth century by Dean William Herbert of England by crossing winter or spring flowering Afrikanders then available, the best known fragrants, arose from a cross between *Gladiolus tristis* and *G. recurvus*. Afterwards, Morgan also got some success. There are some even 40 South African species.

Cultivar Mibloom with Pfitzers' Triumph produced Joli and the former with Mammoth White (non-scented) produced Summer Breeze and Honeysuckle, Perfume was evolved by crossing Joli and Summer Breeze. Yellow Rose and Sachet were evolved by crossing the seedlings of Mibloom × Maid of Orleans (non-fragrant) to Perfume and again to the seedlings of Mibloom × Chance and Mibloom × Incense, all highly fragrant.

Cultivar Rosemarie Pfitzer with Picardy produced reasonably fragrant Gwen. Sweetie, a highly fragrant cultivar, was evolved by crossing Rose Gem and This Is It and Ruflles by Friendship and Beautys' Blush. Spice-fragrant cultivars were evolved by crossing faintly fragrant King David and Party Ruffles, Happy Talk by Party Ruffles and Purple Splender. Little Darling, Indian Princess and Summer Fragrance are reasonably fragrant cultivars evolved by crossing Queen of Bremen, Dream of Beauty and an unknown seedling (Fisher, 1975).

Acacia, one of the best cultivars, was produced by crossing Perfume with Diadem, both fragrant, This Is It by Summer Breeze and New Era, Sweet and Lovely by Paula and Frilled Fragrance, Pink Fragrance by Vista Bonita seedling and Summer Breeze, Sweet Cream by Frilled Fragrance and Perfume, Dark Fragrance and Red Scent by Burma and Cologne, and Lucky Star by Filigree and *Gladiolus callianthus murielae*. Other fragrant cultivars are Progress, Angles Breath, Jimmy Boy, David, Dawn, Gypsy Fire, Carrie Jacob Bond, Lucky Strike, Fragrant Beauty, Sweet Debbie, Sweet Dreams, Cliffie, Fragrant Queen, Bouquet, Royal Fragrance, Golden Fragrance, Incense Smoke, Sweet Rose, Mainliner, Araby, Sweet Lavender, Brenchleyensis, Baron Joseph Hulot, 1910 Rose, Mrs. Dr. Norton, Anna Eberius, Kirkoff's Violet, Aida, Avemaria, Queen of Bremen, Pythia, Mrs. Leon Douglas, Prince of Wales, Lady Eaton, Rosemarie Pfitzer, Arethusia and Camilla. Mrs. Hazel Zeller produced fragrants like Gaytime, Happy Talk, Gypsy Fire and Dawn in 1967 (Fisher, 1975). Gromov of the USSR produced fragrant Ra (Park, 1976). George A. Webster's seedlings S-4, S-69 and S-4 × S-69 are highly rose-fragrant.

Gladiolus dracocephalus produces strange colours like green, brown and slate. Lemoinei hybrids also produced strange colours and with *G. gandavensis* also, apart from *G. dracocephalus*, produced smokies. First smoky produced from Childsii hybrids after 1871 was Prince of India. Lemoinei strain was produced by M. Lemoine of Nancy, France, by crossing *Gladiolus purpureoauratus* with *G. gandavensis* strains. He also used *Gladiolus papilio* with *G. purpureoauratus* which resulted in many of the purples, lavenders and violets (Garrity, 1975).

6.1.0 Some Good Cultivars, According to Flower Colour

6.1.1 White

Althena, Bush Balland, Cotton Blossom, Dream Girl, Eastern Star, Gambier Pearl, Icicle, Jennifer, Lipstick, Lunar Moth, Marjorie Ann, Melissa, Mighty Mite, Moon Frost, Morning Bride, Simplicity, Snowdrop, Snow Dust, Super Star, White Enchantress, White Friendship, White Oak, White Wonder, *etc.*

6.1.2 Green

Armstrong, Green Bay, Green Bird, Green Giant, Green Woodpecker, Green Willow, Emerald Isle, Forest Glade, Lemon Lime, Mint Julip, Oasis, *etc.*

6.1.3 Cream

Ariette, Bonnie, Classmate, Cream Topper, Crinkled Wrinkles, Dairy Queen, Dew Drop, Lady Bountiful, Landmark, Pale Moon, Party Ruffles, Ruffled Lotus, *etc.*

6.1.4 Yellow

Aurora, Brightsides, Chinese Lantern, Fatima, Folksong, Golden Harvest, Golden Peach, Junior Prom, Lemon Ruffles, Limelight, Medusa, Morning Sun, Nugget, Royal Gold, Sweet Fairy, *etc.*

6.1.5 Buff

Adventure, Apricot Delight, Apricot Lustre, Bit O'Honey, Fashion, Happy Birthday, Honeycomb, Illuminator, Mikado, Novelette, Perky, Royal Buff, Royal Glimpse, Sundown, *etc.*

6.1.6 Orange

Autumn Glow, Coral Seas, Fiesta, Fox fire, Gypsy Dancer, Little Mo, Orange Beauty, Orange Chiffon, Setting Sun, Tangerine, *etc.*

6.1.7 Salmon

Atom, Big Daddy, Brigadier Cliffie, Frilled Champion, Goliath, Heritage, Parade, Memento, Salmon Queen, Sister Fortuna, Sugar Babe, Summer Garden, Thunderbird, Tropicana, *etc.*

6.1.8 Pink

America, Dawn Pink, Dresden Doll, Enchantress, FlosFlorium, Friendship, Howard, Kimberly, Legend, Marjayne, Miss America, Miss Salem, Pink Formal, Pink Cheer, Pink Prospector, Pink Triumph, Powder Puff, Spic and Span, Spring Song, Super Ruffles, Sweet Debbie, Tea Party, True Love, Vicki Lin, Windsor, *etc.*

6.1.9 Red

Black Prince, Decathalon, Delicious, Dixieland, Eclipse, Enticement, Fatima, Little Slam, Music Man, Nautilus, Oscar, Red Bantam, Redeem, Sans Souci, Sassy Willie, Sequoia, Shirley Cole, The Barton, *etc.*

6.1.10 Rose

American Beauty, Ecstasy, Mexicali Rose, Patricia Jean, Royal Brocade, Upper Crust, *etc*.

6.1.11 Lavender and Purple

All Aglow, Anniversary, Burgundy Beau, Dawn Mist, Elegance, Lavender Master-piece Memorandum, Purple Giant, Purple Moth, Royal Ruffles, Royal Dame, Shalimar, *etc*.

6.1.12 Violet

Angel Eyes, Blue Bird, Blue Isle, Blue Hawaii, Blue Mist, Blue Ruffles, China Blue, Delphine, Eternal City, Her Majesty, Linda Ruth, Province, Tropic Seas, *etc*.

6.1.13 Tan

Hastings, Little Fawn, Papoose, Table Talk, Tapestry, *etc*.

6.1.14 Smoky

Autumn Charm, Autumn Sensation, Aztec Chief, Blue Smoke, Cosmic, Misty Eyes, Misty Mom, Piccolo, *etc*.

6.1.15 Brown

Autumn Sensation, Brown Beauty, Chocolate Chip, Chocolate Dip, Chocolette, Mystic Glow, Little Tiger, *etc*.

6.2.0 Cultivars Evolved in India

Public sector institutes such as ICAR-IIHR, Bengaluru, IARI, New Delhi, CSIR-NBRI, Lucknow and SAUs *viz.*, PAU Ludhiana, MPKV Pune, BCKV, West Bengal, YSPUHF Solan, SKUAST Srinagar, BHU Varanasi have developed several improved varieties of gladiolus.

Recently ICAR-IIHR, Bengaluru released some outstanding gladiolus varieties suitable for cut flower, bouquet preparation, floral arrangement and garden display. The salient features of these varieties are given below.

Arka Aayush: It is hybrid selection from the cross Gold Medal 412 x Arka Poonam. Its florets are red with yellow border, open faced, thick, slightly ruffled and arranged in double rows. It takes about 76.65 days for flowering. It is resistant to *Fusarium* wilt disease. It has long spike (95.18 cm) with good rachis length (48.81 cm) and bears 17.54 florets per spike.

Arka Manorama: It is hybrid selection from the cross Arka Meera x Picardy. Its florets are red-purple having red-purple streaks, red-purple splash, open faced and medium in texture. It takes about 72.25 days for flowering, has long spike (123.11 cm) with good rachis length (46.77 cm) and bears 12.66 florets per spike.

Arka Amar: It is hybrid selection from the cross Watermelon Pink x Arka Aarti. Its florets are red having red margin and white line on tepals with yellow blotch. It takes about 72 days for flowering, has long spike (101 cm) with good rachis length

Cultivars of Gladiolus

American Beauty

Alexander S

Black Swan

Cardinal

Chandelle Rose

Friendship

Green Star

Jilley

Matchpoint

Cultivars of Gladiolus

Morala

Novalux

Priscilla

Peter Pears

Red Sensation

Summer Sunshine

Sancerre

Tiger Flame

White Friendship

(50 cm) and bears 17 florets per spike. It produces 30.24 spikes/m^2/crop season. It is high yielding and has vase life of 8.5 days. It is resistant to *Fusarium* wilt disease.

Arka Gold: It is hybrid selection from the cross Greenbay x Gold Medal-412. Its florets are yellow having yellow lower lip with red blotch and slightly ruffled. It takes about 74 days for flowering, has long spike (113 cm) with good rachis length (52 cm) and bears 13 florets per spike and it produces 24.80 spikes/m^2/crop season.

Arka Kesar: It is hybrid selection from the cross Vink's Glory x Arka Sagar. Its florets are yellow-orange having orange margin. Blotch orange-red with yellow green lines. It is moderately resistant to *Fusarium* wilt disease. It takes about 61 days for flowering, has long spike (111 cm) with good rachis length (65 cm) and bears 18 florets per spike.

Arka Shobha: It is induced mutant from variety Wild Rose. Its florets are light pink with creamy white throat. It takes about 60 days for flowering, has long spike (105 cm) with good rachis length (62 cm) and bears 18 florets per spike.

Arka Pratham: This variety developed by hybridization and selection. It is early variety, blooms in 61 days, has spike length of 107 cm and produces 1.85 spikes/corm (2.1 lakhs/ha). It produces 1.82 corms per corm and 52.56 cormels per corm. The floret colour is violet with white lines.

Arka Ranjini: This variety developed by hybridization and selection. It blooms in 65 days, has spike length of 104 cm and produces 2 spikes/corm (2.4 lakhs/ha). The florets are in double rows. Floret colour purple with white blotch.

Arka Naveen: It is hybrid selection from the cross Hybrid 74-39-1 x Tropic Sea. Its florets are purple-violet having red-purple margin with yellow-green blotch and wavy. It takes about 66 days for flowering, has long spike (111 cm) with good rachis length (55 cm) and bears 16 florets per spike.

Recently IARI, New Delhi has released seven improved gladiolus varieties namely Pusa **Shubham**, Pusa Kiran, Pusa Manmohak, Pusa Vidushi, Pusa Valentine, Pusa Unnati and Pusa Srijana. The salient features of these varieties are given below.

Pusa Shubham: It is a cross from Lucky Shamrock x Green Lilac Open. It produces 14-16 florets of cream to yellow colour on long sturdy and compact spike. It is an early variety, flowering in about 72 days. The vase life of a spike is 10 days. It produces 1.6-2.3 shoots and more than 2 corms and 20 cormels from each mother corm (Anon., 2011).

Pusa Kiran: It is selection from open pollinated population from cv. Ave. It produces white coloured florets (16-19 in numbers) on long sturdy spikes. It is an early variety flowering in about 75 days. It is a prolific multiplier. The vase life of spikes is about 10 days. It produces 1.9-2.7 shoots and more than 2 corms and 20 cormels from each mother corm (Anon., 2011).

Pusa Manmohak: It is selection from the cross Mayur x Hunting Song. Florets are saffron red with thin whitish stripes on the throat of two oppositely placed lower tepals. It has spike length (>93 cm), rachis length (>55 cm) with 19-21 florets per spike. It flowers in 100 to 105 days with flowering duration of 20-25 days.

Pusa Vidushi: It is selection from the cross Melody x Berlew. Its florets are purplish white with grey purple spots on the base of the throat. First floret opens in 80-85 days. Spikes are straight with good rachis length and have 15-16 florets per spike. The florets are arranged compactly on the spike. At a time about 5 to 6 florets remain open at a time.

Pusa Red Valentine: It is selection from open pollinated seedlings of variety Regency. Florets are brick or blood red in colour with sun ray like small lines on the lower tepals. It flowers in about 95 days. The spikes are straight and long with good rachis length (50-55 cm) and have close arrangement of 18-19 florets on each spike.

Pusa Unnati: It is selection from the cross Berlew x Heady Wine. It has plant height of 158.66 cm, spike length (141.00 cm) with rachis length (71.00 cm).

Pusa Srijana: It is selection from the cross Berlew x Wine. It attains a height of 103.22 cm, having spike length (85.25 cm), rachis length (49.55 cm) with 16.66 florets per spike. PAU, Ludhiana recently released following improved varieties of gladiolus *viz.*, Punjab Glad-1, Punjab Flame, Punjab Glance, Punjab Lemon Delight, Punjab Pink Elegance, Punjab Glad-3, Punjab Glad-2. The salient features of some varieties are given below.

Punjab Glad-1: It is selection from the cross Happy End x True Yellow. Its florets are orange with yellow centre. It has average spike length (85 cm), number of florets per spike (15), flowers in 99 days and vase life 16 days. It is moderately resistant against *Botrytis* blight and *Fusarium* wilt.

Punjab Flame: It is selection from the cross Sylvia x White Prosperity. Its florets are carmine pink with red blotch. It has average spike length (103 cm), number of florets per spike (17), flowers in 114 days and vase life 16 days. It is moderately resistant against *Botrytis* blight.

Punjab Glance: It is selection from the cross Happy End x Yellow Stone. Its florets are orange with yellow centre. It has average spike length (85 cm), number of florets per spike (12), flowers in 78 days, number of corms per corm (2.61) and vase life 13 days. It is moderately resistant against *Botrytis* blight.

Punjab Lemon Delight: It is selection from the cross Jacksonville Gold x White Prosperity. Its florets are lemon yellow colour. It has average spike length (84 cm), number of florets per spike (12), flowers in 80 days, number of corms per corm (2.61) and vase life 13 days. It is moderately resistant against *Botrytis* blight.

Punjab Pink Elegance: It is selection from the cross Suchitra x White Prosperity. Its florets are soft pink with magenta centre. It has average spike length (85 cm), number of florets per spike (18), flowers in 86 days and vase life 17 days. It is moderately resistant against *Botrytis* blight.

MPKV, Rahuri has released four outstanding gladiolus varieties namely, Phule Ganesh, Phule Prerna, Phule Tejas, Phule Neelrekha. The salient fatures aof these varieties are iven below.

Phule Ganesh: Its floret colour is pale yellow, spike length (130-135 cm), number of florets per spike (17-19), with productivity of 80 to 100 thousand spikes.

Phule Prerna: Its floret colour is attractive pink, spike length (115-120 cm), number of florets per spike (16-17), with productivity of 60 to 80 thousand spikes.

Phule Tejas: Its floret colour is attractive mauve, spike length (80-85 cm), number of florets per spike (14-15), with productivity of 60 to 80 thousand spikes.

Phule Neelrekha: Its floret colour is deep purple, spike length (110-115 cm), number of florets per spike (20-21), with productivity of 80 to 100 thousand spikes.

BHU, Varanasi released three improved varieties of gladiolus namely, Malviya Kiran (white florets with red-purple throat on lower petal), Malviya Shatabdi (white florets with pinkish tinge) and Malviya Kundan (light yellow florets). These were developed though gamma irradiation.

ICAR-NBPGR, New Delhi has registered trait specific unique gladiolus germplasm for further use in breeding. The unique traits of registered germplasm are given in Table 2.

Table 2: Unique Gladiolus Germplasm Registered with ICAR-NBPGR, New Delhi

Sl. No.	*National Identity*	*Donor Indetity*	*INGR No.*	*Year*	*Pedigree*	*Novel/ Unique Features*
1.	IC296733	IHBT-G-1	13	2000	Green Woodpeeker x Oscar	Hybrid, with capsicum red (RHS-33A) Flowers, lip petals charetreuse yellow (RHS-2D), petals ruffled, average spike length (138 cm), mean number of florets (18.5) and flower diameter (10.2 cm).
2.	IC296774	IHBT-G-2	14	2000	Vink's Glory x Eurovision	Hybrid, with brick red petals having Dresden yellow on lip petals nicely ruffled, average spike length (115 cm) mean number of floret (18.3) and flower diametr (11.8 cm).
3.	IC296775	IHBT-G-3	15	2000	Snow princess x Her Majesty	Hybrid, with white petals having deep purple violet (RHS80A) edges, average spike length (125.4 cm), mean number of florets (17.1) and flowers (10.5 cm) in diameter.
4.	IC296776	IHBT-G-4	16	2000	Snow princess x Her Majesty	Hybrid, with reddish purple (RHS-74D) petals and dark reddish purple (RH-66A) blotches on the edges of both sides of the petal, lip petal magenta coloured with white stripes.
5	IC0584125	IIHGR-5 or IIHR-84-7-11	10067	2010	Hybrid 74-39-1 x Tropic Seas	Floret colour: (R.H.S. colour chart) Purple-violet (82.C) having Red-Purple (72.B) margin with Yellow-Green (154.D) blotch.
6.	IC0584126	IIHGR-9 or IIHR-84-7-11	10068	2010	Greenbay x Gold Medal-412	Floret colour: (R.H.S. colour chart) Yellow (4.C) having yellow (6.C) lower lip with Red (39.A) blotch.

Sl. No.	*National Identity*	*Donor Indetity*	*INGR No.*	*Year*	*Pedigree*	*Novel/ Unique Features*
7.	IC0584127	IIHRG-10 or IIHR-87-22-1	10069	2010	Watermelon Pink x Aarti	Resistant to *Fusarium* wilt disease cause by *Fusarium oxysporum* f.sp. gladioli race 1. floret colur: (Based on R.H.S. colour chart) Red (46.D) having Red (45.B) margin and white (155.B) line on tepals with Yellow (2.C) blotch.
8.	IC0611879	P-16-1 x Eurovision	14052	2014	A cross betwwen P-16-1 and Eurovision	Early blooming (76-80 days) and florets colour in red group (41 C, RHS colour chart) with dark red stripes on inner two tepals (42A, RHS colour chart), and also red spots on outer throat. Spike length (93 cm) with more than 16 florets. At a time 5-6 floretes remain open on the spike. From each mother corm, it produces more than 2.00 corms.
9.	IC0611878	Melody open	14053	2014	A selection among the open pollinated seedlings of the variety "Melody"	Very early in flowering (74 days) and florets colour in red group (50C, RHS). Two red spots on two central tepals. Number of florets (18.44) and spikes length (>100 cm). Corm multiplications rate is higher and produces more than 2.00 corms from each mother corm.
10.	IC0621473	IIHRG 6	17088	2017	Poonum x Gold medal-412	Floret colour: (RHS color chart). Purple (78.A) middle. Red-purple (72.A) margin with Green Yellow (1.D) blotch. Floret type is open faced and floret placement is in double rows.
11.	IC0621474	IIHRG 12	17089	2017	Junior prom x Poonum Year	Floret colour: (RHS color chart). Purple violet (82.A) having purple (77.A) margin. Green-White (157.C) line on lower lip. Early flowering (61.54 days).
12.	IC0620379	IIHRG-7	20066	2020	Meera x Picardy	Spike with variegated florets. Floret colour [Red-Purple (65.B) having Red-Purple (62.A) streaks with Red-Purple (67.B) splash].
13.	IC0620380	IIHRG-11	20067	2020	Gold Medal 412 x Poonam	Resistant to *Fusarium* wilt disease. Floret colour [Red (41.C) having Red (41.A) margin. Blotch Red (46.B) with yellow (13.C) border].

Source: http://www.nbpgr.ernet.in:8080/ircg/index.htm.

7.0 Genetics and Breeding

7.1.0 Genetics

The modern garden cultivars came from diverse genetic parentages. It has cumulative heterozygosity for many characters inherent with complex genetic constitution. Most members of this genus were heteroploids having very small chromosomes ranging from 2n=30 to 120 (diploid, triploid, tetraploid, pentaploid, hexaploid, octaploid and hyperaneauploid).

Majority of the South African species were diploid; summer-flowering commercial ones tetraploids and central African and Eurasiatic species polyploids (Bamford, 1935; Tamberg, 1967; Lewis *et al.*, 1972). In subgenus *Gladiolus*, *G. atropurpurens* had x = 12 (2n = 24 + 0-5B, 36); *G. scrapiflorus*, *G. gregarious* and *G. pseudospicatus* had x = 11 (2n = 22); *G. unguiculatus* had x = 13 (2n = 26 + 0-2B, also possibly 2n = 24) and *G. actinomorphanthus* had 2n = 28 (Goldblatt *et al.*, 1993). Mensinkai (1939) stated that although the longest known chromosomes were found in the Liliaceae, in gladiolus they were very small. It was the Asiatic group which led to improved cold hardiness.

In gladiolus, even diverse parents crossed together and the cultivars and species that differed widely in chromosome numbers are also cross-fertile. The genealogy of the modern cultivars involved about 20 or more species and even a few allied genera.

The desirable strains obtained in F_1 generation were perpetuated vegetatively without being segregated in the subsequent generations; the present day might be F_2, F_3,., F_7 or so of a particular cross or further involved with some other parents at each stage of flowering and seed set. Thus, they were not allowed to segregate freely privileged to be propagated asexually. The modern cultivars have become so complex that the off-springs do not appear to be similar (Misra, 1975); on repeated selfing for certain generations the inbred lines become difficult to maintain (Risley, 1953; Ryan and Ulman, 1953; Troyer, 1954; Palmer, 1954; Klein, 1972).

The genes located on a given chromosome are said to be linked together and therefore transmitted to the progeny provided crossing-over does not take place at meiosis. Large segments and heavy ruffling, heavy ruffling and short flower heads, blue-violet or red colour and certain pathogenic organisms, large floret size and loose floret attachment, flower colour and corm pigmentation, production of large cormels and late-blooming, *etc.*, are said to be linked characters (Griesbach, 1972). In case of very closely linked genes, a large number of seedlings of a cross are grown to obtain more number of desired combinations.

There are various patterns of pigmentation of the gladiolus florets. The free parts of the petals, the throat area and the in-between portions, even the inner and outer petals may differ in colour because of the individual gene expression. In fact, pigmentation patterns in the florets closely relate to the symmetry or form as often the falls differ in colour from those of central and wing segments (Griesbach, 1972).

Macek *et al.* (1946) reported the presence of both anthocyanins and anthoxanthins in gladiolus flowers. Anthocyanins were responsible for a range of colours from violet to red including shades of lavender, pink and rose and anthoxanthins varied from ivory through cream to rather intense yellow. Hamilton (1963) reported that E.C. Bate-Smith identified a violet coloured pigment in cv. Ravel as malvidin. Delphinidin which is responsible for the colour of truly blue flowers does not appear to be present in gladiolus. The carotenoid pigments were not known to occur in gladiolus even in the yellow cultivars since this colour is governed by anthoxanthins. The green shades occurring in certain basic white and yellow flowers presumably occurred due to the presence of one or both of the chlorophylls in the chloroplasts, clearly seen at the time of bud opening and which faded readily in sunlight. Chloroplasts contained both chlorophylls and carotenoids, whereas the chromoplastids contained only carotenoids which were absent in gladioli (Griesbach, 1972).

For breeding a particular colour or form, gladioli being highly heterozygous, the same type of parents were chosen with much probability for getting alikes. In case where genetics of two given cultivars were known for their recessive whites or yellows, they might produce white or yellow colour if crossed. In certain cases, inhibitor genes were present which inhibited the pigments of the lavender-red-purple anthocyanins present in the stamens and/or inner portions of the petals. In the simultaneous presence of one or more inhibitor genes, the effect of inhibitor genes to a certain degree might be suppressed. The cultivars having inhibitor genes might be pure white, cream or yellow which expressed the inhibited colours under certain temperature and light conditions. Such inhibitor genes might or might not inherit in the offsprings, if inherited, the colour might be suppressed, otherwise the colours in the flowers were expressed.

Plants of 5 species and several hybrids were used by Raamsdonk and de Vries (1989) to assess the extent of variability. Morphometric analyses were conducted using data on 25 characters. Bulk characters were useful for identification and the cluster analysis revealed 5 main groups (3 with PCA). The subspecific status of *G. communis* ssp. *byzantinus* was confirmed. Hybrids of 4 parental combinations were found in addition to a polyploid series (3x, 4x, 6x, 8x, 12x) and aneuploid cytotypes.

Patil *et al.* (2017) found that clustering pattern showed that genotypes of different geographical areas were clubbed in one group and also the genotypes of same geographical area were grouped into same cluster as well as in different cluster indicating formal relationship between geographical diversity and genetic diversity.

Thakur *et al.* (2016) assessed genetic diversity in 20 genotypes. Grouping of genotypes by Torcher's method revealed that all the genotypes were grouped into eight clusters, indicating the presence of considerable divergence and variation for different growth, flowering and corms traits. The genotypes 'Regency' from cluster II, 'Pacifica' and 'Tiger Flame' from cluster VIII, 'Priscilla' 'American Beauty' and 'Praha' from cluster IV and Acidanthera from cluster I had considerable divergence.

Simple correlation studies in gladiolus revealed that spike length had a positive association with days to shoot emergence, leaf breadth, florets per spike, rachis length, internodal length, size of floret and size and weight of daughter corms.

Kumar *et al.* (2015) found positive highly significant correlation of days to heading with days to first floret showing colour and days to opening of first floret was observed. Spike length was also significant and positively correlated with flower diameter and number of cormels/plant. Corm diameter was positive and highly significantly correlated with corm weight and polar diameter.

Rashmi *et al.* (2016) showed that number of spike per plant had positive correlation with plant height, number of leaves, leaf length, leaf width, leaf area, spike length, rachis length, number of florets per spike, vase life, number of corms, weight of corms and diameter of corm. Spike length exhibited positive correlation with plant height, number of leaves, leaf width, leaf area, rachis length, number of florets per spike, diameter of the floret, weight of the spike, length of the floret, vase life, number of corms, number of cormels, weight of corm, diameter of corm.

Momin *et al.* (2017) found that days for sprouting of corms exhibited positive correlation with days for spike initiation and days for floret opening. Vase life was highly significant and positively correlated with rachis length, number of florets/ spike, fresh weight of spike, spike girth and field life.

Nainwal *et al.* (2016) reported that spike length was positively correlated with plant height, leaf length, leaf width and average corm weight except tillers per clump at genotypic level. Number of florets per spike was positively correlated with plant diameter, leaf width, tillers per clump and corms per clump. Taj *et al.* (2015) found that spike length had positive correlation with plant height, days to emergence of spikes, days to opening of first floret, rachis length, floret size, number of florets per spike and number of florets remain open at a time.

Kumar *et al.* (2015) found that number of cormlets per plant had positive correlation with spike per corm, while, numbers of flowers per spike were positively correlated with length of rachis and spike and flower diameter. Zaharia (2018) found positive (diameter of the corm x weight of the corm) and negative (CW × PH) values of the correlation coefficient in hybrids obtained from cross hybridizations and their genitors.

Naresh *et al.* (2019) found that plant height exhibited positive correlation with number of spikes per plant; spike length and number of florets per spike were associated with spike yield, however, association between spike yield and corm weight per plant was highly significant. Swetha *et al.* (2020) reported that number of spikes per plant had positive correlation with number of leaves, number of shoots and number of daughter corms per plant. Spike length exhibited positive association with rachis length, number of florets, vase life and number of cormels. Number of florets exerted positive correlation with vase life.

Number of spikes per corm planted, florets per spike and rachis length had high and positive direct effects on spike length (Mahanta and Paswan, 1994).

Kumar *et al.* (2015) revealed that maximum direct effect on length of rachis due to visibility of spike, while, spikes per corm, opening of first flower, leaf length and weight of corm had direct negative effect on length of rachis. The indirect path revealed maximum indirect effect of number of spikes per corm, weight of corm and diameter of corm via number of cormlets per plant on length of rachis.

Zaharia (2018) reported that weight of the corm has a direct and positive influence on the number of flowers, length of the inflorescence and the height of the plant in hybrids obtained from cross hybridizations and their genitors.

Naresh *et al.* (2019) reported that a dependent traits *viz.,* number of spikes per plant was controlled by several component characters which were inter-related in a complex manner. The number of days taken for sprouting of corms along with number of days taken for spike initiation exhibited a high negative and direct effect on number of spikes per plant. Fourteen independent traits had positive indirect effect of number of days taken for sprouting of corms on spike yield.

Pant *et al.* (1998) indicated that the inheritance of characters like days to sprouting, plant height, leaves/plant, days to flowering and floret size was predominantly controlled by additive and additive epistatic components of genetic variance. Cultivars like Oscar, Friendship and Apple Blossom were good general combiners.

Emna and Faouzi (2015) assessed morphologically 15 accessions Gladiolus in Tunisia. Cluster analysis revealed two major clusters, first one includes the accessions of *Gladiolus italicus* and second consists populations with seeds wing.

Hiremath *et al.* (2020) phenotypically characterised 84 Indian and exotic gladiolus genotypes for 21 traits as per the DUS (Distinctiveness, Uniformity and Stability) guidelines to distinguish and grouped selected genotypes. Among DUS traits, seven were dimorphic and 14 as polymorphic. Characterization information for different traits could be used as reference collection for precise identification and distinction of gladiolus varieties.

Mishra *et al.* (2014) reported highest heritability estimates for vase life. High genetic advance in percentage of mean coupled with high heritability was recorded for number of sprouts, number of spikes per plant, length of spike, vase life, weight of corms per plant, weight of cormels per plant, number of corms per plant, number of cormels per plant provide greater scope for further improvement of these traits in advance generations.

Amrutha *et al.* (2014) noticed that number of leaves per plant, number of spikes per plant and corm weight showing higher values of genotypic coefficient of variation, these characters are the most suitable characters for selection. Kadam *et al.* (2014) found higher values of PCV and GCV were recorded for cormel weight and number of cormels per plant, respectively. Maximum heritability was recorded for days to first floret opening followed by rachis length and days to spike initiation. Highest values for genetic advance were recorded for plant height and spike length.

Hort *et al.* (2014) estimated the variability and the heritability for some hybrid combinations of Gladiolus. High heritability in broad sense was recorded to all the cyclic combination. The heritability was between 0.360-0.677. Kumar *et al.* (2015) found high PCV along with GCV observed for plant height, leaf length, florets/ spike, vase life, corm weight and cormels/plant. High heritability coupled with high genetic advance as per cent of mean was recorded for plant height, leaf length, florets/spike, corm diameter, corm weight and number of cormels/plant.

Rashmi *et al.* (2016) reported that rachis length, number of spikes per plant, weight of spike, number of florets per spike, number of corms per plant, number of cormels per plant, weight of corm, yield of corms kg per hectare and yield of corms tons per hectare exhibited high heritability (>75 per cent) coupled with high genetic advance as per cent mean (>30 per cent) which indicated the prevalence of additive gene action in expression of these traits.

Bhajantri and Patil (2016) studied the genetic variability induced by gamma rays radiation for fifteen vegetative and floral traits in M_1V_2 generation of gladiolus variety Ethyl Cav cole and White Prosperity. High heritability estimates along with moderate genetic advance observed for number of leaves per plant (at 10 Gy), leaf width, days to full spike emergence, days to first floret to show colour, spike length, rachis length, florets per spike, inter floret length and floret diameter were observed in all gamma ray treated mutant populations.

Nainwal *et al.* (2016) found high heritability coupled with high genetic advance and coefficient of variability were observed for plant height, plant diameter, tillers per clump, spike length, florets per spike, corm weight and corms per clump. Momin *et al.* (2017) reported that maximum heritability was also noticed in corm weight, while, high heritability with genetic advance was recorded for corm weight, which indicated possible role of additive gene action.

Ishwarraddy *et al.* (2018) found high PCV and GCV for weight of cormels, number of corms, number of florets, day to 50 per cent flowering and number of leaves at maturity. High heritability estimates coupled with high genetic advance over mean was observed for all the traits except leaf width at maturity, plant height at maturity and number of days for lowest floret to open.

Kaushik *et al.* (2018) reported that PCV was higher than GCV for all traits, indicating that these traits are governed by environmental factor influencing their expression. Heritability ranged from 41.40–88.30 per cent, the highest was in weight of corm. Genetic advance expressed as percentage of mean was high for weight of corm, number of florets per spikes and moderate in number of spike per corm, number of corms per plant, leaf width, yield of corm and cormlets per hectare, length of spikes, diameter of corms, number of cormlets per plants.

Sheikh *et al.* (2006) carried out stability analysis of 10 elite cultivars for important floral traits (florets/spike and spike length) across 5 random locations (environments) revealed the linear regression was non-significant for floret/spike revealing that most of the cultivars except 'Apple Blossom' and 'Big Time Supreme' were average in stability.

Naik *et al.* (2011) evaluated 14 promising cultivars under three environments, *viz.*, shade-house, polyhouse and the open condition. Genotypes x Environment interactions were significant for days to spike emergence and days from first floret to last floret opening. For days taken from first to last floret opening, genotypes Eighth Wonder, Melody, Friendship, Pricilla, Pacifica, White Prosperity and Vedanapoli were stable, with high mean values. The genotype Vedanapoli was stable for number of daughter corms per plant, Pacifica and Eighth Wonder were stable for corm diameter, with high mean values. Though 'Vedanapoli' showed unpredictable performance for average weight of corm, ten cormel weight and number of cormels per plant, it gave the highest corm yield per plant and showed predictable performance for corm diameter, exhibiting a great promise for hybridization with consistent corm yield outcome.

Mahato and Gantait (2020) studied the combining ability in 5 × 5 full diallel mating; found that the genotypes, *viz.*, ACC-15, ACC-11 and ACC-14 were good general combiners for most characters. The cross ACC-04 × ACC-14 had the highest SCA effects for diameter of cormel and can be used in both yield and quality improvement.

Most of crosses showed a negative mid-parent and high-parent heterosis, hybrids expressing the highest MPH and HPH heterosis overall, were ACC-15 × ACC-13, ACC-14 × ACC-15 and ACC-04 × ACC-14.

Azimi *et al.* (2020) generated two new promising high-quality hybrids (OPRC16 and OPRC57) through hybridization between 'Amsterdam' and 'White Prosperity'. The hybrid OPRC16 showed that Hsp value was negative for peduncle length, leaf width, stem diameter, and cormels diameter, but it was positive for other traits. The hybrid OPRC57 showed Hsp value positive for leaf length, bud diameter, vase life, cormlet number, and cormles yield but negative for other traits. When 'Amsterdam' used as the maternal plant in crosses, it has more positive heterosis effects than when it is used as the paternal parent.

7.2.0 Breeding

Plant height, stem types such as slenderness and flexibility to high wind, spike length, bud counts and compactness, flower size and shape, flower placement and facing, texture and frilling, side spikes, corm multiplication and resistance to diseases and pests are the important parameters for improvement of gladioli.

7.2.1 Hybridization

Gladioli being hermaphrodite has both the male and female organs in the same flower, so the desired combinations are made by choosing either of the parents as male or female; however, in certain cases where any of the organs is sterile, that may be used only for the other organs. Flowers start opening after the sunset and open fully in the morning with a few exceptions, *e.g.*, cv. Setting Sun which normally opens by 11.00 am, in cloudy days, it opens even by evening. Certain species also have nyctinastic movements, as *Gladiolus floribundus* closes its segments at night whereas *G. Tristis* opens fully in the evening (Lewis *et al.*, 1972).

With the opening of the flowers, the anthers are matured which may be used for pollinating already emasculated females. Emasculation and pollination, both are done in the morning. The stigma generally becomes receptive on the third day of the opening of the flowers or anther maturity, and that is the reason when the flowers are bagged individually for selfing do not set seeds. If the pollens from the freshly opened flowers of the same plant are applied, the seeds are set provided other conditions remain favourable. Emasculation is carried out at bud stage when these have started swelling for opening. After pollination, the flowers are bagged with perforated butter paper bags and tagged with labels mentioning the parents and the date of pollination.

Capsules mature generally within four to six weeks of pollination, which when start dehiscing are collected, sun-dried, sometimes dewinged and stored in moisture-proof envelopes until sowing.

Magie (1957-58) reported a new cultivar Florida Pink which was vigorous and produced high yield of quality flower. It had considerable resistance to *Fusarium oxysporum*, *Curvularia trifolii* and *Stemphyllum* sp. Resistance to *Botrytis gladiolorum* was as good as that of its pollen parent Spic and Span. Negi *et al.* (1991) tested about 1000 hybrid seedlings from crosses, 5 susceptible commercial varieties with 4 varieties tolerant of *Fusarium oxysporum* f. sp. *gladioli*, 18 selected for field resistance over 4 years were further tested following inoculation. One of the 18 proved resistant and 10 were tolerant. The resistant 82-11-90 and tolerant 82-7-59 and 82-18-16 had good scores for source for various quantitative and qualitative traits. The resistant hybrid was derived from a cross between the susceptible Beauty Spot and resistant Psittacinus Hybrid.

Wilfret (1981) selected Florida Flame from a cross between WBL 047 and Folies Bergere. The maximum spike length is 160 cm with 14-17 florets. The petals are light scarlet-red intensifying to darker throat veins. The cultivar is *Fusarium* tolerant. The flower and corm production in Florida Flame was compared with 5 commercial cultivars (Table 3). Morning Mist cultivar was originated by Wilfret (1993) from a cross between seedling GC354 (Blue Baron x Beverly Ann) and GC143 (China Blue x Beverly Ann). This pale violet-flowered variety was released in 1988. It combines good cut flower characteristics, such as tall spikes, with a floral bud count of 20-24 and the ability to open from the tight bud stage. Morning Mist is tolerant of *Fusarium oxysporum* f. sp. *gladioli*, yielding good quality corms larger than those of Dr. Magie.

Cohen and Barzilay (1991) used pollen of *Gladiolus tristis* (2n=30) to pollinate several commercial cultivars of *G.*x *grandiflorus* (2n=60) for producing miniature gladiolus. From the triploid hybrids (2n=45), 5 new cultivars were selected (female brackets): Adi, Nirit (both Red Cascade), Kinnereth (Yamit Fidelio) and Ronit (Admirant). These new cultivars particularly from Orchidiola group are distinct from the Pixiola group in their flower colour pattern and growth duration. They flower in winter after 70-100 days compared with 110-150 days for the *G.*x *grandiflorus* progenitors.

Crossing of diploid scented *Acidanthera bicolor* var. Murielae with the amphidiploid (allotetraploid) non-scented gladiolus cultivar Filigree produced two scented triploids and a number of tetraploids with gladiolus characters, the latter probably arising through apomixes. Back-cross to both parents and selfings of the triploids produced plants varying from tetraploid to pentaploid and showing a wide range of fragrance between the extremes of the two parents. Aneuploids, arising from the loss of chromosomes by the triploids and pentaploids during egg formation, were less fragrant than the corresponding euploids suggesting that loss of fragrance determining chromosomes at meiosis is a normal feature (Rattenbury, 1966). The parents and some of the hybrids are arranged according to their relative fragrance in Table 4.

Table 3: Flower, Corm and Cormel Yield Comparisons of Florida Flame and 5 Commercial Gladiolus Cultivars grown at Arec-bradenton (spring, 1977) on Unfumigated Land*

Cultivar^z	*Total Spikes*	*Spike Length (cm)*	*Rachis Length (cm)*	*No. of Florets/ Spike*	*No. of Corms*	*Wt. of Corms (g)*	*Wt. of Cormels (g)*
Florida Flame Jessie M.	20.0 ab^w	132.1 a	57.2 a	14.4 c	21-8 ab	1061a	218 a
Conner^Y	16.0 bc	96.0 cd	45.0 c	17.0 b	16.5 c	752 bc	199 a
Friendship^Y	14.5 c	104.4 c	46.5 bc	15.8 bc	14.0 cd	465 de	17 b
Valeria^Y	8.0 d	85.8 d	30.5 d	10.2 d	9.4 d	244 e	3 b
Friendship^X	21.8 a	117.9 b	52.3 ab	17.6 b	25.5 a	999 ab	159 a
Traveler^X	17.8 abc	91.7 d	44.7 c	16.0 bc	15.0 c	675 cd	65 b
Peter Pears^X	18.2 abc	116.8 b	55.6 a	20.0 a	18.8 bc	522 cd	37 b

z: 15 corms (3.5 cm dia) planted per plot with each plot replicated 4 times; y: Corms grown at AREC-Bradenton the previous year and given no postharvest or pre-plant fungicide dip; x: Corms grown on a commercial farm the previous year and given a postharvest fungicide dip of dowicide B. w: Mean separation within columns by Duncan's multiple range test 5 per cent level.

* Wilfret (1981).

Ohri and Khoshoo (1983) reported the results of intra and interploidal crosses (cv. La Palonea x *Acidanthera bicolor*). Only two of the progenies of self and cross combinations involving the tetraploid cultivars were not tetraploid. The progeny essentially showed the same meiotic behaviour of the parents with prominent bivalent formation. The 4x x 2x cross was successful only when a tetraploid was used as the female parent, 28 of 29 plants were triploid. *Acidanthera bicolor* var. Murialae (2n = 30) could be crossed with gladiolus *cvs.* La Palonea and Pacifica (2n=60) when these were used as female parents.

Bylov (1990) studied the hybrids of *Gladiolus hybrid* × *Acidanthera bicolor*. The F_1s showed different capacities for selfing. In the F_2, a wide range of forms segregated for various traits, including flower colour. The F_2s exceeded the parents but not the F_1s in height. When the F_1s were pollinated by various gladiolus varieties, the forms obtained were of high ornamental value. There was variation in chromosome number even within a given cross. More than 45 promising hybrids were selected.

Poon *et al.* (2013) found that among four cross combinations, 'H.S. 88-10-22 × Shobha' recorded 27.63 well-developed seeds per capsule and in cross 'H.S. 84-7-11 × Shobha', it was 27.01. The interaction effect of fresh pollen- cum- 'H.S.88-10-22 × Shobha' produced higher number of well developed seeds/capsule (41.02). In contrast, cryopreserved pollen- cum- 'H.S. 88-10-22 × Shobha' produced the lowest number of well developed seeds/capsule (10.85). Inclusion of fresh pollen to genotype 'Shobha' and cryopreserved pollen of H.S. 88-10-22' revealed better performances for maximum seed set per capsule.

Mahato and Gantait (2017) studied the seed setting in crosses and open pollinated twenty gladiolus genotypes. Five maternal parents *viz.*, ACC-04, ACC-

15, ACC-11, ACC-13 and ACC-14 produced a high percentage of capsule set to capsule harvest and number of harvestable seed per capsule from cross and open pollination. Conversely, the genotype, ACC-05, ACC-16 and ACC-17 were sterile, while ACC-10 and ACC-18 showed varying percentage of pod set after pollination which were shrinkled thereafter and died.

Mahato and Gantait (2020) in 5 × 5 full diallel mating, found that days required for maturity of capsules, harvest percentage of capsule, number of seeds per capsule, hundred seeds weight and seed germination percentage were varied among the crosses. The maximum harvestable capsules was recorded in ACC-13 × ACC-14 and its reciprocal cross while maximum number of seeds per capsule was found in the cross ACC-15 × ACC-13.

Table 4 : Fragrance Scale of Parents and Hybrids*

Plants Arranged in Order of Decreasing Fragrance	*Acidanthera chromosomes*	*Gladiolus chromosomes*
Acidanthera bicolor var. Murielae	30	0
Triploid *Acidanthera*	30	30
Triploid hybrids (fragrant)	15	30
Triploid *Gladiolus*		
Best	15	60
Less fragrant	10+	60
Gladiolus cultivar Filigree	0	60

* Rattenbury (1966).

Szczepaniak *et al.* (2016) discovered putative hybrid plants in Poland. Natural hybridization between *G. palustris* and *G. imbricatus* was confirmed by chloroplast (*psb*A-*trn*H and *rpl32*-*trn*L) DNA and nuclear ribosomal DNA (ITS1) sequences, AFLP markers and macro-, micromorphological and reproductive characters. Based on molecular data, the hybridization events are likely to have occurred relatively recently with *G. palustris* as the maternal species and *G. imbricatus* as the pollen donor in interspecific crosses. The existence of a shared common cpDNA haplotype in all hybrids and *G. palustris* indicates unidirectional hybridization. Analyses of AFLP data and polymorphisms of ITS1 sequences showed additive inheritance of parental genomic fragments in *G.×sulistrovicus*. The hybrids exhibited either morphological similarity to *G. imbricatus* or intermediateness in phenotypic characters. The corm structure of flowering plants and seed capsules clearly distinguish the hybrid. The new taxon is characterized by a lower generative reproduction than the parental species, however, hybrids produce ~50 per cent viable pollen and seeds.

7.2.2 Mutation Breeding

Many gladiolus cultivars are available in commerce by spontaneous mutation (Anon., 1972; Fisher, 1975), and only a few through induced ones. Two such sports have been reported in the cvs. Salman's Sensation (Misra, 1975) and Ratna's Butterfly (Misra, 1982). In many cases changes are so peculiar that flowers become monstrous, and in most cases the mutations are deleterious. The induced changes mostly revert back in further generations because of diplontic selection.

The only record of pollen treatment by physical mutagens from 100 to 200 kR gamma rays has been reported where colour mutants in the seedlings were obtained with 17.23 per cent viable pollen grains (Dryagina, 1970). However, the earliest record for seed treatment available is the *cv.* Lord Selkirk, pure white in colour (1939-40 introduction) produced by J. Twomeys, Winnipeg, Canada by irradiating the seeds of Mrs. Anna Pfitzer × Queen Mary (Anon., 1972). Decreased survival with increased exposure of X-rays on the F_2 seeds of White Star with 12, 24, 48 and 96 seconds was obtained with a number of floral abnormalities (Marek, 1954) and it was suggested not to expose for more than 30 seconds. Seeds of cultivar Vista Bonita exposed to atomic irradiation showed faster germination the same year and in the following season, 40.44 per cent of 136 seedlings eventually bloomed (four times more than the percentage normally occurs in usual cases) from salmon to deep blotched pink via various shades with the seedlings having hood, ranging from 200 to 500 size, and abnormalities appeared as ruffling of petals, unopening of florets, increased floral parts, bicolouration into rose and white, short head, poor texture and attachment and small cormels (Hubbard, 1966). Likewise, gamma radiation has also been found effective for seed treatment on different cultivars (Dryagina and Kazarinov, 1965, 1966 a, b; Dryagina 1970).

X-ray treatment of corms and cormels has also been reported effective (Yamamoto *et al.*, 1958; Jenkins, 1961). In the former case, colour mutation was obtained, whereas in the latter one where eight cultivars were exposed, the sprouting of cormels and plant height were reduced, the lethal dosages were 15000 and 20000 R and none of the cormels sprouted.

^{60}Co gamma radiation has been found equally effective as X-ray treatment (Sheehan, 1959; Sheehan and Sagawa, 1959, 1960; Sheehan and Lorz, 1960; Sedel'nikova and Tarasenko, 1980). White Frosting and *Dixiland gladioli* at 1000 R produced 14 per cent M_1 plants in the former cultivar showing variation in many morphological characters from the controls. White Frosting was comparatively more sensitive, whereas in the latter case the treatment was given by placing the plant material at certain distances from the centre of the gamma source at 33000 R and obtained leaf deformities, about 12 per cent decrease in blooming and partially double flower with no effect on sprouting (Sheehan, 1959). Co-60 at 10 kR sharply inhibited growth and development (Dryagina, 1962), and with X-ray treatment on eight cultivars, 500, 2000 and 5000 R dosages had a slight stimulating effect on sprouting of corms and growth of the plants in the cv. Summer Freuds. Among the four cultivars, Summer Freuds, Vincent Van Gogh, Edith Eddy and Roma, former two were resistant to radiation while other two sensitive (Dryagina, 1964), and higher dosage of 10 kR by single irradiation proved fatal while in a gamma-field irradiation the corms tolerated even a dosage of 74.8 kR, although 5 kR was found most suitable and changes like double and instaminate flowers, branching of flower stalk, colour change of the petals, *etc.*, were the conspicuous features (Dryagina, 1964, 1968, 1970). Effectiveness of gamma radiation is now well known (Meshit-suka *et al.*, 1963; Iba *et al.*, 1964, 1965; Dryagina and Kazarinov, 1965, 1966a) and thus the mutations including the chimaeral ones were produced (Anon., 1967, 1968). Chronic irradiation on a gamma-field from 10 to 28.3 kR on the *cv.* Zommer Froide (Summer Freuds) resulted in a general reduction of the percentage of flowering plants and in the retardation of morphogenesis of the flower stalk (Dryagina and Akhramova, 1962, 1963).

Mutations in terms of flower colour and several other characters were also induced by the use of gamma radiation (Buiatti *et al.*, 1965). A study on the effects of gamma radiation on the mutated sector size revealed that petal in gladiolus is derived from 1-4 initial cells (Scarascia-Mugnozza, 1968). The effect of environment on diplontic selection by implicating the results in relation to Gaul's theory of diplontic selection after growing the gladiolus *cv.* Oscar under two conditions (glasshouse and open field), revealed that faster growth from sprouting to spike emergence led to an increase in mutation frequencies and decrease in mutated sector size (Buiatti *et al.*, 1970). Hawaii, a tetraploid, was treated with 2500, 5000, 10000 and 15000 R gamma rays; in 2500 R chimaeral flower mutants in two per cent of the spikes were obtained, and in 5000 R 8.1 per cent, and entire mutant spike appeared in vM_2 in 23.9 per cent from those treated with 2500 R and 60.98 per cent from those treated with 5000 R (Moes, 1966). The total percentages of mutated spikes among all the progeny plants (corms and cormels) were 4.08 and 19.7, respectively. Some chimaeral mutants disappeared in vM_2 and were reduced to only 1.06 per cent of the spikes. A lot of modifications like spots, specks and sectors which were lighter than the normal colour of the segments were recorded by Grabowska (1972, 1975a,b) and Grabowska and Mynett (1970) by gamma radiation of gladiolus corms. In the latter work, one plant with colour mutation was obtained. Lethal dosages in vM_1 for gladiolus by gamma irradiation was reported from 10000 to 40000 R by Glazurina and Chemagin (1974). Cultivar Hawaii at 4000 R (Seilleur, 1975) delayed the flowering. Seilleur (1977a,b,c) while studying the presence of pigments in these mutants found one-dimensional chromatographic methods inadequate for separating the pigments whereas two-dimen-sional paper chromatography facilitated the separation. He described the methods for separating the anthocyanins and anthocyanidins. Six common anthocyanidins-pelargonidin, peonidin, malvidin, cyanidin, petunidin and delphinidin were found in the cv. Hawaii and most mutants. With regard to anthocyanins, Hawaii contained 18, but the mutants only some or all of these apart from six more anthocyanins. Further he had made quantitative evaluation of pigments separated by thin layer of chromatography in cv. Hawaii and 12 mutants where he found certain modifications due to mutation.

In India, cv. Shobha was evolved from cv. Wild Rose in vM_3 under 1 kR dosage (Raghava *et al.*, 1988). Jo Wagenaar produced double flowers, stunted spikes and leaf variegation (chlorophyll mutation) in 2 and 4 kR of gamma rays (Kaicker, 1975), but in vM_2 these abnormalities disappeared because of the diplontic selection of the tissues. Different types of variations were reported by Misra (1978), Misra and Choudhury (1979) and Misra and Bajpai (1983) by treating several gladiolus cultivars to various dosages of gamma irradiation. Bicolouration of petals, increase in the number of floral parts, formation of twin flowers, russeting of florets, leaf variegation and deformities, stem branching, plant stunting, *etc.*, were observed in the *cvs.* Jo Wagenaar and Picardy, where a yellow flower mutant was also isolated from the latter cultivar (Misra, 1976a).

A dose of 50 Gy had lowest survival rate but was sub-lethal in *cvs.* 'Eurovision' and 'Wine and Roses'. *In vitro* selection by challenging with fusaric acid (1-1.5 mM) and culture filtrate (20 per cent) of *F. oxysporum* f.sp. *gladioli* (FOG) resulted *Fusarium* yellows insensitive mutants at the end of the 3rd selection cycle. They confirmed that fusaric acid and culture filtrate were suitable phytotoxins for selecting optimal resistance against FOG (Pathania and Misra, 2003).

At 6 kR and 7 kR proved injurious as the number of florets and length of spike reduced drastically in cv. American Beauty while plants remained blind in *cvs.* Novalux and Eurovision. Colour variation in florets and whole spike, increase and decrease in the number of floral organs and their fasciations and spike bifurcation were also observed from 4 kR to 7 kR. Three desirable colour mutants obtained from three varieties and one mutant with bifurcated spike, having all similar characters as original except doubled number of florets. Treatments of 5 kR and 6 kR were observed to be good for induction of mutants (Patil and Dhaduk, 2009).

Reduction in survival rate, root length, number of spike/plant, number of floret/spike, days to flower, self life, vase life, floret size, number of corm/plant and weight of corm/plant were observed after irradiation. Abnormal leaves increased with increase in exposure of gamma rays. Cent per cent mortality was recorded in plants treated with 15 KRD gamma rays. It may be concluded that 5 to 10 KRD doses of gamma rays are suitable for induction of somatic mutation in gladiolus (Tiwari *et al.*, 2010).

Percentage of sprouting and survival was affected at 1 kR to 4 kR. LD50 was found to be beyond 7 kR dose for both sprouting and survival. Plants treated with 6 kR and 7 kR did not produced flower spikes. Higher doses reduced plant height, leaf number and leaf size and become narrow and leathery. Colour variations in florets and whole spike were also increased with increase in dose rate along with increase or decrease in number of floral organs. Radiation treatments at higher doses caused delayed in spike initiation with decrease in spike length, number and size of florets, vase life and yield per plot while lower doses responded positively. Three desirable mutants with light colours were isolated from all three varieties in 5 kR whereas one mutant have bifurcated spike at 6 kR from *cv.* American Beauty (Patil, 2014).

Corms of different varieties irradiated with various gamma doses showed narrower and short leaves as gamma doses increased. A stable mutant with different flower colour from variety Tiger Flame was identified at 4.5 kR. Lower petals of mutant were yellow colour (Sisodia *et al.*, 2015). Higher dose 5.5 kr reduces sprouting, at 3.5, 4.5 and 5.5 kr found lethal in *cvs.* Aldebaran, Legend and Praha and failed to produce spike. Early spike emergence was recorded with 1.5 kr while late opening of 1st floret was observed with 2.5 kr. Abnormal plant growth, bunched spike in some plants of cv. Pusa Kiran was observed with 4.5 kr. A creamish white mutant from cv. Jyotsna and white colour florets mutant from cv. Punjab Dawn was recorded at 2.5 kr. A yellow florets coloured mutant was found at 3.5 kr. in *cv.* Tiger Flame (Sisodia and Singh, 2015).

Plants at 55 Gy and 70 Gy showed deleterious effect, maximum number of corms per plant recorded at 25 Gy, while, minimum at 70 Gy, whereas, number of cormels per plant was also reduced with the increase in gamma rays dose. Homeotic mutants were isolated from Yellow Golden, Nathan Red, White Friendship, Purple Flora and Algarve at 55 and 70 Gy (Kumari and Kumar, 2015). Corms of varieties 'Punjab Glance' and 'Sylvia' irradiated with different doses of gamma rays revealed that the pollen fertility, size, stomata number, size and number of chloroplast decreased with the increasing dose of irradiation from 50 to 150 Gy (Dogra *et al.*, 2017).

Maximum reduction in plant height and number of leaves was recorded with highest dose (150 Gy) in *cv.* Punjab Glance. Earlier (13.72 days) corm sprouted with lowest dose at 50 Gy, while, highest dose (150 Gy) delayed (20.67 days), corm number per corm (1.09) and cormels per corm (16.39) was reduced with increased dose of gamma rays (150 Gy). The leaf abnormalities in terms of changes in leaf shape, leaf size, leaf margin and leaf apex were observed (Dogra and Dhatt, 2017).

Irradiation of three cultivars at 15, 30, 45, and 60 Gy showed that lower doses had stimulative effect on sprouting percentage, days taken to sprouting, spike initiation, opening of first florets whereas higher doses were detrimental for vegetative characters (Shukla *et al.*, 2018).

Higher dose 60 Gy observed maximum leaf and morphological abnormalities in cvs. Candyman, American Beauty, Her Majesty during vM_2 generation. Two variants from var. Candyman at 45 Gy and 60Gy and one variant from *var.* Her Majesty at 30 Gy were identified and isolated for changes in floret color whereas one variant was identified and isolated from *var.* American Beauty having variation in floret arrangement (Kashyap *et al.*, 2018).

Eight varieties irradiated at different doses of gamma rays (^{60}Co) evaluated in M_1 and M_2 generations revealed that chlorophyll a, chlorophyll b, total chlorophyll content and carotenoid content was inversely related to gamma rays doses. Percentage of abnormal leaves, blind plants and abnormal spikes were directly related to gamma rays dose and were highest at 70 Gy, irrespective of varieties. In general, morphological changes and abnormalities were reduced in M_2 generation as compared to M_1 generation, which shows diminishing effect of gamma rays in successive generations (Kumari and Kumar, 2019).

Gladiolus variety 'Sylvia' treated with ^{60}Co gamma rays revealed that as the dose was increased from 50 to 150 Gy, reduction in survival rate, days taken to sprouting of corms, plant height, number of leaves, leaf length, width and leaf area, corm and production were also decreased. Earlier sprouting of corms recorded at 50 Gy, while delayed with 150 Gy. Leaf abnormalities were recorded in terms of leaf margin and apex bud (Dogra and Dhatt, 2019). Increased tillers per corm (2.33) and leaves per plant (12.88) recorded at 25 Gy. Flower diameter (78.45 mm) and number of florets (1.86) was slightly increased at 15 Gy treatment as compare to untreated plants. Homeotic mutants were isolated from American Beauty, Dull Queen, Saffron, Candyman and Summer Sunshine at 35, 45 and 55 Gy (Sathyanarayana *et al.*, 2019).

Abdulrahman *et al.* (2019) investigated the effect of electric current severity and shock timing of corms on the vegetative, flowering and corm growth characters of

Gladiolus grandiflorus cvs. Bimbo and Sapporo. The corms were first soaked for 3 h in 1 per cent NaCl solution, and then soaked in fresh water for the same period of 3 h before planting. The corms of *cv.* Bimbo which were exposed to low current for a short period treatment *i.e.* 4 A × 1 min significantly reduced sprouting time of vegetative growth to 12 days and flowering date to 68 days, increased the number of lateral branches (2.75 lateral branches/plant), largest leaf area (191.35 cm^2), chlorophyll content (59.65 SPAD), longest spike flower (12.50 cm), highest number of florets in spike flower (27.85 flowers/spike) and highest dry weight of corms (19.81 g). However, the treatment (6 A × 3 min) significantly increased the plant height (128.95 cm), dry weight of vegetative parts (12.09 g) and diameter of corm (6.13 cm).

Chemical mutagens such as EMS, DEMS, SEMS, DES *etc.* also found effective in inducing mutation in gladiolus. EMS @ 0.5 per cent treated corms cv. American Beauty recorded early sprouting and flowering with maximum number of sprouts, plant height, number of leaves at 6 kR + 0.2 per cent EMS in *cv.* Pink Beauty. The plants could not survive with combined tratement of gamma rays and EMS. It is concluded that 0.2 to 0.6 per cent EMS doses improved leaf area, number of florets per spike, spike length, floret size, blooming period, number of spikes per plant and number of corms. Higher doses of all mutagens adversely affected growth, flowering and yield parameters of gladiolus (Patil *et al.*, 2017).

Corm weight and corm diameter recorded high phenotypic and genotypic coefficient of variation at 0.50 per cent EMS treated population in cvs. Ethyl Cav Cole and White Prosperity. The highest variability for days to spike initiation was at 0.50 per cent EMS treated recorded in plants of Ethyl Cav Cole and White Prosperity (Bhajantri and Patil, 2013).

Bulbs of 'White Prosperity' treated with gamma rays, Dry Ethyl Methane Sulphonate (DEMS) and soaked Ethyl Methane Sulphonate (SEMS) with different concentrations, found that EMS and gamma radiations delayed emergence of sprouts. Maximum number of leaves and shoot per corm were recorded at 2Kr and soaked EMS @ 0.25 per cent, respectively, however, EMS @ 0.25 per cent increased number of spike per plant. It was concluded that 0.25 to 0.50 per cent dry and soaked EMS doses are suitable for inducing mutation in gladiolus (Tirkey and Singh, 2019).

Yadav *et al.* (2016) treated corms of gladiolus *cv.* Pink Beauty with gamma rays, EMS both alone and in combination including control *i.e.* untreated. The days to corm sprouting delayed after treatment with gamma rays and recorded maximum at 6 kR + 0.2 per cent. Duration of sprouting was decreased by EMS and was observed maximum at 3 kR gamma rays. The gladiolus plants could not survive in gamma rays and its combination with EMS. They concluded that 0.2 to 0.6 per cent EMS doses are suitable for inducing mutation in gladiolus.

Survival per cent of DES treated corms was less compared to gamma rays in variety 'Psittacinus Hybrid'. DES produced larger spikes and with more numder of florets as compared to Gamma rays. Plants were more responsive to treatment of DES as compared to gamma rays (Patel *et al.*, 2018) (Tables 5 and 6).

Table 5: Comparative Effect of Gamma Rays and DES on the Vegetative Characters of Gladiolus Variety Psittacinus Hybrid*

Treatment	*Days to Sprouting*	*Sprouting Per cent*	*Survival Per cent*
5.5 kR γ rays	23.40	10.02 (101)	9.94 (99.24)
6.0 kR γ rays	25.67	9.94 (99.24)	9.94 (99.24)
6.5 kR γ rays	30.67	9.85 (97.52)	9.72 (95.04)
7.0 kR γ rays	24.40	9.85 (97.52)	9.76 (95.76)
7.5 kR γ rays	28.13	9.51 (90.75)	9.51 (90.88)
8.0 kR γ rays	48.27	8.53 (73.21)	8.29 (69.17)
0.1 per cent DES	29.07	10.02 (101)	9.85 (97.52)
0.2 per cent DES	26.93	9.94 (99.24)	9.90 (98.57)
0.3 per cent DES	31.40	9.50 (90.75)	9.82 (96.87)
0.4 per cent DES	29.73	9.49 (90.56)	9.28 (86.56)
0.5 per cent DES	26.20	9.68 (94.26)	9.63 (93.29)
Rest	29.44	9.88	9.60
Control	32.60	10.02(101)	10.02(101)
S.Em.±	1.68	0.20	0.25
C.D. at 5 per cent	4.93	0.57	0.72
C.V per cent	9.81	3.50	4.42

* Patel *et al.* (2018).

Table 6: Comparative Effect of Gamma Rays and DES on Flowering Characters of Gladiolus Variety Psittacinus Hybrid*

Treatment Details	*Days to Flowering*	*Spike Duration (Days)*	*Spike Length (cm)*	*No. of Florets/ Spike*	*Vase Life (Days)*	*Floral Abnormalities (Per cent)*
5.5 kR γ rays	71.07	8.00	65.33	12.07	8.90	0.71 (0.99)
6.0 kR γ rays	77.33	8.67	63.60	11.67	8.38	3.03 (9.66)
6.5 kR γ rays	75.00	7.67	69.60	13.33	9.38	2.25 (5.54)
7.0 kR γ rays	83.20	6.67	58.93	11.40	9.00	0.71 (0.99)
7.5 kR γ rays	101.27	5.67	55.80	9.93	7.50	0.71 (0.99)
8.0 kR γ rays	104.00	5.33	50.73	8.80	5.20	0.71 (0.99)
0.1 per cent DES	112.60	8.67	76.13	16.67	11.26	3.58 (13.32)
0.2 per cent DES	90.53	9.67	85.13	17.73	12.26	3.52 (12.91)
0.3 per cent DES	87.27	7.00	67.53	12.13	9.07	2.57 (7.164
0.4 per cent DES	86.80	7.67	78.00	12.53	9.08	0.71 (0.99)
0.5 per cent DES	88.00	6.67	73.13	15.13	10.08	2.67 (7.63)
Rest	88.82	7.42	67.63	12.85	9.10	1.92
Control	89.63	6.33	69.13	12.27	7.00	0.71(0.99)
S.Em.±	1.53	0.53	1.19	0.43	0.49	0.08
C.D. at 5 per cent	4.49	1.55	3.50	1.26	1.43	0.22
C.V per cent	2.98	12.50	3.05	5.83	9.45	7.21

* Patel *et al.* (2018).

7.2.3 Polyploidy

In order to increase the commercial and horticultural value by improving the ornamental traits of gladiolus cv. 'White Prosperity', polyploidy was induced by soaking gladiolus corms in different colchicine concentrations (0.1 per cent, 0.2 per cent and 0.3 per cent) for 24 h. Colchicine at 0.1 per cent improved the ornamental value of gladiolus by increasing vase life, whereas, colchicine at 0.3 per cent was effective in increasing floret diameter. However, the colchicine treated plants exhibited delayed and reduced percentage of flowering corms. Both pollen and stomata size were increased while stomatal density and pollen fertility was reduced in polyploid plants. Induction of polyploidy (mixoploids + octoploids) was at 0.2 per cent and 0.3 per cent of colchicine were effective for producing large number of polyploid plants. Colchicine at 0.1 per cent and 0.3 per cent concentration has effectively enhanced the gladiolus ornamental value by improving its floral value such as vase life and floret diameter (Manzoor *et al.*, 2018).

7.2.4 Screening for Biotic Stress

Tombolato *et al.* (2002) reported that in Brazil during 1986, the breeding program restarted with one of the objectives to select new varieties tolerant or resistant to rust (*Uromyces transversalis*), two new varieties released 'Carmim' and 'Paranapanema' were very rustic and vigorous plants.

Taj *et al.* (2016) tested eight gladiolus genotypes for resistance to *Fusarium* wilt under pot culture. *Psittacinus* hybrid was resistant to *Fusarium* inoculation with no disease incidence and mortality, whereas Arka Naveen was susceptible with 100 per cent disease incidence. They concluded that the genotypes *Psittacinus* hybrid, Arka Amar and IIHRG-12 are superior gladiolus genotypes having best floral characteristics and resistance against *Fusarium* wilt.

Kumari *et al.* (2019) evaluated 30 gladiolus genotypes for reaction to two virulent strains of *Fusarium, viz.* IARI isolate (*F. oxysporum* f. sp. *gladioli – Fog* isolate 1) and Pantnagar isolate (*F. oxysporum* f. sp. *gladioli - Fog* isolate 2). The genotypes Dhanwantri, Pusa Unnati, Swarnima, and hybrid Smokey Lady × Oscar were found tolerant to both the *Fusarium* isolates. The tolerant genotypes had higher levels of leaf total chlorophylls and total phenols.

Singh and Singh (2013) screened fifty varieties of gladiolus in the field under artificial epiphytotic conditions against *Curvularia trifolii* f.sp. gladioli. Nine varieties, namely SJN, Yellow Stone, Plum Tart, Jacksonville Gold, Pink Elegence, Delhi Pink, Shagun, Arka Kesar and Punjab Flame showed very good level of genetic resistance against the disease (disease reaction: = 1.0). Nineteen varieties were found to be moderately resistant (disease reaction: >1.0–2.0); while, eleven varieties were moderately susceptible (disease reaction: >2.0–3.0) to the disease. Eleven varieties, namely Jester, Chandani, Punjab Dawn, Gunjan, Shobha, Solist, Jessica, Aldebaran, Purple Flora, Overture and Alexander the great were observed to be susceptible to the disease (disease reaction: >3.0–4.0). Among the susceptible varieties, Alexander the great and Overture were found to be the most susceptible to leaf spot, having disease reaction of 4.0. The resistant varieties may be used as such or exploited in the breeding programme for development of suitable varieties.

8.0 Growth and Flowering

Genetic constitution of the plant, growing environment, nutrition, soil moisture, *etc.*, are the important factors influencing the growth and flowering of gladiolus. The role of growth substances in the regulation of growth and flowering is also well known. Influence of nutrients and soil moisture has been dealt separately and the other factors are discussed in this chapter as well.

8.1.0 Genetic Factors

The range of variation in gladiolus is very large. Species and cultivars with different growth habits producing various types of flowers of different shapes, sizes and colours are available. Some produce tall plants bearing large flowers while others medium and small, bearing medium and miniature types of flowers. Also, there are early, intermediate and late cultivars suitable for growing in different periods. All these characters are said to be governed by the genetic factors.

8.2.0 Environmental Factors

Flower induction in gladiolus is autonomous (Halevy, 1972). Inflorescence initiation always follows a definite number of leaves, usually 8-10 (Hartsema, 1937). Further development of flower spikes and also of individual florets are affected by environmental factors, mainly light, temperature and water deficit. Growth of corms and production of cormels are also influenced by environment.

8.2.1 Light

8.2.1.1 Photoperiod

Parker and Borthwick (1951) induced earlier flowering in gladiolus with 12 h light but the spikes were shorter and flowers were fewer as compared to 15 h photoperiod. However, artificial light from 30 minutes to one and half hour at mid-night from the date of sprouting showed the effect similar to that of long photoperiod. Experiments were made on the influence of day length, on blindness in gladiolus using supplementary light and artificial shading. The flower bud apparently differentiates as the stem elongates regardless of day length, but the flower often aborts at various stages before the formation of ovules and pollens, when the day is short, which may be due to low light intensity. Lengthening the day with artificial light in the winter months increased flower production. The response to day length differed with cultivars, Kunderd White being influenced more than Hector by short-day conditions (Kosugi, 1953). The cultivar Spotlight when grown under long (24 h) and short (8 h) photoperiods from October 10 until November 6. The difference in length of spike increased with time and blindness occurred in the 8 h photoperiod 2 weeks later. Anatomical studies showed that in 8 h plant cells at the top of the flower spike began to die 3 weeks after the initiation of treatment, and the death of cells proceeded downwards, resulting in complete death of the flower spike 1 week later (Kosugi, 1959). Early, intermediate and late cultivars of gladiolus were exposed for 45 days after planting to 93 h illumination after sunset with 200 W, 500 W or ultraviolet radiation. All treatments increased plant height but retarded flowering, results being more marked with 500 W (Ghisleni,

1956). Yasuda and Yokoyama (1954) also found earlier flowering and increase in the ratio of blind individuals by limiting day light to 10 to 11 h and suggested that 9-hour day length is insufficient to induce flowering. Long day treatment delayed flowering but improved the quality of the blooms, while exposure to 8-10 h of day light in summer increased the number and weight of cormels (Kosugi *et al.*, 1957). Nitsch (1966) reviewed the effects of photoperiodism on the formation of bulbs in different plants including gladioli. Earlier production of cormel and replacement of corm under short day was reported by Cohat (1974). He found that when corms were stored for 40 weeks in darkness at 25-30°C before planting, long duration storage at higher temperature induced early corm development. Interaction between temperature and day length showed the earliest corm formation in plants raised from corms stored at 30°C and grown under short day. Wassink (1965) observed that formation and/or growth of small new corms was stimulated by a 10 h day length but the greatest increase in size of the replacement corm took place in 18 h day length. The ratio of the breadth-length of leaf (taking 18 h day as equal to 100) showed that very short days (½ hour) produced ratio of 48 compared with 22 in total darkness. In an experiment on artificial illumination and source of light, Raikov and Bulykov (1975) reported the highest percentage of flowering, the earliest and largest spike and increase in flower number, when treated for 14 h per day with artificial light from DKsT-10000 provided with quartz bulbs starting from 30 days after planting as compared to that from xenon arc lamp DKsT-10000 beginning from 15, 30 or 65 days after planting. Further, Shillo and Halevy (1976a, b) reported that shortening the length of day-light period (SD) caused a decrease in either flowering percentage or the number of florets per spike or both depending on the cultivar. SD also prevented production of secondary inflorescence. A few plants flowered when short-day treatment was applied from 1- to 5-leaf stage, number of florets per spike was reduced when shaded from 4-leaf stage to anthesis. Floret number and flowering percentage fell when shaded from 4-leaf stage to spike emergence. The adverse effect of short day appeared to result from a reduction of total solar irradiance. In winter, photoperiodic low light intensity, extension of natural day length (LD) delayed flowering and increased the number of florets, the number of secondary inflorescence and sometimes the flowering percentage also. Plant response to long day treatment was similar at all stages of flower development. In subsequent investigations, Shillo and Halevy (1981) observed that plants grown in winter under natural short day condition (10.5 to 12.0 h light) bloomed earlier (after 127 days) with less number of florets per spike whereas those under 14.5 to 16.0 h light (long day) flowered later (after 142 days) with more florets. In an experiment with two cultivars, Eurovision and Spic and Span, on plant density, season of planting and duration of light, Shillo *et al.* (1981) found the highest percentage of flowering in a population of 27 plants/m^2 under long-day condition (14 to17 h light). These treatments enhanced flower quality and number of florets.

McKay *et al.* (198lb), in an experiment on plant density (150000, 300000 and 450000 corms/ha), arrangement of planting (bed, double and single row) and two day lengths, found that a 24 h photoperiod delayed flowering by about 15 days and increased the number of inflorescence harvested from low, medium and high density treatment by 20, 91 and 169 per cent, respectively, compared with inflorescence yield under natural day length (12.3-14.5 h). However, weight of new corms, cormlet weight per plot and per corm were reduced with a 24 h photoperiod by 32, 71 and 63 per cent, respectively. The yield of inflorescence, corms and cormlets and inflorescence quality of winter gladioli (cvs. Prof. Goudrian, Oscar, Bali Hai) were studied under varying levels of illumination (IL) of 12 h day length extension ranging from 0-333 lux. Inflorescence yield was increased by increasing the day length extension IL. Ninety-seven per cent of the maximum inflorescence yield was obtained at a day length extension illuminance of 144 lux, while 97 per cent of the maximum number of florets/spike and other flower quality characteristics were obtained at a day length extension IL of 45 lux. There was no clear relationship between the day length extension IL and number and weight of new corms (McKay *et al.*, 1982).

Effects of day length of 10 and 14-16 h, at different developmental stages on cormel formation, were studied by Asahira *et al.* (1968). They found that short day treatment of 10 h after flowering increased the yield of cormels in cvs. Spotlight and Snow Princess.

Zhang and Gao (1992) reported that under short day (8 h), the flowers of cv. Peter Pears were all blasted but, the development of new corms and root growth increased; *cv.* Rose Charm suffered seriously due to abortion of flower buds and the vegetative growth was markedly inhibited. Long days (>14 h) generally increased stalk and spike elongation and the number of florets per spike.

Hong and Kim (1988) reported that corms produced under short days and cool conditions in a greenhouse produced larger amounts of GA-like substances during the first 40 days of postharvest storage sprouted earlier and flowering percentage was higher in comparison with those obtained from open cultivation.

Feng and Zhang (2003) reported that plants in short day (SD) treatment flowered 100.6 days after planting (DAP) which was significantly earlier than plants in long day (LD) (105.1 DAP) and NI (107.0 DAP). SD treatment resulted in the highest dry matter (DM) and mineral contents in spikes, while, LD treatment resulted in the highest flower, longest spike and the most florets per spike in *cv.* Hunting Song.

Lopéz *et al.* (2006) found that long days had significant effect on flower development and flower quality, resulting in the largest flower, longest spike and the most florets per spike (Table 7).

Table 7: Length of the Floral Spike and the Number of Flower Buds in Plants Treated with different Photoperiod*

Day Length (DL)	*Floral Spike (cm)*		*Number of Flower Buds*	
	1st Sprout	*2nd Sprout*	*1st Sprout*	*2nd Sprout*
SD	18.3 A	12.1 A	4.9	3.6
LD	21.3 B	15.6 B	5.2	3.9
Cold (C)				
0 Weeks	20.4 b	16.8	5.3 b	4.1 b
2 Weeks	21.2 b	15.7	5.3 b	4.3 b

Day Length (DL)	Floral Spike (cm)		Number of Flower Buds	
	1st Sprout	2nd Sprout	1st Sprout	2nd Sprout
3 Weeks	22.9b	13.8	5.5 b	3.7 ab
4 Weeks	20.1 b	13.4	5.2 b	3.8 ab
5 Weeks	16.2 a	12.4	4.0 a	3.2 a
Interac.DL x C	(*)(**)	(ns)	(*)(**)	(ns)

Note: Different capital letters point to significant differences between lightening conditions and small letters show significant differences between the treatments acording to LSD ($p<0.05$). *, ** and 'ns' refers to a level of significance of 0.05, 0.01 and not significant, according to two-way ANOVA.

* Lopéz *et al.* (2006).

8.2.1.2 Light Intensity

Gladioli prefer high light intensities but very high intensities without provision for temperature control adversely affect growth, although low light intensity is the main cause for failure in flowering of gladioli. Low light intensity during winter lead to serious flower bud abortion during greenhouse forcing of gladiolus. Gladiolus plants were grown under four light intensities, *viz.*, 12, 37, 75 and 100 per cent of natural day light. Stem diameter, the number of internodes, the number of vascular bundles and the average length and width of bundles increased markedly with the increase in light intensity (Wassink, 1960; 1963). It also influenced the leaf morphology, leaf area ratio and assimilation rate. In 1965, Wassink found that dry weight of aerial parts of gladiolus at flowering time was nearly 90 per cent of the total. Vegetative and floral growths during flowering season were markedly dependent on photosynthesis. Leaf shape was influenced by light intensity; leaf width increased in relation to length as the light became stronger. He observed increase in total dry weight and stem weight with increase in light intensity, and the leaf area reached its maximum in 37 per cent day light as did stem length, but the highest light intensity produced the shortest stems. Jacoby and Halevy (1970) foundtable 6 that for induction of contractile roots thermal fluctuations and illuminations were necessary and suggested that light stimulus was perceived by leaves and not by the corms. They also reported that high light intensity was presumably necessary not simply for photosynthesis but might be necessary for the translocation of required factors from the leaves to the roots. In the Netherlands, in 1975, the solar radiation was unusually brighter in February and March and had very low light intensity in December and January. There was 14 per cent flower bud blasting with close planting on 1 January whereas no blasting was noticed on those planted on 1 February. With wide spacing (50 rather than 75 corms/m^2) there was no blasting in either case (Groen and Kappelhof, 1975). Total solar radiation is an important factor affecting winter flowering as low light intensity in winter affects flower development directly, whereas short day has a dual effect : one, reducing the total daily solar radiation and the other photoperiodic effect of reducing the time from planting to flowering. Plants at the 4th to 6th leaf stages are most sensitive to prevailing light conditions (Shillo and Halevy, 1975). In another set of experiments, Shillo and Halevy (1976b) provided 20 and 25 per cent shading by hessian screens at different stages of plant growth in the field. They observed reduction in flowering percentage and number of florets when shaded from 4- to 6-leaf stage, decrease in the flowering percentage only when shade was provided from corm sprouting to 4-leaf stage and a reduction in floret number by shading at later stages. Pre-planting treatment of the base of cormels with intermittent pulses of light, stimulated shoot growth and when the light rays were directed on one cormel only, the stimulus was transmitted on to 9 adjacent cormels. Hot-water (35°C) treatment of corm buds resulted in increased cormel size, the increase being greater or smaller (7 mm) than larger (9 mm) buds. Smaller corm buds (5-7 mm) also responded better to light treatment than larger buds (9-13 mm) by producing cormels of larger size (Kukushkin, 1983).

Talia *et al.* (1988) exposed the gladiolus plants to light intensities of 2000 or 4000 lux from 16 h to 22 h in addition to normal day light. This extension of day length increased the production of marketable spikes and improved flower quality compared to the unlighted control. Supplementary light of 5000 lux for 12 h reduced the number of blasted plants and enhanced the number of marketable cut flowers. The growth stage from third leaf to anthesis was most sensitive to light intensity (Zhang, 1995).

8.2.2 Temperature

With cultivars Spotlight and Hector, Iwama and Iwai (1952) found that plant growth was markedly affected by temperature and the plants grown in the warmer regions made better growth. The plants grown from larger cormels formed flower buds earlier than those grown from smaller cormels. In an experiment on 68 cultivars, Kosugi and Sano (1962) found no relationship between the date of flowering and the percentage of corm flowering. Although corms grown at a higher temperature flowered about 1 month earlier than those grown at a lower temperature, no effect of temperature on the percentage of corm flowering was recorded. Cocozza (1971) observed, with 5 cultivars planted between January and early March under glass, flowering occurred a week earlier with ventilation and a temperature range of 15-25°C than with constant relative humidity of 15 per cent and minimum temperature of 15°C. A further delay of 3 days occurred in a glasshouse kept at 10° and 20°C for alternate 12-h period and flowering was delayed by 19 days where 3°C was the minimum temperature. Reduction of day length to 8 hours adversely affected yield under glass, not in the open. In Queensland, McKay *et al.* (1981a) planted corms of 12 cultivars at a 3- or 4-week interval from February to September. Days to flowering varied from 77 to 132 for various cultivars and planting date combination. Corms planted on 28 May took longest time to flower. Temperature was the major factor influencing the number of days to flowering. Yield and quality of inflorescence from February to early May plantings which flowered in winter were lower than from late planting. Shillo and Halevy (1963) exposed gladiolus plants in pots to low night or day temperature of 2°C at various stages of development and reported that 2- and 5 to 6-leaf stages were most sensitive to low night temperature, which reduced the flowering percentage at both stages and the number of florets per spike at 2-leaf stage. In another experiment with low or high temperature under various conditions of day length and light intensity, Shillo and Halevy (1976c) observed increased occurrence of flower blasting under low light intensity in winter at low temperature of 1-4°C. The chilling damage was found at 2 stages, immediately after planting and at the 7th leaf stage when spike emergence began. The plants were tolerant to high temperatures (up to 50°C) so long the air humidity and soil moisture level were optimum.

Variations in reponse to temperature have been reported in cultivars. In Spain, the flowering percentage of gladiolus corms which are planted outdoors in the summer is low for most cultivars, due to mainly very high temperatures during early growth. In a trial of 43 cultivars sent from the Netherlands to Barcelona, the flowering percentage was above 70 per cent in only 9 cultivars (Advance, Carqueiranne, Fond Memory, Monte Negro, Peter Pears, Priscilla, Semarang, Serafijn and Stardust). The flower quality was good or better than average in Carqueiranne, Fond Memory, Monte Negro, Peter Pears and Priscilla; it was also atisfactory in 8 of the cultivars with a low flowering percentage. The corms used were 12-14 cm in circumference and had been stored at 5°C before despatch to Barcelona. They were planted on ridges, at a density of 40 corms/m^2 and at a depth of 6 cm (Buschman and Groen, 1989).

De Donato (1968) observed earlier flowering in both early and medium-late cultivars in heated beds (13.2-19°C) than control (11.3-16.6°C). However, with the latter cultivars soil warming resulted in the production of shorter spikes with rather fewer flowers. With 3 *cvs.* Hunting Song, Peter Pears and Jessica planted under glass, provided with soil heating at 17.4°C and a reduced air temperature of 12°C flowering was more advanced than those grown at 15°C without soil heating. With soil heating at 20.4 and 15°C air temperatures growth was equal to that with 19°C (air) (Groen and Schipper, 1982a). Shoots emerged 3 weeks earlier when plants were grown in soil heated to 70°F (21.1°C) for 9 h/day rather than in unheated (minimum 50°F) (White and Kling, 1991).

In an experiment conducted for three years with 18 gladiolus cultivars at two places in Russia, Tamberg (1978) observed late sprouting but earlier flowering at Krasnodar (latitude 45°N) and in Leningrad (59° 44′ N), and explained that early cultivars had a temperature sum requirement of 1500-1700°C and late ones 2000-2500°C. Flowering stems were longer and number of florets greater at Krasnodar but flower size was similar at 2 locations.

Atom, Hector, Scarlet Princess and Spic and Span proved relatively insensitive to low temperatures and reduced light intensity and duration. Birch Red and the autumn flowering cultivar Nissen were most sensitive to low temperatures and reduced light. Some cultivars were more affected by reduced light than by low temperature but others were more sensitive to temperature changes (Kosugi and Surnimoto, 1956).

Low air temperatures (15 to 20°C) was best for pollen tube growth, fertility, and fruit set in the cross between *G.* × *grandiflora* and *G. tristis* (Takatsu *et al.*, 2001).

The cultivars Snow Princess and American Beauty raised in Phytotron with temperatures treatments 20/18, 26/22 or 30/26 °C (Day/Night) with 14 h light and 60–65 per cent relative humidity. Results showed that 30/26 °C temperature was found with lowest days for corm sprouting. There was no effect of temperature on sprouting percentage of corms, leaf number and number of florets per spike. The taller spike and rachis length was found with 26/22°C in *cv.* American Beauty. Flowering duration reduced with the temperature increase. However, the effect of highest temperature of 30/26°C was more adverse and significant for various vegetative and reproductive growth parameters. Increase in temperature concomitantly decreased flower initiation, hastened sprouting of corms and increased plant height (Kadam *et al.*, 2013).

Muttoni *et al.* (2017) reported that the cardinal temperatures for the gladiolus planting-emergence phase are 5°C, 22.5°C and 35°C for base, optimum and maximum temperature, respectively (Table 8).

Table 8: Duration of the Planting-Emergence Phase (PL-EM) in Gladiolus (*Gladiolus* x *grandiflorus* Hort.) Cultivar 'Amsterdam' under different Temperature with their Respective Normalized Emergence Rates*

Temperature (°C)	*PL-EM (days)*	*Normalized Emergency Rate*
5	76	0
7	64	0.2031
10	30	0.4329
13	20	0.6500
16	18	0.7228
18	15	0.8667
20	14	0.9286
22	13	1.0000
25	13	1.0000
30	15	0.8667
33	19	0.6842
35	30	0

*Muttoni et al. (2017)

8.2.2.1 Effect of Corm Storage Temperature

In Italy, Cocozza Talia and Lucia (1994) observed that storing cormlets at 5°C for 5 months after harvesting, advanced corm development and led to increased yields of large corms. Gonzalez *et al.* (1998) found that corms stored for 3 or 6 weeks at 5°C and 90 per cent RH flowered 20 and 11 days before the control, respectively, while the duration of the flowering period increased by 35 days after 6 weeks of storage. The vegetative and quality parameters of the flowering stem and spike length increased after 6 weeks of cold treatment.

The storage temperature of May-harvested corms of *Gladiolus grandiflorus* when reduced from 33 to 5°C caused severe reduction in flowering whereas similar treatment given to July to September-harvested corms did not show any adverse effect on flowering. A 10 per cent reduction in water content of summer-harvested corms by drying immediately after lifting had no deleterious effect (Shillo and Simchon, 1973).

Corms of *cv.* Schnee Prinzess were stored at 9-15°C (control) or kept for 35 days at 20, 25 or 30°C before planting; over 80 per cent early flowering was obtained from plants whose corms were kept at 25°C for 35 days. The control plants gave only 15.8 per cent early flowering (Lemini and Lemini, 1981). Morphologically developed and undeveloped cormels (with more than 3 and less than 3-leaf primordia in the terminal bud, respectively) of 2 cultivars were stored at 8 or 20-22°C. A high storage temperature stimulated terminal bud development in the morphologically developed cormels whereas a low temperature retarded it. Converse temperature responses were observed in the undeveloped cormels. Cormels of an early flowering *cv.* Leif Fleim were more sensitive to storage temperatures than those of a medium-late *cv.* Oscar, and within the cultivars small cormels were more sensitive than large cormels (Kukushkin, 1983).

The corms subjected to a temperature of –6.6°C (20°F) for 8 h grew well after storage but were weak or did not grow when heated for 16 and 24 h. Growth and flowering were normal when corms were cured at 43.3°C (110°F) for 24 h (Forsberg, 1952). Paulin (1958) recorded early flowering in Snow Princess from the corms stored at 28°C for 60 days as compared to storage at 34 and 10°C. Storage of corms at 21, 26.5 and 32.5°C for 15 and 30 days prior to planting enhanced sprouting, the best results in respect of yield of flowers and corms were obtained from those stored at 26.5°C for 15 days (Jenkins, 1963). Storage of corms at 10°C until four weeks before planting followed by storage at 20°C resulted in higher total weight of corms and more large corms as compared to 10°C throughout the storage period (Groen *et al.*, 1976). From the results of the experiments, Stienstra (1976b) concluded that storing gladiolus cormels (planting stock) at a temperature higher than normal for a few weeks before planting had a beneficial effect on corm growth and storage at 20°C appeared to be better than at 17-18°C. Cormels held at a constant 10°C and then transferred to 20°C for four weeks before planting showed a more beneficial effect on corm growth.

Aoba (1975) treated the corms of cvs. Red Radiance, West Point and Hector at 10, 20 or 25°C for three to seven months, planted in sand vermiculite mixture and were grown in darkness around 15°C. He found enhanced formation of daughter corms by pre-treatment at 10°C. When Red Radiance was pre-treated at 20°C for five months and planted in darkness at 5, 10, 15 or 25°C, he recorded enhanced daughter corm formation at 15 and 25°C, delay at 10°C and no new corm formation at 5°C. In another experiment, corms of several gladiolus species were pre-treated at 5, 10 or 25°C for 90 days. In *G.* x *tubergenii* cv. Charming Beauty chilled at 5° and 10°C new corm was formed directly above the old one. No appreciable effect of chilling on corm formation was observed in *G.* x *colvillei*, *G.* x *nanus* and *G.* x *ramosus*. Storage of corms at 18 and 20°C during the last four weeks before planting, increased subsequent gladiolus corm yield considerably; prolonged high temperature treatment upto 8 weeks had no harmful effect and was probably beneficial (Kruyer and Groen, 1977).

Zalewska and Antkowiak (2009) stored corms of three varieties at four groups of storage temperature and duration *viz.*, group I (12-week storage at 17°C), II (12-week storage at 5°C), III (6-week storage at 17°C and then 6-week storage at 5°C) as well as IV (exposed to preparation treatment at 22-23°C, stored for 4 weeks at

17°C) and grown in glasshouse. It was found that maintaining temperature of 17°C during corm storage was most favorable, as all cultivars flowered at similar dates as those from the corms additionally exposed to preparation. Flowering occurred three weeks earlier than the flowering from the corms stored at 5°C. Irrespective of the cultivar, in the group of corms stored throughout the storage period at 5°C or at 17°C, there were reported shorter stems than those from the corms stored at 17°C and then prepared. The corm storage temperature did not affect the spike length, the number of florets and foliage leaves or the stem diameter.

Corms of six varieties stored at 8°C in a potato store house, at 12°C in a fruit store house and at 27±3°C in a farmer's store house showed that increase in storage temperature concomitantly declined corm weight losses during storage, delayed germination and flower initiation, declined in germination percentage, reduced root, shoot and spike length, number of flowers, corm weight after harvest and number and biomass of cormlets. However, the effect of highest storage temperature of 27±3°C was more adverse for germination and various vegetative and reproductive growth parameters as compared to lower temperatures. Thus, for better germination, plant vegetative and reproductive growth, corms of different varieties of gladiolus should not be stored at 8°C (Riaz *et al.*, 2009).

Corms of varieties Priscilla and Semarang stored in four groups of duration and storage temperature *viz.*, groups I (twelve-week storage at 5°C), II (twelve-week storage at 17°C), III (four-week storage at 5°C and eight-week storage at 17°C) and IV (storage for four weeks at 5°C followed by exposure to preparation treatment for eight weeks at a temperature of 22–23°C). The reaction to corm-storage temperature is typically cultivar-specific. More plants flowered from the group of corms stored only at 17°C than from those in the groups of corms cooled at 5°C in group I and at 5°C and 17°C for four and eight weeks, respectively in group III. Temperature of 17°C, when compared with storage at 5°C accelerated *Gladiolus* flowering to a similar extent (by more than three weeks) as exposing them to preparation treatment (Wony and Antkowiak, 2013) (Table 9).

Table 9: Length of Inflorescence (cm) and the Number of Florets per Spike Depending on the Corm Storage Temperature (A) and the Cultivar (B)*

Cultivar (B)	*Corm-Storage Temperature (A)*									
	5°C	*17°C*	*5/17°C*	*5/22-23°C*	X_B	*5°C*	*17°C*	*5/17°C*	*5/22-23°C*	X_B
	Inflorescence									
	Length					*Number of Florets*				
Priscilla	18.8c	25.3a	23.5ab	23.5a	22.8a	8.2bc	9.4b	7.6c	8.3bc	8.4b*
Semarang	24.3b	16.0c	24.0b	19.8bc	21.0b	10.8c	7.4c	8.9b	9.0b	9.0a
X_A	21.5a	20.7a	23.7a	21.7		8.4b	9.5b	8.4b	8.2ab	–

* Values marked with the same letters for a given cultivar do not differ significantly at α =0.05.

* Wony and Antkowiak (2013).

8.2.3 Interaction of Factors

8.2.3.1 Light and Temperature

Gladiolus cv. Traveler plants were exposed to low light intensity (75 per cent shade) or low temperature conditions (12°/2°C day/night) at various developmental stages. Plants were most sensitive to low light inensity at the 4- to 5-leaf stage. Low light intensity during this period resulted in a decrease in the flowering percentage and the number of florets per flower spike. Low light intensity given at the 6- to 7-leaf stage reduced only the number of florets per spike. The developmental stage, most sensitive to low temperature, was the 6-leaf stage and, at this stage, low temperature reduced flowering percentage under low irradiance. Artificial long day (16 h) lighting was very effective in preventing a decrease in flowering percentage and the number of florets per spike produced (Imanishi and Imae, 1990).

8.3.0 Growth Regulators

Growth regulators have been found to influence the growth and flowering of gladiolus. In an experiment on *cv.* Sylvia, Dua *et al.* (1984) found improved flower quality and better corm multiplication when the plants were sprayed thrice with 100 ppm of GA_3. Arora *et al.* (1992) observed that cormel weight and diameter increased by 239.4 per cent and 59.1 per cent respectively in *cv.* Mayur when treated with 100 mg GA_3/l. Bala *et al.* (1986) observed that for corolla growth on intact spikes endogenous gibberellins from the stamens were necessary until the separation of the outer bract and at the early stages of corolla growth, and exogenous application of GA_3 replaced the promoting effect of stamens.

Treatment with GA_3 increased the weight of cormels and IAA that of corms (Winkler, 1969). In a trial on the effect of soaking of corms in solutions of IAA, IBA, NAA, GA_3 and kinetin, each at 100, 500, 1000 and 2000 mg/l, Jonecki (1979) recorded inhibition of early growth with all the chemicals except GA_3. He (1980) also reported that soaking in GA_3 and IAA solutions hastened differentiation of floral primordia but kinetin retarded corm sprouting and shoot apex differentiation. Awad and Hamied (1985), working with GA_3 (10-500 ppm), kinetin (1-50 ppm) ethephon (1-100 ppm) and gamma irradiation (23-415 R), recorded differentiation of individual flowers and longer flower spikes by treatment with low concentrations. Mohanty *et al.* (1994) observed that GA_3 increased plant height at a concentration of 250 ppm and colour break in the basal florets occurred significantly earlier than the control. A single foliar spray of GA_3 (100 and 200 ppm) enhanced vegetative growth, flowering and number of corms and cormels produced but, adversely affected individual corm weight (Misra *et al.*, 1993). Ravidas *et al.* (1992) observed the greatest plant growth, earliest flowering, largest spike, largest rachis and highest number of florets/spike with GA_3 treatments (50-100 ppm). In a field trial in Kanpur, U.P., Prakash and Jha (1998) observed that GA_3 treatment at 150 ppm improved all the floral traits (time of flowering, inflorescence length, spike length, floret length and number of florets/spike) in gladiolus, *cv.* Friendship. The longest inflorescence and spikes with the highest number of florets/spike was produced with 150 ppm GA_3. In another experiment, 20 ppm GA_3 gave the greatest spike length while 40 ppm GA_3 produced spikes having the longest (16.2 days) life in the field (Pal and

Chowdhury, 1998). Soaking of corms at 100 ppm GA_3 for one hour shortened the time from planting to harvest and increased flower number per unit area, the length of flower stem and spikes, the number of flowers per spike and the diameter of flower stems (Karaguzel *et al.*, 1999).

Kumar *et al.* (2008) reported that GA_3 500 ppm induce early flowering. Ramos-Garcýa *et al.* (2009) dipped corms of var. Blanca Borrego in Biorend 1 at 1.5 per cent before planting resulted in accelerated corm emergence in about 4 days, the number of flowers increased by 2–7 and the vase life extended for 3 days. The number of cormlets was also duplicated.

GA_3 (100, 200 and 300 ppm) was found most effective for enhancing vegetative growth, earliest sprouting, number of leaves, initiation of spike, spike length, rachis length, corm diameter and corm weight (Kumar *et al.*, 2010; Kumar and Misra, 2012; Dogra *et al.*, 2012; Parween *et al.*, 2019).

Khan *et al.* (2013) found that GA_3 (100 ppm) and BA (50 ppm) was found most pronounced on dormancy breaking. Corms treated with GA_3 (75 ppm) and stored for 90 days produced the maximum percentage of spikes, whereas, 90 days stored corms treated with BA (125 ppm) produced the highest number of plants and corms/hill. However, corms treated with GA_3 (100 ppm) and stored for 90 days produced the heaviest and largest corms.

Jadhav *et al.* (2015) observed that pre-harvest foliar spray of GA_3 50 mg/l followed by GA_3 25 mg/l was found most significant in influencing number of spikes per plant and per meter square, number of sprouts per corm, corm and cormel yield. Spike length and vase life were highest with spary of GA_3 50 mg/l followed by GA_3 25 mg/l. Enriched banana pseudostem sap at 2 per cent improved vegetative and flowering parameters.

Baskaran and Abirami (2016) observed early corm sprouting, early spike emergence and maximum duration of spike with GA_3 (500 ppm). While, maximum spike length, rachis length, number of florets per spike, weight of single corm, weight of corms/plant, size of single corm and volume of single corm were recorded by GA_3 (750 ppm).

Foliar spray of GA_3 (150, 200 ppm) produced maximum plant height, leaves per plant, leaf area, spike length, rachis length, spikes per plant, diameter of first floret, number of florets/spike, vase life, earlrier spike emergence and first floret opening, number of corms and cormels per plant (Chaudhary *et al.*, 2016; Sable *et al.*, 2016).

However, early spike emergence and maximum duration of the spike with GA_3 (500 ppm), while, maximum spike length, rachis length, number of florets/spike, floret size and corm weight, corm size, its volume and propagation coefficient was recorded with GA_3 (750 ppm) (Baskaran *et al.*, 2017).

Rahman *et al.* (2020) reported that Gibberellic acid at 200 mg/l recorded maximum sprouting, number of leaves/plant, daughter corms per plant, less number of days to corm sprouting, days to spike emergence and flower opening (Table 10).

Table 10: Effect of Gibberellic Acid (GA_3) on Corm Sprouting, Spike Emergence, Flower Opening, Corm Diameter, Daughter Corms*

Cultivar	*Corm Sprouting (Days)*	*Spike Emergence (Days)*	*Flower Opening (Days)*	*Corm Diameter (cm)*	*Daughter Corms (Nos.)*
White Prosperity	20.2	55.3 a	76.1 a	2.8 b	2.7 b
Red Advance	20.4	53.8 b	74.5 b	3.1 a	3.0 a
F	0.16	32.43	73.24	4.75	6.44
Gibberellic acid					
0	22.8 a	60.5 a	79.2 a	2.47 c	2.4 c
50	21.5 ab	57.0 b	76.6 b	2.84 b	2.5 bc
100	20.1 bc	54.3 c	75.4 c	2.85 b	2.8 b
150	18.8 cd	52.2 d	73.9 d	3.37 a	3.3 a
200	17.9 d	60.5 a	72.7 e	3.25 a	3.5 a
F	11.86	215.98	77.76	16.39	17.5 g
C X G					
F	0.15	1.07	1.62	0.21	0.80
CV per cent	6.89	1.36	0.92	7.36	10.41

Means followed by the different letter are significantly different at 5 per cent level of significance by LSD test.

ns: Non-significant and *: Significant at 5 per cent level of significance.

* Rahman *et al.* (2020).

Corm treatment with BA at 50, 75, 100 ppm recorded more number of shoots per corm, maximum corms, cormels per plant and weight of cormels/plant (Kumar *et al.*, 2010; Baskaran and Abirami, 2016; Baskaran *et al.*, 2017).

Kumar *et al.* (2010) soaked corms with Panchgavya (5 per cent) for 24 hours before planting, recorded maximum leaves/plant, florets/spike, spike diameter, rachis length, corm weight, cormels/plant and propagation coefficient, while, Panchgavya (10 per cent) recorded maximum plant height, leaf breadth, spike length, length of first internode and corm diameter.

Soil drenching with 0.8 per cent CCC increased stem length and number of flowers per spike and slightly advanced the date of anthesis (Halevy and Shillo, 1970). The effect of GA_3 and CCC on growth and assimilate translocation by labelling a single leaf with $^{14}CO_2$ was studied by Ginzburg (1974). He recorded that GA_3 stimulated assimilate movement towards the inflorescence at the expense of the corm, CCC had a similar but smaller effect. However, after the growth of the inflorescence translocation from the labelled leaf was promoted by CCC but inhibited by GA_3. In an experiment on cv. Friendship, Bhattacharjee (1984) observed increased vegetative growth, cormel production and improvement in flowering by using GA_3 at 10 and 100 ppm. Application of SADH, CCC at 1000, 2500 and 5000 ppm each as soil drench and ethrel at 500, 1000 and 2000 ppm increased flower size. Zalewska (1997) studied the effects of B-Nine (daminozide) and Topflor (flurprimidol) on

10 gladiolus cultivars grown from corms planted in a heated greenhouse. The growth retardants reduced the plant height by about 4-9 per cent depending on the treatment. The most cultivars, the growth retardants reduced the percentage of plants which flowered.

Treatment with ethephon increased percentage of sprouting, reduced days to sprouting, and increased fresh weight of corm (Goo *et al.*, 1999). Ethrel at 100 ppm significantly increased leaf area, induced early appearance of flower spike, highest number of florets/spike and largest individual florets (Pal and Choudhury, 1998). They also reported that BA induced early sprouting of corms. Wang *et al.* (1998) reported that methyl jasmonate (30, 60 or 120 mg/l) treatments decreased length of flower stalk and the concentration of chlorophyll in leaves and increased the fresh and dry weight of leaves of gladiolus *cv.* Hunting Song.

Cormels (300 mg) of cultivars Happy End and Apricot were soaked for 60 h in solu-tions of bayleton [triadimefon] (10-5 M), GA_3 (10-4), ABA (10-4), kinetin (10-4), ethrel (10-4) or water, dried for 2 h and then planted in wooden boxes filled with a soil/manure (3:1) mixture. Growth regulators induced early and vigorous sprouting. Bayleton was more effective for Happy End and GA_3 and ABA for Apricot. Growth regulator treatments had positive effects on leaf parameters (length, width and num-ber). Spike appearance only occurred in Happy End. All treatments except ethrel advanced flowering in this cultivar and bayleton doubled the number of spike compared with the control (Ogale *et al.*, 2000).

8.4.0 Growth of Corms

Apart from planting time, corm size, depth of planting and environmental and nutri-tional factors, growth of corms was influenced by leaf number and time of harvest of cut flower and corms.

Mukhopadhyay and Das (1978) studied the effects of removal of flower and foliage on the yield of gladiolus corms and cormels and concluded that removal of spikes at early stages resulted in the increase of corm weight, whereas flower spikes removed along with two leaves had adverse effect. They also opined that on removing the flower buds just before opening but retaining the rachis, higher corm weight was noted, thereby suggesting rachis as a supplementary food manufacturing organ. However, there was no effect on weight or yield of cormel.

In trials with *G. gandavensis cv.* True Love, Hong *et al.* (1989) found that the diameter and weight of daughter corms, number and weight of cormels, and the number of grade 4 cormels increased as the number of leaves left after cutting the flowers increased from zero to all leaves. The minimum number of leaves for new corm and cormel production and flowering in the following years was 4. If only the daughter corm was to be used in the following year, corms could be lifted at 30 days after flowering but the optimum lifting date for corm and cormel production and flowering was 60 days after flowering.

In trials on cultivar Peter Pears, Eurovision, White Friendship, Gold Field, Nicole, Deciso, Red Beauty, Traderhorn and Victor Borg, minicorms (3-4 cm in circumference) were planted at 100 corms/m^2. The percentage of large corms suitable for forcing for cut flower production was greater when corms were harvested on 12 October (Cocozza *et al.*, 1992).

Cultivars Nova Lux, Mascani, White Prosperity and Piricilla, were harvested when the leaves were green, beginning to yellow or when they were drying. Korkut *et al.* (1998) reported the second and third harvests gave the highest number of cormlets, but cormlet diameter and cormlet weight decreased from the first to the third harvest. The cutivar Mascani gave the greatest corm girth and weight. Suh and Kwack (1990) reported that when large cormels were used with early planting dates and late harves-ting dates, formation of good quality corms was promoted.

9.0 Propagation

Gladiolus is propagated both sexually by seed and asexually by corms and cormels. Division of big corms for multiplication, although not very common, is also used. The differentiation of organs and regeneration of plants through tissue culture technique is very effective to get virus-free material and for rapid multiplication.

9.1.0 Seed Propagation

Seed propagation is followed to evolve new cultivars. Gladioli set seeds abundantly and about 60 seeds are normally formed in three chambers of each capsule. The plants raised from seeds require four seasons to come to bloom under ordinary conditions and under best cultural treatments it may be reduced to two seasons (Misra, 1978).

The capsules when start turning brown, *i.e.*, about four to six weeks after flowering are ready for dehiscence and collection. For most of the cultivars the time for collection of capsules is between August and October in the hills and January and March in the plains of India. The winged seeds extracted from the capsules are sown in boxes, pots or directly on raised beds thinly and at 1 cm depth during February-March in the hills and November-December in the plains. The soil should be porous and rich in organic matter. While sowing in boxes, the seed compost should contain 3 parts loamy soil, 2 parts well decayed sterilized leaf mould and small amount of bone meal and some sand. After sowing, the containers are kept in greenhouse at a temperature between 7 and 13°C and seeds germinate in 3 – 5 weeks. Carpenter *et al.* (1991) reported that a constant temperature of 20°C promoted higher total germination (97 per cent) and optimum seed storage condition at 14°C and 26 per cent RH (Table 11).

Seeds on germination strike a single root down and send up single grass-like blade above the ground. When this root is well established it starts contracting and takes the plant down into the soil. The second leaf may appear after 3 weeks. After the seedlings have attained a good growth (in about 2-3 months), these are shifted from greenhouse to a sheltered position at 24°C to grow for whole of the season (Misra, 1978). The seedlings raised in open need not be shifted and may be grown throughout the cycle. The soil should be kept moist and during summer the seedlings are to be fed with liquid manure to help in developing the corms. The moisture in seed bed could be maintained by using some mulch.

Table 11: Germination of Gladiolus 'Dr. Magie' Seed under Alternating Temperature*

Temperature Cycle[z] (°C)	*Germination*		
	G[y]	*T50[x]*	*T90-T10[w]*
15-20	48	7.1	7.3
15-25	81	5.4	5.6
15-30	57	22.2	11.4
15-35	27	27.8	30.9
20-25	96	4.0	4.6
20-30	85	6.9	7.9
20-35	51	11.3	11.7
25-30	96	7.9	9.8
25-35	22	13.9	14.4
Tukey's HSD, 5 per cent	7	2.6	3.4

z: Temperature alternated every 12 h. y: Per cent absolute final germination; x: Days from 10 per cent to 90 per cent relative germination.

*Carpenter *et al.* (1991)

When the foliage starts withering, the cormels are lifted, washed and stored in cool place which are planted 20 cm apart in the following season in the usual time of planting. Some of these cormels may produce flowers in the same year (*i.e.*, in the next season of seed sowing), while the others the next or third year (Misra and Choudhury, 1977). Sometimes, *Curvularia* infection is noticed even when the seeds are sown in sterilized media. The tip of the leaves of seedlings show browning and damping off of the seedlings can be controlled by spraying of 0.2 prt cent dithane M-45 at fortnightly intervals after germination or sprouting (Misra, 1978).

9.2.0 Vegetative Propagation

9.2.1 Corms and Cormels

Corm is a thickened underground perennating structure consisting of a short vertical stem having many ring-like nodes full of buds, covered with tunics all round, and a disc-like root zone at the base. Corm develops at the base of flowering stem (pseudostem) when the old corm is gradually used up. Cormels, produced in clusters on stolons between mother and daughter corms, are important sources for maximizing the multiplication. These vary in sizes, each one enclosed in a hard shell which is less permeable to water and thus prolongs the dormancy. In certain cultivars, tunics on cormels are found cracked at harvesting or during the process of curing, but in certain others the tunics are intact throughout and are impermeable to water. Cormels are selected carefully to prevent spreading of disease. Hot-water treatment or water soaking for four to seven days facilitates early and uniform sprouting (Forsberg, 1961). Some fungicides are found to complement the action. Cormels are dipped in 0.1 per cent benomyl plus 0.18 per cent captan for 30 minutes in warm water at 53-55°C, for 10 minutes in cold running water and then air dried and stored at 2-4°C until planting (Magie, 1971; 1975). Dormancy of larger cormels is usually broken in 4 months of treatment. Swelling of root at the base indicates that cormels are ready for planting. The cormels are planted in fumigated and fertilized soil of pH 5.8-6.5. It should be well drained. Most cormels are planted at a depth of 5 cm and in rows 10 to 13 cm wide furrows, spaced 60-75 cm apart. About 130 large cormels are planted per metre of row. The soil is initially kept moist to facilitate uniform germination. Young gladiolus cormels required 2-3 seasons of vegetative growth before flowering could be induced (Ogale *et al.*, 1995).

Fertilizers should be used every month to obtain higher yield within 5-6 months after planting (Waters, 1965). About 100 corms larger than 1.3 cm in diameter can be obtained per metre of row when larger cormels are planted and cultivated properly. Corms from 1.3 to 2.5 cm in diameter are used for the production of flowering size corms.

In case of planting stock, the temperature for the treatment in fungicide suspension should be about 46°C and the duration of treatment is up to 15 minutes (Magie, 1975). Small corms (<2.5 cm diameter) are planted in one or two rows per bed at depth of 6-8 cm and the number of corms per metre of row varies from 50-80, depending upon corm size. Adequate nutrition and moisture should be provided to promote growth and corm production. Removal of flower spikes improves corm size. About 430000 or more flowering size corms (>2.5 cm diameter) can be obtained per hectare (Magie *et al.*, 1966). The corms should be cleaned and dipped in a fungicide solution within 2 days of digging. The dormant corms require 3- to 4-month cold storage (2-4°C) to break the dormancy. An alternative method of breaking the dormancy is to expose the corms to ethylene chlorohydrin (Denny, 1938). According to the size of the corms and cormels, the North American Gladiolus Council have suggested six grades as below (Wilfret, 1980).

Dilta *et al.* (2004) evaluated the performance of 10 gladiolus varieties with respect to two corm sizes *i.e.* A-grade (> 5.00 to < 6.00cm) and B-grade (> 2.50 to < 3.50cm). A-grade corms increased plant height, ealry flowering and produce more number of florets/spike than B-grade corms.

Kumar and Yadav (2006) found that large sized corms (>5.10 to 6.00 cm) induced early flowering and increased plant height, leaf breadth, length of spike and rachis, florets/spike and cormels/plant with corresponding increase in weight when compared to medium or small sized corms in gladiolus *cv.* Pusa Jyotsena.

Memon *et al.* (2016) reviewed the improved conventional techniques for gladiolus multiplication such as division of the corms, removal of leaf and flower spikes, use of standard corm size, and mechanical removal of sprouts.

In order to promote development of corms and cormels, the spikes are cut when the florets show colour, leaving four to five leaves intact. Cormels if cared well, may flower the same year otherwise in the following year, but spikes should be removed, to promote the growth of corms and cormels (Halevy and Monselise, 1961). Shallow planting, *i.e.*, at 2.5 cm depth (Kolesnikov, 1965) and early planting (Pickell, 1975) help in multiplication.

Grades of Gladiolus Corms Developed by the North American Gladiolus Council

Description		Size (diameter, cm)
Large		
Jumbo		> 5.1
No.1		> 3.8 to 5.1
Medium	Flowering stock	
No.2		> 3.2 to 3.8
No.3		> 2.5 to 3.2
Small		
No. 4		> 1.9 to 2.5
No. 5	Planting stock	> 1.3 to 1.9
No. 6		> 1.0 to 1.3

Memon *et al.* (2009a) reported that clipping three leaves with the flower spike produced the highest total weight of corms and cormels (136.59 g/plant) with 20.16 per cent higher production than the untreated control. This also increased the mean number of corms (27.86 per cent) and cormels (17.47 per cent) per plant.

Memon *et al.* (2009b) planted corms of three different sizes, namely, small (diameter 2.2 to 2.4 cm), medium (diameter 2.7 to 3.0 cm) and large (diameter 3.2 to 3.5 cm) from three different varieties of Gladiolus. Large sized corms produced higher weight of corms per plant, cormels per plant and combined total weight of corms and cormels per plant in all the three varieties of Gladiolus. Cormel production also depicted significant results in response to large sized corms in all the three varieties.

Noor-ul-Amin *et al.* (2013) planted cormels of *cv.* White Friendship of three different sizes (>1.5 cm and <2 cm, >1.0 cm and <1.5 cm and >0.5 cm and <1 cm) and observed that large sized cormels produced maximum diameter of corms, number of cormels per plant and corms weight. Kareem *et al.* (2013) also reported that large sized corms (3 to 3.5 cm) produced the more number of cormels per plant as compared to medium (2 to 2.5 cm) and small (1 to 1.5 cm) sized corms.

9.2.2 Division of Corms

The cultivars which produce less number of corms and cormels are propagated by division to increase the number of planting materials (Kireeva, 1972). The corms are cut 7-10 days before planting. Small corms are divided into 3-4 parts, large into 7-10 parts and very large ones may be divided into 12-15 parts depending on the number of buds. After treatment with charcoal powder or sulphur dust the pieces of corms are placed in boxes in warm well ventilated room for drying of cut-surface. Each division should have a bud and a portion of root zone (Gromov, 1972). Even better results are obtained by quartering corms into radial cuts (La Rue and Curry, 1947), provided the split corms are treated with some fungicides such as dithane M-45 or TMTD or benomyl at 0.2 per cent before planting to check the infection of some of the soil borne pathogens, *e.g., Fusarium, Botrytis, etc.* The splitted and treated corms should better be planted on well prepared, warm and moist sterilized medium. Corms of cultivars Peter Pears and White Goddess when divided in 4 or 8 sections and planted in 50 per cent peat and 50 per cent perlite substrate, produced larger number of grade I corms (4.8 cm diameter), while soaking the corms for 24 hours in 500 ppm GA_3 solutionincreased the cormel production (Lepez Oliveras *et al.*, 1984).

A medium size corm has 6 to 8 dormant and waking buds. In some cases, very large over-grown corms may have up to 12 to 15 such buds (Gromov, 1972). Near the dominating apical or terminal bud, there are one or two more growing buds which also come out together with apical bud and produce multi-shoot plant. Each shoot when flowers simultaneously produces corm.

Cutting of corms markedly increased the growth of the filial corms, the weight of the corms, the number and weight of cormlets in comparison with those from whole corms. It is also evident that in *cv.* Lumumba's Memory weights of corms, the cormlets from cut-corms were more than that obtained in Sans Souci and Hawaii.

Memon *et al.* (2009a) reported rapid propagation of new planting material can be successfully and economically achieved by using half corms instead of whole corms as planting of half corm increased corm and cormel production by 64 per cent in 'Traderhorn', 36 per cent in 'White friendship', and 37 per cent in 'Peter pears' (Table 12).

Table 12: Mean, Standard Error (SE) of 2 Years Corm and Cormel Diameter (cm) and Number per Plant of Three Varieties of Gladiolus (*Gladiolus* spp.) after Various Treatments*

	WC	*SHC*	*LR*	*HCAC*	*LFsR*
Traderhorn					
Corm diameter	6.32±0.15 a	5.62±0.17 bcd	6.06±0.13 ab	5.41±0.11 cd	6.22±0.09 ab
Cormel diameter	0.68±0.04 c	0.60±0.06 c	0.72±0.04 c	0.60±0.05 c	0.73±0.08c
Total no. corms	1.10±0.06 cd	1.00±0.00 d	1.13±0.07 cd	1.00±0.00 d	1.27±0.03 c
Total no. cormels	49.60±3.48 fg	34.60±1.57 i	51.23±0.62 ef	21.10±0.36 k	60.67±3.72 cd
White Friendship					
Corm diameter	5.39±0.11 cd	5.36±0.10 cd	5.07±0.01 d	5.61±0.01 bcd	4.99±0.10 d
Cormel diameter	1.69±0.04 a	1.50±0.07 ab	1.46±0.10 ab	1.42±0.00 ab	1.51±0.23 ab
Total no. corms	1.60±0.10 b	1.13±0.07 cd	1.90±0.06 a	1.03±0.03 d	2.07±0.03 a
Total no. cormels	71.57±1.97 b	43.00±1.53 gh	71.47±1.47 b	57.30±4.01 de	86.63±2.59 a
Peter Pears					
Corm diameter	5.63±0.15 bcd	5.81±0.10 abc	5.20±0.11 cd	5.81±0.59 abc	5.28±0.01 cd
Cormel diameter	1.28±0.16 b	1.26±0.12 b	1.31±0.10 b	1.20±0.06 b	1.15±0.12 b
Total no. corms	1.50±0.12 b	1.20±0.06 cd	1.93±0.03 a	1.03±0.03 d	2.03±0.07 a
Total no. cormels	60.50±5.15 cd	36.37±3.34 hi	52.87±1.85 ef	27.87±1.75 j	66.13±1.13 bc

Note: WC, whole corms; SHC, simple halt corms; LR, clipping of three leaves: HCAC, half corms treated with activated charcoal; LFsR, clipping of three leaves with flower spike.

For each parameter, the values sharing the same letter (s) are not significantly different at the 5 per cent level by Duncan's multiple range test*

* Memon *et al.* (2009a).

Mahasen *et al.* (2015) used cut corm of 4 gladiolus cultivars as planting materials for flowering and corm-cormel production. All flowering and corm-cormel characters recorded were on par between half corm with whole corm. Use of the cut corm did not affect the significantly on the floral and reproductive characters. Therefore, using half corm may be beneficial for farmers by reducing half of the planting materials for gladiolus production.

9.3.0 Micropropagation

Tramier and Andreoli (1965) were able to obtain planting material free from *Fusarium oxysporum* by *in vitro* culture of the apical meristem or of the whole terminal bud plus the upper portion of the corm. Simonsen and Hilderbrandt (1971) regenerated virus-symptom-less plants from cormel tips. It is now possible to produce foundation stocks of virus-indexed gladiolus cormels on a commercial scale. Growing of such cormels demonstrated the production of slightly earlier, uniform and high quality flowers (Holdgate, 1977). The optimum size of gladiolus corm explants was found to be 0.3 mm^3. The survival rate of 0.2 mm3 explants was only 60 per cent and 0.5 mm^3 explants retained some viruses. The virus-free corms were, however, slightly late in sprouting but flowered early and were of better quality (Takatsu, 1982).

Studies on *in vitro* propagation of gladiolus revealed that cormel tips were most suitable explant for culture establishment than leaf and inflorescence segments. MS medium supplemented with 1 mg/l BAP + 0.5 mg/l NAA was optional for shoot multiplication (Misra and Singh, 1999). On the other hand, Wang and Chen (1997) found a culture medium containing 0.1-0.5 mg/l BA to be suitable for micropropagation of gladiolus. Nagaraju *et al.* (1998) observed the maximum rate of shoot multiplication in presence of BA at 0.25-1.0 mg/l depending on cultivar.

In vitro regeneration and transformation studies were conducted on two cultivars of gladiolus (Babu and Chawla, 2000). Cormels of 1.0 to 1.5 cm diameter cut into 2-3 mm thick slices of top, middle and bottom, and *in vitro* derived bisected shoo tips were used as explants on MS medium supplemented with 18.6 µM kinetin for multiple shoot induction. Amongst the cormel slices, the top slice gave better shoot induction response of 89 per cent with an everage of 2.4 shoots per explant over both cultivars.

In vitro derived bisected shoot tips were inoculated on the medium oriented cut-side up, cut-side down and vertically both with and without the cormel base attached. Bisected shoot tips without attached cormel base and inoculated in the cut-side down orientation showed an average of 90 per cent shooting response (Babu and Chawla, 2000).

Micropropagated shoots of gladiolus rooted on MS medium containing 1.0 mg NAA/l (Hegde and Kuruvinashetty, 1998a). Kang and King (1998) reported that iso-citrate dehydrogenase activity during adventitious root formation were the highest in root primordia.

In order to speed up the offer of the new hybrids by a fast propagation system and to produce a stock of virus-free mother plant, Ruffoni *et al.* (2011) established a

Propagation of Gladiolus

Seed Pod and Seeds | Sprouted Seedlings | Sowing in Field

Corms | Corm with Cormels | Sprouting Corms

Micropropagation of Gladiolus

Induction of somatic embryos and regeneration into plantlets of locally cultivated orange color gladiolus (*Gladiolus dalenii*) through corm culture; (a) Corm explants proliferating embryogenic callus on Murashige and Skoog (MS) 7.5 mg/l napthaleneacetic acid culture after 30 days; (b) Highly proliferated embryogenic callus at the same medium culture after 90 days; (c) Prominent embryogenic callus with regeneration state on transferring to medium containing MS 0.5 mg/l benzylaminopurine 0.5 mg/l Kin culture after 30 days; (d). Prominent regeneration of embryogenic callus at the same medium transferring after 45 days; (e) Healthy and strong shoot formation at the same medium transferring after 60 days; (f) Root induction of somatic embryo-derived shoots on half-strength MS 2.5 mg/l indole-3-butyric acid culture after 30 days; (g) In vitro raised plant resumed new growth in the earthen-pot; (h) 4 months old in vitro raised plant at flowering state in the field [Kabir *et al.* (2014).

combined protocol using *in vitro* propagation by simple bioreactors and *in vivo* corm enlargement and multiplication. Memon *et al.* (2016) emphasised that a number of *in vitro* protocols have been developed for regeneration of Gladiolus plantlets using different media by using various explants sources of the plant. However, literature is rather scanty on *in vitro* cormel formation and acclimatization of *in vitro* propagules.

9.3.1 *In vitro* Cormel Formation

Kinetin induced *in vitro* cormel formation, which was inhibited by GA_3; abscisic acid instead of affecting cormel formation inhibited the growth (Ginzberg and Ziv, 1973). Growth retardant also helped in *in vitro* cormlet formation in gladiolus. In liquid medium 10 µM CCC (chlormequat) resulted in higher percentage of cormletformation (65.8 per cent) than with the same concentration of paclobutrazol, ancymidol or ABA. It took nearly 28 days for *in vitro* cormlet production (Kim and Han, 1993). Goo and Kim (1994) reported that *in vitro*cormlet formation from the shoot base of gladiolus *cv.* Topaz was greatest (90 per cent) with 9 per cent sucrose whereas in *cv.* Friendship 96 per cent of the shoots formed cormlet on liquid medium containing 6 per cent sucrose (Dantu and Bhojwani, 1995). Micropropagation by shoot-tip culture, subculturing 4 times and then being transplanted into the greenhouse for 3 months, could yield 10752 cormlets, each weighing 251 mg on average, in a year (Wang and Chen, 1998).

Gladiolus bud explants propagated in agitated liquid medium in the presence of growth retardants like daminozide, ancymidol, paclobutrazol or uniconazol proliferated into massive bud aggregates without leaves. In a bubble bio-reactor the rate of proliferation was higher than in agitated cultures. The buds developed into protocorms and after sub-culture to a hardening agar solidified medium, formed cormlets 8-10 mm in diameter. The corms were not dormant and following transplanting to the greenhouse, developed into plantlets 5-6 months after explant isolation. Transplanted plants produced corms, 20-22 mm in diameter after 10-12 weeks in the greenhouse (Ziv, 1990).

Emek and Erdað (2007) developed an *in vitro* propagation method for *Gladiolus anatolicus* using corm sections. The highest number of shoots per explant was obtained in MS medium containing 0.2 mg/l BA and 2 mg/l NAA (4.7 shoots per explant). Corm formation in the base of shoots was observed in the medium with 0.1 mg/l BA. Additionally, 5-6 cormel per shoot occurred during sub-culturing.

Memon *et al.* (2012) regenerated cormels under *in vitro* conditions, direct organogenesis in cvs. Traderhorn, White Friendship and Peter Pears followed using cormel sprout as an explant. Cormel sprouts obtained by culturing the whole cormels (0.6 g) in a polar position on MS medium supplemented with BAP (4 mg/l). Out of the three varieties, White Friendship obtained more number of shoots (22.07). Cormel production was not affected by any variety of the gladiolus using MS medium supplemented with IBA (1 mg/l) and sucrose (7 per cent).

Jin-ping *et al.* (2012) established regeneration system using petals of *Gladiolus* 'Rose Supreme'. The results showed that the best medium to induce callus was MS+2, 4-D 4.0 mg/l+ 6-BA 0.5 mg/l and the best medium to induce somatic embryogenesis was MS+2,4-D 1.0 mg/l + TDZ 0.3 mg/l. New petals could be formed from petal callus directly.

Memon *et al.* (2014) reported that the *in vitro* propagules initiated the development of cormels producing the highest mean number of cormels (12.06) on MS medium supplemented with 5 per cent sucrose plus IBA (1 mg/l). Middle and Bottom slice of the cormels exhibited statistically similar and more potential results for cormel induction and number of cormels (Table 13).

Table 13: Cormel Induction and Number of Cormels Produced from Callus Cultures of Cormel Slices*

MS + different Concentrations of PGR (mg/l) + Sucrose	*White Friendship*		*Peter Pears*		*Mean*
	Middle Slice	*Bottom Slice*	*Middle Slice*	*Bottom Slice*	
Cormel Induction (per cent)					
MS+ 1 IBA+ 3 per cent sucrose	66.67	46.67	36.67	30.00	43.00 B
MS+ 1 IBA+ 5 per cent sucrose	90.00	73.33	60.00	76.67	75.00 A
MS+ 1 IBA+ 7 per cent sucrose	76.57	70.00	73.33	50.00	70.00 A
MS+ 1 IBA+ 1 BAP	20.00	30.00	0.00	36.67	21.67 C
MS+ 1 IBA+ 1KIN	80.00	70.00	76.67	70.00	74.17 A
Mean	66.67 A	58.00 AB	49.33 B	54.67 B	
Number of cormels/culture					
MS+ 1 IBA+ 3 per cent sucrose	5.53 gh	5.67 gh	2.13 ij	3.20 hi	4.13 C
MS+ 1 IBA+ 5 per cent sucrose	8.47 ef	13.30 a	13.93 a	12.53 ab	12.06 A
MS+ 1 IBA+ 7 per cent sucrose	9.33 c-f	8.80 ef	11.80 abc	6.87 fg	9.20 B
MS+ 1 IBA+ 1 BAP	3.07 i	2.13 ij	0.00 j	2.80 i	2.00 D
MS+ 1 IBA+ 1KIN	11.50 a-d	10.07 b-e	9.20 def	5.60 gh	9.12 B
Mean	7.50 A	7.99 A	7.41 A	6.20 B	

* Memon *et al.* (2014).

Bera *et al.* (2015) developed an efficient protocol for mass scale *in vitro* corm production of Gladiolus *cv.* White Friendship using liquid culture and coconut coir explored as a matrix. The high frequency of *in vitro* corms was initiated in liquid MS media supplemented with 0.5 mg/l NAA and 6 per cent sucrose in coir matrix.

Dutta and Gantait (2016) use cormel tip as explant for *in-vitro* regeneration. Maximum (55.0 per cent) and earliest (56.0 days) formation of cormels was observed on MS medium supplemented with 4.0 mg/l NAA elevated with 6 per cent sucrose with good cormel size of up to 7.0 mm diameter. It was proved that metabolic and enzymatic activity during *in-vitro* morphogenesis help gladiolus for rapid organogenesis and multiplication through cormel formation.

Kumar *et al.* (2018) reported that the BAP efficiently produced multiple shoots in gladiolus on Gamborg medium supplemented with 0 to 4.0 mg/l BAP and MS media. After 30 days of incubation, the length of *in vitro* developed multiple shoots varied from 2.2 to 3.8 cm in Gamborg medium supplemented with 0 to 4.0 mg/l

BAP media and 1.1 to 2.9 cm in MS media. Also a monomorphic banding profile was obtained using Randomly Amplified Polymorphic DNA (RAPD) markers indicating that there was no genetic variation in *in vitro* raised plants with respect to the mother plant when *in vitro* regeneration was carried out. Hence, *in vitro* regeneration could be suggested for more efficient and cost effective mass propagation of Gladiolus.

9.3.2 Acclimatization

During the hardening phase, addition of paclobutrazol to the medium resulted in the formation of cormlets with 100 per cent survival following transfer to the greenhouse. The plantlets had a survival rate of 58 per cent when hardening was done without paclobutrazol (Ziv, 1991). Hegde and Kuruvinashetty (1998b) observed that the survival of micropropagated plantlets was the maximum in vermiculite.

10.0 Dormancy of Corm and Cormels

10.1.0 Biochemical and Physiological Changes in Corms

Investigations on the biochemical and physiological changes during dormancy and sprouting revealed that sucrose tended to increase and starch to decrease in gladiolus corms stored at 5°C for 20 to 30 days but in some cases the trend was reverse (Yasuda and Yokoyama, 1955). Storage of corms at different temperatures showed a decrease in reducing sugar at high temperature and an increase at low temperature. The reducing sugar content was positively correlated with sprouting percentage (Tsukamoto and Uneno, 1957). The rate of respiration rose in cormels with broken or removed husks to a level as high as 16 times that of cormels with intact husks, with the disappearance of dormancy the rate of respiration did not rise but sometimes decreased (Apte, 1962).

The endogenous growth substances were found to influence dormancy and germination of corms and cormels. Auxin was present in corms during growth, it disappeared after harvest and appeared again when stored at room temperature for 2 months. The inhibitor content in growing corms was low, began to increase at harvest and reached a maximum after a month storage, after which it gradually declined. The fluctuation of auxin and inhibitor was related to the degree of dormancy (Tsukamoto and Yagi, 1959).

Tsukamoto and Konoshima (1972) also reported a rapid decrease in endogenous inhibitor activity and increase in growth promoting substances when dark grown corms were given cold treatment. In light grown corms without cold treatment the inhibitor activity was consistent even after 2 months. The earliest sprouting was noted in the dark grown corms given cold treatment and the latest in light grown corms without cold treatment. Tsukamoto (1975) also assayed the extracts of gladiolus corms for growth inhibiting activities. Linolenic, linoleic, stearic and palmitic acids and a compound thought to be ferulic acid were isolated as the growth inhibitor in gladiolus. The outer scales of large and small cormels contained growth stimulants and inhibitors of phenolic nature. In the scales of large cormels the stimulants were more active whereas in those of small cormels the inhibitors were more active. Soaking of cormels in water for one day leached out the inhibitors and improved sprouting (Sarkisyan *et al.*, 1975).

ABA (abscisic acid) was found to be the major endogenous growth inhibitor controlling sprouting of cormels. The ABA level in dormant cormels was 5 – 10 times higher than in non-dormant ones, and it was higher in *cv.* Texas with pronounced dormancy than that of Friendship with shallow dormancy (Ginzburg, 1973). Physiological changes on bulbing and dormancy breaking of gladiolus were studied. Suh and Kwach (1992) reported that the ABA content in *G. gandavensis* (cultivars True Love and Topaz) corms was negligible at the start of corm development but increased gradually to high levels at maturity. Cytokinin levels were also high at corm maturity. In contrast, corm gibberellin content was high at the beginning of development then decreased with maturity. In another investigation, Ginzburg (1981) recorded similar phospho-enolpyruvate (PEP) carboxylase activity in dormant and non-dormant cormels of Eurovision but PEP carboxykinase activity was higher in non-dormant than in dormant cormels. Malate utilization was slower in dormant than in non-dormant cormels, and citrate and glutamine accumulated in dormant cormels. Imbibition of water promoted dark fixation cormels by non-dormant but had little effect on dormant ones.

10.2.0 Breaking of Dormancy

Extensive investigations have been carried out to break the dormancy of gladiolus corms by exposing to temperature and using chemicals. Cultivars could be classified into 3 groups depending on dormancy status : (i) shallow dormancy status as shown by Madame Vahldiek and True Love; (ii) moderate dormancy status as shown by Nova Lux, Red Beauty, Spic and Span, Topaz and White Friendship; and (iii) deep dormancy status as shown by Calista, Hunting Song and White Prosperity (Goo *et al.*, 1998).

Gladioli corms are stored between 2 and 10°C, preferably at 4.5°C (Crossley and Arrowsmith, 1972) in a well ventilated place having 75 per cent relative humidity. At this temperature sprouting is not encouraged. The place where corms and cormels are stored, should be opened for one hour once or twice a week to prevent building up of toxic gases generated from respiration, rotting and sprouting. About one to three weeks prior to planting, cold-stored corms and cormels are brought to a warmer condition to break dormancy and to encourage root swelling.

Piya *et al.* (2012) reported that corms treated with moist sand rooted earlier. At the 47th day lab experiment, highest root initiation was from chilling treatment and at 62nd day, maximum shoot was from chilling treatment. From field experiment, maximum flower size was recorded from chilling treatment followed by scales removal, however, highest corm yield was recorded from scales removal treatment. They concluded that chilling corms in moist sand is the best and easy method to get early root and shoot.

10.2.1 Effect of Storage Temperature

Denny (1936) in an experiment reported that when corms of 8 cultivars were subjected to a storage period of 6 weeks at 3 and 10°C starting from 10 weeks after harvest, all the corms germinated much earlier than those stored at 30 and 35°C. Variation was recorded in response to the storage temperature and its duration to germinate in different cultivars. In a similar investigation, Tsukamoto and Asahira (1956) observed that treatment of corms at 35°C for 10 days followed by cold storage (0°C) for 10-15 days was the most effective way of breaking dormancy in all the cultivars used. Tsukamoto and Yagi (1960) also reported that high temperature treatment at 35°C followed by a low temperature of 0-5°C for 25 days at 15-day intervals, resulted in sprouting of corms when treated early but later treatment inhibited sprouting. A pre-planting storage temperature of 3-9°C, after 4 months at 5-10°C, was found to promote germination in cormels of 6 cultivars. Sprouting was reduced at 15°C and was significantly reduced at 20°C. At 23-30°C sprouting was stimulated in some cultivars but not in others (Apte, 1959). Apte (1962) in another experiment, found that dormancy in cormels disappeared more quickly during storage at 6 or 10°C than at 15 or 20°C. With several cultivars the removal of husk replaced the low temperature requirement of the cormels. In moist soils, the dormancy was not broken when kept at constant low temperature but an interruption of 66 days at 6°C during storage at 10 or 20°C, or a diurnal cyclic alternation of 10 and 22°C could break the dormancy. The dormancy in cormels was deeper than that of corms. He also reported that the corms and cormelsgrown at higher temperature (23°C, 25°C) or in long day (16 h light) were slower to sprout after harvest than those grown at 14 or 16°C or in short day (8 h light). Alternate heating at 35°C and cooling at 4°C had a better effect than cooling alone on sprouting of summer corms which showed delayed and uneven sprouting (Evanri *et al.*, 1950).

In an experiment on breaking of dormancy in corms, Ryan (1955) found that the corms which reached maturity with soil temperature above 15°C had somewhat deeper rest period than those exposed intermittently to temperature at 10°C or below during the final few weeks before harvest. Winter grown corms sprouted more promptly after curing at 35°C than at 21°C curing. Tsukamoto (1954) reported that the most successful treatment for breaking the dormancy was dry storage at 35-38°C for 10-15 days and then at 1°C for 10 days.

In a study on 15 cultivars, the natural period of dormancy lasted from 17 to 27 weeks. Early cultivars tended to have short dormant period and late cultivars have long ones. The length of dormancy also varied with grade-size of corms. A 40-day treatment at 5°C successfully broke the dormancy (Hsieh and Huang, 1970). Cohat (1971) reported earlier sprouting of corms stored at 5°C than those stored at 20°C.

In vitro sprouting of corm tips took 7 days in cv. Topaz when corms were stored at 5°C for 20 days. After 110 days of storage (at 5°C or at 10/20°C day/night), the time taken by *cv.* Topaz to sprout *in vitro* was reduced to 4 days (Kim and Goo, 1991). Goo *et al.* (1998) reported that the storage of *in vitro* raised cormlets at 4°C for 60-90 days was most effective in breaking dormancy.

Cormels (5-6 mm diameter) were harvested in October and stored in a dry place at room temperature until planting. They were planted from November to April and forced at 25°C. Planting of newly harvested cormels did not sprout upto 4 months but late planting resulted in 80 per cent sprouting in 6 weeks (Imanishi, 1981a). Imanishi (1981b) also reported improvement in sprouting by storing alternately under dry and humid conditions (wet sphagnum) weekly for 3 cycles (*i.e.*, 6 weeks of wetting and drying) before planting. Imanishi (1982) also found that the cormels stored in a dry place at room temperature for 17 months and planted in the next spring sprouted well but dormant cormels kept in moist condition remained dormant for a year or longer.

Amingad *et al.* (2013) studied the effect of storage temperature (4°C and room temperature 27±2°C) and length of storage (50, 70 and 90 days) on dormancy of corms in three gladiolus cultivars. Corms stored at 4°C resulted in early sprouting, early spike emergence, early opening of first floret and highest sprouting percentage.

Karimi *et al.* (2014) reported that pre-storage curing with one or two months of cold treatment 5°C resulted in the best dormancy release and faster bolting. These corms produced better plant and cut flower quality. The elongation of flowering shoot and flowering delayed when the corms stored at 5°C for the whole storage period, but produced good quality cut flowers, showing interruption of dormancy release. In the other hand, the corms stored during whole storage period at 13°C had lower and longer plant emergency and produced lower quality plants and flowers.

10.2.2 Effects of Chemicals

Sprouting of cormels in 5 cultivars of gladiolus was hastened by the use of 3 – 5 ml of 40 per cent ethylene chlorohydrin per litre of air space within a closed container for 3 to 5 days. The time required to break the dormancy varied with the cultivars (Denny and Miller, 1934). In an another experiment, Denny (1938) observed that the corms whose dormancy had been maintained for 7 months in soil at room temperature when removed and treated with ethylene chlorohydrin germinated promptly and produced blooms before the untreated control corms showed any emergence of sprouts. In Israel, Evanri *et al.* (1950) reported delayed and uneven sprouting of summer corms, while the winter corms showed little or no rest period. In experiments with summer corms of several cultivars, pre-cooling at 4°C for 20 days or 3-day treatment with either 0.3 per cent solution of ethylene chlorohydrin or 4 per cent solution of thiourea shortened the rest period, stimulated sprouting and flowering and increased the average number of flowers per inflorescence. Nakasone (1953) suggested that ethylene chlorohydrin was more effective in shortening the rest period than cold storage when the total time required for treatment and a desirable percentage of sprouting were considered. El-Gamassy (1957) observed 87 per cent sprouting in corms of cv. Valeria 75 days after planting by treatment with ethylene chlorohydrin vapour, while the untreated corms and those treated with thiourea and potassium thiocyanate showed 0 per cent, 3 per cent and 18 per cent respectively. A mixture of thiourea and KSCN was as effective as the sum of the effect of each chemical used separately. The larger corms responded more readily to the chemical treatment than the small ones. He (1960) further reported that the corms of 3 cultivars in the deepest dormancy showed prompt sprouting after treatment with a mixture of ethylene chlorohydrin, ethylene dichloride and carbon tetrachloride (4 : 1: 1) and proved more effective than ethylene chlorohydrin alone. Bautista and Cadiz (1964) observed early and high percentage of sprouting by exposure to acetylene and smoke and dipping in ethyl alcohol. The low percentage of sprouting, with ethylene chlorohyd-rin treatment could have been due to increased infection by *Fusarium* sp. Soaking of cormels in 0.1 per cent solution of $KMnO_4$ for 1 hour improved sprouting by 15.2 – 25.9 per cent in comparison with the untreated cormels (Germanyan and Ambartsumyan, 1976).

Treatment with ethrel at 1000 ppm enhanced sprouting when the corms were stored at high temperature or stored for a short period at 5°C. Treatment of cold stored corms at higher concentration delayed sprouting. Ethrel promoted sprouting in dormant but inhibited in non-dormant cormels. It increased corm splitting, delayed flowering and slightly reduced the length of flower spikes. The most important effect of ethrel was the marked improvement in health and survival of treated plants infected with latent *Fusarium* sp. (Halevy *et al.*, 1970). Bylov *et al.* (1988) observed that ethephon based preparations like hydrel, flordimex and camposan (all applied for 30 min at 0.04 per cent) were the best among several substances tested as corm treatments. In addition to breaking dormancy, the treatments improved cut flower production and ornamental quality. Ganyushkin (1993) recommended treatment of corms with 0.04 per cent camposan (50 per cent ethephon) to reduce the dormancy period. The response was rather best when camposan was applied 10 days after lifting the corms; later application inhibited metabolism.

Treatment of gladiolus cormels *cv.* Red Beauty with smoke for 1 h/day for 3 days prior to planting increased percentage of sprouting and weight of cormels produced (Suh, 1989). While treatment with 200 ppm ethephon for 6 h increased sprouting but not corm weight.

Hong and Goo (1991) in an experiment on cultivars Spic and Span and True Love found that corm sprouting percentage was increased by smoking once a day for 3 days, cracking the outer shell of the corm or treating with 400 ppm ethrel [ethephon] for 24 h. Cracking led to the production of larger, heavier corms and treatment in hot water at 28-40°C for 12 h resulted in high weights of new corms and cormels. Treat-ment with 400 ppm ethephon for 24 h gave the highest total weight of corms and high grade corms and so was considered to be the best treatment for corm and cormel production.

Soaking of corms in 20 ppm BA solution for 24 hours, followed 3 days later by resoaking in a GA_3 solution at 100 ppm resulted in breaking of dormancy and good shoot and root growth after sprouting (Tsukamoto, 1974). Goo *et al.* (1998) reported that dipping of *in vitro* developed cormlets in 10 ppm BA for 24 hours was more effective in breaking the dormancy as compared to GA_3 or ABA treatments. The effects of growth substances and storage temperature (25° or 6°C) on cormel dormancy of several gladiolus cultivars were studied by Ginzburg (1973). Cold storage and BA treatment promoted cormel germination but ABA or GA_3 inhibited it.

Foliar application of NAA (10, 50, 100 mg/l) at weekly intervals during growth, however, inhibited sprouting when treated with higher concentration. Soaking of corms in NAA in the same concentration for 14 hours also inhibited sprouting (Tsukamoto and Yagi, 1958). El-Gamassy (1957) observed retention of rest period by using MH, methyl ester or sodium salt of NAA and 2,4,5-T. A higher percentage of cormels sprouted when soaked in H_2O_2 solution (3 per cent) for 3, 12 and 24 h (42, 62, 58 per cent respectively) compared with other soaking treatments and untreated (24 per cent) and cracked (40 per cent) cormels (Imanishi, 1981b).

Hosoki and Kubara (1989) reported that the dormant period of corms in spring flowering gladiolus (*Gladiolus ×tubergenii cv.* Charm) was shortened by a 4 h treatment with methyl disulphide (MeS_2). Flowering rate of treated corms with or without 10 days of storage at 30°C was greatly enhanced compared to the untreated (control) corms. The most beneficial effect of 4 h MeS_2 treatment was an enhancement in the cut flowers on the second shoots which did not flower on control corms. Hosoki (1995) studied the changes in endogenous phytohormones in dormant gladiolus corms treated with methyl disulphide. The concentration of endogenous promoters in the corm tissue increased and inhibitors decreased within 24 h of treatments. Corms treated with MeS_2 sprouted 30 days earlier than the untreated corms which contained high concentrations of inhibitors.

Ram *et al.* (2012) found salicylic acid (100 ppm) was found optimum for increasing number of leaves, leaf length, days to opening of first floret and visibility of first spike, spike length, and number of florets per spike.

10.3.0 Tetrazolium Test

Tetrazolium (triphenyltetrazolium chloride) test has been a reliable indicator for measuring the depth of dormancy in gladiolus corms. A positive correlation was found between the degree of tetrazolium (TTC) colouration and germination potential of the cormels (Roistacher *et al.*, 1953, 1957). It was also observed that cormels stored at 40°F (4.44°C) showed an increase in tetrazolium reddening over a period of 5 months and this increase was positively correlated with increased germination of cormelsplanted at various intervals. Cormels with outer husk cracked had a higher degree of tetrazolium reddening and a correspondingly higher germination than uncracked cormels. The tetrazolium test may also assist in determining the tolerance to heat treatment against pathogens. Tsukamoto (1954) found the strongest reaction to tetrazolium when the corms received the most effective storage treatment. El-Gamassy and Shedeed (1961) reported that the most suitable test conditions was using 1 per cent tetrazolium salt solution at 40°C for 2 hours. The test was considered superior in several respects to respiration test as an indicator of dormancy. The level of dihydrogenase activity in corms and cormels during the various stages of dormancy by tetrazolium method was determined. A negative correlation between the depth of dormancy and the rate of TTC reduction was found (Simchon *et al.*, 1972).

11.0 Open Cultivation

Gladiolus is grown in all parts of the world. In temperate countries, the most important period is summer, in tropical and subtropical climate, it is grown in winter and spring.

11.1.0 Soil and Climate

11.1.1 Soil

Gladiolus can be grown in a wide range of soils from a light sandy to a clay loam but deep (at least 30 cm), well drained, friable, soils rich in organic matter and nutrients, are preferable. For best results they require a slightly acidic soil of pH 5.5 to 6.5 where most of the nutrients become available to the plants. Jenkins (1958) reported that *cv.* Snow Princess was insensitive when grown in soil with low pH *i.e.*, 4.8 to 6.5, but by the application of sulphur, corm yield declined because of the reduction in the soil pH to 4.4 or 4.5, however, maximum flower and corm yields obtained at soil pH values between 5.5 and 6.0. With a little care gladioli may be grown to their perfection even upto pH 7.5. Liu *et al.* (1998) found that cormel germination was suppressed when soil pH was more than 7.8. Plant growth was stunted when soil EC was more than 2 dS/m and/or when chloride content in the soil was more than 110 mg/l. The extent of damage caused by salt varied with cultivar and corm size planted. Salt tolerance increased with increase in cormel size. They observed that application of pig manure (50 t/ha) to the soil decreased the soil pH and increased the yield of corms and cormels. If the soil is light and sandy or heavy, adequate amount of well rotten organic matter should be applied and in very heavy soils, coarse sand may be added to improve the texture.

11.1.2 Climate

The effects of environmental factors like temperature and light on growth and flowering have been discussed under growth and development.

Gladioli can be grown in a wide range of climate but during the growth and development, the climatic factors play a very important role. The temperature in particular, is the major factor, though, some deviations may also be attributed to light conditions, humidity and water regime. For the successful cultivation of this crop, mild climate is ideal while very hot and too cold atmospheric conditions are harmful. The day temperature should range between 15 and 20°C. Temperature falling below 6°C may cause frost injury to the plant. At the time of planting, the soil temperature should not be less than 10°C (Anon., 1976).

Gladiolus prefers sunny situation and requires at least 80 per cent of total sunlight for their proper growth and flowering. During winter, lack of sufficient light may result in abortion of flower spikes (Anon., 1976). Constant humid atmosphere is not favourable as it attracts various pathogens.

11.2.0 Selection of Site

The gladiolus is a sun-loving plant throughout its growing period, so the place where it is to be planted should be sunny. The soil should have sufficient moisture and there should be some protection from strong winds so that plants may not break. At shady sites, plants become leggy and flower quality is adversely affected. Therefore, gladioli should never be planted under shade of trees or any other location where a large part of the day is not sunny.

11.3.0 Preparation of Soil

It is better to give land a rest and, therefore, green manuring should also be done. For this purpose, sunnhemp, *dhaincha, moong*, lupins, winter rye or other leguminous cover crops can be grown. The green manure crop should be turned down into the soil at least two months before planting of corms. If this is not possible, application of compost or peat moss has been suggested. Animal manure is not recommended as it is a source of pathogens (Morey and Staples, 1972).

The first ploughing is done to a depth of 30 cm two months before planting of gladioli and then left as such, about two to three weeks before planting, second ploughing is done followed by rolling for a fine tilth and at this time weeds are thoroughly removed.

11.4.0 Selection of Cultivars

The selection of cultivar is very important for achieving success in growing gladiolus either for garden decoration or for commercial cultivation. Gladiolus has many cultivars and wide range of colour. The size of plant may vary from miniature to giant which grows up to 2 metres and the size of floret may be as large as 16 cm.

In Israel, Chavagnat (1965) reported that cultivars Nouvelle Europe, Princesse de Neiges, Sans Souci and Jo Wagenaar were suitable for cut flower and corm production. Wilfret (1970) critically evaluated twenty gladiolus cultivars of autumn and winter plantings based on flower quality and quantity and corm production, Wild Rose was rated the best.

Of 34 cultivars, tested in the Maresme region, 3 red cultivars (Red Beauty, Cordula and Hunting Song), 2 pink cultivars (Wild Rose and Happy End), 2 yellow cultivars (Charlou and Yellow Emperor) and one white cultivar (May Bride) gave promising results (Varela Nieto *et al.*, 1980). Swarup and Raghava (1972) in Delhi assessed the performance of many cultivars and George Mazure, Patricia, Snow Princess, Jo Wagenaar, Apple Blossom and Ratna's Butterfly were found to be very promising. Laeser (1981) in an experiment on 44 cultivars under polythene cover, recommended Eyecatcher, H. v.d Mark, Continental, Traderhorn in red cultivars, My Love, Stuttgartia, Applause, Deciso in pink cultivars, Sancerre in white and Mr. W. Cobley in lavender colour.

Among cultivars in the large-flowered, primulinus and butterfly types of gladiolus Anitra, Atom, Bon Voyage, Cambome, Cordula, H. v. d Mark, Jessica, Memorial Day, Peter Pears and White Friendship were found very suitable for forcing and early flowering in heated glasshouse (Groen and Vander Lans, 1980).

11.5.0 Selection of Corms and Planting

The size of the corms markedly influences the growth and development of gladiolus including flower and corm production. If the flowers are to be used in the home, medium size corms will produce satisfactory bloom and also corms and cormels for future use. In most cultivars large corms will give maximum bud count and hold most florets open. Exhibitors prefer at least a number 1 size corms or jumbos. Small size corms, though most will bloom, produce smaller spikes than medium and large corms. Small-flowered cultivars, however, produce good blooms from any size bulb. Each cultivar requires has a fairly definite number of days to bloom, ranging from 60-69 days, 70-74 days, 75-79 days, 80-84 days, 91-99 days and 100 days or more for very early, early, early mid-season, mid-season, late and very late cultivars, respectively. These growing periods are applicable only to large corms, medium size corms take about 10 days more and smaller corms about 10 days longer than the medium. If flowers are desired over a long blooming period either planting the same cultivar at intervals or one planting of different sized corms will extend the blooming period, the smaller corms taking longer to bloom.

Corms should be checked before planting and only healthy, disease-free corms should be planted. Each corm has about 6 eyes or potential shoot, 2-3 of these eyes will produce medium bloom. To obtain one large bloom, all the eyes except the strongest and closest to the centre should be removed. This de-eying is done with a sharp-pointed knife or a potato peeler. It is suggested to apply a complete fertilizer 5-10-10 (NPK) at the time of soil preparation. If fertilizer has not been applied to the soil prior to planting, it is recommended that 3-5 lb superphosphate per 100 ft (33.3 m) of row to be spread in the row under the bulbs. Nitrogenous fertilizer should not be used under the bulbs as it damages the root. The corms are dusted with a fungicide (phaltan or benlate) and a pesticide. The corms are planted in the row base down, and they are covered as soon as possible (Morey and Staples, 1972).

11.5.1 Time of Planting

In India, optimum time for corm planting of gladioli is March-April in the hills and September-October in the plains, whereas in uniformly mild climate it may be planted throughout the year. It is normally a seven-month crop from planting up to lifting of corms. Most of the experiments carried out in the plains in India clearly indicated that planting of corms between mid-September and mid-November produced the best spikes and florets depending on the climate and cultivar (Kalasareddy *et al.*, 1998; Dhankar *et al.*, 1999).

In Japan, planting of cormels in March in the hills was found to promote plant growth and improve flowering, while August planting in warmer regions showed good results by using large cormels (Kosugi and Kondo, 1959). In Italy, planting is usually done in July-August. When corms were planted at fortnightly intervals from mid-July to mid-October, in most cases, the plants did not flower when planted after mid-September (Cocozza, 1971). Shimizu and Uchibori (1976) observed that cormlets required 90-150 days from planting to flowering, 90 to 100 days for cormels and 80 to 100 days for corms. With cormlets, successful flowering was obtained only from early planting. Plots planted after June, except with the *cv.* Travela, failed to flower. With corms 80-100 per cent flowering was obtained on plots planted after July. They also suggested that the temperature sum requirement from planting to flowering was 1800-2200°C for cormlets, 1500-1700°C for cormels and 1400-1500°C for corms. The time limit for planting was mid-June for cormels and corms of cvs. Oscar and Firebrand but in case of *cv.* Travela, it was possible to delay planting till late June or early July. By planting cormels around 20 March and further on two consecutive dates at 18 days interval, Grabowska (1978) observed the largest number of flowering plants and more number of corms in earliest planting. Vita (1981) found that the selection of cultivar was more important for yield and corm size, as Peter Pears and Victor Borg showed the best result in December planting, Gold Field and Spic and Span in late January and White Prosperity from 10 January to 20 February.

11.5.2 Size of Corms

Weight of the corms between 2 and 40 g each influenced the production and size of the corms, plant height, number of leaves, length of spikes and number of flowers per spike (Gill *et al.*, 1978). Khanna and Gill (1983), in an experiment with 6 cultivars, planted small, medium and large corms in mid-August, mid-September and mid-October, in Ludhiana (India). They observed earliest colour break of the basal florets from large corms planted early, highest cormel production from large and medium corms planted in mid-October and King Lear was the most productive cultivar. Increase in yield of flowers and planting materials, with the use of large corms, was also recorded by Mukhopadhyay and Yadav (1984).

The effect of 9 different corm sizes (from jumbo to 0.6 cm in diameter) on flowering and corm production was studied by Misra *et al.* (1985). The number of florets did not vary significantly upto 3.5 cm corm diameter. Number and weight of corms and cormels increased with the increase in corm size. The productivity of corms > 8, 6-8, 4-6, 2-4 and >2 cm in circumference was studied by Vinceljak-Toplak (1990). In general, shoots from larger corms started to emerge earlier than those from smaller corms. The planting-to-emergence period was similar for both cultivars, but Peter Pears started to flower after 75 days and produced mature corms (after 127 days) earlier than Oscar (89 and 149 days, respectively). Oscar yielded 25 (4-14 cm) corms/m^2 and Peter Pears only 20. However, the yield of small cormels (< 4 cm) and total corm yield (up to 14 cm) were similar for the 2 cultivars. Oscar produced about 14 per cent higher yield of corms suitable for cut flower production (8-14 cm) than Peter Pears.

The results of the investigation on the effect of corm size revealed that large corms were superior in terms of number of shoots per corm, plant height, spike length, number of spikes, number of florets per spike and the diameter of corm produced (Syamal *et al.*, 1987; Singh, 1996; Kalasareddi *et al.*, 1998).

In an experiment on cvs. Prof. Goudrian, Oscar and Golden Bog by using corms varying between 13 and 50 mm diameter, McKay *et al.* (1981a) reported that planting of whole large corms produced larger and better quality blooms. Flowering percentage was increased by cutting large corms but was reduced by cutting smaller corms and flower quality from cut-corms was poorer than from whole corms. Similarly, corm production was more by cutting large corms but was reduced by cutting smaller corms. They also observed that the method of cutting corms influenced the subsequent plant development. Cutting to remove storage tissue and damaging the root initials, but not the buds, stimulated multiple shooting, increased inflorescence yield, decreased inflorescence quality and substantially increased the number and weight of new corms. Removal of buds when cutting reduced the inflorescence number and corm and cormlet yields.

The effects of cormlet size and protected cultivation (plastic tunnel) on the growth of cormlet and growth size *cv.* Qin-gu-hong were investigated by Zhang (1998). Large cormlets were better than small ones in terms of plant growth, development and flower size. The best growth and development were observed for cormlets grown in plastic tunnels. Cormlets required 1-2 years before flowering.

Flowering behaviour and final corm yields from corms of 6 different sizes (<1-35 g), at different stages of developmental maturity, were studied in cultivars Happy End and Apricot (Ogale *et al.*, 1995). In both cultivars there was a direct correlation between corm size, flower production and final corm yield. Hong *et al.* (1989) reported that the number of daughter corms and flowering ability increased with increasing corm size upto 4-5 cm diameter, but there were no further increases for corms >5 cm in diameter.

11.5.3 Corm Size and Planting Time

In an experiment on corm size and planting dates, cultivars Hunting Song and Spic and Span corms having circumference of 10-12 cm, 8-10 cm and 6-8 cm were planted on 19 May, 17 June and 15 July at an elevation of 600 m to investigate flowering and cut flower production in the highlands. The percentage of plants flowering from the first 2 planting dates was 95 per cent but <90 per cent from planting on 15 July. Early planting with larger corms produced longer stems and longer spikes and cut flower weight, number of florets, floret height and diameter and the percentage of high quality flowers were increased. Grade 4 corms resulted in particularly poor flower quality (Ko *et al.*, 1994). Suh and Kwack (1990) reported that when large cormels were used in early plantation of good quality corms was promoted.

Zubair *et al.* (2006) planted corms of 8 cultivars on first day of November, December and January showed variation with respect to their flowering and flower quality traits. Fatihullah and Bostan (2009) found that planting of corms of on 10th February observed maximum sprouting percentage, corm weight, number of cormels/plant, plant height, corm diameter and number of leaves/plant.

Schwab *et al.* (2018) concluded that it is possible to cultivate gladiolus throughout the year, but the duration of gladiolus cycle is higher when planting is carried out in periods of low temperatures. Extreme temperatures (above 35°C and below 0°C) reduce the commercial flowers quality and even cause the death of plants.

11.5.4 Depth of Planting

Depth of planting depends more on the size of the planting materials. The medium and smaller-sized corms are generally planted up to 7 cm deep while large or jumbo corms to a depth of up to 15 cm. The planting depth also varies due to soil texture, as in light soils the depth is more as compared to heavy soils. Generally, the depth varies from 10 to 15 cm. Kofranek (1951) reported marked decrease in the production of spikes and corms when planted 15 cm deep. In general, planting at depths of 5 and 10 cm resulted in significantly greater production of spikes and corms than planting 15 cm deep but the largest corm by weight was produced at 15 cm depth. An increase in the depth of planting from 7 cm to 12-18 cm delayed flowering by 10-11 days and propagation coefficient of corms and cormels also decreased (Kolesnikov, 1965). Astvatsatryan and Sarkisyan (1974a) planted corms at depths from 1 to 10 cm and observed that 9-10 cm deep planting produced the largest cormels whereas those planted 1-2 cm deep produced the largest number of vegetative buds on the corms. Misra (1980) also found low corm and cormel production by increasing the planting depth from 4 to 10 cm.

At a very shallow planting, elongation and thickening of the contractile roots were very poor and at deeper planting of 15 and 30 cm these roots elongated slightly but showed little thickening, whereas at 5 cm depth they elongated and thickened satisfactorily (Iziro and Hori, 1983). Konoshima (1980) reported that the daughter corms from deeper planted corms (20 or 30 cm) sprouted earlier than the shallower planted corms and the leaves and corms derived from the corms planted deeply contained less amount of inhibitors than those planted at 15 cm or less depth.

11.5.5 Planting Density

Planting distance varies with the cultivar, corm size and the purpose for which the plants are grown. Increasing the planting density of cv. Monsoer from 40 to 67 per m^2 increased the number of flowering shoots from 40.7 to 65 but reduced flower number per stem and duration of flowering (Cocozza, 1971). Fernandes *et al.* (1975) found reduction of spike length, number of cormels and the weight of the corms by increasing the number of corms from 200000 to 600000 per hectare. Cirrito and Vita (1980) by using three planting densities, *i.e.*, 100, 150 and 200 corms per m^2 did not find any relation between planting density and cormel production but corms of optimum size were formed under the highest plant density.

Corms are spaced in furrows at 13-17 per metre in row. Single rows spaced 0.76 m apart are adequate in heavier soils whereas sandy soils normally require wider spacing. In Florida, growers use either single row on 1.4 m centres or two rows per bed spaced 0.5 m apart with beds on 1.8 m centres. Corm densities for single rows on 0.76 and 1.4 m centres are 173000 and 96000/ha, respectively. Double rows in beds can hold about 136000 corms/ha. Large corms are planted 15 to 23 cm deep and medium corms 13 to 16 cm deep (Wilfret, 1980).

The number of harvested corms increased with increase in plant density from 461000 to 738000/ha. The corm size and cormel production were reduced with the increasing plant density but the number and quality of flowers improved (Armitage, 1973). Grabowska (1980b) planted corms of *cv.* Kopemik at 60, 80 or 100/m^2 in heated plastic tunnels (24 × 4.5 m) and he reported that high planting density delayed flowering and reduced the number of flowering plants, plant height and spike length. He suggested 80 plants/m^2 as the optimum planting density. By surveying commercial Dutch holdings, Vroomen and Rotten (1981) found that planting density greatly increased over 10 years which consequently increased the number of corms produced per hectare but the percentage reduction was only partly related to planting density. White and Kling (1991) in an experiment on plant density concluded that a minimum of 96.9 m^2 or 72.7/m^2 for single-row spacing or 53.8/m^2 for double-row spacing appeared to be necessary.

A planting depth of 5 to 10 cm and spacing of 18.5 or 37 cm delayed the appearance of the sprouts but the extent of the delay varied with the cultivar and corm size whereas shallow planting with wide spacing produced tallest plants. Larger corms gave rise to taller and stronger plants. Wide spacing and large corms resulted in significant increase in the weight of spikes and number of florets. Deep planting, wide spacing and planting of large corms produced heavier corms than shallow planting, close spacing or planting of small corms. Shallow planting with wide spacing resulted in the heaviest cormel production and larger corm yielded more cormels than small corms (El-Gamassy and El-Gendy, 1962). El-Gamassy*et al.* (1977a) also reported that deep planting and wide spacing promoted early flowering, improved flower quality and corm multiplication, whereas shallow and close planting increased cormlet production. They recommended a spacing of 20 cm between the plants in furrows, 50 cm apart.

In greenhouse Klasman *et al.* (1995) obtained best commercial quality spikes in terms of stem length and number of flowers per stem in a density 25 plants/m^2. The number of cormels produced was also the highest (102.85/plant) in this treatment.

Bahar and Korkut (1998) observed significant effects of the planting density on number of new corms, new corm circumference, new corm weight, cormels weight and total number of cormels, the best planting distance, however, was concluded to be 15 × 15 cm. In Italy, cormels of circumference < 2 cm or 2.25 cm planted at densities 400, 600 and 800/m^2 produced corms for cut flower production and propagation material even at the highest planting density. The yield of corms for cut flowers was greater when larger cormlets were used (Cocozza Talia and Lucia, 1994). In a 2- year trial in Italy, *cv.* Peter Pears cormels were planted at densities of 75, 100, 125 or 150/m^2 and at depths of 2, 4, 8, 16 or 20 cm (Incalcaterra, 1992). Increasing the planting density increased corm yields but reduced corm quality. Planting depths of 16 and 20 cm gave markedly lower yields than the shallower planting depths. The best compromise for yield and quality was planting at a density of 125 cormels/m^2 and at a depth of 8 cm.

Kumar and Yadav (2006) found that planting of corms at wider spacing (30 × 30 and 40 × 30 cm) increased leaf length, plant height, flowering duration, length of spike and rachis, florets/spike, number of corms and cormels and their weight in gladiolus *cv.* Pusa Jyotsena. Dogra *et al.* (2012) reported that corms planted at a spacing of 40 × 40 cm exhibited highest plant height, rachis length, number of florets per spike, floret diameter, number of corms per plant, corm diameter, corm and cormel weight. Ram *et al.* (2012) found optimum spacing 20 × 20 cm with salicylic acid (100 ppm) for number of leaves, leaf length, days to opening of 1st floret and visibility of first spike, spike length, and number of florets per spike.

Maniruzzaman *et al.* (2017) reported that highest yield of flowers (2.92 lac/3650 USD) was recorded from combination of 1/2 cut corm and closer spacing (25 x 15 cm). The highest yields of corm (32.33 t/ha) was obtained from the combination of whole corm and closest spacing (25 x 10 cm) and highest yields of cormel (1.13 t/ha) from the treatment combination of 1/2 cut corm and the closest (25 x 10 cm) spacing.

Bohra and Nautiyal (2019) reviewed the effect of plant density and INM on vegetative, floral and yield attributes of gladiolus for better understanding and to develop future strategies for quality and quantity of production of gladiolus spikes and corms.

11.6.0 Irrigation

The soil should have sufficient moisture at the time of planting of corms so that no watering is required till sprouting. Gladioli need plenty of water during dry months but not wet feet. The frequency of irrigation depends largely on the type of soil and prevailing weather conditions. During warm weather, watering should be done twice a week sufficiently to wet the roots. In an investigation on the sensitivity of gladiolus to drought in various stages of its development, Halevy (1962) observed that the critical soil moisture content (CM), which decreased yield, differed at various stages of development and according to the condition whether the crop is grown for flower or corm. When growing for corms, the CM corresponds to the tension of 0.6 to 0.8 atm for the periods from the emergence of the first to the full extension of fifth leaf and from anthesis to the end. In regard to the number of flowers per stem a CM corresponding to 0.3 to 0.6 atm is operative from the sprouting to the full extension of the second leaf only. To attain maximum stem elongation till full extension of the fifth leaf, CM is the same as for corm production, while later on, till anthesis, it is necessary to keep very high soil moisture as CM corresponds to tension even lower than 0.3 atm.

Halevy and Richter (1962) found shorter plants with smaller corms when grown in dry soil than those grown under adequate soil moisture. Soaking of corms in borax solution did not increase drought resistance. Halevy (1972) found two stages of growth sensitive to water stress, (i) the very early growth stage when flower initiation takes place and (ii) from the 4th leaf stage, through flower stem elongation. Grinenko *et al.* (1973) studied the flowers, leaves and corms of cultivars susceptible and resistant to drought which were dehydrated in hypertonic sucrose solutions of different concentrations (suction force 10, 19 or 34 atm). The water retaining capacity of the cells showed a close relationship with the degree of drought resistance of the cultivars in hot and dry weather. The corms of the drought susceptible cultivars when subjected to heat treatment after storage showed greater resistance to dehydration.

Robinson (1983) found that water deficit in gladioli reduced the assimilate mobilizing ability of the inflorescence, increased that of the corm and delayed translocation from leaves. By external application of ethephon, he found variation in assimilate distribution and growth similar to that caused by water stress but no relation to ethylene production and water status of plants could be established. Robinson *et al.* (1983) studied the distribution patterns of photosynthates derived from $^{14}CO_2$ and of applied ^{14}C-sucrose in flowering gladiolus cv. Eurovision watered normally or kept under stress. The water stress decreased the mobilizing ability of inflorescence and increased that of the corms, reduced $^{14}CO_2$ fixation and slightly delayed assimilate translocation from the source leaves. They opined that under water deficit conditions the water potential of the main sink (inflorescence) decreased and its turgor dependant growth was slowed down. Oszkinisowa and Haber (1968) found the best growth and development of gladiolus at 60, 75 or 90 per cent total soil moisture capacity. In an irrigation experiment flower production in cultivar Monsoer responded markedly when water was applied to soil capacity each time the content fell to 24 per cent (pF 3) and almost as well when the minimum was allowed to reach 19 per cent (pF 3.5). Flower quality was the best at the higher irrigation rate. It was concluded that high flower quality could be maintained by irrigation when soil moisture tension at the deepest part of the root zone reached 2 atm (Cocozza, 1971).

Cultivation of Gladiolus

Planting in Open

Watering

Intercropping in Gladiolus

Cultivation in Polyhouse

Flowering in Open

Flowering in Polyhouse

Shillo and Halevy (1964) investigated the effect of watering gladioli at different degrees of stomatal aperture as determined by the absorption of fluid through stomata and they found that flower yield was more affected than yield of corms by soil water stress. An absorption fluid is suggested on the basis of the results for use as an indicator of irrigation time for gladiolus grown for cut flowers. In another experiment, Halevy and Monselise (1963) studied the leaves from dusk to dawn for water content on fresh and dry weight basis, water saturation deficit, water weight per leaf area unit and ratio of dry weight to leaf area, where they recorded maximum water content and minimal dry weight at midnight followed by sharp fall in water content and a corresponding rise in dry weight for about 2 hours and then a further reversal.

Kosugi and Kondo (1961), by watering gladioli every 1, 3, 6 and 9 days, recorded 15.5 and 51 per cent blind plants when watered on the sixth day and ninth day, respectively. They found higher resistance to drought at the early stages than at the later stages of growth, and the day length and nutrition were also important factors causing blindness. In an experiment in Egypt, the yield and quality of flowers were best when the plants were watered at an interval of 12 or 17 days, and a minimum soil moisture content of 58 per cent field capacity for successful gladiolus culture in sandy clay soil was recommended by El-Gamassy *et al.* (1977b). Sasso (1962) maintained soil moisture ranges between 20 and 60, 50 and 90 and 80 and 100 per cent of field capacity with the help of Bouyoucos soil moisture meters. He recorded nearly the same number of corms in all the treatments whereas the number of cormels was slightly increased by increasing soil moisture treatment. The increase in soil moisture content increased the corm diameter and weight.

In furrow-irrigated gladioli soil, water potentials over 1.0 bar resulted in better plant growth than at lower values and at 15.0 bar good corm formation occurred. The ratio of actual to potential evapotranspiration was determined (Klar *et al.*, 1974). Dierickx (1977) stated that at critical soil moisture tension irrigation became necessary in gladioli. Rodriguez *et al.* (1983) grew gladioli in pots with minimum soil water potentials of -15.0, -0.8, -0.2 or -0.09 where control plants were watered daily. With the decrease in the soil moisture, the area and fresh and dry matter yields of aerial parts also decreased but soil moisture had no effect on plant height or on the soluble carbohydrate contents in the aerial parts. Rodriguez *et al.* (1980) also found the highest transpiration rates with 26, 29 and 32 per cent soil moisture at the vegetative stage (four weeks before the beginning of reproductive phase) when gladioli were grown in pots with 20 to 32 per cent soil moisture content. They recorded relative water content positively correlated and the refractive index of sap negatively correlated with soil moisture content.

Maggio *et al.* (1993) planted corms of gladiolus cv. Jester and Spic and Span and irrigated at 3 rates. Flower weight and length were the greatest when about 1500 m^3 water/ha were applied, but application of 700 m^3 water/ha was sufficient to achieve standard yields. They concluded that irrigation rate had no effect on the number and weight of corms. D'Andria *et al.* (1996) observed the highest yield of cut flowers due to irrigation when the soil water potential reached 0.05 Mpa at a depth of 0.15 m.

Yadav *et al.* (2020) reported that 60 per cent of cumulative pan evaporation (CPE) water and 80 per cent water solubilizing fertilizer (WSF) through drip recorded maximum yield of Gladiolus (110119 spikes/ha) and (110119 corms/ha), which is 29 per cent more over conventional irrigation. Water used by the crop was maximum in the conventional method (454.4 mm), water use efficiency was recorded maximum in 60 per cent CPE water by drip (72.52 per cent) (Tables 14 and 15).

Table 14: Effect of Irrigation and Fertigation Cost Economics of Gladiolus*

Irrigation Levels	*Spike No. (Yield/ha)*	*Corm No. (Yield/ha)*	*Benefit-Cost Ratio*
T_1	1101194	110119.4	2.51
T_2	101190.7	101190.7	2.09
T_3	101190.7	101190.7	1.97
T_4	105148.0	105148.0	2.35
T_5	96220.23	96220.23	1.94
T_6	95238.09	95238.09	1.81
T_7	10166.67	104166.67	2.32
T_8	96220.23	96220.23	1.94
T_9	95238.09	95238.09	1.80
Control	86.309.52	86309.52	1.46

Note: T_1= I1F1, T_2= I1F2, T_3= I1F3, T_4= I2F1, T_5= I2F2, T_6= I2F3, T_7= I3F1, T_8= I3F2, T_9= I3F3); I_1–Irrigation of 0.6 CPE by drip (Cumulative Pan Evaporation), I_2–Irrigation of 0.8 CPE by drip, I_3–Irrigation of 1.0 CPE by drip, control –Surface Irrigation at CPE; F_1–80 per cent RDF (Recommended dose of fertilizer), F_2–100 per cent RDF, F_3–120 per cent RDF.

* Yadav *et al.* (2020).

Table 15: Water Use Efficiency under different Treatment Combinations*

Treatment	*Fertilizer Use Efficiency (kg/kg)*	*WUE (per cent)*
I_1F_1	6.38	72.52
I_1F_2	4.78	66.64
I_1F_3	3.89	66.64
I_2F_1	6.12	51.93
I_2F_2	4.65	47.52
I_2F_3	3.87	47.04
I_3F_1	6.06	41.16
I_3F_2	4.58	38.02
I_3F_3	3.61	37.63
Control	3.68	22.79

Note: T_1= I1F1, T_2= I1F2, T_3= I1F3, T_4= I2F1, T_5= I2F2, T_6= I2F3, T_7= I3F1, T_8= I3F2, T_9= I3F3); I_1–Irrigation of 0.6 CPE by drip (Cumulative Pan Evaporation), I_2–Irrigation of 0.8 CPE by drip, I_3–Irrigation of 1.0 CPE by drip, control –Surface Irrigation at CPE; F_1–80 per cent RDF (Recommended dose of fertilizer), F_2–100 per cent RDF, F_3–120 per cent RDF.

* Yadav *et al.* (2020).

Porto *et al.* (2014) evaluated the effect of different water replacement levels in the soil and nitrogen doses on crop growth and production in a greenhouse. The management of water irrigation at replacement levels above 75 per cent of field capacity allowed for the length of the floral stems to be classified for commercial purposes, and for the number of flowers, water replacement levels above 100 per cent provided a better commercial classification.

Pereira *et al.* (2016) found that number of flowers and days to flowering was separately influenced by either doses of wood ash or soil moisture, which ranged from 8 – 12 g/dm^3 wood ash, and 33 – 35 per cent soil moisture. The ash may be used as both soil corrective and fertilizer for greenhouse gladiolus production, besides improving soil water retention capacity. The ideal volumetric soil moisture for flower production is 33 per cent.

11.7.0 Manuring and Fertilization

Soil fertility is an important factor influencing the fertilizer requirement of gladiolus. It also depends on cultivar, size and chemical composition of corms or cormels. Cultivars showing rapid growth and developing large plants and large flower spikes respond more to fertilizer than those with low vigour and producing smaller plants and spikes. Cormels and smaller corms require more fertilizer than larger corms mainly due to their stored reserve and partly to greater uptake ability of the extensive root system produced by large corms (Woltz, 1972).

The moisture content showed a marked effect on the rate of absorption and quantity of absorbed N, P_2O_5 and K_2O and also on the distribution of nutrients in the plants. At alow moisture content (20 per cent total moisture capacity) gladiolus plant contained 0.1029 g N, 0.0264 g P_2O_5 and 0.0687 g K_2O and most of them were accumulated in corms and roots. At a high moisture content (90 per cent of total moisture capacity), the plant contained 0.4995 g N, 0.1295 g P_2O_5 and 0.5294 g K_2O, mostly concentrated in leaves and inflorescences (Baber, 1968).

Woltz (1955a, 1955b) carried out extensive investigations in sand culture on the nutritional requirement of gladiolus using commercial size, corms of cvs. Elizabeth the Queen and Picardy. Omission of N was a factor in causing blindness and also reduced the number of florets per spike. In order of decreasing effect N, K, P and Ca deficiencies lowered the average floret count and K, Mg, N and Ca deficiencies resulted in shorter spikes while K and Mg appeared to delay flowering. The weight of spikes was reduced by the omission of N, Ca, P and Mg, while leaf production was most affected by N, K and P deficiencies. Deficiency of N, K and Ca also caused marked reduction of leaf nutrient contents. Cheal (1963) in another experiment in sand culture used Long Ashton nutrient (LAN) and LAN minus N, P, K, Mg or Ca. Leaf tips of plants receiving no nutrient, no N or no Ca became scorched and died back, magnesium-deficient plants did not flower and flowering of N-deficient plants was severely affected.

11.7.1 Nitrogen

Nigrogen is one of the most important nutrients producing growth and yield responses in gladiolus. Leaf analysis indicates that the leaves should contain, on a dry weight basis, 2.5 to 3.0 per cent nitrogen or more for optimum yield. Nitrogen deficiency results in pale green colouration of foliage and reduction in the number of florets per spike and number of spikes per corm. The utilization of food reserve partly in old corms is also hampered by nitrogen deficiency. The form in which nitrogen is supplied in important. The quality of flower is improved when nitrogen is applied partly as nitrogen and partly as ammonium than when the nutrient is applied entirely from either source. Gladiolus is sometimes damaged by large amount of ammonium nitrogen in the soil, especially when nitrogen is at low level. Excess nitrogen especially in the absence of adequate potassium and calcium is unfavourable to produce quality flower due to nutritional imbalance (Woltz, 1972, 1976).

Effect of nitrogen application on growth, flowering and corm production studied by Brewer (1953) on *cvs.* Valeria and Snow Princess revealed that higher rates of N from 20 lb to 60 lb/acre resulted in 20 per cent more florets per spike, 50 per cent increase in the weight of spikes and also increase in weight and size of corms. Further increase in N rate of 180 b/acre caused higher percentage of Rhizoctonia neck rot infection and reduced the number of harvested corms. Gladioli responded most readily to N, especially when this was applied as side dressings during the development of the 2nd and 5th leaves. The effect of N was enhanced when K was applied (Mantrova, 1956).

Kosugi and Sano (1961) by growing small corms of the *cv.* Spotlight in sand and using N (l00 ppm) twice a week recorded greater plant height, weight of tops, percentage of flowering and also N content in the dry matter of the top than those grown without N. Skalska (1970) reported best corm production when N was applied in 2 or 3 split doses. Fernandez *et al.* (1974a) obtained better growth when N was applied 5 cm deep and 10 cm on each side of a plant row. Of the two sources of N, ammonium sulphate showed better result than sodium nitrate. Corm and cormel production in *cv.* Friendship after removal of flower spikes and by treatment with different rates of nitrogen was reported by Misra and Negi (1977). The weight of corm and cormel increased with higher rates of N and the effect was more pronounced in plant from which the flower spike was removed. The number of cormels per plant, however, did not show any variation due to the treatments. The beneficial effect of higher rates of N on flower and cormel production was also reported by Deswal *et al.* (1983). In field trials with 5 cultivars N was applied at 0, 20, 40 or 60 g/m^2, with half the dose being applied at the 2-leaf stage and the remainder at the 4-leaf stage. With regard to spike length the cutivars Snow Princess, Melody, Green Meadow and Suchitra showed good response to 20 g/m^2 but the cultivar Mayur did not show any response to N application (Sidhu and Arora, 1989).

Singh (1996) reported that large cormels produced larger corms with more cormels per plant than small cormels. Different rates of N significantly influenced corm weight and number of cormels per plant; the largest corms with the highest number of cormels per plant were obtained from the 250 kg N/ha treatment. In another investigation cv. Pink Friendship received 3 N rates (100, 200 or 300 kg/ha, as urea) in 3 or 4 equal split doses : (i) at planting, 60 days after planting (DAP) and 90 DAP; (ii) at 30, 60 and 90 DAP; (iii) at planting and at 30, 60 and 90 DAP; and (iv) at 30, 60, 90 and 120 DAP. Effects on corm per cormel yield and yield parameters were assessed. The highest N rate produced significantly more large-sized corms ($\geq$ 3.81 cm in diameter), a higher total number of corms per plot and higher average corm diameter and weight, compared with the lower rates (Singh, 1998).

The effect of treatment with N in previous year was studied by Kosugi and Kondo (1960). They observed that where N had been supplied there were increases in the percentage of germination and flowering, the number of florets, weight of corms and the number of cormels.

Devecchi and Barni (1997) studied the effect of ammonium nitrate (40 or 80 g N/m^2) on the colour of gladiolus spikes and observed that fertilizer application affected flower colour but not in a systematic manner. They concluded that colour variation could be minimised and spike colour could be adjusted to meet customer demand by manipulating the fertilizer regime. Similarly, Hatibarua and Misra (1999) reported that the corms, those received CAN and ammonium sulfate, in the previous year performed better as compared to urea. Application of N at 15, 30 and 40 days after germination of corms increased the weight of corms and cormels and the earliest flowering was obtained with sodium nitrate in surface bands on both sides of the row (Puccinelli and Fernandez, 1974).

Role of nitrogen on incidence of *Fusarium* corm rot has been investigated. Woltz (1955b) observed increased incidence of *Fusarium* corm rot by increasing the N level. Groen (1990) also found that despite the fungicide treatment there was a high incidence of 30, 36, 45 and 40 per cent fusarium infection and the levels were clearly related to the previous year's N treatment with 0, 75, 150 or 225 kg/ha, respectively and recommended soil sampling for N by the grower or advisory service before topdressing with N, to avoid wasteful overdosing and increasing the risk of infection.

Kumar *et al.* (2003) reported that 60 g N m^2 recorded maximum leaves per shoot, leaf area per plant, plant height, diameter of first, durability of first floret and whole spike, florets/spike, spike length, rachis length and useful life of the spike. However, lower N dose at 20 g/m^2 registered maximum number of florets open at first, earliest 50 per cent heading and first floret showing colour, while, higher dose of nitrogen (80 g/m^2) resulted maximum corms per plant, corm size, corm weight, cormel weight, cormels/plant and propagation coefficient.

Dhakal *et al.* (2017) also found that higher doses of nitrogen produced a significant difference on the length of flower spikes, rachis length, spike diameter and floret width, however, did not affect number of florets on the spike.

11.7.2 Phosphorus

The quantity of phosphorus required by gladiolus is about one-tenth of the nitrogen expressed in terms of foliar analysis. The symptoms of phosphorus deficiency are dark green foliage and purple colouration in the lower leaves. The new plant readily utilizes the phosphorus in the corms and it is an important source of the nutrient. It is likely that the plant can take up enough phosphorus at the early stage of growth to last the whole growing period, if the level of available phosphorus declines towards the end of the season (Woltz, 1972).

In an experiment with ^{32}P, applied at various stages of growth in different sets of plants, Sheehan and Joiner (1966) reported that phosphorus was absorbed and translocated with equal ease and rapidity to all parts of the plant, it was increased in all tissues for about 16 weeks after planting and then decreased. The percentage of total P was stable throughout the experiment in all tissues. Phosphorus fixation is a problem in clay soils having high iron and aluminium contents. The problem is met by liming and proper placement of fertilizer. It is not a major problem in sandy soils, low in iron and aluminium. Seasonal application of P_2O_5 should vary from 40 to 160 lb per acre depending on soil fertility (Woltz, 1976).

11.7.3 Potassium

Gladiolus requires about 3 – 4 per cent potassium in the leaves on a dry weight basis for the best yield and quality of flowers. Potassium deficiency results in shortening of spike length, reduction in the number of florets, general yellowing of older leaves and yellowing between the veins of younger leaves. In case of severe deficiency the older leaves show marginal leaf burn (Woltz, 1972). Increased K application was more beneficial to spring-planted than to autumn-planted corms (Woltz, 1954).

Trials in North Carolina on a sandy loam soil showed increase in corm yields and length of spike following side dressings of potassium but no response of N, Mn or Fe fertilizers was recorded. Application of 50 lb available K/acre was recommended when the plants began to send up flower spikes (Jenkins, 1961). Anstetta (1958) suggested that the fertilization with K and N should be done about a month before flowering for better flower production. Application of muriate of potash should be avoided when soil or irrigation water contains excess chlorides (Woltz, 1972).

The effects of applied K (0, 100, 200 or 300 kg/ha) on *cv.* Pink Friendship were investigated by Singh *et al.* (1997). Significantly higher corm weight and lower corm rot incidence occurred when K was applied at 100 kg/ha compared with controls. The number and weight of cormels/plant were significantly higher when K was applied at 200 kg/ha. Corm weight in 4 cultivars increased with increasing rates of K but corm number was not affected. Spike removal promoted corm growth, with the best time to remove spikes being at spike emergence (Das, 1998).

Devi *et al.* (2020) reported that K_2O at 210 kg/ha recoreded maximum number of leaves, plant height, length of the leaves, corm weight, corm diameter, number of cormels, corm yield per ha in tones and highest benefit-cost ratio.

Yasin (2018) evaluated the role of potassium (K) on plant growth and amelioration of cadmium (Cd) stress in *Gladiolus grandiflora* under greenhouse conditions. Supplementation of 60 mg/Kg K in treated plants (C+K, Cd50+K and Cd100+K) improved quantity of total soluble protein and proline (Pro) along with activity of antioxidant enzymes including superoxide dismutase (SOD), catalase (CAT), peroxidase (POD) and ascorbate peroxidase (APX) under Cd stress. Potassium supplemented plants exhibited better vegetative and reproductive growth. The improved stress tolerance in K supplemented plants was attributed to the reduced quantity of MDA and H_2O_2, enhanced synthesis of protein, proline, phenols, flavonides and improved activity of antioxidant enzymes.

11.7.4 Nitrogen and Phosphorus

Singh and Ha (1990) while studying the influence of N and P in phosphorus deficient soils observed a marked improvement in corm production with N at 40 g/m^2 or P_2O_5 at 20 g/m^2 but the combined N + P application had no significant effect on corm yield. The best cut flowers in *cv.* Debonair in terms of both quantity and quality were obtained with the highest N, P rate of 40 g/m^2 (Gowda *et al.*, 1998).

Kumar *et al.* (2003) observed that $N_{30}P_{30}$ g/m^2 increased plant height, florets/spike, spike length and rachis length, while, leaf length, leaf width, floret length, corms/plant and cormel's weight/plant increased with $N_{40}P_{20}$ g/m^2. However, corm size, cormel's weight/plant, floret width and propagation coefficient were maximum with higher doses of $N_{40}P_{30}$ g/m^2.

Kumar and Misra (2011) found that $N_{80}P_{20}$ g/m^2 produced maximum leaves/shoots, leaf area/plant, corm size, weight of corm and cormel per mother corm, cormels/plant and propagation coefficient *cv.* Jester Gold. They further reported that $N_{80}P_{20}$ g/m^2 recorded maximum nitrogen uptake at spike emergence (77.03 kg/ha) and at harvest (159.62 kg/ha), phosphorus content at spike emergence (0.38 per cent) and at harvest (0.55 per cent). Higher dose of $N_{80}P_{25}$ g/m^2 noted maximum potassium content at spike emergence (4.38 per cent) and at harvest (4.53 per cent). However, earliest 50 per cent heading was recorded under $N_{20}P_5$ g/m^2.

Chandana and Dorajeerao (2014) reported that nitrogen 300 kg/ha, phosphorus 200 kg/ha both individually and in combination improved spike yield. Nitrogen 400 kg/ha plus phosphorus at 200 kg/ha recorded maximum total weight of corm per plant, while, highest dry weight of whole plant and spike was registered with nitrogen 400 kg/ha plus phosphorus 250 kg/ha. However, the increase over 300 kg/ha of nitrogen and 200 kg/ha of phosphorus was not effective. They further reported that uptake of nitrogen and phosphorus was highest with nitrogen at 300 kg/ha plus phosphorus at 200 kg/ha.

Dhakal *et al.* (2017) observed that combination 150 kg/ha nitrogen and 100 kg/ha phosphorous produced highest plant height, spike length, rachis length and combination of 100 kg nitrogen/ha with 75 kg phosphorus/ha recorded longest vase life.

11.7.5 Nitrogen and Potassium

In a fertilizer experiment, *cvs.* Rosa Van Lima and June Bells showed greater respon-ses to N than Picardy and Valeria, which might be partly due to high incidence of *Fusarium* in the latter two cultivars. Small corms and cormels were benefited from application upto 160 lb N per acre. The general requirement of K was met by applying 160-240 lb K per acre during the growing period. When N was applied at 160 lb/acre and K was increased from 160 lb to 320 lb per acre flower production was increased (Woltz, 1956). Increasing levels of N and K applied to *cvs.* Picardy and Valeria resulted in increased average percentages of N and K, respectively, in corms except at the highest K level of 320 lb/acre (Woltz, 1956-57). Flowering size Valeria corms needed only a moderate N supply for best production of flowers, Rosa Van Lima and June Bells grown in the spring did not show much response to N probably owing to sub-optimal water supply and nematode attack, but showed a greater response to K than the autumn grown crop. Picardy and Valeria, however did not show response to increased K. It was concluded that small corms required more fertilizers, especially N, than large corms (Woltz, 1955c).

In another experiment with potassium chloride at 12 g K_2O/linear metre and/or ammonium sulphate at 6 g N/m applied at different days after sprouting, Fernandez *et al.* (1974b) reported largest number of corms from application of N and K at 45 days after sprouting. They also found that N had beneficial effect on the number of corms and K on corm weight.

11.7.6 Nitrogen, Phosphorus and Potassium

In 1950, Stuart and McClellan reported the highest flower production in Florida by treatment with 800 lb per acre of 5-10-5 NPK fertilizer in row before planting and 400 lb per acre of 5-10-5 in 2 side dressings. The treatment also produced high yield of corms.

In an experiment on cormels, higher rates of nitrogen delayed flowering. The percen-tage of flowering was highest with 100 ppm N, 50 ppm P and 80 ppm K (4N2P2K) and lowest with 200 ppm N, 25 ppm P and 40 ppm K (4N1P1K). The percentage of blind stalk was 11.8 in the 4N2P1K treatment, in the 4N1P1K and 4N2P1K treatments 5.9 per cent of the stalks had no flowers. The percentage of undifferentiated plants was highest in the control (100 per cent) and lowest with 4N2P1K (35.2 per cent). No significant difference was observed in the number of florets per spike due to the treatments. The number of cormels decreased with increasing N while 2N produced the heaviest cormels (Kosugi and Kondo, 1961).

The beneficial effect of fertilizer application on Morgenrot gladioli was reported by Lemini and Lemini (1965). Treatment with 600 kg superphosphate, 300 kg muriate of potash and 400 kg ammonium nitrate resulted in greatest plant height, flower number and flower size. In an experiment on a soil, very poor in P and K, Cirrito (1976) suggested that the amount of N, P and K optimum for corm enlargement depended on cultivar. For *cv.* Eurovision the dosages were 50 : 328 : 58 kg/ha and for Tequendama 60 : 240 : 42 kg. Application of N, P and K at 60, 50 and 60 kg/ha, respectively, was found optimum for cormel formation in cv. Spic and Span. Skalska (1968) observed that higher rates of N, regardless of the ratio of other nutrients, increased the weight of corms and cormels. The best results were obtained with 7.5 kg ammonium sulphate, 10 kg superphosphate and 10 kg muriate of potash per 100 sq m. Fertilization with N, NP or NPK resulted in increase in number, size and weight of corms and cormels (Shoushan *et al.*, 1980a). In another experiment, treatment with NK proved very effective for the development of corms (Shoushan *et al.*, 1980b). Increasing rates of N on cv. Friendship advanced the time of flowering and increased the spike length, weight and size of corms and number of cormels. Higher rates of P and K tended to improve flower quality, cormel growth and corm production (Bhattacharjee, 1981b).

In an investigation on changes in N, P and K contents in different parts of gladiolus, Mega and Tamura (1957) found that total amount of N in the above ground parts reached its maximum in mid-growth and total N in the whole plant was abundant, especially in the period of flower spike elongation and during the growth of new corms but at flowering time a decrease was noticeable. As regards P, the total amount in the whole plant decreased during the earlier and later stages of growth. The P content in non-fertilized plants was only half than that in fertilized plants. It was also recorded that K occurred in greater abundance than N and P throughout the growing period.

Low N resulted in low dry matter production and poor utilization of P and K. The uptake of P was comparatively small and bulk of it accumulated in new corms. All parts of the plant accumulated considerable amount of K and N. Based on the results of nutrient uptake the quantities recommended were 90, 4 and 85 kg/ha N, P and K, respectively (Barbagallo and Silvia, 1964). Analysis of plants at 75 days showed the following nutrient contents in mg/plant-586.1 K, 362.6 N, 92.4 Ca, 78.4 P, 27.9 Mg and 35.9 S (Hagg*et al.*, 1970). Optimum corm yield was obtained by treatment with 30 lb each of N and K_2O/acre applied 7 times during the growing period. The chemical composition of the corms associated with optimum yield was 1.5-2.0 per cent N, 1.25-1.5 per cent K, 0.37 per cent Ca, 0.15 per cent Mg and 0.36 per cent P (Waters, 1965).

Gamal (1974) recorded increase in the number and size of flowers by 3 fortnightly applications of growth promoting fertilizer poly-crescal followed by applications of flower stimulating poly-fertesal at 0.4 or 0.8 per cent. The treatments also increased the corm size. Cocozza (1971) reported that cultivar Monsoer responded equally and positively in yield to manuring with FYM at 1000 q/ha or NPK (13:13:20) at 20 q/ha or to a combination of both, but NPK alone was unfavourable to quality, FYM applied alone had no effect on quality, flowering was delayed by NPK with or without FYM.

In pot and field experiments with N, P_2O_5 and K_2O in various combinations, Amen and Kovanci (1990) found that leaf and flower spike uptake of N significantly influenced flower quality. The rates of P and K applied did not show any significant effects, although quality appeared to be improved by increasing rates of P in the field experiments.

Sharma (2006) found that N40, P20, K20 g/m^2 increased spike length, number of florets per spike, diameter of the first floret on the third day of opening, useful life of spike as well as durability of the whole spike in field. However, N50, P25, K25 g/m^2 increased daughter corms, corm size and weight as well as number of cormels.

Kumar *et al.* (2008) found that longevity of first floret (4.36 days) increased by $N_{80}P_5$ g/m^2. A dose $N_{60}K_{20}$ and $N_{80}P_{10}$ g/m^2 recorded bigger first and third fully opened floret, respectively. A dose of $N_{80}K_{20}$ g/m^2 resulted into maximum fully opened florets and total florets/spike in the vase. Effective useful life, vase life, water uptake on third day and at senescence in the vase was increased at $N_{60}P_{20}$ g/m^2. Maximum increment of spike length at senescence, fresh weight of spike at harvest at third and at fifth day in the vase was recorded with $N_{80}P_{20}$ g/m^2.

Durga *et al.* (2018) recorded that nutrient consortium, FYM 10 t/ha and 300, 120 and 120 kg/ha NPK found to be most effective in enhancing plant height, number of leaves, spike length, rachis length, florets per spike and spike yield/m^2.

11.7.7 Calcium

Calcium is an important nutrient element for producing good quality flowers. A mild deficiency will result in the disorder 'topple' which causes breaking over of gladiolus in the vase after most of the florets have opened. In North Carolina, a high proportion of spikes of several cultivars was found to break and topple over after they had been harvested. Spraying plants with 2 per cent calcium nitrate at 54 gal per acre reduced the percentage from 67 to 18.7 (Fink, 1953). Severe calcium deficiency resulted in symptoms similar to physiological bud-rot, the affected florets did not open normally, petals curved inwards and there might be breakdown of petal tissue with water soaking apparent (Woltz, 1972). In severe deficiency, the leaves emerged from sheathing leaves in a soft, gelatinous condition, initially light-coloured but later turned brown. Some horizontal cracking also appeared (Woltz, 1976). The spikes should contain on dry weight basis 0.2 or preferably 0.3 per cent calcium to ensure its optimum concentration. Calcium is not translocated freely from leaves to spikes and it is, therefore, necessary to make the element available in adequate quantities during the development of spikes. Ordinary superphosphate is a good source of calcium. To avoid calcium deficiency side dressing with calcium nitrate which contains soluble calcium or spraying two to three times between spike emergence and flowering have been suggested (Woltz, 1972). High Mg level aggravated Ca deficiency symptoms. Studies on the effect of N, P and Mg on cv. Rosa Van Lima at varying levels of Ca showed that flower production was decreased by high levels of the three elements at low Ca supply (Woltz, 1954). Experiments in sand culture showed that flowering size gladiolus corms did not contain sufficient reserves of N and K to produce normal foliage and flowers while the reserve for Ca was adequate for normal production of foliage but inadequate for flowers. High levels of N, K and Mg accentuated Ca deficiency (Woltz, 1955b).

Treatment with ordinary lime or dolomitic lime at the highest rate (4.5 tonnes/ha) increased the number of corms and cormels without affecting their average weight. Liming did not, however, show any effect on earliness or flower number/spike (Fernandez and Lima, 1974).

Kumar *et al.* (2003) found that sparying of 0.5 per cent $CaSO_4$ at 3-leaf stage recorded maximum corm weight in gladiolus *var.* Sylvia. Qureshi *et al.* (2015) reported that four sprays of calcium nitrate and cormel grade 2.5-3.0 cm were best for most of the vegetative and corm growth in gladiolus cv. Priscilla.

11.7.8 Magnesium

Symptom of magnesium deficiency in gladiolus is yellowing between veins which appeared first in the older leaves. Deficiency of magnesium also causes delay in flowering although not as much as with potassium deficiency. For normal growth, the leaves should contain about 0.2 – 0.4 per cent magnesium. Excessive supply of this element interferes with calcium nutrition (Woltz, 1972).

11.7.9 Iron

Iron deficiency is manifested first by yellowing between the veins of new leaves and as the deficiency becomes more severe, the whole leaf becomes yellow. Iron deficiency may not be due to lack of iron only but due to a number of complex factors. Therefore, total iron content in leaves is not a reliable indication of the deficiency of this element; moreover, much of the iron may be in the plant in an insoluble form and not usable. High soil pH and higher levels of copper, manganese and zinc are common causes. The quickest method of correcting iron deficiency is the side dressing of ten to twenty pounds of chelated iron per acre in moist soil near the root zone (Woltz, 1972).

The leaves showed iron deficiency symptoms for several years in the Sanford district in Florida, chiefly in soils with high Cu residues (Magie, 1951).

Fe deficiency can also be corrected more easily by spraying Fe-EDTA at 0.25 ppm than by ferrous sulphate or ferric citrate and greening was apparent in 4 days after treatment. At 3-10 ppm, Fe-EDTA produced brown blotches on the leaves (Woltz, 1955b). Among the different sources of Fe-fertilizers used by Chen *et al.* (1982), Fe-enriched peat ('bar-teva 55'), Fe-EPDHA (sequestrene-138 and $FeSO_4$+L-77) effectively prevented chlorosis. Treatment with 'bar-teva 55' resulted in higher percentage of flowering, and markedly increased the number of corms and cormels as compared to other treatments.

Singh *et al.* (2016) observed that foliar spary of $FeSO_4$ 0.4 per cent +$ZnSO_4$ 0.2 per cent recorded maximum diameter of third floret, while, $FeSO_4$ 0.4 per cent prolonged the days to opening of the first floret followed by $FeSO_4$ 0.2 per cent and $ZnSO_4$ 0.2 per cent.

11.7.10 Boron

Gladiolus has a moderately high requirement of boron. Boron deficiency caused reduction in the size and thickening of young leaves, and edges sometimes became thin, cracked and colourless. In older plants the cracking occurred near the base of larger leaves. It was corrected by using 4 lb boric oxide or 10 lb borax per acre or by spraying at an interval of two weeks until flowering with 1 lb borax in 100 gallons of water (Magie, 1951). Woltz (1972) reported that deficiency of boron caused horizontal cracking, starting from the margin of the leaf and extending towards the midrib. In some cultivars, there was less cracking but translucent areas develop between the veins. Boron toxicity caused scorching on the margin of older leaves. The toxicity increased and affected younger leaves when the boron content rose to 400 ppm or more. Foliar sprays of borax, boric acid or solubor at one pound per 100 gallon were beneficial for correcting boron deficiency.

In an experiment in sand culture, Woltz (1955a) suggested that 10-15 ppm B in recently matured leaves was probably the critical level. Deficiencies of N, P, K and Ca reduced the B content of the older leaves. He also observed that by increasing the level of K in substrate, the B content of lower leaves was increased and toxicity symptoms in the older leaves occurred when B content reached 150-200 ppm. Kofranek *et al.* (1957) found an increase in B content in the plant from 150 to 1700 ppm with a change of B from 0.3 to 10.3 ppm in the nutrient solution.

Kumar *et al.* (2003) reported that spike length, number of florets per spike, width of floret, size of corm and weight of cormels increased with borax 0.2 per cent + $CaSO_4$ 0.5 per cent in var. Sylvia.

11.7.11 Zinc

Sharova *et al.* (1977) reported improvement in plant growth including leaf size and number of corms by treatment of corms and cormlets with micronutrient solution; Zn was the most effective among the nutrients tested.

Availability of zinc is markedly decreased under neutral or slightly alkaline soil condition. When gladiolus is grown in such soils it is suggested to apply zinc either in soil or as foliar spray to prevent the zinc deficiency (Woltz, 1972).

Kumar *et al.* (2003) found that spray of $ZnSO_4$ at 0.75 per cent at 3-leaf stage on var. Sylvia induced earlier flowering and increased the number of corms. Singh *et al.* (2015) reported that Zn 0.4 per cent + Cu 0.2 per cent recorded earliest spike emergence, spike showing colour, opening of the first, third, fifth and seventh floret and durability of the florets, however, Zn 0.2 per cent recorded maximum duration of flowering. Maximum zinc accumulation was recorded with Zn 0.4 per cent followed by Zn 0.4 per cent + Cu 0.4 per cent and Zn 0.2 per cent, whereas, Zn 0.4 per cent + Cu 0.4 per cent recorded maximum copper accumulation which was on par with Zn 0.2 per cent + Cu 0.4 per cent.

Singh *et al.* (2016) found that foliar spary of $ZnSO_4$ 0.4 per cent was most effective for enhanced length of rachis and duration of flowering. Maximum vase-life was observed with $ZnSO_4$ 0.4 per cent followed by $FeSO_4$ 0.2 per cent + $ZnSO_4$ 0.4 per cent. Saeed *et al.* (2013) observed that zinc at 6 mg/kg rendered the highest impact for increasing the leaf area, spike length, flower size, fresh and dry biomass weight, whereas, earlier flowering and higher number of florets per spike was recorded with 8 mg Zn/kg. Chlorophyll and protein contents were highest at 6 mg Zn/kg; whereas, Zn contents were highest with 10 mg Zn/kg.

Ingle *et al.* (2015) found that zinc at 1.0 per cent and iron at 1.0 per cent improved the yield, quality, content and uptake of N, P, K, Zn and Fe of gladiolus flower and plant. Zinc at 1.0 per cent increased the total content of plant N (1.962 per cent), P (0.314 per cent), K (2.74 per cent), Zn (9.47 per cent) and Fe (11.52 per cent).

11.7.12 Manganese

In the case of Mn deficiency the leaf veins tended to remain green whereas the leaf was yellow. When Mn is used in a fertilizer, a soluble Fe-salt should be added to avoid upsetting of Fe-Mn balance (Magie, 1951). Woltz (1972) also reported that yellowing between veins of new leaves is due to Mn deficiency. The principal cause of the deficiency is the low availability of this element under high soil pH. The preventive measures include the use of manganese sulphate in small amount in mixed fertilizer or foliar application of fungicides which contain manganese.

11.7.13 Copper

A disorder caused due to Cu deficiency of gladiolus in Florida was characterized, by soft spikes, weak, soft stem and flaccid leaves, which was corrected by spraying, with a copper compound and scattering 45 lb $CuSO_4$ per acre before planting. This condition occurred only on new land, in plants from corms, grown two years in succession on this land. The cultivar Valeria and Snow Princess were more severely affected (Magie, 1951). Some cultivars are more susceptible than others. Application of copper sulphate @ 20 lb or more/acre is suggested as a preventive treatment. Copper toxicity also is a problem with gladiolus in older soils where the element has been accumulated from past application. Toxicity of copper is manifested first as root damage and second as copper induced iron chlorosis of the leaves. Copper toxicity to the roots is reduced by liming to a pH of about 6.5 (Woltz, 1972).

Chaturvedi *et al.* (1986) applied micronutrients and nitrogen in single or double foliar sprays of Agromin (containing Zn, Mn, Cu, Mg, Fe, B and Mo + 2 per cent N) at 0, 1000, 2000, 3000 and 4000 ppm to cv. Sylvia. Application of 3000 ppm of Agromin improved plant height, increased number of leaves and floret size but delayed spike formation. Two sprays at a concentration of 4000 ppm enhanced the duration of flowering with the longest spikes but delayed the emergence and opening of the first floret.

11.7.14 INM and Biofertilizers

Jha *et al.* (2020a) reviewed the effect of organic and inorganic fertilizer in gladiolus with special reference to growth, yield and flower quality. Fertilizer requirements in gladiolus has vital role in growth, quality, corm and cormel production. Major nutrients like nitrogen, phosphorus, potassium along with zinc noticeably increase the number of flowers, florets/spike, length of spike and flowering stem of gladiolus. The microorganisms in biofertilizers restore the soils natural nutrient cycle and build soil organic matter. Through the use of biofertilizers, healthy plants can be grown while enhancing the sustainability and the health of soil.

Basoli *et al.* (2015) found that ¾ N, P and K+ *Azotobacter* +PSB + KSB was found most effective in enhancing plant height, bud emergence, days taken to spike emergence, size of florets, length of spikes, rachis length, chlorophyll content, photosynthetic rate and N content in leaves of gladiolus. Maximum leaf area, number of spikes/plot and number of florets/spike, and number of florets opened/ spike, P and K content of the leaves were recorded with ½ N, P and K+ *Azotobacter* + PSB + KSB. However, ¾ N, P and K+ PSB + KSB resulted in maximum stomatal conductance of spikes.

Sathyanarayana *et al.* (2017a) reported that 100 per cent RDF + FYM @ 7.5 t/ha + Azotobacter + PSB + KMB + 1 per cent foliar spray of Nauroji Novel Organic Liquid Fertilizer found most effective in plant height, leaves/plant, early spike initiation, days to spike harvesting, spike length, rachis length, florets/spike, diameter of 2nd floret, vase life, spikes/plant in cv. American Beauty. They (2017b) further observed that same treatment recorded maximum number of corms per plant, corm and cormels weight per plant, corm size, nitrogen, phosphorus and potash contents in leaf, minimum soil EC with highest available nitrogen, available phosphorus, available potash, soil organic carbon and microbial population.

Jha *et al.* (2020b) found that 75 per cent RDF + FYM 10 t/ha performed better for days to sprout, numbers of sprouts, leaves per plant, girth of plant base, width of leaf, plant height, days to spike emergence, corm diameter, weight and number of corms per plant. However, 75 per cent RDF + vermicompost 3 t/ha recorded maximum spike length, florets per spikes and vase life in *cv.* Candyman (Table 16).

Table 16: Effect of Integrated Nutrient Management on Sprouting, Flowering, Quality and Yield of Gladiolus cv. Candyman*

Treatment	Days to Corm Sprouting	Days to Spike Emergence	Spike Length (cm)	Diameter of Corm (cm)	Corm Weight (g)
T_1	9.00	72.50	80.65	2.92	30.10
T_2	7.70	68.50	90.85	4.20	43.27
T_3	7.50	68.00	96.65	4.25	47.57
T_4	6.70	67.50	99.12	4.35	51.35
T_5	8.00	69.00	86.97	3.95	39.80
T_6	6.70	65.50	100.18	4.65	56.95
T_7	5.20	63.50	101.80	5.82	65.07
T_8	8.00	69.20	87.15	3.62	39.50
T_9	6.70	66.00	100.20	4.57	53.37
T_{10}	6.50	63.25	101.88	5.75	60.02
S.Em±	0.411	1.665	3.970	0.393	2.952
C.D (p=0.05)	1.193	4.833	11.521	1.142	8.568

T_1: Control, T_2: 50 per cent RDF, T_3: 75 per cent RDF, T_4: 100 per cent RDF, T_5: FYM 10t/ha, T_6: 50 per cent RDF + FYM 10t/ha, T_7: 75 per cent RDF + FYM 10t/ha, T_8: Vermicompost Only 3t/ha, T_9: 50 per cent RDF+ Vermicompost 3t/ha, T_{10}: 75 per cent RDF+ Vermicompost 3t/ha; RDF: 120:80:60 kg/ha.

* Jha *et al.* (2020b).

Devi *et al.* (2019) reported that vermicompost (0.5 kg/m^2) + bio-inoculant treated corm + bio-control agents (*Pseudomonas fluorescens* + *Trichoderma* @ 0.5 kg/m^2 each) + nitrogenous bio-fertilizers (Azotobacter + Azospirilum mixture @ 0.5 kg/m^2 each) + phosphatic bio-fertilizer (Phosphate solubilizing bacteria @ 0.5 kg/m^2) recorded tallest plants and maximum leaves/plant, earliest spike emergence and spike length, florets/spike, vase-life, weight and diameter of corms.

Dhakar *et al.* (2020) revealed that organic, inorganic and biofertilizers under INM system showed significant effect on vegetative, flowering and yield of corm parameters. Maximum sprouting, number of florets per spike and early flowering were recorded at 75 per cent RDF + vermicompost at 10 t/ha + *Azospirillum* + PSB, whereas, maximum plant height, spike length, number of corm per plant and number of cormels per plant were observed at 75 per cent RDF + FYM @ 5 t/ha + Vermicompost at 5 t/ha + *Azospirillum* + PSB.

Baruati *et al.* (2018) reported that soil enzymes fluorescein di-acetate (FDA), phosphomonoesterase (PMEase), dehydrogenase (DH), arylesterase (ARE) and arylsulphatase (ARS) involved in energy flow and nutrient cycling showed higher activities under different organic inputs. Enriched compost 5 t/ha demonstrated clear increase in FDA (8.35 µg fluorescein g soil/ha), PMEase (63.76 µg p-nitrophenol g soil/ha), DH (125.42 µTPF/g/24 h), ARS (20.79 µg p-nitrophenol/g soil/h), ARE

(147.97μg p-nitrophenol/g soil/h), MBC (522.90 μg/g soil) and microbial population in gladiolus.

Karagoz *et al.* (2019) indicated that autoclaved vermicompost (60 kg/ha) + PGPB formulation (*Bacillus subtilis* RK-1977 + *Bacillus megaterium* RK-1978 + *Pseudomonas fluorescens* RK-1979) promoted flower diameter, leaves per plant, florets per spike, stem diameter, spike length, fresh and dry weight of flowers, number and diameter of corm.

Ahmad *et al.* (2013) found that three applications of humic acid and NPK, applied at planting, 3-leaf, and 6-leaf stages of plant development, followed by two applications of humic acid and NPK, applied at planting and 3-leaf stage, proved best for early and uniform sprouting, more foliage growth per plant, greater leaf area, and total leaf chlorophyll contents, earlier spike emergence, greater number of florets per spike, longer stems and spikes, and greater diameter of a spike, higher flower quality, longer vase life, higher number of cormels per clump, and cormel diameter and weight in gladiolus cvs 'Eminence', 'Cantate', 'Essential', 'Corveira' and 'Fado'.

Bashir *et al.* (2016) found that soil application of humic acid (3.0 mL) at planting and 3rd leaf stage in combination with NPK (17:17:17) produced higher sprouting, plant height, leaves/plant, leaf area, spike length, spike diameter, stem length, vase life, number of cormels/plant, diameter of cormels and weight of cormels/plant.

Keisam *et al.* (2012) revealed that vermicompost @ 2.5 t/ha + VAM + humic acid @ 0.2 per cent (FS) registered the maximum plant height, number of leaves, leaf length, leaf width and early sixth leaf stage, earlier spike emergence and first flower open, and maximum spike length, rachis length, florets per spike, flower weight and the number of marketablespike per plot. The same treatment also recorded longer vase life.

Sankari *et al.* (2015) revealed that foliar spary humic acid (0.2 per cent) led to early spike emergence, highest spike length, rachis length and florets per spike, spike yield/plant and per m^2, and longevity of cut spike. The same treatment also increased corm weight, corm diameter and number of cormels in gladiolus *cv.* White Prosperity.

Pansuriya (2018) reported that humic acid 0.2 per cent with Azotobacter 2.5 ml/plant + PSB 2.5 ml/plant + KSB 2.5 ml/plant sprayed at 30, 45 andt 60 days after planting and soil application of biofertilizers at the time of planting and two month after planting recorded higher corm, cormels and spike yield in gladiolus *cv.* American Beauty.

11.8.0 Interculture and Aftercare

Once the corms are planted, the interculture starts with the appearance of the sprouts through the soil, although this practice differs depending on the success from previous experiences. However, after the corms have sprouted well, watering should be done, if necessary. When the shoots are about 20 cm high they are hilled up to a height of 10 – 15 cm. This enables the plants to grow erect despite high winds and rains and suppresses weed growth. Hilling is a must in case of light soils. This also holds the prims, butterflys and pixiolas even if they are not staked. In case where spikes grow longer or stems are not strong enough to bear the lodging or mild stroke of wind, they are supported with about 1.5 metre strong stakes. Strings instead of stakes may be used at the time of the appearance of the spikes. Strings are stretched between the stakes alongside each row to provide easy and adequate support which does not look as bad as the stakes.

11.9.0 Mulching

Mulching is important for conserving the moisture and reducing the weed population and its effectiveness depends on the type of material and thickness of the mulch used. Mulching is done between and across the rows. Care should be taken to mulch the beds only when there is sufficient moisture, otherwise it will not serve any purpose. Fresh manure, chopped straw, dried grass, clippings, saw dust, peat, husk, bark and strips of black polythene may effectively be used as mulching materials. Dobos (1964) used 3 cm thick layer of half-rotten farmyard manure and observed better results as compared to compost, bog soil, peat or river sand.

Wisniewska-Grzeszkiewicz (1978) used composted pine bark, fresh pine bark, moss peat and straw as 5 cm deep mulches and observed more number of cormels in moss peat and composted bark and also increase in plant size and floret number. All materials except straw inhibited weed growth, bark being better than peat. Mulching with fresh or composted bark also resulted in longer flower spikes than other treatments. In another experiment by using composted, fresh or old (carbonated bark undisturbed for 5 years) bark, used as a soil amendment and source of organic fertilizer at 20 or 40 litres per square metre soil, of which about 85 per cent was derived from Scotch pine and 15 per cent from birch. All three amendments improved growth, flower and cormlet yield and quality, and composted bark gave better results than others (Grzeszkiewicz, 1978a, b).

Gladioli mulching with slit mulch was found equally good as intact polythene mulching (Smith, 1973). Application of bitumen emulsion at 3 or 6 tonnes/ha as a complete cover or in bands 15, 25 or 35 cm wide increased plant emergence by 5-20 per cent depending on corm size and corm production upto 25 per cent. The higher rate was generally superior (Benkenstein, 1974; Benkenstein *et al.*, 1975). Early sprouting, flowering and improvement in production of corms and flowers by the use of polythene mulch have been reported by several workers (Montesi and Lazzereschi, 1966; Lepori, 1967; Beck *et al.*, 1970). Improvement in flower quality due to polythene mulch was also reported. Planting of corms through holes of perforated plastic sheet increased the corm production by more than 50 per cent as compared to control, and variation in response of different cultivars to growth and flowering due to mulching was also evident. Fodor (1974) recorded better quality flower and increased corm multiplication by mulching the field with plastic film shavings. Plastic mulching advanced flowering, increased the number of flowering plants and the yield of corms of suitable size for export (Benkenstein *et al.*, 1976). Covering of gladiolus field with perforated plastic sheet till plants attained 15-20 cm height resulted in better emergence, longer stems and earlier flowering (Anon., 1980). Elia and Accati (1964), by using perforated black plastic sheet and planting corms through the holes, found earlier emergence, flowering and more number of longer spikes. In 4 cultivars of gladioli, Hunting Song, Peter Pears, and White Friendship, covering with perforated plastic film advanced harvest of flowers and protected from frost damage (Bondt, 1982). Reflective film mulching increased the percentage of flowering and improved flower quality in cultivars Hunting Song and Pink Glory (Huh *et al.*, 1996).

Gantait (2015) reported that Gladiolus plants of *cv.* Pusa Suhagin mulched with black polyethylene sheet established superior plant and yield followed by rice straw mulches than plants mulched with coloured polyethylene sheet and white transparent mulches.

11.10.0 Weed Control

Weed control is an important practice in commercial cultivation of gladiolus. In India, it is generally done manually, and if not properly done, damages the plants. Swaroop *et al.* (2015) reviewed the integrated weed management in gladiolus. Weed is one of the most important yield limiting factors and significantly reduces the yield. However, the most critical period for crop weed competition is first six weeks after planting of crop. During this critical period, weedings essentially required by chemical, nonchemical means or in integrated manner. Therefore, timely weed control at early stage is imperative for realizing desirable level of productivity. The use of herbicides offers selective and economic control of weeds right from the beginning, giving crop an advantage of good start and competitive superiority. Pre-and post emergence herbicides may also be viable option to control the weeds. Manual weed control is effective, if done frequently, but, it is very expensive. Hence, a successful integrated weed management program utilizes mulching or a combination of chemical measures, taking into consideration labour costs and availability of materials is the best way.

Results of experiments on chemical weed control in gladiolus have revealed that several chemicals are very effective in eradicating weed flora without affecting the growth and flowering of gladiolus. These herbicides are applied either as a pre-plant or pre/post-emergence of weeds.

The kind of chemicals, their concentration, time and method of application vary according to soil types and prevalent weed population, only one chemical can not be used universally.

Chenopodium album, *Amaranthus retroflexus* and *Digitaria sanguinalis* were controlled by spraying terbacil (3-t-butyl-5-chloro-6-methyluracil) and DuPont 733 (3-t-butyl-5-bromo-6-methyluracil) each at 0.9 kg/acre and monuron at 0.9 kg, triflu-ralin at 1.8 kg, linuron at 1.8 kg and chlorthal-dimethyl at 3.6 kg + 2, 4-DES at 1.8 kg/acre, although first two herbicides caused delayed killing of the gladiolus foliage (Welker, 1967). Diuron at 1.13 kg or trifluralin at 0.9 kg/acre was found very effective and safe for broad leaved and grassy weeds in gladiolus whereas dimethylamine salt of 2,4-D sprayed at 0.225-0.45 kg/acre prior to spike formation controlled late season weeds (Jenkins *et al.*, 1968). Monaco and Jenkins (1969, 1970) recommended diuron incorporation at 0.54 kg and trifluralin 0.45-0.9 kg/acre with a follow up spraying of 2, 4-D amine at 0.225-0.45 kg/acre at the base when the gladioli plants were 30 cm high; incorporation of nitralin at 0.9 kg/acre, dichlobenil and diphenamid each at 2.7 kg/acre, and metobromuron at 1.35 kg/acre immediately after planting; and para-quat, chloramben and SD-15418 (2-4-chloro-6-ethylamino-1, 3, 5- triazin-2-ylamino-2-methylpropionitrile) in screening trials.

Bing (1978, 1979) reported alachlor or napropamide at 0.9-3.6 kg/acre was less effective as compared to diuron at 0.225 kg + chlorthal at 5.4 kg/acre. However, napropamideupto 7.2 kg and prodiamine at 0.9-1.8 kg/acre were found safe while oxadiazon at 1.8-3.6 kg and oxyflurofen at 0.9-1.8 kg/acre, and alachlor at 1.8-3.6 kg/acre proved injurious to the crop grown from cormels. Fetvadzhieva and Dinova (1982) recommended napropamide, lenacil, alachlor, metabromuron and chloridazon as surface application at the time of planting of cormels, however, application of 1.5kg napropamide + 0.8 kg lenacil/ha on moist soil controlled *Echinochloa crusgalli, Chenopodium album, Amaranthus retrojlexus, Portulaca oleracea* and *Sinapsis arvensis* effectively. Use of metoxuron at 3 kg a.i/ha as pre-emergence application produced quality blooms whereas perfluidone proved phytotoxic (Koutepas, 1982).

Wilfret and Burgis (1977) used EPTC and napropamide each at 2.7 kg/acre and MBR-8251 (perfluidone) at 1.8 kg/acre at one month intervals for three times, last being before four weeks of planting which controlled the weeds effectively. Koster and Rooy (1981) found fervin (alloxydim) to be toxic, as post-emergence application, and metoxuron to be effective. Pitt *et al.* (1981) controlled the weeds with linuron at 1 kg/ha applied 24 days after planting. Groen and Vander Lans (1980) sprayed the plants grown from cormels at 1 kg linuron in 100 l of water per hectare on 1 June, 1 July, 1 August and 1 September, where only August spraying was found the best while others caused flower aberrations and malformations.

Kwon *et al.* (1996) found 16 species of weeds in the gladiolus field out of which *Digitarias anguinalis* and *Portulaca oleracea* were dominant. The most effective herbicides for controlling grass weeds were alachlor, simazine, napropamide, linuron and pendimethalin. However, in vinyl house culture, the incidence of abnormal plants increased and slight injury to flowers was observed at high rates of simazine, linuron and oxadiazon + pendimethalin; no injury was observed with napropamide or alachlor + pendimethalin application.

The most effective herbicide in controlling the weeds in gladiolus field and promoting the growth and flowering of the crop, under West Bengal (India) conditions, was found to be oxyfluorfen applied at the rate of 0.5 kg a.i. per hectare; higher concentration (1.0 kg), however, markedly reduced the growth and yield of spikes and florets. Pre-plant application of this herbicide markedly reduced weed population and gave the highest yield of spikes as compared to the other treatments (Yadav and Bose, 1987). Bing *et al.* (1988) reported that pre-emergence spraying of oryzalin, prodiamine and pendimethalin at 2.2 and 4.5 kg/ha did not adversely affect vigour or corm yield of gladiolus cv. Promenade. In the hills, Laskar and Jana (1995) indicated that the weed control was the best with higher doses of herbicides, 2 l/ha oxyfluorfen being the most effective and comparable to the hand weeding treatment. Widaryanto *et al.* (1995) reported that mulching and herbicides were effective in suppressing weed growth for first 30 days after planting, especially the broad leaved species such as *Portulaca oleracea, Ageratum conyzoides* and *Convolvulus arvensis* whereas mulching and weeding suppressed weed growth for first 60 days after planting. *Galinsoga parviflora, Cyperus rotundata, Echinochloa colonum, E. crusgalli* and *Digitaria sangunalis* were prominent ones among the weeds in the gladiolus field (Chahal *et al.*,1994). They recommended use of atrazine (1-2 kg/ha) which was found to be more effective than pendimethalin (1-2 kg/ha) during the first period (up to 95 days after planting).

Kadam *et al.* (2014) found *Ageratum conyzoides, Chenopodium album, Convolvulus arvensis, Cynodon dactylon, Cyprus rotundas, Digera arvensis, Digitaria adscendance, Euphobia hirta, Parthenium hysterophorus* and *Portulaca* are major weeds infesting gladiolus.

Jankiram *et al.* (2014) reported that herbicide atrazine 1.0 kg/ha as pre-emergence followed by rice residue at 5 tonnes/ha at 2 days after atrazine caused the greatest reduction in density and dry weight of weeds. Tank-mix pre-emergence is usually more effective than the single herbicide application. The tank-mix pre-emergence of pendimethalin 0.75 kg/ha + metribuzin 0.3 kg/ha resulted in greater plant height, cut-flower yield, corm yield and net returns compared to weedy check.

Kadam *et al.* (2014) found pre-emergence use of Atrazine (1.0 and 1.5 kg a.i./ha) recorded low weed count, fresh weight and dry weight and also recorded maximum corms/plant, fresh weight and diameter of corms, however, Atrazine (1.5 kg a.i./ha) recorded maximum weed control efficiency (WCE). Pendimethalin (1.0 and 0.75 kg a.i./ha) as pre-emergence had superior effect on the plant height, spike length, rachis length and number of florets (Tabel 17).

Table 17: Effect of Herbicides on Vegetative and Floral Characters in Gladiolus var. White Prosperity*

Treatment	*Plant Height (cm)*	*Spike Length (cm)*	*Rachis Length (cm)*	*Days to First Floret Opening*	*Second Floret Diameter (mm)*	*Florets/ Spike*	*Florets Open at a Time (Nos.)*	*Leaves/ Plant*
Atrazin 1.0 kg a.i./ha	95.7	79.1	43.1	122.0	97.8	12.6	5.2	6.2
Atrazin 1.5 kg a.i./ha	106.4	90.7	52.3	121.1	97.2	11.5	5.1	6.3
Metribuzin 0.25 kg a.i./ha	103.0	93.2	49.3	122.0	93.9	12.0	5.4	6.4
Metribuzin 0.50 kg a.i./ha	87.8	75.6	38.0	126.7	96.0	9.8	5.2	6.3
Butachlor 1.0 kg a.i./ha	98.4	83.8	37.9	124.9	97.2	11.0	4.9	6.5
Butachlor 1.5 kg a.i./ha	88.4	75.4	40.5	126.6	90.4	9.6	5.3	6.2
Pendimethalin 0.75 kg a.i./ha	107.4	94.8	52.9	125.6	96.9	12.1	5.3	6.6
Pendimethalin 1.0 kg a.i./ha	112.9	98.3	58.8	122.7	92.5	14.1	5.2	6.3

Treatment	Plant Height (cm)	Spike Length (cm)	Rachis Length (cm)	Days to First Floret Opening	Second Floret Diameter (mm)	Florets/ Spike	Florets Open at a Time (Nos.)	Leaves/ Plant
Control (Unweeded)	76.7	79.3	36.3	122.5	99.0	9.1	4.8	6.3
Control (weed free)	111.5	91.9	57.1	117.1	93.8	14.2	5.2	6.5
CD (P=0.05)	22.3	15.1	14.4	4.1	NS	4.3	0.5	NS

* Kadam *et al.* (2014).

Dhakar *et al.* (2015) observed that reuduction in uptake of nitrogen, phosphorus and potassium by weeds was 4.1, 1.2 and 6.6 kg/ha, respectively by use of metribuzin 0.4 kg/ha pre-emergence + residue (dry grass 5 t/ha). The yield of corms, cormels and number of marketable spikes/ha, net profit and benefit cost ratio in cv. Pusa Srijna was received maximum with metribuzin 0.4 kg/ha pre-emergence + residue.

Dhakar *et al.* (2016) observed that atrazine 0.75 kg/ha as pre-emergence + carfentrazone 0.030 kg/ha post-emergence at 40 days after planting recorded least number of monocot and dicot weeds, their fresh and dry weight, however, metribuzin 0.4 kg/ha pre-emergence + residue (dry grass 5.0 tonnes/ha) recorded maximum corm and cormel yield (89.48 q/ha), and marketable spikes (1.43 lakh/ ha), net profit (Rs.3.48 lakh/ha) and benefit : cost ratio (1.99).

Qadeer *et al.* (2016) observed that Pendimethalin at 12 ml/litre was found to be the most effective in controling all weeds except *Cyperus rotundus*. S metolachlore efficiently controlled (91 per cent) *Cyperus rotundus.*

Swaroop *et al.* (2017) found that atrazine @.75 kg/ha as pre-emergence + carfentrazone (0.03 kg/ha) post-emergence at 40 days after sowing recorded minimum number of dicot weeds, whereas, minimum monocot weeds were recorded with pendimethalin (1.0 kg/ha) as pre-emergence+dry grass residue (5.0 tonnes/ ha). Pendimethalin 0.75 kg/ha + metribuzin 0.3 kg/ha (tank-mix) pre-emergence recorded maximum plant height, spike length, number of corms/ha (1.20 lakh/ha) and estimated net returns (Rs. 2.63 lakh/ha), however, metribuzin 0.4 kg/ha pre-emergence + dry grass residue 5.0 tonnes/ha recorded maximum marketable spikes.

Rawat (2017) found that atrazine 1.50 kg a.i./ha recorded minimum count, fresh weight of weeds and earlier spike emergence, whereas, atrazine 1.00 kg a.i./ha recorded maximum plant height, spike length and florets/spike, while, pendimethalin 0.75 kg a.i./ha increased fresh weight of spike and duration of flowering.

11.11.0 Intercropping

Singh and Datta (2006) observed that intercropping of marigold with gladiolus paired system recorded an additional yield than the pure cropping of gladiolus paired system. The net income due to the intercropping was almost two fold higher than the pure cropping of conventional practice.

Prakash *et al.* (2011) reported that gladiolus is remunerative intercrop with sugarcane as gross higher monetary return in sugarcane + gladiolus intercropping system was found in comparison to cucumber, okra and French bean.

Kumar *et al.* (2012) observed that gladiolus as intercrop with litchi plantation was forund ideal. The highest net profit was recorded with the Okra - Gladiolus intercrop rotation. Inter-cropping of Okra-Gladiolus with litchi fetched better cost : benefit ratio.

Singh and Singh (2014) observed that gladiolus yield attributes and yield were improved under intercropping with Coriander, Fenugreek and Soya than as sole crop.

Singh *et al.* (2015) found that individual intercropping of coriander, fenugreek and soya with gladiolus *i.e.* gladiolus + coriander, gladiolus + fenugreek and gladiolus + soy improved seed germination per cent, plant height, number of leaves and branches/plant. Gladiolus crop performed better for growth, flowering and yield with intercropping of gladiolus + fenugreek.

Ghosh *et al.* (2017) reported that introduction of gladiolus as intercrop recorded additional flower production into the coconut orchard by 1.50 lakh gladiolus/ha/year. Growing of gerbera and gladiolus in existing coconut orchard may be ecologically beneficial as well as remunerative. Singh *et al.* (2018) indicated that in the *Rabi* season, gladiolus recorded the maximum gross and net returns as intercrops (Rs.112400 and Rs.56820, respectively) and B: C ratio of 1.02.

12.0 Greenhouse Cultivation

Gladiolus being a very popular flower is in constant demand throughout the year. However, in cold countries it is difficult to grow the crop outside in the field during the winter months. Several investigations have revealed that gladiolus can, successfully be grown in different structures like glasshouse, plastic tunnels, hot beds, *etc*. The advantages of glasshouse are manifold as it provides protection against unfavourable conditions. It ensures desired temperature, light and humidity. The carbon dioxide content of the air is also enhanced, assuring better photosynthetic activity.

12.1.0 Greenhouse

Hoogeterp and Mohlmann (1965) suggested that for growing in heated glasshouse, the corms should be stored at 9-17°C during November and the first half of December and then for 6 weeks at 20°C. Soil temperature should not exceed 10-12°C during the first 6 weeks after which 12-14°C should be maintained until flower spikes appear. In Poland, Grabowska (1975b) found that Karmazyn, Kopemik and Sarmata are suitable for forcing in heated glasshouse. Corms of Karmazyn and Sarmata stored at 5-8°C were ready for planting in early February while those at 20-24°C were ready for planting in late February. Plants from corms stored at high temperature flowered earlier than those from corms stored at low temperature.

In Yugoslavia, Oscar and Sans Souci grown in greenhouse had the largest leaf area and flowered about 26 days earlier than those grown outdoors (Prijic, 1980).

Keszi-Hermath (1960-61, 1962), however, observed that gladioli took longer time to flower under glasshouse than in the open and earlier the planting the greater the number of days required for flowering. In an experiment in glasshouse, corms of cv. Rosalind were planted on 4 February and together with those of Hunting Song on 5 March. They flowered between 22 April and 5 May in case of first planting, and 3-27 March in case of second. The glasshouse temperature was maintained at 13°C until March, it was raised to 14°C in March and in April to 16°C. The energy saving by increasing the planting density to 128/m^2 was estimated at 35-40 per cent (Groen and Schipper, 1982a). A saving of 27 per cent energy with an early forced gladiolus crop was obtained from the quick growing cv. Hunting Song planted on 25 February compared with Peter Pears planted on 1 February, both the cultivars flowered on 19 May (Groen and Schipper, 1982b).

In an experiment in Holland, Pasterkamp (1981) found Caravelle, Love Letter, Kim's Delight and Wine and Roses suitable for forcing in warm house. Love Letter was the earliest and had the best flower quality. The initial growing temperature was 12°C in late January and rose to 16°C in April. In another experiment, corms of 4 gladiolus cultivars were planted under glass in early February and those of 3 quick growing cultivars in early March, after storing the corms at 20°C for 6 weeks before planting. The cultivar Appledoorm was most satisfactory of February-planted cultivars with least bud blasting and best quality. With March planting, Love Letter and Come Back gave better results than Applause, all flowered in mid-May, about a week after the February-planted cultivars (Pasterkamp, 1982). Working with some large flowered cultivars Andes (cream-flowered), Breakdance (pink with a red centre) and Vedi Napoli (pink), he (1989) reported that the flowers were harvested in mid to late-May, Peter Pears being the earliest, followed by Breakdance, Andes and Vedi Napoli a few days later. Flower quality was good in all cultivars, particularly, Vedi Napoli. Vase life was excellent in Vedi Napoli and Peter Pears, good in Andes and fairly good in Breakdance.

Substratum in greenhouse was found to influence growth and flowering of *Gladiolus nanus* and large flowered gladiolus hybrids. Flower quality in *G. nanus* was similar with Einheitserde, rockwool and foam, but significantly lower with expanded clay. Flower quality in the hybrids was similar with rockwool, perlite and expanded clay. Covering the planted crop with black plastic film had little effect. Covering with the substrate was necessary for good growth (Leinfelder and Rober, 1989).

Singh *et al.* (2019) found that shade net house was significantly superior to the open field condition in respect to percentage of sprouting, leaf area, plant height, days to spike initiation, basal floret opening and last floret opening.

12.1.1 Plastic Tunnel

Selection of cultivars suitable for cultivation in plastic tunnels is very important. Eight cultivars were grown under plastic tunnel. Corms were planted at 64/m^2 in January and flowers were harvested in April-May. The maximum/minimum temperatures were 25-30°C/4°C in January-February. The cultivars Hunting Song (1 flower spike/corm), INRA 67-23-3 and INRA 68-45-1 (both with 2 flower spikes per corm) were outstanding for cultivation inside plastic tunnel. Quality was better in INRA 67-23-3, Peter Pears, Cordula, Hunting Song, Vander Mark and INRA 68-45-1 (Vidalie, 1978). In another investigation on 15 commercial cultivars grown in a large plastic tunnel with 200 hours more sunshine during the culture period, Vidalie and Laffair (1983) observed increased earliness and production of shorter plants bearing shorter flower spikes with fewer flowers. Among the 7 cultivars grown in unheated plastic tunnel Kopernik, Karmazyn and Bird of Dawning showed the shortest growing period and was earliest to flower (Grabowska, 1980a). Fodor (1974) recorded good or very good flower quality by growing 13 cultivars in a heated Dutch frame over which plastic structure was erected. In China, Zhang (1998) observed best growth and development of cormlets grown in plastic tunnels.

13.0 Diseases, Pests and Disorders

Weather, pests and diseases are three common hazards of crop production where relative importance of these is not certain. The inter-relationship among these factors is very complicated. Unfavourable weather alone may invite more pests and diseases. The following diseases and pests attack gladioli. Disorders caused by fluoride and sulphur dioxide have also been discussed.

13.1.0 Diseases

Baiswar *et al.* (2007) reviewed the status of gladiolus diseases and their management In India. Diseases caused by fungi, bacteria and viruses result in considerable loss of flowers and corms every year. Through certain cultural practices, attack of many diseases as well as pests can be prevented. Corms are required to be thoroughly checked at the time of planting and those showing any sign of abnormality are to be discarded and destroyed. Corms are to be planted under full sun and in well-drained soil. The same soil should not normally be used for more than once in four years for gladiolus planting, otherwise it should be fumigated. Field should be dry at the time of cutting the spikes, and the cutting knives should also be dipped in alcohol frequently to check transmission of the diseases, especially viruses. At the time of lifting, care should be taken not to bruise the corms, and after lifting corms they should thoroughly be cured, cleaned and treated with fungicides like benomyl and captan at the time of storage. Plant debris should be burnt. If these practices are thoroughly adopted, many diseases can be eliminated. The important diseases reported on gladioli with their suitable control measures are given hereunder.

Fungal Diseases

Bhagat *et al.* (2018) reported that gladiolus was infected with different fungal diseases namely wilt (*Fusarium oxysporum* f. sp. gladioli), corm rot (*Rhizoctonia solani*) and leaf spot (*Alternaria alternata*).

13.1.1 Fusarium Rot and Yellow

Commonly known as 'vascular disease', 'dry rot' or 'core rot' is caused by *Fusarium* sp. McCulloch (1944) described *Fusarium orthoceras* var. *gladioli* attacking vascular bundles of gladioli in the USA, and Nelson (1948) described different diseases attacking gladioli with their suitable control measures. Harrison (1962) reported the history, economic importance and symptoms of *Fusarium oxysporum* f. sp. *gladioli*. Michail and Elarosi (1965) when cultured two isolates from wilted gladioli corms and inoculated on new corms, found distinct strains of *Fusarium oxysporum* f. sp. *gladioli*. Singh (1969) reported *Fusarium oxysporum* f. sp. *gladioli* from India. In France, Albony (1966) found the leaf hoppers *Euscelis plebejus* and *Macrosteles quadripunctulatus* transmitting fusarium disease in gladioli. Georgeva and Peikova (1976) in Bulgaria recorded *Fusarium oxysporum* f. *gladioli*, *F. moniliforme* (*Gibberella fugikuroi*) var. *subglutinasis*, *F. heterosporum* (*Gibberella cyanea*) and *F. sporotrichiella* (*F. sporotrichioides*) var. *tricinctum* causing brown rot of gladiolus corms in field and storage. Ashour and El-Din (1964) isolated *Fusarium oxysporum* f. sp. *gladioli* from diseased corms of gladiolus in Egypt. Albores *et al.* (2014) identified three fungal species namely *F. oxysporum*, *F. solani* and *F. Proliferatum* from basal stems and damaged corms taken from field plants. Molecular analyzes corroborated the species identified. *Fusarium solanii* has also been found to infect the corms in storage in India (Tandon and Bhargava, 1963). The vascular disease described by McCulloch (1954) on gladiolus proved to be pathogenic to the plants of *Babiana, Crocus, Freesia, Iris* (bulbous), *Ixia, Sparaxis, Streptanthera, Tritonia* and *Watsonia* (McClellan, 1945).

In India, *Fusarium oxysporum* f. sp. *gladioli* has been found to be very serious on gladioli. It causes curving, bending, arching, stunting, yellowing or drying of leaves associated with root and corm rot in field as well as in storage. The disease is characterized by dry rot in storage, often restricted to the corm base. Corms when split into halves, show radiating dark-coloured streaks extending from the corm base through the flesh and in severe cases the centre of entire corm is black and rotten. On the corms red-brown lesions, having clear and concentric markings with well-defined margins, are found. Initial infection of the corms comes either through the soil or by latent corm infections of previous year which in adverse weather conditions develops bright yellow interveinal streaks on leaves. Plants bending geotropically with leaf tip burn are a conspicuous feature, bending always starting from the site of the corms showing rot. Leaf yellowing developing from older ones spreads over the entire foliage. *Fusarium solanii* causes dry rot and mummifies the corms.

Cruz *et al.* (2019) characterized the wound healing in corms of *G. grandiflora* stored under refrigeration and how it affects the entry and establishment of *F. oxysporum* f. sp. gladioli infection. Corms were wounded and stored at 12±4°C and relative humidity of 90±5 per cent. Wounds on corms caused trans-epidermal damage with collapse and cell death. Physiological (increased loss of mass and respiration) and biochemical changes (deposition of lignin and suberin, enzymatic activity) occurred in the cells neighboring those deaths by the injury. The injury caused gradual darkening of the tissue, injured and neighbor. *Fusarium oxysporum* infection decreased with wound healing. The healing of injured *G. grandiflora* corms stored at 12°C occurs from the 3rd day after injury by the accumulation of suberin, lignin, and melanin, inhibiting *F. oxysporum* f. sp. gladioli infection.

Bhagat *et al.* (2018) observed that initiation of wilt (*Fusarium oxysporum* f.sp. gladioli) symptoms occurred during 26th meteorological week with an incidence 1.25 per cent. Higher disease incidence of 43.12 per cent was recorded during 46th

meteorological week. Maximum temperature showed positive correlation with wilt infection, whereas, negative and significant correlation was established by rainfall, minimum temperature and relative humidity.

Chandel and Deepika (2010) reported that management practices generally employed for Fusarium wilt control include resistant cultivars, chemical applications, cultural practices and biotechnological approaches. However, incorporation of integrated management provides a better opportunity to manage this disease. Magie (1956) suggested water soaking at 57.2°C for cormels upto 30 minutes provided they had been lifted in mid-April or early summer and treated 6 to 10 weeks after harvest, and Roistacher *et al.* (1957) reported that this fungus died between 53.3° and 56.7°C while the cormels were destroyed above 56.7°C.

Magie (1987a) studied the effect of fungicides as postharvest treatment on corms to control *Fusarium* rot disease and reported that dowicide B, mertect, benlate and busan were effective. He also found that dowicide B and mertect were as effective as or more effective than benlate in flower and corm production. D'Aulerio *et al.* (1994) evaluated various fungicides for their effectiveness against *Fusarium oxysporum*f. sp. *gladioli* and found that immersion of corms in flutriafol at 0.5 or 1.25 g a.i./l of water for 30 min before planting gave effective results.

Raj and Upmanyu (2006) recommended that wilt of gladiolus can be effectively controlled by planting corms treated with Quintal (carbcndazim + iprodione) (0.25 per cent) in a field which is amended with cabbage leaves (10 tonnes/ha), irrigated and solarized with transparent polyethylene mulch (25 μ*m*) for 40 days in summer months. Baiswar *et al.* (2008) reported that Local isolates of *Trichoderma roseum* (52.5 per cent) followed by *Trichoderma* spp. (Local isolate B) (46.2 per cent) were found to be highly effective against *Fusarium oxysporum* f.sp. gladioli.

Kadam *et al.* (2014) found that fungicides Difenconazole (0.1 per cent) and Propiconazole (0.1 per cent) completely inhibited the mycelial growth of *Fusarium oxysporum* f.sp. gladioli followed by Thirum (85.22 per cent), Captan (82.77 per cent), Carbendazim (65.33 per cent) and Copper oxychloride (50.77 per cent) tested under tested *in vitro* were found effective in inhibiting the mycelial growth of *Fusarium oxysporum* f.sp. gladioli. Among the different plant extract tested against the same fungus, complete inhibition of mycelial growth was achieved due to 10 per cent Jatropha followed by Soapnut (37.00 per cent), Neem seed kernel extract (35.00 per cent), Cassia (26.44 per cent), Neem leaf extract (22.00 per cent), bulb extract of Garlic (17.22 per cent).

Khan *et al.* (2017) found that corm dressing with *T. harzianum* or carbendazim checked the corm rot and yellows and increased the plant growth and flowering of gladioli. Chandel and Kumar (2017) found that soil beds amended with cabbage leaves residue showed the minimum Fusarium disease incidence (11.97 per cent) followed by the mustard (19.02 per cent) compared to control (42.06 per cent).

Yasmin *et al.* (2018) observed that corms treated with Bavistin (0.1 per cent) for 15 minutes + Poultry refuse 5 t/ha in soil application 25 days before corm sowing + Bavistin @ 0.1 per cent as soil drenching at 45 days after corm sowing found best integrated management option for reducing Fusarium wilt of gladiolus. Besides, integration of Bavistin (0.1 per cent) as corm treatment for 15 minutes + Mustard oil cake 600 kg/ha in soil application 25 days before corm sowing + Bavistin (0.1 per cent) as soil drenching at 45 days after corm sowing was also better option for combating Fusarium wilt of gladiolus. The alternate option was integration of Bavistin (0.1 per cent) as corm treatment for 15 minutes + Bio-pesticide in soil application 7 days before corm sowing + Bavistin (0.1 per cent) as soil drenching at 45 days after corm sowing was effective against the disease incidence(Table 18).

Table 18: Effect of Integrated Management Options on Flower Production of Gladiolus*

Treatment Combination	*Spike Length (cm)*	*Rachis Length (cm)*	*Florets Spike^{-1}*	*Flower Stick Weight (g)*	*Flower Sticks Plot^{-1}*	*Flower Sticks^{-1}*
$A_1 = T_1+T_4+T_7$	77.23 a	45.18 a	12.73 a	68.71 a	40.75 a	97000 a
$A_2 = T_1+T_5+T_7$	76.00 ab	44.57 ab	12.25 ab	68.13 a	40.00 ab	95250 ab
$A_3 = T_1+T_6+T_7$	75.40 ab	44.00 abc	12.11 bc	67.88 a	39.75 ab	94750 ab
$A_4 = T_2+T_4+T_7$	75.25 ab	43.50 bc	12.36 ab	67.88 a	39.25 ab	93250 ab
$A_5 = T_2+T_5+T_7$	74.54 bc	43.00 cd	12.40 ab	67.50 a	39.00 b	92750 b
$A_6 = T_2+T_6+T_7$	74.25 bcd	42.75 cde	12.23 ab	67.00 a	38.75 b	92250 b
$A_7 = T_3+T_4+T_7$	72.63 cde	42.13 de	11.93 bc	64.45 b	36.50 c	87000 c
$A_8 = T_3+T_5+T_7$	72.25 cde	41.50 e	12.34 ab	64.13 b	36.25 c	86000 c
$A_9 = T_3+T_6+T_7$	71.75 de	41.63 e	11.85 bc	63.88 b	36.00 c	85000 c
A_{10}= Control	70.25 e	38.75 f	11.60 c	61.00 c	33.00 d	78000 d
CV per cent	2.16	1.94	2.84	2.51	2.65	2.65

Means followed by the same letter do not differ significantly at the 5 per cent level by DMRT.

T_1 = Corm treated with Bavistin @ 0.1 per cent for 15 minutes, T_2 = Corm treated with hot water @ 54°C for 5 minutes, T_3 = Corm treated with hot water @ 52°C for 10 minutes, T_4 = Poultry refuse @ 5 t/ha, T_5 = Mustard oil cake @ 600 kg/ha, T_6 = Bio-pesticide @ 64 kg/ha, T_7 = Bavistin @ 0.1 per cent as soil drenching.

* Yasmin *et al.* (2018).

Khan *et al.* (2017) reported that treatment with carbendazim-nemacur mixture or *P. fluorescens* effectively controlled the *Fusarium oxysporum* f.sp. gladioli (Fog) + *Meloidogyne incognita* (Mi) disease complex leading to decrease in the corm rot and yellows (4-43 per cent and 5-21 per cent) and root-knot indices (9-33 per cent and 4-40 per cent).

13.1.2 Core or Spongy Rot

Also known as soft corm rot, grey mould, neck rot, floral rot or botrytis blight is caused by *Botrytis gladiolorum*. It is a soil-borne highly destructive fungus which affects the plants during cool and humid weather, most favourable temperature being 13 to 15°C. It was first recorded by Timmermans (1941) in Holland as a new species, different from *Botrytis gladioli* by Schmidt (1949) in Austria and by Agarwala *et al.* (1965) in India. Yet another species, *Botrytis elliptica* has been recorded in Washing-ton by Maclean (1948). Timmermans (1941) mentioned that the disease starts as small spots which develop into large spots covered with a mould, soon killing the leaves. Characteristic symptoms include small red-bordered and rusty-coloured specks on leaves and stems and colourless spots on the flowers. Through the vascular bundles of stem and leaves the fungus reaches the corms and causes core rot. Disease incidence is less frequent on the corms lifted early. During storage at a higher temperature (25.5-30°C) the corms dry quickly and the disease hardly spreads. McClellan *et al.* (1949) found that its infection to gladioli foliage is severe at 12.8° to 18.3°C and the optimum temperature for infection on freshly harvested corms is 17°C. Groen *et al.* (1979) when stored the corms at low temperature for 8 to 12 months, corms rotted due to attack of *Botrytis gladiolorum* (*Sclerotinia draytonii*) and they suggested storage at 5°C after six weeks of lifting.

Botrytis survives in corms, soil and in plant debris. During cool and humid weather and after frost-injury, it becomes infectious and affects various parts of the plant. Air-borne spores from infected plants spread the disease rapidly through petals and buds during dew and rainfall in case of neck rot while brown rot develops at soil level, and on dead tissues grey mould is seen which is most common when the flowers are cut. Spongy rot is caused by this fungus on the corms which becomes soft. Rotting of the corms start from the top of the base which on scrapping shows browning. This causes spotting and rotting of all parts of the plants. The fungus may appear as very small groups of the spores in grape-like bunches. The leaf spots may be large, round to oval and brown to pale brown with reddish brown margins or very minute rusty-brown spots on the exposed side of the leaf and after a long spell of humid weather the floral bases rot. Translucent water soaked areas on the petals are seen early in the day and during transit these infected flowers rot due to its spread.

Soft rot of the corms is very common in storage. The rot is primarily restricted to core and/or vascular tissues and later on the infected corms become light in weight, shrivelled and mummified. Stalk or neck rot caused by this pathogen has abundant dark brown irregular spots on the corms beneath the tunics. The effective fungicides for controlling *Botrytis* are vinclozolin, iprodione, and benlate (Magie, 1987c). Maneb and ferbam sprays also control its infection.

Baiswar *et al.* (2008) reported that Local isolate B (57.3 per cent) and A (54.3 per cent) of *Trichoderma* spp. were found highly effective at 25°C against *Botrytis gladiolorum*. These isolates were also effective at 20°C, indicating that *Trichoderma* spp. can be used at lower temperature also.

13.1.3 Dry or Neck Rot

It is also known as root rot, caused by *Stromatinia gladioli* or *Sclerotinia gladioli*. Dry rot of gladiolus corms caused by *Sclerotinia gladioli* is seen on the stored corms as small, dark, more or less superficial spots or lesions. The pathogen may cause collar rot, killing the plants or its delayed attack may only harm the new corm for carrying the disease to the next season. Agarwala *et al.* (1965) reported *Sclerotinia gladioli* infecting gladioli in India and Vigodsky (1969) from Israel. Jeves and Coley-Smith (1976) found high proportion of buried sclerotia surviving for five years in the field and its spread was favoured by close planting.

Dry rot is a wide spread disease attacking gladioli plants in the field and corms in storage. It is more severe during humid conditions in the field where premature yellowing occurs ultimately causing death of the plants. The leaves turn brown from the tips downwards and at bases they decay (neck rot), but corms remain attached to the stems firmly. Later, the sheath become shredded and covered with numerous round and black resting bodies just large enough to see. These tiny bodies and the absence of grey spore masses differentiate this disease from botrytis stem rot. Newly rotted neck tissues are yellow-brown with mouldy odour.

Diseased corms show numerous round, black and small lesions (first reddish brown) varying from merely a dot to 0.6 cm in diameter with slightly raised edges which may coalesce together forming irregular area. Often the small lesions are found in the ring along the leaf scars and then black resting bodies are seen on the decayed bases and tunics. These spots on the corms are more or less circular and sunken but can be distinguished from scab which produces shiny or gummy exudation. In moist storage conditions these black lesions coalesce together to make the corms mummified but in dry season it is not so serious. The tunics of these infected corms look snuffy and turn brittle. Diseased tissues are characteristically corky and usually shrivel. *Sclerotinia gladioli* are a soil-borne fungus and its longevity in soil has been found to be 10 years or more.

Control measures are the same as for *Fusarium*. Cormels may be cured by soaking them first in cool water for 24 hours and then for 30 minutes in hot water at 54.5°C. Vasilevskjj (1959) recommended the use of 0.1 per cent potassium permanganate or 0.2 per cent granosan for 20 minutes and spraying with copper solution twice to thrice during the growing season at fortnightly intervals. Haasis (1954) suggested immersion of corms shortly after harvesting for five minutes in ceresan-M, 4.5 g per gallon, for effective control during storage. Magie (1987c) reported that dicloran, rizolex are effective fungicides.

Rhizoctonia solani attacks cormels and small corm plants causing neck rot, poor growth and 'damping-off', and in large plants, it causes root rot, brown corm spots and a shredded neck. In case of root rot, contractile roots become yellowish to dark brown. This fungus is most active during warm weather, and to attack gladiolus the fungus requires partially decayed organic matter (Magie, 1987b). He also advocated that undecomposed crop residues should not be left in the field and the field should be sprayed with demosan and botran in open furrow after planting the corms and cormels. Applications of chlorothalonil (bravo), benlate or botran also control this fungus. Mantel and Groen (1988) recommended hot water treatment for cormlets (0.5 h at 53-55°C) plus a soak or dry dressing with sumisclex (procymidone) to control the spread of dry rot within the planting stock most effectively.

13.1.4 Curvularia Blight

It is caused by *Curvularia trifolii* f. sp. *gladioli*. *Curvularia lunata* was, however, reported first in the USA by Magie (1948) and in the Philippines by Mendiola-Ela (1952). *Curvularia trifolii* f. sp. *gladioli* was recorded in New South Wales in 1960 (Anon., 1962), in France by Grouet (1965) and in India by Singh (1968). Kore and Bhide (1976) reported *Curvularia brachyspora* infecting gladiolus from Parbhani

(India). Torres *et al.* (2013) reported for the first time incidence of leaf spot caused by *Curvularia gladioli* in *Gladiolus grandiflorus* in Brazil. Symptoms include oval to round leaf spots brown with a dark border surrounded by a halo of yellow tissue.

This is also a soil-borne pathogen which survives in the soil for three years or more. This disease may be very serious for the plants grown from seeds and cormels which are often attacked at or below soil level. Leaf infection by this fungus leads to the infection of corms also. The disease is characterized by firm, sunken, black lesions and corking out. In corms, the red spots are small to medium in size and dark brown in colour, rounded, elongated or irregularly shaped surrounded by a reddish brown ring, mostly shallow. All plant parts including leaves, stems, flowers, corms and roots are infected. Aerial parts are more prone to infection causing unmarketable spikes. Young leaves are very susceptible. This disease becomes serious during warm (18.3-32.2°C) and humid weathers; infected plants or soils which are moist, provide a source of infection for the next crop. But this can be reduced by digging the corms in dry weather and curing immediately after harvest.

The first appearance of symptoms on the leaves is seen as severe yellowing and brow-ning of leaf tips, initially with circular to oval spots which in favourable conditions become irregular and extend lengthwise along the leaf blades to the extent of about 12 cm. These spots start as a point which when coalesce extend to about 20 cm in length with 1 cm breadth. These pale yellow spots in later stages turn to tan or dark brown surrounded by a darker reddish brown border. Soon the central area is filled with black spores and in the later stages the entire leaf is invaded causing distortion and breaking of leaves. The spots develop severely on stems and corms also. Distorted flowers with brown oval spots on petals and young leaves are seen in abundance. These water-soaked spots are colourless which later on turn brown to black after spore production.

Infected soil from the previous crop causes rotting of leafy stalks below the soil, also of the corms, and sometimes the seedling plants damp off without showing any symptom on the leaves. The hard sunken spots on the corms can be separated easily from healthy tissues if the whole corm is not dead. The dark bordered spots in the corms elongate towards tunic juncture, and the infection spreads to the roots which blacken and rot.

This disease is controlled by the same fungicides used for the control of the *Botrytis*. Maneb is very effective in controlling this disease. Dithane M-45 (mancozeb) and miltox (copper oxychloride + zineb) each at 0.2 per cent, sprayed at fortnightly intervals control this disease.

13.1.5 Storage Rot

During storage, the corms are severely affected due to the attack of certain fungi. Jackson (1962) stated that infection of *Penicillium* sp. occurs through wounded callused basal core tissue or through infected spike stubs. Core rot caused by *Penicillium funiculosum* and storage rot caused by *P. gladioli* have been reported in India by Singh (1970).

Penicillium gladioli causes corky corm rot with large reddish brown sunken lesions and numerous cream and pinkish resting bodies are seen on the corms, particularly in the centre of the lesions. At low temperature the lesions are full of greenish mouldy growths which pose a problem when corms are piled in heaps without treatment, subsequent to lifting from the field in cool weather and moist storage, spreading to corms and vascular tissues. The infection on the corms initially occurs on the tunic scars and spreads rapidly. In field, it causes rot by latent infection of corms. *Penicillium funiculosum* is mainly responsible for field rot, and is of minor importance to storage rot. This disease is not serious where corms have been cured for a week at 29-32°C immediately after harvest. Benomyl dips advocated for the control of *Fusarium* is also effective against *Penicillium gladioli.*

Rhizopus arrhizus develops soft rot and covers with black 'bread moulds'. It attacks when corms are injured by artificial heat or sun scald after lifting and during curing by which it causes watery rot of the corms (McClellan, 1956). He also found that corms heat-treated at 29-34.5°C became liable to infection with this fungus which produces light brown soft rot. Sanitary measures should be adopted and injury to the corms should be avoided for prevention of this fungus.

13.1.6 Leaf Spots

Alternaria fasciculata (syn. *Macrosporium fasciculatum*) has been reported in India by Rao (1964a) from Bombay. It causes dirty dark brown to black irregular spots in long patches mostly on apical and marginal parts on the leaves. In a very few cases, it causes corm rot. *Alternaria tenuis* also infects gladioli in India.

Septoria gladioli, first reported from India by Singh (1965), is confined only to the temperate regions of the country. This is of minor importance and although it may cause corm rot in its serious infection but in India it has been found confined only to the foliage and that too in the plants grown from cormels and seeds and sparingly in those grown from corms. This fungus invades even internodes of spikes and florets. The lesions are circular, brown to purple-brown, medium in size with black spore bodies in the centre of older spots. In wet weather, spores ooze out and cause rapid spread of disease. The disease can be effectively controlled by spraying 0.2 per cent dithane M-45 or miltox. Mancozeb or daconil, dyrene or kemate are also effective.

Magie (1948) reported that *Stemphylium* sp. causes leaf spots which may kill the leaves before flowering, causing total loss. It disappears in summer and autumn and reappears in humid winters on older leaves. Schenk (1955) in Germany reported *Stemphylium botryosum* causing leaf spots on gladioli. The disease causes small reddish brown spots surrounded by ring of translucent pale yellow tissues. Treatment with maneb will also control this disease as in case of *Botrytis* or *Curvularia.*

Bertus (1974) reported *Drechslerabicolor* from Australia causing leaf spot of gladioli. Rangaswami and Seshadri (1970) recorded *Colletotrichum* sp. and *Phyllosticta gladioloides*, Rao (1964b) also reported them from Dharwar along with a *Phyllosticta* sp. from Coorg and Kapoor and Gill (1961) recorded *Mycosphaerella minimaepuncta* (syn. *Sphaerella minimaepuncta*) from Kalimpong, causing leaf spots of gladiolus. *Macrosporium candidum* also causes leaf spots of gladioli (Shumilenko, 1973). All these minor diseases can be controlled by the use of 0.2 per cent dithane-M-45.

13.1.7 Gladiolus Rust and Smut

Collingwood (1970) observed *Uromyces transversalis* in Africa and Viennot-Bourgin (1978) in Morocco, causing rust on the leaves. Langdon and Herbert (1944), in Australia, recorded *Urocystis gladioli* causing smut on corms, stems and leaves. Maclean (1951) recorded smut disease in the USA, causing blistering, shredding and necrosis of stem and leaf tissues and the death of severely infected seedlings. Ceruti-Scurti (1955) described another pathogen *Tuburcinia gladioli* causing smut on gladioli. All these diseases are controlled by hot water treatment and disinfection before planting. Smut spores in the soil may be treated with formalin, chlorobromopropene, trapex (methylisothiocyanate) and mylone (Schenk, 1961). Dipping the corms in water for up to 30 minutes at 43.3°C kills the pathogen (Langdon and Herbert, 1944). Ferriera and Nevill (1989) reported that weekly applications of bitertanol or triadimefon gave significantly better control of gladiolus rust caused by *Uromyces transversalis* than fenarimol. There was a direct relationship between the number of applications and the incidence of rust. Corm yield was adversely affected by rust, and weekly applications of biteranol or triadimefon were necessary to ensure a reasonable yield.

Bacterial Diseases

13.1.8 Bacterial Scab Blight and Spots

Also known as 'neck rot' is caused by *Pseudomonas marginata*, and 'bacterial blight and spot' is caused by *Xanthomonas gummisudans* which results in neck rot and spots on leaves during the rains. The disease is much serious on the plants grown from cormels and seeds and in over-crowded plots. Rainio (1936) described *Pseudomonas marginata, P. gummisudans* (*Xanthomonas gummisudans*), *Bacillus omnivorus* and *B. variegatus* attacking gladioli in Finland, and Abo-El-Dahab *et al.* (1972) obtained seven virulent isolates in Egypt which were pathogenic to freesia, *etc.* In India, bacterial soft rot caused by *Erwinia carotovora* and neck rot caused by *Pseudomonas marginata*were reported by Misra (1976b). *Pseudomonas marginata* appears as red-brown specks on leaves which become dark brown in the later stages. On leaves the rough spots run together destroying parenchymatous tissues, and leaves start drying from tip when the attack is severe, and tissues at the base decay in wet weather. On corm husks, scab appears as elongated lesions with black rough margins, which sometimes penetrate entirely through husks. Warm and humid weathers are favourable for this disease. McKnight (1948) found this disease infecting gladioli which was controlled by dipping corms in 1:1000 corrosive sublimate solution for 12 hours before planting. Bald (1952) described that under cool and humid weather where soil is warm and moist, drops of water exude through the stomata of the leaves which during high temperature and low humidity retract and draw bacteria from the leaf surface into substomatal cavities and intercellular spaces. El-Goorani *et al.* (1974) found that presence of *Meloidogyne javanica* increased the severity of *Pseudomonas marginata* attack. Koch (1950) in Canada found *Erwinia carotovora* causing soft rot in gladioli similar to vegetables, and Saaltink (1969) in Holland found *Xanthomonas gummisudans* attacking gladioli. Dipping of corms in mercuric chloride at the time of planting and storage gives a better crop stand.

Viral Diseases

Viral diseases may cause serious losses to the growers as they spread rapidly and are very difficult to control. Once a plant is infected it usually remains infected for life even though the symptoms may not become marked. Brierley (1963) reported that gladiolus cormels may develop into virus-free plants if the corms are harvested before the virus moved into the cormels. Symptoms of virus vary depending on virus and host and include stunting and mottling, change of flower colour, twisting of flowers, blotching of leaves, misshapen and discoloured corms, multiple sprouts. In gladiolus, viruses are mostly spread by a few specific insects, while some viruses can infect healthy plants through mechanical means such as cutting of flower spike (Brierley, 1962).

13.1.9 Cucumber Mosaic Virus

Cucumber mosaic virus (CMV) is now one of the major diseases of gladiolus. The infection is expressed as white discontinuous streaks on the foliage like that of severe attack by thrips except that the streaks are more angular in outline. In cultivars with coloured flowers there is usually a streaking or colour break in the flowers and it is noticed in the bud stage as the petals come through the sheath. Some cultivars with white or yellow flowers show distortion of flowers. CMV symptoms also appear as wrinkling or pitting of the corms and discolouration of vascular bundles in the corms. It also affects the plant size and the number of spikes. Many species of aphid including common green peach aphid, potato aphid and cabbage aphid transmit CMV from the infected to healthy plants (Bing, 1972). Lee *et al.* (1983) reported that aphid, *Myzus persicae* could transmit the CMV and BYMV (bean yellow mosaic virus). The use of systemic insecticides, dimethoate and phorate, did not significantly reduce the spread of the virus.

The only method successful in repelling aphids from gladiolus plants is the use of reflective surface (aluminium foils) around the plants (Johnson *et al.*, 1967).

13.1.10 Aster Yellows

This virus is common on many plants of the Compositae family. Two strains of aster yellows, eastern and western, occur in the United States, the western strains being more virulent. Its infection causes precocious sprouting and grassy top, weak shoots, affected root system, stunting, corkscrew malformation and spiral bending of spikes, straw-yellowing of younger leaves and spikes and virescent flowers (Freitag and Tompkins, 1963). Such plants die from the tip towards the base. Due to attack of this virus, the vascular tissues at the base of the corms may show degeneration or discolouration. Rihne *et al.* (2019) observed about 8 per cent to 96 per cent disease incidence of leaf yellowing and streaks in eleven varieties of gladiolus. The presence of mixed infection of 16SrI-D subgroup phytoplasmas and CMV associated with leaf yellowing and streaks symptoms in gladiolus varieties. Groen and Slogteren (1980) described its symptoms and achieved success in transmitting the virus through six spotted leaf hoppers, *Macrosteles sexnotatus*.

Lee *et al.* (1983) observed that BYMV was transmitted through *Myzus persicae* from gladiolus to gladiolus and gladiolus to beans. It can also be transmitted mechanically. Its symptoms appear as light and dark green angular mottling usually on younger leaves and variable streaking of lighter or darker tones in flowers which are insufficient in reducing the flower quality.

It is difficult to control aster yellows but careful roguing and control of leaf-hopper will be helpful.

13.1.11 Tobacco Ring Spot Virus

Tobacco ring spot virus is transmitted by mechanical processes (Brierley and Smith, 1960), certain aphids and grasshoppers but it is not a serious disease. The symptom may appear from symptomless to brilliant chlorotic ring spots on leaves. Its occurrence has not yet been established in India.

13.1.12 Tobacco Rattle Virus (TRV)

TRV has successfully been transmitted from diseased to healthy plants of gladioli using root and leaf extracts (Cremer and Schenk, 1967). They also found that *Trichodorus pachydermus* and *T. similis* also transmit this virus.

The infection of this virus causes constriction on leaves with or without holes of different sizes. The most common symptom is obvious indenting on the margin of the leaves.

13.1.13 Tomato Ring Spot Virus (Tom R)

Tom R has been found to be transmitted by *Xiphinema* nematodes. Its infection causes stunting of plants and flower spikes without any obvious mottling or colour break (Bozarth and Corbett, 1957). The occurrence of this virus in India could not be established.

All these viruses to some extent can be controlled by planting the gladiolus crop quite away from beans, cowpea, tobacco, tomato, cucurbits and the members of Compo-sitae. The control of vectors by the use of systemic insecticides like malathion, metasystox, metacid 50, dimecron, nuvan, nurvacron, *etc.* from 0.02 to 0.2 per cent concentrations controls the viruses. Mineral oil spraying and use of aluminium foils repels the vectors, and flower cutting tools dipped in 10 per cent trisodium orthophosphate or 25 per cent methyl alcohol will check the spread of the viruses by cutting tools (Bing, 1972). The infected plants should also be uprooted and destroyed.

Groen and Asjes (1989) obtained four virus-free strains (733, 736, 740 and 743) of the widely grown cultivar Rose Charm, obtained through meristem culture, were compared, for cut flowers under glass, with standard material. The virus-free strains were more vigorous, and bore one or two more florets per stem, although flower colour and size were similar to those of the standards.

13.1.14 Bean Yellow Mosaic Virus (BYMV, Potyvirus, Potyviridae)

It causes serious disease in *Gladiolus* spp. Thermotherapy combined with meristem culture can greatly improve BYMV elimination efficiency from infected gladiolus corms, resulting in the production of BYMV free gladiolus plants. Thermotherapy, meristemtip culture and combination of both techniques on infected corms/meristem-tip explants (0.5–1 mm in length) resulted in BYMV elimination up to 15.38, 78.04 and 86.66 per cent, respectively, as determined by double antibody sandwich-enzyme-linked immunosorbent assay (DAS-ELISA) and reverse transcription-polymerase chain reaction (IC-RT-PCR). BYMV elimination in gladiolus corms was significantly affected by thermotherapy treatment of infected corms (Parisa *et al.*, 2015).

13.2.0 Pests

Many insects and pests attack gladioli at different stages of growth and development, and even rats and porcupines relish its corms. The most commonly occurring insect pests attacking gladioli are described below.

13.2.1 Seedcorn Maggot

Hylemya cilicura (syn. *Phorbia platura*) maggots feed on the seeds of gladioli by entering the capsules in the field or by webbing the seeds in storage. It was found harmful for gladiolus shoots and leaves and was controlled with 4 per cent wettable chlordane powder diluted in water at 1.35 kg per 100 gallon (Pritchard, 1950). Audemard (1969) reported good control by dipping the corms for one hour in thiram + insecticide solution 24 hours before planting.

13.2.2 Aphids

Aphis gossypii suck the sap of young plants. *Myzus persicae, Macrosiphum solanifolii, M. gossypii, etc.*, damage plants by extracting the sap from the foliage and by transmitting viral diseases (Brierley and Smith, 1947; Smith, 1948; Smith and Brierley, 1953). These aphids may well be controlled with 0.2 per cent methyl parathion sprayings used for the control of cut worms and by using aluminium foils.

13.2.3 Thrips

Taeniothrips simplex feed on flowers and floral-sheaths. The yellow nymphs and dark brown adults rasp the foliage and flowers and cause silvering and whitish streaks. Flower thrips (*Frankliniella* spp.) often invade gladiolus and damage the florets. Thrips also thrive on corms in storage. At the time of spike emergence the plants may be sprayed with methyl parathion at 0.04 per cent to control the pest. Labanowski and Mynett (1990) recommended dipping of corms in cypermethrin at 0.05 per cent and spraying of methamidophos at a concentration of 0.1 per cent to control thrips. Conijn and Groen (1990) reported that storage of corms at 2°C for 6 or 8 weeks killed larvae, pupae, adults of the thrips. A predatory mite (*Amblyseins barkeri*) gave promising results as a biological control. Acetate and omethoate proved very effective when sprayed 2-3 times during the growing season.

13.2.4 Cut Worms

In India, the larvae of *Agrotis segetum* (Misra, 1975) and *Feltia subterranea* and *Prodenia dolichos* in the USA, feed on different parts of the plants and may sever plants at base particularly when young (Magie and Poe, 1972). Since the insects are nocturnal in habit, during day time they hide in the soil surrounding the plants. It may be controlled by spraying the crop with 0.2 per cent methyl parathion at fortnightly interval. Fumigation of infested land has also been suggested (Kelsheimer, 1956).

13.2.5 Mites

Mites (*Rhizoglyphus echinopus, Tetranychus urticae, T. telarius, T. bimaculatus*, etc.), have been found feeding on gladiolus leaves under warm and shady conditions. White specks appear on leaves which consequently results in bronzing or silvering of the affected plants (Magie and Poe, 1972). Careful observation may reveal webbing on the plants provided these mites have good population. Spraying with acaricide will control the mites.

Bulb mites were reported to cause severe damage of bulbs in Taiwan. Liu (1995) investigated on chemical control of bulb mites including *Rhizoglyphus robini* and *R. setosus* on bulbs of gladiolus. He suggested dipping of the bulbs prior to planting and applying acaricide solution (prothiophos) twice onto the soils between 7-28 days after planting. Granular acaricides (such as 10 per cent terbufos G) applied in the planting ditches at 60 kg/ha also gave good control and was stated to be labour saving and more economical. Liu *et al.* (1998) suggested sterilization of loosened soil with steam at 60°C for 30 min or 80°C for 20 min for 100 per cent mortality of bulb mites.

13.2.6 Loopers

Several species including *Trichoplusi ni* and *Pseudoplusia includens* have been found feeding primarily on leaves but occasionally attacking florets or spikes. Loopers and army worms are readily killed by commercial preparation of *Bacillus thuringensis* by causing intestinal paralysis and death of the larvae.

13.2.7 Corn Ear Worms

Kelsheimer (1956) reported in the USA that larvae of *Heliothis zea* feed on young spikes and florets and by boring the stalk they sometimes destroy it. The control measures are similar to that for cut worms.

13.2.8 Other Lepidopterous Larvae

Army worms, *Laphygma frugiperda, Prodenia ornithogalli, Spodoptera eridania, S. exigua* and other species feed on leaves and spikes and may completely destroy these parts of the plants (Magie and Poe, 1972).

Pan'ko (1973) found 57 to 100 per cent success in controlling the eggs of cabbage moth (*Mamestra brassicae*) by releasing 30000 per hectare of *Trichogramma evanescens* parasite. Ciampolini and Zangheri (1975/1976) recorded that *Sasamiano nagrioides* causes 30 – 90 per cent loss which they controlled with carbaryl, mevinphos or azinphosmethyl applied every 7 to 10 days.

Magie (1987c) suggested a number of insecticides for effective control of some common pests of gladioli, (i) lannate or nudrin, dipel and orthene, for chewing insects (worm) (ii) sevin and orthene for thrips, (iii) cygon and dursban for thrips and aphids, malathion for aphids, (iv) vendex for spider mites, and (v) lindane for wire worms, dipel for loopers.

Diseases and Pests of Gladiolus

Fusarium Wilt

Rust

Carvularia Blight

Iris Leaf Spots

Spongy or Core Rot

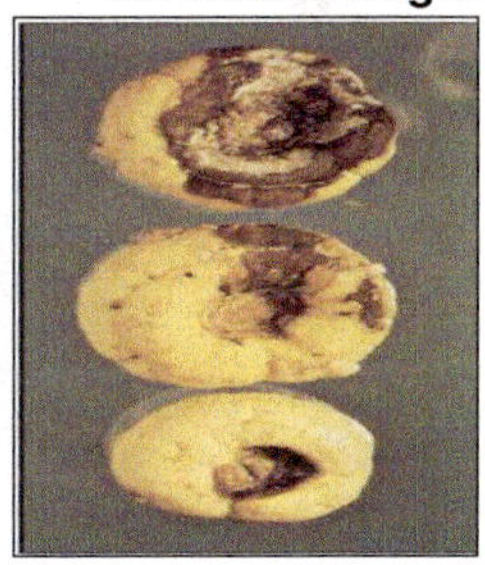

Corm Rot

Cucumber Mosaic Virus (CMV)

Stem Borer

Aphid Infestation

Thrips Infestation

Mites Infestation

Nematode Infestation

13.2.9 Slugs and Snails

Slugs and snails are disastrous in the hills with the gladioli plantings at a screened situation or near the wild plants and grasses, especially during cloudy and humid weathers. These are nocturnal in habit. They climb the plants and feed on the leaves battering them completely. While feeding, they leave a trail of slimy reflective substance. During day time, in cloudy weather, these may be seen feeding on the underside of leaves.

These may be light trapped or with the help of poison bait these may be killed. Poison bait may be prepared by using 250 g wheat bran + 50 g jaggery and 6-8 g metaldehyde.

13.2.10 Nematodes

Nematodes damage the roots of gladioli in well drained loamy and sandy soils. Damage is less likely in heavy clay soils in cool climates. *Meloidogyne* and *Trichodorus* attack gladioli corms and roots (Kelsheimer, 1954). Reddy *et al.* (1979) reported *Meloidogyne incognita* causing root-gall formation in gladioli in Bangalore (India). Abbas and Waliullah (2010) reported that root-knot nematode *Meloidogyne, Helicotylenchus* and *Pratylenchus* spp., were associated with gladiolus roots.

Assumi *et al.* (2017) reported that gladiolus crop severely infested with *M. incognita* showed significant reduction in plant growth parameters and flower production. In severe infestation, no flower was produced. The histopathology of plant roots revealed that 2nd stage juveniles penetrated into root cortex and moved along the cortical layer of cells and started feeding. During feeding process, they produced some hypertrophied cells in vascular cylinder and hyperplasia in pericycle regions. Though the site of infection was vascular tissue but due to the nematode development and giant cell formation, the entire nematode complex and giant cells appeared irregular in root section. The nematodes damaged complete thickness of roots and this condition lead to the destruction of feeder roots which ultimately affected plant growth by reducing water and nutrient absorption from soil.

Pandey *et al.* (2018) investigated biochemical changes in gladiolus varieties inoculated with root-knot nematode (*Meloidogyne incognita*). Reduced percentage of total chlorophyll contents were observed in inoculated samples than the healthy counterparts. However, an increase in amount of total protein, phenol, proline and total sugar contents was observed in the diseased tissues.

Soil fumigation and treatment of dormant planting stock in hot water (53°C) for 30 minutes kills the nematodes. Khan *et al.* (2017) found that corm treatment with nemacur or *P. fluorescens* suppressed the galling and egg mass production of *M. incognita*.

13.3.0 Disorders

Fluoride injury has been observed in gladioli. Woltz (1963, 1964, 1965) and Woltz *et al.* (1953) reported leaf scorch of gladioli in Florida due to the presence of fluorine compounds in the atmosphere which was accumulated on the tips of leaves. Woltz (1963) found that the toxicity of sodium fluoride sprays was reduced by applying calcium sulphate to the soil as well as salts of calcium, manganese and magnesium as foliar spray. He also stated that fluoride injury was associated with heavy application of superphosphate. In 1965, he found that leaves imbibed with fluoride decreased the photosynthetic activities. Hitchcock *et al.* (1962) made the comparative studies of different cultivars by fumigating the plants with hydrofluoric acid (HF) where he found that the plants grown in fumigated area developed more tip burns with more fluoride accumulation, the injury being more severe on the middle-aged leaves whereas fluoride accumulation was more in older leaves. Brewer *et al.* (1966) found deteriorated flowers in terms of both quality and quantity by fluoride injury under controlled greenhouse conditions. Spierings (1967) while working on the *cv.* Snow Princess which is very sensitive to fluoride found a positive correlation between fluorine content of the atmosphere, the extent of leaf tip burn and the fluorine content of the leaf tips. In further experiments, he (1969) found that 10 cm cut leaves of gladioli absorbed fluorine readily from fluoride-treated tap water which was accumulated mainly in the apical 4 cm part, whereas the cut-spikes placed in fluoride-treated water accumulated fluorine in the tips of the flower bracts. Water used for cut flowers should also be low in soluble salts (Waters, 1966, 1968) and free of dissolved fluorides (Waters, 1968; Marousky and Woltz, 1971) because even as little as 0.25 ppm fluoride is detrimental for certain cultivars.

The effects of hydrogen fluoride on Snow Princess and Elizabeth the Queen were studied by growing the plants in sand culture in complete nutrient solution and also under deficiencies of K, Ca, P, N, Mn or Fe and using hydrogen fluoride fumigations after 6 and 7 weeks of planting (McCune and others, 1966). Increased tip burn was associated with K and P deficiencies in Snow Princess and with Mg deficiency in Elizabeth the Queen and decreased tip burn with Ca and N deficiencies in Snow Princess; no effect of Fe or Mn deficiency on tip burn was found in either cultivar.

Singh *et al.* (1990) reported reduction in chlorophyll content, shoot length, phyto mass production in gladiolus when exposed to 2µg SO_2/l for 2 h daily. The SO_2 sensitivity varied in 5 different cultivars.

14.0 Harvesting and Yield

14.1.0 Harvesting of Flower Spikes

For cut flower purpose the spikes are inspected regularly in the field for their growing style, placement of the flowers on spikes or for any other possible mishaps. Poor placements may be corrected by the judicious use of fingers everyday. Care should be taken that spikes must grow straight, therefore, stakes are to be placed in time. Any hand manipulation in the spikes should be effected only when the temperature is above 24°C, otherwise there are chances that spikes or flowers may be damaged. The harvesting should be done at the tight bud stage with at least four leaves left on the plant and from one to five buds showing colour so that these may easily open indoors one by one. The leaves left on the plant are needed for development of corms and cormels (Compton, 1960; Hussein *et al.*, 1962; Wilfret and Raulston, 1974). The cut-spikes are immediately placed in water indoors at 6.0°C for the local market. The spikes cut well in advance are refrigerated at 2.3 to 2.8°C if they are to be stored for a week whereas for one or two days these may be stored at 4.5°C. Refrigerated spikes are frequently taken out to facilitate opening of flowers.

14.2.0 Yield of Flowers

The flower spike yield in gladiolus varies according to the cultivar, corm size, planting density and management practices, *etc*. Generally, one plant produces single marketable spike and plantable corm. High plant density with proper management have been found to give higher yield per unit area. Approximate yield of flower spike would be around 2,00,000 per hectare.

14.3.0 Lifting of Corms and Cormels

Corms are matured when 25 per cent cormels become brown which generally takes 30 to 45 days from flowering, moreover the leaves also start yellowing (Hockings, 1972; Webster, 1972). Late harvesting of corms results in increased spike length and floret number in the following season whereas earlier flowering is obtained from corms lifted earlier. As soon as corms mature these should be gently lifted because a little can be gained by delaying it. During mild winters, corms survive in the soil provided the soil remains dry throughout, but in severe winters the corms are infected by a number of diseases like *Botrytis*, *Erwinia*, *etc*., and also by certain saprophytic fungi. When corms severely damaged by frost were planted, sprouting was noticed in one sixth of the number of corms planted (Gigante, 1973a,b). It was also observed that very low temperature caused browning of the whole or part of the corm surface and discolouration of the underlying tissues resulting in rotting. In Russia, Astvatsatryan and Sarkisyan (1974b) noticed that autumn planting of cormels less than 7 mm in size was unsuitable to overwinter until they were covered with a 10 cm layer of saw dust in frames. Certain cultivars are, however, quite hardy and can tolerate winter without any cold damage (Misra and Choudhury, 1977).

The cormels may be gathered by pulling the plant gently as they are attached to the corms. In developed countries lifting is carried out by well designed machines but in India lifting is done with the help of spades and khurpi (a handy tool used for loosening the soil and interculture operation). However, by using a frame of iron or nylon mesh, the corms and cormels are cleaned more rapidly. This is very helpful if the soil adhering the corms and cormels is loose and dry. At the time of lifting, the dried stems are twisted off and the withered mother corms attached to the bottom of daughter corms are taken out easily by pushing with the thumb, otherwise the corms may be injured while removing dry and hard mother corms (Webster, 1972). Care should be taken to prevent damage of the corms which may be the source of infection with pathogenic as well as saprophytic fungi. At the time of lifting, corms may also be checked for any disease infection and the affected corms are discarded. Dried stems and leaves, and other withered remains may be collected and burnt. In India, lifting in the hills starts from September to November and in the plains April to May.

14.4.0 Yield of Corm and Cormels

The yield of gladiolus corm and cormels is influenced by cultivar, corm size and other factors, Singh (1996) reported that large cormels produced larger corms with more cormels/plant. Bose (1984) obtained 33.1 t/ha of corm and cormels of *cv.* Oscar when planted at 10 x 30 cm. The same cultivar produced 41.3 t/ha when sufficient nutrition was applied.

15.0 Postharvest Management

The postharvest handling is the most important aspect is cut flower industry because about 20 per cent losses due to improper handling. The vase life of cut flowers of gladiolus depends on the genotype of crop and also influenced by different chemical and sucrose, added to the vase solution (Kumar and Awasthi, 2012). Usually, stems are harvested at the bud stage and need water and carbohydrates to support inflorescence development and to ensure flower quality. Vase solutions can supply these elements. However, sugars may benefit bacterial proliferation, which eventually can block xylem vessels. Bactericides potentially inhibit microorganism proliferation and new compounds are being tested constantly (Casares *et al.*, 2017).

Kumar *et al.* (2015) provided the initial evidence that exogenously applied ABA (100 μM) accelerate the senescence process by enhancing the activity of several senescence processes such as lipid peroxidation and lipoxygenase activity. On the contrary, the antioxidant enzymatic activity was reduced, *viz.*, superoxide dismutase, catalase, peroxidase, ascorbate peroxidase and glutathione reductase, in petal tissue. Supplementation of GA_3 in the vase solution having ABA reversed various senescence related parameters, which were down or upregulated by ABA proving its role in senescence of gladiolus flowers.

Costa *et al.* (2016) expressed that the gladiolus flower is classified as insensitive to ethylene. Abscisic acid is involved in the induction of senescence-related events in gladiolus flowers, such as high loss of membrane stability and abnormal flower opening. GA_3 regulates the action of ABA in the maintenance of cell membrane and opening of gladiolus flowers.

15.1.0 Flowers

15.1.1 Grading

Gladiolus spikes are graded into following five grades based on overall quality, length of spike and number of florets in each spike (Wilfret, 1980).

Grade	Spike Length (cm)	Number of Florets (minimum)
Fancy	> 107	16
Special	> 96 to 107	15
Standard	> 81 to 96	12
Utility	81	10

15.1.2 Packaging

For local markets spikes may be taken submerged in water but for distant markets these may be carried dry in cardboard or wooden boxes or in trunks and this way these can easily be retained for 12 hours whether it is for interstate or to be lifted by air for international markets. These rectangular boxes may be prepared having 1.2 metre length, 60 cm width and 30 cm height, perforated at several places for circulation of air. Care should be taken to avoid bruising the flowers while packing. Between the spikes, inside the bundles, absorbent cotton may also be placed near the swollen flowers to avoid them being flattened by inadvertent pressure. The spikes are arranged head to tail alternately, and tied with cotton wool so that they remain in position. The spikes carried dry should immediately be placed in water after these have reached the destinations. A 2.5 cm diagonal cut is made at the bottom of each stem so that capillary tissues are exposed for maximum water intake. These are kept straight in fresh water until petals and spikes become fully turgid. These spikes are then placed in a floral preservative, containing at least a source of carbohydrate and a bactericide and by the use of floral preservatives the vase life may be prolonged from three to five days (Kofranek and Paul, 1974; Marousky, 1971).

Bundles of 50 to 100 spikes are prepared for air lifting in the perforated and light-proof cardboard boxes. The upper portion of the spikes may bend downwards if cardboards are not kept vertically. Refrigerated vans are also used to transport gladioli spikes to distant markets. After the spikes are put in flower vases they continue to blooming for 10 to 15 days. One floret ordinarily lasts four days and generally one floret from bottom to top opens every day. While opening indoors deep colours become lighter.

Beura and Singh (2003) found that cellophane and butter paper packed flowers stored at 4°C packed in cellophane for 2 days, in cellophane or butter paper for 4 days and in butter paper for 6 days were proved to be the best to increase the postharvest life of cv. Her Majesty.

Grover *et al.* (2005) observed that polypropylene of 100 gauge (25 μ) packages maintained modified atmospheres with low levels of O_2 and high levels of CO_2 throughout the duration of the storage. These packages can be held in transit for 15 days with post-storage vase life of more than 6 days. Sisodia *et al.* (2018) revealed that cellophane paper packaging recorded maximum diameter, days to withering of first and third florets, weight of spike at third, sixth and ninth day, and spike packed in news paper recorded maximum shelf life and length of first floret.

Kumawat *et al.* (2018) observed that the maximum diameter of first floret and length of third and fifth floret with spike packed in brown paper. However, maximum weight and shelf life were recorded with spikes stored in cellophane paper.

Chand *et al.* (2019) reported that spikes *cv.* White Prosperity pulsed for 12h, wrapped with cellophane and stored for 7 day recorded least number of florets was opened. The percentage opening of florets decreased with increase in storage duration and also the postharvest quality deteriorated with increase in storage duration.

15.1.3 Storage

Gladiolus spikes of *cv.* White Friendship, wrapped in polythene or kraft paper placed in cartons, were subjected to simulated condition at 4.4°C (40°F) or 10°C

(50°F) for 3 days. They were then held at 23.3°C (74°F) either in water or in 400 ppm hydro-xyquinoline citrate (8-HQC)+3 per cent sucrose. Spikes held in preservative continued to produce open florets and remained turgid for 8.5 to 10 days whereas spikes held in water wilted after 4.5 to 6 days. Spikes wrapped in kraft paper lost about 10 per cent moisture but this did not affect floret opening (Marousky, 1974). Holmes *et al.* (1957-58) found that cut gladioli lost moisture about 8 times faster at 28.3°C (89°F) than at 10°C (50°F). Buds of spikes cut early in the morning remained tighter in storage than those cut in mid-morning. Satisfactory keeping quality is obtained when spikes are stored at 1.6-4.4°C (35-40°F) for 6-9 days, 4.6-7.3°C (40-45°F) for 4-6 days or 10-29.9°C (50-80°F) for 2-4 days. Spike turgor and the appearance and opening ability of flowers, covered with moisture proof wrapping or stored in boxes lined with wrapping are superior to those in paper container (Waters, 1966). Patil *et al.* (1994) observed the longest (7.75 days) vase life when the cut spikes were stored at 4-4.5°C for 48 h.

Floret opening in gladiolus improves by controlled atmospheric storage with 1 per cent O_2 and 5 per cent CO_2 for 6-8 days. The respiration rate in CA stored flowers was 20-30 per cent lower than those stored in normal air (Hardenberg *et al.*, 1969). Meir *et al.* (1995) sealed groups of 25 spikes of mini-gladiolus cv. Adi in individual poly-thene liners, in which a MA composed of 4-7 per cent CO_2 and 10-14 per cent O_2 was created. It significantly improved floret opening and longevity following dry storage for 14 days at 2°C, and retarded bract and leaf yellowing. MA packaging retarded leaf and bract senescence presumably due to high CO_2 concentration accumulated within the package.The storage of flowers can also be extended to over 30 days by low pressure storage techniques (Staby *et al.*, 1978). However, this method is not economically feasible.

Belwal *et al.* (2019) found enhanced storage life of gladiolus cv. Nova Lux with sucrose 3 per cent. Kumawat *et al.* (2018) revealed that spikes stored in cool chamber condition found higher vase life than the ambient condition. Slow opening of florets, diameter of florets and percentage of opened florets were found maximum in cool chamber condition (16°C).

Avneet *et al.* (2016) observed that increase in storage duration cause decrease in size of floret and water absorption in spikes. Sisodia *et al.* (2018) revealed that there was no significant difference between ambient (22 ± 2°C) and cool chamber condition (16°C) on weight of spike at third, sixth and ninth day, days to withering of first and third floret and shelf life.

Bhat and Sheikh (2015) reported that shorter storage durations (0 days) recorded higher vase life, longevity of open floret and total water absorption, while, 21 days storage found superior for days to open the basal floret and floret size. The cut spikes stored for 28 days did not open either pulsed with chemical solutions or in control.

da Costa *et al.* (2017) evaluated the vase life and the rehydration capacity three gladiolus varieties after dry storage at low temperature (5 ± 1°C) and relative humidity (85 per cent) for 12, 24, 36, and 48 h and then kept in water at room temperature. In dry cold storage conditions, for up to 36 h, the vase life was not affected. Rehydration capacity of the stem is linked to the staggered opening of flowers along the inflorescence.

15.1.4 Senescence of Flowers and Vase Life

Investigations for extending the vase life have been successfully applied to keep the flower of gladiolus fresh for a long period. Bravdo *et al.* (1974) found that sucrose improved water balance in cut-gladioli, however, this cannot account for the great increase in water uptake in sucrose-treated gladioli. Marousky (1968, 1969) found 600 ppm 8-hydroxyquinoline citrate (8-HQC) with 4 per cent sucrose effective for increasing the water uptake and decreasing vascular blockage in gladiolus cultivar White Friendship, whereas Rameshwar (1974) found Al_2SO_4 (0.1 per cent) equally effective as 8-HQC (600 ppm).

Mayak *et al.* (1973) found that 20 per cent sugar for 20 hours at 20°C improved the opening, size and durability of gladiolus flowers, whereas Kofranek and Halevy (1976) suggested the use of 20 per cent sugar with 1000-1200 ppm $AgNO_3$ before storage for 7 or 10 days at 21°C for pulsing of the gladiolus stem which consequently resulted in greater floret opening and longer life of the flowers and might be used for the flowers prior to shipping or storage of gladiolus for long periods. Lukaszewska (1978) reported improvement in flower opening and vase life by placing the spikes in solution containing 5 per cent sucrose and 8-hydroxyquinoline sulphate (8-HQS), $AgNO_3$, or $NaNO_3$. Beneficial effect of 3.5 per cent sucrose and 200 ppm 8-HQS solution in improving the vase life was also observed by Riest (1977).

The vase life was considered best when inflorescences at green bud stage were stored at 4 to 6°C for 3 weeks and pulsed over 5 h in a solution containing sucrose + silver salts + pyrosulfurous acid. High ABA content in the petals was associated with flower senescence (Kondrat'eva and Belynskaya, 1989).

Zhou *et al.* (1994) also used a chemical solution including $CoCl_2$ that extended the vase life of cut gladiolus flowers from 8 days in control to 16 days and also increased the quality of cut flower. Similar investigations using sucrose, STS and 8-HQS also proved effective in extending the vase life of gladiolus (Song *et al.*, 1992; Hwang and Kim, 1995). Zhou *et al.* (1995) also recorded the maximum vase life of 12 days in the holding solution containing sucrose + silver nitrate + 8-HQS + organic acids.

Preservation was enhanced by lowering the pH to 3.5 with citric acid. Water uptake by the stems increased as conditions for prolonging the vase life improved. In an experiment on cv. Firebrand, the vase-life of spikes increased by dipping stem bases in $AgNO_3$ at 1000 ppm for 15 minutes or 2000 ppm for 5 minutes. Sucrose at 3 to 6 per cent in a preservative solution of pH 3 also extended it (Choi and Roh, 1980).

Merodio and Plaza (1989) evaluated the effects of postharvest ethylene removal treatments on cut flower stems of gladiolus using three different absorbants : Ethysorb, New Green Pack and Green Keeper. Ethylene removal from storage atmosphere contributed to a reduction in the respiration rate and in weight loss of fresh spikes. They observed a correlation between ethylene content and stem rigidification. Ethylene removal was effective in retarding ethylene evolution, colour development of petals and the enhancements of more intensive flower opening in both 'white' and 'pink' gladiolus cultivars.

Spikes harvested at 1-2 basal florets show colour had a higher vase life, lowest transpirational loss, highest florets diameter, least decrease percentage of lipids and proteins White Prosperity (Kumar *et al.*, 2015). Spikes of *cv.* Nova Lux harvested at basal pair opened bud stage recorded maximum vase life with continuous increase in spike length, absorption of vase solution, first increase thereafter decreased and continuous florets opening, maximum spike weight and moderate florets drooping (Dwivedi *et al.*, 2018).

Evaluation of 10 gladiolus varieties for postharvest life revealed that maximum changes in fresh weight (7.21 g) on day 2 was recorded in Pusa Sagun and minimum in Super Star (1.86 g), while at senescence stage maximum water loss in fresh weight was observed in Sancerre (-16.86 g) and minimum in Yellow (-4.36g). Maximum water uptake on days 2 (41.00 ml), 4 days (59.50 ml), 6 day (73.00 ml) and on senescence (75.50 ml) was measured in Pusa Suchitra. However, maximum numbers of opened florets were recorded in Sancerre (8.00), Pusa Sagun (8.00) and peter pears (8.00) at day 6 in vase. Maximum increase in spike length at senescence was measured in Pusa Suchitra (6.65 cm) whereas, longest vase life was found in Pusa sagun (8.65 days) followed by Psittacinus Hybrid (8.00 days) (Kumar *et al.*, 2009).

Among varieties evlautaed for vase life, Novalux, Hunting Song, Princess Margarate Rose, Blue Sky and Precillia recorded greater vase life (7.11-6.11 days). However, Charisma, Grandpics, Ocillia, Novalux, Princess Margarate Rose performed better to per cent opening of floret, while, Purple Flora, White Prosperity, Victor Berge, Friendship White, Oscar and NG-35 recorded bigger florets with more length and width (Patra and Mohanty, 2015). Out of 12 gladiolus genotypes evaluated for vase life, maximum vase life and number of florets remaining open at a time was observed in cvs. Darshan (17.56 days) and Hunting Song (6.78 florets), respectively (Singh *et al.*, 2017) (Table 19).

Table 19: Performance of Gladiolus Varieties with Respect to Vase Life, Number of Florets Open at a Time, Days to Wilting of Basal Floret and Number of Florets Remaining Unopened*

Genotype	*Vase Life of the Spike (Days)*	*Floret Remaining Open at Time*	*Days to Wilting of Basal Floret*	*Floret Remaining Unopened*
African Star	15.89	5.11	5.89	5.89
Arti	14.33	5.56	5.78	1.11
Darshan	17.56	4.22	4.89	0.22
Friendship	15.44	6.11	7.00	1.67
Huntaing Song	16.11	6.78	5.89	1.22
Legend	11.33	5.89	6.11	2.78
Pusa Srijana	14.44	4.78	5.78	3.89
Pusa Kiran	15.22	5.44	6.44	0.33
Snow Princess	15.44	4.33	5.33	3.78
Sunaina	13.78	5.11	6.11	3.67
Trader Horn	14.89	5.44	6.00	0.44

Genotype	Vase Life of the Spike (Days)	Floret Remaining Open at Time	Days to Wilting of Basal Floret	Floret Remaining Unopened
Urmi	15.67	5.00	6.22	0.00
Mean	15.00	5.31	5.95	1.83
SEm±	0.75	0.67	0.51	0.41
CD (0.05)	1.57	1.32	NS	0.84

* Singh *et al.* (2017).

Pre-storage pulsing with sucrose (20 per cent)+$Al_2(SO_4)_4.16H_2O$ (400 ppm) vase life, opening of florets, floret size, number of florets opening at one time and decreased per cent loss of spike weight in storage (Grover *et al.*, 2005).

Pulsing cut flowering stems of *cv.* White prosperity in 20 per cent sucrose solution for 20h before holding them in 2 per cent sucrose preservative solution under room temperature (21±1°C) improved the change per cent in fresh weight of inflorescences, water uptake, opening flower, flower diameter, longevity, photosynthetic pigments content in the leaves, as well as total carbohydrates per cent in the leaves and petals till the end of vase life (Darwish *et al.*, 2018). The pulsing with 5 per cent sucrose recorded maximum vase life and number of florets opened, proved best pulsing treatment for prolonging the vase of life *cv.* American Beauty (Nijasure *et al.*, 2004).

Pulsing cut spikes of *cv.* White Prosperity in 20 per cent sucrose solution for 20 h found the best before transferring them to the preservative solution, as it significantly improved the change per cent in fresh weight of inflorescences, water uptake, opening flower, flower diameter, longevity, photosynthetic pigments content in the leaves, as well as total carbohydrates per cent in the petals (Sarhan *et al.*, 2018).

Gladiolus *cv.* White Prosperity spikes pulsed with 20 per cent pulsing solution for 24 hours recoreded maximum vase-life with continuous opening of florets, increase in spike length, absorption of vase solution (Gupta and Kumar, 2018). Treatment of cut spikes with 12 h with 20 per cent sucrose solution prolongs vase life, greater number of opened flowers and marketable spikes (Ramos *et al.*, 2015; Babaji *et al.*, 2014). Salicylic acid (SA) found helpful in extending the vase life of gladiolus cut flower (Tariq *et al.* (2016).

Vase solution containing $AgNO_3$ 25ppm, $AgNO_3$ 25ppm + sucrose 2 per cent and sucrose 2 per cent influenced the vase life, water uptake, flower diameter, rachis length, floret length and florets open in cv. Yellow Stone (Tariq *et al.*, 2016), Na-DCC plus sucrose (4 per cent) improved water balance, increased open flowers, reduced wilting of flowers and fresh mass loss, therefore, enhance postharvest quality of cut gladiolus inflorescences in cv. White Friendship (Casares *et al.*, 2017), aluminium sulphate to be the best preservative for maximum floret opening, increase in spike length and longer vase life in cvs. White Prosperity and Red Beauty (Gupta *et al.*, 2020). However, holding cut spikes in 2 per cent sucrose preservative solution at the room temperature after pre-treatment with 20 per cent sucrose pulsing gavesbest results (Sarhan *et al.*, 2018). Sharma (2006) found that sucrose (4 per cent) + 8-HQC

(200 ppm) recorded longest vase life and total blooming period. Similarly, the maximum days taken for opening of basal floret of gladiolus spike, floret size, maximum open florets at a time, and vase solution uptake were also obtained with sucrose (4 per cent) + 8-HQC (200 ppm). However, the maximum longevity of floret and increase in spike length were noted in sucrose (4 per cent) + $Al_2(SO_4)_3$ (300 ppm).

Kumar *et al.* (2009) observed that 4 per cent sucrose + 8-HQC (250 ppm), improved the diameter of 1st and 3rd floret, longevity of first florets, effective useful life and vase life. However, minimum loss in fresh weight of spike at senescence was observed with 4 per cent sucrose + acetyl salicylic acid (200 ppm) in gladiolus *cv.* Pusa Jyotsana. Kumar *et al.* (2010) found that 4 per cent sucrose +250 ppm 8-HQC increased days to basal floret opening, floral size fifth and second floret, however, 4 per cent sucrose + 300 ppm $Al_2(SO_4)_3$ recorded maximum spike length, vase life, vase solution uptake and longevity of first five florets in *cv.* White Prosperity. Mehraj *et al.* (2013) reported that sucrose (100 ppm) + lemon juice solution recorded maximum days taken for floret senescence, number of floret opened during floret senescence, highest longevity with minimum fresh weight loss, whereas, minimum days required for floret senescence in 100-ppm lemon juice solution (Table 20).

Table 20: Assesment of Vase Life of Cut Gladiolus (*Gladiolus hybridus* Hort.) to different Solution in Terms of Various Attributes*

Treatment	*12 Days after Treating*		
	Spike Diameter (mm)	*Fresh Weight Loss (Per cent)*	*Floret Opening (Per cent)*
Control	8.3 c	39.7 a	50.0 e
Distilled water	8.5 b	36.6 c	58.0 d
Sucrose solution (100 ppm)	8.3 c	35.4 d	66.7 c
Lemon juice solution in distilled water (100 ppm)	8.2 d	39.4 b	75.0 b
Sucrose+ lemon juice solution in distilled water (100 ppm)	8.6 a	28.2 e	91.7 a
CV per cent	0.11	0.01	0.01
LSD (0.05)	0.02	0.01	0.01

Note: In a column means having similar letter(s) are significantly identical and those having dissimilar letter (s) differ significantly as per 0.05 level of probability.

* Mehraj *et al.* (2013).

Uddin *et al.* (2016) found that preservative solution 3 per cent sucrose + 100 ppm STS recorded maximum spike diameter and floret opening, days to basal floret senescence, number of floret open during basal floret senescence, solution absorption, vase life with minimum fresh weight loss and floret wilting at 12th days after treatment. Belwal *et al.* (2019) reported that 8-HQC 300 ppm and sucrose 6 per cent as floral preservatives was found optimum in improving floral floret diameter, longevity, spike length, preservative uptake, floret opening percentage, blooming period in *cv.* Nova Lux. Sharma *et al.* (2017) found that $AgNO_3$ (300 ppm) with 5 per cent sucrose was best for vase life, solution uptake, number of basal florets open at a time, numbers of florets open at senescence of basal florets, in six varieties of gladiolus. Dwivedi *et al.* (2016) found that sodium nitroprusside (100 ppm/l) increased vase life by increasing the scavenging mechanism of reactive oxygen species (ROS) in terms of antioxidants activity, membrane stability and down-regulation of GgCyP1expression in gladiolus cut flowers of cv. Snow Princess.

Higher temperature, there was significant reduction in fresh weight of cut spike of gladiolus at harvest and senescence was accelerated by increased temperature (Kadam and Singh (2009). Boric acid (30 to 120 ppm) in vase solution led to increase the florets diameter, duration period and inhibit the growth of microorganisms in the vase solution, while, 5-salfosalicylic acid (100-200 ppm) gave a fast opening of the florets, increased the number of the opened florets, decreased the number of the non-opened florets per spike and increased absorbed vase solution in cv. White Prosperity (Khattab *et al.,* 2017).

Putrescine (300 mg/l) in preservative mixture increase flower diameter, whereas, water uptake was elevated with Putrescine increased to 300 mg/l. Salicylic acid (300 mg/l) had positive and significant effect on soluble solids (Mohammadi *et al.,* 2014). Ethylene does not influence the flower opening or vase life of the gladiolus cultivars 'Amsterdam', 'Blue Frost', 'Gold Field', 'Green Star', 'Jester', 'Lavender', 'Red Beauty', 'Rose Supreme', 'Traderhorn', and 'Verônica'. The vase life differed between cultivars regardless of ethylene action. The maintenance of flower opening is attributed to total soluble sugars content, which, in turn, is a major determinant of the vase life (Da Costa, 2016). Addition of calcium chloride in vase solution resulted in the maximum days to 1st floret opening, 1st floret fading and full spike fading, fresh and senesced floret weight as well as the least weight loss florets. Calcium chloride was the best source among various sources of calcium (Table 21) when applied in holding solution (Ahmad and Rab, 2019).

Table 21: Fresh Floret Weight (FFW), Senesced Floret Weight (SFW) and Weight Loss (WL) of Gladiolus Spike as ASffected by Various Sources of Calcium and Application Methods*

Treatments	*Details*	*Physical Quality Attributes in Gladiolus Floret and Spike*		
		FFW (g)	*SFW(g)*	*WL (per cent)*
Calcium sources (S)	Distilled water	7.341 d	3.450 e	36.966 a
	Tap water	7.549 d	3.667 d	34.820 b
	Calcium chloride	8.860 a	6.798 a	8.229 e
	Calcium sulphate	7.913 bc	6.146 c	10.328 c
	Calcium nitrate	7.877 c	6.037 c	10.580 c
	Calcium gluconate	8.115 b	6.394 b	9.682 d
	$LSD_{0.05}$	0.211	0.173	0.285
Application methods (M)	Holding solution	8.105	5.757	17.674
	Foliar application	7.780	5.073	19.194
	Significance	—	—	—

Treatments	*Details*	*Physical Quality Attributes in Gladiolus Floret and Spike*		
		FFW (g)	*SFW(g)*	*WL (per cent)*
Year	Year 1	7.889	5.369	18.381
	Year 2	7.996	5.461	18.348
	Significance	NS	NS	NS
Interaction	S X M	**	**	**

NS = Non-significant, *= Significant at $P \leq 0.05$, **= Significant at $P \leq 0.01$

Means followed by different letters in respective colums are significantly different from each other at $P \leq 0.05$

* Ahmad and Rab (2019).

15.1.5 Physiology of Senescence

The opening and senescence of gladiolus florets was accompanied by climacteric and non-climacteric pattern of respiration and ethylene production depending on cultivar. A climacteric pattern of ethylene production by the youngest buds on the spike was stimulated by cold storage (2°C), and was not affected by holding the spike in a preservative solution containing sucrose. Ethylene treatment had no effect on senescence of florets. Pulsing treatment of the spikes with silver thiosulphate (STS) improved floret opening but not the life of individual florets.

Sucrose and STS had similar but not synergistic effects on floret opening, suggesting that STS improved flower opening in gladiolus by overcoming the effects of carbohydrate depletion (Serek *et al.*, 1994). Yamane and Ogata (1995) reported that gladiolus cv. Fujinoyuki spikes in a vase containing a solution of 300 μM cycloheximide (CHI), an inhibitor of protein synthesis, decreased perianth length, fresh weight, dry weight, total soluble sugar and protein contents. CHI did not affect perianth acid invertase activity but it inhibited other protein synthesis and lowered other perianth total soluble sugar content. CHI treatments of excised, fully opened florets prevented fresh weight and dry weight losses and the decrease of total soluble solid and protein in the perianths. CHI also inhibited an increase of ion leakage from perianth tissues and delayed the onset of wilting. Yamane *et al.* (1999) observed that treatment of cut spikes with sodium benzoate and n-propyl gallate, as free radical scavengers, slightly delayed the wilting of perianths in detached gladiolus florets. Superoxide dismutase (SOD) activity of the perianths sharply decreased within two days after their full unfolding, when they commenced to wilt. The decrease of SOD activity was alleviated by treatments with 300 μM cycloheximide which prevented perianths from wilting.

Vase solution of 4 per cent sucrose + 4 ppm PbSNPs (*Piper betle* mediated silver nanoparticles) maintained spikes fresh and dry weight, reduce the vascular blockage, improve the antioxidative defense, and stabilize the membrane integrity that leads to delay senescence. PbSNPs in vase solution might have played a significant role for scavenging the reactive oxygen species by inducing antioxidant enzyme system at the time of senescence, leading to a decrease in lipid peroxidation and increase in the membrane stability (Maity *et al.*, 2019).

Tree bark extracts (major forest plant residues), as an ecofriendly and natural antioxidant preservative in holding solutions, is a novel tool for extending flower longevity. *Magnolia acuminata* and *Taxus cuspidata* bark extracts was used as additives in holding solutions. *G. grandiflorus* subjected to bark extracts as well as catechin and protocatechuic acid (main phenols) displayed significant increased longevity (up to 18 days), an increased number of open florets, and increased floret fresh weight. Higher antioxidant activities were detected and associated with increased superoxide dismutase and catalase enzyme activities and reduced H_2O_2 accumulation. The bark extract treatments associated with reduced expression of GgCyP1 (produces cysteine protease) and increased expression of both GgDAD1 (defends against apoptotic activity) and GgEXPA1 (regulates petal expansion). Several mechanisms were implicated in these effects, including maintenance of water content, enhanced management of reactive oxygen species (ROS), increased sugar and protein composition, and control of microbial growth. Thus, bark extracts and isolated phenols could be developed as an ecofriendly, non-toxic, and cost-effective natural preservative for cut gladiolus flowers (Elansary, 2020).

Calcium at 200 mM through vase holding solution extended 1st floret and full spike fading, the higher fresh floret weight and senesced floret weight, retention of highest reducing and non-reducing sugars, least electrolyte leakage and extended vase life, and least loss of protein and highest total phenols in florets of gladiolus spikes. Therefore, application of 200 mM calcium through holding solution was most effective in retaining the quality of gladiolus florets and spikes over extended period and enhanced the vase life of gladiolus spikes (Ahmad and Rab, 2020).

Zinc being an activator of certain enzymes, regulates antioxidant activity; therefore, it could enhance the shelf life of cut flowers. Vase quality parameters like percent florets opened, vase life and fresh weight change were greater with 8 mg Zn/kg, and the least membrane leakage was also ensured at this rate. Antioxidant enzymes, *viz.*, SOD, CAT, POD and free radical scavenging activities in cut flowers remained at the highest with 6 mg Zn/kg. They concluded that Zn applied at 6–8 mg/kg imparts greater beneficial effects vase life quality, and antioxidative activities in gladiolus cut flower (Saeed *et al.*, 2013).

It is observed that fresh weight, vase solution uptake and vase life of gladiolus flowers kept in Sodium Nitroprussside, which is a source of nitric oxide, 100 ppm solution remained higher in comparison to control. Vase solution containing nitric oxide (100 ppm) maintained high membrane stability index in comparison to control, at all stages of flower development. The expression of senescence associated genes (SAGs) *viz.*, GgCyP1, GgERS1a and GgERS1b were downgraded while GgDAD1 was upregulated by nitric oxide treatment during the course of flower development. As nitric oxide delays the senescence, it can be used to prolong the vase life of gladiolus (Anonymous, 2020b).

Foliar spray of salicylic acid (SA) (0.5 mM) on harvested flowers before or after simulated transport to local market showed increased flower vase life, as well as membrane stability, anthocyanin and flavonoid concentrations, and activities of catalase and superoxide dismutase enzymes in the flower petals of gladiolus *cv.* Amsterdam. Significant reduction was observed in lipid peroxidation of the SA-treated flowers. It is concluded that foliar application of SA (0.5 mM), improves gladiolus vase life through modulating its antioxidative system and reducing lipid peroxidation (Rahmani *et al.*, 2015).

1-methylcyclopropene (1-MCP) or STS extended vase life and minimized the weight loss of gladiolus spikes. 1-MCP) and STS enhanced the relative water content (RWC) of leaves and maintained chlorophyll and carbohydrate contents compared with the control values, which were decreased. Ethylene production was increased in florets of untreated spikes and membrane stability was reduced while 1-MCP or STS treatments minimized ethylene production and retained membrane stability. An increase in floret antioxidant enzyme activities (CAT, SOD and POX) was observed in 1-MCP or STS treated spikes (Hassan and Ali, 2014).

Li *et al.* (2017) reported for the first time, the effects of nano-silver (NS) pulses on cut spikes of gladiolus 'Eerde' towards reducing bacterial colonization of and biofilm formation on their stems. The NS treatments enhanced floret 'opening rate' and 'daily ornamental value'. NS (25 mg/l) pulse treatment tended to give the longest vase life and the best 'display quality'. All NS pulse treatments significantly improved water uptake by and reduced water loss from flowering spikes, thereby delaying the loss of water balance and maintaining relative fresh weight. NS (25 mg/l) pulse treatment effectively inhibited bacterial colonization and biofilm formation on the stem-end cut surface and in the xylem vessels, respectively.

15.2.0 Corms and Cormels

15.2.1 Curing

After lifting and removing the adhering soil, the corms and cormels of each cultivar are kept in trays in a shady but well ventilated place for about a fortnight (Hockings, 1972). If the corms and cormels are washed after topping, these should be dried quickly with the aid of warm dry air or blower but the temperature should never exceed 30°C (Stienstra, 1976a). Magie and Myers (1975) noticed earlier flowering and better corm production at 24 – 33°C with low air velocity for six days before cleaning.

Proper and effective curing of the corms are very important for storing them successfully. For curing, the layers of corms should not exceed three which may be cured for five weeks at 21°C. If the old withered corms have not been removed after lifting and during the course of curing (suberization) these become very hard to remove and if forced the corms may be damaged. It should be done either at lifting or whenever the old corms can be detached easily in the process of curing. In an investigation on *cvs.* Valeria and Queen Elizabeth, Magie (1987b) concluded that six days of curing at temperature regimes 25.5 – 34.5°C and 35 -16.6°C with corms protected against rapid weight loss resulted in 37 per cent greater flower production compared with 6 days in open trays in the same air streams. Six days of curing in the 35°C (95°F) cabinet gave 49 per cent increase in flowering for protected corms compared to those in open trays.

Harvesting and Marketing of Gladiolus

Harvesting Stage

Harvesting of Spikes

Harvested Spikes

Cut Flowers in Market

Packing with Polythene

Harvesting of Corms from Field

Storage of Corms

Gladiolus Corms for Sale

15.2.2 Cleaning and Grading

After the corms are fully cured, these are cleaned and diseased ones are discarded. The old withered corms are taken out and dirts are cleaned. Treatment with 0.02 per cent or 0.2 per cent captan 15 days before storage is very useful. Further curing for seven more days at 27°C will help to form callus at the bases and the corms become more resistant to diseases. Pre-storage fungicidal dip is given instead of dusting, after which the corms are dried by blowing warm air for 36 hours.

After cleaning, the corms and cormels are graded in different grade sizes as formulated by the North American Gladiolus Council, USA (Wilfret, 1980).

15.2.3 Storage of Corm

Proper storage of corm and cormels is very important, as otherwise storage rot of corm may occur due to fungal infection. Corms are stored in a single layer in wooden trays having a wire bottom. The scales over the corms are not removed during storage. Before storage corm/cormels should be treated with fungicides, than air dried and stored. Crossley and Arrowssmith (1972) reported that gladioli corms are stored between 2 – 10°C, preferably at 4.5°C in a well ventilated place having 75 per cent relative humidity. About 1-3 weeks prior to planting, cold-stored corms and cormels are brought to a warmer condition to encourage root swelling. Imanishi (1981) stored the October harvested cormels (5-6 mm diameter) in a cool dry place at room temperature until planting.

16.0 Biotechnology

Investigations have been carried out on various aspects of biotechnology of gladiolus and many of the results have been successfully applied for its improvement. A comprehensive review on biotechnology on gladiolus by Nagaraju and Parthasarathy (2000) will provide valuable information on the subject.

Certain important aspects of biotechnology of gladiolus have been summarized below:

16.1.0 Somatic Embryogenesis

The greatest potential for clonal multiplication was through somatic embryogenesis where a single isolated cell could be induced to produce first an embryo then a complete plant. Friable embryogenic callus and somatic embryos were obtained from young leaf base and corm slices of gladiolus when MS medium was supplemented with 2,4-D or NAA. Somatic embryos transferred to MS hormone-free medium regenerated into plantlets (Stefaniak, 1994). The embryogenic friable callus was maintained for more than 2 years in culture and retained very high regeneration ability. Tomotsune *et al.* (1995) obtained suspension callus from MS medium supplemented with NAA. The highest rate of somatic embryogenesis (224.1 somatic embryos/g) was obtained when callus induced in medium containing 5 mg NAA/l was placed on growth regulator-free medium, but callus induced in medium containing 10 mg NAA/l when transferred to medium containing 0.1 mg BA/l also gave good results. Somatic embryogenesis from stem tips of cormels

(MS + NAA 5 mg/l) resulted in flower colour variations among regenerated plants (Kasumi *et al.*, 1999a).

In vitro shoot multiplication in gladiolus by using leaf disc as explants revealed that BAP and Kinetin are the best shoot induction hormones. The highest percentage of shoot regeneration (91.66 per cent) and the highest shoot length was 4.24 cm with a minimum number of days (13.33) was observed in the MS medium supplemented with 3.0 mg/l BAP and 0.5 mg/l Kn (Shaheenuzzaman *et al.*, 2011).

Rakosy-tican *et al.* (2012) studied somatic embryogenesis of *Gladiolus imbricatus. In vitro* cultures started from wounded seeds or excised mature embryos. MS solid medium with 30 g/l sucrose and 1 mg/l N-6-benzyl adenine (BA) used for multiplication. Increased concentrations of sucrose promote corm development. Two-step culture protocol, on solid and liquid shaken MS based media revealed that acetic acid + sucrose 9 per cent, alone or in combination with ancymidol stimulated corm formation and yield.

Meristem tips were multiplied in MS medium containing 2iP (2.98 µM). After 3 months of shoot proliferation, the explants were enlarged onto a medium containing IBA (4.90 µM). Cormels have been then transferred in vessels for Temporary Immersion Culture. Three minutes of flooding every three hours were set up and the vessels were cultured at 23°C at a 16 h photoperiod at 35 µE/m/s (Ruffoni *et al.*, 2008). MS medium supplemented with BA (2.0-3.0 mg/l) in combination with 0.5 mg/l NAA, 30.0 g/l sucrose and 7.5 g/l agar exhibited higher shoot proliferating efficiency, number of shoot(s) per explant and shoot of higher length. Higher root proliferating efficiency was documented on rooting medium MS + 0.5 mg/l IBA + 0.5 mg/l Kinetin + 15.0 g/l sucrose + 7.5 g/l agar. (Vidhyashankar, 2017).

Devi *et al.* (2019) reported that *in vitro* regeneration of *cvs.* Sylvia, White Prosperity and Amsterdam were achieved using shoot bud of cormels as an explant. The highest callus induction rate of 76 per cent explants was recorded on MS medium supplemented with 2 mg/l 2,4-D, after 28 days.

Arya and Kumar (2012) reported that direct organogenesis was derived from roots of *in vitro* plantlets as an explant in *Gladiolus* sp. MS medium found best for initiation of shoots. Highest number of shoots regenerated in 1x MS with 2.0 mg/l BAP. The shoots produced roots in MS with 1.0 mg/l IBA and 6 per cent sucrose. After five weeks, rooted plantlets were transferred for initiation of cormels. The protocol enabled to harvest more than 50,000 cormels within 150 days starting from a single *in vitro* root explant.

Rakosy-tican *et al.* (2012) reported that for medium-term conservation maintenance of the *in vitro* cultures in low temperature in the dark proved to be the best in reducing the growth rate of the shoots after 3, 6 and 12 months in *Gladiolus imbricatus.*

16.1.1 Callus Culture

Explants from inflorescence stalks of gladiolus when cultured *in vitro* regenerated new plants within 6-7 weeks. Regeneration started by the formation of a thin layer of callus and root primordia on the basal end. The regeneration of the various organs from the explants was found to be polarized and depended on the levels of growth substances added to the basal medium (Zivet *al.*, 1970). Bajaj *et al.* (1983) obtained the best callus growth from segments of flower stalk cultured on a basal MS medium supplemented with NAA and kinetin. The callus mostly underwent rhizogenesis and occasionally differentiated some shoots. They regenerated complete plants from *in vitro* cultured cormels, cormel tips and axillary buds. Nagaraju and Parthasarathy (1995) obtained the maximum culture weight and callus + axillary bud weight in MS medium supplemented with 1 mg BA/l which was found to be better than kinetin. Kim and Lee (1993) used 10 mg 2,4-D/l to induce callus in corm tissue of gladiolus. Similar results were obtained by Kang *et al.* (1998). The calluses were subcultured on media containing NAA or 2,4-D either alone or in combination. They observed adventitious bud formation from calluses on media supplemented with low level of BA (0.01-0.1 mg/l) or without BA, regardless of auxin type during subculture of calli. Adventitious bud formation increased as the volume of media and ammonium nitrate increased. Kasumi *et al.* (1998) obtained callus from the lower part of the perianth and ovary. Among ovaries at different developmental stages, those cultured 2-3 days before anthesis were the most efficient to form callus and regenerate shoots. The medium supplemented with 5 mg NAA/l and BA induced callusing. Shoots regenerated most efficiently from the callus on MS medium containing 2 mg BA/l.

Kshirsagar (2009) found that the callus was regenerated from slices of cormel sprouts of *Gladiolus primulinus* cv. Golden Wave, cultured on MS medium with 4.0 mg/l NAA, 2.0 mg/l BAP or 1.0 mg/l 2,4-D. Shoots were regenerated from the callus on MS medium with 0.5–2.0 mg/l Kn or/and 0.5–2.0 mg/l BAP. Highest number of shoots was regenerated with 2.0 mg/l BAP from one-month-old callus which was initiated in the medium with NAA. Shoot regeneration capacity of callus decreased with the age and after fifth month no shoot was regenerated.

16.2.0 Somaclonal Variation

Somatic embryogenesis from the stem tips of cormels resulted in flower colour variation among regenerated plants of three gladiolus cultivars. The highest frequency of somatic embryos (51.1 per cent) was induced on MS medium supplemented with NAA at 5 mg/l. Varietal differences in somatic embryogenesis occurred; Topaz and Traveler formed more somatic embryos than Hector. Flower colour variants originated among regenerated plants grown in field, but no chimeric changes occurred in the floret and flower (Kasumi *et al.*, 1999a). Kasumi *et al.* (1999b) obtained a mutated sectorial chimeric flower colour variant by gamma irradiation of cormels of gladiolus *cv.* Traveler. The variant had two flower stalks in which the florets of one had pink perianths, whereas the florets of other had pale pink perianths. Ovary culture was carried out in order to isolate plants with a fixed flower colour. Regenerated plants from ovaries of pink florets produced pink flowers. Thirtyfive plants out of 52 regenerated plants (67.3 per cent) from pale pink florets produced pale pink flowers, but 17 plants (32.7 per cent) reverted to the original pink colour. None of the regenerated pink or pale pink had sectorial chimeras, and no changes in flower colour occurred among 63 vegetatively propagated plants obtained from pale pink plants. It is concluded that shoot regeneration from callus tissue of ovaries is effective in isolating plants of fixed flower colour from sectorial chimeras of gladiolus *cv.* Traveler.

16.3.0 Biochemical and Molecular Markers

Choi and Kim (1997) reported the possibility of using peroxidase activity as a biochemical marker for organogenic ability of gladiolus callus. Callus derived from cormel tissues of Nova Lux, White Prosperity and White Friendship turned brown and showed high peroxidase activity. Callus of Spic and Span, True Love, Red Beauty and Hunting Song regenerated spontaneously during subculture and it was difficult to maintain the cultures as callus. Topaz which recorded lowest peroxidase activity could be maintained and proliferated. Further, the callus subcultured in media containing NAA at 0.5 mg/l showed higher enzyme activity than in the presence of 2,4-D at the same concentration and formed organs during subculture. Protein contents and peroxidase activities in non-organogenic callus were lower than in organogenic callus. The pattern of peroxidase isoenzymes was determined in roots, leaves, corms and organogenic and non-organogenic calli of Topaz by native PAGE and found ten isoenzyme bands. Non-organogenic callus had the lowest number and amount of peroxidase isoenzymes.

Directed amplification of minisatellite DNA (DAMD) markers were used to investigate the relationships among 50 Gladiolus cultivars. DAMD analysis using 10 primers produced 120 polymorphic bands with 80 per cent polymorphism showing polymorphic information content (PIC = 0.28), Marker index (MI = 3.37), Nei's gene diversity (h = 0.267), and Shannon's information index (I = 0.407). Based on the study, they identified diverse cultivars of Gladiolus that could be the potential source as breeding material for further genetic improvement (Singh *et al.*, 2018).

RAPDs was used to identified and characterize 54 *Gladiolus* cultivars and to estimate the genetic relationships between the selected genotypes. Principal component analysis based on RAPD data showed *cv.* 'Pusa Lohit' is quite different from other genotypes (Pragya *et al.*, 2010).

Memon *et al.* (2012) assessed the *in vitro* produced cormels of different gladiolus cultivars for their clonal fidelity using RAPD and ISSR markers. In comparison, ISSR primers produced higher percentage of similarity matrix than RAPD. The ISSR cluster analysis for genetic similarity between mother and *in vitro* propagated cormels had varied degree of differences detected 90 per cent in Traderhorn, 96 per cent in White Friendship and 85 per cent in Peter Pears.

ISSR markers were used for molecular characterization of *Fusarium* wilt in resistant along with susceptible gladiolus (*Gladiolus* spp) genotypes. Primer UBC 861 has produced specific band of size 350 bp for both resistant genotypes, *viz.*, Arka Amar and Arka Aayush. UBC Primer 891 primer produced band of size 250 bp for both susceptible genotypes, *i.e.* Pink Friendship and IIHRG-12. The results suggest that ISSR markers are useful for detection of resistant bands and also assessing genetic relationship in gladiolus varieties. The identified ISSR marker will help in screening the gladiolus population for *Fusarium* wilt resistance (Kumari *et al.* 2016).

Geeta *et al.* (2016) determined genetic diversity of 15 gladiolus varieties using Sequence Related Amplified Polymorphism (SRAP), a new molecular marker technology developed based on PCR. SRAP analysis carried out using twenty five primer combinations out of which, eleven primer combinations were polymorphic and revealed a total of 80 scorable bands, among which 68 were polymorphic with an average of 6.18 polymorphic bands per pair of primers. The similarity matrix coefficient ranged from 31 – 73 percent, suggesting a higher genetic variation within gladiolus varieties.

Rashmi *et al.* (1016) studied diversity and genetic relationship of 20 genotypes through Sequence Related Amplified Polymorphism (SRAP) analysis. The percentage of polymorphism ranged from a maximum of 93.33 per cent in Me 3 + Em 3 primer combination to a minimum of 50.00 per cent by Me 2 + Em 6. The similarity matrix coefficient ranged from 18 to 79 percent, suggesting a higher genetic variation within gladiolus genotypes. Shobha a mutant genotype was identified quite distinct from other genotypes.

Singh *et al.* (2016) demonstrated the efficiency and discriminatory potential of an individual as well as cumulative analysis of RAPD, DAMD and ISSR markers to estimate the genetic variability and inter-relationships among indigenously developed Gladiolus cultivars. A total of 30 primers (10-RAPD, 10-DAMD and 10-ISSR) resulted into 230 DNA fragments, of which 177 (75.8 per cent) were polymorphic in nature. The polymorphic information content (PIC = 0.18), resolving power (RP = 11.2) and marker index (MI = 11.4) was found to be highest in ISSR markers.

Singh *et al.* (2017a) used directed amplification of minisatellite DNA (DAMD) markers to investigate the relationships among 50 Gladiolus cultivars. DAMD analysis using 10 primers produced 120 polymorphic bands with 80 per cent polymorphism showing polymorphic information content (PIC = 0.28), Marker index (MI = 3.37), Nei's gene diversity (h = 0.267), and Shannon's information index (I = 0.407).

Singh (2017b) developed sequence characterised amplified region (SCAR) markers from RAPD and ISSR fragments for identification and authentication of Gladiolus germplasm. The SCAR markers, *viz.* ScG12, ScG34, and ScG36, are specific to the DNA from all 62 Gladiolus cultivars, as they did not amplify the DNA of other taxa of the family Iridaceae, including Iris, Amaryllis, and Narcissus. All three SCAR markers distinguished Gladiolus from other taxa of the family Iridaceae, whereas marker ScAm was specific to the 'Amethyst' cultivar.

Niraj *et al.* (2018) analyzed genetic variability and relatedness in 62 Gladiolus cultivars using directed amplification of minisatellite DNA (DAMD) and inter simple sequence repeats (ISSR). The cumulative analysis carried out with DAMD and ISSR markers showed 83.32 per cent polymorphism revealed that amount of genetic variability in the Gladiolus genome is relatively high. They suggested that that DAMD and ISSR are useful markers to elucidate the genetic variability and relationships amongst Gladiolus cultivars, and is a prelude for further utilization of promising and genetically divergent materials in the breeding programmes.

Daco *et al.* (2019) studied the genetic population structure of populations of the endangered *Gladiolus palustris* in the western limit of its distribution area using AFLP markers and clarified the taxonomic status of populations where hybridization with the closely related *G. imbricatus* was suspected based on morphology by sequencing the ITS region of the nuclear ribosomal DNA as well as two regions of the chloroplast DNA. The analysis of the genetic population structure of *G. palustris* revealed a strong differentiation among geographical regions, which was much higher than the differentiation among populations within regions. The genetic analyses indicated that some French populations of *G. palustris* are of hybrid origin.

16.4.0 Genetic Transformation

Genetic engineering in gladiolus is being attempted to transfer genes for flower colour, for resistance to viral and fungal diseases and herbicides.

Graves and Goldman (1987) reported gladiolus transformation by A. *tumefaciens*. Slices of gladiolus corms were inoculated with A. *tumefaciens* and evidence of transformation was based upon the presence of nopaline and octopine in the resulting tumors cultured on growth regulator-free medium.

Kamo (1997) studied factors affecting *Agrobacterium tumefaciens* mediated *gus A* expression and opined synthesis in gladiolus using five tumorigenic strains of *A. tumefaciens* on corms, cormels of plants grown *in vitro* and seed-derived seedlings of several cultivars of gladiolus. Tumor formation was observed on 12 per cent of the plant tissues inoculated and 1 per cent of these tumors synthesized either octopine or nopaline. *A. tumefaciens* showed 0.5 and 35 per cent GUS expression for plants grown *in vitro* and regenerable callus respectively. Maximum GUS expression of 40-61 per cent was recorded when *A. tumefaciens* was incubated with leaf explants obtained from seedlings grown in the dark, whereas leaves from seedlings grown under 16-h light showed non-GUS indicating the significant effect of etiolation on transient GUS expression mediated by *A. tumefaciens.*

Particle bombardment (PB) has also been reported by Kamo *et al.* (1995a) who obtained more than 100 transgenic plants using suspension cells and callus. Stably transformed calli were selected on medium containing either phosphinothricin (PPT) or bialaphos followed by transfer to a regeneration medium to recover transgenic plants. Stable transformation was confirmed by detection of the *pat* gene by DNA gel blot analysis and by enzymatic assays to measure GUS activity. In general, particle bombardment of regenerable suspension cells rather than callus resulted in large number of transformants. The rate of expression for *gus* in PPT resistant plants was high (approximately 70 per cent). Promoters that are typically more efficient in dicotyledonous plants were very active in gladiolus, a monocot. This protocol would now allow the introduction of transgenes to confer resistance to the viral and fungal pathogens that decrease gladiolus production. Stable transformation via particle gun bombardment has been applied to a variety of tissues (Kamo *et al.*, 1995b).

Bowers *et al.* (1994) established regenerable callus and suspension cell cultures of the gladiolus *cv.* Jenny Lee. Transient beta-glucuronidase (EC-3.2.1.31) gene (*gus*) expression levels from the *uidA* gene were measured following particle bombardment. Gene promoters which are typically efficient in dicots were more active in gladiolus cell cultures than the monocot-derived promoters tested (*e.g.*, rice *Actl*) which usually express more highly in transformed monocots. Phosphinothricin resistant gladiolus calli were recovered following bombardment with a phosphinothricin-acetyltransferase (PAT) expressing plasmid and selection on PPT containing medium. Transformation was confirmed by the detection of GUS activity and a PAT gene-specific polymerase chain reaction product in transformed calli. After PPT resistant calli were moved to a regeneration medium, large numbers of transgenic plantlets were obtained. Creation of an efficient transformation system for gladiolus would allow the introduction of new phenotypic traits for commercial production of ornamental plants. Kamo *et al.* (1994) stably transformed over 100 commercially important cultivars by particle bombardment on phosphothricin-acetyltransferase gene (confe-rringphosphinothricin-resistance) and cotransformed with either virucide genes or with the *uidA* reporter gene encoding beta-glucuronidase (EC-3.2.1.31). Transgenic plants were regenerated following selection on concentrations of phosphinothricinwhich varied with the type of tissue bombarded. Foreign DNA integration was confir-med by the polymerase chain reaction.

Embryogenic cell lines of *Gladiolus* were bombarded with the *bar-uidA* fusion gene under the cauliflower mosaic caulimovirus (CaMV) 35S promoter (pDM327) or cobombarded with *uidA* under the CaMV 35S promoter (pBCG) and *bar* under the CaMV 35S promoter (pDM307). Over 500 cell lines were isolated for either the fusion gene or cobombarded cells following selection on MS medium supplemented with 2 mg (9 µM) 2,4-D and 6 mg phosphinothricin/1. Three times as many cell lines (72 per cent of the cell lines) containing the *bar-uidA* fusion gene expressed *gus* as compared to cobombarded cell lines (23 per cent) following histological staining. *Gus* expression ceased after 1 year in culture for 5 per cent of the cell lines containing the fusion gene and 3 per cent of the cobombarded cell lines. The bifunctionality and utility of the *bar-uidA* fusion gene were demonstrated, accompanied by enhanced *gus*expression (Kamo *et al.*, 2000).

Yamada *et al.* (2004) isolated a homolog of the potential anti-apoptotic gene, defender against apoptotic death (DAD1) from gladiolus petals as full-length cDNA (GlDAD1), and investigated the relationship between its expression and the execution processes of programmed cell death (PCD) in senescing petals. RNA gel blotting showed that GlDAD1 expression in petals was drastically reduced, considerably before the first visible senescence symptom (petal wilting).

Kamo *et al.* (2012) transformed plants of Gladiolus 'Peter Pears' or 'Jenny Lee' with single-chain variable fragments (scFv) to Cucumber mosaic virus (CMV) subgroup I or II. The eight plant lines that showed resistance and two susceptible lines expressed the transgene as determined by Northern hybridization and reverse transcriptase PCR.

Kamo *et al.* (2014) compared the growth of transgenic Gladiolus *cvs.* 'Peter Pears' and 'Jenny Lee' with non-transformed plants either regenerated from embryogenic callus or micropropagated *in vitro*. Micropropagated and regenerated plants of 'Peter Pears' showed similar sprouting percentage of corms *in vitro* and daughter corm. Transgenic plants of 'Peter Pears' had similar corm weights to those from regenerated plants, but sprouting percentage, leaf length, and daughter corm production was less than that of regenerated plants.

Kamo *et al.* (2015) transformed Gladiolus cv. Peter Pears plants by particle bombardment with plasmid DNA containing a 90 bp D4E1 gene that was under the control of the duplicated CaMV 35Spromoter. Transgenic plants were tested *in vitro* for resistance to *F. oxysporum*, and several lines appeared to be more resistant than the control plants that lacked D4E1. Cell extracts from the five transgenic lines showed either an inhibition of *F. oxysporum* mycelial colony formation or less fungal hyphae were observed to infect their roots as compared to non-transformed Gladiolus plants.

Wu *et al.* (2016) reported that AFPs (ABA insensitive five binding proteins) acts as a negative regulator in ABA signaling and is related to dormancy. They isolated and characterized a novel AFP family member from Gladiolus dormant cormels, GhAFP-like, containing three conserved domains of the AFP family. Quantitative PCR analysis revealed that GhAFP-like was expressed in dormant organs and its expression was down-regulated along with corm storage.

Seng *et al.* (2016, 2017) expressed that ADP-glucose pyrophosphorylase (AGPase) plays a key role in regulating starch biosynthesis in storage organs and is likely one of the most important determinant of sink strength. They isolated and characterized the large subunit of AGPase gene GhAGPL1 which was highly expressed in sink organs (cormels and corms). The expression of GhAGPL1 was induced by glucose, sucrose, and mannitol, and it was repressed by abscisic acid (ABA). Silencing GhAGPL1 in gladiolus decreased the transcript level of GhAGPL1 and GhSus, and resulted in the reduction of AGPase activity and starch content in gladiolus corm and cormel. Meanwhile, sucrose content was higher in GhAGPL1-silenced corm. Surprisingly, silencing GhAGPL1 in gladiolus produced smaller corms and fewer numbers of cormels. Overall, GhAGPL1 contributed to the quality and quantity of gladiolus corms and cormels.

Jian *et al.* (2018) isolated an ABA biosynthesis gene, GhNCED, from the transcriptome database of corm dormancy release (CDR), and characterized its negative role in regulating CDR. GhTCP19 is dramatically induced by exogenous cytokinins (CKs) and is induced during CDR. Silencing of GhTCP19 in dormant cormels delayed CDR, resulting in higher expression of GhNCED and ABA levels. Meanwhile, endogenous CK biosynthesis and signaling were inhibited in GhTCP19-silenced cormels. Results revealed that GhTCP19 is a positive regulator of the CDR process by repressing expression of an ABA biosynthesis gene (GhNCED), promoting CK biosynthesis (GhIPT) and signal transduction (GhARR) as well as inducing cyclin genes.

Jian *et al.* (2019) revealed that ABA (abscisic acid) signaling regulator, GhPP2C1 (protein phosphatase 2C1) promotes ABA signaling and synthesis, and inhibits cytokinin biosynthesis pathways, thereby inhibiting corm dormancy release. GhNAC83 regulates the ABA and CK pathways, and therefore controls corm dormancy.

Exhibitions of Gladiolus

17.0 References

Abbas, A. and Waliullah, M.I.S. (2010) *Indian J. Nematol.*, **40**(2) : 179-183.

Abdulrahman, Y.A. (2019) *Res. Crop.*, **20**(4) : 774-781.

Abo-E1-Dahab, M.K., El-Goorani, M. A. and El-Banoby, P. E. (1972) *Phytopathologia Mediterranea*, **13**: 131-135.

Agarwala, R. K., Kaicker, U. S. and Arora, K. N. (1965) *Indian Hort.*, **9** : 31-33.

Ahmad, I., Saquib, R.U., Qasim, M., Saleem, M., Khan, A.S. and Yaseen, M. (2013) *Chil. J. Agric. Res.*, **73**(4) : 339-344.

Ahmad, M. and Rab, A. (2019) *Biosci. Res.*, **16**(2) : 1647-1659.

Ahmad, M. and Rab, A. (2020) *Pak. J. Bot.*, **52**(1), doi: 10.30848/PJB2020-1(28).

Albony, J. (1966) *Epiphyt.*, **17** : 81-91.

Albores, L.C., Baños, S.B., Herrera, J.M., Necha, L.B., López, M.H. and Hernández, A.C. (2014) *Afr. J. Microbiol. Res.*, **8**(8) : 724-733.

Amen, T.A. and Kovanci, I. (1990) *Fen Bilimleri Enstitusu Dergisi*, **1** : 191-195.

Amingad, V.S., Rao, T.M., Venugopalan, R., Kumar, D.P., Dhananjaya, M.V. and Bhanuprakash, K. (2013) *J. Hortic. Sci.*, **8**(1) : 114-117.

Amrutha, N., Madhavi, K. and Suchitra, V. (2014) *Trends Biosci.*, **7**(12) : 1227-1229.

Anonymous (1962) *Agric. Gaz. N. S. W.*, **3** : 125-126.

Anonymous (1967) *Ann. Fac. Agr. Univ. Pisa*, **27** : 93-100.

Anonymous (1968) *Ann. Fac. Agr. Univ. Pisa*, **29** : 109-129

Anonymous (1972) *Splash*, pp. 45-47.

Anonymous (1976) *Min. Agric. Fish.*, New Zealand, pp. 9.

Anonymous (1980) *Verbond. voor de Belgis. Siert.*, **24** : 811- 813.

Anonymous (2011) *IARI News*, 27(3) : 2-3.

Anonymous (2020a) https://www.zauba.com/exportanalysis-fresh+cut+flower+gladiolus-report.html.

Anonymous (2020b) https://www.krishisewa.com/articles/pht/290-gladiolus-vase-life.html.

Anstetta, A. (1958) *Proc. 15th Int. Hort. Cong.*, Nice, pp. 493-499.

Aoba, T. (1975) *J. Yamagata Agric. Forest Soc.*, **32** : 48-53.

Apte, S. S. (1958) *Gladiolus*, **33** : 31-46.

Apte, S. S. (1959) *Meded. Dir. Tuinb.*, **22** : 207-211.

Apte, S. S. (1962) *Meded. Land Hogesch*, Wageningen, **62**: 1-47.

Armitage, M. S. (1973) *Misc. Rpt. Malkerns Res. Sta.*, Swaziland, **91** : 21.

Arora, J.S., Singh, K. and Grewal, N.S. (1992) *Indian J. Plant Physiol.*, **35** : 202-206.

Arya, A. and Kumar, S. (2012) *Trends Biosci.*, **5**(3) : 168-170.

Asahira, T., Imanishi, H. and Tsukamoto, Y. (1968) *Mem. Coll. Agric. Kyoto Univ.*, **93** : 21-34.

Ashour, W.A. and El-Din, I.R.G. (1964) *Ann. Agric. Sci., Cairo*, **9** : 205-224.

Assumi, B., Gogoi, B.B., Mahanta, B., Bora, L.C. and Borgohain, N. (2017) *Appl. Biol. Res.*, **19**(1) : 105-106.

Astvatsatryan, Z. A. and Sarkisyan, E. D. (1974a) *Referativayi Zhurnal*, 4.55 826.

Astvatsatryan, Z. A. and Sarkisyan, E. D. (1974b) *Biologicheskii Zhurnal Armenii*, **27** : 89-91.

Audemard, H. (1969) *Phytiatrie-Phytopharcie*, **18** : 207-211.

Avneet, K.; Kushal, S. and Nirmaljit, K. (2016). *Indian J. Ecol.*, **43**(1) : 219-223.

Awad, A. R. E. and Hamied, A. A. (1985) *Acta Hortic.*, **167** : 177-185.

Azimi, M.H. (2020) *Ornam. Hortic.*, **26**(2) : 177-189.

Babaji, J. P.; Bharat, P. N.; Suresh, D. S. and Davabhai, J. P. (2014) *Ann. Biol. Res.*, **5**(8) : 13-17.

Babu, P. and Chawla, H.S. (2000) *J. Hort. Sci. Biotech.*, **75** : 400-404.

Bahar, S.N.G. and Korkut, A.B. (1998) *Turkish J. Agri. Forestry*, **22** : 51-58.

Baiswar, P., Chandra, S. and Kumar, R. (2007) *J. Ornam. Hortic.*,**10**(4) : 209-214.

Baiswar, P., Chandra, S. and Kumar, R. (2008) *J. Ornam. Hortic.*,**11**(1) : 79-80.

Bajaj, Y. P. S., Sidhu, M. M. S. and Gill, A. P. S. (1983) *Sci. Hortic.*, **18** : 269-275.

Bala, R., Rao, I. V. R. and Ram, H. Y. M. (1986) *J. Pl. Physiol.*, **122** : 87-92.

Bald, J. G. (1952) *Amer. J. Bot.*, **39** : 97-99.

Bamford, R. (1935) *J. Agric. Res.*, **31** : 945-950.

Barbagallo, L. and Silvia, A. (1964) *Tech. Agric.*, **16** : 63-71.

Baruati, D., Talukdar, M.C., Nath, D. and Bhuyan, T. (2018) *Int. J. Curr. Microbiol. Appl. Sci.*, **7**(10) : 1477-1482.

Bashir M., Qadri R.W.K., Khan I., Zain M., Rasool A. and Ashraf U. (2016) *Pak. J. Sci.*, **68**(2) : 121-127.

Baskaran, V. and Abirami, K (2016) *Indian J. Agric. Sci.*, **86**(2) : 216-219.

Baskaran, V., Abirami, K. and Simhachalam, P. (2017) *Indian J. Agric. Sci.*, **87**(4) : 551–554.

Basoli, M., Kumar, P. and Kumar, S. (2015) *Indian J. Agric. Sci.*, **85**(5) : 649-654.

Bautista, O. D. K. and Cadiz, T. G. (1964) *Philipp. Agric.*, **48** : 288-306.

Beck, A. R., Peterson, L. E. and Robbins, M. L. (1970) *Ia. Fm. Sci.*, **25** : 39-40.

Belwal, S., Bhuj, B.D., Srivastava, R., Shahi, N.C. and Chand, S. (2019) *Prog. Hortic.*, **51**(2) : 196-200.

Benkenstein, H. (1974) *Archiv. fur Gartenbau.*, **22** : 419-427.

Benkenstein, H., Hallwig, J. and Buder, A. (1976) *Archiv. fur Gartenbau.*, **24** : 251-258.

Benkenstein, K., Kruger, W. and Nehl, H. (1975) *Archiv. fur Gartenbau.*, **23** : 71-80.

Bentham, G. and Hooker, J. D. (1883) *Genera Plantarum 3*, London.

Bera, A.K., Maity, T.R., Samanta, A., Dolai, A., Saha B. and Datta S. (2015) *J. Appl. Hortic.*, **17**(3) : 222-224.

Bertus, A. L. (1974) *Phytopath. Z.*, **80** : 181-185.

Beura, S. and Singh, R. (2003) *J. Ornam. Hortic.*, **6**(4) : 322-327.

Bhagat, S.S., Ingle, S.T. and Gupta, V. (2018) *Int. J. Chem. Stud.*, **6**(5) : 1554-1558.

Bhagat, S.S., Ingle, S.T. and Gupta, V. (2018) *Int. J. Curr. Microbiol. Appl. Sci.*, 7(6) : 707-714.

Bhajantri, A. and Patil, V.S. (2013) *J. Farm Sci.*, **26**(3) : 403-407.

Bhajantri, A. and Patil, V.S. (2016) *Green Farming*, **7**(2) : 388-391.

Bhat, Z.A. and Sheikh, M.Q. (2015) *Sci. Res. Essays.*, **10**(24) : 655-658.

Bhattacharjee, S. K. (1981a) *Singapore J. Primary Ind.*, **9** : 18-22.

Bhattacharjee, S. K. (1984) *Gartenbauwiss.*, **49** : 103-106.

Bhattacharjee, S.K. (1981b) *Singapore J. Primary Ind.*, **9** : 23-27.

Bing, A. (1972) *The World of Gladiolus*, NAGC, USA, pp. 182-191.

Bing, A. (1978) *Proc. N-eastern Weed Sci. Soc.*, **32**: 284-287.

Bing, A. (1979) *Proc. N-eastern Weed Sci.* Soc., **33**: 264-269.

Bing, A., Senesac, A.F. and Macksel, M. (1988) *Proc. 42nd annual meetings of the North eastern Weed Science Soc.*, pp.137-140.

Bohra, M. and Nautiyal, B.P. (2019) *Agric. Rev.*, **40**(1) : 45-52.

Bondt, K. De (1982) *Verbond. voor de Belgis. Siert.*, **26** : 144-145.

Bose, T.K. (1984) *Indian Agriculturist*, **28** : 53-72.

Bowers, A.D., Kamo, K., Smith, F., Van-Eck, J. and Serlemitsos, K. (1994) *J. Cell. Biochem.*, Suppl., 18A, 99.

Bozarth, R. F. and Corbett, M. K. (1957) *Plant Dis. Reptr.*, **42** : 217-221.

Bravdo, S., Mayak, S. and Gravrieli, Y. (1974) *Canadian J. Bot.*, **52** : 1271-1281.

Brewer, R. F. (1953) *Gladio Grams, NAGC Bull.*, **34** : 17, 19.

Brewer, R. F., Guillemet, F. B. and Sutherland, F. H. (1966) *Proc. Amer. Soc. Hort. Sci.*, **88** : 631-634.

Brierley, P. (1962) *Plant Dis. Reptr.*, **46** : 505.

Brierley, P. (1963) *Plant Dis. Reptr.*, **47** : 863.

Brierley, P. and Smith, F. F. (1947) *Flor. Rev.*, Feb. 6.

Brierley, P. and Smith, F. F. (1960) *Plant Dis. Reptr.*, **44** : 463-464.

Buell, C. D. (1972) *The World of the Gladiolus*, NAGC, USA, pp. 138-140.

Buiatti, M., Baroncelli, S., Tesi, R. and Boscariol, P. (1970) *Rad. Bot.*, **10**: 531-538.

Buiatti, M., Ragazzini, R. and Tognoni, F. (1965) *Rad. Bot.*, **5** : 77-98.

Buschman, J.C.M. and Groen, N.P.A. (1989) *Bloembollencultuur*, **100** : 18-19.

Bylov, V. N., Raikov, N. I., Mokhamed, D. S. and Ganyushkin, E. V. (1988) *Introduktsiya, Izucheniei Osnovy Selektsii Dekorativnykh Rastenii*, pp. 171-178.

Bylov, V.N. (1990) *Selekfsiyai Semenovodstvo* (*Moskva*), **6** : 17-21.

Carpenter, W.J., Wilfret, G.J. and Cornell, J.A. (1991) *HortScience*, **26** : 1054-1057.

Casares, M.C., Mattiuz, B.H., Muniz, A.C. and Mattiuz, C. (2017) *Ornam. Hortic.*, **23**(1) : 79-87.

Ceruti-Scurti, J. (1955) *Ann. Sper. agrar.*, **9** : cvii-cvi.

Chahal, D.S., Sehgal, O.P. and Sing, K.K. (1994) *Annals Biol.*, **10** : 245-249.

Chand, S., Panwar, P., Rawat, A.S., Srivastava, R., Bhuj, B.D. and Sharma, R. (2019) *Int. J. Chem. Stud.*, **7**(3) : 3148-3151.

Chandana, K. and Dorajeerao, A.V.D. (2014) *Plant Arch.*, **14**(1) : 197-204.

Chandel, S. and Deepika, R. (2010) *J. Fruit Ornam. Plant Res.*, **18**(2) : 361-380.

Chandel, S. and Kumar, V. (2017) *Int. J. For. Crop Improv.*, **2**(8) : 121-124.

Chaturvedi, O.P., Shukla, I. N. and Singh, A.R. (1986) *Prog. Hort.*, **18** : 196-199

Chaudhary, M.B., Patel, R.B., Chawla, S.L., Patil, S., Patel, K.J., Chaudhari, B.B. (2016) *J. Ornam. Hortic.*, **19**(1 and 2) : 44-47.

Chavagnat, A. (1965) *Rev. Hort.*, Paris, **137** : 807-810.

Cheal, W. F. (1963) *Exp. Hort.*, **9** : 98-100.

Chen, Y., Steiniz, B., Cohen, A. and Elber, Y. (1982) *Sci. Hortic.*, **18** : 169-175.

Choi, H.R. and Roh, S. M. (1980) *J. Korean Soc. Hort. Sci.*, **22** : 176-184.

Choi, J.D. and Kim, K.W. (1997) *J. Korean Soc. Hort. Sci.*, **38** : 581-587.

Ciampolini, M. and Zangheri, S. (1975/76) *Bolletina di ZoologiaAgraria e di Bachico1tura*, II-**13** : 37-47.

Cirrito, M. (1976) *Ann. Dell' Inst. Sper. F1ori.*, **7** : 91-120.

Cirrito, M., Vita, M. De. and Zizza, G. (1980) *Ann. Dell' Inst. Sper. Flori.*, **11** : 123-148.

Cocozza Talia, M. A. and Lucia, B. de. (1994) *Colture Protette*, **23** : 81-85.

Cocozza Talia, M.A., Ferrari, I. and Costanza, I. (1992) *Acta Hortic.*, **325** : 369-372.

Cocozza, M. (1971) *Pepinieristes Horticulteurs Maraichers*, **119** : 31-36.

Cohat, J. (1971) *Pepinieristes Horticulteurs Maraichers*, **122** : 23-29.

Cohat, J. (1974) *Ann. Amel Plantes*, **24** : 13-24.

Cohen, A. and Barzilay, A. (1991) *HortScience*, **26** : 216.

Collingwood, E. F. (1970) *FAO Plant Prot. Bull.*, **18** : 69-70.

Compton, O. C. (1960) *Proc. Amer. Soc. Hort. Sci.*, **75** : 688-692.

Conijn, C.G.M. and Groen, N.P.A. (1990) *Acta Hortic.*, **266** : 539-546.

Costa, L.C.D., Araujo, F.F.D., Lima, P.C.C., Pereira, A.M. and Finger, F.L. (2016) *Bragantia*, **75**(4) : 377-385.

Cremer, M. C. and Schenk, P. K. (1967) *Netherlands J. Plant. Pathol.*, **73** : 33-48.

Crossley, J. H. and Arrowsmith, S. (1972) *Gladiolus Growing*, Canada Deptt. Agriculture, p. 19.

Cruz, R.R.P., Ribeiro, W.S., Silva, S.M., Finger, F.L., Zanuncio, J.C., Corrêa, E.B., Bruno, R.L.A., Fugate, K.K., da Costa, F.B. and Railene. (2019) *Plant Signal. Behav.*, **14**(10), doi: 10.1080/15592324.2019.1652520.

D, Aulerio, A.Z.; Dallavalle, E. and Marchetti, L. (1994) *Informatore Agrario*, **50** : 63-66.

D'Andria, R., Mori, M. and Morelli, G. (1996) *Irrigazione e Drenaggio*, **43** : 22-27.

Da Costa, L.C. and Finger, F.L. (2016) *Ornam. Hortic.*, **22**(2) : 147-153.

Daco, L., Maurice, T., Muller, S., Rossa, J. and Colling, G. (2019) *Conserv. Genet.*, **20** : 1339-1354.

Dantu, P. K. and Bhojwani, S. S. (1995) *Sci. Hortic.*, **61** : 115-129.

Darwish, M.A., Sarhan, A.M.Z., Emam, K.A.M. and Alm-Eldeen, R.E.A. (2018) *Biosci. Res.*, **15**(3) : 2041-2048.

Das, T.K. (1998) *Indian J. Agron.*, **43** : 756-761.

De Donato, M. (1968) *Coltiv. Giorn. Vinic. ital.*, **5** : 12.

De Meyer, Jr. (1982) *Family Food Garden*, August-September issue, pp. 46-49.

Denny, F. E. (1936) *Contr. Boyce Thompson Inst.*, **8** : 137-140.

Denny, F. E. (1938) *Contr. Boyce Thompson Inst.*, **9** : 403-408.

Denny, F. E. and Miller, L. P. (1934) *Contr. Boyce Thompson Inst.*, **6** : 31-38.

Deswal, K. S., Patil, V. K. and Anserwadekar, K. W. (1983) *Indian J. Hort.*, **40** : 254-259.

Devecchi, M. and Barni, E. (1997) *Colture Protette*, **26** : 79-82.

Devi, K.L., Maitra, S., Bhattacharya, P.M. (2019) *Int. J. Curr. Microbiol. Appl. Sci.*, **8**(10), doi: 10.20546/ijcmas.2019.810.208.

Devi, L.T., Singh, U.C., Devi, A.B., Athokpam, H.S. and Singh, N.G. (2020) *Pharm. Innov. J.*, **9**(3) : 01-04.

Devi, P., Kumar, P., Sengar, R.S., Yadav, M.K., Kumar, M., Singh, S.K. and Singh, S. (2019) *Int. J. Curr. Microbiol. Appl. Sci.*, **8**(7) : 1345-1350.

Dhakal, K., Khanal, D., Ayer, D.K., Khanal, A.P., Pandey, L., Pant, S.S., Upadhyay, K., Magar, S.S.B., Pandey, L., Prasad, M.L., Joshi, S., Ashmit, K.C. (2017) *Res. Rev.: J. Agric. Sci. Technol.*, **6**(3): 1-7.

Dhakar, S., Swaroop K., Singh, K.P., Sharma, V.K., Singh, S. and Singh, N. (2020) *Indian J. Agric. Sci.*, **90**(8) : 1460-1466.

Dhakar, S., Swaroop, K., Singh, K.P., Das, T.K., Kumar, P. and Singh, N. (2015) *J. Ornam. Hortic.*, **18**(3-4) : 56-60.

Dhakar, S., Swaroop, K., Singh. K.P., Das. T.K., Kumar, P. and Singh, N. (2016) *Indian J. Hortic.*, **73**(4) : 570-575.

Dhankhar, D.S., Singh, R., and Rana, J.S. (1999) *Sci. Hortic.*, **6** : 141-146.

Dierickx, W. (1977) *Boer en de Tuinder*, **83** : 15.

Dilta, B.S., Badiyala, S.D., Sharma, Y.D. and Verma, V.K. (2004) *J. Ornam. Hortic.*, **7**(2) : 153-158.

Dobos, M. (1964) *Kert. szal. Foisk. Kozlem.*, **28** : 267-274.

Dogra, N. and Dhatt, K.K. (2017) *Int. J. Agric. Sci.*, **9**(53) : 4905-4908.

Dogra, N. and Dhatt, K.K. (2018) *J. Hill Agric.*, **9**(1) : 88-93.

Dogra, N., Dhatt, K.K. and Kaur, N. (2017) *Int. J. Curr. Res. Aca. Rev.*, **5**(9) : 44-50.

Dogra. S., Pandey, R.K. and Bhat, D.J. (2012) *Int. J. Pharm. Bio. Sci.*, **3**(4) : 1083-1090.

Dryagina, I. V. (1962) *Soveschchanie po MarfogenezuRast. Trudy*, **1** : 173-176.

Dryagina, I. V. (1964) *Moscow Univ. Vestnik Ser. VI : Biol. Pchv.*, **19** : 46-56.

Dryagina, I. V. and Akhramova, V. F. (1962) *Nauch. Dok. Vysshei. Shkoly. Biol. Nauki*, **4** : 98-102.

Dryagina, I. V. and Akhramova, V. F. (1963) *Experimental Morphogenesis*, Moscow Univ., Moscow, pp. 409-415.

Dryagina, I. V. and Kazarinov, G. E. (1965) *Nauch. Dok. Vysshei. Shkoly. Biol. Nauki*, **1**: 91-94.

Dryagina, I. V. and Kazarinov, G. E. (1966a) *Trudy Moscowskowo Obszaczestwa Ispytatieliej Prirody*, **23** : 272-279.

Dryagina, I. V. and Kazarinov, G. E. (1966b) *NSA*, **20** : 352-390.

Dua, I.S. SehgaI, O. P. and Chork, K. S. (1984) *Gartenbauwiss.*, **49** : 91-94.

Durga, M.L., Raju, D.V.S., Pandey, R.N., Pandey, R., Kumar, P., Singh, K.P. and Krishnan, S.G. (2018) *Indian J. Hortic.*, **75**(1) : 119-123.

Dutta, K. and Gantait, S.S. (2016) *Indian J. Agric. Sci.*, **86**(1) : 120–126.

Dwivedi, N., Deen, B. and Kumar, A. (2018) *Pharm. Innov.*, **7**(2) : 114-116.

Dwivedi, S.K., Arora, A., Singh, V.P., Saira, R. and Bhattacharya R.C. (2016) *Sci. Hortic.*, **210** : 158-165.

Elansary, H.O. (2020) *Processes*, **8**(1), 71, doi:10.3390/pr8010071.

El-Gamassy, A. M. (1957) *Ann. Agric. Sci., Cairo*, **2** : 239-246.

El-Gamassy, A. M. (1960) *Ann. Agric. Sci., Cairo*, **5** : 101-8.

El-Gamassy, A. M. and El-Gendy, S. A. (1962) *Ann. Agric. Sci., Cairo*, **7** : 353-385.

El-Gamassy, A. M. and Shedeed, M. R. (1961) *Ann. Agric. Sci., Cairo*, **6** : 205-218.

El-Gamassy, A. M., El-Barkousky, M. and Hashem, M. (1977a) *Tech. Bull. Qubba Botanic Garden*, **1** : 17.

El-Gamassy, A. M., El-Barkousky, M. and Hashem, M. (1977b) *Tech. Bull. Qubba Botanic Garden*, pp. 18-19.

El-Goorani, M. A., Abo-El-Dahab, M. K. and Mehiar, F. F. (1974) *Phytopath.*, **64** : 271-272.

Elia, P. and Accati, E. (1964) *Fruitticoltura*, **26** : 249-256.

Emek, Y. and Erdað, B. (2007) *Pak. J. Bot.*, **39**(1) : 23-30.

Emna, El Chaieb and Faouzi, H. (2015) *Int. J. Agron. Agric. Res.*, **6**(1) : 54-62.

Engler, A. (1908) *Die PflanzenbeltAfrikas*, In :*Vegetation der Erde*, **9** : 2.

Evanri, M., Konis, E. and Zirkin, D. (1950) *Palestine J. Bot.*, **5** : 32-45.

Fatihullah and Bostan, N. (2018) *Int. J. Environ. Sci. Nat. Res.*, **9**(1), doi: 10.19080/IJESNR.2018.08.555753.

Fen, J. and Zhang, Q. (2003) *Acta Hortic.*, **624** : 265-269.

Fernandez, P. D. and Lima Filho, S. A. (1974) *Cientifica*, **1** : 33-41.

Fernandez, P. D., Fysink, A. J. J., Bisogni, A. C. and Oliveira Filho, J. J. De. (1975) *Cientifica*, **3** : 42-47.

Fernandez, P. D., Haag, H. P. and Oliveira, G. D. (1974a) *Anais da Escola Superior deAgricultura "Luiz de Queiroz,"* **31** : 621-634.

Fernandez, P. D., Haag, H. P., Simao, S. and Mattos, J. R. De (1974b) *Anais da Escola Superior de Agricultura "Luiz de Queiroz, "* **31** : 635-644.

Ferreira, J. F. and Nevill, W. G. (1989) *Plant Dis.*, **73** : 987-990.

Fetvadzhieva, N. and Dinova, M. (1982) *Gradinarskai Lozarska Nauka*, **19** : 82-88.

Fink, H. C. (1953) *Plant Dis. Reptr.*, **37**: 285-286.

Fisher, S. N. (1975) *Gladiolus Variety Parentages* (Rev. ed.), NAGC, USA.

Fodor, B. (1974) *Kerteszeti Egyetem Kozlemenyei*, **37** : 59-69.

Forsberg, J. L. (1952) *Flor. Rev.*, **111**: 33-44, 45.

Forsberg, J. L. (1961) *Nat. Hist. Surv., Biol. Notes*, **43** : 1-12.

Freitag, J. B. and Tompkins, C. M. (1963) *Plant Dis. Reptr.*, **47** : 617-621.

Gamal, Taha Mousa (1974) *Kerteszeti Egyetem Kozlemenyei*, **38** : 223.

Gantait, S.S. (2015) *Indian Hortic. J.*, **5**(1-2) : 28-31.

Ganyushkin, E.V. (1993) *Byulleten' Glavnogo Botanicheskogo Sada*, **167** : 102-107.

Garrity, J. B. (1975) *Gladioli for Everyone, Devon and Charles*, London.

Geeta, S.V., Shirol, A.M., Nishani, S., Shiragur, M. and Varuna, K.J. (2014) *Res. J. Agric. Sci.*, **5**(4) : 658-661.

Georgeva, M. and Peikova, E. (1976) *B'lgarski Plodove, Zalenchutsii Konservi*, **8** : 22-29.

Germanyan, N. M. and Ambartsumyan, S. A. (1976) *Izvestiya S.-Kh. Nauk*, **6** : 50-55.

Ghisleni, P. L. (1956) *Riv. Ortoflorofruttic*. Ital., **46** : 134-44.

Ghosh, D.K., Bandyopadhyay A., Maheswarappa, H.P., Sahu, P.K., Chakrabarti, K. and Biswas, B. (2017) *The Bioscan.*, **12**(3) : 1653-1659.

Gigante, R. (1973a) *Sementi Elette*, **19** : 29-35.

Gigante, R. (1973b) *Sementi Elette*, **19** : 37-42.

Gill, C. M., Minami, K. and Demetrio, C. G. B. (1978) *Solo*, **70** : 66-68.

Ginzburg, C. (1973) *J. Exp. Bot.*, **24** : 558-66.

Ginzburg, C. (1974) *J. Exp. Bot.*, **25** : 995-1003.

Ginzburg, C. (1981) *J. Exp. Bot.*, **32** : 1105-1109.

Ginzburg, C. and Ziv, M. (1973) *Ann. Bot.*, **37** : 219-224.

Glazurina, A. N. and Chemagin, N. C. (1974) *Bulleten Gasudarstvennogo Nikitskogo Botanicheskogo Sada*, **1** : 51-54.

Goldblatt, P. (1971) *J. S. Africa Bot.*, **37** : 317-460.

Goldblatt, P. (2008) *Novon.*, **18**(2) : 164-167.

Goldblatt, P., Manning, J.C., Wightman, N. and Bytebier, B. (2019) *S. Afr. J. Bot.*, **123** : 293-295.

Goldblatt, P., Takei, M. and Razzaq, Z.A. (1993) *Annals of the Missouri BotanicalGarden*, **80** : 461-470.

Gonzalez, A.; Banon, S.; Fernandez, J.A.; Franco, J. A; Casas, J.L. and Ochoa, J. (1998) *Sci. Hortic.*, **74** : 279-284.

Goo, D. H. and Kim, K.W. (1994) *J. Korean Soc. Hortic. Sci.*, **35** : 400-405.

Goo, D.H., Cho, H.R., Huh, K.Ya., Ko, J.Y., Kim, K.W. (1999) *J. Korean Soc. Hortic. Sci.*, **40** : 385-388.

Goo, D.H., Park, I.S., and Kim, K.W. (1998) *J. Korean Soc. Hortic. Sci.*, **39** : 479-482.

Gowda, J.V.N.; Jayanthi, R. and Raju, B. (1998) *Curr. Res.*, Univ. of Agric. Sciences, Bangalore, **17** : 80-81.

Grabowska, B. (1972) *Biul. Inst. Hodowli Aklimat. Rosl*, **5-6** : 267-268.

Grabowska, B. (1975b) *Prace Instytutu Sadownictwa Ser. B. Rosliny Ozdobnych.*, **1** : 41-50.

Grabowska, B. (1978) *Prace Instytutu Sadownictwai Kwiaciarstwa w Skierniewicach*, B. **3** : 15-22.

Grabowska, B. (1980a) *Prace Instytutu Sadownictwai Kwiaciarstwa w Skierniewicach*, B. **5** : 41-47.

Grabowska, B. (1980b) *Prace Instytutu Sadownictwai Kwiaciarstwa Skierniewicach*, B. **5** : 49-54.

Grabowska, B. and Mynett, K. (1970) *Biul. Inst. Hodowli Aklimat. Rosl.* (IHAR), **1-2** : 85-88.

Grabowska, B. (1975a) *Prace Instytutu Sadownictwa w Skierniewicach*, B.1 pp 27- 40.

Graves, A.C.F. and Goldman, S.L. (1987) *J. Bacteriol.*, **169** : 1745-1746.

Griesbach, R. A. (1972) *The World of the Gladiolus*, NAGC, USA, pp. 138-140.

Grinenko, V. V., Pospelova, Yu. S. and Chirva, Z. F. (1973) *Bjull. glav bot. Sada*, **87** : 87-93.

Groen, N. P. A. and Kappelhof, N. J. (1975) *Bloembollencultuur*, **86** : 62.

Groen, N. P. A. and Schipper, J. A. (1982a) *Bloembollencultuur*, **93** : 15.

Groen, N. P. A. and Schipper, J. A. (1982b) *Bloembollencultuur*, **93** : 526-527.

Groen, N. P. A. and Slogteren, D. H. M. Van (1980) *Bloembollencultuur*, **91** : 327-328.

Groen, N. P. A. and Vander Lans, A. M. (1980) *Bloembollencultuur*, **91** : 443.

Groen, N. P. A., Kruijer, C. J. and Ruyter, J. P. (1976) *Bloembollencultuur*, **86** : 679.

Groen, N. P. A., Muller, P. J. and Kappelhof, N. J. (1979) *Bloembollencultuur*, **90** : 572-573.

Groen, N.P.A. (1990) *Bloembollencultuur*, **101** : 35.

Gromov, A. N. (1972) *The World of the Gladiolus*, NAGC, USA, pp. 98-102.

Groren, N.P.A. and Asjes, C.J. (1989) *Bloembollencultuur*, **100** : 26-27.

Grouet, D. (1965) C. *R. Acad. Agric. France*, **51** : 588-594.

Grover, J.K., Singh, K., Gupta, A.K. and Kumar, A. (2005) *Indian J. Plant Physiol.*, **10**(3) : 248-253.

Grzeszkiewicz, H. (1978a) *Acta Hortic.*, **82** : 31-36.

Grzeszkiewicz, H. (1978b) *Acta Hortic.*, **82** : 37-41.

Guillarmod. A. J. (1977) *Gladiolus Annual*, pp. 44-45.

Gupta, S. and Kumar, A. (2018) *Int. J. Curr. Microbiol. App. Sci.*, **7**(9) : 330-334

Gupta, S., Kumar, A., Pathak, S. and Maurya, N. (2020) *Pharma Innov. J.*, **9**(5) : 274-277.

Haag, A., Oliviera. P., De, G. D. and Mathos, J. R. (1970) *Anais da Escola Superior de Agricultura*, **27** : 125-141.

Haasis, F. A. (1954) *Plant Dis. Reptr.*, **38** : 518-520.

Haber. Z. (1968) *Biul. Inst. Hodowli Aklimat. Rost.*, **3-4** : 107-113.

Halevy, A. H. (1962) *Advances in Hort. Sci. and their Applications*, pp. 484-492.

Halevy, A. H. (1972) *HortScience*, **7** : 113-114.

Halevy, A. H. and Monselise, S. P. (1961) *Gladiolus*, **36** : 71-76.

Halevy, A. H. and Richter, R. (1962) *Proc. 16th Int. Hort. Cong.*, Brussels, **1** : 350.

Halevy, A. H. and Shillo. R. (1970) *Physiol. Plant.*, **23** : 820-827.

Halevy, A. H., Shil1o. R. and Simchon, S. (1970) *J. Hort. Sci.*, **45** : 427-434.

Halevy. A. H. and Monselise, S. P. (1963) *Bot. Gaz.*, **124** : 343-346.

Hamilton, A. P. (1963) *The Gladiolus Annual*, British Gladiolus Society, pp. 46-50.

Hardenberg, R. E., Uota, M. and Parson, C. S. (1969) *Proc. 12th. Int. Cong. Refrig.*, **3** : 339-347.

Harrison, D. E. (1962) *J. Agric. Vict.*, **60** : 377-387, 389.

Hartsema, A. M. (1937) *Verli. Akad. Ulet. Amst.*, **36** : 1-35.

Hassan, F.A.S. and Ali, E. (2018) *Int. J. Adv. Res.*, **2**(10) : 188-199.

Hatibarua, P. and Misra, R.L. (1999) *J. Ornam. Hort.*, **2** : 111-114.

Hegde, N. and Kuruvinashetty, M. S. (1998a) *Karnataka J. Agri. Sci.*, **11** : 223-224.

Hegde, N. and Kuruvinashetty, M. S. (1998b) *Karnataka J. Agri. Sci.*, **11**: 225-226.

Heidmann, I., Schade-Kampmann, G., Lambalk, J., Ottiger, M. and Di Berardino, M. (2016) *PLoS ONE* **11**(11). doi:10.1371/journal.pone.0165531.

Hiremath, V.M., Singh, K.P., Swaroop, K. and Panwar, S. (2020) *Indian J. Agric. Sci.*, **90**(6) : 1180-1184.

Hitchcock, A. E., Zimmerman, P. W. and Coe, R. R. (1962) *Contr. Boyce ThompsonInst.*, **21** : 303-344.

Hockings, F. D. (1972) *Qd. Agric. J.*, **98** : 425-434.

Holdgate, D. P. (1977) In: *Plant Cell, Tissue and Organ Culture* (Eds. J. Reinert and Y. P. S. Bajaj), Springer Verlag.

Holmes, E. S. Myers, J. M. and Magie, R. 0. (1957-58) *A. R. Fla. Agric. Exp. Sta.*, pp. 33-42.

Hong, Y.P. and Goo, D.H. (1991) *Res. Rep. Rural Develop. Admin. Hort.*, **3** : 125-130.

Hong, Y.P. and Kim, Y.K. (1988) *Res. Reports of the Rural dev.Admin., Hort.*, Korea Republic, **30** : 99-103.

Hong, Y.P., Goo, D.H. and Hano, I.S. (1989) *Res. Rep. Rural Develop. Admin. Hort.*, **31** : 54-58.

Hongbo, Li., Hongmei, Li., Jiping, Liu., Zhihong, Luo., Daryl, Joyce. and Shenggen, He. (2017) *Postharvest Biol. Technol.*, **123** : 102-111.

Hoogeterp, P. and Mohlmann, J. (1965) *Weekbl. Bloembollcult.*, **76** : 491.

Horþ, D.A., Cantor, M. and Buta, E. (2014) *Bull. UASVM Hortic.*, **71**(1), doi: http://dx.doi.org/10.15835/buasvmcn-hort:9860

Hosoki, T. (1995) *HortScience*, **30** : 1251-1252.

Hosoki, T. and Kubara, T. (1989) *HortScience*, **24** : 968-971.

Hsieh, K. C. and Huang, S. T. (1970) *Taiwan Agric. Quart.*, **6** : 92-103.

Huh, K.Y., Cho, H.Y., Bang, C.S., Song, J.S., Goo, D.H., Kim, Y.J. and Kim, B.H. (1996) *RDA J. Agric. Sci., Hort.*, **38** : 503-507.

Hussein, M. F., El-Gamassy, A. M. and Serry, G. A. (1962) *Agric. Res. Rev.* Cairo, **40** : 1-9.

Hutchinson, J. (1932) *The Families of Flowering Plants. II-Monocotyledons.*

Hwang, M.J. and Kim, K. S. (1995) *J. Korean Soc. Hort. Sci.,* **36** : 410-419.

Iba, S., Matsubara, H., Inokoshi, Y. Oka, M. and Meshitsuka, G. (1964) *TokyoMetropol. Isop. Res. Cent.* Japan, A. Rep., **3** : 109-13.

Iba,S., Matsubara, H., Oka, M., Meshitsuka. G., Matsumato, H. and Kudo, I. (1965) *Tokyo Metropol. Isop. Res. Cent.,* Japan, A. Rep., **4** : 85-88.

Imanishi, H. (1981a) *J. Jap. Soc. Hort. Sci.,* **50** : 92-99.

Imanishi, H. (1982) *J. Jap. Soc. Hort. Sci.,* **50** : 503-510.

Imanishi, H. and Imae (1990) *Acta Hortic.,* **266** : 189-196.

Imanishi. H. (1981b) *J. Jap. Soc. Hort. Sci.,* **50** : 355-362.

Incalcaterra, G. (1992) *Colture Protette,* **21** : 83-94.

Ingle, S.N., Kuchanwar, O.D., Choudhari, P.L., Gurav, P.P., Pawar, Y.S. and Patangray, A.J. (2015) *Trends in Biosci.,* **8**(23) : 6605-6608.

Ingram, C. (1967) *J. Roy. Hort. Soc.,* **92** : 309-311.

Innes, C. (1985) *The World of lridaceae-a comprehensive record,* Holly Gate International Ltd., England.

Ishwarraddy, Kandpal, K., Hugar, A., Ramesh, G. and Amaregouda, A. (2018) *Int. J. Curr. Microbiol. App. Sci.,* **7**(11): 2566-2573.

Iwama, S. and Iwai, S. (1952) *J. Hort. Ass. Japan,* **21** : 183-188.

Iziro, Y. and Hori, Y. (1983) *J. Jap. Soc. Hort. Sci.,* **52** : 51-55.

Jackson, C. R. (1962) *Phytopathol.,* **52** : 974-977.

Jadhav, P.B., Mangave, B.D., Singh, A., Dekhane, S.S. and Patel, D.J. (2015) *Int. J. Curr. Res.,* **7**(7) : 17674-17676.

Jakoby, B. and Halevy, A. H. (1970) *Bot. Gaz.,* **131** : 74-77.

Jankiram T., Swaroop, K. and Das, T.K. (2014) *Indian J. Agric. Sci.,* **84**(8) : 951-56.

Jenkins, J. M. Jr. (1958) *Flor. Exch.,* **131** : 11.

Jenkins, J. M. Jr. (1961) *Gladiolus,* **36** : 67- 70.

Jenkins, J. M. Jr. (1963) *Res. Fmg.,* N. C. **22** : 14.

Jenkins, J. M. Jr., Chambers, E. E. and McGee, F. G. (1968) *Weed Sci.,* **16** : 86-88.

Jeves, T. M. and Coley-Smith. J. R. (1976) *Trans. British Mycol. Soc.,* **67** : 419-425.

Jha, S., Kaushik, M., Pariha, B.S. and Sinha, P.K. (2020a) *J. Pharmacogn. Phytochem.,* **9**(3) : 826-830.

Jha, S., Kaushik, M., Parihar, B.S. and Sinha, P.K. (2020b) *J. Pharmacogn. Phytochem.,* **9**(3): 831-833.

Jian, W., Wenjing, W., Jiahui, L., Yujie, J. and Gazz. (2018) *Plant Cell Physiol.*, **60**(1), doi: 10.1093/pcp/pcy186.

Jin-ping, F., Shu-fang, G. and Dai-di, C. (2012) *J. Northeast Agric. Univ.*, **19**(3) : 38-42.

Johnson, G. V. A., Bing, A. and Smith, F. F. (1967) *J. Econ. Entomol.*, **60** : 16-18.

Joneki, J. (1979) *Acta Hortic.*, **91** : 201-206.

Joneki, J. (1980) *Acta Hortic.*, **109** : 347-355.

Judith, O., Saffudin, D., Catherine, L. and Abiy, Y. (2014) *J. Pharm. Res.*, **8**(9) : 1218-1222.

Kabir, M.D., Mamun, A.N.K., Yasmin, F. and Subramaniam, S. (2016) *J. Phytology.*, **6**: 40-45

Kadam, G. B. and Singh, K. P. (2009). *J. Ornam. Hortic.*, **12**(3) : 175-181.

Kadam, G.B., Kumar, G., Saha, T.N., Kumar, R., Tiwari, A.K. and Kumar, R. (2014) *Indian J. Agric. Sci.*, **84**(12) : 1546-1549.

Kadam, G.B., Kumar, G., Saha, T.N., Tiwari, A.K. and Kumar, R. (2014) *Indian J. Hort.*, **71**(3) : 379-384.

Kadam, G.B., Singh, K.P. and Singh, M.P. *Indian J. Plant Physiol.*, **18** : 49-54.

Kadam, J.J., Agale, R.C., Rite, S.C., Pandav, S.M. (2014) *Discovery Agric.*, **2**(9) : 61-64.

Kaiker, U. S. (1975) *NAGC Bull.*,**122** : 73-76.

Kalasareddi, P.T., Reddy, B.S., Patil, S.R., Ryagi, Y.H. and Gangadharappa, P.M. (1998) *Karnataka J. Agric. Sci.*,**11** : 1133-1135.

Kamo, K. (1997) *Plant Cell Rep.*, **16** : 389-392.

Kamo, K., Aebig, J.A., Guaragna, M.A., James, C., Hsu, H.T. and Jordan, R.L. (2012) *Plant Cell Tiss. Organ Cult.*, **110**(1), doi: 10.1007/s11240-012-0124-y.

Kamo, K., Blowers, A., Smith, F. and Van-Eck, J. (1994) *HortScience*, **29** : 529.

Kamo, K., Blowers, A., Smith, F. and Van-Eck, J. (1995b) *Plant Sci.*, **110** : 105-111.

Kamo, K., Lakshman, D., Bauchan, G., Rajasekaran, K., Cary, J. and Jaynes, J. (2015) *Plant Cell Tiss. Organ Cult.*, **121** : 459–467.

Kamo, K., McElroy, D. and Chamberlain, D. (2000) *In Vitro Cellular and Developmental Biology - Plant*, **36** : 182-187.

Kamo, K., O'Connor, A. and Gelvin, S. (1995a) *Plant Physiol.*, **108** : 152.

Kamo, K., Rajasekaran, K. and Cary, J. (2014) *J. Appl. Hortic.*, **16**(3) : 193-198.

Kang, M., Choi, J.D. and Kim, K.W. (1998) *J. Korean Soc. Hort. Sci.*, **39** : 338-342.

Kang, Misuk and Kim, KiuWeon. (1998) *J. Korean Soc. Hort. Sci.*, **39** : 487-490.

Kapoor J. N. and Gil1. H. S. (1961) *Indian Phytopath.*,**14** : 149-153.

Karagoz, F.P., Dursun, A., Tekiner, N., Kul, R. and Kotan, R. (2019) *Ornam. Hortic.*, **25**(2) : 180-188.

Karaguzel, O., Alian, S., Doran, I. and Sogut, Z. (1999) In: *Improved Crop quality by nutrient management* (Eds. Anag, D.; Martin-Prevel, P.) Kluwer Academic Publishers, pp. 250-262.

Kareem, A., Khan, M.A., Shoib-ur-Rehman and Afzal, I. (2013) *Adv. Zool. Bot.*, **4** : 86-91.

Karimi, M., Kalantari, S., Naderi, R. and Shafiei, M. (2014) *Iranian J. Hortic. Sci.*, **45**(1) : 55-66.

Kashyap, M., Sharma, G., Tirkey, T. and Sharma, D. (2018) *Int. J. Curr. Microbiol. Appl. Sci.*, **7**(10) : 1832-1838.

Kasumi, M., Takatsu, Y., Tomotsune, H. and Sakuma, F. (1998) *J. Japanese Soc. Hort. Sci.*, **67** : 951-957.

Kasumi, M., Takatsu, Y., Tomotsune, H. and Sakuma, F. (1999a) *J. Japanese Soc. Hort. Sci.*, **68** : 168-175.

Kasumi, M., Takatsu, Y., Tomotsune, H., Sakuma, F. and Lida, S. (1999b) *J. Japanese Soc. Hort. Sci.*, **68** : 195-197.

Kaur, S. and Dhatt, K.K. (2019) *Int. J. Curr. Microbiol. Appl. Sci.*, **8**(2) : 1625-1631.

Kaushik, A.K., Singh, B., Dhaiya, R. and Goswami, A. (2018) *Ann. Hortic.*, **11**(1) : 46-51.

Keisam L., Pradeep, K., Manivannan and Kumar S.R. (2014) *Asian J. Hortic.*, **9**(2) : 416-420.

Kelsheimer, E. G. (1954) *Proc. Fla. Sta. Hort. Soc.*, **67** : 297-300.

Kelsheimer, E. G. (1956) *Fla. Agr. Exp. Sta. Circular*, S-91.

Keszi-Hermath (1960-61 and 62) *Lucr. Sti. Inst. Cerc. horti. vitic.*, **4** : 811-816.

Khan, F.N., Rahman, M.M. and Hossain, M.M. (2013) *J. Ornam. Hortic. Plants.*, **3**(1) : 59-71.

Khan, G., Kapoor, V. P. and Farooqi, M. I. H. (1980) *Indian J. Hort.*, **37** : 182-184.

Khan, M.R., Shahid, S., Mohidin, F.A., Mustafa, U. (2017) *Biol. Control.*, **115**. doi: 10.1016/j.biocontrol.2017.09.010.

Khanna, K. and Gill, A. P. S. (1983) *Punjab Hort. J.*, **23** : 116-120.

Khattab, M., El-Torky, M., Torabeih, Abd-El-H. and Rashed, H. (2017) *Alexandria Sci. Exch. J.*, **38**(3) : 588-598.

Kim, K W. and Han, S.Y. (1993) *J. Korean Soc. for Hort. Sci.*, **34** : 136-144.

Kim, K.W. and Goo, D.H. (1991) *J. Korean Soc. for Hort. Sci.*, **32** : 388-393.

Kim, K.W. and Lee,J.S. (1993) *J. Korean Soc. for Hort. Sci.*, **34** : 301-307.

Kireeva, M. F. (1972) *Referativnyi Zhurnal*, **3** : 55,758.

Klar, A. E., Mattos, J. R. and Ranzani, G. (1974) *Turrialba*, **24** : 285-97.

Klasman, R., Molinari, J. and Benedetto, A. de. (1995) *Horticultura Argentina*,**14** : 65-68.

Klein, L. G. (1972) *Splash*, pp. 11-12.

Ko, J.Y., Kim, S.K., Um, N.Y., Han, J.S. and Lee, K.K. (1994) *RDA J. Agri. Sci. Hort.*, **36** : 430-434.

Koch, L. W. (1950) *Sci. Agric.*, **30** : 483-487.

Kofranek, A. M. (1951) *Flor. Exch.*, **116** : 12.

Kofranek, A. M. and Halevy, A. H. (1976) *HortScience*, **11** : 572-573.

Kofranek, A. M. and Paul, J. L. (1974) *Acta Hortic.*, **41** : 199-206.

Kofranek, A. M., Lunt, O. R. and Kohl, H. C. Jr. (1957) *Proc. Amer. Soc. Hort. Sci.*,**69** : 556-560.

Kolesnikov, B. M. (1965) *Sadovostva Vinogradastvai Vinodeliya Mald.*, **5** : 47.

Kondrat'eva, V.V. and Belynskaya, E.V. (1989) *Byulleten' GlavanogoBotanicheskogoSada*, **152** : 25-28.

Konoshima, H. (1980) *J. Jap. Soc. Hort. Sci.*, **49** : 403-408.

Kore, S. S. and Bhide, v. P. (1976) *Curr. Sci.*, **45** : 74.

Korkut, A.B., Butt, S.J. and Dozalan, E. (1998) *Pakistan J. Sci. and Indust. Res.*, **41** : 199-202.

Koster, A. T. J. and Rooy, M. De. (1981) *Bloembollencultuur*, **91** : 1184.

Kosugi, K. (1953) *Tech. Bull. Kagawa Agri. Coll.*, **5** : 137-143.

Kosugi, K. (1959) *J. Hort. Ass., Japan*, **28** : 188-192.

Kosugi, K. and Kondo, M. (1959) *Tech. Bull. Fac. Agri. Kagawa Univ.*, **11** : 173-176.

Kosugi, K. and Kondo, M. (1960) *J. Hort. Ass.*, Japan, **29** : 163-168.

Kosugi, K. and Kondo, M. (1961b) *J. Japanese Soc. Hort. Sci.*,**30** : 366-370.

Kosugi, K. and Sano, Y. (1961a) *J. Japanese Soc. Hort. Sci.* **30** : 259-262.

Kosugi, K. and Sano, Y. (1962) *Tech. Bull. Fac. Agric. Kagawa Univ.*, **13** : 141-145.

Kosugi, K. and Sumimoto, M. (1956) *Tech. Bull. Kagawa Agric. Coll.*, **7** : 113-122.

Kosugi, K., Sumimoto, A. and Katagiri, T. (1957) *Tech. Bull. Fac. Agric. Kagawa Univ.*, **9** : 59-65.

Koutepas, N. G. (1982) *Zizaniologia*, **1** : 39-42.

Kruyer, C. and Groen, N. P. A. (1977) *Bloembol1encultuur*, **87** : 699.

Krvavych, A.S., Konechna. R.T., Petrina, R.O., Kyrka, M.S., Zayarnuk, N.L., Gulko, R.M., and Stadnytskane, Novikov, V.P. (2014) *Res. J. Pharm. Biol. Chem. Sci.*, **5**(1) : 240-246.

Kshirsagar, A.B. (2009) *Bioinfolet*, **6**(3) : 224-226.

Kukushkin, V. A. (1983) *Byulliton Glownogo Botanicheskego Sada*, **129** : 76-79.

Kumar A., Kumar A., Sharma V., Mishra A., Singh S. and Kumar P. (2018) *Int. J. Curr. Microbiol. Appl. Sci.*, **7**(10) : 2900-2909.

Kumar, A., Pandey, S.D. and Rai, R.R. (2012) *Veg. Sci.*, **39**(2) : 128-131.

Kumar, B.S., Kameswari, P.L., Pratap, M. and Venkateswarrao, P. (2015) *J. Agric. Res.*, **49** : 71-76.

Kumar, M., Kumar, S. and Chaudhary, P. (2015) *Int. J. Agric. Stat. Sci.*, **11**(2): 527-532.

Kumar, M., Singh, V.P., Arora, A, Raju, D. and Sakhare, A. (2015) *Indian J. Hortic.*, **72**(3) : 397-401.

Kumar, P. and Awasthi, A.K. (2012) *Ann. Hortic.*, **5**(2) : 234-239.

Kumar, P.N. and Misra, R.L. (2012) *Indian J. Agric. Sci.*, **82**(7) : 632-637.

Kumar, P.S., Kumar, R., Choudhary, V.K. and Kanwat, M. (2015) *Indian J. Agric. Sci.*, **85**(6) : 845–849.

Kumar, R. and Chaturvedi, O.P. and Misra, R. (2003) *J. Ornam. Hortic.*, **6** : 100-103.

Kumar, R. and Misra, R.L. (2003) *J. Ornam. Hortic.*, **6**(2) : 95-99.

Kumar, R. and Misra, R.L. (2011) *Indian J. Hortic.*, **68**(4) : 535-539.

Kumar, R. and Misra, R.L. and Singh, S.K. (2008) *Indian J. Hortic.*, **65**(1) : 83-86.

Kumar, R. and Yadav, D.S. (2006) *J. Ornam. Hortic.*, **9**(1) : 33-36.

Kumar, R., De, L.C., Roy, A.R. and Verma, M.R. (2009) *Indian J. Agric. Sci.*, **79**(8) : 604-607.

Kumar, R., Roy, A.R. and Patel, R.S. (2009) *Environ. Ecol.*, **27**(2) : 607-610.

Kumar, R., Singh, G.N. and Misra, R.L. (2003) *J. Ornam. Hort.*, **6**(2) : 104-106.

Kumar, R., Deka, B.C. and Roy, A.R. (2010) *J. Ornam. Hortic.*, **13**(1) : 35-40.

Kumar, S., Kumar, A. and Chandra, S. (2010) *Asian J. Hortic.*, **5**(1) : 44-48.

Kumari, K. and Kumar, S. (2015) *Int. J. Agric. Environ. Biotechnol.*, **8**(2) : 413-422.

Kumari, K. and Kumar, S. (2019) *Inter. Arch. Appl. Sci. Technol.*, **10**(4) : 71-80.

Kumari, M., Swaroop, K., Singh, S.K., Kamil, D., Singh, B. and Bharadwaj, C. (2019) *Indian J. Agric. Sci.*, **89**(2) : 268-278.

Kumari, P., Rao, T.M., Sane, A., Kumar, R. and Dhananjaya, M.V. (2016) *Indian J. Agric. Sci.*, **86**(7) : 849–853.

Kumawat, P., Sisodia, A., Singh, A.K., Padhi, M. and Barman, K. (2018) *Int. J. Curr. Microbiol. Appl. Sci.*, **7**(8) : 870-874.

Kwon, O.D., Kim, H.J., Kim, J.K. and Han, K.P. (1996) *RDA J. Agric. Sci., Crop Protection*, **38** : 495-500.

La Rue, C. D. and Curry, La V. (1947) *Amer. J. Bot.*, **34** : 586.

Labanowski, G and Mynett, M. (1990) *Acta Hortic.*, **266** : 547-551.

Laeser, H. (1981) *Deutscher Gartenbau*, **36** : 38-41.

Langdon. R. F. and Herbert, D. A. (1944) *Univ. Qd. Pap. Dep. Biol.*, **2** : 4.

Laskar, M.A.; Jana B.K. (1995) *Indian Agriculturist*, **39** : 43-50.

Laurie, A., Kiplinger, D. C. and Nelson, K. S. (1958) *Commercial Flower Forcing*, McGraw Hill Book Co. Inc., New York.

Lee, S. H., Kim, J. S. and Choi, Y. M. (1983) *Korean J. Pl. Protection*, **22** : 198-202.

Leinfelder, J. and Rober, R. (1989) *Gartenerborse Gartenwelt*, **89** : 946, 948-953.

Lemini, C. and Lemini, V. (1965) *Grad. Via Liv.*, **14** : 66-68.

Lemini, C. and Lemini, V. (1981) *Productia Vegetala, Horticultura*, **30** : 39-51.

Lepez Oliveras, A. M., Lopez Perez, D. and Pages Pallares, M. (1984) *Anales del Inst. Nacinal de Investigaciones Agrarias, Agricola*, **27** : 29-45.

Lepori, G. (1967) *Inf. Agrar.*, Verona, **23** : 684.

Lewis, G. J., Obermeyer, A. A. and Barnard, T. T. (1972) *Gladiolus-a revision of the African species*, Purnell, C. P.

Liu T. (1995) *Bull. Taichung Dist. Agric. Improv. Sta.*, **49** : 33-40.

Liu, T.S., Huang, Y.C., Twu, J.S. and Hsieh, J.H. (1998) *Plant Protection Bull.* (Taipei), **40** : 241-249.

Liu, W.C., Wu, G.D., Yao, T.P., Chen, F. and Chen, W.H. (1998) *Taiwan Sugar*, **45** : 20-25.

López, J., González, A., Fernández, J.A., Franco, J.A. and Bañón, S. (2006) *Acta Hortic.*, **711** : 241-246.

Lucas Cavalcante da Costa, Fernanda Ferreira de Araújo, Mirelle Nayana de Sousa Santos, Paula Cristina Carvalho Lima, Ariana Mota Pereira, Fernando Luiz Finger (2017) *Cienc. Rural.*, **47**(2), http://dx.doi.org/10.1590/0103-8478cr20160139.

Lukaszewska, A. (1978) *PraceInstytutu Sadownictwai Kuraciarstiva w Skierniewicach*, B., **3** : 69-79.

Macek, L., Robinson, G. M. and Robinson, R. (1946) *Nature*, **158** : 342.

Maclean, N. A. (1948) *N. W. Sci.*, **12** : 17.

Maclean, N. A. (1951) *Phytopath.*, **41** : 941.

Maggio, A., D" Andria, R. and Morelli, G. (1993) *ColtureProtette*, **22** : 63-69.

Magie, R. O. (1948) *Plant Dis. Reptr.*, **32** : 11-13.

Magie, R. O. (1951) *Gladiolus Mag.*, **15** : 18-21.

Magie, R. O. (1956) *Phytopath.*, **46** : 19.

Magie, R. O. (1957-58) *A. R. Fla. Agric. Exp. Sta.*, pp. 317.

Magie, R. O. (1971) *Plant Dis. Reptr.*, **55** : 82-85.

Magie, R. O. (1975) *Gladio Grams, NAGC*, **17** : 4-6.

Magie, R. O. (1987a) *Gladio Grams, NAGC*, **66** : 6.

Magie, R. O. (1987b) *Gladio Grams, NAGC*, **66** : 10-11.

Magie, R. O. (1987c) *G1adio Grams, NAGC*, **66** : 4.

Magie, R. O. and Myers, J. M. (1975) *Proc. Fla. Sta. Hort. Soc.*, **87** : 478-482.

Magie, R. O., Overman, A. J. and Waters, W. B. (1966) *Bul1. Fla. Agric. Exp. Sta.*, **644** : 42.

Mahanta, P. and Paswan, L. (1994) *J. Agric. Sci. Soc. North East India*, **7** : 165-168.

Mahasen, M., Ona, A. F., Taufique, T., Mehraj, H. and Jamal Uddin, A. F. M. (2015). *J. Biosci. Agric. Res.*, **4**(1) : 10-19.

Mahato, S.K. and Gantait, S.S. (2017) *Indian Hortic. J.*, **7**(1) : 72-75.

Mahato, S.K. and Gantait, S.S. (2020) *Indian J. Agric. Sci.*, **90**(5) : 938–942.

Maniruzzaman, S., Rabbani, M.G., Akter, T., Barua, R., Brua, J.L. and Haque, M.M. (2017) *Am-Euras. J. Agric. and Environ. Sci.*, **17**(6) : 519-529.

Manning, J.C., Mannheimer, C. and Goldblatt, P. (2011) *S. Afr. J. Bot.*, **77**(3) : 790-794.

Mantel, P. and Groen, N. P. A. (1988) *Bloembollencultuur*, **99**: 36-37.

Mantrova, E. Z. (1956) *Bju1. glav. bot. Sada*, **24** : 64-70.

Manzoor, A., Ahmad, T., Bashir, M.A., Baig, M.M.Q., Quresh, A.A., Shah, M.K.N. and Hafiz, I.A. (2018) *Folia Hortic.*, **30**(2) : 307-319.

Marek, J. J. (1954) *Gladiolus*, **29** : 138-143.

Marousky, F. J. (1968) *Proc. Fla Sta. Hort. Soc.*, **81** : 415-419.

Marousky, F. J. (1969) *Proc. Fla. Sta. Hort. Soc.*, **82** : 411-414.

Marousky, F. J. (1971). *Proc. Trop. Reg. Sta. Hort. Sci.*, **15** : 216-222.

Marousky, F. J. (1974) *Agric. Res.*, **20** : 15.

Marousky, F. J. and Woltz, S. S. (1971) *Proc. Fla. Sta. Hort. Soc.*, **84** : 375-380.

Matraszek-Gawron, R., Chwil, M., Terlecka, P and Skoczylas, M. (2019) *Pharm.*, **12**(4), doi: 10.3390/ph12040172.

Mayak, S., Bravdo, B., Gvilli, A. and Ha Jevy, A. H. (1973) *Sci. Hortic.*, **1** : 357-365.

McClellan, W. D. (1945) *Phytopath.*, **35** : 921-930.

McClellan, W. D. (1956) *Phytopath.*, **46** : 687-689.

McClellan, W. D., Baker, K. F. and Gould, C. J. (1949) *Phytopath.*, **39** : 260-271.

McCulloch, L. (1944) *Phytopath.*, **34** : 263-287.

McCune, D. C. and others. (1966) *Contr. Boyce Thompson Inst.*, **23** : 1-11.

McKay, M. E., Byth, D. E. and Tommerup, J. A. (1981a) *Aust. J. Exp. Agric. Anim. Husb.*, **21** : 343-348.

McKay, M. E., Byth, D. E. and Tommerup, J. A. (1981b) *Sci. Hort.*, **14** : 77-92.

McKay, M. E., Hisse, B. J. and Mulder, J. C. (1982) *Sci. Hort.*, **17** : 277-279.

McKnight, T. (1948) *Qd. Agric. J.*, **66** : 104-105.

Mega, K. and Tamura, J. (1957) *Stud. Inst. Hort. Kyoto*, **8** : 145-150.

Mehraj, H., Mahasen, M., Taufique, T., Shiam, I.H. and Uddin, A.F.M.J. (2013) *J. Expt. Biosci.*, **4**(2) : 23-26.

Meir, S., Philoshoph-Hadas, S., Michaeli, R., Davidson, H., Fogelman, M. and Shaffer, A. (1995) *Acta Hortic.*, **405** : 335-342.

Memon N., Jasakni, M.J., Qasim, M. and Sharif, N. (2014) *Pak. J. Agri. Sci.*, **51**(2) : 475-482.

Memon, N., Qasim, M., Jaskani, M.J., Ahmad, R. and Anwar, R. (2009b) *Pak. J. Agric. Sci.*, **46** : 13-19.

Memon, N., Qasim, M., Jaskani, M.J., Ahmad, R., Ahmad I. (2009a) *N. Z. J. Crop Hortic. Sci.*, **37**(4), doi: 10.1080/01140671.2009.9687586.

Memon, N.U.N., Qasim, M., Jaskani, M.J., Awan, F.S., Khan, A.I., Sadia, B. and Hussain, Z. (2012) *Pak. J. Bot.*, **44**(2) : 769-776.

Memon, N.U.N., Wahocho, N.A., Miano, T.F. and Leghari, M.H. (2016) *Afr. J. Biotechnol.*, **15**(32) : 1600-1710.

Mendiola-Ela, V. (1952) *Philipp. Agric.*, **35** : 517-533.

Mensinkai, S. W. (1939) *Cytologia*, **10** : 59-72.

Merodio, C. and Plaza, J.L. De La (1989) *Acta Hortic.*, **246** :327-333.

Meshitsuka, G., Oka, M., Iba, S., Matsubara, H. and Inokishi, Y. (1963) *TokyoMetrop. Isot. Res. Cent. Japan*, A. Rep., **2** : 169-173.

Messar, Y., Srivastava, R., Sharma, R., Kumar, A., Bhuj, B.D. and Pal, S. (2016) *Ecol. Environ. Conserv.*, **22**(4) : 1795-1798.

Michai1, S. H. and Elarosi, H. (1965) *Trans. Brit. Mycol. Soc.*, **48** : 377-379.

Mishra, P., Singh, A.K. and Singh, O.P. (2014) *J. Agric. Vet. Sci.*, **7**(7) : 23-26.

Mishra, S. and Singh, R. (1999) *J. Ornam Hortic.*, **2** : 67-70.

Misra, A. K. and Negi, S. S. (1977) *South Indian Hortic.*, **25** : 88-89.

Misra, R. L. (1976a) *NAGC Bu11.* **125** : 80, 85.

Misra, R. L. (1976b) *Prog. Hort.*, **8** : 35-36.

Misra, R. L. (1977) *Delhi Garden Mag.*, pp. 20-25.

Misra, R. L. (1978) *Prog. Hort.*, **9** : 55-59.

Misra, R. L. (1980) *Haryana J. Hort. Sci.*, **9** : 152-153.

Misra, R. L. (1982) *Haryana J. Hort. Sci.*, **11** : 207-210.

Misra, R. L. and Bajpai, P. N. (1983) *Haryana J. Hortic. Sci.*, **12** : 1-6.

Misra, R. L. and Choudhury, B. (1977) *Indian Hortic.*, **22** : 19-20, 26.

Misra, R. L. and Choudhury, B. (1979) *Indian Hortic.*, **23** : 21.

Misra, R. L., Verma, T. S., Kumar, R., Singh, A. and Singh, B. (1985) *Indian J. Hortic.*, **42** : 290-295.

Misra, R.L. (1975) *Glad. Br. Assn. Newsletter*, **12** : 2-5.

Misra, R.L., Tripathi, D.K. and Chaturvedi, O.P. (1993) *Prog. Hortic.*, **25** : 147-150.

Moes, A. (1966) *Bull. Rech. Agron. Gembloux, N. S.*, **1** : 76-95.

Mohammadi, G.A., Sardoei A.S. and Shahdadneghad, M. (2014) *Int. J. Adv. Biol. Biomed. Res.*, **2**(2) : 417-426.

Mohanty, C.R., Sena, D.K. and Das, R.C. (1994) *Orissa J. Hort.*, **22** : 1-4.

Momin, B.C., Kumar, S., Dewan, N., Nongrum, H.B. and Marwein, B. (2017) *Indian J. Agric. Sci.*, **87**(6) : 825–833.

Monaco, T. J. and Jenkins, J. M. (1969) *Rep. South. Nurserymens' Assn. N. C. Sta.Univ.*, Raleigh, pp. 69-70.

Monaco, T. J. and Jenkins, J. M. (1970) *Proc. 23rd A. Meet. South. Weed Sci. Soc.*, pp. 218 (abstr.).

Montesi, M. and Lazzereschi, F. (1966) *Atti. Conv. naz. Forz. F1oric. Sett. aff.*, Geneva, pp. 37-44.

Morey, D. and Staples, D. (1972) *The Wor1d of the Gladiolus*, NAGC, USA.

Mukhopadhaya, T. P. and Yadava, L. P. (1984) *Haryana J. Hortic. Sci.*, **13** : 95-99.

Mukhopadhyay, A. and Das, P. (1978) *Orissa J. Hort.*, **6** : 1-5.

Muttoni, M., Alberto, C.M., Bartz, A.C., Uhlman, L.O., Tartaglia, V. de L. and Streck, N.A. (2017) Ciência Rural, 47(10). https://doi.org/10.1590/0103-8478cr20160824.

Nagaraju, V. and Parthasarathy, V.A. (1995) *Folia Hortic.*, **7** : 93-100.

Nagaraju, V. and Parthasarathy, V.A. (2000) In : *Biotechnology of Horticultural Crops*, Vol. 3, Naya Prokash, Calcutta, India, pp. 108-132.

Nagaraju, V., Bhowmik, G. and Parthasarathy, V. A. (1998) *Int. J. Tropical Agric.*, **14** : 131-139.

Naik, K., Nataraj, S.K., Kulkarni, B.S. and Reddy, B.S. (2011) *J. Hortic. Sci.*, **6**(1) : 41-45.

Nainwal, R.C., Katiyar, R.S., Singh, D. and Tewari, S.K. (2016) *Indian J. Hortic.*, **73**(3) : 396-399.

Nakasone, H. Y. (1953) *Circ. Hawaii Agric. Exp. Sta.*, **41** : 10.

Naresh S., Rao, A.V.D.D., Krishna, K.U. and Rao, B.B. (2019) *Bull. Env. Pharmacol. Life Sci.*, **8**(4) : 87-92.

Naresh, S., Rao, A.V.D.D., Vani, C.H.S., Vanaja, L., Aditya, G.and Tejaswi, R. (2019) *Bull. Env. Pharmacol. Life Sci.*, **8**(4): 83-86.

Negi, S.S., Raghava, S.P.S., Chacko, T.I., Rao, T.M. and Ramachandran, N. (1991) *Current Plant Science and Biotechnology in Agriculture*, Vo. 12.

Nelson, R. (1948) *Spec. Bull. Mich. Agric. Exp. Sta.*, **350** : 63.

Neto A.J.S., Grossi J.A.S., Finger F.L. and Ferreira S.O. (2009) *Acta Hortic.*, **847** : 309-312.

Ngoupaye, G.T., Bum, E.N., Taiwe, G.S., Moto, F.C., and Talla, E. (2014). *Afr. J. Tradit. Complement Altern. Med.*, **11**(1): 53-61.

Nijasure, S.N.; Ranpise, S.A. and Gondhali, B.V. (2004) *J. Ornam. Hort.* (3-4) : 381-385.

Nitsch, J. P. (1966) *Bull. Soc. Franc. Physiol. Veg.*, **12** : 233-246.

Noor-ul-Amin, Khattak, A.M., Ahmad, I., Ara, N., Alam, A., Ali, M. and Ali, I. (2013) *Pak. J. Bot.*, **4** : 1407-1409.

Ogale, V.K. ; Rode, V.A. and Mishra, S.D. (1995) *Indian J. Plant Physiol.*, **38** : 241-243.

Ogale, V.K., Venu-Babu, P. and Mishra, S.D. (2000) *J. Ornam. Hortic.*, **3** : 1-5.

Ohri, D. and Khoshoo, T. N. (1983) *Zeitscrift fur Pflanzenzuchtung*, **91** : 46-60.

Oszkinisowa, K. and Haber, Z. (1968) *Biul. Inst. Hodowli Aklimat. Rosl.*, **3-4** : 101-106.

Oyetayo F.L., Odeniyi I.A., Olaseigbe O.E. and Jegede F.O. (2020) *FASEB J.*, **34**(S1). doi: 10.1096/fasebj.2020.34.s1.00601.

Pal, P. and Chowdhury, T. (1998) *Hort. J.*, **11** : 69-77.

Palmer, E. F. (1954) *Pollen, Pistil and Pad, NAGC*, **2** : 2-4.

Pan'ko, N. A. (1973) *Vestsi Akademii Navul BSSR Biyalogichnykh Navuk*, **6** : 59-62.

Pandey, R.K., Nayak, D.K. and Mohalik, R.P. (2018) *Indian J. Nematol.*, **48**(1) : 68-72.

Pansuriya, P.B., Varu, D.K. and Viradia, R.R. (2018) *Int. J. Chem. Stud.*, **6**(2) : 2659-2663.

Pant, C.C., Lal, S.D. and Shah, D. (1998) *J. Ornam. Hortic.*, **1** : 32-36.

Parisa, S.N., Mina, K.H., Akbar, D. and Siamak, K. (2015) *J. Crop. Prot.*, **4**: *533-543.*

Park, R. (1976) *A Handbook of the World's Gladioli*, DeBeMiniglads, UK.

Parker, N. W. and Borthwick, H. A. (1951) *Flor. Exch.*, **117** : 38-40.

Parween, N., Mishra, S., Adil, A., Pal, A. and Jha, K.K. (2019) *Int. J. Curr. Microbiol. Appl. Sci.*, **8**(6) : 1739-1750.

Parween, N., Mishra, S., Adil, A., Pal, A. and Jha, K.K. (2019) *Int. J. Curr. Microbiol. Appl. Sci.*, **8**(7) : 2840-2857.

Pasterkamp, H. P. (1981) *Bloembollencultuur*,**92** : 397.

Pasterkamp, H. P. (1982) *Bloembollencultuur*, **93** : 484-485.

Pasterkamp, H.P. (1989) *Bloemballencultuur*, **100** : 21.

Patel, D., Patil, S., More, S.J. and Dohiya, T.P. (2018) *Int. J. Curr. Microbiol. Appl. Sci.*, **7**(1) : 645-652.

Pathania, N.S. and Misra, R.L. (2003). *Acta Hortic.*, **624** : 487-494.

Patil, M.T., Katwate, S.M., Singh, B.R., Kale, P.N. and Deokar, A.B. (1994) *J. Maharashtra Agric. Univ.*, **19** : 142-143.

Patil, S., Chawla, S.L. and Chaudhary, P. (2017) *Int. J. Chem. Stud.*, **5**(5) : 2305-2308.

Patil, S., Chawla, S.L. and Patel, A.I. (2017) *Indian J. Agric. Sci.*, **87**(7) : 953–957.

Patil, S.D. (2014) *Int. J. Adv. Res. Biol. Sci.*, **1**(6) : 15-20.

Patil, S.D. and Dhaduk, B.K. (2009) *J. Ornam. Hortic.*, **12**(4) : 232-238.

Patra, S.K. and Mohanty, C.R. (2015) *Int. J. Agric. Sci. Res.*, **5**(2) : 27-32.

Paulin, A. (1958) *Rev. Hort. Paris*, **130** : 1921-1923.

Pereira, M.T.J., da Silva, T.J.A., Bonfim-Silva, E.M. and Mazzini-Guedes, R.B. (2016) *Aust. J. Crop Sci.*, **10**(3) : 393-401.

Pickell, C. A. (1975) *NAGC Bull.* **121** : 51-53.

Pitt, H. B., Hiwale, B. G., Worke, D. C. and Patil, V. K. (1981) *Ann. Conf. Indian Soc. Weed Sci.*, p.3.

Piya, S., Bajracharya, A.S.R., Mandal, J.L. and Choudhary, B.P. (2012) *J. Hortic. For.*, **4**(3) : 54-60.

Poon, T.B., Rao, T.M., Rajshekharan, P.E., Aswath, C. and Kumar, D.P. (2013) *Nepal J. Sci. Technol.*, **11** : 47-50.

Poon, T.B., Rao, T.M., Rajshekharan, P.E., Aswath, C. and Kumar, D.P. (2013) *Nepal J. Sci. Technol.*, **14**(1) : 31-34.

Porto, R.A., Koetz, M., Silva, E.M.B., Polizel, A.C. and Da Silva T.J. (2014) *Agric. Water Manag.*, **131**(1) : 50-56.

Pragya, Bhat, K.V., Misra, R.L. and Ranjan, J.K. (2010) *Indian J. Agric. Sci.*, **80**(9) : 766-772.

Prakash, S., Yadav, K.G. and Kumar, A. (2011) *J. Agric. Sci.*, **7**(1) : 77-79.

Prakash, V. and Jha, K. K. (1998) *J. Appl. Biol.*, **8** : 24-28.

Prijic, L. (1980) *Archen. Za Poljoprivredne Nauka*, **41** : 425-435.

Pritchard, A. E. (1950) *Calif. Agric.*,**4** : 13.

Puccinelli, L. F. R. and Fernandez, P. D. (1974) *Cientifica*, **1** : 24-32.

Qadeer, A., Ali, Z., Qasam, M., Toor, S. and Ahmad, H. (2016) *Plant Gene and Trait.*, **1**(9). doi: 10.5376/pgt.2016.07.0006

Qureshi, A., Qadri, Z.A. and Nazki, I.T. (2015) *J. Ornam. Hortic.*, **18**(3-4) : 126-134.

Raamsdonk, L.W.D. van and Vries, T. de (1989) *Syst. Evol.*, **165** : 189-198.

Raghava, S. P. S., Negi, S. S. and Sharma, T. V. R. S. (1988) *Indian Hortic.*, **26** : 2-3.

Rahman, A., Hussain, I., Nabi, G. and Ziaullah, Z. (2020) *Ornam. Hortic.*, **26**(2) : 244-250.

Rahmani, I., Ahmadi, N., Ghanati, F., Sadeghi, M. (2015) *N. Z. J. Crop Hortic. Sci.*, **43**(4) : 294-305.

Raikov, N. J. and Bulykov, V. I. (1975) *Bjull. glav. bot. Sada*, **98** : 65-68.

Rainio, A. J. (1936) *Valt. Maatalousk. Julk.*, **84** : 102.

Raj, H. and Upmanyu, S. (2006) *Indian J. Agric. Sci.*, **76**(5) : 307-311.

Rakosy-tican, E., Bors, B. and Szatmari, A.M. (2012) *Afr. J. Biotechnol.*, **11**(81) : 14703-14712.

Ram, M., Pal, V., Singh, M.K. and Kumar, M. (2012) *HortFlora Res. Spectrum.*, **1**(3) : 270-273.

Rameshwar, A. (1974) *Indian J. Hort.*, **31** : 95-96.

Ramos, P.A.S.; Megguer, C.A. and Finger, F.L. (2015) *Acta Hortic.*, 1060153-157.

Ramos-Garcýa, M., Ortega-Centeno, S., Hernandez-Lauzardo, A.N., Alia-Tejacal, I., Bosquez-Molina, E. and Bautista-Banos, S. (2009) *Sci. Hortic.*, **121** : 480–484.

Rangaswami, G. and Seshadri, V. S. (1970) *Fungi of South India*, pp. 29, 63-66.

Rao, V. G. (1964a) *Sydowia ann. Mycol.*, **18** : 46-64.

Rao, V. G. (1964b) *Mycopathologia Mycol. appl.*, **22** : 157-167.

Rashmi, R., Chandrashekar, S.Y., Hemanth, K.P and Geeta, S.V. (2016a) *Green Farming*, **7**(2) : 383-387.

Rashmi, R., Lakshmanareddy, D.C., Chandrashekar, S.Y. and Sukanya, T.S. (2016) *Green Farming*, **7**(6) : 1346-1351.

Rashmi, R., Lakshmanreddy, D.C. and Chandrashekar, S.Y. (2016b) *Ecol., Environ. Conserv.*, **22** : 137-142.

Rattenbury, J. (1966) *The Gladiolus Annual*, British Gladiolus Society UK, pp. 32-44.

Ravidas, L.; Rajeevan, P.K. and Valsalakumari, P.K. (1992) *South Indian Hort.*, **40** : 329-335.

Rawat, R., Bhuj, B.D., Srivastava, R. and Chand, S. (2017) *Prog. Hortic.*, **49**(2) : 164-167.

Reddy, P. P., Singh, D. B. and Rao, V. R. (1979) *Curr. Sci.*, **48** : 82.

Riaz, T., Khan, S.N. and Javaid, A. (2009) *Int. J. Agric. Biol.*, **11** : 498–500.

Riest, A. (1977) *Revue Suisce de Vite. d.' Arboric. et hort.* **9** : 221-225.

Rihne, T., Kumar, M., Shreenath, Y.S., Pant, R.P., Taloh, A., Swaroop, K. and Rao, G.P. (2019) *Phytopathogenic Mollicutes*, **9**(1). doi: 10.5958/2249-4677.2019.00075.6.

Risley, E. B. (1953) *Gladiolus*, pp. 121-128.

Robinson, M. (1983) *Ann. Bot.*, **51** : 779-785.

Robinson, M., Halevy, A. H., Galili, D. and Plaut, Z. (1983) *Ann. Bot.*, **51** : 461-468.

Rodriguez, J. D., Klar, A. E., Pedras, J. F., Rodriguez, S. D. and Pinho, S. Z. (1980) *Phyton*, Argentina, **39** : 57-76.

Rodriguez, J. D., Klar, A. E., Pedras, J. F., Rodriguez, S. D. and Delachiave, M. E. S. (1983) *Phyton*, Argentina, **43** : 81-95.

Roistacher, C. N., Bald, J. G. and Baker, K. F. (1953) *Science*, **118** : 186-187.

Roistacher. C. N., Baker, K. F. and Bald, J. G. (1957) *Hilgardia*, **26** : 659-684.

Ruffoni B., Pamato M. and Brea M. (2011) *Acta Hortic.*, **886** : 219-224.

Ruffoni B., Pamato M., Giovannini A. and Brea M. (2008) *Propag. Ornam. Plants.*, **8**(2) : 102-104.

Ryan, G. F. (1955) *Proc. Amer. Soc. Hort. Sci.*, **65** : 463-471.

Ryan, R. W. and Ulman, P. (1953) *Rept. Joint Res. Committee*, NAGC, USA.

Saaltink, G. J. (1969) *Tuirbouw Meded.*, **32** : 274-285.

Sable, P.B., Ransingh, U.R. and Waskar, D.P. (2016) *J. Soil. Crop.*, **26**(2) : 249-252.

Saeed, T., Hassan, I., Abbasi, N.A. and Jilani, G. (2014) *Plant Growth Regul.*, **72**(1) : 89-95.

Saeed, T., Hassan, I., Jilani, G. and Abbasi, N.A. (2013) *Sci. Hortic.*, **164** : 124-129.

Sagiroglu, M. and Akgül, G. (2014) *Turk. J. Bot.*, **38**(1) : 31-36.

Sankari, A., Anand, M. and Arulmozhiyan, R. (2015) *Asian J. Hortic.*, **10**(1) : 86-94.

Sarhan, A.M.Z., Alm-Eldeen, R.E.A., Emam, K.A.M. and Darwish, M.A. (2018) *Biosci. Res.*, **15**(3) : 2041-2048.

Sarkisyan, E. D., Azatyan, S. A. and Avundzhyan, E. S. (1975) *Biologicheskii Zhurnal Armenii*, **28** : 72-76.

Sasso, G. (1962) *Riv. Ortoflorofruttic.* Ital.,**46** : 245-254.

Sathyanarayana, E., Patil S. and Chawla, S.L. (2017a) *Int. J. Chem. Stud.*, **5**(5) : 317-319.

Sathyanarayana, E., Patil, S., Chawla, S.L. and Patel, D.K. (2017b) *Int. J. Curr. Microbiol. Appl. Sci.*, **6**(8) : 379-386.

Sathyanarayana, E., Sharma, G., Tirkey, T., Das, B.K., Divya, K. and Kumar, J. (2019) *J. Pharmacogn. Phytochem.*, **8**(5) : 227-230.

Scarascia-Mugnozza, G. T. (1968) *Gamma Field Symposia*, **7** : 111-121.

Schenk, P. K. (1955) *Tijdschr. Plziekt.*, **61** : 154-158.

Schenk, P. K. (1961) *Tijdschr. Plziekt.*, **67** : 313-316.

Schmidt, T. (1949) *PfiSch. Ber. Wien.*, **3** : 97-111.

Schulze, A., Roberts, R. and Pietersen, G. (2017) *Plant Dis.*, **101**(5): 846, doi: 10.1094/PDIS-10-16-1446-PDN.

Schwab, N.T., Streck, N.A., Uhlmann, L.O., Becker, C.C., Ribeiro, B.S.M.R., Langner, J.A. and Tomiozzo, R. (2018) *Ornam. Hortic.*, **24**(2) : 163-173.

Sedel'nikova, L. L. and Tarasenko, N. D. (1980) *Bju1l. glav. bot. Sada*, **116** : 37-42.

Seilleur, P. (1975) *Bu1leten des Researches Agronomiques de Gembloux*, **10** : 291-306.

Seilleur, P. (1977a) *Bu1leten des Researches Agronomiques de Gembloux*, **12** : 97-106.

Seilleur, P. (1977b) *Bu1leten des Researches Agronomiques de Gembloux*, **12** : 107-120.

Seilleur, P. (1977c) *Bulleten des Researches Agronomiques de Gembloux*. **12** : 121-134.

Seng, S., Wu, J., Liang J., Zhang, F., Yang Q., He, J. and Yi, M. (2017) *J. Am. Soc. Hortic. Sci.*, **142**(2) : 119-125.

Seng, S., Wu, J., Sui, J., Wu, C., Zhong, X., Liu, C., Liu, C., Gong, B., Zhang, F., He, J. and Yi, M. (2016) *Biochem. Biophys. Res. Commun.*, **474**(1) : 206-212.

Serek, M. Jones, R.B. and Reid, M.S. (1994) *J. Amer. Soc. Hort. Sci.*, **119** : 1014-1019.

Shaheenuzzaman M., Haque M., Karim M. and Noor Z. (2011) *J. Bangladesh Agric. Univ.*, **9**(1) : 21-26.

Sharma, G. (2006) Thesis. http://krishikosh.egranth.ac.in/handle/1/88827.

Sharma, S., Mishra, A., Moond, S.K., Vikee and Sharma, A.K. (2017) *J. Pharmacogn. Phytochem.*, **6**(6) : 498-501.

Sharova, N. C., Rybak, Yu. G. and Marina, N. E. (1977) *VyvashchivanieTevetochno-DekoratRast. V Moldavii Kishinev*, pp. 11-17.

Sheehan, T. J. (1959) *A. R. Fla. Agr. Exp. Sta.*, pp. 130-131.

Sheehan, T. J. and Joiner, J. N. (1966) *Proc. Fla. Sta. Hort. Soc.*, **79** : 443-446.

Sheehan, T. J. and Lorz, A. P. (1960) *A. R. Fla. Agr. Exp. Sta.*, pp. 126, 167.

Sheehan, T. J. and Sagawa, Y. (1959) *Proc. Fla. Sta. Hort. Soc.*, **72** : 388-391.

Sheehan, T. J. and Sagawa, Y. (1960) *Flor. Exch.*, **134** : 15.

Sheikh, M.Q., Paul, T.M. and Ahmad, M. (2006) *Indian J. Agric. Sci.*, **76**(10) : 307-311.

Shillo, R, and Halevy, A. H. (1981) *Scientia Hortic.*, **15** : 187-196.

Shillo, R. and Halevy, A. H. (1963) *Israel J. Agric. Res.*, **13** : 141-145.

Shillo, R. and Halevy, A. H. (1964) *Israel J. Agric. Res.*, **14** : 89-95.

Shillo, R. and Halevy, A. H. (1975) *Acta Hortic.*, **47** : 277-285.

Shillo, R. and Halevy, A. H. (1976a) *Scientia Hortic.*,**4** : 139-146.

Shillo, R. and Halevy, A. H. (1976b) *Scientia Hortic.*, **4** : 131-137.

Shillo, R. and Halevy, A. H. (1976c) *Scientia Hortic.*, **4** : 147-155.

Shillo, R. and Simchon, S. (1973) *Scientia Hortic.*, **1** : 57-62.

Shillo, R., Valis, R. G. and Halevy, A. H. (1981) *Sci. Hortic.*, **14** : 367-3 75.

Shimizu, K. and Uchibori, K. (1976) *Bu1l. Yamanishi Agric. Expt. Sta*, **18** : 69-79.

Shoushan, A. M., Dahab, A. M.A., El-Bagoury, H. M., El-Dahb, R. and El-Khateeb (1980b) *Res. Bull. Fac. Agric. Ain. Shams Univ.*, **1343** : 15.

Shoushan, A. M., El-Bagoury, H. M., Fahmy, G. E., Dahab, A. M. A., El-Dabh, R. S. and El- Khateeb, M. A. (1980a) *Res. Bull. Fac. Agric.*, Ain Shams Univ., **1342** : 13.

Shukla, A., Sharma, G., Kashya, S. and Netam, M. (2018) *Int. J. Curr. Microbiol. Appl. Sci.*, **7**(11) : 2493-2499.

Shumilenko, E. P. (1973) *Khimiya v Sel 'skomKhozyaistve*, **11** : 45-47.

Sidhu, G.S. and Arora, J.S. (1989) *Indian J. Hortic.*, **46** : 250-254.

Simchon, S., Shillo, R. and Halevy, A. H. (1972) *J. Hortic. Sci.*, **47** : 505-508.

Simonsen, J. and Hilderbrandt, A. C. (1971) *Canadian J. Bot.*, **49** : 1817-1819.

Singh, A.K., Asmita, Sisodia, A. and Hembrom, R. (2015) *Indian J. Agric. Sci.*, **85**(7) : 955–959.

Singh, A.K., Hembrom, R., Sisodia, A., Pal, A.K. and Asmita (2016) *Indian J. Agric. Sci.*, **86**(10) : 1316–1319.

Singh, B., Salam, P.K., Singh, K.P. and Patel, R. (2018) *Adv. Biores.*, **9**(4) : 152-154.

Singh, D. and Singh, K.M.P. (2015) *HortFlora Res. Spectrum*, **4**(1) : 76-78.

Singh, D. V. (1965) *Indian J. Microbiol.*, **5** : 87.

Singh, D., Mishra, A., Meena, B., Khatik, V.K. and Meena, B. (2017) *Chem. Sci. Rev. Lett.*, **6**(22) : 1129-1132.

Singh, H. and Singh, P.J. (2013) *Plant Dis. Res.*, **28**(2) : 152-156.

Singh, K.M.P. and Singh, D. (2014) *J. AgriSearch*, **1**(4) : 246-250.

Singh, K.P. (1996) *Advances in Plant Sci.*, **9** : 241-243.

Singh, K.P. and Ha, K.S. (1990) *South Indian Hortic.*, **38** : 208-210.

Singh, K.P., Ramachandran, N. and Uma, S. (1997) *Indian J. Agric. Sci.*, **67** : 404-406.

Singh, N., Mahar, K.S., Verma, S., Meena, B., Roy, R.K., Tewari, S.K., Goel, A.K., Rana, T.S. (2018) *Indian J. Biotechnol.*, **17**(1) : 118-127.

Singh, N., Pal, A.K., Meena, B., Roy, R.K., Tamta, S. and Rana, T.S. (2017b) *J. Hortic. Sci. Biotechnol.*, https://doi.org/10.1080/14620316.2017.1309995.

Singh, N., Pal, A.K., Roy, R.K., Tewari, S.K., Tamta, S. and Rana, T.S. (2016) *Nucleus*, doi : 10.1007/s13237-016-0181-4.

Singh, N., Pal, A.K., Roy, R.K., Tewari, S.K., Tamta, S. and Rana, T.S. (2017a) *Biochem Genet.*, https://doi.org/10.1007/s10528-017-9835-4.

Singh, N., Pal, A.K., Roy, R.K., Tewari, S.K., Tamta, S. and Rana, T.S. (2018) *Biochem Genet.*, **56**(1-2) : 128-148.

Singh, N., Singh, S.N., Srivastava, K., Yunus, M., Ahmad, K.J., Sharma, S.C. and Sharga, A.N. (1990) *Ann. Bot.*, **65** : 41-44.

Singh, R. N. (1968) *Plant Dis. Reptr.*, **52** : 662.

Singh, R. N. (1969) *Indian Phytopath.*, **22** : 402-403.

Singh, R. N. (1970) *Plant and Soil*, **33** : 249-250.

Singh, S. (1998) *J. Ornam. Hortic.*, **1** : 79-80.

Singh, S. and Datta, S.K. (2006) *J. Ornam. Hortic.*, **9**(1) : 37-39.

Singh, S., Sehrawat, S.K. and Sharma, S. (2019) *Int. J. Curr. Microbiol. Appl. Sci.*, **8**(11) : 656-662.

Singh, V, Meena, B., Pal, A.K., Roy, R.K., Tewari, S.K., Tamta, S. and Rana, T.S. (2017) *J. Genet.*, **96**(1), doi: 10.1007/s12041-017-0755-1/.

Sisodia, A. and Singh, A.K. (2015) *Indian J. Agric. Sci.*, **85**(1) : 79-86.

Sisodia, A., Kumawat, P., Padhi, M. and Singh, A.K. (2018) *Int. J. Curr. Microbiol. Appl. Sci.*, **7**(8) : 889-893.

Sisodia, A., Singh, A.K. and Sisodia, V. (2015) *Indian J. Agric. Sci.*, **85**(8) : 1059-1064.

Skalska, E. (1968) *Acta Pruhon.*, **17** : 29-47.

Skalska, E. (1970) *Vedecke Prace Vuzkumneho Ustava Okrasneho Zahradnictvi v Pruhonicich,* **5** : 261-270.

Smith, F. F. (1948) *Phytopath.,* **38** : 581-583.

Smith, F. F. and Brierley, P. (1953) *Plant Dis. Reptr.,* **37** : 547-552.

Smith, N.J. (1973) *Amer. Veg. Gr.,* **21** : 13-14.

Song, C.Y., Shin, D. G., Rho, T.H. and Lee, J.S. (1992) *J. Korean Soc. Hort. Sci.,* **33** : 209 - 219.

Spierings, F. (1969) *Tuinbouwmededelingen,* **32** : 110-114.

Spierings, F. H. F. G. (1967) *Nethelands J. Plant Pathol.,* **73** : 25-28.

Staby, G. L., Robertson. J. S., Kiplinger, D. C. and Conover, C. A. (1978) *Chain of Life,* Ohio State University, Columbus, Ohio, USA.

Stefaniak, B. (1994) *Plant Cell Rep.,* **13** : 386-389.

Stienstra, A. E. (1976a) *Bloembo1lencultuur,* **86** : 720.

Stienstra, A. E. (1976b) *Bloembol1encultuur,* **86** : 757.

Stuart, N.W. and McClellan, W.E. (1950) *Flor. Exch.,* **114** : 11, 47.

Suh, J.K. (1989) Abst. of Communicated Papers (Hort. Abst.), *J. Korean Soc. Hort. Sci.,* **7** : 158-159.

Suh, J.K. and Kwach, B.H. (1992) *J. Korean Soc. Hortic. Sci.,* **33** : 466-470.

Suh, J.K. and Kwack, B.H. (1990) *J. Korean Soc. Hortic. Sci.,* **31** : 294-299.

Swaroop, K., Raju, D.V.S., Das, T.K., Sharma, V.K. and Dhaker, S. (2017) *J. Ornam. Hortic.,* **20**(1-2) : 61-68.

Swaroop, K., Singh, K.P., Dhakar, S., and Kumar, P. (2015) *J. Ornam. Hortic.,* **18**(3-4) : 79-86.

Swarup, V. and Raghava, S. P. S. (1972) *Indian Hortic.,* **17** : 27-29.

Swetha, S., Kulkarni, B.S., Shiragur, M., Kulkarni, M.S., Naik, M.B.N., Mulge, R. and Hegde, L. (2020) *Int. J. Curr. Microbiol. App. Sci.,* **9**(2) : 525-531.

Syamal, M.M. ; Rajput, C.B. S. ; Singh, S.P. (1987) *Res. Development Reports,* **4** : 10-12.

Szczepaniak, M., Kamiñski, R., Kuta, E., S³omka, A., Heise, W. and Cieœlak, E. (2016) *Preslia.,* **88** : 137–161.

Taj, A., Rao, T.M., Ansar, H., Kumar, R. and Kumar, V.B.S. (2015) *Trends Biosci.,* **8**(23) : 6412-6415.

Taj, A., Rao, T.M., Ansar, H., Sriram, S. and Kumar, R. (2016) *Indian J. Agric. Sci.,* **86**(3) : 331-339.

Takatsu, I. (1982) *Bul1. Ibaraki. Ken Hort. Exp. Sta.,* **10** : 11-13.

Takatsu, Y., Kasumi, M., Manabe, T. and Hayashi, M. (2001) *HortScience,* **36**(2) : 341-343.

Takatsu, Y., Manabe, T., Kasumi, M., Yamada, T., Aoki, R., Inoue, E., Morinaka, Y., Marubashi, W. and Hayashi, M. (2002) *Breed. Res.*, **4**(2) : 87-94.

Talia, M.C., Vendola, D. and Ferrari, I. (1988) *Colture Protette*, **17** : 61-68.

Tamberg, T. G. (1967) *Genetika*, **3** : 123-128.

Tamberg, T. G. (1978) *Trudy po Prikladnov Botanike Genetikei Selektsii*, **62** : 115-121.

Tandon, R. N. and Bhargava. S. N. N. (1963) *Curr. Sci.*, **32** : 377.

Tariq, S.; Imran, H.; Abbasi, N. A.; Ghulam, J. (2016) *Sci. Hortic.*, **210** : 236-241.

Thakur, N., Bhuj, B.D., Srivastava, R. and Chand, S. (2016) *Int. J. Basic Appl. Agric. Res.*, **14**(3) : 285-287.

Tilak Raj Maity, Aveek Samanta, Babita Saha, Siraj Datta (2019) *Bull. Natl Res. Cent.*, **43**(1) : 1-11.

Timmermans, A. S. (1941) *Meded. Inst. Phytopath. Lab. Bloembol. Lisse*, **67** : 32.

Tirkey, P. and Singh, D. (2019) *Int. J. Curr. Microbiol. App. Sci*, **8**(11) : 1510-1516.

Tiwari A.K., Srivastava R.M., Kumar V., Yadav L.B. and Misra S.K. (2010) *Prog. Agric.*, **10**(3) : 75-82.

Tombolato, A.F.C.T., de Castro, J.L. and Matthes, L.A.F. (2002) *Acta Hortic.*, **570** : 219-224.

Tomiozzo, R., Uhlmann, L.O., Becker, C.C., Schwab, N.T., Streck, N.A. and Balest, D.S. (2019) *Ornam. Hortic.*, **25**(3), https://doi.org/10.1590/2447-536x.v25i3.2048.

Tomotsune, H., Kasumi, M. and Takatsu, Y. (1995) In: *IPPS Japan Potential region.* First annual meeting, 19-22 Sep, 1994. International Plant Propagators' Society: Combined Proceedings, **44** : 239-244.

Torres, D.P., Silva, M.A., Pinho, D.B., Pereira, O.L. and Furtado, G.Q. (2013) *Plant Dis.*, **97**(6) : 847. https://doi.org/10.1094/PDIS-12-12-1118-PDN.

Tramier, R. and Andreoli, C. (1965) *C. R. Acad. Agric. France*, **51** : 918-922.

Tripathi, M. K., Ram, K.M., Vidhyashankar, M. and Patel, R.P. (2017) *Int. J. Agric. Technol.*, **13**(4) : 583-599.

Troyer, B. I. (1954) *Gladiolus*, pp. 133-137.

Tsukamoto, Y. (1954) *J. Hort. Ass., Japan*, **23** : 16-20.

Tsukamoto, Y. (1974) *Acta Hortic.*,**43(II)** : 393-404.

Tsukamoto, Y. (1975) *Proc. 14th Int. Hortic. Cong.*, 1974, **4** : 293-306.

Tsukamoto, Y. and Asahira, T. (1956) *J. Hortic. Ass. Japan*, **15** : 133-40.

Tsukamoto, Y. and Konoshima, H. (1972) *Physiol. Plant*, **26** : 244-249.

Tsukamoto, Y. and Uneno, Y. (1957) *J. Hortic. Ass., Japan*, **26** : 137-140.

Tsukamoto, Y. and Yagi, M. (1958) *J. Hortic. Ass., Japan*, **27** : 144-148.

Tsukamoto, Y. and Yagi, M. (1959) *J. Hortic. Ass., Japan*, **28** : 59-64.

Tsukamoto, Y. and Yagi, M. (1960) *Plant Cel1 Physiol.*, **1** : 221-230.

Uddin, A.F.M.J., Shamsuzzoh, M., Nusrat, A., Taufique, T. and Mehraj, H. (2016) *J. Biosci. Agric. Res.*, **10**(01) : 837-842.

Uphof, J. C. Th. (1955) *Dictionary of Economic Plants, Engelman*, pp. 170.

Valente, L.M., Savolainen, V., Manning, J.C., Goldblatt, P. and Vargas, P. (2011) *Glob. Ecol. Biogeogr.*, **20**(6) : 881- 892.

Van Der Burgt, X.M., Konomou, G., Haba, P.M. and Magassouba, S. (2019) *Willdenowia*, **49**(1) : 117-126.

Varela Neito, F., Matal1ana Gonzalez, A. and Lopez Perez, M. D. (1980) *Investigationes Agrarias Producion Vegetal*, **12** : 115-132.

Vasil'evskij, A, P. (1959) *Svetovodstvo*, **2** : 20.

Vidalie, H. (1978) *Pepinieristes Horticulteurs Maraichers*, **192** : 13-18.

Vidalie, H. and Laffaire, M. (1983) *P. H. M.-Revue Horticole*, **233** : 35-39.

Viennot-Bourgin, M. G. (1978) *C. R. Acad. Agric. France*, **64** : 880-885.

Vigodsky, H. (1969) *Israel J. Agric. Res.*, **19** : 205-207.

Vinceljak-Toplak, M. (1990) *Poljoprivredna Znanstvena Smotra*, **55** : 379-392.

Vita, M. De (1981) *Ann. dell' Inst. Speri. Flor.*, **12** : 91-105.

Vroomen, C. O., N. De. and Rotten, L. A. J. M. van de. (1981) *Bloembol1encultuur*, **91** : 1332-1333.

Wang, D.R., L.F.Q., L.F.X. and Bin, J.H. (1998) *Advances in Hortic.*, **2** : 719-722.

Wang, T.Y. and Chen, M.M. (1998) *Special Publication - Taichung District Agricultural Improvement Station*, **40** : 39-48.

Wassink, E. C. (1960) *Proc. 3rd lnt. Congr. Photobiol.*, Copenhagen, pp. 371-378.

Wassink, E. C. (1963) *Meded. Landb Hogesch.*, Wageningen, **63** : 1-8.

Wassink, E. C. (1965) *Meded. Landb Hogesch.*, Wageningen, **65** : 32.

Waters, W. E. (1965) *Proc. 18th South Weed Control Conf.*, pp. 237-245.

Waters, W. E. (1966) *Proc. Fla. Sta. Hort. Soc.*, **79** : 452-456. ;

Waters, W. E. (1968) *Proc. Amer. Soc. Hort. Sci.*, **92** : 633-640.

Webster, G. A. (1972) *The World of Gladiolus*, NAGC, USA, pp. 521-559.

Welker, W. V. (1967) *Proc. 21st N-East Weed Control Conf.*, pp. 182-183.

White, J.W. and Kling, G.L. (1991) *Bulletin Pennsylvania Flower Growers*, **405** : 1-6.

Widaryanto, E., Soetopo, L. and Andajani, R.D. (1995) *Agrivita*, **18** : 14-20.

Wilfret, G. J. (1970) *Proc. F1a. Sta. Hort. Soc.*, **83** : 423-427.

Wilfret, G. J. (1980) In: *Introduction to Floriculture* (ed. R. A. Larson), Academic Press Inc., pp. 166-181.

Wilfret, G. J. (1981) *HortScience*, **16** : 787-788.

Wilfret, G. J. and Burgis, D. S. (1977) *Proc. 30th Ann. Meet. South Weed Sci.*, pp. 177-204.

Wilfret, G. J. and Raulston, J. C. (1974) *J. Amer. Soc. Hort. Sci.*, **99** : 38-40.

Wilfret, G.J. (1993) *HortScience*, **28** : 752-753.

Winkler, G. (1969) *Arch. Gartenb.*, **17** : 325-340.

Wisniewska-Grzeszkiewicz, H. (1978) *The Gladiolus Annual*, British Gladiolus Society, pp. 30-33.

Woltz, S. S. (1954) *A. R. Fla. Agric. Exp. Sta.*, pp. 259-260.

Woltz, S. S. (1955a) *Proc. Fla. Sta. Hort. Soc.*, **68** : 358-362.

Woltz, S. S. (1955b) *Proc. Amer. Soc. Hort. Sci.*, **65** : 427-435.

Woltz, S. S. (1955c) *Proc. F1a. Sta. Hort. Soc.*, **67** : 330-334.

Woltz, S. S. (1956) *Proc. Fla. Sta. Hort. Soc.*, **69** : 347-351.

Woltz, S. S. (1956-57) *A. R. Fla. Agric. Expo Sta.*, pp. 302-303.

Woltz, S. S. (1963) *Proc. Fla. Sta. Hort. Soc.*, **75** : 469-471.

Woltz, S. S. (1964, 1965) *Proc. Fla. Sta. Hort. Soc.*,**77** : 516-517.

Woltz, S. S. (1972) *The World of the Gladiolus*, NAGC, USA, pp. 143-149.

Woltz, S. S. (1976) *Gladio Grams*, NAGC, **21** : 1-5.

Woltz, S. S., Magie, R. O. and Geraldson, C. M. (1953) *Proc. Fla. Sta. Hort. Soc.*, pp. 306-309.

Wony, A. and Antkowiak, M. (2013) *Ann. Warsaw Univ. Life Sci., Hortic. Landscape Architect.*, **34** : 35–43.

Wright, J. (1971) *Amer. Hort. Mag.*, **50** : 79-83.

Wu, J., Jin, Y., Liu, C., Vonapartis, E., Liang, J., Wu, W., Gazzarrini, S., Httporcid, O., He, J., Yi, M. (2019) *J. Exp. Bot.*, **70**(4) : 1221-1237.

Wu, J., Seng, S., Carianopol, C., Sui, J., Yang, Q., Zhang, F., Jiang, H., He, J. and Yi, M. (2016) *Biochem. Biophys. Res. Commun.*, **471**(1) : 198-204.

Yadav, L. P. and Bose, T. K. (1987) *Acta Horic.*, **205** : 177-185.

Yadav, M., Rajput, V., Saharan, R.S., Sehrawat, S.K. and Ravika (2016) *Asian J. Hortic.*, **11**(1) : 114-118.

Yadav, U., Agrawal, N., Katre, P., Tamrakar, S.K. and Tripathi M.P. (2020) *Int. J. Curr. Microbiol. App. Sci.*, **9**(6) : 2913-2923.

Yamada, T., Takatsu, Y., Kasumi, M., Marubashi, W. and Ichimura, K. (2004) *J. Plant Physiol.*, **161**(11) : 1281-1283.

Yamamoto, M., Matsuoka, K., Otani, S. and Kondo, N. (1958) *Japan J. Br.*, **8** : 201-202.

Yamane, K. ; Kawabata, S. and Fujishige, N. (1999) *J. Japanese Soc. Hort. Sci.*, **68** : 798-802.

Yamane, K. and Ogata, R. (1995) *J. Japanese Soc. Hort. Sci.*, **64** : 411-416.

Yasin, N.A. (2018) *Int. J. Phytoremediation*, **20**(3) : 274-283.

Yasmin, L., Ali, M.A. and Khan, F.N. (2018) *Bangladesh J. Agric. Res.*, **43**(1) : 13-23.

Yasuda, I. and Yokoyama, N. (1955) *J. Hort. Ass. Japan*, **24** : 207-210.

Yasuda, J. and Yokoyama, J. (1954) *Sci. Rept. Fac. Agric. Okayama*, **4** : 12-22.

Zaharia, A., Cantor, M., Borsai, O., Erzsebet, B. and Jucan, D. (2018) *Rom. Biotechnol. Lett.*, **23**(4) : 13723- 13727.

Zalewska, M. (1997) In : *Zeszyty Problemowe Postepbw Na. Rdniczych*, **449** : 237-245.

Zalewska, M. and Antkowiak, M. (2009) *Electron. J. Pol. Agric.*, **12**(1) : #03.,http://www.ejpau.media.pl/volume12/issue1/art-03.html.

Zhang, X.P. (1995) *Acta Hortic. Sinicia*, **22** : 73-76.

Zhang, X.P. (1998) *Advances in Hortic.*, **2** : 716-718.

Zhang, X.P. and Gao, Y. (1992) *Acta Hortic. Sinicia*, **19** : 367-370.

Zhang, Z., Wang, D., Zhang, D. and Yuan, X. (2007). *J. Radiat. Res. Radiat. Processing*, **25**(4): 201-205.

Zhou W., Li B., Wang, Y., Zhang, C. and Xu, M. (1995) *J. Jiangsu Agric. College*, **16** : 27-30.,

Zhou, Y. You, Z.S. Yu, Y. H. and Zhang, W.J. (1994) *Acta Hortic. Sinicia*, **21** : 189-192.

Zhygalova, S.L., Futorna, O.A. and Levanets, A. (2014) *Environ. Socio.-econ. Stud.*, **2**(4) : 21-27.

Ziv, M. (1990) *Acta Hortic.*, **280** : 204-207.

Ziv, M. (1991) *Proc. 14th int. conf. on plant growth substances*, Amsterdam, 21-26 July Agric.Univ., **19** : 142-143.

Ziv, M., Halevy, A. H. and Shillo, R. (1970) *Ann. Bot.*, **34** : 671-676.

Zubair, M., Wazir, F.K., Akhtar, S. and Ayub, G. (2006) *Pak. J. Biol. Sci.*, **9** : 1669-1676.

6

TUBEROSE

T. Usha Bharathi, L.P. Yadav, R.G. Maity, T.K. Bose and Debmala Mukherjee

1.0 Introduction

The tuberose (*Polianthes tuberose* L.) occupies a very selective and special position. Among the ornamental bulbous plants which are valued for the beauty and fragrance. It has a great economic potential for cut flower trade and essential oil industry (Sadhu and Bose, 1973). Due to their great demand, it is currently cultivated in most of the tropical and sub-tropical countries of the world (Jowkar and Salehi, 2006; Benschop, 1993).

2.0 Importance and Uses

2.1.0 Aesthetic Uses

Tuberose has gained considerable importance and it is cultivated commercially for its varied uses. It can successfully be grown in pots, beds and borders. The flower remains fresh for pretty long time and stands long distance transportation and fills a useful place in the flower market (Desai, 1957). It was grown as an ornamental and medicinal plant by the Aztecs more than 400 years ago (Gonzalez *et al.*, 2012). It is used for artistic garlands, floral ornaments, bouquets and buttonholes. The long flower spikes are excellent as cutflowers for floral decoration when arranged in bowls and vases. The variegated type with golden striped leaf margins is very attractive and suitable for the beautification of gardens.

2.2.0 Therapeutic and other Uses

The flowers emit a delightful fragrance and are the source of tuberose oil. The flower oil of tuberose remains today as an expensive raw material for the perfume industry. The common constituents of tuberose concrete are geraniol, nerol, benzyl alcohol, methyl benzoate, methyl salicylate, eugenol, benzyl benzoate and methyl anthranilate (Hussain, 1986). The fragrance is also added with stimulants or sedatives to the favourite beverage prepared from chocolate and served either cold or hot as desired (Trueblood, 1973). It is commercially grown for its fragrant cut flowers and perfume industry in India, New Zealand, Japan, France and Mexico (Gonzalez *et al.*, 2012). The fragrant oil, almost all of which comes from India, is one of the most expensive of the fragrant oils used in perfumes, valued more than $2,000.00 per pound (Hodges, 2010). The components of tuberose oil such as geraniol, indole and methyl anthranilate are known to have antifungal activity (Nidiry and Babu, 2005).

The bulbs are reported to contain an alkaloid lycorine, which causes vomiting. Two steroidal sapogenins namely, hecogenin and a small amount of tigogenin, a poly fructosan have been isolated from the bulbs. Presence of a transfructosidase has also been observed (Srinivasan and Bhatia, 1954; Bhatia and Srinivasan, 1954). The bulbs are considered diuretic and emitic. They are rubbed with turmeric and butter and applied as a paste over red pimples of infants (Kirtikar and Basu, 1935). Dried tuberose bulbs in the powdered form are used as a remedy for gonorrhoea.

3.0 Origin and History

Tuberose is a native of Mexico from where it spread to the different parts of the world during the 16th century. The name tuberose is derived from tuberosa,

Uses of Tuberose

In Border

In Pot

In Bed

In Flower Vase

In Interior Decoration

Essential Oil from Tuberose

In Flower Arrangement

this plant being the tuberous hyacinth as distinguished from the bulbous hyacinth. The name, therefore, is tuber-ose, not tube-rose. Clusius, a famous botanist of the 16th century called it as *Hyacinthus indicus tuberosa radice*. Several other botanists of 16th and 17th centuries, including Caspar Bauhin (1560-1624) and Parkinson (1507-1650) referred to the plant as *Hyacinthus*. By the middle of the 18th century the tuberose was still referred to as *Hyacinthus indicus tuberosus*. In 4th edition of the *Gardeners Dictionary* printed in 1743 and written by Philip Miller, two species were cited the first, *Hyacinthus indicus ruberosus flore*, the Indian tuberose, and a second, *Hyacinthus indicus tuberosus flore pleno*, the Indian tuberous-rooted hyacinth with a double flower, commonly called the double tuberose.

With publication of Linnaeus *Species Plantarum* in 1753 the classification of plants based on the structure of the flower parts and the morphology of the plant became established, the nomenclature simplified and the use of long polynomial phrases gave way to a consistent binomial system. Originally, Linnaeus has used the name *Polianthus floribus alternis* in his *Hortus Cliffortianus* (1738). Later, in *Species Plantarum* he grouped this and *Hyacinthus indicus tuberosus* of Caspar Bauhin and Carlos Clusius under the genus *Polianthes* and the species *tuberosa*.

In India tuberose is commonly known as *gulchari* and *gulshabbo* in Hindi, *rajanigandha* in Bengali, *sukandaraji* and *nelasampengi* in Telegu, *nilasampangi* in Tamil and as *sugandhraja* in Kannada.

The tuberose, which was cherished and cultivated in Mexico before the conquest in 1522, is one of the many flowers which have come from the ancient culture of the Nahuatl-speaking people. These intensely fragrant cream flowers were known to the Aztecs as omixochitl or bone flower, a name which in Spanish language became flor de hueso. In Mexico today one also hears the names nardo, azucena, amole and amiga de noche applied to the white tuberoses sold in the flower markets in Mexico city, neighbouring towns and especially in and near Guadalajara where these flowers are cultivated extensively.

The first of these, nardo has a long history which dates back to biblical times. It is referred to in the bible as nard, an ingredient in a precious ointment. In the 12th century Moses Maimonides (1135-1204) mentioned nardo (*Valeriana tuberosa*), which was used in an unfermented beverages by the Arab Muslims. The early Spanish in Mexico associated the fragrant omixochitl or flor de hueso with nardo which was known later as the tuberose. The same plant they called azucena. The Nahuatl names of amole (soap) and amolixochitl (soap flower) were applied to plants of Mexico with roots containing saponiferous substance. The high concentrations of sapogenin are reported in the rhizomes and roots of tuberose, many species especially *P. geminiflora*, *P. graminifolia* Rose and *P. tuberosa* have been used as substitutes of soap.

Following the introduction of tuberose into Europe, the bulbs were distributed to England, the Netherlands and France. Propagation from bulbs began commercially in southern France where the mild winter permitted storage without danger of freezing.

The texts of the 16th century are the earliest reference to the different species of tuberose. About twelve species were discovered in the cool mountain valleys of western and northwestern Mexico and the highest number of species occurs in the state of Jalisco (Gonzalez *et al.*, 2012). These species range in colour from white, orange-red, red to striped. Among those botanists whose studies were concerned with the genus *Polianthes* are Cyrus G. Pringle (1838-1928), Edward Palmer (1831-1911), Christopher C. Parry (1823-1890) and J.N. Rose (1862-1928).

In 1878, Christopher C. Parry on a collecting trip with Edward Palmer in the mountains of San Luis Potasi found out *Polianthes sessiliflora*. Two years later Hemsley referred to the species as *Bravoa sessiliflora*, but Rose gave it the final name *P. sessiliflora*. This was a white variety of tuberose as was also *Polianthes nelsonii*, which was identified by E.E. Nelson on August 1, 1898, and referred to by Rose as a new species.

The most extensive contributions to new species of *Polianthes* were made by Rose (1903-05) on various collecting trips to Mexico. Of the twelve species reported in his studies on Mexican and Central American plants, eight were discovered by him.

Thus the fragrance of the omixochitl, cultivated in the gardens of the Nahuatl-speaking people of Mexico 400 years ago, is still cherished today by people of diverse cultures wherever fine perfumes are appreciated. While less cultivated in American gardens of today, these cream white flowers are still to be seen in gardens in tropical and subtropical regions of the world.

Most of the *Polianthes* species have white flowers, *P. geminiflora* (Lex.) Rose, has reddish-orange flowers, and *P. densiflora* (B.L.Rob and Fernald) Shinners, yellow ones. Hawkmoth pollination has been suggested for most species except for *P. geminiflora*, which has been considered to be pollinated by humming birds (Rocha *et al.*, 2006).

4.0 Distribution

The species of *Polianthes* were recorded in 18 Mexican states, where most occur in pine forest, oak forest, grasslands and few species are found in tropical dry or semi- deciduous forests (Solano, 2000). Distribution of the genus *Polianthes* was highest in the southersn Sierra Madre Occidental and northern side of the Trans Mexican Volconic Belt (Solano and Feria, 2007). Trans Mexican Volconic Belt is considered as biographic barrier for *Polianthes* species dispersal and which was the reason two species such as *P. bicolour and P. oaxacana* Garcia-Mend. and E. Solano occurring in Oaxaca (Solano, 2000).

Polianthes geminiflora has the broadest geographical and ecological distribution (Solano, 2000) whereas the other five species *P. densiflora*, *P. howardi* Verh-Will. *P. longiflora* Rose, *P. palustris* Rose and *P. platyphylla* Rose are listed as rare and are under the category of special protection as per Mexican law (DOF 2002). Ferria *et al.* (2010) also confirmed that the *P. howardii* is in risk of extinction, *P. longiflora* and *P. platyphylla* are endangered and *P. palustris* is probable extinct.

Tuberose is cultivated on large scale in France, Italy, South Africa, North Carolina, USA and in many tropical and sub-tropical areas, including India. Guenther (1952) reports that for centuries the chief centre of cultivation of *Polianthes* for the perfume industry has been the fertile valley of the Siagne River not far from Cannes and Grasse. Formerly, as much as 75 metric tonnes of flowers were harvested annually and processed by the flower oil manufacturers of Grasse. More recently the need for land for comestibles to supply the growing population of France has led to a reduction of the acreage devoted to flower culture and to the expansion of cultivation of the tuberose in Morocco.

It is commercially cultivated in India in an area of about 16,190 ha, with a loose flower production of 107910 metric tonnes and cut flower production of 89.29 lakh numbers (Anon., 2016). The commercial cultivation of tuberose is confined mainly to the following states of India such as West Bengal, Tamil Nadu, Karnataka, Andra Pradesh, Chhattisgarh, Maharashtra, Madhya Pradesh, Assam, Madhya Pradesh, Haryana, Uttarakhand, Telangana, Manipur and Meghalaya.

5.0 Morphology and Classification

Tuberose is a half-hardy, bulbous perennial perpetuating itself through the bulblets. Bulbs are made up of scales and leaf bases and the stem is a condensed structure which remains concealed within scales. Roots are mainly adventitious and shallow. The leaves are long, narrow, linear, grass-like, light green, and arise in rosette. The flowers have a funnel shaped perianth and are fragrant, waxy white, about 30-50 mm long, single or double and borne in a spike. The terminal flower spikes arising from the bulb produce flowers for a number of days. Stamens are six in number, anthers dorsifixed in the middle; ovary 3 and 4 locular, ovules numerous and fruits are capsule. The embryo sac follows a monosporic pattern of the polygonum type and starts its development from the chalazal megaspore. At maturity, the embryo sac is of a pyriform shape with a chalazal haustorial tube where the antipodals are located, just below the hypostase, which connects the embryo sac with the nucellar tissue of the ovule. The central cell nucleus shows a high polarity, being located at the chalazal extreme of the embryo sac and the endosperm was classified as of the helobial type (Gutierrez and Garay, 2016).

Floral biology of 2 cultivars (Single and Double) was investigated by Srivastava and Sridhara (1999). Time of anthesis and dehiscence were similar. Flowers of Single cultivar usually had a functional pistil. In Double cultivar, most flowers had rudimentary pistils, with 0.1 per cent of flowers having a functional pistil. Pollination on only the first day was not sufficient for seed set and pollination for 2-3 days was required for optimum seed set. Double cultivar produced fewer seeds than Single. Seeds from Single exhibited better germination than those from Double.

Hutchinson (1960), however, grouped the genera *Polianthes* and *Prochnyanthes*, *Pseudobravoa* in tribe Poliantheae under the family Agavaceae. He characterized this tribe as having a stem rising from tuberous rhizome with narrow basal leaves. This classification as proposed by Hutchinson which placed the genus *Polianthes* in family Agavaceae rather than in the Amaryllidaceae was supported by cytological studies of Sato (1938). Rose (1903-05) and Bailey (1939) placed the genus under family Amaryllidaceae and it was accepted by most of the taxonomists. The genus *Polianthes* (Agavaceae) is endemic to Mexico, and includes 14 species, three varieties, and two cultivars (Solano, 2000). Most species in this genus are used for ornamental and ceremonial purposes. The best-known taxon is *Polianthes tuberosa*, which has been cultivated and used for medicinal, ornamental, and ceremonial practices since prehispanic times. Other species (*e.g. P. bicolor, P. geminiflora, P. longiflora*) are important for their medicinal properties (Solano, 2000). At present the genus *Polianthes* belongs to Asparagaceae, subfamily Agavoideae (APG III, 2009).

6.0 Species and Cultivars

6.1.0 Species

Polianthes was considered as a monotypic genus and closely related to *Bravoa* from which it differs in having a perianth widened upwards. According to Rose (1903-05), the genus contains about a dozen of species but they are not clearly distinct. Nine of the species described by him has white flowers, one was white and red and two were red. All were found growing wild with the exception of *P. tuberosa* which has never been found anywhere except under cultivation. The only close relative of *P. tuberosa* is believed to be *Polianthes gracilis* which was described by Link and Otto in 1821.

The botanical exploration carried out in its geographical range where several new species were discovered and according to Solano *et al*. (2019) *Polianthes* consists of 19 species.

Castro *et al.* (2015) reported that *Polianthes* Linnaeus (1753) is a genus that comprises 20 species (Table 1), arranged in two sub-genera: *Polianthes* and *Bravoa* (La Llave and Lexarza); Roemer (1847) (Verhoek-Williams, 1975, Eguiarte *et al.*, 2000, Solano 2000). *Polianthes* subgenus *Polianthes* includes 12 species with geminate flowers, sessile or with pedicels up to 6 mm long. The perigone is erect or curved almost to mid portion and then infundibular, with an expanded throat between the curvature of the tube and the base of the lobes, white or pink when mature, with perpendicular or reflexed lobes relative to the perigone tube axis, scented, and pollinated by nocturnal moths. The stamens are inserted 2–3 mm below the perigone lobes. In contrast, *Polianthes* subgenus *Bravoa* includes eight species with geminate flowers, or solitary in *P. howardi* (Verhoek-Williams, 1976), and with pedicels 1.1–3.5 cm long. The perigone is pendulous, tubular, with perpendicular lobes or extended at the same plane as the perigone axis, usually red, yellow, pink, or exceptionally bicolor greenish orange as in *P. bicolor* (Solano and Mendoza, 1998), scentless, but scented in *P. multicolor* (Solano and Aranda, 2003), and pollinated by humming birds. Lastly, the stamens are inserted below the middle of the perigone tube.

Table 1: Diversity of the Genus *Polianthes* (Asparagaceae, Agavoideae) in Mexico*

Sl.No.	*Species*
	Polianthes subgenus *Bravoa* (Llave and Lex.) M.J.Roem.
1.	*P. bicolour* E.Solano and García-Mend.
2.	*P. cernua* Art.Castro, J.G. González and Aarón Rodr. sp. nov.
3.	*P. geminiflora* (Lex.) Rose var. *geminiflora*
	P. geminiflora var. *clivicola* McVaugh
	P. geminiflora var. *pueblensis* E.Solano and García-Mend.

Sl.No.	Species
4.	*P. graminifolia* Rose
5.	*P. howardii* Verh.-Will.
6.	*P. multicolor* E.Solano and Dávila
7.	*P. oaxacana* García-Mend. and E.Solano
8.	*P. zapopanensis* E.Solano and Ríos-Gómez *Polianthes* subgenus *Polianthes*
9.	*P. alboaustralis* E.Solano and Ríos-Gómez
10.	*P. densiflora* (B.L.Rob. and Fernald) Shinners
11.	*P. durangensis* Rose
12.	*P. elongata* Rose
13.	*P. longiflora* Rose
14.	*P. montana* Rose
15.	*P. nelsonii* Rose
16.	*P. palustris* Rose
17.	*P. platyphylla* Rose
18.	*P. pringlei* Rose
19.	*P. sessiliflora* (Hemsl.) Rose
20.	*P. tuberosa* L.**

* Castro *et al.* (2015).

** Concerning *P. tuberosa*, its origin and natural distribution in Mexico is unknown, but it is widely cultivated in the country and in the tropical and sub-tropical areas of the world.

The description of important twenty species is given below.

Polianthes tuberosa

An erect herb, 60-120 cm high with stout and short bulbs; leaves basal, 6-9 in number, 30-45 cm long, about 1.3 cm wide, linear, bright green, reddish near the base, flowers funnel shaped, waxy white, the tube bent only near the base, filaments attached on upper part of corolla, fragrant, in long terminal racemes.

P. palustris

This species was collected in swamps on the western foothills of the Sierra Madre, Territorio de Tepic in 1897. This white flowering species was described as having bulbs which are oval to oblong; stems approximately 38 cm high, erect and glabrous, several basal leaves, narrow with parallel veins, stems with leaves or bracts reduced in size near the terminal end of the stem; flowers in 3 to 5 pairs arising from single bracts; perianth slightly curved with segments short and spreading filaments short near top of the tube, anthers not extended; ovary free at tip.

P. durangensis

It was collected in 1897 on the west slope of the last range of the Sierra Madre in the state of Durango. The bulbs are small in this species and the flowers arranged in one to six pairs are all sessile, erect at first but becoming curved and purplish with age.

P. montana

This is another species with white flowers which was collected in 1897 in a deep canyon near Santa Teresa in the Sierra Madre. In this species, the bulbs are oblong, the stem long and slender and possessing 12 pairs of flowers all with pedicels. The flowers are short with lobes small, erect and rounded.

P. longiflora

In August 1901, Rose obtained this species which he saw for sale in great abundance in the market at Guadalajara. This is a tall species with flowers in three to five pairs. Perianth is white tinged with purple colour.

P. platyphylla

This was found by Rose in 1897 near the borders between the states of Jalisco and Zacatecas. Flowers are white tinged with red. The florets are arranged in four to seven pairs with tube of the floret bent at the base and with short, rounded lobes. The presence of the red colour, the short lobes and the curvature at the base of the perianth suggest the natural crossing of the wild red and white species of tuberose.

P. graminifolia

This deep red species was very common in Mexico and was found by Rose in 1897. The local names were amole de flor and amoli sochitl (soap flower). The species closely resembles to *P. geminiflora* which has also saponaceous roots and were used as soap substitutes. The bulbs are long and tuberous. Stems are slender, densely hairy below and glabrous above. Leaves are long and grass-like. The flowers are in 8 to 15 pairs, the lower ones often on peduncles; deep red, bent downward near the base and the filaments are slender and attached low down in the tube.

P. geminiflora

This was originally described by Lexington in 1824 as *Bravoa geminiflora*. In 1901, Rose found it near Tultenango and stated that it was very rarely collected although native in the Central Mexico. He observed the bulbs in the Central Mexico, shallow and giving off several thickened roots. Stems are smooth throughout with several basal leaves. Flowers are light orange-red in colour and arranged in pairs of 6 or more. The perianth is similar to that of the species *P. graminifolia* and bent down sharply at the base. Lobes are short, broad and rounded at the tip.

P. gracilis (P. tuberosa var. *gracilis)*

This species is supposed to be Mexican and is distinguished by slender habit and narrower leaves. Perianth tubes are long and slender. Possibly this was the original form of *P. tuberosa*.

P. blissii

This is a garden hybrid between *Bravoa geminiflora* and *P. tuberosa* and was grown by Bliss in 1905.

P. pringlei

Plants of this species have small bulbs and smooth stems. Flowers are sessile, in pairs of 3 or 4, curved, fragrant and white in colour which changes to slightly purplish on drying.

P. sessiliflora

This was collected from the mountains of San Luis Potosi in 1878. Plants produce flowers of white colour.

P. nelsonii

This species was first reported by W. Nelson in 1898. Flowers are white in colour. Howard (1985, 1986) made efforts to collect *Polianthes*, many of which are becoming endangered. The trips to Mexico were very successful and several of the species mentioned above had been collected.

P. howardii

This species was observed with restricted distribution in Colima, Jalisco of Mexico and listed under endangered category (Arroyo *et al.*, 2010). The flowers are reddish purple, red in the base and gradually green in the lobes without any fragrance, produces during July to September.

P. bicolour

This species was located at Oaxaca, Mexico and the flowers are orange greenish with green lobes, non fragrant and produced during June to August.

P. oaxacana

The species was found at Oaxaca, Mexico and the flowers are pink outside and yellow inside, without fragrant and produced during Septemebr to October.

P. zapopanensis

This species was first reported by Solano and Gomez (2011) found in Jalisco. The leaves of the flowering plants are not in a rosette, and this species differs from the others in the genus by its thick infundibuliform flowers and by its usually branched inflorescence, with one or two branches 6–24 cm long between the fifth and eighth nodes. The colour of the flower is orangish pink colour, non fragrant and produced during July to September.

P. multicolour

This species was located by Solano and Aranda (2003) from Guanajuato, Mexico. This species is related to *Polianthes geminiflora* (Lexarza) Rose var. *geminiflora* and *Polianthes geminiflora* var. *clivicola* McVaugh. It is differs from the above species in the shape of the perianth tube, spreading lobes, the presence of bulliform cells in the foliar epidermis, and the absence of tannins in the mesophyll.

P. densiflora

This species is located at Chihuahua, Mexico and is classified under endangered category (Ferria *et al.*, 2010). The yellow colour flowers without any fragrance are produced during June-August.

P. venustuflora

This new species was first reported by Solano *et al.* (2019) is endemic to the north of Michoacán, Mexico and is morphologically related to *P. montana* and *P. platyphylla*, differing from these by presenting 3 to 7 leaves in the rosette, occasionally 9, leaves from 12 to 35, rarely 40 cm in length, with papillose to regularly denticulated margin; length of the inflorescence is 24 to 68 cm and number of floral nodes from 3 to 7. It is assigned the category of risk in Critically Endangered (CR).

6.2.0 Cultivars

There are two types of tuberose such as single with one row of corolla segments, and double having more than two to three rows of corolla segments. Appreciable variations in the morphological characters of plants including the flower have been recorded in the population of single and double types and selection of large flowered and small flowered types have been done in both the cases. Named cultivars are few, except a double petalled cultivar called the Pearl or Dwarf Pearl Excelsior, one or more variegated forms and a single form usually called Mexican Single but sometimes Mexican Ever-blooming. Often the florets of the double tuberose fail to open completely which results in the reduction of its fragrance. Single flowered type is more fragrant and is widely cultivated than the other types.

Sometimes, cultivar names are given after places like Calcutta Single and Calcutta Double (Nambisan and Krishnan, 1983). Calcutta Single and Mexican Single have been identified as promising ones for large scale cultivation (Irulappan *et al.*, 1980).

The double-flowered cultivar, commonly known as Calcutta Double, Hyderabad Double, Pune Double with 3-5 whorls of perianth segments commonly being grown for the last two decades, is an excellent and popular item in cut flower trade. This is available almost round the year. In winter months, the length of the flower may be up to 5 cm, each spike bearing 30-40 flowers over a length of 25-35 cm. This cultivar shows marked variation in flowering, length of spike and number, size and colour of flowers in response to temperature. During autumn and winter, with the decrease in temperature, the spikes are longer and stouter, flower size is large, and with brighter perianth segments, opening much better than those produced during summer and rainy seasons. The spike produced during higher temperature is shorter and the flowers are smaller, white in colour, opening imperfectly. In winter, the outer perianth develops a light pink colour on the external surface and flowers are more attractive. The yield of spikes in winter is, however, reduced (Bose *et al.*, 1999).

The double cultivars are not normally used for making floral ornaments and the price of spikes is higher than those of single cultivar. The bulbs of the single cultivar are smaller in size and they multiply more rapidly compared to the double cultivar. The double type previously known as Pearl does not open well and is not in commercial cultivation now.

Species and Cultivars of Tuberose

Polianthes tuberosa

Polianthes tuberosa* *'Pink Sapphire'

Polianthes tuberosa* *'Yellow Baby'

Wild *Polianthes* Species

(A) and (E) **P. geminiflora* var. *clivicola*; (B): *P. geminiflora* var. *graminifolia*; (C) and (P) *P. geminiflora*; (D) and (Q) *P. geminiflora* var. *geminiflora*; (F) *P. graminifolia*; (G) *P. howardii*; (H) *P. longiflora*; (I) *P. multicolor*; (J) *P. nelsonii*; (K) and (S) *P. palustris*; (L) *P. platyphylla*; (M) *P. pringlei*; (N) *P. sessiliflora*; (O) *P. montana; (R) *P. multicolour*. [Source: Gonzlez *et al.* (2012) *Mexican GI*, Mexican Geophytes I. The Genus *Polianthes*].**

Cultivars of Tuberose

Arka Nirantara

Calcutta Single

Calcutta Double

Phule Rajani

Arka Prajwal

Rajat Rekha

Arka Srinagar

Arka Suvasini

Arka Vaibhav

Two mutants, named Rajat Rekha and Swarna Rekha were released from the National Botanical Research Institute, Lucknow. Rajat Rekha is single-flowered type with silvery white streaks along the middle of leaf blade, while in Swarna Rekha, the flowers are double and the leaf margin is streaked with golden yellow.

ICAR-Indian Institute of Horticultural Research, Bengaluru is a premier institute has released six promising cultivars and is named as Arka Shringar, Arka Suvasini, Arka Prajwal, Arka Vaibhav, Arka Nirantara and Arka Sugandhi (Bharathi *et al.*, 2016). The brief description of these varieties is given below:

Arka Shringar: It is a single type, bearing flowers on strong, medium spikes. The flower bud has pinkish tinge. The individual flower is bigger in size than that of the usual Single. Its loose flowers are ideal for making garland while spikes can be used as cut flowers. Yield of loose flowers is about 15,000 kg/ha per year, which is 40 per cent higher than 'Calcutta or Mexican Single' and the concrete content of the hybrid is at par with Mexican Single. Shringar is preferred by farmers and perfumery industries. This hybrid is tolerant to root knot nematodes (*Meloidogyne incognita*).

Arka Suvasini: It is a multi-whorled cultivar with bold, big, pure white flowers borne on long spikes. The number of flowers per spike is more and flower opening is uniform in this hybrid as compared to the local 'Double' cultivar. Spike yield is 26 percent more compared to the local Pearl Double 'Double' cultivar. Spikes are best suited for cut flower purpose. This tuberose hybrid is multi-whorled with bold, large, pure white fragrant flowers borne on long spikes in contrast to off-white flowers of local cv. 'Double'.

Arka Prajwal: It is a single floret type variety, flowers of which are bold with pinkish tinge on buds has great potential and demand in market as loose flower, cut flower, concrete extraction and the flower and yield ranges from 20-22 tonnes/ha/year.

Arka Vaibhav: It is a double floret type variety with medium spikes exhibits early flowering. The flower buds are greenish in colour and flowers are White and ideal as cut flower. The spike yield ranges from 1.0-1.5 lakh spikes/acre/year.

Arka Nirantara: It is hybrid with single type flowers and gives high spike yield. It exhibits early flowering and prolonged blooming period from 6-8 months. It continues to flower when other varieties enter dormancy and ideal for loose flower purpose. It possesses higher concrete content of 0.132 per cent and is tolerant to nematodes (*Meloidogyne incognita*). The yield of loose flowers is 20 t/ha/year.

Arka Sugandhi: It exhibits early flowering and flowers continuously for 180 days. The spikes have upward facing florets and the flowers are almost straight, very attractive dwarf type with green tinged buds and the florets are compactly arranged on the stalk. The flowers are with protruding stigma which is very attractive feature of this variety and it is ideal for planting in beds in the landscape and also for pot culture.

Mahatma Phule Krishi Vidyapeeth, Rahuri, Maharashtra has released two single and one double tuberose cutivars named as Phulae Rajani, GK-TC-4 and Phulae Rajat, respectively.

Bidhan Chandra Krishi Viswavidyalaya, Kolkata has released three single tuberose cultivars namely Bidhan Snigdha, Bidhan Ujwal and Bidhan Jyoti and two double tuberose cultivars Bidhan Rajani-24 and Bidhan Rajani-19.

Bidhan Snigdha: It is having single type flower 40-45 number of florets per spike. This cultivar is well suited in medium to low land cultivation and tolerant to nematode.

Bidhan Ujjwal: This is also a single type flowering cultivar and is mainly used as cut flower. Florets are purely white in colour in case of bud stage also. Flower production is 8-10 spikes per lant per year with 45-50 number of florets per spike. It is tolerant to nematode.

7.0 Genetics and Breeding

7.1.0 Genetics

According to Whitakar (1934) and Joshi and Pantulu (1941), the haploid number of chromosomes in *Polianthes tuberosa* is 30, of which 5 are large and 25 small. In the diploid set two pairs of chromosomes are associated with the organization of the nucleous. The small bivalents during I and II metaphase show secondary pairing. Groups of 2, 3 or rarely 4 bivalents are observed. The two nucleoli formed during telophase are of unequal size. Also in the resting nuclei or during the prophase, when there are two nucleoli, these differ in their size. The two pairs of chromosomes associated with the formation of the nucleoli are also not exactly similar. They remain mostly away from each other. Cytological investigations made by Schiva and Lanteri (1983) showed chromosome numbers of 60 in the Single cultivar which was fertile and used in perfumery and 50 in the infertile Double used for cut flowers.

Most of the studies on cytology of *P. tuberosa* reported the chromosome number of single-flowered forms to be $2n = 60$, while the double-flowered forms have been reported to have chromosome numbers of $2n = 50, 54, 60$ and 120 (Whitaker, 1934; Sato, 1938; Joshi and Pantulu, 1940; Sharma and Ghosh, 1956; Vij *et al.*, 1982; Schiva and Lanteri, 1984; Jayasree *et al.*, 2001; Lin and Shen, 2004). Gupta *et al.* (1984) studied cytomorphology of two gamma ray induced mutant (Rajat Rekha and Swana Rekha) of *P. tuberosa* and their respective mother line. Cytological investigations showed chromosome number of 60 in the single cultivars and 50 in the double cultivars. It has been suggested that the double forms are derived from the single by loss of 10 chromosomes due to abnormal mitotic or meiotic division (Schiva and Lanteri 1983).

Ayangar (1963) on the basis of cytological analysis reported that multiple perianth variety was an autotetraploid having a compliment of 120 chromosomes of which 20 chromosomes were long and 100 were short. Cytological basis of sterility and interrelationship in *P. tuberose* have been determined by Sharma and Bal (1956), Sharma and Bhattacharya (1956), and Rangaswamy (1963). Different types of chromosomal aberrations have been detected when Laxmi *et al.* (1984) analysed cytomorphological characters of control and gamma ray induced mutants of tuberose. Seed setting in tuberose is quite erratic in the single-flowered cultivar and is not observed in the double-flowered one. The variegated type, however, has high degree of seed setting compared to others. Shen *et al.* (1987) found that double cultivar is fertile in early flowering stage, when the female parent in 2–3 days after anthesis can be used as both pollen and seed parents. Rosalind *et al.* (2018) recorded 72 per cent of seed germination in tuberose cross IIHR-6 x Mexican Single. Arka Nirantara recorded 21.77 per cent of fruit set upon open pollination (Hemanta *et al.*, 2016).

The exact cause of sterility in tuberose is not known. Certain attempts have been made in this respect and investigations carried out by Joshi and Pantulu (1941) showed that sterility is not due to any defects or deformation in the formation of the pollen grains or development of the embryo sac. Krishnamurthy and Srinivas (2005) studied histological changes during ovule development after crossing Mexican Single and Pearl Double and reported that reproductive compatibility stimulates the metabolic activities in the associated tissues leading the normal development of fruit seed formation. Unlike many other flowers, there is no attractive colour variation in tuberose flowers.

7.1.1 Self Incompatibility

Self-incompatibility of various degrees has also been reported in some cultivars (Shen *et al.*, 1987; Datta, 2017). Joshi and Pantulu (1941) reported that sterility is not due to any defects or deformation in the formation of the pollen grains or development of embro sac and the exact cause of sterility is not known. Uma and Gowda (1999) tried to overcome self-incompatibility through some horticultural manipulations. They studied bud pollination, used IBA and IAA at the time of pollination to the ovaries and also used gamma irradiated (0.5kR) pollen. Bud pollination failed to induce fruiting. Irradiated pollen pollinated two days after anthesis resulted in fruit set upto 4.78 per cent and the seed viability was as high as 48.23 per cent. Fruit drop after seven days of pollination was noticed in tuberose genotypes Calcutta Single, Hyderabad Single, Kahikuchi Single, Mexican Single, Navsari Local, Phule Rajani, Prajwal, Pune Single, Shringar and Variegated Single (Ranchana and Kannan, 2016). Self compatible tuberose lines are identified by ICAR-IIHR and selfed progenies were raised and evaluated. Bundrant (1985) developed three hybrids: *P.* x *blissii*, *P.*x *bundrantii* (*P.tuberosa* x *P. howardii*) and *P. tuberosa*, *P.* (Manfreda) *maculosa*. *P. tuberosa* has the characters of dichogamy and self-incompatibility. Shen *et al.* (1987) reported that there was no seed production in both selfed single varieties and selfed double varieties due to self-incompatibility.

7.1.2 Morphological and Molecular Characterization

Characterizations of plant genetic resources based on molecular markers superior than the morphological and biochemical based markers and DNA-based markers identify polymorphism at DNA sequence level, therefore, they are independent of environmental influence, and furthermore they sample the whole genome (Ovesna *et al.*, 2002). Sarkar *et al.* (2010) reported molecular characterization of both single and double type of tuberose, 12 RAPD markers were used and 157 RAPD bands were generated. Genotypes Guwahati Double and Swarna Rekha showed good morphological similarity revealing high similarity coefficient (0.90) suggesting them to be very closely related. Genotypes, Arka Vaibhav and Pune

Single were found to have least pair-wise similarity although they had a greater morphological similarity with each other.

The wild species of tuberose along with cultivated species were characterized using the amplified fragment length polymorphism (AFLP) markers. The average similarity was 0.75, and this value indicates the existence of a good level of diversity in the genus which can be used to generate new diversity through intergeneric crosses. Individuals from *P. tuberose* 'Double' showed the highest similarity values (0.95) who indicate the homogeneity of this asexually propagated cultivated variety (Gonzalez *et al*, 2013). Lalitha *et al.* (2014) subjected seven single type tuberose genotypes for molecular characterization. Six ISSR primers were used and 62 bands were produced of which 53 bands were polymorphic, accounting for 85.48 per cent polymorphism. The ISSR dendrogram grouped the genotypes into two major clusters. Among the seven genotypes, Hyderabad Single and Phule Rajani are divergent from other genotypes *viz.*, Arka Shringar, GK-T-C-4, Arka Nirantara, Arka Prajwal and Calcutta Single. Ten tuberose varieties were characterized by using RAPD and ISSR markers and the study revealed 53 per cent and 73 per cent polymorphism, respectively. The dendrogram (UPGMA), based on Jaccards co-efficient as similarity index for RAPD and ISSR, grouped ten varieties into two major clusters, and, combined RAPD-ISSR cluster analysis formed three major clusters based on their genetic relatedness/variation (Khandagale *et al.*, 2014). Sirohi *et al.* (2017) studied the genetic variation among 21 tuberose genotypes evaluated using seven randomly amplified polymorphic DNA markers. The results revealed that the RAPD markers produced much better reproducible bands and were more efficient in grouping germplasm. The dendogram generated by RAPD markers revealed four major groups and noted considerable amount of genetic variation among the genotypes. Genotypes namely Mexican White Double and Swarna Rekha cultivar were found to be more diverse in molecular analysis.

7.2.0 Breeding

7.2.1 Hybridization

Bundrant (1985) summarized the efforts on hybridization in tuberose and its allied genera. The genus *Polianthes* now includes, according to Shinners (1966), not only those species originally included in *Polianthes*, but all those formerly placed in the genera *Bravoa, Pseudobravoa, Manfreda, Prochnyanthes, Runyonia* and the herbaceous species of *Agave*. In 1899, the first hybrid in this group was produced using *Polianthes* (*Bravoa*) *geminiflora* and *P.* (*Prochnyanthes*) *bulliana*, but the first cross involving the tuberose was reported in 1911 as *Polianthes* × *blissii*, a cross between *P. geminiflora* and *P. tuberosa*. After many years Verhock-Williams (1975) reported having crossed *P.* (*Manfreda*) *virginica* with *P. tuberosa*, obtaining self-fertile plants that looked much more like the seed parent, *P. virginica* than *P. tuberosa*. While they were intensely fragrant, these hybrids were not thought to have much value as horticultural introductions.

Howard (1977, 1978) reported two hybrids, a repeat of *P.* × *blissii* and a new cross, *P.* × *bundrantii* (*P. tuberosa* × *P. howardii*). Since *P. geminiflora* occurs in two forms, one with slightly glaucous, red, thin-walled flowers, and another with thick-walled, bright orange flowers, clones of *P.* × *blissii* are variable in both substance and colour. Flowers vary in colour from reddish pink without, pinkish white within, to salmon or peach outside and whitish yellow inside and are displayed on a loose raceme in 12 to 16 pairs. To obtain this cross, all flowers on the scapes of *P. geminiflora* were dusted with pollen of *P. tuberosa. Polianthes howardii*, one parent of *P.* × *bundrantii*, has bright green, succulent leaves in a more or less flattened rosette. The flowers are single at the nodes and rather widely spaced along the stem to form a raceme of up to 48 flowers. The colour of these flowers is most interesting, being red on the lower part of the tube, blending into green on the upper side and the outside of the sepals. The inside of the flower is dark maroon in colour, appears almost black. Crossing this species with the tuberose produced a plant intermediate in foliage and flower size, but the flowers were coloured pink on the outside and mauve to rich purple on the face. Scape height and number of flowers approximate both parents. The flowers are usually paired at the nodes, but the lower ones may be solitary (Bundrant, 1985).

Both the above crosses are between the genera *Polianthes* and *Bravoa* and in them the traits of genus *Bravoa* are easily discerned, but of the following two the same can not be said. The parent of the first of these crosses with *P. tuberosa* is a small plant with narrow glaucescent leaves and *Bravoa*-like flowers, with bright orange-red exterior and orange-yellow interior. It carries the Howard collection number 74-95. While it appears to be a member of the subgenus *Bravoa*, it has very short filaments attached near the mouth of the tube, as in subgenus *Polianthes*. When crossed with the tuberose, a relatively small plant with yellow flowers streaked red on the outside was obtained. These open well and are quite striking in their delicate beauty. Other equally pretty clones had deep pink flowers. The second parent, Howard collection number 73-35, is much larger in all parts than 74-95, has flowers with a coral exterior and yellow interior, and a perianth shape resembling subgenus *Prochnyanthes*, *i.e.*, the lower part of the tube is small and bends abruptly at mid-point, becoming bulbously inflated beyond. The foliage is bright green and deeply channelled like *P. tuberosa*. Members of the subgenus *Prochnyanthes* have much broader, leathery, dark green leaves. When crossed with *P. tuberosa*, 73-35 produced flowers coloured like hybrids of 74-95, but the flowers were larger. Marcia Wilson dubbed at *Polianthes* hybrids having yellow flowers streaked red, 'Sunset tuberoses', and lumped all the coloured hybrids, including *P. blissii* and *P. bundrantii*, into a single group which she called 'Rainbow tuberoses'.

In addition to those tuberose hybrids mentioned, Bundrant (1985) succeeded in getting another which deserves mention, *i.e.*, *P. tuberosa*, *P.* (*Manfreda*) *maculosa*. While these hybrids are intermediate, it seems the second parent contributes more. The plants are robust, the white flowers open well, and since both parents are very fragrant, the hybrids are also fragrant. All of the above tuberose hybrids are self-fertile, and the F_2 generation looks just like the F_1 but not many F_2 s have been observed. The next experiment required to be performed is crossing of the existing hybrids with each other, to get back to the size and form of the commercial tuberose. In spite of the interesting beauty of the hybrids, none are as large as the parent. Other experiments need to be made using the double and variegated forms of tuberose.

Intergeneric hybrid named × *Polifreda* 'Lindstrom' was developed between *P. tuberosa* × *Manfreda virginica* by Ritchie and Lindstrom (2014). It inherited desirable characters from *P. tuberosa* include two flowers per node, a pleasant fragrance, heat and drought tolerance and suited for perennial borders, rock gardens, situated in full sun or partial shade. Crosses and back crosses among *P. tuberosa* 'Single', *P. tuberosa* 'Double', *P.* × *howardii* and *P.* × *blissii* were made and several hybrids showing pink, reddish-purple, purple, orange and yellow flower colors were selected. However, the long spikes of flowers in these colored hybrids were only suitable for cut flowers. Four hybrids showing a dwarf plant types were re-selected for use as pot and/or bedding plants (Shen *et al.*, 2003). The intergeneric and interspecific crosses performed by Gonzalez *et al.* (2013) indicated that *P. geminiflora* var. *clivicola* was successful both as a receiver of pollen as well a donor with different *Polianthes* species and with *Prochnyanthes mexicana*. However, this species was not compatible with tuberose to produce viable seeds from 105 crosses. At the same time, *P. howardii* was the better pollen donor but was not a good pollen receptor even with *P. tuberose* 'Double' from 153 crosses which produced 4 capsules with seeds. Also, *P. tuberose* 'Double' was more effective as a pollen receptor than pollen donor, because the pollen in these plants was sterile. Huang *et al.* (2001) studied the flower pigment constitutions of nine hybrids with pink, reddish purple, purple, orange and yellow fowers obtained from the crossings and back-crossings using *Polianthes tuberosa* 'Single' and 'Double' and *Polianthes howardii* were analysed. Among the parents and hybrids, two white flowered cultivars of *P. tuberosa* had neither carotenoids nor anthocyanins, one hybrid had only carotenoids, four hybrids had only anthocyanins and an additional other four hybrids along with *P. howardii* had both carotenoids and anthocyanins in their petals. The main anthocyanidin in the petals containing cyanidin with and some hybrids also contained delphinidin. Hybridization carried out between single and double tuberose cultivars resulted in complete single tuberose population without any double tuberose and wide range of variations in the progenies were observed (Bharathi and Kirthishree, 2019).

7.2.2 Breeding for different Flower Colours

Polianthes howardii, also native to Mexico (Howard, 1985) and single flowered, has reddish purple flowers. It was introduced to Taiwan from Mexico in 1985 (Shen *et al.*, 1993), and Shen *et al.* (1997) have started the breeding to bring *P. howardii's* flower colours into tuberose (*P. tuberosa*) by crossing and backcrossing *P. tuberosa* 'Single', 'Double' and *P. howardii*, and until now more than 50 hybrids having pink, reddish-purple, purple, orange and yellow flower colours have been selected. The process and the results of their breeding program in tuberose and some morphological characters of the hybrids have already been reported in detail (Shen *et al.*, 1997).

Huang *et al.* (2001) studied the flower pigment constitutions of nine hybrids with pink, reddish purple, purple, orange and yellow fowers obtained from the crossings and back-crossings using *Polianthes tuberosa* 'Single' and 'Double' and *Polianthes howardii* were analysed. Among the parents and hybrids, two white flowered cultivars of *P. tuberosa* had neither carotenoids nor anthocyanins, one hybrid had only carotenoids, four hybrids had only anthocyanins and an additional other four hybrids along with *P. howardii* had both carotenoids and anthocyanins in their petals. The main anthocyanidin in the petals containing cyanidin with and some hybrids also contained delphinidin. Hybridization carried out between single and double tuberose cultivars resulted in complete single tuberose population without any double tuberose and wide range of variations in the progenies were observed (Bharathi and Kirthishree, 2019).

Investigations of various light conditions on colour of the florets by Muriithi *et al.* (2009) confirmed the full sun conditions enhanced colour development more that all the other light conditions. Application of soil amendments such as ammonium sulphate and magnesium nitrate lowered the soil pH but that lightness, chroma, hue, and anthocyanins in the florets were not significantly linked to the soil pH. For temperature and magnesium concentration, it was determined that magnesium was not required for colour to develop at low temperature but when temperatures increased magnesium appeared to contribute to the maintenance of the developed colour.

Taipei University, Taiwan developed colour tuberose varieties and licensed to the breeding company, Ludwig and Co, Netherlands. The company has introduced number of coloured varieties like Pink Sensation, Yellow Baby, Cinderella, Double Pink and Super Gold. Huang from Taiwan patented the tuberose variety Chia Nong Pink Sapphire (USPP26460P2) and CNGH (USPP26461P2) and received US plant patent in the year 2016.

7.2.3 Mutation Breeding

Mutation breeding can be utilized to develop improved strains and some attempts have already been made in this respect. Several mutagens like radiation, ultraviolet light and a variety of chemicals have been utilized for this purpose. Abnormalities like trifurcated spikes, three flowers at one place instead of two were recorded in bulbs treated with 1.5 kR gamma rays (Patil *et al.*, 1975). By treating the bulbs with gamma rays and fast neutrons, several mutants of ornamental value have been obtained by Abraham and Desai (1976a, b). However, they all showed colour variation in leaves and not in flowers. Similarly, two gamma induced mutants, the single-flowered cultivar Rajat Rekha and the double-flowered cultivar Swarna Rekha were developed at NBRI, Lucknow, India (Gupta *et al.*, 1984). Irradiation with gamma rays at 10 Gy or above reduced growth rate, and percentage of bulblet survival. All irradiated plants had leaf chimaeras but no flower colour mutations were found (Khasaechai, 1992).

Younis and Borham (1975) studied the genetical and morphological variability induced by gamma irradiation using 500 to 3000 rad and LD_{50} was determined as 2.5 krad. Abraham and Desai (1976a) used fast neutrons and gamma rays and studied their effect on vegetative and flowering characters of single and double type tuberose. They reported that bulb size is not very critical in determining lethality in both the varieties. In a separate experiment, Abraham and Desai (1976b) used X or gamma ray and fast neutrons separately and in combination for induction of mutation in single and double tuberose. Double cultivar was found to be more sensitive to all radiations. They detected leaf colour mutants in single and the double cultivar. One large flower mutant of tuberose was induced by 0.5 Krad gamma ray

(Patil *et al.*, 1975). Gupta *et al*. (1974) developed two Chlorophyll variegated mutants using 2.5 Krad gamma rays introduced onto the market.

Mutation studies in tuberose recorded anatomical changes in the flower stalk and stomatal features on the leaves, variation in morphological characters such as variegated and albino leaf mutants of tuberose while using gamma rays (Abraham and Desai, 1976; Younis and Abraham, 1975; Khasaechai, 1992; Shukla and Datta, 1993; Basaldua *et al.*, 2011; Navabi *et al.*, 2016; Abhangrao *et al.*, 2019; Kayalvizhi *et al.*, 2017a and 2017b; Sharavani *et al.*, 2019a, 2019b: Jyothi *et al.*, 2019).), EMS (Singh *et al.*, 2013; Singh and Sadhukhan, 2019), X-rays (Kaintura and Srivastava, 2015; Kaintura *et al.*, 2016) and UV radiation (Pohare *et al.*, 2012) as mutagen. Kutty *et al.* (2019) found 9.64 Gy. of gamma rays as Lethal dose (LD_{50}) and reported that the floral volatile emission rate was increased in the MV_2 variant line treated with 10 Gy of gamma radiation and further low-dose application of γ-irradiation in *P. tuberosa* bulbs leading to enhanced plant growth and improved floral scent profile.

8.0 Growth and Flowering

Tuberose bulb goes through a characteristic cycle of development beginning with its initiation as a growing point. In the vegetative stage the bulblets grow and attain their maximum size. The subsequent reproductive stage includes the initiation of flower differentiation of floral parts, elongation of the flowering shoot and finally flowering and sometimes seed production.

Tuberose normally begins to flower in 80 to 95 days after sprouting. It flowers profusely during summer and rains (April-September) in the plains of eastern part of India, in the hills it flowers in May-August, while in milder climate this flowers well throughout the year (Sadhu and Bose, 1973). Flower initiation in tuberose was studied in bulbs (2.5-3.0 cm diameter at planting) dug up at 10-day intervals and sectioned longitudinally (Bankar, 1991). Flower initiation was first observed 40 days after planting (4.76 per cent), increased sharply at 110 days (from 7.14 per cent to 49.20 per cent) and increased further (to 87.20 per cent) at 126 days after planting.

The sugar content of tuberose bulbs (cv. Excelsior) in 3 size classes was analyzed by Gonzalez *et al.* (1995) with the aim of identifying biochemical indicators of flower initiation. Fructose and sucrose contents did not vary throughout bulb development; the contents of these 2 sugars were higher than glucose content. Glucose content decreased as bulb size increased, suggesting that glucose is being metabolized to starch from early stages of bulb development.

Ding *et al.* (1999) studied the correlation of free IAA with flower development of tuberose cultivar double and the level of free IAA in vegetative corms is correlated with leaf differentiation, and that the early floral initiation stage is correlated with a reduction in free IAA and an increase in IAA conjugates in the corms. According to Su *et al.* (2002) dramatic reduction in the free ABA content in bulb tissues are correlated with floral initiation and flower development in *P. tuberosa* (Table 2). At the vegetative stage, the corms contained approximately twenty times more free ABA per unit mass as compared with corms at the early floral initiation and flower differentiation stages. In contrast, bound ABA contents were an order of magnitude lower than free ABA and were not clearly affected by different growth stages.

Table 2: Endogenous ABA in *P. tuberosa* Bulbs at the Vegetative, Early Floral Initiation and Floral Development Stages. Values are ± SD (n=2)*

Growth Stage	*ABA Content (ng g^{-1} fresh weight)*	
	Free	*Bound*
Vegetative growth	252.9 ± 76.3	29.5 ± 5.4
Early flower initiation	12.8 ± 1.5	48.9 ± 14.6
Flower development	14.8 ± 3.0	24.8 ± 0.9

* Su *et al.* (2002).

Huang *et al.* (2004) studied the changes in polyamine pattern during floral initiation and development of tuberose cultivar double and reported that the decrease in putrescine and spermine and increase in spermidine and cadaverine content in the bulbs during early floral initiation stage. Cadaverine content in corms disappeared at the flower development stage. Changes in the levels of endogenous polyamines were determined in tuberose bulbs during their dormant periods by Sood and Nagar (2005). High free putrescine and low spermine and spermidine levels were associated during initial stages of dormancy and contrary to that high spermine and spermidine levels were observed with dormancy release (Table 3). The conjugated putrescine level increased during the period with an increase in conjugated spermine and spermidine levels.

Table 3: Variation in Putrescine to Spermidine Plus Spermine (Put[Spd+Spm]) Ration in the Tuberose Bulbs*

Days after Harvest	*Put[Spd+Spm]*		
	Free	*Conjugated*	*Bound*
15	1.29	0.14	0.24
30	1.28	0.13	0.21
45	1.04	0.2	0.17
60	0.35	0.23	0.12
75	0.22	0.25	0.1
90	0.17	0.34	0.07

* Sood and Nagar (2005).

Jayanthi *et al.* (2018) sequenced the RNA obtained from tuberose and generated 11,100 transcripts which represent a total of 7876 unigenes that were considered for downstream analysis and 21 flowering genes have been identified.

8.1.0 Environmental Factors

8.1.1 Temperature

According to Hartsema (1961), temperature is the most important factor affecting growth and initiation and subsequent flowers development of bulbous plants. High (40°C) as well as low temperature (10°C) reduce spike length, weight and quality of tuberose flowers. The optimum temperature range for growth and development is 20 to 30°C (Sadhu and Bose, 1973). Kosugi and Kimura (1961) observed that a temperature of 19.3°C was optimum for the differentiation and development of flower bud.

Storage temperature also influences the sprouting of bulbs after planting and subsequent growth and flower production. Preplant storage of bulbs at 10°C for a period of 30 days was found to improve the plant growth and increase yield of spikes and flowers. This treatment also caused early emergence of flower spikes and improved the quality of flowers (Dhua *et al.*, 1987). The bulbs remain dormant during winter months in places where the temperature is low.

Storage of tuberose bulbs at 25°C was found to be ideal, though which increased the loss of water content in the bulbs, hastened the sprouting and flowering after planting and enhanced the cut flower quality as compared to low temperature treatments such 5,10 and 15°C. Storing the tuberose bulbs at 5°C shielded the bulbs from drying up and delayed sprouting and flowering and deteriorated the cut flower quality (Huang and Okubo, 1995).

Table 4: Effect of Storage Temperature on Flowering of Tuberose Bulbs*

Initial Storage		*Final Storage at 25°C*	*Days to Sprouting*	*Days to Flowering*
Temperature (°C)	*Weeks*			
25	20	0	3.6	86.0
20	4	16	5.2	87.4
	8	12	4.4	87.7
	12	8	4.5	88.5
	16	4	5.3	88.8
	20	0	5.5	91.4
15	4	16	4.3	85.3
	8	12	5.8	85.8
	12	8	4.8	88.1
	16	4	5.2	91.1
	20	0	10	96.5
10	4	16	4.3	86.3
	8	12	4.4	87.2
	12	8	5.3	89.8
	16	4	7.7	93.0
	20	0	15.3	100.0
5	4	16	4.7	87.6
	8	12	4.9	90.1
	12	8	5.8	92.1
	16	4	6.8	90.6
	20	0	14.5	99.0

Mean separation within columns by Duncan's multiple range test, 5 per cent.

* Huang and Okuba (1995)

Watako and Ngamau (2013) revealed that the days to sprouting were significantly earlier (14.9 days) when tuberose bulbs were pretreated at 10°C followed by 20°C subsequent temperature storage for 6 weeks compared to 51.1 days to sprouting at 5°C pretreatment with no subsequent temperature storage (Table 5). The highest percent sprouting (99.2 per cent) was obtained with 10°C pretreatment followed by 20°C thawing for 6 weeks. The number of days to flowering were significantly reduced (110.8 days) at 10°C pre treatment followed by 20°C subsequent thawing for 6 weeks compared to 143.1 days at 5°C pretreatment with no thawing, respectively. Stem length of inflorescences significantly improved to 106.8 cm at 10°C with thawing at 20°C for 6 weeks compared to 98.2 cm at 5°C pretreatment and no thawing respectively. Number of florets per spike also significantly increased to 42.4 compared to 34.9 for similar treatments.

Table 5: Effect of Storage Temperature on Days to Sprouting, Flowering and Flower Quality of Tuberose*

Pre-Treatment for 3 Months (°C)	*Subsequent Treatment*		*Sprouting*		*Days taken for Flowering*	*Stem Length*	*No.of Florets (Pairs)*	*No. of Stem per Bulb*
	Temp (°C)	*Time (Weeks)*	*Per cent*	*Days*				
5	20	0	69.3a	51.1e	143.1f	98.2a	34.9a	0.09a
		2	70.2a	49.0e	140.2e	99.7a	35.2a	1.05a
		4	85.3b	38.1d	137.8d	100.1a	36.1ab	1.09a
		6	94.1c	24.2b	129.1b	102.8b	36.8b	1.08a
10	20	0	88.3b	39.2d	138.3d	97.3a	33.8a	1.02a
		2	97.2cd	31.3c	121.2c	102.9b	41.2c	1.09a
		4	98.1d	20.8ab	119.3ab	104.3c	43.5cd	2.01a
		6	99.2d	14.9a	110.8a	106.8d	42.4d	2.10a

Mean separation within columns by Duncan's multiple range tests, at $p \leq 0.05$. Different letters per column indicate significant differences.

* Watako and Ngamau (2013).

8.1.2 Light

Although tuberose is not strictly photosensitive, long-day exposure promotes vegetative growth as well as early emergence of the first flower spike, and also increases the length of the flower spike. A day length of about 16 hours promoted growth and flowering (Sadhu and Bose, 1973). Chang *et al.* (1998) investigated the effects of day length (8 or 16 light) on floret initiation and development of flower. Bulb remained vegetative at 8-h day length when plant height was 0-5 cm. Floret initiation was achieved with 16-h day length. However, no significant difference in floret initiation or development was found between 8 and 16-h day lengths when plants were > 5 cm in height.

Though the influence of light on growth and development of tuberose has not been methodically studied, reduction in light intensity promotes the leaf length whereas increase in the intensity and duration of light result in production of quality blooms provided there is optimum temperature. Low light intensity adversely affects flowering, therefore, in short-day conditions, *i.e.*, winter growing, supplementary light may help producing better blooms.

Chang *et al.* (1998) studied the effect of day length on protein pattern and flowering of tuberose and reported that though tuberose is a day-neutral species floral development was slightly inhibited in short day of 8h day length at the early stage of growth as compared to long day conditions of 16h day length. Long day conditions showed 20 newly synthesized proteins and significant increase of 4 other polypeptides as compared to 8 h of daylength.

Colour tuberose line '77A05' developed by Huang *et al.* (2002) showed reddish purple flowers when cultivated at 20°C under natural light (Table 6), whereas, the color was white at 30°C. It was almost white at 25°C with 45 per cent shading of natural light, but pale with 25 per cent shading. Cultivation of '77A05' in open fields was carried out at four different altitudes, 25, 75, 500 and 1,200 m in Taiwan. The anthocyanin content at 500 and 1,200 m was approximately two and three times higher than that at 25 m.

Table 6: Effect of Temperature and Shading on Anthocyanin Content and Anthocyanin Constituent in Tuberose Line '77A05'*

Treatment		*Anthocyanin Content*[y]	*Anthocyanidin Constituent (per cent)*	
Temperature (°C)	*Shading (Per cent)*		*Cyanidin*	*Delphinidin*
20	0	3.3	81.1	18.9
	25	3.7	80.4	19.6
	45	3.9	81.8	18.2
25	0	3.7	86.0	14.0
	25	3.3	85.2	14.8
	45	2.9	83.8	16.2
30	0	0.9	88.0	12.0
	25	0.7	89.6	10.4
	45	0.5	88.6	11.4

y: Absorbance at 530nm.

* Huang *et al.* (2002)

8.2.0 Effect of Growth Regulators and other Chemicals

Growth and developmental behaviour of bulbous plants is also regulated either by a single or by an interaction of several endogenous growth hormones like gibberellin, auxin, cytokinin and abscisic acid (ABA). They play a major role in directing the movement of organic metabolites and in establishing sinks (Rees, 1972).

Application of certain growth regulators has been found to influence the growth and flowering of tuberose. Soaking of sprouting bulbs for one hour in solutions of 100 ppm GA_3 or 400 ppm CCC advanced the flowering by 17 and 15 days, respectively. Flower yield was greater by treatment with CCC at 1000 ppm (Ramaswamy *et al.*, 1979). Treatment of plants with CCC at higher concentration (3000 ppm), however, was found to delay the flowering (Hassan and Agina, 1980). Soaking of bulbs in GA_3, ethrel, kinetin and thiourea solutions before planting improved the growth and flowering of tuberose. Among the different chemicals used GA_3 and thiourea proved more effective than others. Thiourea promoted plant height and leaf number while GA_3 improved flowering (Table 2). Treatment with GA_3 at 200 mg/l caused earliest flowering and gave the maximum yield of spikes and flowers (Dhua *et al.*, 1987). Pathak *et al.* (1980), on the other hand, reported that treatment of bulbs with GA_3 at 300 ppm had no effect while kinetin at 300 ppm and ABA at 10 or 50 ppm was especially effective in stimulating sprouts in smaller bulbs. GA_3 and kinetin inhibited flowering whereas ABA stimulated it, especially in larger bulbs. Biswas *et al.* (1983) found that leaf emergence was quickest (11 days) in bulbs treated with IAA at 10 mg/l. The first flower spike emerged earliest (87 days) in bulbs soaked in CCC at 2.0 ml/l. The highest number of flower spikes (6/clump) was obtained after foliar application of GA_3 at 1000 mg/l, CCC at 0.2 ml/l and the largest number of flowers per spike (46) with GA_3 at 100 mg/l. Investigations carried out by Basu and Bose (1980) revealed that application of ancymidol at 200 ppm at 5-6 leaf stage increased the number of flowers per spike from 35.4 in the control to 44 and reduced the number of days to first flower bud appearance from 96 to 87. Jana and Biswas (1982) reported that the shortest time to flower opening (97) occurred in plants treated with 10 ppm GA_3, and the highest number of flowers per spike (35.5) was on plants treated with 1000 ppm SADH.

In experiments on growth regulators, Reddy *et al.* (1997c) summarized that plants treated with GA_3 at 200 ppm were early to flower, application of GA, BA and phosphon increased spike length, while CCC and B-9 reduced its size. CCC at 1000 ppm resulted in the production of higher number of spikes per plant. The highest flower yield was obtained with the application of 1500 ppm CCC or B-9. Vase life was longest (7.67 days) for flowers from BA and GA treated plants, while those from phosphon treated plants lasted for the shortest time (6.0 days at the 1500 ppm rate). Treatment with GA, BA, CCC and B-9 increased the number of shoots, plant height, number of leaves per plant, leaf length and leaf chlorophyll content (Reddy *et al.*, 1997a). Sanap *et al.* (2000) obtained the highest flower yield in tuberose Single (27.5 t/ha) when 150 ppm GA_3 was used.

Foliar application of tricontanol and GA_3 75 ppm on plants of cv. Calcutta Double 60 days after planting and 30 days after first spraying increased the spike quality and yield. Maximum spike length was obtained with GA_3 (Ghosh and Pal, 1999).

100 ppm GA_3 (Wagh *et al.*, 2012; Thakre *et al.*, 2019; Rupa *et al.*, 2017), 150 ppm GA_3 (Maheswari and Sivasanjeevi, 2019), 200 ppm GA_3 (Akand *et al.*, 2016; Amin *et al.*, 2017) and 250 ppm (Suseela *et al.*, 2018) as foliar spray improves the spike yield, quality and longevity of tuberose. Foliar application of 100 and 200 ppm GA_3 at 30 and 60 days after sprouting significantly increased spike quality and yield of tuberose cv. Arka Prajwal. Tuberose bulbs pretreated with 150 ppm GA_3 resulted

maximum vegetative growth in terms of plant height, number of leaves, leaf length, leaf width, increased spike length, rachis length, number of florets per spike and floret length (Rani and Singh, 2013). Sultana *et al.* (2016) recorded the highest spike (3,50,000/ha) and bulb yield (21.72 t/ha) by applying poultry litter @ 20 t/ha with 200 ppm GA_3. Lakshmi *et al.* (2020) reported that the foliar application of 200 ppm GA_3 resulted in maximum number of spikes/ha (1.862 lakhs) as compared to 500 ppm Ethephon (Table 7).

Table 7: Effect of Chemicals on Flowering and Yield Parameters of Tuberose cv. Arka Prajwal*

Treatments	*Days to Spike Emergence*	*Spike Length (cm)*	*Rachis Length (cm)*	*No. of Florets per Spike*	*No. of Spikes per Clump*	*No. of Spikes per ha (Lakhs)*
GA_3 100 ppm	130.98	109.21	39.25	51.82	2.056	1.830
GA_3 200 ppm	128.66	111.16	39.95	52.75	2.092	1.862
Salicylic acid 50 ppm	134.13	106.68	38.34	50.62	2.008	1.787
Salicylic acid 100 ppm	133.39	107.26	38.55	50.90	2.019	1.797
BA 50 ppm	137.17	104.34	37.50	49.51	1.964	1.748
BA 100 ppm	135.89	105.31	37.85	49.97	1.982	1.764
Ethephon 250 ppm	141.17	101.41	36.45	48.12	1.909	1.699
Ethephon 500 ppm	146.88	97.96	35.20	47.66	1.876	1.667
Spermidine 50 ppm	139.81	102.39	36.80	48.59	1.927	1.715
Spermidine 100 ppm	138.48	103.36	37.15	49.05	1.946	1.732
Control (water spray)	142.56	100.11	36.10	46.27	1.891	1.683
Mean	136.66	104.50	37.56	49.57	1.970	1.753
S.Em	1.56	0.87	0.32	0.17	0.007	0.006
CD (5 per cent)	4.61	2.58	0.93	0.49	0.020	0.017

* Lakshmi *et al.* (2020).

Dipping of bulbs in 100 ppm GA_3 for 12 h resulted in early sprouting, improved growth and yield parameters of tuberose cultivar Double (Ganesh *et al.*, 2013). Dhumal *et al.* (2018) applied 160 ppm GA_3 as dip treatment of tuberose bulbs of cv. Phulae Rajani for 24 hrs before planting and as a spray 30 day after planting significantly improved growth, flowering and yield parameters. Dip treatment of 100 ppm IAA in tuberose cultivars Culcutta Single and Culcutta Double enhanced the vase-life of spike. Application of 250 ppm GA_3 as dip enhanced the number of bulbs and bulblets per plant in both of the cultivars (Sarkar *et al.*, 2009). Enriched sap of banana pseudostem at 15,000 ppm found to improve growth and flowering parameters of tuberose cultivar Arka Prajwal (Alpeshkumar *et al.*, 2018).

The dormancy can successfully be broken by dipping the bulbs in 4 per cent thiourea solution for one hour (Hakim *et al.*, 1958; Sadhu and Bose, 1973). Ethylene chlorohydrin was effective for large bulbs but did not stimulate sprouting in small bulbs. Soaking in carbon tetrachloride was only partially effective for large bulbs and was harmful to small bulbs.

A high concentration of free ABA was detected by Nagar (1995) in tuberose bulbs during their dormant period. With the release of dormancy the concentration decreased suggesting that ABA is involved in the induction and maintenance of dormancy in bulbs.

Foliar application of 100 ppm of paclobutrazol at 85 DAP had given maximum number of medium, large and total number of bulbs (Nellipalli and Pal, 2018). Foliar application of 1500 ppm CCC recorded the maximum number of spike per plant, per plot and per hectare, number of florets per spike, yield of florets per plant, per plot and per hectare (Gawai *et al.*, 2020).

9.0 Propagation

Tuberose is mainly propagated vegetatively by means of bulbs. Seeds, if produced, are difficult to germinate and are seldom used for raising plants. Although not very common, propagation may also be done by the division of bulbs. To get virus-free material or for a very rapid multiplication, the propagation is done through tissue culture.

9.1.0 Seed Propagation

Under favourable climatic conditions, seed setting is observed recurrently in the single flowered cultivar and rarely in double flowered cultivar. Seeds are sown in well prepared growing medium containing leaf mould and garden soil in equal proportion. Moisture and temperature have a marked effect on germination and soil temperature of 26-30°C is optimum. The bed for raising seedling should thoroughly be prepared by digging and sufficient quantity of organic matter is to be mixed before sowing. The seeds are sown in rows 10 cm apart and 1.5 cm deep in heavy soil and 2.0 cm in light soil. Moisture can be maintained by spreading a mulch on the bed. Seeds start germinating within 10 to 15 days after sowing. After the seedlings have attained a good growth, these are transplanted in pots or ground where they are allowed to grow.

According to Rosalind *et al.* (2018) growing environmental condition also determines the germination rate of the seeds and the maximum number of days taken for seed germination was observed in the seeds of Arka Nirantara (46.50 days) when crossed with IIHR-6 as the male parent and the seeds of IIHR-6 x Arka Nirantara recorded to germinate the earliest (32.00 days). Ranchana and Kannan (2016) recorded 53.64 days for seeds germination of Single x Variegated but Seetharamu (1993) recorded lesser days for seed germination (17.00 days). Seed germination as recorded by Raja *et al.* (2003) started at 22 days after sowing and completed at 55 days after sowing.

The highest mean seed germination of 45.87 per cent was observed when Arka Sugandhi was used as seed parent (Rosalind *et al.*, 2018). Seed germination of 90 per cent was recorded by Seetharamu (1993), whereas, Krishnamoorthy (2014) recorded the highest germination percentage of 16.50 per cent in the cross Arka Shringar x Mexican Single. Hemanta (2015) had also recorded a maximum of 26 per cent seed germination in tuberose and maximum germination of 66 per cent was noticed when fresh seeds of tuberose were sown (Raja *et al.*, 2003).

9.2.0 Vegetative Propagation

9.2.1 Propagation by Bulbs

This is the most common method practiced commercially for the multiplication of tuberose. However, care should be taken in the selection of suitable bulbs. Spindle-shaped bulbs free from diseases and having an average diameter of 1.5 cm or above should always be preferred to conical or flat bulbs (Sharga, 1977). The maximum production of bulbs can be obtained by planting them early and to grow as late as possible. In areas with low temperature in winter, bulb goes to dormancy and in case of early planting dormancy can be successfully broken by dipping the bulbs in 4 per cent Thiourea solution for one hour. Ethylene chlorohydrin can also be used for breaking of dormancy. Proper selection of bulbs is very important for successful cultivation. In general, spindle -shaped diseases free bulbs having diameter between 1.5 and 3.0 cm are suitable for planting. About 1.25 - 1.5 lakh bulbs (8 to 9 tons of bulbs) are required for planting one hectare.

9.2.1.1 By Division

Tuberose also responds to this method of propagation, *i.e.*, by using bulb segments. The success, however, depends on the size of bulbs and it was reported that only the segments from large bulbs (2 cm or more in diameter) regenerated well (Sharga, 1980). Bulbs are cut into 2-3 vertical sections, each containing a bud and a part of the basal plate. Each of these sections is treated with fungicide and planted vertically in a rooting medium with their tips just showing above the surface. A moderately warm temperature, slightly higher than for mature bulbs should be maintained. New bulblets along with roots develop from the basal plate. At this time they are transferred to the ground to continue growth.

9.3.0 Micropropagation

The conventional method of propagation through bulbs is rather slow to meet the growing demand and, therefore, clonal propagation through tissue culture may be utilized for rapid and large scale multiplication. Muralidhar and Mehta (1982) studied *in vitro* propagation potentiality and reported that 800 regenerated plants could be obtained from a single bulb. The method suggested by him consists of radial cutting of bulbs into segments (9-12/bulb) and transferring them sequentially to shoot inducing and then root inducing media. The media for shoot and root induction consisted of MS + 3 per cent sucrose, 2.0 mg/l BAP + 0.5 mg/l kinetin + 0.2 mg/l NAA and MS +1 per cent sucrose + 2.0 mg/l IBA + 0.2 mg/l kinetin, respectively. The culture is to be kept under continuous flourescent illumination of

Propagation of Tuberose

Pods

Dried Pods

Seeds

Emergence of Seedlings

Growth of Seedlings

Harvesting of Bulbs

Separation of Bulbs

Separation of Bulbs

Planting of Bulbs

3500 lux in a culture room at 25 ± 2°C. Bose and Jana (1987) reported regeneration of plantlets from callus of scale-stem by using IAA and kinetin in MS medium.

Direct organogenesis from bulb explants of tuberose (Krishnamurthy *et al.*, 2001; Panigrahi *et al.*, 2013; Kumari and Pal, 2016; Sangavi and Chellapandi, 2008), axillary buds (Kanchana *et al.*, 2018), stem disc explants (Gajbhiye *et al.*, 2011; Raghuvanshi *et al.*, 2013; leaf segments (Datta *et al.*, 2013), immature inflorescence segments (Krishnan *et al.*, 2003; Mendoza *et al.*, 2013) were standardized. MS medium supplemented with 6.0 mg/l BAP+0.5 mg/l NAA+0.7 mg/l 2,4-D+0.5 mg/l TDZ for both petals (4.0) and immature flower bud (4.33) found to be best medium for shoot multiplication (Kadam *et al.*, 2010). Naz *et al.* (2012) used outer and inner scales of tuberose bulbs as explants in the MS medium containing BAP 3.0 mg/l. Multiple shoots were observed when regenerated shoots were transferred to MS medium supplemented with 2.5 mg/l BAP combined with 0.5 mg/l NAA and 0.1 mg/l kinetin. Regeneration protocol for tuberose using the following explants axillary buds, nodal segments of rachis, stem scale, immature flower buds and petal segments were standardized by Surendranath *et al.* (2016). Proliferation of the highest number of axillary buds was observed on MS medium fortified with BAP 4 mg/l and NAA 0.5 mg/l and shoot elongation occurred on MS medium supplemented with MS + BAP (4 mg/l) + NAA (0.5 mg/l) + TDZ (0.5 mg/l). Among different plant growth hormones so far studied, benzylaminopurine (BAP) is considered to be more important for in vitro propagation of these plants through tissue culture techniques (Naz *et al.*, 2012).

Peduncle was found to be the best explant for embryogenic callus induction in tuberose by Kahrizi *et al.* (2008). The highest embryogenic callus from peduncle explants was obtained on MS medium supplemented with 2 mg/l IAA and 0.5 mg/l BAP. Rapid propagation and regeneration of the plantlets via somatic embryogenesis pathway using leaf, stem and flower bud as explants was reported by Abdullah (2012). A combination of 2.0 mg/l Kin with 2.0 mg/l NAA yield the best shoot regeneration from somatic embryos, producing 26.23±0.74 number of microshoot. MS media supplemented with 0.5 mg/l Kin and 2.0 mg/l NAA is the most suitable media for root formation with 4.23±0.40 number of roots formation.

9.3.1 Anther Culture

Anthers with microspores at the uninucleate stage from a *P. tuberosa* variety bearing single flowers were cultured (Gi and Tsay, 1989) on half-strength MS medium supplemented with 1 ppm 2,4-D, 3 per cent sucrose and 0.8 per cent phytagar. The resulting embryogenic callus differentiated into embryoids and plantlets after transfer to medium containing 0.2 ppm NAA and 2 ppm kinetin. Among the 35 plantlets regenerated in this way, 8 reached the flowering stage and revealed considerable variation in floral traits, *viz.*, inflorescence length, number of flower buds, and flower size and colour.

10.0 Soil and Climate

10.1.0 Soil

The tuberose grows in a wide range of soils. Its cultivation can also be extended economically in almost unproductive soils affected by salinity and alkalinity with good agronomic practices. Loam and sandy loam soils having a pH range from 6.5 to 7.5 with good aeration and drainage are considered suitable for its cultivation (Sharga, 1977). The soil should be rich in organic matter and retain sufficient moisture for proper growth (Sadhu and Bose, 1973). According to Rameshwar (1976), heavy soils, where rice is cultivated, should be preferred to light red soils. For cultivation in pots, a mixture of garden soil, farm- yard manure and leaf mould in the proportion of 2:1:1 should be used.

10.2.0 Climate

Tuberose grows in mild climate without extremes of high or low temperatures even though it can be grown under a wide range of climatic conditions. In India, the commercial cultivation of tuberose is mainly confined in warm, humid areas with temperature range between 15 and 35°C. For its luxuriant growth, it requires high humidity and a temperature around 30°C (Sharga, 1977). Temperature above 35°C reduces the spike length and quality of the flowers. Very low temperature and frost also damage the plants and flowers (Sadhu and Bose, 1973). Tuberose grows well in sunny situation.

11.0 Cultivation

11.1.0 Selection of Site

For optimum growth and high yield of flowers, it is preferable to choose a place having plenty of sunlight. A little shade towards the later part of the afternoon is desirable in hot summer. If there is too much shade, the plants grow tall and lanky and the flowering is adversely affected. Tuberose is sensitive to wet feet, hence drainage should be good, water logging even for a short period damages the root system and affects the growth and flowering.

11.2.0 Preparation of Soil

Land preparation is very important for the cultivation of tuberose. The field should be ploughed deep to a good tilth and properly manured. If the soil is not thoroughly prepared and contains soil clods or undecomposed organic matter, bulb production is affected and it also reduces the yield and quality of flowers. Soil should be atleast 45 cm deep, well drained and friable enough. Well-rotten farmyard manure, at the rate of 20-30 tonnes per hectare, should be incorporated with the soil about a month before planting. Care should be taken to make the land free from weeds.

11.3.0 Selection of Bulbs and Planting

Selection of good planting material is necessary for obtaining high yield of good quality of flowers. It has been observed that fresh bulbs if planted, lead only to profuse vegetative growth but less flowering, and a storage for few weeks in well ventilated place is essential for better growth and flower production. Dormance of bulbs can be broken down by dipping 4 per cent Thiourea solution. Propagation of Tuberose should be done through mature bulbs but multiplication of growing stock can also be done with division of bulbs. Bulbs having size of 2.1 cm or more in diameter are suitable for planting. The bulbs are cut into 2 to 3 vertical sections, each containing a bud and part of the basal and treated well with fungicide and planted vertically in a rooting medium with their Tips showing above the surface. Clumps as such should never be planted as the plants grow vegetatively rather than produce flowering spikes. They should carefully be placed upright by hand and covered with soil. For pot cultivation, 2-3 bulbs are planted in the centre of a 20 cm pot.

11.3.1 Bulb Size

The size of bulb plays an important role on growth and flowering of tuberose. It influences the sprouting of bulbs and the time required is inversely proportional to size of the bulb (Sadhu and Das, 1978; Pathak *et al.*, 1980). Larger bulbs were found to take more time for sprouting (Yadav *et al.*, 1984; Mahanta and Paswan, 1995). This might be due to the presence of more layers of membranous scales which interferred exchange of gases and inhibited metabolic process (Kamerbeek, 1962). Plant height and number of leaves per bulb also showed gradual increase with the increase in bulb size up to a certain limit (Yadav *et al.*, 1984; Dhua *et al.*, 1987).

Bulb size also influences flowering. Larger bulbs cause early flowering and give higher yield of spikes and flowers (Sadhu and Das, 1978; Pathak *et al.*, 1980; Curir *et al.*, 1984; Yadav *et al.*, 1984; Reddy *et al.*, 1994; Mahanta and Paswan, 1995). The number of flowers per spike, flower quality, daughter bulb production *etc.*, were also found to be related to bulb size (Kale and Bhujbal, 1972; Ramaswamy and Chockalingam, 1977; Yadav *et al.*, 1984; Dhua *et al.*, 1987). Planting of large size bulbs of 3-4 cm diameter (Ahmed *et al.*, 2009; Khoshkam and Salari, 2018; Singh *et al.*, 2017) and >4 cm (Nagaraju *et al.*, 2012) produced best spike and bulb production in tuberose.

The effect of bulb size on the flowering of tuberose was investigated by Singh and Reddy (1994). Flower spikes from large bulbs emerged and flowered earlier than those from small bulbs. The duration of flowering was shortest for plants from small bulbs. The numbers of spikes/plant and spikes/ha increased with increasing bulb size. Flower spikes from large bulbs had more florets than those from small bulbs. Mahanta *et al.* (1998) also reported improvement in growth and flowering by using larger bulbs at planting.

In general, bulbs having diameter between 2.0 and 3.0 cm are suitable for planting. Yadav *et al.* (1984) studied the effect of four bulb sizes (1.5-2.0, 2.1-2.5, 2.6-3.0 and 3.1-3.5 cm in diameter) on growth and flower production in tuberose cv. Single for a period of three years and recorded that in the plant crop large sized bulbs (3.1-3.5 cm) significantly improved the yield of spikes and flowers, but in ratoon crops bulbs with 2.6 to 3.0 cm diameter showed better result in these respects. Considering the total production over three years, planting of bulbs having 2.6-3.0 cm recorded the highest yield of spikes (15.1 lakhs/ha) and flowers (30.1 tonnes/ha). Nambisan and Krishnan (1983) reported that bulbs weighing 30 to 40 g gave a yield of 37,970 kg flowers/ha which was 101.0 per cent higher than the yield from smallest bulbs weighing 6 to 10 g.

Chattopadhyay *et al.* (2000) observed that larger bulbs (>2.7 cm diameter) of *cv.* Calcutta Double produced early flowering. Medium sized bulb (1.4-2.0 cm in diameter) recorded higher yield/ha. Large bulbs (2.0-2.7 cm in diameter) produced maximum proportion of quality grade flowers.

Misra *et al.* (2000) reported that the bulb size influenced the initiation of spikes in both single and double cultivars. The bulb size (2.60-3.00 cm) produced the highest number of spikes/plant for both cultivars. The large bulb size of 3.0 cm diameter recorded significant increase in number of spikes per plant, the length and girth of spike, keeping quality of the spike at room temperature (Mane *et al.*, 2007).

Chaudhury *et al.* (2010) reported that the large bulbs (2.0–2.5 cm) registered maximum weight of individual bulb (45.48 g), maximum number of bulblets per plant (14.0), and maximum yield of bulb (9.10 t/ha) and bulblets (11.44 t/ha) in tuberose.

The extra large bulbs (<3.0 cm diameter) of tuberose variety Arka Prajwal recorded early sprouting (7.96 days) and recorded significant increase in vegetative parameters like plant height, number of leaves and plant spread in tuberose *cv.* Arka Prajwal (Kumari, 2014) (Table 8).

Table 8: Effect of Bulb Size on Growth of Tuberose*

Grading of Bulb (Dia in cm)	*Days to Sprouting*	*Plant Spread (cm)*	*Number of Leaves per Plant*	*Number of Tillers*	*Leaf Area (cm^2)*
Smaller bulb <1.5 cm	12.60	24.93	76.57	6.90	65.83
Medium bulb 1.5-2.0 cm	10.53	26.75	85.86	7.60	61.51
Large bulb 2.1-3.0 cm	10.53	27.03	97.35	8.50	80.50
Extra large bulb >3.0cm	7.96	27.48	99.23	9.20	74.43
S.Em ±	0.10	0.41	4.42	0.28	3.19
CD @ 5 per cent	0.29	1.91	12.76	0.81	NS

* Kumari (2014).

Kumari (2014) also stated that the bulbs of 1.5-2.0 cm diameter of cv. Arka Prajwal recorded maximum flower production in terms of cut and loose flowers than the larger size bulbs of <30 cm diameter (Table 9).

Tuberose cv. Arka Prajwal bulbs having diameter (2.6-3.5 cm) had significant effect on vegetative growth, quality and quantity production of tuberose spikes, bulbs and bulblets as compared to bulbs having diameter 1.5-2.5 cm (Wagh *et al.*, 2012). Suseela *et al.*, (2016) reported that bulbs with more than 3.0 cm diameter recorded significant increase in growth, flowering and spike yield in tuberose cv. Arka Suvasini. The bulb size of greater than 2.0 cm increased the vase life of cut flowers and yielded highest bulbs and bulblets in the tuberose *cv.* Hyderabad Double (Vishwakarma and Kumar, 2018).

Table 9: Effect of Bulb Size on Flowering and Yield Parameters of Tuberose cv. Arka Prajwal*

Grading of Bulb (Dia in cm)	*Days to First Spike Emergence*	*Days to 50 per cent Flowering*	*Number of Spikes per Plant*	*Length of Spike (cm)*	*Rachis Length*	*Number of Florets per Spike*	*Number of Bulbs*
Smaller bulb<1.5 cm	174.90	177.21	1.95	104.05	23.95	46.63	4.74
Medium bulb1.5-2.0 cm	155.86	146.43	3.01	108.70	26.74	51.28	5.66
Large bulb2.1-3.0 cm	147.68	142.88	3.14	110.31	27.80	52.30	6.23
Extra large bulb>3.0cm	132.94	132.84	2.61	113.31	29.84	54.70	7.41
S.Em ±	0.78	3.82	0.025	0.58	0.43	0.35	0.10
CD at 5 per cent	2.25	11.05	0.073	1.69	1.25	1.03	0.30

* Kumari (2014).

11.3.2 Depth of Planting

Planting depth also markedly influences growth and production of flowers (Bhattacharjee *et al.*, 1979; Yadav *et al.*, 1984). It may vary from 4.0 to 7.0 cm depending on the size of bulb, nature of soil as well as the growing region. The bigger is the bulb, the more is the depth. Similarly, in sandy soil, planting is generally deeper as compared to clay soil. Planting of bulbs at greater depth delays the appearance of shoots. In sandy loam soil, planting of bulbs at the depth of 6.0 cm is recommended (Yadav *et al.*, 1984). In the experiment on tuber size (0.5-3.5 cm diameter) and depth of planting (2-6 cm), Rao *et al.* (1991) reported that large tubers significantly increased flower yield and tuber yield and advanced flowering. Shallow planting resulted in higher flower yield, but generally planting depth had no marked effects on other growth and flower indices. Shallow planting of large tubers is recommended.

Mane *et al.* (2007) observed that planting the bulbs at shallow depth (3 cm) resulted in the earliest spike emergence, but had no identical difference with medium depth (5 cm) of planting. The length of spike, number of spikes per plant and longevity of spike in field increased significantly at 7 cm depth. Singh and Pal (2019) stated that the maximum number of sprouts per clump, number of leaves per plant and length of longest leaf are found where the bulbs were planted at the depth of 5.5 cm.

Suseela *et al.* (2016) found that 6.0 cm depth of planting were found best to achieve higher spike yield per plot (107.83) and per hectare (2.72 lakh/ha) in tuberose *cv.* Arka Suvasini. The effect of planting depth on flowering of tuberose cv. Single is presented in Table 10.

Table 10: Effect of Depth of Planting on Flowering in Tuberose cv. Arka Suvasini*

Planting Depth (cm)	*Days to Spike Emergence*	*Days to 50 per cent Flowering*	*Spike Length*	*Number of Florets per Spike*	*Number of Spikes per Plot*
4.0	143.54	166.84	69.18	39.04	55.42
6.0	160.18	190.64	85.22	41.59	107.38
S.Ed	0.57	0.24	0.35	0.16	0.29
CD at 5 per cent	1.17	0.51	0.73	0.35	0.60

* Suseela *et al.* (2016).

11.3.3 Planting Time

In India tuberose is generally planted in February-April in the plains and in April-May in the hills (Sadhu and Bose, 1973).

In southern parts of India, however, Nambisan and Krishnan (1983) suggested to plant the bulb in the month of July-August. Investigations on effect of time of planting on growth and flowering in cv. Single made by Yadav and Bose (1988) revealed that planting of bulbs in the month of April recorded the highest yield of spikes and flowers. Larger number of spikes from April or May planting was also recorded by Mukhopadhyay and Banker (1981). In a two-year field trial on a medium fertile clay soil, the best time for planting in Italy was reported between 14 and 29 June and the percentage of saleable bulb produced was about 85 (Cirrito *et al.*, 1982). The results of the experiments by Zizzo *et al.* (1999) revealed that for all planting dates, the highest flower yield was obtained with the larger bulbs, but flower quality did not vary with bulb size. The 4 planting dates led to flower production continuously from December to May, and it was concluded that the use of the unheated greenhouse system was successful in extending flower production beyond that achieved in outdoor cultivation.

Misra (1999) recorded that planting bulbs on 7 March followed by 22 March produced plants with a large number of long leaves/clump, thick and big spikes, and a high number of florets and spikes/unit area. Sequential planting may be practised to obtain flowers almost throughout the year, if the temperature is not very low in the winter months. Replanting is necessary after 3 years. In an experiment at Meerut, Uttar Pradesh, India, tuberose planted on 3 dates (2nd week of October, March or June) at 3 spacings (15 × 20, 20 × 20 or 25 × 20 cm) and at 2 planting depths (5 or 8 cm), Singh *et al.* (1999) reported the longest flower spikes and the highest number of spikes/clump from June planting and the lowest plant density. Shallow planting gave more spikes/clump. The number of bulbs/plant was highest for June planting, and this planting date also gave the largest bulbs. As plant density increased, the number of bulbs/plant decreased. Deep planting also gave more bulbs/plant than shallow planting.

Kumar (2008) studied staggered planting in tuberose with seven planting times from last week of February to last week of May and reported that 10th April planting was found to be better and recorded the maximum plant height, number of leaves per plant and average length of leaves, duration of flowering, number of spikes per plant, number of florets per spike, number of bulbs per plant and weight of bulbs per plant.

Meena *et al.* (2018) studied the influence of staggered planting dates *viz.* 15th April, 30th April, 15th May, 30th May, 15th June and 30th June at 15 days interval on growth, flowering and yield of tuberose cv. Phule Rajani at sub-humid southern plains of Rajasthan. Planting on 15th April were found significantly superior for leaves per plant (31.09), plant height (66.81 cm), spike length (59.94 cm), days to spike emergence (111.03), days to flower bud break (122.03), floret/spike (35.72), durability of spike (9.36 days), spike/plant (3.78), spike weight (72.12 g), bulb weight/plant (173.89 g), bulbs and bulblets per plant (6.87, 16.43). Staggered planting of tuberose from 15th April to 30th May can provide constant supply of flowers to market and ensure sustainable income to the farmers. Vaid *et al.* (2019) subjected tuberose *cv.* Sikkim selection to three different planting dates *i.e.* 20th May, 15th June and 10th July and reported that the 20th May planting gave the best results in terms of all the growth and flowering parameters.

11.3.4 Planting Distance

Plant density markedly influences the yield and quality of flowers and bulbs obtained per unit area. Higher plant density has been found to produce greater yield of spikes, flowers and bulbs (Bhujbal and Whaval, 1978; Cirrito and Zizzo, 1980; Patil *et al.*, 1980). The planting distance varies in different regions. Similarly, the quantity of planting material required for an area is also variable. In a three-year experiment at Palermo in Italy, tuberose *cv.* Pearl was grown at planting rates of 24, 36 and 48 bulbs/m^2 on high or medium ridges or on flat soil. The production of bulbs was maximum at the highest planting rate, the planting system did not affect production (Cirrito and Zizzo, 1980). Sadhu and Bose (1973) in West Bengal suggested spacing of 10-15 cm between the bulbs and 25 cm between the rows. Experiments conducted at Bidhan Chandra Krishi Viswavidyalaya, Nadia, West Bengal showed that planting of bulbs at a distance of 20 × 20 cm with a population of 2,50,000 plants per hectare gave the highest yield of spikes, flowers and bulbs over a period of 3 years (Yadav, 1983). Similar planting distance was also recommended by Nambisan and Krishnan (1983) for tuberose cultivation in South India. But Rameshwar (1976) suggested a spacing of 30 to 38 cm between rows and 15 to 20 cm between bulbs, accommodating 40,000 to 50,000 bulbs per acre for cultivation of tuberose in Bangalore. The highest yield of flowers and bulbs per plant was obtained by Bhattacharjee *et al.* (1979) when planting was done at a distance of 20 × 25 cm. Sharga (1977) recommended a spacing of 30 cm apart. Planting of tuberose carried out at a spacing of 25 × 25 cm on flat beds, or on medium (12 cm) or high (18 cm) ridges resulted in the highest yield of spikes and flowers on flat beds (Singh and Sujatha, 1990). Sharma *et al.* (1995) suggested that the 2 close spacings (5 × 10 cm and 10 × 15 cm) should be used for growing tuberose for oil and loose flower production, and that the 2 wide spacings (15 × 20 and 20 × 25 cm) be used for producing cut flower for vase and bulbs.

Misra *et al.* (2000) recommended 30 × 30 cm spacing to achieve higher spike yield per plant and also reported that the plants planted in closer spacing took much time to produce spikes than the wider spacing. Similarly, Nagaraju *et al.* (2012) also reported that wider spacing of 30 × 30 cm as optimum to obtain the maximum growth, flowering and yield of tuberose cv. Double. Wider spacing of 30 × 45 cm was found to be best to obtain maximum number of spikes per plant in tuberose *cv.* Single (Patel *et al.*, 2006). The bigger sized bulbs of 3 cm diameter planted at widest spacing (20 × 25 cm) and at 5 cm depth of planting was found best to get maximum commercial advantage for achieving higher flower yield and quality of tuberose (Mane *et al.*, 2007). Planting tuberose at optimum spacing of 30x30 cm found to be best to get maximum spike yield in tuberose cv. Arka Suvasini (Suseela *et al.*, 2016). Spacing of 30 × 30 cm promoted the number and yield of bulb and bulblets per plant and per hectare (Vishwakarma and Kumar, 2018). Singh and Pal (2019) suggested that the spacing 20 × 20 cm is adequate for tuberose cv. Double to achieve better vegetative, growth and yield of tuberose. Kumari (2014) reported that the bulbs of tuberose variety Arka Prajwal planted at wider spacing of 30x30 cm distance recorded early spike emergence, maximum plant spread, number of spikes per plant, spike length, rachis length, number of florets per spike, weight of hundred flower buds and number of bulbs per plant (Table 11).

Table 11: Effect of different Spacing on Flowering Parameters of Tuberose cv. Arka Prajwal*

Spacing	*Days to Spike Emergence*	*Number of Spikes per Plant*	*Length of Spike*	*Rachis Length*	*No. of Florets per Spike*	*Weight of Hundred Flower Buds*
30 x 30 cm	174.90	1.95	104.05	23.95	46.63	168.17
25 x 25 cm	155.86	3.01	108.70	26.74	51.28	177.44
20 x 20 cm	147.68	3.14	110.31	27.80	52.30	181.79
15 x 15 cm	132.94	2.61	113.31	29.84	54.70	189.15
S.Em ±	0.78	0.025	0.58	0.43	0.35	2.61
CD at 5 per cent	2.25	0.073	1.69	1.25	1.03	7.53

* Kumari (2014).

11.4.0 Irrigation

Soil moisture is an important factor effecting growth and flowering in tuberose. It is suggested to irrigate before planting to provide optimum moisture for sprouting and further irrigation should be avoided until the bulbs have sprouted. Too much moisture in the soil at the time of sprouting results in the rotting of bulbs and similar condition during flowering adversely affects the development of spikes and flowers. The frequency of irrigation depends upon soil type, stage of growth and weather conditions. The tuberose fields should be irrigated at an interval of 10-15 days, if the weather is dry (Sadhu and Bose, 1973). In Lucknow (North India), Sharga (1977) suggested that during summer months (April to June) the crop should be irrigated at weekly interval and during winter at 10-days interval whereas in South India, the crop requires irrigation twice a week during dry months and at weekly interval during other months except in the rainy season (Nambisan and Krishnan, 1983). In an investigation on the effect of irrigation frequency on the growth and flowering, El-Roumi (1981) observed that plants of *cv.* Pearl irrigated at 40 l/m^2 every 7 days were optimum, flowered earlier and produced large bulbs than those irrigated at every 14 days. Cirrito *et al.* (1981) studied the effect of different watering rates and mulching on tuberose for bulb enlargement and reported that drip irrigation gave the best yield of (30-32 bulbs/m^2) bulbs of planting size. Mulching with black polythene did not reduce the water requirement.

The effects of 6000, 8000 or 10000 ppm NaCl or $CaCl_2$ to the irrigation water either separately or in combination, were assessed by Malini and Khader (1989). Flowering in the control occurred at 122 days and it was delayed to 165 days at the highest salinity level. The flower yield, which was 24.7 g/plant in the control, declined to 9.8 g/plant at the highest salinity level.

Irrigation at IW/CPE of 1.2 resulted in significantly higher dry matter accumulation and increased spike length, rachis length and number of flowers per spike in tuberose (Ashok *et al.*, 1995).

Phenophase based irrigation and nutrient scheduling in tuberose cv. Arka Prajwal was studied at IIHR, Bangalore and the results revealed that irrigation scheduling at 0.8 ER during vegetative, 1.0 ER during flowering and 0.6 ER during dormancy phase in combination with nutrients at the rate of 200:200:200 kg NPK/ha/year applied in split doses of 30:25:25 per cent NPK (vegetative phase), 30:40:40 per cent NPK (flowering), 10:10:10 per cent NPK (Dormancy), 30:25:25 per cent NPK (Ratoon vegetative stage, 35:25:25 per cent NPK (Ratoon flowering phase) produced the maximum number of spikes per plant, 100 bud weight, yield of flowers and spikes in the main crop. This schedule accounted for 35 per cent and 57 per cent in yield of florets and spikes, respectively over the conventional practice in the main crop of tuberose *cv.* Arka Prajwal. This irrigation and nutrient schedule increase flowering and yield parameters of first and second year ratoon crop of tuberose (Anon, 2019).

Bahadoran and Salehi (2015) determined the interaction effects of salinity and irrigation intervals on growth and flowering of two cultivars 'Mahallati' and 'Dezfuli' using four irrigation intervals 2, 4, 6, and 8 days and four salinity treatments with the EC values of 0.7 (control), 1.9, 3.1, and 4.3 dS/m. They concluded that tuberose is sensitive to water and salinity stress. Both physiological and biochemical processes can be affected by salinity and drought stress and their interaction.

Irrigation with higher moisture schedule at IW/CPE 1.0 showed maximum spike and bulb yields of plant and it was comparative with moderate deficit irrigation schedule of IW/CPE 0.8 (Banik *et al.*, 2018). Gurjar *et al.* (2018) assessed the impact of waste water irrigation on bulb productivity and water use efficiency in tuberose under varied wastewater-groundwater irrigation regimes. They further reported that the tuberose crop can be grown successfully under waste water irrigation without an adverse impact on its bulb yield. The optimum irrigation requirement of tuberose crop is 1350 mm. Higher water use efficiency and bulb productivity of tuberose can be attained with application of wastewater irrigation at 1.2 ID/CPE in sandy loam soils as compared to groundwater irrigation.

Pal *et al.* (2019) studied the water requirement of tuberose cvs. Arka Prajwal, Calcutta Single, Calcutta Double (Table 12). Irrespective of cultivars, the maximum spike per plot was recorded at 1.0 IW/CPE which gave about 33.15 number of spike per plot, which was superior to 0.8 IW/CPE (32.25) and 0.4 IW/CPE (30.57).

Table 12: Water Use and Water Use Efficiency of Tuberose under different Irrigation Regimes*

Treatment	*Profile Contribution (mm)*	*Irrigation (mm)*	*Rainfall (mm)*	*Total Water Use (mm)*	*Economic Yield (Spike/plot)*	*Water Use Efficiency (Spike/plot/cm)*
0.4 IW/CPE	22.70	240	1319	1581.7	30.57	0.193
0.8 IW/CPE	32.10	510	1319	1861.1	32.25	0.173
1.0 IW/CPE	35.70	630	1319	1984.7	33.15	0.167

* Pal *et al.* (2019).

11.5.0 Manuring and Fertilization

The requirement of manures and fertilizers for tuberose vary with climatic conditions and soil types. During the preparation of soil a basal application of leaf mould, farm yard manure or cattle manure at the rate of 20 to 30 tonnes per hectare depending on climatic conditions and soil type should be done to ensure better growth and flowering (Nambisan and Krishnan, 1983).

Karim *et al.* (2017) subjected tuberose variety Arka Prajwal into different organic manures individually such as vermicompost (2000 g/m^2), neem cake (400 g/m^2) poultry manure (1000 g/m^2) and FYM (4000 g/m^2) and all the treatments were repeated with the combination of treatments of seaweed. The Seaweed+vermicompost @ 2000 g/m^2 recorded maximum plant height, spikes length, number of florets, weight of ten florets, length of florets, rachis length, vase life and the flower yield (Table 13).

The application of 75 per cent recommended dose of fertilizer along with *Azotobacter* @ 2 kg/ha and PSB @ 2 kg/ha was beneficial in increasing the growth attributes and flower yield of tuberose (Sudhagar *et al.*, 2020). Application of 400 kg/ha nitrogen and 200 kg P_2O_5/ha recorded significantly the highest vegetative growth, rachis length and number of florets/spike (Patel *et al.*, 2006). Foliar application of 50 per cent RDF @ 50 : 50 : 50 kg NPK/ha + vermicompost @ 5 t/ha along with foliar application of humic acid @ 0.2 per cent + boron @ 1 per cent at 60 days interval from 60 DAP onwards was found better in increasing plant spread, leaf area, chlorophyll content and dry matter production of tuberose cv. Arka Prajwal (Priya *et al.*, 2020).

Nitrogen, phosphorus and potassium greatly influence growth and flower production. In an experiment on the effect of high and low doses of N, P and K in sand culture, Jana *et al.* (1974) found that high dose of N and P markedly promoted leaf and bulb formation in tuberose. These treatments also improved flowering. The effect of K was less pronounced than N and P in these respects. The plants treated with low dose of N and P showed growth inhibition and failed to flower.

Table 13: Effects of Organic Manure on Flowering Parameters of Tuberose cv. Arka Prajwal*

Treatments	*Length of Spike (cm)*	*Weight of Spike (g)*	*Rachis Length (cm)*	*No. of Florets per Spike*	*Floret Length (cm)*	*Vase Life (Days)*	*No. of Spikes per Plot per Year*
Control (No organic manure)	93.33	98.33	19.00	27.33	6.00	7.67	24.33
Poultry Manure @1000g/sq.m	101.83	113.00	25.67	28.33	6.97	10.00	27.00
Vermicompost @ 1500g/sq m	101.22	118.33	31.33	34.00	6.93	8.34	27.33
Farm Yard Manure @4000g/sq m	102.11	116.67	29.67	28.33	6.50	9.34	26.33
Neem cake @400g/sq m	102.33	115.17	32.00	31.00	6.90	8.67	27.00
Sea weed + poultry manure 1000g/sq m	105.33	126.00	27.33	37.33	6.77	11.34	30.33
Seaweed + Vermicompost 1500g/sq m)	106.33	128.00	32.67	40.33	7.17	11.67	33.00
Seaweed + Farmyard manure 4000g/sq m)	104.22	126.00	31.33	35.67	6.67	11.00	30.67
Seaweed + Neem cake 400g/sq m)	102.99	130.00	33.33	39.67	6.90	11.67	30.67
C.D. at 5 per cent	4.294	16.582	4.945	2.974	0.594	2.222	1.521
SE(m) ±	1.42	5.484	1.635	0.934	0.196	0.735	0.503

* Karim *et al.* (2017).

Though Militiu *et al.* (1970) emphasized the need of higher dose N than P and K, excessive N tends to cause the flower spike quite tall and soft, making it vulnerable to wind and the plants become more susceptible to diseases and pests. The flower quality is also adversely affected (Sadhu and Bose, 1973). Nitrogen deficiency can be manifested as reduction in number of spikes and number of flowers per spike as well as the typical pale green foliage. Symptoms of phosphorus deficiency are dark green upper leaves and purple colouration in lower leaves. Reduction in growth and flowering due to the deficiencies of these nutrients was also reported by Bhattacharjee *et al.* (1979). The deficiency of calcium causes cracking of the spike and acute deficiency results in bud rot. Magnesium deficiency causes interveinal chlorosis of older leaves whereas iron deficiency is manifested by interveinal chlorosis of new leaves. Deficiency of boron causes cracking of leaf margins, deformed leaves and stunted inflorescence. Manganese deficiency is evidenced by yellowing between the veins of lower leaves. Ailincai (1960) found that granular superphosphate at 60 g/m^2 combined with phosphobacter in was very useful to increase the number of flowers per inflorescence and also the life of flowers. As regards the bulb production, experiments conducted by Cirrito (1975) showed that both number of saleable bulbs and individual bulb weight were maximum when 20 kg N, 40 kg P_2O_5 and 60 kg K_2O per 1000 m^2 were applied at planting. In another trial on alkaline sandy soil, Cirrito and Vita (1981) obtained the highest production (144/m^2) of new lateral bulbs measuring 4-6 cm in circumference and ready for flowering (92 per cent) with the application of 800 kg $(NH_4)_2SO_4$, 1000 kg rock phosphate and 180 kg K_2SO_4 per hectare incorporated in the soil after planting of bulbs.

In India, Singh (1971) recommended N and P_2O_5 at 40 and 60 kg/ha respectively in alkaline soil of Uttar Pradesh whereas Sadhu and Bose (1973) from West Bengal suggested 10 g urea, 10 g superphosphate and 8 g muriate of potash per m^2 in two split doses. In a neutral clay soil, having high amount of potassium, the best nutrient combination was found to be 200 kg N and 60 kg P_2O_5 per hectare, K_2O, did not show any response. However, for tuberose cultivation in soils low in K, application of this nutrient at the rate of 50 kg per hectare was suggested (Nanjan *et al.*, 1980; Nambisan and Krishnan, 1983). In an investigation carried out by Yadav *et al.* (1985) application of 300 kg N in two split doses and 200 kg P_2O_5 per hectare per year proved very beneficial in increasing the plant growth and yield of flowers and bulbs in crops grown for three years after planting.

Bankar and Mukhopadhyay (1990) also reported that N application advanced flowering and improved growth. Leaf N content was positively correlated but leaf P and K contents were negatively correlated with number of flower spikes. The highest number of flower spikes/m^2 (20.09) was obtained with the highest N rate. Fertilization of tuberose with N : P_2O_5 : K_2O at 20 : 20 : 20 g/m^2 was recommended. In another experiment both N and K_2O significantly influenced the number of days required for flower spike emergence. Increasing P and K_2O rates resulted in a greater number of flower spikes and number of flowers/spike. The highest yield of flowers (40.20/spike), the longest spikes (81.28 cm) and the longest duration of flowering (29.75 days) were obtained with 200 kg N + 75 kg P_2O_5 + 125 kg K_2O/ha (Gowda *et al.*, 1991). Parthiban and Khader (1991) found that application of 100 kg N + 75 kg P + 62.5 kg K/ha resulted in the highest number of spikes/plant (1.72), number of flowers/spike (39.67) and the highest flower yield (3578.6 kg/ha). The effects of 7 NPK rates were compared on tuberose grown for cut flowers (Baruah *et al.*, 1998). Spike yields were highest with 80 g N + 40 g P + 60 g K/m^2, floret size, and vase life were also greatest in this treatment.

Tuberose treated with 325 kg N, 125 kg P_2O_5 and 125 kg K_2O/ha and planted at 30 × 20 cm resulted in the highest N, P and K uptake, both at 50 per cent flowering stage and harvesting stage (Mohanty *et al.*, 1999). Treatment with 120 : 65 : 62.5

kg NPK/ha + *Azospirillium* resulted in the highest mean spike length, number of flowers per spike, flower weight, number of tubers, and tuber weight per plant, and yield of flowers (3.08 and 2.75 t/ha, for the first and second year, respectively) (Swaminathan *et al.*, 1999).

Planting of 6 tubers per hill resulted in the highest number of shoots and leaves and leaf area per plant, while 3 tubers per hill resulted in the highest plant growth, flower yields and number of spikes. Application of 250 : 200 : 200 kg NPK/ha on plants from 3 tubers per hill resulted in the highest flower yield (7.86 t/ha) and plant growth (43.72 cm leaf length) (Patil *et al.*, 1999). Bhuyan *et al.* (1996) reported that the number and weight of spikes, floret size, shelf life and vase life and bulb production increased as K rate increased up to 60 g/m^2.

Mostafa *et al.* (1996) studied the effects of P (0 or 300 ppm P_2O_5), B (0, 90 or 270 ppm boric acid) and method of B application (pre-planting application to bulb, foliar spray or soil drench) on vegetative growth, flower quality, corm production and essential oil yield and components of tuberose (*cv.* Double) and found that foliar application of B in combination with 300 ppm P_2O_5 increased the number of florets/spike and the percentage of eugenol and farnesol in the essential oil.

Table 14: Response of different Level of NPK on Flowering and Yield of Tuberose cv. Arka Shringar*

Treatment (Levels of NPK kg/ha)	*Days to Flowering*	*Spike Length*	*Rachis Length*	*No. of Florets per Spike*	*No.of Spikes per Clump*	*Yield of Spikes/ ha*	*Yield of Bulbs (t/ha)*
Control	99.19	96.96	21.82	26.44	2.20	296680	10.52
20:40:20	100.78	97.08	22.95	28.68	2.45	331570	10.73
40:80:40	100.97	98.38	23.43	28.76	2.62	354840	10.90
60:120:60	106.32	100.04	24.12	29.39	2.88	389510	13.15
80:160:80	108.85	102.01	25.26	30.91	3.27	442100	13.34
100:200:100	107.14	101.16	25.01	30.85	3.92	529060	15.20
120:240:120	109.68	105.88	25.43	31.21	3.62	488610	13.96
14:280:140	111.54	110.19	26.21	31.41	4.18	564190	17.03
160:320:160	112.07	111.94	27.27	32.17	4.47	604930	17.18
180:360:180	112.94	115.46	28.74	32.85	5.56	749980	18.03
200:400:200	114.56	119.72	32.55	35.31	4.78	645310	17.75
220:440:220	113.20	117.80	31.26	33.58	4.69	633970	17.32
CD. at 5 per cent	0.03	0.04	0.29	0.05	0.19	11.68	

N: Urea, P: Single Super Phophate (SSP), K: Muriate of Potash (MOP).

* Singh *et al.* (2014).

The combined application of 150 kg N and 80 kg P_2O_5/ha had significant impact on gorwth and flowering of tuberose and it resulted in highest number of florets per spike, flower yield per clump, flower yield per hectare, longevity of spikes in the plant, vase life of spikes, number of bulbs and dry matter production (Senthilnathan and Manivannan, 2019). Singh *et al.* (2014) subjected tuberose cv. Arka Shringar to different levels of NPK and found that the maximum number of spike/clump (5.56), yield of spike/ha (749.98) and yield of bulbs (t/ha) were produced by the N: P: K in a ratio of 180:360:180 kg/ha (Table 14).

Jagtap *et al.* (2017) recommended the soil application of 300 kg N, 250 kg P_2O_5 and 300 kg K_2O/ha + FYM 20 t/ha to *cv.* Phulae Rajani for higher number of spikes/ha with increased spike length, rachis length, florets/stalk, number of spikes/ha number of bulbs/ha and maintained nutrient status in respect of available NPK in soil.

Table 15: Effect of Graded Levels of NPK on Flower and Bulb Yield Parameters of Tuberose cv. Phulae Rajani*

Treatments	*No. of Florets per Stalk*	*No.of Spikes per ha (Lakhs)*	*No. of Bulbs per ha (Lakhs)*
200:150:200 NPK kg/ha	33.00	6.42	7.75
200:150:250 NPK kg/ha	36.33	6.69	8.08
200:150:300 NPK kg/ha	37.67	6.90	8.33
200:200:200 NPK kg/ha	38.17	7.17	8.66
200:200:250 NPK kg/ha	39.83	7.41	8.95
200:200:300 NPK kg/ha	41.67	7.65	9.24
250:150:200 NPK kg/ha	39.00	6.95	8.40
250:150:250 NPK kg/ha	41.83	6.94	8.38
250:150:300 NPK kg/ha	42.67	6.76	8.16
250:200:200 NPK kg/ha	40.33	7.15	8.63
250:200:250 NPK kg/ha	41.83	7.61	9.19
250:200:250 NPK kg/ha	46.17	7.95	9.60
300:150:200 NPK kg/ha	43.67	7.23	8.73
300:150:250 NPK kg/ha	44.50	7.25	8.76
300:150:300 NPK kg/ha	45.00	7.28	8.79
300:200:200 NPK kg/ha	44.50	7.55	9.12
300:200:250 NPK kg/ha	45.17	8.34	10.07
300:200:300 NPK kg/ha	54.17	8.90	10.75
SE±	0.88	0.11	0.13
CD. at 5 per cent	2.50	0.31	0.37

* Jagtap *et al.* (2017).

Application of 25t/ha FYM + recommended dose of 200:200:200 kg/ha NPK + Zinc sulphate @ 0.50 per cent + Borax @ 0.50 per cent on 30, 60 and 90 days after planting improved the vegetative, flower and bulb yield parameters (Karuppaiah, 2019). The combination of three micronutrients (B, Zn and Fe) along with GA_3 200 ppm was effective in increasing plant height, number of flower per spike, total leaf area, number of spikes per clump, spike length and yield per plot (Ganesh *et al.*, 2013) (Table 16).

Table 16: Effect of Micronutrients on Growth and Flowering of Tuberose cv. Arka Prajwal

Treatment (Foliar spray)	*Plant Height (cm)*	*Number of Leaves per Plant*	*Days taken for First Flowering*	*Spike Length (cm)*	*No. of Florets per Spike*	*Flower Length (cm)*	*Flower Diameter (cm)*
Control	28.06	16.72	108.59	70.29	32.87	5.85	5.85
H_3BO_3 @ 0.1 per cent	32.35	18.69	106.30	70.59	34.99	6.02	6.02
$ZnSO_4$ @ 0.5 per cent	26.95	18.57	106.76	70.48	35.59	6.03	6.03
$FeSO_4$ @ 0.2 per cent	29.61	20.30	106.64	73.27	37.25	6.15	6.15
H_3BO_3 @ 0.1 per cent + $ZnSO_4$ @ 0.5 per cent	32.61	19.42	97.14	75.01	36.49	5.96	5.96
H_3BO_3 @ 0.1 per cent + $FeSO_4$ @ 0.2 per cent	34.60	22.75	97.04	73.96	36.24	6.26	6.26
$ZnSO_4$ @ 0.5 per cent + $FeSO_4$ @ 0.2 per cent	29.65	18.47	103.15	79.41	37.89	6.19	6.19
H_3BO_3 @ 0.1 per cent + $ZnSO_4$ @ 0.5 per cent + $FeSO_4$ @ 0.2 @ 0.2 per cent	40.04	20.20	100.48	82.61	40.49	6.47	6.47
Mean	31.73	19.39	103.26	74.45	36.48	6.12	6.12
S.E±	2.65	0.75	0.72	1.11	1.13	0.12	0.07
CD at 5 per cent	5.35	1.52	1.46	2.25	2.28	0.24	0.14

* Ganesh *et al.* (2013).

11.6.0 Weed Control

Weeds become a serious problem in the cultivation of tuberose as fertilizer and irrigation requirements of this crop create conducive condition for the growth of various types of weeds. If these weeds are not removed in time, growth and productivity will be markedly reduced. Manual weeding is effective if done frequently. But this method is very laborious and increases the cost of cultivation. Generally, after each irrigation, weeding is done. Hoeing between plants is also necessary to loose the soil and destroy the weeds.

Certain chemicals have proved effective in controlling both monocot and dicot weeds (Christensson, 1982). Chiapparini (1960) reported that pre-emergence applications of diuron (80 per cent) at 2.8 kg a.i. per hectare effectively controlled weeds in tuberose field. Cirrito *et al.* (1980) observed that treatment with treflan at 2 kg per hectare emulsified in 20 l of water controlled most weeds except *Cyperus longus*. Pre-plant application of atrazine (2-chloro-4-ethylamino-6-isopropylamino,

Cultivation of Tuberose

Open Cultivation

Mulching in Tuberose

Drip irrigation

Flowering in Bed

Bud Stage

Full Bloom Stage

1,3,5-triazine) at 3.0 kg a.i. per hectare was also found very useful for the reduction of weed population in tuberose field resulting in the highest yield (10.2 tonnes/ha) of good quality flowers. The corresponding yield from untreated control plants was only 5.8 tonnes per hectare (Yadav and Bose, 1987).

Murthy and Gowda (1993) observed that metolachlor at 2.0 kg and pendimethalin at 1.25 kg/ha resulted in the lowest weed DM at harvest and the greatest flower yields, respectively. Application of gramoxone at 3-4 l/ha 7 days prior to planting of bulbs or 3 times at an interval of 40 days after planting caused maximum reduction of weed biomass and resulted in the highest yield of spikes and flower (Pal and Das, 1990).

Rathod (2015) studied the effect of different pre and post emergent weedicides on tuberose *cv.* Arka Prajwal. The maximum flower yield (t/ha) was recorded in treatment weed free check, followed by Pendimethalin 30 EC @ 1 kg a.i/ha (3.67 t/ha) and Alachlor 50 EC @ 1.5 kg a.i/ha (3.54 t/ha). The higher weed control efficiency (97.23 per cent) was observed in weed free check treatment on 120 DAP. Among the herbicides, Pendimathalin 30 EC @ 1 kg a.i/ha and Alachlor 50 EC @ 1.5 kg a.i/ha were the best which exhibited 84.30 and 83.83 per cent weed control efficiency, respectively. Application of Atrazine (1.0 kg/ha) + rice residue (5t/ha) resulted in better weed control, higher growth, flowering and bulb yield in tuberose cv. Arka Prajwal (Jain *et al.*, 2015).

Jeevan *et al.* (2016) reported that highest plant height (48.11 cm), number of leaves per plant (38.26), SPAD value (64.50), spike length (44.10 cm), floret length (6.01cm), flower yield (7215 kg/ha) and weed control efficiency (97.58 per cent) was observed in plot mulched with black polythene in tuberose cv. Hyderabad Single. Whereas, highest leaf area (63.75 cm^2) was observed in paddy straw mulched plot. Minimum number of days for flowering and maximum number of flowers per plant obtained in pendimethalin followed by hand weeding (Table 17).

Pre-emergence application of pendimethalin @ 0.75 kg a.i./ha + 1 hand weeding (HW) at 40 days after planting (DAP) reported maximum weed control efficiency (93.10 per cent) as compare to weedy check. However, 3 hand weeding at 30, 60 and 90 days after planting had superior effect on the plant height, number of leaves per plant, spike length, spike emergence, days to first flowering, number of floret per spike, number of spikes per plant, flowering duration, Number of bulbs and bulblets per plant and hectare (Mahavar *et al.*, 2019).

11.7.0 Mulching

Prakash *et al.* (2011) tried six different locally available mulch materials *viz.*, paddy straw, dry weed, dry sugarcane leave, transparent poly sheet, black poly sheet and control as without mulch in tuberose cv. Double. Mulching with dry weed was found better than the other mulching treatments and produced longest spike (86.50 cm), rachis (48.42 cm.), maximum weight of bulbs (33.26 g) and highest number of bulblets (24.56). Water hyacinth as mulch in tuberose significantly increased the flower yield (Barman *et al.*, 2015).

Table 17 : Effect of herbicides on growth and flowering of tuberose*

Treatments	*Days to Spike Emergence*	*Number of Flowers per Plant*	*Spike Length (cm)*	*Floret Length (cm)*	*Flower Yield (kg/ha)*
Oxyfluorfen @0.15 kg a.i./ ha + Quizalofop ethyl @ 50 g a.i./ha + 30 DAP	124.38	87.00	31.54	3.56	5020
Atrazine @ 11.0 kg a.i./ha + Quizalofop ethyl @ 50 g a.i.ha + 30 DAP	122.25	90.03	33.77	3.90	5199
Oxyfluorfen @ 0.15 kg *a.i.*/ ha + hand weeding at 30 DAP	122.88	96.19	36.43	5.29	5550
Atrazine @ 1.0 kg *a.i.*/ha + hand weeding at 30 DAP	112.08	104.00	37.10	5.40	6001
Pendimethalin (pre-emergent) @ 0.75 kg *a.i.* / ha + quizalofop ethyl (post emergent) @ 50 g a.i./ha	118.13	91.00	33.77	4.02	5233
Pendimethalin (pre emergent) @ 0.75 kg a.i./ha + one hand weeding at 30 DAT	109.07	115.01	39.00	5.71	6787
Black polyethylene sheet mulch 30 micron thickness	121.08	106.00	44.10	6.01	7215
Paddy straw mulch	116.03	98.08	41.23	5.26	5658
Hand weeding @ 20, 40 and 60 DAP	114.110	112.50	42.83	5.14	6792
Unweeded control	140.07	42.00	20.90	2.69	2150
SEm	2.81	5.68	0.94	0.45	166
CD (5 per cent)	8.36	16.87	2.80	1.34	493

* Jeevan *et al.* (2016).

Deka and Talukdar *et al.* (2017) studied the effect of mulching on growth and flowering of tuberose *cv.* Double and the results revealed that the black polythene mulch significantly influenced the growth, floweing and yield parameters.

Shiombing and Handayati (2017) reported that the mulch improved the flower production in tuberose and it resulted 3.8 stalks/clump, whereas, 2.47 stalks/ clump in without mulch. The damage intensity caused by mealy bug infestation on the mulch treatment was 15.21 per cent, while that without mulch was 22.17 per cent. The use of mulch reduced the weeding cost up to 66.67 per cent and the total production cost up to 20.86 per cent. Black plastic mulch showed best results in various growth and flowering parameters of tuberose like early sprouting of bulb, earliest spike emergence, maximum spike length, maximum rachis length, maximum number of florets per spike, maximum fresh weight of spike, maximum duration of flowering, largest sized bulb (Vaid *et al.*, 2019). Plastic mulching of 30 micron thickness in tuberose was recorded to increase the yield upto 18-22 per cent and reduced weed management cost by 80 per cent (Indhumathi *et al.*, 2020) and Dharmapuri district of Tamil Nadu reported 200 ha under plastic mulching.

12.0 Diseases and Pests

Tuberose is quite hardy and there are no serious diseases and pests posing problems for the cultivation of this crop. However, it is attacked by some diseases and pests which are described below:

12.1.0 Diseases

Fungal Diseases

12.1.1 Stem Rot

The disease symptom is preceded by the appearance of prominent coarse mycelial masses on leaf surface at or near the soil level. Soon the infected spots loose green colour due to rotting which extends and covers the whole leaf. The infected leaves fall on the soil surface and become detached from the plant. More or less round sclerotia, brown in colour, are formed on and around the infected leaves. As a result the infected plants become weak and send out few or none of the flowering shoots in case of severe damage (Das, 1961). The disease can be controlled by soil application of brassicol (quintozene) or zineb at 3-week interval (Dutta, 1975). Based on the results obtained from the experiments conducted in the All India Coordinated Floriculture Improvement Project, dusting of brassicol (20 per cent) at the rate of 30 kg per hectare has been recommended (Malik *et al.*, 1984). Mercuric chloride at 0.1 per cent and 0.2 per cent commercial formalin also gave excellent control (Roy, 1985). Infection by *Sclerotium rolfsii* can be minimized by reducing soil moisture or planting at wider spacing. Destruction of infected plant debris is also important because it harbours the fungus. Treating the collar portion and the soil with benlate, carbendazim, iprodione or copper oxychloride @ 2 g/litre of water is effective. Bharathi and Narayanaswamy (2018) reported that the combinations of *Trichoderma viride* (10g/l) + Press mud (500g/m^2) + Carbendazim (0.1 per cent) recoded a lesser incidence (10.66 per cent), among fungicides the treatments Hexaconazole @ 0.1 per cent recorded minimum disease incidence of 19.33 per cent.

12.1.2 Botrytis Spots and Blight

This disease, caused by *Botrytis elliptica*, when it has not caused serious damage to the leaves, can be controlled by spraying the plants with mancozeb (0.2 per cent) or benjimidazol (0.2 per cent) or mixture of the two fungicides. Priyadarshini *et al.* (2019) reported that the *Botrytis cinerea* infects tuberose crop in Tamil Nadu and causes heavy yield loss by affecting the foliage and flower.

12.1.3 Leaf Spot

Tzeng and Tung (1998) reported a leaf spot disease in Taiwan caused by *Fusarium equiseti*. The initial symptoms of this disease were water-soaked spots which turned into elliptical or round, brown spots. *F. equiseti* could infect leaves, stems and flowers.

As it is a soil borne disease clean cultivation and soil sterilization are suggested as preventive measures. Application of mancozeb is effective to control the disease.

12.1.4 Phoma Blight

Phoma blight is caused by *Phoma* sp. and the disease was first reported in West Bengal, India by Panja *et al.* (2016) and they designated the species as *P. mondouriensis*. The disease appears during wet weather and reddish brown spots appear on leaves and coalesce together. The dark brown or reddish patches appear on stalks which reduces the flower production. Removal of affected leaves followed by application of carbendazim @ 2 g/l of water or Propiconazole @ 1 ml/litre of water will control the disease.

12.1.5 Alternaria Blight

The leaf blight disease caused by *Alternaria polianthi* is considered as a severe problem causing growth reduction and yield loss. The leaf spot starts from leaf tip and progresses gradually inwards and downwards with wavy margin. Conidiospores are brown, septate formed in fascicles. Conidia are subhyaline to olivaceous, obpyriform with 4-6 septa and 1-2 longitudinal septa. This has been reported from Tamil Nadu and Chandigarh. Spraying with mancozeb at 0.2 per cent will reduce the disease incidence. Tebuconazole @ 0.1 per cent could be used for the management of leaf blight. Four sprays taken with Tebuconazole @ 0.1 per cent at 15 days interval starting from onset of disease proved to be the most effective treatment and resulted in lowest percent disease index of 10.80 (Ravikumar *et al.*, 2020). Kakade *et al.* (2016) stated that the leaf blight disease caused by *Alternaria polyanthi* accounting 15-20 per cent losses in yield and quality of tuberose and becomes the major threat. Six sprays of iprodine +carbendazim 0.1 per cent at 10 days interval starting the first spray at disease appearance was found effective in controlling the leaf blight.

12.1.6 Anthracnose

Colletotrichum truncatum causing anthracnose on tuberose in India was reported by Mahadevakumar *et al.* (2019). This disease is characterized by the dark concentric sunken necrotic lesions with spore mass in the acervuli on leaves and peduncles.

Bacterial Diseases

12.1.7 Flower Bud Rot

It is caused by *Erwinia* sp. The disease appears mainly on young flower buds and results in dry rotting of the buds with brown scorched necrotic discolouration of peduncles. At later stage, buds shrivel and become dry. Application of streptocycline (500 ppm) is recommended to control the malady.

Besides the above diseases, a contagious disease known as bunchy top causes malformation of the inflorescence like a ball and the flower spike grows to only 1/3rd of its normal height. Control of thrips which is associated with this disease, checks the malady (Rameshwar, 1976).

Peduncle blight caused by *Lasiodiplodia theobromae* in tuberose in Tamil Nadu was reported by Durgadevi and Sankaralingam (2012). The disease exhibits blossom blight followed by peduncle dieback starting from the tip, and leaf blight at the tips. The microbes such as *Pseudomnas flourescences, Bacillus subtilis Trichoderma viridae* inoculated bulblets showed sustained, timely induction and accumulation of these defense enzymes and PR-proteins which enhances the resistance in tuberose against peduncle blight disease and also improved the vegetative growth, physiological attributes, PR-proteins and phenol contents (Durgadevi *et al.,* 2018).

Viral Diseases

During a survey for ornamental plant virus diseases in 1994, foliar mild mosaic symptoms were noticed on most tuberose plants in Taiwan. A virus was isolated from infected plants and it was subsequently identified as a newly recognized potyvirus, designated as tuberose mild mosaic virus (TMMV), based on particle morphology, aphid transmissibility and the cylindrical inclusions it induced in infected cells. Tuberose was the only host identified so far for TMMV in the inoculation experiment. Based on the result of several serological tests, TMMV was found serologically distinct to 22 other known potyviruses tested (Chang and Chen, 1998, 1999).

Tuberose mild mottle virus from (TuMMoV) from India (Kulshrestha *et al.,* 2005: Krishnareddy *et al.,* 2007), New Zealand (Pearson and Horner (1986) and Tuberose mild mosaic virus (TuMMV) from Taiwan have been reported earlier (Chen and Chang, 1998 and Chen *et al.,* 2002) and China (Lin *et al.,* 2004).

Tuberose Mild Mosaic potyvirus in India was detected molecularly by reverse-transcription polymerase chain reaction (RT-PCR) and serologically by Ouchterlony's double diffusion test from a tuberose plant with mild mosaic disease (Raj *et al.,* 2009).

12.2.0 Pests

12.2.1 Thrips

Thrips feed on leaves, flower stalk and flowers. They suck sap from these parts and ultimately damage the whole plant. Sometimes they are associated with a contagious disease known as bunchy top where the inflorescence is malformed (Rameshwar, 1976). Spraying of oxydemeton methyl 25 EC @2 ml/litre or dimethoate 30 EC @2 ml/litre or acephate 75 SP @1 g/litre along with 0.5 per cent pongamia oil at 15 days interval is recommended followed by drenching the soil with chlorpyriphos 20 EC @5 ml/litre.

12.2.2 Mealy Bugs

Mealy bugs occur on the crown of the plant and feed on the bulbs, which secretes white waxy material that covers their bodies. This insect may move down to below the ground and inhabit roots during the dey weather. This mealy bug is favoured by dry weather. There are many records referring to heavy attacks following periods of prolonged drought (Mani *et al.,* 2011). Hemalatha *et al.* (2018) found that the application of 5 ml/l crude toxin of *Xenorhabdus nematophilus* bacterial isolate MDUStBa15 isolated from the nematode *Steinernema carpocapsae*

Diseases of Tuberose

Stem Rot

Botrytis Spots and Blight

Leaf Spots

Leaf Blight

Anthracnose

Flower Bud Rot

Insect Pests of Tuberose

Aphids

Mealy Bugs

Grasshopper

Thrips Infestation

Red Spider Mite

Foliar Nematode

along with biocontrol agent *Cryptolaemus montrouzieri* registered the 90.55 per cent mortality in crawlers and 73.60 per cent mortality in adults of two tailed mealybug *Ferrisia virgata* at 7 days after spraying. Dipping the basal portion of the planting material in chlorpyriphos 0.05 per cent as a prophylactic measure and drenching of chlopyriphos 20 EC @ 5 ml/litre and Neem cake ground application at the time of planting can effectively control root mealy bugs. Followed by soil drenching with botanicals like nimbicidine (1 per cent) and neem seed kernel extract (NSKE 5 per cent) at monthly intervals. Soil drenching with the chemicals chlorpyriphos 20EC @ 5 ml/litre, thiamethoxam 25WG @ 0.5 g/litre at monthly intervals is recommended for controlling the root mealy bugs in the field.

12.2.3 Grasshopper

The insect feeds on young leaves and flower buds. Spraying the plants thoroughly at fortnightly interval with 0.1 per cent chlorpyriphos keeps away the pest (Sadhu and Bose, 1973). These can slo be controlled by spraying 2 ml/l of Malathion @ 0.1 per cent or Quinalphos @ 0.05 per cent.

12.2.4 Weevil

The adult weevils which become active in darkness damage the shoots and feed the leaves. Generally, they feed at the edge of the leaf producing a characteristic notched effect. The larvae feed on roots and tunnel into the bulbs. Larvae can be controlled by dressing the soil with 1.0 per cent chlorpyriphos before planting and working it into the top soil. *Scypophorus acupunctatus,* a black weevil is reported attacking *Polianthes tuberosa* in Morelos, Mexico with 51 per cent of infestation by Lavin *et al.* (2002).

12.2.5 Aphids

Aphids, feed on the flower buds and growing points. These may be controlled spraying NSKE 5 per cent @ 2.0 ml/l or 2 ml/l of Malathion @ 0.1 per cent or imidacloprid 17.8 SL @ 0.25 ml/l.

12.2.6 Red Spider Mites

Mites suck sap and yellow stripes and streaks appear on the foliage, which spread until the leaves become yellow, silvery or bronze and distorted. Spraying with dicofol 18.5 EC @ 2.5 ml/l or wettable sulphur @ 5 g/l or Profenophos 5 EC @ 2 ml/l has been found very effective to control the mites.

12.2.7 Nematodes

12.2.7.1 Greasy Streak or Foliar Nematode

This is caused by *Aphelencoides besseyi* and believed to be the first report of a foliar nematode on tuberose (Holtzmann, 1968). Khan (2004) evaluated six tuberose cultivars for foliar nematode resistance and found that Calcutta Single, Calcutta Double, Suarna Rekha, Rajat Rekha and Arka Shringar were susceptible and Arka Suvasini was immune to *A. besseyi*. Observation on population fluctuation of *A. besseyi* in tuberose Calcutta single in naturally infested field revealed that population of *A. besseyi* was dependent on relative humidity and atmospheric temperature.

12.2.7.2 Root Galling

It is caused by nematodes and a survey revealed the association of *Meloidogyne incognita* and *M. javanica* with *Polianthes tuberosa*. The plants showed yellowing, drying up of leaves and retarded growth. In severely affected plants the emergence of spikes was suppressed resulting in an absolute loss of flower (Rajendran and Muthukrishnan, 1980). Emergence of side shoots from bulbs were also affected and the numbers were conspicuously less (Jayaraman *et al.*, 1975). Studies on pathogenicity of four root knot nematode species showed that *Meloidogyne arenaria* and *M. acritata* were found to cause the greatest reductions in top and root weight, respectively (Johnson, 1970).

Sundara Babu and Vadivelu (1988) reported that *M. arenaria* caused the maximum reduction in length and weight of shoot and root of tuberose cv. Single, followed by *M. incognita* and *M. javanica* in pot experiments in which 500 nematodes were inoculated per plant. Damage to the plants was directly correlated with the initial inoculum level. Granular application of curbofuran or thimet at 1.5 kg a.i/ha successfully controls the endo- or ecto-parasitic nematode. Incorporation of marigold plant residues at 100 kg/ha (Akhtar and Alam, 1990) or chopped leaves at 40 and 80 g/kg soil) (Walia and Gupta, 1997) inhibits the population of *M. incognita*.

Gandhi *et al.* (2018) screened twenty seven genotypes of tuberose for their resistance to *Meloidogyne incognita*. Suarna Rekha recorded the least gall index and exhibited resistant reaction towards *M. incognita*. Histological characterization of select genotypes revealed active giant cells with dense cytoplasm and multi nucleated condition in RKN susceptible ones. Whereas in the tolerant ones, the giant cell formation was disrupted with higher number of phenolic cells and cells with lignified cell wall that prevented the successful establishment of nematode feeding site. The resistant and tolerant genotypes recorded significantly higher accumulation of phenols and defence enzymes *viz.*, polyphenol oxidase, peroxidase and phenyl alanine ammonia lyase compared to the susceptible genotypes.

Root knot nematode menace in tuberose are widespread and majority of the tuberose fields in North and South India are heavily damaged by this nematode (Rao *et al.*, 2001). They widely occur in subtropical and tropical regions and are reported to cause 10-14 per cent reduction in flower number and spike weight (Khan and Reddy, 1992).

According to Rao *et al.* (2015) treating tuberose bulbs before planting with bio-pesticides *Pseudomonas fluorescens* or *Trichoderma harzianum* @ 10 g/kg found to be effective. Field application at the time of planting with enriched farm yard manure (1 ton) or neem cake with two kg each of the bio-pesticides, *Trichoderma harzianum*, *Pseudomonas fluorescens* and *Paecilomyces lilacinus* under shade for 2 to 3 weeks at 25-30 per cent moisture was also found to be very effective. Periodical application of biopesticide enriched neem cake mixture in water at the ratio of 1: 10 through drip or as soil drench was also found to be very effective in controlling the root knot nematode in tuberose.

13.0 Harvesting and Yield

13.1.0 Harvesting of Flowers

Tuberose is harvested by cutting the spikes from the base for table decoration or the individual flower is picked from the spike for making garlands and other floral ornaments. Picking of flowers should be done in cool hours of the day either in the morning or evening. Harvesting of flowers, on the previous evening and marketing in the next morning leads to a weight loss of about 40 per cent (Rameswar, 1976). Four to five persons can harvest about 60 kg flowers in 2 to 2.5 hours.

The flowers spike for vases should be cut when the first pair of flowers fully open, with a sharp knife and placed immediately in water. The small clasping leaves on the flower stalk should be retained for a longer life of the flowers.

The effects of stem length and harvesting stage on vase life and cut flower quality of tuberose *cv.* Single were studied by De and Barman (1998a). Results indicated that stem length of 75 cm, harvesting when spikes had creamy white buds with one floret opened and using 8 per cent sucrose as the holding solution produced high quality, long-lasting cut flowers. Spike length of 70 cm and stages of harvest of tuberose spikes with 1-2 florets in *cv.* Arka Prajwal retained highest RWC (127.5 per cent) resulting in higher vase life (Khongwir *et al.*, 2015).

13.2.0 Yield of Flowers

Flower production varies with cultivars and depends upon bulb size at planting, time and density of planting and cultural practices adopted. In a 3-year experiment with tuberose *cv.* Single, Nambisan and Krishnan (1983) recorded a flower yield of 12,000 kg/ha, by using farmyard manure alone while fertilization with N and P increased the yield to 29,951 kg/ha. Yields of 4.8 – 9.6 tonnes in the plant crop, 8.9 to 12.1 tonnes in the first ratoon crop and 4.2 – 5.4 tonnes in the second ratoon crop per hectare per year depending on the size of bulbs were recorded in cultivar Single (Yadav *et al.*, 1984). Kumari (2014) registered the highest floret yield of 54.73 t/ha per year and 10.98 lakh spikes/ha per year in cv. Arka Prajwal.

13.3.0 Harvesting, Yield and Storage of Bulbs

Harvesting of tuberose bulbs at proper stage of maturity is important for storage of bulbs and their subsequent growth after planting The bulbs reach maturity when the flowering is over and plant growth ceases, depending on the environmental condition and soil moisture. At this stage, the old leaves become dry and bulbs are almost dormant. Irrigation is withheld and soil is allowed to dry before digging out the bulbs. The leaves are cut at the ground level and bulbs are taken out of the soil.

The bulb yield varies with cultivar, initial bulb size at planting and cultural practices adopted. Sharga (1977) obtained about 80 quintal bulbs per acre. Raja and Palanisamy (2000) found that dry and fresh weight of bulb increased to a peak in month 11. The number of bulblets produced increased with the development of bulbs. Investigations on cv. Single, revealed that bulbs, having a diameter between 2.6 and 3.0 cm, planted 6.0 cm deep at a distance of 20 x 20 cm produced 21.3 tonne bulb when harvested after a period of three years from planting (Yadav, 1983).

Nagaraja *et al.* (1998) found that the number of bulbs produced in tuberose increased with increasing bulb size (1- 4 cm diameter) and spacing (20 x 20, 30 x 20 or 30 x 30 cm). Large bulbs planted in June at 30 x 30 cm produced the highest bulb yield (109.78 g/plant). The large size bulbs of 2.1 to 3.0 cm diameter planted at wider spacing of 30 x 30 cm produced the maximum number of bulbs 8.97 per plant and bulblets 18.62 per plant in *cv.* Arka Prajwal (Kumari, 2014).

Watako and Ngamau (2013) studied the effects of subsequent warm temperature storage after low temperature treatment of tuberose bulbs on growth, sprouting and flower quality. Storage of tuberose bulbs at low temperatures (10°C and 5°C for three months) followed by warm subsequent storage for 2, 4 or 6 weeks at 20°C improved the sprouting and quality of flowers and also could enhance the availability of planting materials for crop production. Weight loss increased with 10°C pretratment and subsequent storage at 20°C for 6 weeks (12.61 per cent) and its recorded low in 5°C pretreatment and subsequent storage at 20°C for 2 weeks (1.87 per cent). Sprouting of bulbs was observed at 10°C pre-treatment but not at 5°C pre-treatment.

Toma *et al.* (2015) reported that the tuberose has specific temperature requirement for its growth and resting of bulbs. The harvested clump of bulbs were stored in two different temperatures 8-10°C and 20-22°C and subjected to macroscopic and microscopic observations. Atmospheric humidity was maintained as 55-60 per cent for the cold climate and 80-85 per cent for the warm climate. The observations revealed that the bulbs stored at 20-22°C of 80 -85 per cent humidity recorded slight increase in size of small bulbs and emergence of new bulblets and central bud is out of the state of latency, with a more advanced development. Cell divisions were noticed in the central bud and formation of the leaf primordia in the bulbs stored at 20-22°C of 80-85 per cent humidity. In contrary to that the bulbs stored at 8-10°C with 55-60 per cent humidity does not show and developments and the central bud is at latent phase.

14.0 Extraction of Concrete and Absolute

The tuberose concrete and absolute is commercially extracted by solvent extraction of fresh flowers and 1200 kg of buds provides 200 g of absolute, which makes tuberose one of the most expensive natural raw materials to use. Tuberose concrete is light to dark brown in colour and waxy solid, soluble in oil. Tuberose absolute is a powerful modifier in floral accords that blends especially well with other exotic scents. Tuberose concrete has 34 to 44 per cent of absolute and it is in solid state and soluble in oil. The absolute is a liquid with dark, red-amber colour, the specific gravity varies from 0.915 to 0.969 and soluble in oil. It has antiseptic and has healing properties that make it a holistic natural skin tonic and finest skin tincture. It also reported to have slightly spicy, heavy, sweet fragrance and narcotic properties (Elumalai *et al.*, 2010). The approximate cost of one kg of absolute is about Rs. 4.5 lakhs. Tuberose flower wax is also produced as byproduct which serves as an excellent skin care product.

The number and type of volatiles emitted were different according to altitude and temperature, thus the effect of altitude and temperature have to be taken into account when growing tuberose for the volatiles (Muriithi *et al.*, 2014). The concrete content varies according to the season and Srinivas *et al.* (1996) reported that the flowers harvested during October yielded a higher concrete content than those harvested during March. The yield of tuberose concrete varies with type of flower such as single types tuberose yielded a higher concrete (0.134–0.136 per cent) than the double types (0.107–0.108 per cent) (Srinivas *et al.*, 1996; Ranchana *et al.*, 2015; Chaudhary and Kumar, 2017) also stated that the yield of tuberose concrete was in the range of 0.11 to 0.16 per cent in single types and 0.06-0.09 per cent in double types.

The major chemical component detected in single type was 9-(4-Aminobenzo-18-crown-6) methyl anthracene and double type was 2-propoxyl-4-phenylthiophene. Gas chromatographic analysis showed the presence of Methyl iso-eugenol (40.37 per cent) and Indole (1.88 per cent) as major constituents in the absolute obtained from the cultivar of Single petalled florets Shringar, whereas Methyl Palmitate (24.46 per cent) was found to be the major component in the absolute extracted from Double petalled florets of cultivar Vaibhav (Singh *et al.*, 2009). Volatile compound profiling of tuberose cultivars Shringar, Hyderabad Single, Phule Rajani, Bidhan Ujjwal, GKTC-4, Prajwal, STR-505 and Bidhan Snigdha indicated that ethyl benzene, ethyl benzoate, methyl benzoate, indole, farnesene, *p*-xylene, benzyl benzoate, n-octadecane, benzyl salicylate, 1, 8- cineole were the major fractions besides minor fractions. Ahmadian *et al.* (2018) studies the variation in concrete content at four developmental stages of flowers. Fully opened flowers recorded the highest total amount of sesquiterpenoids and half-opened stage registered the highest monoterpenoids such as 1, 8-cineole in all methods of extraction. Unopened bud and nearly slight opened bud stages showed the least amount of volatile compounds. Chaudhary and Kumar (2017) reported that the time pf harvesting is also influences the yield of tuberose concrete, tuberose flower of *cv.* Arka Prajwal harvested in the morning gave maximum concrete and absolute recovery of 0.161 - 0.047 per cent. Major differences in essential oil constituents between control and genetically modified plants were reported by Elumalai *et al.* (2010). Further, gas chromatography study revealed four peaks in tuberose each of control and genetically modified plants, the highest peak percentage was observed in the peak number four of GM plants and peak number one of control plants.

15.0 Postharvest Management

15.1.0 Handling and Packaging of Flowers

Loose flowers are transported in polythene bags to the nearby wholesale market where they are sold by weight. The flower spikes are graded according to length of spike, length of flowering zone and quality of individual flower and then bunched in round bundles each having about 50-100 sticks. It is standard practice of Mexico to pack tuberose spikes while still in the field. For this purpose, a blanket is usually placed on the ground whereupon 12 rows of 12–15 spikes are alternately placed until a group of 100–120 spikes is collected (Perez-Arias *et al.*, 2019). Such groupings are then stacked and transported in pickup trucks to the retail locations of choice. For the most part, spikes are transported dry; however, they can also be placed in pulsing solutions (usually inside 20 L containers) and wrapped in newspaper if they are intended for export. The stem portion of the bundles is wrapped in wet newsprint sheets. To avoid damage of the flowers and buds, the whole bundle should be wrapped in soft, white tissue paper or polythene. These bundles may be packed in the card-board boxes for long distance transportation.

In the postharvest management of tuberose flowers, overpacking, rough handling, and inappropriate conditions during transport often cause mechanical damage due to compression and/or vibration (Pérez-Arias *et al.*, 2016). Exemplifying this are the results of experiments by Pérez- Arias *et al.* (2016) where the rates of respiration and of ethylene production, as well as the relative percentages of mass and water consumption, are calculated for tuberose spikes subjected for 2 h to pressures of 70, 120, and 210 Pa and compared to controls where no such pressures were applied. In this case, higher rates of respiration, ethylene production, abscission, and floral opening were observed in treated samples, demonstrating the negative effects of compression in the postharvest life of these flowers. It is generally recognized that proper packing is an important aspect to consider during the storage and transport of cut flower crops as it helps preserve product freshness, adds economic value, and reduces mechanical damage (Kumar *et al.*, 2010). In India, the effect of various packing materials on the postharvest life of tuberose spikes (cv. Double) was evaluated by Varu and Barad (2007, 2008). In their study, damage from handling was kept to a minimum when 50 µm polyethylene sheets or metallic paper were used as packing materials. Also minimized were weight loss and wilting, with higher rates of water uptake and minimal water loss both being responsible for the decreased presence of the latter symptom (*i.e.* the wilting). Flower color was additionally well-preserved, and when considered along with the aforementioned results, the postharvest life of spikes could be extended a further 3–4 days compared to unwrapped controls. Similarly, Kumar *et al.* (2010) reported longer postharvest lives and a greater number of open flowers when newspaper sheets were used to cushion the spikes of cv. 'Shringar'. If, on the other hand, polyethylene sheets replaced newspaper, greater flower diameters (at full bloom) and increased rates of water uptake were instead obtained. It was therefore, concluded that both the newspaper and polyethylene induced positive changes, either in terms of relative humidity (RH) or in terms of the relative concentrations of O_2 and CO_2, along with the expected protection against mechanical damage.

Singh *et al.* (2010) reported that tuberose spikes of cv. Arka Shringar wrapped in news paper stored for 24 hours at 4°C had maximum opened florets, per cent opened florets, and longest vase life of 7.00 days whereas the spikes wrapped in polythene sheet and stored for 24 hours at 4°C temperature resulted in enhanced floral diameter and water uptake at senescence. Shelf life of tuberose florets packed in polythene bags (200 gauge) and stored at 5°C temperature was enhanced up to 11th days with 17.57 per cent reduction in flower weight (Singh *et al.*, 2014).

Khongwir (2015) recommended polyethylene packaging comprising of 100µ with 0 per cent ventilation for tuberose cv. Mexican Single, Arka prajwal and Sikkim Selection which retained floret weight, relative water content, membrane stability index with increased shelf life (5 days), retained highest freshness index for market acceptance (Table 18).

Table 18: Effect of Packaging and Rate of Ventilation on Postharvest Parameters of Tuberose Flowers

Treatments	*Floret Diameter (mm)*	*Floret Weight (g)*	*Relative Water Content (per cent)*	*Membrane Stability Index (Per cent)*	*Total Sample Weight (g)*	*Physiological Loss in Weight (g)*	*Freshness Index*	*Shelf Life (Days)*
Control	3.80	0.54	55.01	60.76	120.93	95.90	1	3.0
50 micron PE @ 0 per cent ventilation	4.45	0.62	81.04	75.84	148.75	4.37	5	5.0
50 micron PE @ 1 per cent ventilation	4.42	0.64	63.96	70.97	136.20	20.96	3	4.0
50 micron PE @ 2 per cent ventilation	4.56	0.66	68.42	70.46	124.72	44.89	1	4.0
75 micron PE @ 1 per cent ventilation	4.49	0.65	72.98	74.96	169.98	4.54	5	5.0
75 micron PE @ 1 per cent ventilation	5.00	0.69	69.14	75.36	146.91	22.50	3	4.0
75 micron PE @ 2 per cent ventilation	4.64	0.67	68.01	73.70	134.58	40.18	1	4.0
100 micron PE @ 0 per cent ventilation	4.98	0.70	82.97	77.32	156.99	1.75	5	6.0
100 micron PE @ 1 per cent ventilation	4.10	0.67	76.07	73.40	144.07	14.01	3	5.0
100 micron PE @ 2 per cent ventilation	4.68	0.68	72.10	68.60	127.21	34.99	1	5.0
CD (5 per cent)	0.80	0.11	12.78	7.98	16.87	19.89	-	0.001
SE±m	0.40	0.05	6.34	3.96	8.45	9.96	-	0.001

PE: Polyethylene.

* Khongwir (2015).

15.2.0 Storage of Cut Flowers

Low temperature (10°C) storage was found to be effective in opening of florets in unopened spikes as well as in those with lower florets opened. With increase in the duration of storage at low temperature (10°C) florets opening was gradually reduced at all harvest stages of spikes. Low temperature storage, however, reduced the wilting of the florets than at room temperature (Chattopadhyay and Dhua, 2000). Spikes wrapped in cellophane stored at 4°C for 48 hours and held in a solution of 2 per cent sucrose and 200 ppm 8-hydroxyquinolinecitrate (HQC) showed the longest vase life (Katwate *et al.*, 1995).

The optimal temperature for dry storage of cut tuberose spikes was 0°C to 5°C although vase life and floret opening were markedly decreased by even short periods of storage at 2°C. Fresh-cut blooms had a vase life of more than 9 days, and 44 per cent of their buds opened during the vase-life period. Three days' storage halved floret opening, and reduced overall vase life by 30 per cent. The vase life of flowers has not shown any remarkable variation upon wet or dry storage. The cut spikes stored in cold storage had problems in flower opening such as partial flower opening, early abscission of tight or developing floret buds and drying of upper part of the spike (Reid, 1996). Precooling of Arka Prajwal buds at 4°C resulted in delayed wilting of fifty per cent of flowers (6.07 days), maximum colour retention and minimum physiological loss in weight during storage at room temperature (Kumari, 2014).

Roychowdhury *et al.* (2011) studied the influence of storage temperature, storage duration and subsequent physiological implications on tuberose cv. Calcutta Double. The study revealed that spikes wrapped with banana leaf exhibited maximum vase life with maximum floret opening and least amount of floret wilting. Further, spikes stored for 24 hours, especially at 10°C exhibited least floret wilting with better turgidity throughout the vase life. Retarded rate of respiration, transpiration, ethylene production and entire metabolism of the flower tissue as a whole was facilitated by low temperature during storage which exhibited by better performance of flowers in vase.

15.3.0 Pulsing

The term "pulsing" means briefly placing freshly harvested flowers in a specially formulated solution meant to increase their storage and vase life. Pulsing can thus last a few seconds to several hours (Reid, 2009). Different studies have tried to determine the effect of pulsing on the postharvest life of cut tuberose. For instance, Naidu and Reid (1989) report an increase in shelf-life when spikes are placed in 20 per cent sucrose after storage for 3 days at 2 °C. Discouragingly, however, the flowers opened more slowly than those of untreated controls.

Cut flower stalks of tuberose were pulsed with 5, 10, 15 or 20 per cent sucrose for 2, 4 or 6 h and then kept in tap water which was changed every day and the flower stalk shortened by 1.5 cm. Vase life was 7 days. Stalk fresh weight was generally highest in the tap water control. Pulsing for 2 h tended to increase flower opening (Gawade *et al.*, 1994).

Waithaka *et al.* (2001) also applied a solution of 20 per cent sucrose to the cut spikes of tuberose flowers in addition to a preservative solution consisting of 250 mg/l of hydroxyquinoline citrate (HQC) + 2 per cent sucrose was also used. In this case, a substantial increase in vase life was reported (10.5–11.3 days vs. 8.8 days in controls), along with a decrease in the negative effects of storage at low temperature. Pulsing also prolonged the duration of maximum weight in cut spikes, extended the period at which they exceeded their initial mass, and inhibited the abscission of young flowers (Waithaka *et al.*, 2001).

A solution of 10 per cent sucrose for 24 h also extends by 4 days the postharvest life of cut tuberose (Hutchinson *et al.*, 2003); this is in addition to a 13 per cent increase in floral opening with respect to controls. However, pulsing for 1 h with 20 mM silver thiosulfate (STS) yields better results as it extends postharvest life by 7 d. It also increases floral opening by 25 per cent compared to controls. Similar results are reported by Choudhury and Barooah (2011) when spikes are pulsed with STS for 1 h (5 days postharvest life; 35.2 per cent floral opening). However, they also report 5-8 mL/d rates of water uptake-values that are significantly lower than those of spikes treated with preservative solutions. Interestingly, application of 10 per cent sucrose directly onto flowers for 10 h reduces both weight loss and the percentage of damaged and wilted blossoms (Varu and Barad, 2007). Freshness and color also preserve much better. Similarly, a 6 h application of 20 per cent sucrose to both flowers and spikes extends postharvest life and the duration of floral opening compared to untreated controls (Jeenbuntug *et al.*, 2007). In addition, the application of sucrose significantly diminishes ethylene production and the rate of respiration, both of which extend the vase life of cut flowers (Jeenbuntug *et al.*, 2007). These results were replicated by Pérez-Arias *et al.* (2015) when they pulsed the spikes of cv. 'Perla' with 20 per cent sucrose for 24 h. However in their case, higher rates of water uptake, an increase in fresh weight, and an adequate appearance for 6 d, were simultaneously observed.

Uthairatanakij *et al.* (2007) investigated the physiological changes of cut tuberose flower pulsed with sucrose and various concentrations of thidiazuron (TDZ). The result showed that the cut spikes pulsed with 20 per cent sucrose combined with TDZ significantly induced percentage of bud opening and vase life. Pulsing with sodium thiosulphate for 1 hr found to increase the vase life of tuberose cv. Calcutta Double (Talukdar and Barooah, 2011). Prestorage pulsing with 20 per cent sucrose and wet storage in 400 ppm citric acid increased the vase life of cut spikes (Kumari, 2014).

As a general rule, application of 10 per cent or 20 per cent sucrose increases storage life in tuberose flowers. For this reason, sucrose is proposed to act either as a respiratory substrate or as an osmolyte (Hutchinson *et al.*, 2003). As an osmolyte, added sucrose would improve water balance by increasing its concentration in petal tissues, thus helping to preserve the integrity of both mitochondrial and plasma membranes. This would then maintain tissue turgor and delay wilting (Hutchinson *et al.*, 2003). The use of silver compounds also has had positive effects on the vase life of spikes (Hutchinson *et al.*, 2003; Bahrehmand *et al.*, 2014).

15.4.0 Vase Life

Investigations have been carried out to prolong the vase life of tuberose spikes.

15.4.1 Effect of Chemicals

Preservative solutions typically consist of a bactericide and a dissolved sugar. The bactericide helps keep the water free of germs, while the dissolved sugar (usually sucrose or fructose at a low concentration) provides a readily-available energy source that the flowers can use to drive metabolic processes. Hydrating solutions, on the other hand, help improve water uptake by drastically lowering pH (generally to values between 3.0 and 3.5).

A preservative solution containing benzimidazole, SADH, ascorbic acid, sucrose and 8-HQC in appropriate proportion was found useful for improving the opening of flowers on tuberose spikes and enhancing the vase life of flowers (Pathak, 1981). Vase solution containing sucrose (1 or 2 per cent) and aluminium sulphate (200 or 400 ppm) increased the freshness of flowers (Gowda, 1990). Naidu and Reid (1989) found that a sugar-containing vase preservative (1.5 per cent) and/or pre-treatment with sugar (20 per cent sucrose for 15-20 h) improved display life of the stems before or after storage. The presence of open flowers on the spike increased solution uptake and the response to sucrose pulse. Solution containing sulphate or chloride of Al, Ca, Mg and Co at various concentration increased vase life, the longest was observed in 1.0 mM Al (Balakrishna *et al.*, 1989). Aluminium sulfate in combination with sucrose significantly enhanced the vase life and quality of spikes by increasing the water uptake and maintaining a better water balance and a higher fresh weight for longer period. Optimum concentrations for the combined treatment were 0.50 mM aluminium sulfate and 4 per cent sucrose (Reddy and Singh, 1996). Saini *et al.* (1994) reported that solutions of silver nitrate, aluminium sulphate or sucrose increased vase life of tuberose flowers cv. Single compared with distilled water (controls).

Among the chemicals used in the holding solution (sucrose, B-9, silver nitrate, copper sulphate, boric and MH and aluminium sulphate at 0.2 per cent and 0.4 per cent, Das and Barman (1990) recorded longest vase life of tuberose (12.6 days) in 0.4 per cent sucrose solution. All treatments increased vase life compared with the control (6.0 days). The highest percentage of fully opened flowers (79.2 per cent) was obtained in 0.4 per cent MH solution compared with 58.9 per cent in the control. De and Barman (1998b) also reported that boric acid (250 ppm), $Al_2(SO_4)_3$ (50 ppm), $CaCl_2$ (1000 ppm), $AgNO_3$ (50 ppm), $CaCl_2$ (25 ppm) and citric acid (400 ppm) proved effective in floret opening and increased flower diameter and vase life. The vase life of tuberose was best with 1 per cent NaCl, which gave a significantly better flower colour and freshness together with stem colour and firmness (Harode *et al.*, 1993).

BA at 100 ppm in vase water delayed flower senescence (Bhaskar and Rao, 1998). Reddy *et al.* (1997c) observed that spikes kept in a solution containing 4 per cent sucrose + 2 mM 8-HQS had the longest vase life (17 days). In another experiment, Reddy *et al.* (1997d) reported the highest number of opened and fully opened florets using 4 per cent sucrose + cobalt, 8-HQS or nickel + 8-HQS or citric acid + 8-HQS treatments. The highest vase life (12.33 days) was observed in 4 per cent sucrose

Harvest and Postharvest Management of Tuberose

Harvesting Stage

Pre-cooling

Harvesting of Bulbs

Cut Flowers

Loose Flowers

Tuberose Jewellery

Tuberose Veni

Garland

Home Decoration

Packaging and Marketing of Tuberose

With Polythene

Packing from Field

Tuberose Garland in Local Market

Cut Flower in Local Market

Garland in Local Market

+ 8-HQS + nickel. Vase life was shortest (6.5 days) for flowers in distilled water. Solution containing HQS (400 ppm) + citric acid (100 ppm) + 4 per cent sucrose also increased the vase life (Reddy *et al.*, 1995). Cobalt along with sucrose enhanced the vase life of tuberose spikes by increasing the water uptake, maintaining better water balance and by maintaining higher fresh weight throughout the vase period. The vase life was extended from 5 days in distilled water to 12.7 days in solution containing 1mM cobalt sulfate and 4 per cent sugar (Reddy *et al.*, 1996). Maximum vase life (16.3 days) was recorded for spikes held in a vase solution containing 4 per cent sucrose + 1 mM nickel, which was significantly superior to all other treatments (Reddy *et al.*, 1997b).

Kumari (2014) stated that the flower buds of tuberose cv. Arka Prajwal soaked in 4 per cent Boric acid for two hours and air dried before packing increased the shelf life up to 6 days (Table 19). Epicuticular wax content of flower petals was found to influence the vase life and postharvest water loss of tuberose flowers. Maximum vase life of 8 days was recorded in double type cultivar Arka Vaibhav which was estimated to have higher (73.52µg/g) epicuticular wax, whereas it was minimum of 4 days with single type cultivar Arka Sugandhi with minimum epicuticular wax content (37.90 µg/g) in the florets.

Table 19: Epicuticular Wax (µg/g) Content, Vase Life (Days) and Water Loss (g) of Tuberose Genotypes*

Genotypes	*Epicuticular Wax (µg/g)*	*Vase Life (Days)*	*Water Loss (g)*
Arka Vaibhav	73.52	8	0.34
Arka Suvasini	51.82	7	1.11
Arka Shringar	54.11	7	0.94
Arka Prajwal	68.37	7	0.76
Arka Nirantara	67.87	7	0.34
Arka Sugandhi	37.90	4	1.65
Phule Rajani	42.54	5	1.17
Hyderabad Single	53.61	6	1.20
Mexican Single	50.58	5	1.28
GK-TC-4	64.35	5	0.87

* Kumari (2014).

Some unusual ingredients for preservative solutions have also been reported. For instance, ethanolic extracts of Piper betle leaves (Piper betle leaf extract or PBLE) are reported to extend the postharvest life of cut tuberose to 10.8 d; this is in stark contrast to controls, which last only 6.6 days (Raj-Maity *et al.*, 2014). Greater freshness and higher rates of water uptake are also observed, which combined lead to a greater content of water. PBLE-treated spikes are also purported to have higher contents of soluble proteins and of chlorophyll, as well as higher activities of the enzymes catalase and superoxide dismutase. These beneficial effects are attributed to higher amounts of sugars in the tuberose petals, which would help improve the balance of water in the spikes. Such natural preservatives are important alternatives to the use of other chemicals that could have negative effects on the environment.

15.5.0 Physiology for Flower Senescence

Changes in ethylene production and respiratory rate of the florets during their different stages of development including senescence were studied by Chattopadhyay and Dhua (2000). Ethylene production by florets continued in all stages of development leading to full opening and senescence. While ethylene production attained a peak level at full-open stage of florets, the respiration rate was low at bud stage, increased markedly to a peak before full opening and declined thereafter. The respiratory rate, however, increased again after full opening *i.e.*, before senescence, but at a lower magnitude than the initial increase and the second hike in respiration was considered to be the onset of senescence phase (Table 20).

Table 20: Changes in Respiration and Ethylene Production at different Stages of Floret Development*

Stage of Floret	*Respiration mg CO_2/kg/h*	*Ethylene Production nM/kg/h*
S1 Tight bud stage	193.11	14.5
S2 Floret at the time of opening	207.53	14.78
S3 Fully open floret	195.07	22.56
S4 Senescing floret	200.44	44.33

Prevailing room temperature and relative humidity (RH per cent) in laboratory during the experiment.

Respiration : Max. temp. 30°C, Min. temp. 29°C, RH 73 per cent.

Ethylene production: Max. temp. 27°C, Min. temp. 22.5°C, RH 60 per cent.

* Chattopadhyay and Dhua (2000).

Citric acid prepared with sterilized distilled water had a desirable effect on the vase life of cut tuberose flowers. This effect increased with the increment of the acid up to 450 mg/l (Jowkar and salehi, 2005). Citric acid, especially in higher concentrations help to increase the turgidity and the number of opened florets compared to other treatments. This may be due to reducing the solution viscosity and the microorganism growth (Reddy *et al.*, 1995; van Doorn and Peirik, 1990) via lowering the pH of vase solution. Nasibi *et al.* (2014) stated that tuberose flowers placed in the vase solution containing arginine, cysteine and 5-sulfosalicylic acid recorded lower polyphenol oxidase (PPO) activity, higher phenylalanine ammonialyase (PAL) activity, prolonged membrane integrity and increased vase life The extended vase life in treated flowers was associated with increase in flower opening and decrease in flower abscission and lipid peroxidation.

16.0 Biotechnology

16.1.0 Micropropagation

Tuberose can be propagated by seed as well as vegetatively by bulbs and bulblets. However, the requirement of planting material is more in this method because only one plant can be obtained from a bulb and this method is rather slow. To meet the growing demand, massive *in vitro* propagation through tissue culture is the only option. Furthermore, to increase the oil content, flower colour, quality and increased vase life of tuberose, modern molecular biology and genetic engineering techniques need to be exploited fully for tailoring transgenic plants. For achieving per se goals, an efficient and reproductive regeneration system from explant cultures must be available. In tuberose, various explants have been tried to produce regenerable cultures via *in vitro* morphogenesis are shoot tips (Hutchinson *et al.*, 2004), bulb scale (Muralidhar and Mehta,1982; Bose *et al.*, 1987; Khan *et al.*, 2000; Rajasekaran *et al.*, 2000; Nazneen *et al.*, 2003; Mishra *et al.*, 2006), rhizome (Sangavai and Chellapandi, 2008), leaf disc (Bindhani *et al.*, 2004), root (Narayanaswamy and Prabhudesai, 1979), stem disc (Gajbhiye *et al.*, 2011), flower bud (Hernández-Mendoza *et al.*, 2014) and anther (Gi and Tsay, 1989). However, the regeneration frequencies in all above experiments were found considerably moderate to high.

The recent study of Khanchana *et al.* (2019) led to the inference that micropropagation of tuberose (both single and double types) using axillary buds as explants could be a viable and rapid means of producing disease free true-to-type plants. It has been observed that around 425 plants can be obtained from one mother plant in a time period of approximately seven months. This multiplication ratio is 17 times higher than that which can be obtained through conventional propagation using bulbs.

16.2.0 Molecular Markers

Morphological traits have long been used to estimate systematic relationships in crops and ornamentals. Though, simple and irreplaceable, these descriptors suffer from many drawbacks and may be influenced of environment variation. Therefore, molecular based genetic diversity is extremely useful for more accuracy, intellectual property protection particularly in the determination of essential derivation. These markers are highly polymorphic and have been successfully used in many bulbous flowering crops. Among the available molecular marker techniques, RAPD is widely exploited by horticulturists as well as floriculturists, because results obtained from this technique are quickly and fairly inexpensive (Marrow, 2005; Chen *et al.*, 2004). RAPD has been used successfully for genotype identification in several ornamental plants including Lilium species, alstromeria, heliconia, gladiolus, and roses. Only few studies are available with regard to diversity and genetic relationship in tuberose using DNA markers (Sarkar *et al.*, 2010; Kameswari *et al.*, 2014; Khandagale *et al.*, 2014; Bharti *et al.*, 2016; Sirohi *et al.*, 2017).

For identification of simple sequence repeats (SSRs), the tuberose transcripts were searched with perl script MISA and a total of 1,656 SSRs in 1,280 transcripts with frequency of one SSR per 9.33 kb of the sequence were identified (Madhavan *et al.*, 2019). In 7876 EST sequence of tuberose, they have identified 87 copies of miRNA. Theses SSR markers can be used for genetic diversity analysis of tuberose and can be used to select tuberose varieties with required traits. The study will help in better understanding of the possible role of the transcribed microsatellites and miRNA. Further the data will provide deeper knowledge of genetic mechanisms, evolutionary relationships between tuberose and other plants.

16.3.0 Transcriptome Analysis

The floral scent volatiles of *Polianthes tuberosa* is mainly consist of terpenes and benzenoids that emit a charming fragrance. However, the molecular mechanism responsible for the floral scent of *P. tuberosa* is limited. *De novo* transcriptome analysis has become a useful tool in the discovery of genes involved in various metabolic pathways, including the determination of sequence and expression patterns without the need for a reference genome. Using transcriptome sequencing and de novo assembly, a total of 228,706,703 high quality reads were obtained (Fan *et al.*, 2018), which resulted in the identification of 96,906 unigenes (SRA Accession Number SRP126470, TSA Acc. No. GGEA00000000). Approximately 41.85 per cent of the unigenes were functionally annotated using public databases. A total of 4,694 differentially expressed genes (DEGs) were discovered during flowering. Gas chromatography-mass spectrometry analysis revealed that the majority of the volatiles comprised benzenoids and small amounts of terpenes. Homology analysis identified 13 and 17 candidate genes associated with terpene and benzenoid biosynthesis, respectively. Among these, PtTPS1, PtDAHPSs, PtPAL1, and PtBCMT2 might play important roles in regulating the formation of floral volatiles. The data generated by transcriptome sequencing provide a critical resource for exploring concrete characteristics as well as for supporting functional genomics studies. The study also laid the foundation for the elucidation of the molecular mechanism underlying the regulation of floral scents in monocots.

17.0 References

Abdullah, S.B. (2012) *M.Sc Thesis*, Institute of Biological Sciences, University of Malaya, Kualalumber.

Abhangrao, A.K., Yadlod, S.S. and Ghormade, G.N. (2019) *Int. J. of Chem. Stud.*, **7**(4) : 11-15.

Abraham, V. and Desai, B.M. (1976a) *Indian J. Genet. and Plant Breed.*, **36** : 230-337.

Abraham, V. and Desai, B.M. (1976b) *Indian J. Genet. and Plant Breed.*, **36** : 328-331.

Ahmed, I., Ahmad, T., Asif, M., Saleem, M. and Akram, A. (2009) *Sarhad J. Agric.*, **25**(3) : 391-397.

Ailincai, N. (1960) *Lucr. Sti. Inst. Agron. Iasi.*, 355-360.

Akand, M.S.H., Patwary, N.H., Sultana, Z., Khatun, M.M. and Amin, R. (2016) *Int. J. of Appl. Res.*, **2**(3) : 103-107.

Akhtar, M. and Alam, M. M. (1990) *Pak. J. Nemato.*, **8** : 25-28.

Alpeshkumar, D.S., Patel, B., Naik, P.R. and Patel, V. (2018) *Int. J. Chem. Stud.*, **6**(6) : 1076-1079.

Amin, M.R., Sultana, S., Hossain, M.I. and Sarkar, M.D. (2017) *Int. J. Bio-resource Stress Manag.*, **8**(4) : 553-555.

Anjum, A.K., Naveed, F., Shakeel, F. and Amin, S. (2001) *J. Res.*, **12**(1) : 01-07.

Anonymous. (2016) National Horticulture Board, *Indian Horticulture Database*. Ministry of Agriculture, Government of India.

Anonymus. (2019) *Technical Bulletin* No.04/2019. ICAR-IIHR, Bengaluru.

APG III. Angiosperm Phylogeny Group. (2009) *Bot. J. Linn. Soc.*, **161**: 105–121.

Arias, G.A.P., Tejacal, I.A., León, M.T.C., Aispuro, M.J.S. and Vargas, J.E.A. (2015) *Acta Agrícola Y Pecuaria*, **1**(1) : 29-36.

Ark, P.A. and Maclean, N.A. (1951) *Plant Dis. Reptr.*, **36** : 45-46.

Asgari, M., Azimi, M. H., Hamzehi, Z., Mortazavi, S. N. and Khodabandelu, F. (2013) *J. Agron. Plant Prod.*, **4**(4) : 680-387.

Ashok, S.H., Sujatha, K. and Prabhakar, M. (1995) *Indian J. Hortic.*, **52**(1) : 74-76.

Ayangar, R.K. (1963). *Proc. 50th Indian Sci. cong. Part III section Vi (Bot.) Abst. No.* **247** : 444-445.

Bahadoran, M. and Salehi, H. (2015) *J. Agr. Sci. Tech.*, **17** : 415-426.

Bailey, L.H. (1939) *The Standard Cyclopedia of Horticulture*, The Macmillan Co., N.Y.

Balakrishna, H.V., Reddy, T.V. and Rai, B.G.M. (1989) *Mysore J. Agric. Sci.*, **23** : 344-348.

Banik, M., Ghatak, P., Ray, R. and Patra, S.K. (2018) *Int. J. Curr. Microbiol. App. Sci.*, **7**(04) : 922-930.

Bankar, G.J. (1991) *Prog. Hort.*, **23** : 119-121.

Bankar, G.J. and Mukhopadhyay, A. (1990) *Indian J. Hort.*, **47** : 120-126.

Banker, G.J. (1988) *Prog. Hort.*, **20** : 49-52.

Barman, A., Abdullah, A.H.M., Hossen, A., Asrafuzzaman, M. and Rahman, K.M.H. (2015) *Int. J. Res. Rev.*, **2**(6) : 301-306.

Baruah, N., Talukdar, M. C., Pradip, Mahanta., Baruah, N. and Mahanta, P. (1998) *Indian J. Agric. Sci.*, **68** : 629.

Basaldua, A.A.E., Santos, M.E.P., Torres, E.C., Palacios, A.M., Romero, C.S. and García, J.L.M. (2011) *Rev. Mex. Cienc. Agríc. Pub. Esp.*, **31**: 445-458.

Basu, T.K. and Bose, T.K. (1980) *Sci. and Cult.*, **46** : 323-325.

Benschop, M. (1993) *The Physiology of Flower Bulbs*, Elsvier, Amsterdam, pp. 589-602.

Bharathi, D.A.R. and Narayanaswamy, H. (2018) *J. Pharmacogn. Phytochem.*, **7**(3) : 1282-1283.

Bharathi, T.U and Kirthishree, S.P. (2019) *Int. J. of Chem. Stud.*, **7**(1) : 189-193.

Bharathi, T.U., Srinivas, M., Nair, S. A. and Dhanajaya, M. V. (2016) *Kisan World*, 35-36.

Bharti, H., Singh, K. P., Singh, R., Kumar, R. and Singh, M.C. (2016) *Indian J. Hort.*, **73**: 238-244.

Bhaskar, V.V., Rao, P.V. and Reddy, Y.N. (2000) *J. Ornam. Hortic.*, **3** : 6-11.

Bhatia, J.S. and Srinivasan, M. (1954) *J. Sci. Industr. Res.*, **13** : 373.

Bhattacharjee, S.K., Yadav, L.P. and Mukhopadhyay, T. (1979) *The Lal Baugh*, **24** : 24-29.

Bhujbal, B.G. and Whaval (1978) *Indian J. Ornam. Hort.*, **2** : 27-28.

Bhuyan, B., Paswan, L. and Mahanta, P. (1996) *J. Agric. Sci. Soc. North-East-India*, **9** : 119-122.

Bindhani, B.K., Dalal, A.K. and Behara, B. (2004) *Indian J. Genet.*, **64**:173-174.

Biswas, J., Bose, T.K. and Maiti, R.G. (1983) *South Indian Hort.*, **31** : 129-132.

Borah, S.K. (1993) *South Indian Hort.*, **41** : 162-165.

Bose, T.K. and Jana, B.K. (1987) *Indian J. Hort.*, **42** : 100-101.

Bose, T.K., Jana, B.K. and Moulik, S. (1987) *Indian J. Hort.*, **44**: 100-101.

Bose, T.K., Maiti, R.S., Dhua, R.S and Das, P. (1999) *Floriculture and Landscaping*, Naya Prokash Calcutta.

Bundrant, L.A. (1985) *Herbertia*, **41** : 55-60.

Castro, A.C., Gallegos, J.G.G. and Rodriguez, A. (2015) *Phytotaxa*, **201**(2) : 140-148.

Chang, C.A and Chen, C.C. (1999) *Plant Path. Bull.*, **8** : 89-94.

Chang, C.A. and Chen, C. C. (1998) *Plant Disease*, **82** : 45-49.

Chang, Y. P., Ding, S. F., Chou, C. C., Du, B. S. and Chen, W. S. (1998) *Bot. Bull. Acad. Sin.*, **39** : 199-203.

Chattapadhyay, M., Pal, P. and Dhua, R.S. (2000) *Nat. Sym. Comm. Flor. in India*, B.C.K.V. West Bengal, 54-63.

Chaudhary, V. and Kumar, M. (2017) *Int. J. of Chemical Studies*, **5**(4): 1416-1420.

Chen, C.C, Hsiang, T., Chiang, F.L. and Chang, C.A. (2002) *Bot. Bull. Acad. Sin. (Taiwan)* **43**: 13–20.

Chen, C.C. and Chang, C.A. (1998). *Plant Disease*, **82**: 45–49.

Chen, J., Devanand, P. S., Norman, D. J., Henny, R. J. and Chao, C. T. (2004) *Ann. Bot.*, **93**: 157–166.

Chiapparini, L. (1960) *Not. Mal. Piante*, No. **53**, **54**, pp. 155-173.

Choudhury, S., Amin, M. R., Uddin, A. K. M. and Uddain, J. (2010) *Int. J. Bio-resource Stress Manag.*, **1**(3): 180-183.

Christensson, H. (1982) *Proc. 23rd Swedish Weed Conf.*, Upsala, **1** : 58-67.

Cirrito, M. (1975) *Annali dell'Istituto Sperimentale per la Floricoltura*, **6** : 27-43.

Cirrito, M. and Vita, M. de (1981) *Annali dell'Istituto Sperimentale per la Floricoltura*, **12** : 41-51.

Cirrito, M. and Zizzo, G.V. (1980) *Annali dell'Istituto Sperimentale per la Floricoltura*, **11** : 149-168.

Cirrito, M. vita, M. de and Zizzo, G.V. (1981) *Annali dell'Istituto Sperimentale per la Floricoltura*, **12** : 53-65.

Cirrito, M., Vita, M. de and Zizzo, G.V. (1980) *Annali dell'Istituto Sperimentale per la Floricoltura*, **11** : 123-148.

Cirrito, M., Zizzo, G.V. and Vita, M. de (1982) *Annali dell'Istituto Sperimentale per la Floricoltura*, **13** : 33-34.

Curir, P., Fusconi, A.S., Sulis, S. and Parisi, L. (1984) *Annali dell'Istituto Sperimentale per la Floricoltura*, **15** : 73-88.

Das, A.C. (1961) *Sci. and Cult.*, **27** : 549.

Das, S.N. and Barman, D. (1990) *Punjab Hort. J.*, **30** : 203-205.

Datta, S. K. (2017) *Curr. Sci.*, **113** : 1255–1263.

Datta, S.K., Misra, P., Mandal, A.K.A. and Chakrabarty, D. (2013) *Israel J. of Plant Sci.*, **50** : 287–291.

De, L. C. and Barman, D. (1998) *J. Ornam. Hortic.*,**1** : 66-68.

De, L.C. and Barman, D. (1998) *Orissa J. Hort.*, **26** : 66-69.

Deka, M. and Talukdar, M.C. (2017) *Res. on crops*, **18**(1) : 129-132.

Desai, B.L. (1957) *Indian Farming*, **7** : 10.

Dhua, R.S., Ghosh, S.K., Mitra, S.K., Yadav, L.P. and Bose, T.K. (1987) *Acta Hortic.*, **205** : 21-28.

Dhumal, S.S., Kaur, M., Dalave, P., Garande, V.K., Pawar, R.D. and Ambad, S.S. (2018) *Int. J. Curr. Microbiol. App. Sci.*,**7**(9) : 1622-1626.

Ding, S.F., Chen, W.S., Sul, C.L., Du, B.S., Twitchin, B. and Bhaskar, V.K. (1999) *Plant Physiol. Biochem.*, **37** (2) : 161-165.

DOI: 10.17660/ActaHortic.2005.669.54

Durgadevi D. and Sankaralingam A. (2012) *New Dis. Rep.*, **26** : 5.

Durgadevi D., Srivignesh, S. and Sankaralingam A. (2018) *Int. J. Bio-resource Stress Manag.*, **9**(4) : 510-517.

Dutta, A.K. (1975) *Sci. and Cult.*, **41** : 424.

Eguiarte, L.E., Souza, V. and Silva-Montellano, A. (2000) *Boletín de la Sociedad Botánica de México*, **66** : 131–150.

El-Roumi, M.K. (1981) *Agric. Res. Rev.*, **59** : 305-312.

Elumalai, S., Murugesan, R. and Chinnasamy, N. (2010) *Biomed. Pharmacol. J.*, **3**(1) : 63-72.

Fan, Ronghui, Chen, Yiquan, Ye, Xiuxian, Wu, Jianshe, Lin, Bing, Zhong, Huaiqin (2018) *PLoS ONE*, **13**: e0199261. https://doi.org/10.1371/journal. pone.0199261

Ferria, A.T.P., Solano, E. and Mendoza, A.G. (2010) *Acta Botanica Mexicana*, **92** : 11-28.

Gajbhiye, S.S., Tripathi, M.K., Vidya Shankar, M., Singh, M., Baghel, B.S. and Tiwari, S. (2011) *J. of Agri. Tech.*, **7**(3) : 695-709.

Gandhi, D.P., Bharathi, T.U., Umamaheswari, R., Kalaivanan, D. and Prathiba, S. (2018). *J. Env. Biol.*, **40** : 1151-1158.

Ganesh, S., Soorianathasundaram, K. and Kannan, M. (2013) *Asian J. Hort.*, **8**(2) : 696-700.

Gawade, B.J., Patil, M.T., Ranpise, S.A. and Jadhav, M.S. (1994) *J. Maharashtra Agri. Univ.*, **19** : 470-471.

Gawai, Y.R., Bayaskar, S., Tayade, M. and Davhale, P.N. (2020) *Ind. J. Pure App. Biosci.*, **8**(1) : 179-182.

Ghosh, P. and Pal, P. (1999) *Indian Perfumer*, **43** : 202-206

Gi, H.S. and Tsay, H.S. (1989) *J. Agric. Res., China*, **38** : 346-352.

Gonzalez, B.G, A., Sanchez, E. E., Fernandez, H. J., Arias, S.B and Gonzalez, I. M. (1995) *Horti. Revista de Hortalizas, Flores-y-Plantas orn.*, **107**(13) : 15-16, 18-19.

Gonzalez, R.B., Dominguez, J.M.R., Saucedo, Cruz, A.D.L.C., Banuelos, T.Y.L. and Campos, E.T. (2013) *In Proc. VIIth IS on New Floricultural Crops, Acta Hortic.*, **1000** : 505-510.

Gonzalez, R.B., Dominguez, J.M.R., Saucedo, M.C.C., Rodriguez, A., Van Tuyl, J.M and Campos, E.T. (2012) *Flori. and Ornam. Biotech.*, **6**(1) : 122-128.

Gowda, J.V.N. (1990) *Curr. Res. Univ. Agric. Sci.*, Bangalore, **19** : 14-16.

Gowda, J.V.N., Jacob, S. and Huddar, A.G. (1991) *Indian Perfumer*, **35** : 100-101.

Guenther, E. (1952) *The Essential Oils*, Vol. I-IV, Van Nostrand Co. New York.

Gupta, M.N., Laxmi, V. and Data, S.K. (1984) *J. Nuclear Agric. Biol.*, **13** : 140-141.

Gupta, M.N., Sumiran, R.and Shukla, R. (1974) In : Symposium, Use of radiations and radioisotopes in studies of plant productivity, Pantnagar : 169-179.

Gurjar, D. S., Kaur, R. and Singh, K.P. (2018) *Indian J. of Agric. Sci.*, **88**(12) : 1893–1896.

Gutierrez, A.G.G. and Garay, B.R. (2016) *ÿþSpringerPlus*, **5** : 1804-1817.

Hakim, S. El-Gamassy, A. and Shedid, M. R. (1958) *Ann. Agric. Sci.Cairo*, **3** : 101-113.

Harode, S.M., Kotave, V.G. and Zend, J.P. (1993) *J. Maharashtra Agric. Univ.*, **18** : 448-450.

Hartsema, A. M. (1961) In : *Encyclopedia of Plant Physiology* (ed. W. Ruhland), Springer-Verlag, Berlin.

Hassan, A. M. and Agina, E. A. (1980) *Ann. Agric. Sci.*, **13** : 145-152.

Hemalatha, D., Prabhu, S., Rani, W.B. and Anandham, R. (2018) *Toxicon*, **146** : 42-49.

Hemanta, L. (2015). *Ph.D. Thesis*, GB Pant Univ. Agri. Technol., Pantnagar, India.

Hemanta, L., Srivastava, R., Kumar, A. and Chand, S. (2016) *J. Agri. Technol.*, **3**(1) : 71.

Hernández-Mendoza, Fanny, Carrillo-Castañeda, Guillermo, Pedraza-Santos, Martha E., Eulogio de la Cruz-Torres, Mendoza-Castillo Ma. del Carmen (2014) *Revista ElectrónicaNova Scientia*, **7:** 32-47.

Hodges L. (2010) *The cut flower*, **22** : 55-59.

Holtzmann, O. V. (1968) *Plant Dis. Reptr.*, **52** : 56.

Howard, T. M. (1977) *Plant Life*, **33** : 82-84.

Howard, T. M. (1978) *Plant Life*, **34** : 126-129.

Howard, T. M. (1985) *Herbertia*, **41** : 98-117.

Howard, T. M. (1986) *Herbertia*, **42** : 84-86.

Huang, C.K., Chang, B.S., Wang, K.C., Her, S.J., Chen, T.W., Chen, Y.A., Cho, C.L., Liao, L.J., Huang, K.L., Chen, W.S. and Liu, Z.H. (2004) *J. Plant Physiol.*, **161**(6) : 709–713.

Huang, K., L. I. H.O. Miyajima, T.M. Shen and. T.S. Huang. (2001) *Sci. Hortic.*, **88** (3,4) : 235-241.

Huang, K.L., Miyajima, I., Okubo, H., Shen, T.M. and Huang, T.S. (2002) *Acta Hortci.*, **570** : 367-371.

Huang, K.L.and Okubo, L. (1995) *J. Fat. Agr., Kyushu Univ.*, **39**(3 and 4) : 115-124.

Hussain, A. (1986) In : *Ornamental Horticulture in India*, ICAR, New Delhi.

Hutchinson, J. (1960) *The Families of Flowering Plants*, Clarendon Pres, Oxford.

Hutchinson, M.J., Onamu, R. and Obukosia, S. (2004) *JAGST*, **6**:48-59.

Indhumathi, K., Shanmugam, P. S. and Sangeetha, M. (2020) *Asian J. Agric. Ext., Econ. and Sociol.*, **38**(8) : 78-86.

Irulappan, I., Doraipandian, A. and Muthuswami, S. (1980) *Nat. Sem. Prodn. Tech. Comm. Flower Crops*, Tamil Nadu Agricultural University, Coimbatore, pp. 69-70.

Jagtap, P.G., Chalakand, S.U. and Nale, S.S. (2017) *Trends in Biosci.*, **10**(13) : 2395-2413.

Jain, R., Janakiram, T., Das, T.K. and Kumawat, G.L. (2015) *Current Hortic.*, **3**(1) : 57-60.

Jana, B. K. and Biswas, S. (1982) *South Indian Hortic.*, **30** : 62-65.

Jana, B. K., Roy, S. S. and Bose, T. K. (1974) *Indian J. Hortic.*, **31** : 182-185.

Jayaraman, V., Rajendran, G. and Muthukrishnan, T. S. (1975) *Indian J. Nematol*, **5** : 101-102.

Jayasree, P. L., Sreenivas, M. and Krishnan, R. (2001) *Nucleus*, **44** : 196–201.

Jeevan, U., Vathamma, P.A.S., Deepak, G.N. and Bhagya, H.P. (2016) *The Ecoscan*, **10** (1 and 2) : 177-180.

Johnson, A. W. (1970) *J. Nematol.*, **2** : 191-192.

Joshi, A. C. and Pantulu, J. V. (1941) *Indian Bot. Soc.*, **20** : 37-69.

Jowkar, M.M. and Salehi, H. (2005) *Acta Hortic.*, **669** : 411-416.

Jowkar, M.M. and Salehi, H. (2006) *J. Sci. Tech. Agri. Natural Resour.*, **10** : 306-309.

Jyothi, B., Singh, K.P., Mohapatra, T. and Kumar, K.N. (2019) *Int. J. Curr. Microbiol. App. Sci.*, **8**(8) : 1815-1824.

Kadam, G.B., Singh, K.P., Singh, A.K. and Jyoti, R. (2010) *Indian J. Hortic.*, **67**(1) : 76-80.

Kahrizi, D., Barzegar, B. and Azadi, P. (2008) *J. Biotechnol.*,**136**S : S147–S169.

Kaintura, P, Srivastava, R. and Kapoor, M. (2016) *Int. J. of Agri. Sci.*, **8**(15) : 1257-1260.

Kaintura, P. and Srivastava, R. (2015) *Tropical Agri. Res.*, **26**(4) : 721-732.

Kakade, D.S., Jadhav, S.B. and Katwate, S.M. (2016) *Int. J. Pure App. Biosci.*, **4**(5) : 78-80.

Kale, P. N. and Bhujbal, B. G. (1972) *Indian J. Ornam. Hort.*, **3** : 101-103.

Kamerbeek, G. A. (1962) *Acta. Bot. Neerl.*, **11** : 331-410.

Kameswari, P. L., Girwani, A. and Radh-Rani, K. (2014) *Elect. J. Plant Breed.*, **5**: 52-57.

Kanchana, K., Kannan, M., Hemaprabha, K. and Ganga, M. (2018) *Int. J. Chem. Stud.*, **7**(1) : 236-241.

Karim, K.B., Kumar, N.V., Raghupati, B. and Pal, A.K. (2017) *Int. J. Curr. Microbiol. App. Sci.*, **6**(6) : 1-8.

Karuppaiah, P. (2019) *Hort. Int. J.*, **3**(1) : 7-11.

Katwate, S. M., Patil, M. T. and Singh, B. R. (1995) *J. Maharashtra Agric. Univ.*, **20** : 289-290.

Kayalvizhi, K., Kannan, M. and Ganga, M. (2017a) *Int. J. Curr. Microbiol. App. Sci.*, **6**(4) : 2492-2499.

Kayalvizhi, K., Kannan, M. and Ganga, M. (2017b) *Bull. Env. Pharmacol. Life Sci.*, **6** (1) : 422-429.

Khan, M.R. (2004) *Annals of Plant Protection Sciences*, **12**(1) : 106-109.

Khan, N.H., Zaidi, N., Jabeen, S. and Javaid, I. (2000) *Pak. J. Sci. Ind. Res.*, **43**: 118-122.

Khan, R.M. and Reddy, P.P. (1992) *In Nematode pest of crops*, CBS Publishers and Distributers, Delhi, pp. 250-257.

Khanchana, K, Kannan, M, Hemaprabha, K and Ganga, M (2019) *International Journal of Chemical Studies*, **7**: 236-241.

Khandagale, K., Padmakar, B., Lakshmana Reddy, D. C., Sane, A. and Aswath, C. (2014) *J. Hort. Sci.*, **9** : 5-11.

Khasaechai, A. (1992) *J. Natural Sci.*, **26** : 6-11.

Khobragade, R.I., Damke, M.M., Jadhao, B.J. and Hedau, C.V. (1997) *PKV Research J.*, **21** : 44-47.

Khoshkam, S. and Salari, A. (2018) *Int. J. Farm Alli Sci.*, **7**(1) : 25-30.

Kirtikar, K. R. and Basu, B. D. (1935) *Indian Medicinal Plants*, Lalit Mohan Basu, Allahabad.

Kosugi, K. and Kimura, Y. (1961) *Tech. Bull. Fac. Agric., Kagawa*, **12** : 230-234.

Krishnamoorthy, V. (2014) *Asian J. Hortic.*, **9**(2) : 515-517.

Krishnamurthy, K. B., Mythili, J. B. and Srinivas, M. (2001) *J. Appl. Hortic.*, **3** : 82-84.

Krishnamurthy, K.B. and Srinivas, M. (2005) *J. Ornam. Hortic.*, **8**(2) : 81-85.

Krishnan, G. (2003) *Ph.D. Thesis*, College of Horticulture, Vellanikara, Thrissur.

Krishnareddy, M., Smitha, R., Devaraju. and Jalali, S. (2007) *Indian Phytopathol.*, **60** : 251–258.

Kulshrestha, S., Mehra, A., Hallan, V., Raikhy, G., Ram, R. and Zaidi, A.A. (2005) *Cur. Sci.*, **89** : 870–872.

Kumar, A. and Deen, B. (2017) *J. Pharmacogn. Phytochem.*, **6**(4) : 1233-1236.

Kumar, S. (2008) *M.Sc Thesis*, Chaudhary Charan Singh Haryana Agricultural University, Hisar, India.

Kumari, C.G. (2014) *M.Sc. Thesis* submitted to University of Horticultural Sciences, Bagalkot.

Kumari, N. and Pal, A.K. (2016) *The Bioscan*, **11**(3) :1407-1410.

La Llave, P. and Lexarza, J. (1824) *Novorum vegetabilium descriptiones* I. M. Rivera, Mexico city, 32 p.

Lakshmi, R.N., Reddy, M.L, Bhagavan, B. V. K., Rao, A. V.D.D., Subbaramamma, P. and Uma Krishna, K.U. (2020) *Int. J. Curr. Microbiol. App. Sci.*, **9**(1) : 253-269.

Lalitha, K.P., Girwani, A. and Radha, R.K. (2014) *Electronic J. Plant Breed.*, **5** (1) : 52-57.

Lavin, M.C., Gomez, V.R.C., Brito, R.F., Lianos, L.A. and Estarda, M,E,V. (2012) *Florida Entomologist*, **85**(2) : 392.

Laxmi, V., M.N. Gupta and S.K. Datta. (1984) *J. Nuclear Agric. Biol.*, **13** : 140-141.

Lin, L., Zheng, H.Y., Chen, J., Chen, J.P., Zhang, Q.Y., Zhao, M.F. and Antoniw, J.F. (2004) *Arch. Virol.*, **149** : 1107–1116.

Lin, T.S. and Shen, T.M. (2004) *L. J. Agric. For. NCYU*, **1** : 1–12.

Madhavan, Jayanthi, Jayaswal, Pawan, Singh, Kanchan BM, Thakur, Prasoon Kumar and Rao, Uma (2019) *Indian J. Hort.*, **76**: 141-147.

Mahadevakumar, S., Chandana, C. and Janardhan, G.R. (2019) *Crop Protection*, **118** : 1-5.

Mahanta, P. and Paswan, L. (1995) *Hortic. J.*, **8** : 75-83.

Mahanta, P., Paswan, L. and Siddique, A.B. (1998) *Annals Agri. Bio. Res.*, **3** : 35-38.

Mahawar, T.C., Mahawar, L.N., Bairwa, H.L. and Meena, R.H. (2019) *J. Pharmacogn. Phytochem.*, **8**(6) : 2372-2375.

Maheswari, T.U. and Sivasanjeevi, K. (2019) *Ann. Plant Soil Res.*, **21**(1) : 48-50.

Malik, R. S., Singh, B. and Dadlani, N. K. (1984) *Indian Hort.*, **29** : 9-15 and 22.

Malini, G. and Khader, M.A. (1989) *South Indian Hort.*, **37** : 357-358.

Mane, P.K., Bankar, G.J. and Makne, S.S. (2007) *Indian J. Agric. Res.*, **41**(1) : 71-74.

Mani, M., Krishnamoorthy, A. and Shivarajan, C. (2011) *J. Hortic. Sci.*, **6**(2) : 85-100.

Marrow, A. W. (2005) *Acta Hortic.* **683**, 43-62.

Meena, R.K., Mahawer, L.N. and Bairwa, H.L. (2018) *Indian J. Hortic.*, **75**(4) : 662-666.

Mendoza, A., Silva-Montellano, A., Gaut, S.B., Souza, V., Eguiarte, L.E. (2006) *Agave Aliso.*, **22** : 329-344.

Mendoza, F.H., Castaneda, G.C., Santoa, E.P., Torres, E.C and Castillo, M.C.M. (2013) *Revista Electrónica Nova Scientia*, **7**(1) : 32-47.

Militiu, A., Vierasu, E. and Iliescu, A. F. (1970) *Lucr. Sti. Inst. Agron. N. Balescu, B.*, **13** : 115-120.

Mishra, A., Pandey, R.K. and Gupta, R.K. (2006) *Progressive Horticulture*, **37**: 226-236.

Mishra, H.P. (1999) *J. Appl. Hortic.*, **1** : 145-148.

Misra, H., Singh, A. K. and Singh, O.P. (2000) *Adv. Plant Sci.*, **13**(2) : 563-566.

Mohanty, B. K., Sankar, C. R. and Dayanand, T. (1999) *South Indian Hortic.*, **47**(1-6) : 327-330.

Mostafa, M. M., El Fadl, IAA. and Hussein, E. H. (1996) *Alexandria J. Agric. Res.*, **41** : 93-107.

Mukhopadhyay, A. and Banker, G. J. (1981) *Indian Agric.*, **25** : 131-134.

Muralidhar, C.E. and Mehta, A. R. (1982) In: *Plant Tissue Culture*, Jap. Assoc. Pl. Tissue Cult. Tokyo, pp. 693-694.

Muriithi, A. N., Wamocho, L. S. and Nioroge, J. B. M. (2009) In: 9th African Crop Science, Conference Proceedings (eds Tenywa, J. S. *et al.*), Cape Town, South Africa, pp. 227–234.

Muriithi, A.N. (2014) *Ph.D Thesis,* Jomo Kenyatta University of Agriculture and Technology, Kenya.

Murthy, G.M.A. and Gowda, J.V.N. (1993) *Crop Res. Hisar*, **6** : 176-178.

Nagar, P.K. (1995) *Sci. Hortic.*, **63** : 77-82.

Nagaraja, G. S., Gowda, J. V. N. and Farooqui, A. A. (1998) *Karnataka J. Agric. Sci.*, **11** : 1145-1147.

Nagaraja, G. S., Gowda, J. V. N. and Farooqui, A. A. (1999b) *Karnataka J. Agric. Sci.*, **12** : 236-238.

Nagaraju, H.T., Sreenivas, K.N., Anna, M.R Raj. and Venkatesha, J. (2012) *J. Ornam. Hortic.*, **7**(3 and 4) : 177-181.

Naidu, S.N. and Reid, M.S. (1989) *Acta Hortic.*, **261** : 313-317.

Nambisan, K.M.P. and Krishnan, B.M. (1983) *Indian Hortic.*, **28** : 17 and 20.

Nanjan, K., Nambisan, K.M.P., Veeraragavathatham, D. and Krishnan, B.N. (1980) In: *Nat. Sem. Prodn. Tech. Comm. Flower Crops*, Tamil Nadu Agricultural University, Coimbatore, pp. 76-78.

Narayanswamy, S. and Prabhudesai, V.R. (1979) *Indian J. Exp. Biol.*, **17**: 873-875.

Nasibi, F., Farahmand, H., Kamyab, A. and Alipour, S. (2014) *Agric.Comm.*, **2**(2) : 35-41.

Navabi, Y., Norouzi, M., Arab, M., Daylami, S.D. (2016) *Electronic J. Biol.*, **12**(2) : 168-172.

Naz, S., Aslam, F., Ilyas, S., Shahzadi and Tariq, A. (2012) *J. Med. Plants Res.*, **6**(24) : 4107-4112.

Nazneen, S., Mussarat, J. and Ilahi, I. (2003) *Pak. J. Bot.*, **35**: 17-25.

Nellipalli, V.K. and Pal, A.K. (2018) *The Pharma Innov. J.*, **7**(12) : 34-36.

Nidiry E.S.J. and Babu, C.S.G. (2005) *Phytotherapy Res.*, **19** : 447-449.

Ovesna, J., Pola´kova´, K., Leisova´, L. (2002) *Czech J. Genet. Plant Breed.*, **38** : 29–40.

Pal *et al.* (2019) AICRP on water management, BCKV, pp. 40-46.

Pal, A.K. and Das, S.N. (1990) *South Indian Hortic.*, **38** : 143-149.

Panigrahi, J., Rana, V.D. and Patel, I.C. (2013) *Indian J. Res.*, **3**(5) :12-14.

Panja, B.N., De, A., Mandal, M. and Saha, J. (2016) *J. Mycopathol. Res.*, **54**(1) : 171-174.

Parthiban, S. and Khader, M.A. (1991) *South Indian Hortic.*, **39** : 363-367.

Patel, B. (2013) *Ph.D. Thesis*, Navsari Agricultural University, Navsari, India.

Patel, M.M., Parmar, P.B. and Parmar B.R. (2006) *J. Ornam. Hortic.*, **9**(4) : 286-289.

Pathak, S. (1981) *Ph.D. thesis*, University of Burdwan, West Bengal.

Pathak, S., Choudhuri, M.A. and Chatterjee, S.K. (1980) *Indian J. Plant Physiol.*, **23** : 47-54.

Patil, A.V., Kale, P.N. and Kaulgud, S.N. (1975) *Curr. Sci.*, **44** : 406-407.

Patil, B.A., Bhore, D.P.,Patil, J.D. and Patil, A.V. (1980) In: *Nat. Sem. Prodn. Tech. Comm. Flower Crops*, Tamil Nadu Agricultural University, Coimbatore, pp. 131-134.

Patil, P. R., Reddy, B. S., Patil, S. R. and Kulkarni, B. S. (1999) *South Indian Hortic.*, **47**(1-6) : 335-338.

Pearson, M.N. and Horner, M.B. (1986) *Australasian Plant Pathol.*, **15** : 39.

Pohare, M.B., Rathod, H.P., Shahakar, S.B., Kelatkar, S.K. and Suryawanshi, P.P. (2012) *Res. J. Agric. Sci.*, **3**(6) : 1307-1308.

Prakash, S., Shripal, Arya, J.K. and Kumar, S. (2011) *Int. J. of Agric. Sci.*, **7**(1) : 169-170.

Priya, G.S., Sureshkumar, R., Rajkumar, M., Sendhilnathan, R. and Barathkumar, T.R. (2020) *Plant Arch.*, **20**(1) : 941-944.

Priyadarshini, S., Kamalakannan, A. and Paranidharan, V. (2019) *J. Pharmacogn. Phytochem.*, **8**(3) : 4709-4712.

Raghuvanshi, S., Tripathi, M.K., Sankar, M.V. and Singh, O.P. (2013) *Plant Cell Biotech. Mol. Biol.*, **14**(1 and 2) : 49-59.

Raj, S.K., Snehi, S.K., Kumar, S., Ram, T. and Goel, A.K. (2009) *Australasian Plant Dis. Notes*, **4** : 93–95.

Raja, K. and Palanisamy, V. (1999a) *Orissa J. Hort.*, **27** : 53-56.

Raja, K. and Palanisamy, V. (2000) *Madras Agric. J.*, **86**(1-3) : 165-166.

Raja, K., Palanisamy, V., Selveraju, P. (2003) *Prog. Hort.*, **35**(2) : 233-233.

Rajasekharan, V., Haripriya, K., Arumugam, S. and Shakila, A. (2000) In: *Centennial conference on spices and aromatic plants: challenges and opportunities in the new century*, pp.86-88.

Rajendran, G. and Muthukrishnan, T.S. (1980) In: *Nat. Sem. Prodn. Tech. Comm. Flower Crops*, Tamil Nadu Agricultural University, Coimbatore, pp. 111-112.

Ramaswamy, N. and Chockalingam, P. (1977) *Prog. Hortic.*, **8** : 39-41.

Ramaswamy, N., Paulraj, C. and Chockalingam, P. (1979) *Annamalai Univ. Agric. Res. Annual*, **7/8** : 29-33.

Rameshwar, A. (1976) *Indian Hortic.*, **21** : 22-23.

Ranchana, P. and Kannan, M. (2015) *Trends Biosci.*, **8**(6) : 1435-1438.

Ranchana, P. and Kannan, M. (2016) *The Bioscan*, **11**(1) : 33-36.

Ranchana, P., Kanan, M. and Jawaharlal, M. (2015) *Prog. Hortic.*, **47**(2) : 267-270.

Rangaswami, A. (1963). Proceedings of 50th Indian Science Congress Abs. Part II. Section VI. Botany Abs. No. 247.

Rani, P. and Singh, P. (2013) *J. Trop. Plant Physiol.*, **5** : 33-41.

Rao, D.V.R., Reddy, K.B., Naidu, L.N., Suryanarayana, V., Bhaskara Reddy, K. and Naram Naidu, L. (1991) *South Indian Hortic.*, **39** : 143-145.

Rao, M.S., Reddy, P.P. and Wallia, R.K. (2001) In National Nematology congress-centenary celebrations, New Delhi, India.

Rao, M.S., Umamaheshwari, R., Priti, K., Grace, G.N., Rajinikanth, R., Prabu, P., Kamalnath, M. and Narayanaswamy, B. (2015) *IIHR Technical Bulletin No. 46*. ICAR-IIHR, Bengaluru.

Rathod, A. (2015) *M.Sc. Thesis*, University of Agricultural Sciences, Dharwad.

Ravikumar, M.R., Harish, D.K., Kumara, B.H. and Kumar, A. (2020) *Current J. Appl. Sci. Tech.*, **39**(8) : 12-17.

Reddy, B., Singh, K., Gangadharappa, P.M. and Singh, K (1997d) *Karnataka J. Agric. Sci.*, **10** : 731-737.

Reddy, B.S. and Singh, K. (1996) *J. Maharashtra Agri Univ.*, **21** : 201-203.

Reddy, B.S., Singh, K. and Reddy, B.S. (1994) *Adv. Agric. Res.* **2** : 123-130.

Reddy, B.S., Singh, K. and Singh, A. (1995) *Advan. Agr. Res. India* **3** : 161-167.

Reddy, B.S., Singh, K., Gangadharappa, P.M. and Singh, K. (1998) *Adv. Agric. Res.*, **9** : 59-63.

Reddy, B.S., Singh, K., Gangadharappa, P.M., Singh, K. and Reddy, B.S (1997a) *Karnataka J. Agric. Sci.*, **10** : 1049-1054.

Reddy, B.S., Singh, K., Gangadharappa, P.M., Singh, K. and Reddy, B.S. (1997b) *Karnataka J. Agric. Sci.*, **10** : 1223-1227.

Reddy, B.S., Singh, K., Gangadharappa, P.M., Singh, K. and Reddy, B.S. (1997c) *Karnataka J. Agric. Sci.*, **10** : 1055-1059.

Reddy, B.S., Singh, K., Singh. A., Sathyanarayana Reddy, B., Singh, K. and Singh, A. (1995) *Adv. Agric. Res.* **3** : 161-167.

Reddy, S., Singh, B.K., Singh, A. and Saini, A. (1996) *Adv. Agric. Res.*, **6** : 39-45.

Rees, A.R. (1972) *The Growth of Bulbs*, Vol. I., Academic Press, London and New York.

Reid, M. (1996) *Perishables Handling Newsletter*, 88.21.

Ritchie, W. and Lindstrom, J.T. (2014) *J. Environ. Hort.*, **32**(1) : 51–53.

Rocha, M., Good-Avila, S., Molina-Freaner, F., Arita, T.H., Castillo, A., Garcia

Roemer, M.J. (1847) *IV. Ensatae*. Landes-Industrie-Comptor, Weimar, p. 314.

Rosalind, L., Bharathi, T.U., Kulkarni, S.B., Dhanajaya, M.V., Nair, S.A. and Munikrishnappa, P.M. (2018) *J. Pharmacogn. Phytochem.*, **7**(6) : 23-25.

Rose, J.N. (1903-05) *Studies of Mexican and Central American Plants*, No. 3, Cont. U.S. Nat. Herbarium, **8** : 1-55.

Roy, A.K. (1985) *Indian Hort.*, **30** : 19-21.

Roychowdhury, N., Chakrabarty (Das), S. and Munsi, P. (2011) *Acta Hortic.*, **886** : 359-364.

Rupa, R., Misra, S. and Prasad K.K. (2017) *Int. J. Agric. Sci.*, **9**(42) : 4667-4670.

Sadhu, M.K. and Bose, T.K. (1973) *Indian Hortic.*, **18** : 17-21.

Sadhu, M.K. and Das, P.K. (1978) *Indian J. Hortic.*, **35** : 147-150.

Saini, R.S., Yamdagni, R. and Sharma, S.K. (1994) *South Indian Hortic.*, **42** : 376-378

Sambandamurthi, S. and Appavu, K. (1980) *Nat. Sem. Prodn. Tech. Comm. Flower Crops*, Tamil Nadu Agricultural University, Coimbatore, pp. 73-75.

Sanap, P. B., Patil, B. A. and Gondhali, B. V. (2000) *Orissa J. Hortic.*, **28** : 68-72.

Sarkar, J., Misra, R. L., Bhat, K. V., Singh, A. and Singh, S. K. (2010) *Indian J. Genet.*, **70**: 182-188.

Sarkar, J., Misra, R.L., Bhar, K.V., Singh, S. and Singh, S.K. (2010) *Indian J. Genet.*, **70**(2) : 182-188

Sarkar, J., Misra, R.L., Sanjay, K., Singh, Prasad, K.V. and Ajay, A. (2009) *Indian J. Hortic.*, **66**(4) : 502-507.

Sato, D. (1938) *Cytologia*, **9** : 203-242.

Schiva, T. and Lanteri, S. (1983) *Annali dell'Istituto Sperimentale per la Floricoltura*, **14** : 1-18.

Schiva, T. and Lanteri, S. (1984) *Ann. Ist. Sper. Flor.*, **15** : 1–18.

Seetharamu, G. (1993) *M.Sc. Thesis*, Univ. Agri. Sci., GKVK, Bangalore, India.

Sendhilnathan, R. and Manivannan, K. (2019) *Annals of Plant and Soil Res.*, **21**(3) : 261-264

Sharavani, C.S.R., Kode, S.L., Priya, B.T., Bharathi, T.U., Sekhar, M.R., Ruth, C. and Ramakrishna, M. (2019a) *Adv. Res.*, **10**(1) : 109-113.

Sharavani, C.S.R., Priya, B.T., Bharathi, T.U., Sekhar, M.R., Ruth, C. and Ramakrishna, M. (2019b) *Bull. Env. Pharmacol. Life Sci.*, **8** : 105-109.

Sharga, A.N. (1980) *Prog. Hortic.*, **12** : 71-77.

Sharma, A. K. and Ghosh, C. (1956) *Genetica*, **28** : 99–111.

Sharma, A.K. and Bal, A.K. (1956) *J. of University of Ganhati*, **10** : 65-76.

Sharma, A.K. and Bhattacharya, U.C. (1956) *Genetica*, **28**(1-2) : 99-111.

Sharma, C.K., Barman, D., Singh, I.P. and De L.C. (1995) *J. Hill Res.*, **8** : 271-273

Shen, T. M., Huang, K.L., Shen, R.S. and Du, B.S. (2003) *Acta Hortic.*, **624** : 73-76.

Shen, T.M., Huang, K.L. and Huang, T.S. (1987) *Acta Hortic.*, **205** : 71-74.

Shinners, L.M. (1966) *Sida*, **4** : 333-338.

Shiombing, D. and Handayati, W. (2017) *2nd International Conference on Sustainable Agriculture and Food Security: A Comprehensive Approach*, KnE Life Sciences, 579–586.

Shukla, R. and Datta, S.K. (1993) *J. Ornam. Hortic.*, **1**(2) : 36-41.

Singh, D., Singh, V.K., Kumari, S., Pandey, S.K. and Singh, D. (2014) *New Agriculturist*, **25**(2) : 1–5.

Singh, K. and Reddy, B.S. (1994) *Adv. Agri. Res.*, **2** : 123-130.

Singh, K.P. and Sujatha, K. (1990) *South Indian Hortic.*, **38** : 291-292.

Singh, K.P., Jain, R., Panwar, S., Kumari, P. and Kumar, V.P. (2014) *J. Ornam. Hortic.*, **17**(1 and 2) : 54-58.

Singh, K.P., Poonam, S., Jitendra, M. and Singh, M.C. (2009) *Prog. Hortic.*, **41**(2) : 145-147.

Singh, K.P., Saha, T.N. and Hollajer, P. (2017) *HortFlora Res. Spectrum*, **6**(2) : 128-130.

Singh, L.B. (1971) *Econ. Bot.*, **26** : 361-363.

Singh, O. and Pal, V. (2019) *Ann. Hortic.*, **12**(1) : 67-69.

Singh, P.K. and Sadhukhan, R. (2019) *J. Crop Weed*, **15**(2) : 40-45.

Singh, P.K., Sadhukhan, R., Roy, K. and Sarkar, H.K. (2013) *Floric. Ornam. Biotech.*, **7**(1) : 103-105.

Singh, P.V., Kumar, M. and Kumar, M. (1999) *J. Ornam. Hortic.*, **2** : 127-130.

Singh, R.S., Motial, V.S. and Singh, L.B. (1974) *Plant Sci.*, **6** : 80-88.

Sirohi, U., Kumar, M., Chauhan, P., Singh, M.K., Kumar, N., Prakash, S., Chand, Pooran, Sharma, V.R. and Chaudhary, V. (2017) *Int. J. Agric. Stat. Sci.*, **13** (2) : 519-526.

Solano, C.E. (2000) *Ph.D. Thesis*, UNAM. Facultad de Ciencias, Mexico, 291.

Solano, C.E. (2000) *Tesis de doctorado*. Facultad de Ciencias, Universidad Nacional Autónoma de México. México, D.F., 291 pp.

Solano, C.E. and Aranda, P.D.D. (2003) *Novon*, **13** : 119–122.

Solano, C.E. and Mendoza, G.A. (1998) *Sida*, **18** : 473–477.

Solano, E. and Feria, T.L. (2007) *Biodivers. Conserv.*, **16** : 1885-1900.

Solano, E. and Gomez, R.R. (2011) *Brittonia*, **63** : 70-74.

Solano, E., Mendoza, A.B. and Gomez, R.R. (2019) *Acta Botanica Mexicana*, **126** : 1-10.

Sood, S. and Nagar, P.K. (2005) *Sci. Hortic.*, **105** : 483–490.

Srinivas M., Murthy N. and Chandravadana, M.V. (1996) *J. Essential Oil Res.*, **8**(5) : 541-542.

Srinivasan, M. and Bhatia, J.S. (1954) *Curr. Sci.*, **23** : 192.

Srivastava, H. C. and Sridhara, C. J. (1999) *J. Med. Aromat. Plant Sci.*, **21** : 934-936.

Su, W.R., Huang, K.L., Shen, R.S. and Chen, W.S. (2002) *J. Plant Physiol.*, **159** : 557–559.

Sudagar, I. P., Sankarnarayannan, R. and Aruna, P. (2009) *Asian J. Hortic.*, **4**(2) : 421-423.

Sudhagar, R., Rajaselvam, M., Kamalakannan, S., Kumar, S. and Maheswari, T. U. (2020) *Plant Arch.*, **20**(1) : 2415-2418.

Sultana, S., Amin, M.R., Hossain, M.I. and Sarkar, M.D. (2016) *Int. J. Sci. Res. Pub.*, **6**(12) : 384-388.

Sundara Babu, R. and Vadivelu, S. (1988) *Indian J. Nematol.*, **18** : 146-148.

Surendranath, S., Ganga, M. and Jawaharlal, M. (2016) *Environ. Ecol.*, **34**(4D) : 2556-2560.

Suseela, T., Bhaskar, V.V. and Chandrasekhar, R. (2018) *Int. J. Curr. Microbiol. App. Sci.*, **7**(2) : 3617-3625.

Suseela, T., Chandrasekhar, R., Bhaskar, V.V., Suneetha, D.R.S. and Umakrishna, K. (2016) *The Bioscan*, **11**(4) : 2715-2720.

Swaminathan, V., Ramaswamy, N. and Pillai, O. A. A. (1999) *South Indian Hortic.*, **47** : 1-6, 331-334.

Thakre, S., Panchbhai, D.M., Mahadik, M.K. and Badge, S.A. (2019) *Int. J. Chem. Stud.*, **7**(3) : 153-156.

Toma, F., Georgescu, M.I., Petra, S., Dobrescu, E. (2015) *Agric. Sci. Procedia,* **6** : 179 – 183.

Tzeng, S. L. and Tung, B. K. (1998) *Plant Protection Bull. Taipei,* **40** : 145-152.

Uma, S. and J. V. N. Gowda. (1999) *Crop Research Hisar,* 18(3) : 450-453.

Uthairatanakij, A., Jeenbuntug, J., Buanong, M. and Kanlayanarat, S. (2007) *Acta Hortic.*, **755** : 477-481.

Vaid, N., Chaudhary, S.V.S., Sharma, B.P., Gupta, Y.C. and Chauhan, G. (2019) *Int. J. Curr. Microbiol. App. Sci.*, **8**(5) : 199-206.

van Doorn, W.G. and Peirik, R.R.J. (1990) *J. Amer. Soc. Hort. Sci.* **115** : 979-981.

Verhoek-Williams, S. (1975) *Ph.D. Thesis*, Cornell University. Ithaca, 405 pp.

Verhoek-Williams, S.E. (1976) *Phytologia,* **34** : 365–368.

Vij, S.P., Sharma, M.and Chaudhary, J.D. (1982) Cytologia, **47** : 649-663.

Vishwakarma, S.K. and Kumar, A. (2018) *Plant Arch.*, **18**(1) : 271-274.

Wagh, V.K., Chawla, S.L., Gaikwad, A.R. and Parolekar, S.S. (2012) *Prog. Hortic.*, **44** : 27-31.

Walia, K. K. and Gupta, D. C. (1997) *Indian J. Nematol.*, **27** : 18-23.

Watako, A.O. and Ngamau,K. (2013) *Jomo Kenyatta University of Agric. and Tech.*, **15**(1) : 5-14.

Whitaker, T.W. (1934) *J. Arnold Arb.*, **15** : 135-143.

Yadav, L.P. (1983) *Ph.D. Thesis*, B.C.K.V., Kalyani, West Bengal, India.

Yadav, L.P. and Bose, T.K. (1987) *Acta Hortic.*, **205** : 177-185.

Yadav, L.P. and Bose, T.K. (1988) *Farmer and Parliament*, **23** : 15-16.

Yadav, L.P., Bose, T.K. and Maiti, R.G. (1984) *Prog. Hortic.*, **16** : 209-213.

Yadav, L.P., Bose, T.K. and Maiti, R.G. (1985) *Prog. Hortic.*, **17** : 83-86.

Younis, S.E.A. and Borham, J.H. (1975) *Egyptian J. Bot.* **18**(1 and 3) : 205-217.

Zizzo, G.V., Fascella, G. and Agnello, S. (1999) *Colture Protette*, **28** : 71-74.

7

DAHLIA

Sangeetha Priya S., Rajiv Kumar,
S.K. Bhattacharjee, Swami Vinayananda,
P. Pal and Subhrajyoti Chatterjee

1.0 Introduction

Dahlia is one of the most important garden plants. Multitude of colours, great variation in sizes (ranging from miniature, less than 2.5 cm across to giant over 40 cm in diameter), attractive shapes, many forms, profusion of flowering and easy cultivation have made them immensely popular. It was Abbe' Cavanilles who gave the genus the name *Dahlia* in 1791 (Smith, 1963).

As per Mexican Association of Dalia (AMD), the main growing states are Mexico City, Puebla and Mexico State. Since 2007, the National Day of Dahlia is celebrated in Mexico on each August 4th (Saar *et al.*, 2003). At present, there are 41 international associations concerned with the cultivation and spreading of dahlia; among them are the National Dahlia Society in the UK, the American Dahlia Society (ADS), other groups of New Zealand in European countries, France and Germany (Saar *et al.*, 2003).

2.0 Importance and Uses

2.1.0 Aesthetic Uses

Dahlia unquestionably occupies a place of pride in any garden anywhere. They are easy to grow both in ground and pot, and are extensively used for exhibition, garden display and home decoration. For exhibition and garden display all types of dahlias are used. Dwarf growing types are suitable for borders, beds or even in mixed borders. Large flowering dahlias in pots are popular for terrace garden or verandah display. The long-stemmed flowers of various forms and colour are used in flower arrangement. Cut flowers of pompon and miniature types stay fresh in flower vases for many days and also make moderately good garlands. Dahlia as potential crop in Cuban floriculture attributed to its perennial, busy growth habit, intense lilac petals and easier propagation by cuttings (Liudmila, 2015).

2.2.0 Therapeutic and other Uses

There are certain medicinal and nutritional uses of dahlia. Tubers of this plant contain significant amount of inulin and fructose and small quantities of medicinally active compounds, such as phytin and benzoic acid (Whitley, 1985). Nitsch (1992) reported that an inulin extract from tubers of dahlia is used in diagnosis of renal function. Legorreta *et al.* (2016) found that inulin extracted from wild species *Dahlia coccinea* was significantly better. Furthermore, cultivation of wild Dahlias significantly increased inulin and fructan contents of their roots. Tuberous roots of Merck Dahlia and 6 varieties of *D. variabilis* possessed 11.9-66.5 per cent inulin, 0.833-1.173 per cent tannins, 0.118-1.151 per cent essential oils, 0.190-0.213 per cent organic acids, 6.14-11.37 per cent and 0.20-0.24 per cent ascorbic acid. It could be used as an alternative source of Jerusalem artichoke inulin (Pupykina *et al.*, 2015).

Cruz-Alvarez *et al.* (2019) reported that tuberous roots of Dahlia wild species *viz.*, *D. campanulata*, *D. coccinea* and *D. brevis* showed good nutraceutical and antioxidant properties which could be used as food as well as a source of selection of traits of nutraceutic interest for genetic improvement. Among the different wild species,

D. campanulata performed better for Vitamin C (0.05 mg/100 g), antioxidant capacity (1.88 mg VCEAC/g) and inulin (72.25 per cent). However, its inulin content was similar to *D. coccinea* (66.17 per cent) which had high total phenols (1.74 mg GA/g).

Flavonone 3-hydroxylase, which catalyses the conversion of flavanones to dihydro-flavonols, was detected in flower extracts of cyanic strains of dahlia (Forkmann and Stotz, 1984). Dye can also be extracted from dahlia flowers which could be used for dyeing wool yarns, in carpet industries. Good fastness properties of dyed yarn confirm its utility in textile industries for eco-dyeing and can potentially replace phenolphthalein as indicator. Pradeep *et al.* (2012) observed that ultrasound assisted extraction (UAE) is more efficient than solvent extraction. Change in colour with pH of extracted dye is observed due to bathochromic shift from blue green (pH 8.9) to deep red (pH 2.0).

Seeds of dahlia are a good source of fats and proteins. It contained >16 per cent oil and 20.9-47.1 per cent protein (Nigam and Misra, 1992). Amendments of chopped green shoots of *Dahlia pinnata* in pot mixture for raising seedlings of cauliflower and cabbage gave a considerable reduction in the population of *Tylenchorhynchus brassicae* with no phytotoxicity (Tiyagi and Wani, 1991). *Hoplolaimus indicus, Tylenchus filiformis, Helicotylenchus indicus, Meloidogyne incognita* and *Tylenchorhynchus brassicae* were exposed to various concentrations of roots exudates of *Dahlia pinnata*. Exudate was nematotoxic, the mortality of the nematode increasing with increased concentration of exudates and exposure period (Wani and Ansari, 1993).

Feeding inulin derived from dahlia tuber powder and extract significantly increased lactic acid bacteria population, daily body weight gain but decreased *E. coli* number, intestinal pH and rate of passage in crossbred native chickens. It was also mentioned that feeding inulin of dahlia tuber powder at 0.9 per cent was optimum (Krismiyanto *et al.*, 2014).

DmAMP1, an antimicrobial peptide isolated from the seeds of *Dahlia merckii* was found to have antifungal properties against wild type of *Saccharomyces cerevisiae* (Thevissen *et al.*, 2000). Mukhopadhyay *et al.* (2013) assessed three different solvent extracts from *Dahlia pinnata* for their analgesic potential by acetic acid induced writhing assay in Swiss albino mice. The methanol extract was found to be most potent analgesic followed by chloroform, and petroleum ether extracts.

Dahlia variabilis has many beneficial effects to human health as antibacterial and antifungal agent. Saryono *et al.* (2015) investigated the potency of endophytic microbes isolated from Dahlia as antimicrobial agents. LBKURCC41 (*Aureobasidium* spp.) showed most significant inhibition zones against *Escherchia coli* (19.33 ±1.15 mm) and *Staphylococcus aureus* (17.00±1.00 mm). Whereas, the inhibition zone of amoxsan and ketoconazole were 12.55±0.38 and 17.67±0.57 mm, respectively. On further genetical characterization, the isolate showed 99 per cent similarity with that of *Fusarium oxysporum*.

Saryono *et al.* (2017) revealed that n-hexane and methanol extracts and fractions of red-coloured dahlia tubers exhibited antifungal activity against pathogenic skin fungi *Candida albicans* and *Microsporum gypseum*. Maximum activity was observed with extracts and fractions of methanol. The minimum inhibitory concentration (MIC) for n-hexane and methanol fraction against *C. albicans* was 1.5 and 0.5 per

Uses of Dahlia

In Garden

In Border

In Beds

In Pot

As Cut Flowers

In Flower Arrangement

In Flower Vase

cent, respectively. While, both the fractions resulted in MIC of 0.5 per cent against *M. gypseum*. Thus, the methanol extract could be used as a medicinal drug for candidiasis and other fungal skin infections.

Suraya *et al.* (2019) isolated endophytic fungus LBKURCC40 from dahlia tubers and molecular identified using ITS4 and ITS5 primers followed by testing of homology level using BLAST program. Results revealed that the isolate was identified as *Trichoderma asperellum* with 99 per cent homology level. Closest kinship of this isolate with *T. asperellum* strain Tr266B was obtained from phylogenetic analysis using N-J bootstrap test with 10000 repetitions.

3.0 Origin, History and Morphology

The native land of dahlia is known to be Mexico (Willis, 1966). The species introduced into the Old World were *Dahlia imperialis*, *D. coccinea*, *D. merckii* and *D. juarezii*. The later is actually a hybrid of *D. variabilis* from which most of the garden types have been developed. Continuous crossing among the species and cultivars and subsequent selection resulted into the present-day hybrids.

Dahlia was originally discovered by Aztecs who called it *cocoxochitl*. Early reference revealed that the Aztecs grew dahlia as medicinal plant and probably also as a flower for worship as early as 1570. The plant was first mentioned by Francisco Hernandez, physician of Philip II, in 1651. Vincente Cervantes, Director of the Mexico City Botanic Garden, sent seeds to Abbe' Cavanilles, Director of the Madrid Botanic Garden in 1789 and this was the first time dahlia came out of Mexico. In Europe, the plants flowered in 1791 and were named Dahlia in honour of Dr. Andreas Dahl (1751-1789), a Swedish botanist and pupil of Linnaeus. It quickly spread to other countries of Europe and became popular. Within 15 years of its introduction into England, dahlia had around 300 distinct cultivars developed from a few types and cultivars only. In 1826, 60 cultivars were grown by the Royal Horticultural Society, London. In 1841, one English dealer had over 1200 cultivars. Caledonian Horticultural Society offered 50,000 francs in 1864 to the raiser of a blue dahlia, which does not exist now (Pizzetti and Cocker, 1975). Dahlia was introduced to India as early as 1857 under the auspices of the Agri Horticultural Society of India (formerly, Royal Agri-Horticultural Society of India), Calcutta. Apart from cultivation in ground, dahlias in pots are very popular in most parts of the country.

Dahlias are half-hardy perennials with tuberous roots. Stems are mostly erect, branched, glabrous or scabrous. The leaves 1-3 pinnate, with slightly serrated margin, are produced opposite each node on the stem. They bear flowers on a long, stiff stems well above the foliage. Ray-florets are neutral or pistillate and disc florets perfect. They belong to the family Asteraceae.

4.0 Species, Classification and Cultivars

4.1.0 Species

The most comprehensive dahlia taxonomy till date is that of Sorensen (1969). He, using numerous chromosome counts and extensive field collections and

Species of Dahlia

Dahlia australis

Dahlia coccinea

Dahlia imperialis

Dahlia merckii

Dahlia pinnata

Dahlia variabilis

Dahlia tenuicaulis

Dahlia tenuis

Dahlia sherffii

observations, recognized 27 species and 4 varieties. According to Willis (1966), however, there are about 20 species. The description of the important species is given below.

Dahlia coccinea

It is a highly variable taxon possessing both diploid (n=16) and tetraploid (n=32) races. It is believed that localized hybridization is occurring periodically either at the tetraploid or the diploid level with subsequent polyploidization. Despite its morphological variability, this species is a distinct taxon because of its orange, scarlet or yellow rays and is unlikely to be mistaken for any other species. Plants are slender, 0.4 – 2 m tall, stems glaucous, pinnae arrangement on rachilla is opposite.

D. excelsa

Erect, up to 6 m tall or more, glaucous, woody; leaves 2-pinnate, ovate, slender-pointed, toothed; flower heads 10 cm across; rays 8, pale purple; disc pinkish. Var. *anemonaeflora* : disc-florets lilac or yellow, rays flat.

D. gracilis

Bushy, about 1.2 m tall; leaves 2-pinnate, ternate, glabrous; ovate, coarsely toothed; flowers brilliant orange-scarlet; outer scales almost circular.

D. imperialis

Commonly called as tree dahlia. It is a suffrutescent diploid (n=16) species. Plants 2 – 6 m tall; stems woody; leaves 2-3 pinnate, arrangement on rachilla is opposite; flowers single, 10 – 18 cm across, white, red tinge near base; rays lanceolate, acute, not toothed.

D. juarezii

Cactus Dahlia. Flower heads bright scarlet; florets length irregular; rays long narrow, margins recurved. Has been crossed with *D. rosea* and a distinct race of garden varieties with intermediates.

D. merckii

Plant 60 – 100 cm tall, glabrous; leaves 2-pinnate; flower heads lilac; rays short, roundish, acute; peduncles branched, long, wiry, raising flower heads much above foliage.

D. pinnata

This is a tetraploid (n = 32) species. Plants 60-90 cm tall, glabrous; leaves 5-foliolate, roundish teeth; flowers large, solitary; rays more than one series, margins incurved, bluish-red.

D. rosea syn. *D. variabilis*

Plant 100 – 150 cm tall; leaves pinnate, or sometimes 2-pinnate; flower heads various colours; rays margins straight; florets cupped in double forms. The chief parent of the garden dahlia.

D. scapigera

Plants 60 cm tall; leaves pinnate, ovate to narrow-linear, toothed; flower heads about 5 cm across, various colours; peduncles long, glabrous, often in pairs or threes; disc yellow; rays ovate, 3-toothed at apex; outer involucral bracts rather spreading, not reflexed.

4.2.0 Classification and Cultivars

All advanced dahlia-growing countries have their own system of classification. The scheme of classification in a country also changes from time to time. There is, however, a conscious attempt in different countries to have an identical method of classification. National Dahlia Society of England is one of the oldest societies and their scheme of classification is probably the oldest, which is usually revised every alternate year. According to the classification of this society all cultivars are classified into ten groups. But this scheme of classification does not recognize fimbriated and water lily as separate groups though these are highly popular in most of the countries. Decorative, cactus, and semi-cactus, and also fimbriated are further divided into 5 sub-groups according to the size (diameter) of the normal flowers. These sub-groups are : (i) giant-flowered over 260 mm (over 10.2 inch); (ii) large-flowered between 220 mm and 260 mm (8.6 to 10.2 inch); (iii) medium-flowered between 170 mm and 220 mm (6.7 to 8.6 inch); (iv) small-flowered between 115 mm and 170 mm (4.5 to 6.7 inch); and (v) miniature-flowered upto 115 mm (4.5 inch).

Number of dahlia cultivars is very large. The Tentative Classified List and International Register of Dahlia Names, published in 1969, contains the names of around 20,000 cultivars. However, it is not known whether all or most of these cultivars are available or existing. According to Swami Vinayananda (1985), about two thousand cultivars are expected to be under commercial cultivation all over the world. Development of new cultivars is in progress and the number is rapidly increasing every year.

Dahlia is the unique plant that has the largest number of varieties among all plant species and more than 50,000 has been registered in the Royal Horticultural Society of England (Bye and Linares, 2008). According to the rules of International Code of Nomenclature for Cultivated Plants, 6 new cultivars of Dahlia published *viz.*, 'Meska Slapukë', 'Tumë', 'Dalelytë', 'Giedrës Kasos', 'Lapë', and 'Orija' (Balsevièius *et al.*, 2016).

Below is given a list of different groups, the first ten based on the Classified Directory of Dahlias (1983) published by the National Dahlia Society of England.

4.2.1 Single-flowered

Flowers up to 10 cm across with only a single row of petals. Disc apparent. Suitable for bedding purposes, *e.g.*, Bambino, Little Dorrit, Pinnochio and Yellow Hammer.

4.2.2 Anemone-flowered

Flowers with one or more rows of petals surrounding a dense group of long tubular disc florets. Fully double. Good for flower arrangement, *e.g.*, Comet, Scarlet Comet and Guinea.

4.2.3 Collerette

Flowers with one row of normal petals and one or more rows of small petals (the collar). Disc apparent. Very good for flower arrangement, *e.g.*, Choh, Chimborazo, Clair de Lune, Easter Sunday, Sincerity and Thais.

4.2.4 Paeony-flowered

Flowers have two or more rows of generally flattened petals. Disc apparent. Not much popular and the number of cultivars belonging to this group is extremely small, *e.g.*, Bishop of Llandaff and Fascination.

4.2.5 Decorative

Decorative dahlias have fully double flowers. The petals are broad, more or less flat or slightly twisty and wavy. The tips of the petals may either be rounded or pointed. Where the petals are revolute it should be less than one-quarter of the length of any petal. The back petals should recurve towards the stem. The decorative group covers a wide range of forms, from almost globular bloom with the petals more flat than the pompon or ball to less than half revolute petalled bloom. In many countries, decorative dahlias are again divided into two sub-groups: informal and formal. Decorative is the largest group among dahlias, and as mentioned earlier, this group is further divided according to size into following five sub-groups.

4.2.5.1 Giant Decorative

This is the most popular group among all dahlias. Giants have no maximum limit and so the natural tendency is to grow them as large as possible. There are cultivars producing blooms of 45 cm, *e.g.*, African Queen, Barbara Elaine, Bonaventure, Croydon Ace, Edge of Gold, Hamari Girl, Kelvin, Kenya Yellow, Nobbys Light, Shinsui, White Alvas and Walter Hardisty.

4.2.5.2 Large Decorative

Large decoratives are moderately popular and the number of cultivars under this sub-group is quite large. In India, they are very popular, and one of the most famous cultivars is Black Out. Few others in the sub-group are Arthur Hambly, Black Out Sport, Inca Metropolitan, Polyand, Silver City and Xantine.

4.2.5.3 Medium Decorative

Medium decoratives are highly popular and the bicolour ones among them are still more popular. They are very good for garden display and the bicolour ones are also good for flower arrangement, *e.g.*, April Dawn, Alloway Cottage, Bhikkus Mother, Daleko National, Evelyn Foster, Inca Matchless, King Soccer, Suffolk Spectacular.

4.2.5.4 Small Decorative

Small decorative cultivars are numerous and extensively used for garden display and flower arrangement. Bicolour ones are more popular for flower arrangement, *e.g.*, Corton Linda, Disneyland, Hamari Fiesta, Lady Linda, Nina Chester, Swami Madhavananda, Swami Lokeswarananda and White Linda.

4.2.5.5 Miniature Decorative

One of the most popular types for flower arrangement is the miniature decorative. Vase life of these flowers is also very good. They are also very useful for garden display, *e.g.*, Abridge Taffy, Christine Hammett, Elizabeth Hammett, Eastwin, Margaret Anne, Mistill Delight and Robert Walker.

4.2.6 Ball

Ball dahlias have fully double flowers. They are ball shaped or slightly flattened. Petals are blunt or rounded at the tips and they are involute for more than half their length. Many balls look very similar to pompons but when attain size they are larger than the latter group. Balls are subdivided into small ball (10 – 15 cm) and miniature ball dahlias (5 – 10 cm), *e.g.*, Alltamy Cherry, Crichton Honey, Nijinsky and Risca Miner are small ball dahlias and Connoisseur's Choice, Nettie, Rothsay Superb and Wootton Cupid are miniature ball dahlias.

4.2.7 Pompon

Pompon dahlias have fully double blooms, globular in appearance, like table tennis balls. Petals are involute for whole of their length and in this formation an individual petal looks somewhat like the cell of a honeycomb. The size limit for pompon is 50 mm. These are extremely popular for flower arrangement and have the longest vase life among all the dahlias, *e.g.*, Diana Gregory, Hallmark, Honeycomb, Little Willo, Mark Lockwood, Noreen, Small World, William John and Yellow Baby.

4.2.8 Cactus

Cactus dahlias have fully double flowers. Outer petals are narrow, preferably with revolute edges overlapping from the tips for at least two-thirds of their length. The petals may be either incurving or straight and involute instead of being revolute. This is a very large group and as mentioned earlier, is further divided into five sub-groups.

4.2.8.1 Giant Cactus

Giant cactus sub-group contains only few cultivars, *e.g.*, Donna Huston, Polar Sight and Yvonne Marie.

4.2.8.2 Large Cactus

Large cactus dahlias are very good exhibition items but the number of cultivars in the sub-group is not a large one, *e.g.*, Eastwood Snow, Golden Crown and Wootton Court.

4.2.8.3 Medium Cactus

Medium cactus is a large sub-group. The cultivars of this group are good for garden display and also for flower arrangement, *e.g.*, Arthur Lashlie, Camano Classic, Sioux and Sunset.

4.2.8.4 Small Cactus

Small cactus dahlias are very popular and suitable for garden display. The vase life of these cultivars is also good, *e.g.*, Alvas Doris, Border Princess, Dandy, Doris

Day, Klankstad Kerkrade, Lavender Athalie, Majestic Kerkrade, Park Princess, Shirley Alliance, and Tanjoh.

4.2.8.5 Miniature Cactus

Miniature cactus dahlias are more popular as cut flowers for flower arrangement than as exhibition flowers. The total number of cultivars under this sub-group is, however, small, *e.g.*, Frank Soetan, Little Glenfern, Rokesley Mini and Wee Willy.

4.2.9 Semi-cactus

Semi-cactus dahlias have fully double flowers. Like decorative dahlias, the half base of the petals is broad and flat. The other half of the outer petal is revolute from the pointed tip for more than one-quarter but less than half of the full length of the petal. The petals are either straight or slightly incurving. Semi-cactus is a large group and as mentioned above, this is divided into five sub-groups.

4.2.9.1 Giant Semi-cactus

In popularity, the giant semi-cactus dahlias are second only to giant decoratives. It is very good for exhibition. Since there is no upper size limit for any giants, these cultivars are also grown for producing largest possible flower for a plant. Examples, Alfred C., Carol Channing, Daleko Jupiter, Daleko Polonia, Gateway, Inca Dambuster, Le Vonne Splinter, Rene, Super and Vantage.

4.2.9.2 Large Semi-cactus

Large semi-cactus sub-group is also popular but not as the giant semi-cactus, *e.g.*, De Sarsate, Hamari Katrina, Hamari Princess, Neveric, Reginald Keene, Salmon Keene, *etc.*

4.2.9.3 Medium Semi-cactus

The cultivars of this sub-group are particularly useful for garden display and for flower arrangement *e.g.*, Amelisweerd, Daleko Venus, Davenport Sunlight, De-la-Haye, Hamari Bride and Lavender Symbol, *etc.*

4.2.9.4 Small Semi-cactus

Small semi-cactus dahlias are highly popular for cut flowers and garden display. A good number of quality cultivars are available under this sub-group, *e.g.*, Cefm Glow, Cryfield Bryn, Kimono, Match, Mt. St. Helens and Shandy, *etc.*

4.2.9.5 Miniature Semi-cactus

This sub-group does not contain large number of cultivars but it is very popular as cut flowers, *e.g.*, Douwa, Jean Wanless, Lynne Bartholomew, Mary Jo, Snip, Snow Queen and Twinkle Toes, *etc.*

4.2.10 Miscellaneous

This group consists of dahlia cultivars which do not fall into any of the above mentioned 9 groups. Many of the cultivars of this group are very good for flower arrangement, *e.g.*, Golden Star, Jescot Julie, Pink Giraffe and White Orchid, *etc.*

4.3.0 Fimbriated

Flowers fully double. Petals fimbriated from the tips, preferably for at least 10 mm.

If fimbriated factor is not taken into consideration, the cultivars of this group can conform very well to the above mentioned different groups. Thus, it can be a fimbriated semi-cactus cultivar or a fimbriated decorative cultivar and so on. As mentioned earlier, fimbriated dahlias are divided into four sub-groups.

4.3.1 Giant Fimbriated

Cultivar – Frontispiece.

4.3.2 Large Fimbriated

Cultivars – Louise McKelvey and Zangiku.

4.3.3 Medium Fimbriated

Cultivars – Frilly Dilly, Komplement, Valley St.Lambert.

4.3.4 Small Fimbriated

Cultivars – Cheynne, Sultan, Zetsubi.

4.3.5 Water Lily

The blooms are fully double. Outer petals are broad and slightly cupped with rounded ends. From the side view, a bloom looks like a saucer. The rows of petals are openly spaced to give the bloom a delicate and light appearance. The flowers resemble water lily (Nymphaea) flowers in appearance. Water lily dahlias are becoming very popular as cut flowers and a good number of cultivars are available these days, *e.g.*, Christopher Taylor, Erin Ann, Ferne Irene, Figurine, Jescot Lyn, Porcelain, Ruthie G., Scaur Princess, Yamabiko, Yelno Petite Glory, *etc.*

Both fimbriated and water lily are recognized as independent groups in some countries and these two groups are becoming increasingly popular.

4.3.6 Star-flowered

Small cupid-shaped flowers having two or three rows of pointed petals which overlap very slightly. These are now seldom cultivated, *e.g.*, White Star.

Thus, a large number of cultivars are available to suit the requirement of the growers. However, selection for commercial cultivation should be made considering several factors, particularly on the basis of their performance and demand in the area. Performance of several cultivars was studied by Gardner (1963) and he recommended 12 cultivars for the production of cut flowers. Similarly, Moe (1978) suggested cultivars suitable for pot cultivation. Mastalerz (1976) also listed dahlia cultivars which can be grown in different seasons of the year and used as a new cut flower crop. In a series of investigations under various weather conditions Mehlis (1976a, b; 1978) tested a large number of cultivars and suggested the promising ones. Cultivars Bodensceperle, Kurdirektor, Dikkmann, Rote Spinne, Hanna Hofbauer and Wolstad were found to be outstanding.

Groups of Dahlia

Grp 1 Dahlia 'Moonfire'

Grp 2 Dahlia 'Mexico'

Grp 3 Dahlia 'Mars'

Grp 4 Dahlia
'Keith H'

Grp 5 Dahlia
'Pink Kenya'

Grp 6 Dahlia
'Carstone Valiant'

Grp 7 Dahlia 'Bowen'

Grp 8 Dahlia 'Apricot Beauty'

Grp 9 Dahlia 'Amy K'

Grp 10 Dahlia
'Jescot Julie'

Grp 11 Dahlia
'Hannah Baker'

Grp 12 Dahlia
'Midnight Star'

Cultivars of Dahlia

Avoca Amanda

Badger

Bishop

Black Torch

Bode

Cherwell Linnet

Croydon Monarch

Devon Eclipse

Edinburgh

Eternal Snow

Honka Red

Jescot Julie

Cultivars of Dahlia

John Hill

Kelvin Floodlight

Mistil Delight

Nenekazi

Pineland Princess

Pink Jupitor

Pop Willo

Prime Minister

Rycroft Brenda

Sophie Taylor

Tahoma Hope

White Lace

In Lucknow, Sharga (1975) evaluated 22 different cultivars and found that Axford Triumph, Kelvin Rose, Mrs. John Ralph and Nirmal Chandra were the most promising and suitable cultivars for outdoor cultivation under North Indian conditions. Performance of a large number of cultivars was studied by Bhattacharyya *et al.* (1976) in the plains of West Bengal and those suitable for cultivation were recommended.

5.0 Genetics and Breeding

5.1.0 Genetics

Today's dahlias are not just hybrids but hybrids of hybrids. Origin of these cultivars can be traced back to a sizeable number of wild species which are still available in Mexico, Central America and northernmost South America. There are different opinions about the combinations of species that influenced the present dahlias most. Darlington (1973) considered it to be *D. imperialis* x *D. coccinea* while Sorensen (1969) suggested it to be *D. coccinea* x *D. pinnata*. Beside these three species, there are other species also which have contributed to the evolution of the present-day dahlias.

The chromosome number in dahlia has been found to vary. According to Darlington (1973), the wild species from which *D. variabilis* (n = 64) is supposed to have arisen are tetraploids with 32 chromosomes. The union of the wild species (*D. imperialis* and *D. coccinea*) resulted in octaploid. The octaploid retained free pairing among the 4 chromosomes each of 16 kinds. Giannasi (1975) stated that cultivated dahlias are tetraploid (n=32) and different species have different chromosome numbers. The races of *D. coccinea* are either with 16 or 32 chromosomes, *D. imperialis* with 16, *D. merckii* with 18, *D. linearis* with 17 and *D. pinnata* with 32. According to Fedorov (1979), *D. coccinea* has 32 or 64 chromosomes; *D. coronata*, 32; *D. dissecta*, 36; *D. imperialis*, 32; *D. juarezii*, 64; *D. maxonii*, 32; *D. merckii*, 36 and *D. variabilis*, 32 or 64. In each cell of *D. variabilis*, there are eight sets of chromosomes with eight individual chromosomes in each set, hence 64 chromosomes in all.

Darlington (1973) suggested that octaploid *D. variabilis* appears to be strictly self-incompatible. The incompatibility accelerated the very process which was necessary to produce recombination at a time when breeders would use natural open-pollinated seeds. The great variability in dahlia plants was found to be due to the prevailing self-incompatibility which maintained in all progeny, a rather high degree of heterozygosity. Swami Vinayananda (1986), however, stated that all dahlias are not self-incompatible and more than 25 per cent cultivars are self-compatible. Hybrids raised from *Dahlia scapigera* × *D. coccinea* were also found to be self-compatible (Sorensen, 1969). The self-fertilized seeds or resultant plants also retain the heterozygous nature which is common to all double dahlias.

From the pollen-fertility point of view, there are fertile and sterile cultivars. Easy seed-bearing but pollen-sterile cultivars are very useful for induced bee pollination. Both pollen-sterile and self-incompatible cultivars are much in use for hand pollination in the Western countries.

According to seed-bearing habit, dahlia cultivars can be divided into three groups : (i) easy seed-bearing, which forms seeds both in the disc and ray florets, (ii) those which form seed only in the disc floret and the ray florets are sterile, and (iii) cultivars in which both disc and ray florets are sterile. Whether there are cultivars with ray floret fertile and disc floret sterile is not known. Ray florets are used more frequently in hand pollination than disc florets.

Characters like plant height, growth habit, appearance, position and colour of flowers, *etc.*, are governed by genes. Only a few genes may be required to determine the length of the stem, while there may be several genes which are collectively responsible for the colour of a specific cultivar (Paul Hale, 1983).

The flowers of *D. variabilis cv.* Scarlet Star contain relatively high activity of (+)-dihydroflavonol (DHF) 4-reductase (Fischer *et al.*, 1988). The yellow pigments, the 4´-malonylsophoroside and 4´-malonylglucoside of butein, were characterized in the petals of *Dahlia variabilis* cv. Rocquencourt and *D. coccinea*. They also occur with malonylated anthocyanins in these flowers. Their presence in the Mexican *D. coccinea* supports the view that the cultivated *D. variabilis* could have arisen from hybridization involving this wild species occurring in plants with malonyl attachment (Harborne *et al.*, 1990). Stich *et al.* (1994) also reported that isoliquiritigenin 4´-glucoside and butein 4´-glucoside are responsible for yellow colour in *D. variabilis*.

Bicolour flowering dahlias are generally characterized by a coloured basal part and a white tip. However, they produce single-coloured petals frequently. This petal colour lability prevents the uniform cut/pot flower production and reduces the economic value of bicolour cultivars. Ohno *et al.* (2016) studied this mechanism in red-white bicolour flowering cultivar, 'Yuino' and observed a strong relationship between inflorescence colour and leaf phenotype. Red petals-producing plants accumulated flavonoids in leaves, whereas bicoloured flowering plants failed for flavonoid accumulation which suggests that petal colour is associated with flavonoid synthetic potential in shoot. Among flavonoid biosynthetic genes including four chalcone synthase (CHS)-like genes (DvCHS1, DvCHS2, DvCHS3 and DvCHS4), DvCHS1 and DvCHS2 had significantly lower expression levels in the white part of bicolour petals than in red petals, while DvCHS3, DvCHS4 and other genes had almost the same expression levels. Small RNAs from the white part of a bicolour petal were mapped onto DvCHS1 and 2, whereas small RNAs from red single coloured petal were not mapped on any of the 4 CHS genes. The expression level of DvCHS2 was down-regulated in flavonoid-poor leaves and small RNAs from these leaves were mapped onto DvCHS2 suggesting the possibility of down-regulation post-transcriptionally. Genomic analysis also suggested DvCHS2 as the key gene involved in bicolour formation via post-transcriptional gene silencing (Ohno *et al.*, 2018).

Deguchi *et al.* (2016) investigated the mechanism of black flower colour of dahlias which might be attributed to high accumulation of cyaniding (Cy)-based anthocyanins. Four major anthocyanins accumulated in dahlia petals: 3,5-diglucoside (3,5diG) and 3-(6″-malonl glucoside)-5-glucoside (3MG5G) of Pelargonidin and cyaniding were purified and their colours were evaluated *in vitro* at various pH and various concentrations. At pH 5 which is approximately same as that of dahlia petals and at pH 3 at which anthocyanins are relatively stable, the Petal lightenes (L*) and chroma (C*) of Cy 3,5diG were similar to or higher than that of Pg 3,5diG suggesting that Cy3, 5diG did not contribute more to black flowering than Pg3,5diG. On the other hand, the L* and C* of Cy3MG5G were significantly lower than that of Pg3MG5G suggesting the main contribution of Cy3MG5G for black colour than Pg3MG5G. It was concluded that because of predominant accumulation of 3MG5G type anthocyanins in petals and higher contribution of Cy-type in lowering L* and C*, the high accumulation of Cy-based anthocyanins was found critical for black flower colour of dahlias.

A correlation coefficient at genotypic and phenotypic levels among the seven traits studied by Bhattacharjee and Wahi (1982) showed the same trend, and genotypic correlations were on the higher side. The estimates of heritability and genetic advance were quite high for all the characters except the number of leaves per plant. Path analysis of thirty nine different genotypes indicated that the flower diameter, plant height, longevity of flowers and number of branches were important component characters for the number of flowers per plant. Vikas *et al.* (2011) found that higher PCV and GCV were obtained for stalk length, flowering duration, number of ray florets per flower and individual flower weight. High heritability with high genetic advance was recorded for vase life, flower diameter, flower stalk length and individual flower weight.

The nature and magnitude of genetic diversity, as measured by Mahalanabis's D^2 statistics was studied by a set of 7 quantitative traits in 39 indigenous and exotic cultivars of *Dahlia variabilis*. The population were grouped into 9 different clusters. The principal component analysis showed that the major portion of variability in the population was explained by the first 2 canonical roots (Suman *et al.*, 1984). On the basis of D2 analysis of 9 characters, Misra *et al.* (1990) grouped the 23 *D. variabilis* varieties into 10 clusters. Intercluster values suggested that very little domestication had occurred. Three clusters comprised of only one variety each (Arthur Godfrey, African Queen and Pioneer) and these varieties were recommended for the exploitation of hybrid vigour.

Based on the cluster analysis, Nimbalker *et al.* (2006) revealed that out of 10 clusters cluster I, II and III were largest with 5 genotypes in each cluster, cluster IV, V and VI having 2 genotypes in each cluster and VII, VIII, IX and X were monogenotypic clusters. The inter-cluster distance was between the cluster VI and DC. It was also found that number of ray florets per flower, fresh flower weight, days to full bloom from bud break to complete withering, freshness from bud break to withering and number of disc florets/flower contributed considerably to divergence.

Devi *et al.* (2020) reported that flower yield per plant had positive genotypic correlation with leaf area index, crop duration, flowering duration, tuber weight and change in fresh weight at day 3. Path-coefficient analysis revealed a positive direct effect of crop duration, flowering duration, flower diameter, vase life, total chlorophyll content and change in fresh weight at day 3 on flower yield per plant proving the direct selection of these traits for yield improvement in dahlia.

5.2.0 Breeding

Breeding programmes have focussed on improving various characteristics to enhance ornamental values, including flower colour, size and form, and production quality. Although desirable traits have been introduced by classical breeding, there are limitations to this technique. Firstly, distant crosses may be limited by incompatibility or differences in ploidy level that is very common in dahlias. Secondly, characteristics such as uniform growth and synchronous flowering are polygenic. Thirdly, several viruses are known to infect dahlia. Alternative breeding methods will provide faster procedures. In the past, classical breeding approaches like introduction, hybridization, composite crossing, multiline, and backcross breeding were utilized for this purpose. However, each of these methods has advantages and disadvantages. Recent developments in plant biotechnology such as directed mutation, genomics and recombinant DNA technology were adapted by breeders to develop more improved cultivars of dahlias.

5.2.1 Selection and Hybridization

Breeding in dahlia should be done to create something new and extremely beautiful. One must be able to observe the desirable characteristics in the parents or at least should have evidence that the parents carry those characteristics. The parents are to be selected for testing, initially by analyzing, the characters which they possess. A potential parent should be observed for several years before the test crosses are made. After hybridization, the progeny is to be analyzed. There is a difference between the performance of a cultivar as a plant and as a parent. By progeny testing, a number of superior parents can be identified and used in the subsequent combinations. Cross pollination can be effected by planting parents and combinations of parents in isolation from other dahlias. Insects carry out the cross pollination during the flowering season, and bees are important dahlia pollinators. Dahlia hybridization has so far been done by the commercial dahlia growers and amateures in different parts of the world, mostly in America, New Zealand, Holland and England. In India, Swami Vinayananda, a monk of the Ramakrishna Order, has done a good deal of breeding work using hand-pollination technique. Some of his most popular hybrids are Bhikkus Mother, Bhikkus Vivek and Swami Lokeswarananda.

Gatt *et al.* (2000) successfully produced a large number of artificial hybrids between *Dahlia* species based on x=16 and studied their genomic relationships. Chromosomal behaviour in these artificial hybrids was similar to that observed in parental species suggesting that there was a considerable degree of homology between the genomes of these *Dahlia* species. Using genomic *in situ* hybridization (*GISH*) demonstrated that in these hybrids bivalent formation involved pairing only between parental genomes. The ability of *GISH* to differentiate between parental genomes in artificial hybrids was variable, indicating that molecular divergence of highly repeated sequences accompanied the evolution of these species. However, the extent of chromosome pairing and chiasma formation in the hybrids did not reflect the differences that can be detected by *GISH*.

The vase life of cut dahlia (*Dahlia variabilis*) flowers is very short and genetic improvement of this trait is desirable. Onozaki and Azuma (2019) carried out conventional cross-breeding techniques and found large significant differences among 24 dahlia cultivars. Out of which, 9 cultivars, Syukuhai, Rinka and Micchan had long vase life; 8 cultivars *viz.*, Kamakura, Agitate, Benifusya had normal vase life and 7 cultivars *viz.*, Gin-Ei, Port Light Pair Beauty, Yumesuiren had short vase life. Twenty two cultivars were repeatedly crossed and selected for promising offspring with longer vase life. Two cycles of selection and crossing led to a 1.7 day increase in vase life from first to third generation proving that this approach could extend the vase life of dahlia flowers. The mean vase life of 'Kamakura' was 5.0-6.2, 6.0-6.8 and 6.0-7.6 days, whereas, it was 6.2-12.0, 6.6-10.2 and 9.4-13.6 days for six finally selected lines in distilled water, isothizolinic antibacterial agent and GLA solution (10 g/l glucose, 0.5 ml/l CMIT/MIT and 50 mg/l aluminium sulphate), respectively. Moreover, the selected second generation line 606-46 showed a stable longer vase life that 'Kamakura'. 'Micchan' which had longer vase life was a common progenitor used for breeding of parental lines in cross combinations with long vase life and the six selected lines were all found as progeny of 'Micchan'. This strongly suggested that 'Micchan' had heritable genes related to long flower vase life.

5.2.2 Mutation Breeding

Mutation is a natural or artificially induced change of the genetic information contained in a cell. It occurs when the order of gene is changed or damaged. This can happen sexually in the ovule through the chance mating of dissimilar chromosome. However, mutants can be induced by irradiation or chemical interference. The cultivar Juanita provides a good example of natural mutation. Dohare *et al.* (1974) located a sectorial chimera in White Pearl which has a deep magenta streaked flower and named as Manali. There are some cultivars which have resulted from spontaneous mutation and these are listed by Robertson (1982) as follows : Reginald Keene, a bronze blend sported pink blend (Candy Keene) and salmon pink blend (Salmon Keene). The cultivar Frank Hornsey, a bronze and yellow blend changed to pink (Pink Frank Hornsey), yellow with bronze (Yellow Hornsey), yellow (Lemon Hornsey), pink and white blend (Pearl Hornsey) and yellow with rose flush (Rose Hornsey). Klankstad Kerkrade, a cream flowered dahlia sported white (White Kerkrade) and cream and pink blend (Majestic Kerkrade). The cream coloured cultivars like Alva's Supreme and Rustig sported to White Alva's and White Rustig. The orange and yellow blend Daleko Jupiter sported Pink Jupiter. Another orange cultivar, Shirley Alliance changed to pink (Pink Shirley Alliance and Rosemary Clare). The Bronze Symbol produced dark pink (Pink Symbol) and lavender and yellow blend (Lavender Symbol). Lemon Chester and Yellow Chester are yellow coloured sports from the bronze coloured Paul Chester. Work on improvement was done by Broentjes and Ballego in Holland in the 1960's in which multiple flower mutations were obtained through tuber irradiation. The best known cultivars were Rotonde and Ornamental Rays obtained from Salmon Rays. A similar work, however, did not produce any desired result (Robertson, 1981). In an irradiation experiment on the tubers of 14 dahlia cultivars with gamma rays at doses up to 8 kR, Das *et al.* (1978) recorded 19 mutants, the optimal dosage for mutation induction being 2 to 3 kR. Tubers of 9 cultivars of dahlia were treated with 60Co gamma radiation (doses were in the range 0-4.5 kR) and were evaluated. After 0.5 kR treatment the 4 dahlia cultivars showed increased plant height (Misra, 1990).

5.2.3 Transgenic Breeding

Suzuki *et al.* (2002) identified and isolated a cDNA coding for 3-glucoside-specific malonyl transferase for anthocyanins *i.e.* malonyl-coenzyme A: anthocyanidin 3-O-glucoside-6″-O-malonyl transferase (DvMaT) from dahlia flowers. Dv3MaT coded for a protein of 460 amino acids, Quantitative real-time PCR analyses of Dv3MaT showed that the transcript was present in accordance with the distribution of 3MaT activities and the anthocyanin accumulation pattern in the dahlia plant. This cDNA was expressed in *E. coli* and introduced into petunia whose red colour is exclusively provided by cyanidin 3-O-glucoside and 3,5-O-diglucoside. Thirteen transgenic petunia lines were found to produce malonylated products of these anthocyanins (11-63 mol per cent of total anthocyanins in the flower). The spectral stability of Dv3MaT was significantly higher than that of cyaniding 3-O glucoside. Moreover, 6″-O-malonylation of this effectively prevented the anthocyanin from glucoside attack. This insisted that malonylation should serve as a strategy for pigment stabilization in the flowers.

Otani *et al.* (2013) established an efficient system for Agrobacterium-mediated transformation for *Dahlia x pinnata* 'Yamatohime'. Mass of shoot primordia (MSP) induced on MS medium supplemented with 10 mg/l TDZ were inoculated with *Agrobacterium tumefaciens* strain EHA101 (pIG121-Hm) harbouring both β-glucuronidase (GUS) and hygromycin resistant genes. After 2 days of co-cultivation, the MSP were transferred to a selection medium containing hygromycin with meropenem for bacterial elimination. Shoots were successfully regenerated from survived MSP on hormone-free medium without hygromycin and rooted on hormone-free medium containing hygromycin. Hygromycin resistant plants showed histochemical blue staining for GUS. Transformation of plants was confirmed by PCR and Southern blot analysis. Transgenic 'Kokucho' were also produced by using the same transformation procedure, suggesting wide applicability of this transformation method for other dahlia cultivars.

Thevissen *et al.* (2000) identified a *Saccharomyces cerevisiae* mutant that is markedly more resistant than wild-type to DmAMP1. A genomic library of DmAMP1-sensitive wild type yeast was introduced into DmAMP1-resistant yeast mutant and screened for restored sensitivity. The gene determining sensitivity of the yeast was reported as IPT1, a gene encoding an enzyme involved in the last step of synthesis of sphingolipid-(mannose-inositol-phosphate) 2-ceramide. Strains with noon-functional IPT1 allele lacked the above sphingolipid in their plasma membranes and bound with significantly less DmAMP1 and were resistant. However, these deviations could be restored by introducing functional IPT1 gene.

6.0 Growth and Flowering

6.1.0 Environmental Factors

6.1.1 Temperature and Light

Dahlia plants are responsive to day and night temperatures and also to day length. Plants of several dahlia cultivars grown under glass at different seasons initiated flower buds at all temperatures tested, in the ranges 50-70°F (10-21°C) at night in winter and 60-80°F (15.6-26.7°C) in summer. Bud development was accelerated and plants flowered sooner as temperature increased. Plant age or size did not influence flower initiation but day length was the regulating factor. A screening programme of cultivars showed 2 groups of plants : (i) flower initiation and development could not be prevented by continuous long day, (ii) plants remained vegetative in 16 h or longer days and initiated flowers in day length shorter than 14 h. When vegetative plants from group (ii) were given 9 h light until the buds opened, the stems were very short and the floret number was greatly reduced. However, normal stem length and flower shape were obtained by subjecting vegetative plants to 9 h light for 14 days followed by long days (18-24 h) until the buds opened (Mastalerz, 1976).

In an experiment with *cvs.* Kolchelsee and Park Princess, Durso and Hertogh (1977) observed that the best quality plants were obtained when grown in 25°C day and 16°C night temperatures. Flowering was delayed at 24°C (day temperature) and 12°C (night temperature) and although flowering was accelerated at 28°/17°C and 29°/20°C day/night temperatures, plant quality was adversely affected (Table 1).

Table 1: Effect of Day/Night Temperature Regimes on the Forcing of the Tuberous-Rooted Dahlia 'Kolchelsee'*

Plant Organ Evaluated	*Planting Date*	*(°C)z*			
		24/12	*25/16*	*28/17*	*29/20*
Days to flower					
First flower	Jan 30	81a/d	71b/d	59c/d	55c/d
	Feb 20	84a/d	75b/d	57c/d	54c/d
Second flower	Jan 30	88a/d	76b/d	63c/d	62c/d
	Feb 20	91a/d	82a/d	61b/d	59b/d
Third flower	Jan 30	94a/d	81b/d	67c/d	66c/d
	Fed 20	96l/d	84b/d	66c/d	59c/d
Plant height (cm)					
First flower	Jan 30	37.6a/d	31.6a/d	34.4a/d	19.5b/d
	Feb 20	38.1a/d	30.7b/d	32.1ab/d	29.4b/e
Second flower	Jan 30	42.0a/d	32.7b/d	38.7b/d	24.7c/d
	Feb 20	47.3a/e	37.5bc/e	39.3b/d	31.5c/e
Third flower	Jan 30	45.2a/d	35.6bc/d	40.7ab/d	29.5c/d
	Feb 20	50.4a/d	37.5b/d	39.4b/d	36.7b/e
Flower longevity (days)					
First flower	Jan 30	5.8a/d	6.1a/d	6.5a/d	5.9a/d
	Feb 20	6.3a/d	6.0a/d	6.0a/d	5.7a/d
Flower diameter (cm)					
First flower	Jan 30	6.0a/d	6.0a/d	6.3a/d	5.7a/d
	Feb 20	6.9a/d	7.3a/d	6.2a/d	5.3a/d

Plant Organ Evaluated	*Planting Date*	*(°C)z*			
		24/12	*25/16*	*28/17*	*29/20*
No. of shoots/plant					
Total shoots	Jan 30	2.3a/d	2.5a/d	1.9a/d	2.5a/dd
	Feb 20	1.9a/d	2.0a/d	2.1a/d	1.9a/d

z Means followed by the same letter (a,b,c) preceding the slash(/) are not significantly different as determined by Tukey's test at the 5 per cent level for each row. Each organ for each date and parameter was analysed separately. Means followed by the same letter (d,e) after the slash (/) are not significantly different at the 5 per cent level for each column. Each organ per treatment and parameter was determined separately.

* Durso and Hertogh (1977).

Brondum and Heins (1993) studied the effects of temperature and photoperiod on growth rates and morphological development of *Dahlia pinnata* cv. Royal Dahlietta Yellow. Potted plants were grown under day temperature (DT) and night temperature (NT) ranging from 10 to 30°C and photoperiods from 10 to 24 h/day. Photoperiod influenced vegetative development more than reproductive development as plants flowered in all photoperiods. Temperature interacted with photoperiod to greatly increase tuberous root formation as temperature decreased from 25 to 15°C. Flower count and diameter decreased as average daily temperature increased. The minimum, optimum and maximum temperatures for leaf-pair unfolding rate (0.29 leaf pair/day, 5.5, 24.6 and 34.9°C, respectively), flower development rate from pinching to visible bud (0.07 flower/day, 2.4, 22.4 and 31.1°C, respectively) and flower development rate from visible bud to full bloom (0.054 flowers/day, 5.2, 24.4 and 31.1°C, respectively). It is concluded that a photoperiod of 12-14 h and a temperature of 20°C are needed to produce fast-developing plants of appropriate height and with many large flower buds which are suitable for the pot plant market.

Heins *et al.* (1995) treated dahlia seedling plugs to different combinations of light (0, 1 or 5 μmol/m^2/s from fluorescent tubes) and temperature (0, 2.5, 5, 7.5, 10 or 12.5°C) during storage for up to 6 weeks. Plant quality was better in the light than in darkness as storage temperature and duration increased. The optimum storage temperature was 5°C. The major problem with plug storage was the increased incidence of botrytis under conditions of low vapour pressure deficit (VPD) and the requirement for frequent irrigation under conditions of high VPD.

Zimmer (1989) reported that dahlia is a facultative SD to day neutral plant. Natural photoperiod during spring which increased from 10 to 14 h was optimal for forcing (Durso and Hertogh, 1977). Long days given either as 16 h photoperiod or as 4 h night-break slightly delayed flowering. Dahlias grown under 8 h photoperiod flowered earliest but all plants did not flower (Table 2). High light intensities were necessary during forcing. Under 50 per cent shading, plants were too tall for pot plants. Canham (1969) also studied the effect of night-break lighting on the production of cuttings from dahlia tubers. He reported that cvs. Tornado and Jescot Jim produced 37 per cent and Brandaris 61 per cent more cuttings under 2-night-break using warm white fluorescent tubes than the untreated ones. Positive response to the night-break was also recorded by Hall (1969).

Table 2: Effect of Various Photoperiods on Forcing of the Tuberous-Rooted Dahlias, 'Kolchelsee' and 'Park Princess'*

Cultivar and Treatments[w]	*Days to Flower*[y]	*Plant Height (cm)*	*Flower Longevity (Days)*	*Flower Diameter (cm)*	*Number of Shoots*
Kolchelsee					
Natural daylength (ND)	65ab[x]	31.3a	5.8a	5.9a	1.7ab
8 h	(0 per cent)	–	–	–	1.7ab
16 h	67ab	32.6a	6.3ab	6.0a	1.5ab
Night break (4 h)	72b	32.2a	7.6b	6.2a	1.7ab
4 h/ND	69ab	36.3a	6.9ab	5.9a	1.7ab
16 h/ND	64a	35.4a	6.0a	6.1a	1.8a
QD/4 h	64a	35.1a	6.0a	6.1a	1.6ab
QD/16 h	65ab	33.4a	5.6ab	6.3a	1.1b
Park Princess					
Natural day length (ND)	68a	30.8a	8.5a	12.5ab	1.5a
8 h	56b (78 per cent)	34.0a	5.8b	12.0a	1.4a
16 h	70a	29.1a	9.0a	13.7b	2.2b
Night break (4 h)	71a	32.8a	9.4a	13.4b	2.1b
4 h/ND	71a	30.7a	8.2a	13.5b	1.9ab
16 h/ND	72a	30.9a	8.9a	13.4b	1.9ab
ND/4 h	69a	31.7a	9.0a	13.2ab	1.8ab
ND/16 h	71a	32.9a	8.4a	13.1ab	1.7ab

y: Maximum per cent of the plants flowering in parentheses.

x: Mean separation in columns within cultivars by Tukey's test, 5 per cent level.

w: Combination treatments were changed 5 weeks after planting.

* Durso and Hertogh (1977).

Hasegawa *et al.* (2009) studied the effect of planting time and short-day treatment on growth and flowering of *Dahlia imperialis*. The stems grew slowly during June-August at high temperatures, whereas, elongated rapidly from September at low night temperatures. The stem elongation was completed stopped during flowering. It was reported that stem length, number of flowers produced and flowering duration were better when planted early. However, plants cultivated in glasshouse produced very few flowers. The plants were also treated with 10 hrs light (short day) and planted in Wagner pot. Results showed that plants undergoing early short-day treatment hastened flowering and maintained a prolonged flowering period.

Okada *et al.* (2020) studied the effect of low temperature on flower colour fading of dahlia *cv.* 'Nessho' ray florets. Solid ray florets of this variety frequently exhibited colour fading from autumn to spring. The plants that produced original red-coloured capitulum in February were maintained as relatively low sensitive (RL) to low temperature plants and two plants that showed colour fading in June as relatively high sensitive (RH). Cuttings obtained from three RL (RL 1, 2, 3) and two RH (RH 1, 2) plants were evaluated. RL-2 line plants exhibited a significantly lower extent of colour fading index and significantly higher number of completely red capitulum than all RH-line plants when the minimum temperature was below 10° C. It was suggested that the RH-lines had greater sensitivity and could decrease productivity of red coloured capitulum. From June-July, A RL-2 line produced colour-faded capitulum which needed elimination of RH plants for controlling colour fading in 'Nessho'.

Ciobanu *et al.* (2016) reported that forcing of dahlia tubers affected the accumulation of soluble dry matter content of dahlia tubers. The highest soluble dry matter content was obtained with forced tubers of 'Star Favourite' (25.47 per cent), unforced tubers of 'Tsuki Yori No Sisha' (24.80 per cent) and 'Kennemerland' (24.27 per cent), forced tubers of 'Hayley Jane' (23.97 per cent) and 'Tsuki Yori No Sisha' (22.57 per cent). These dahlia cultivars could be utilized for inulin extraction owing to their higher soluble dry matter content.

6.2.0 Growth Regulators

Synthetic growth substances have proved very effective in the regulation of growth and flowering of dahlia. A broad range of effects, both morphological and physiological, was observed by the application of growth substances by various workers. Different chemicals displayed a remarkable variation on growth, flowering and tuberous root formation of dahlia (Bhattacharjee *et al.*, 1978; Kumar *et al.*, 1978; Kannangara and Booth, 1978; Maiko and Yashchenko, 1980).

6.2.1 Effect on Vegetative Growth

Retardation of plant height in different cultivars of dahlia with the application of SADH (Table 3) and ineffectiveness of CCC in this regard was reported by Bhattacharjee *et al.* (1971). Suppression of plant height by the application of MH and ethrel and stimulation of more number of branches by CCC and ethrel were observed by Pappiah and Muthuswamy (1974). Foliar application of cycocel at 4000 ppm reduced plant height of decorative dahlia cvs. Prime Minister and Thelma Davidson but increased in cv. Vigour (Hossain *et al.*, 1999). Treatment with IAA and GA_3 at 100 and 1000 ppm showed marked promotion in shoot length of *Dahlia variabilis* cv. Masterpiece (Bhattacharjee *et al.*, 1976). Similarly, TIBA at 2000 ppm or MH at 5000 ppm markedly reduced plant height compared with the control (Bhattacharjee, 1984).

Runkova (1985) studied the effect of cytokinins on plant growth. Plants were treated with kinetin, BA or kartolin each at 20-40 ppm at the beginning of growth. Plant height stimulation with kinetin and BA was slight while kartolin had practically no influence and in some cases it had slightly inhibited growth. BA and kinetin appreciably increased the number of lateral branches on the main stem. *D. variabilis* plants were sprayed with GA_3 or alar (daminozide) at 25 and at 35 days after transplanting (20 November). GA_3 treated plants were significantly taller. Alar at 150 ppm produced shorter plant, largest leaf number, the widest stem diameter, the maximum number of branches per plant, the most flowers per plant, a longer flowering period and the highest seed weight per plant, compared with controls and most other treatments (Singh *et al.*, 1994).

Table 3: Effect of SADH on Growth and Flowering in some Cultivars of Dahlia*

Cultivar	*Treatments*							
	Control		*SADH*					
	Height of Plant (cm)	*Number of Flowers*	*Per cent of Retardation Over Control*			*No. of Flowers per Plant*		
			2500 ppm	*5000 ppm*	*10000 ppm*	*2500 ppm*	*5000 ppm*	*10000 ppm*
Dorothy Stanton	89	5.6	5	28	33	7.0	6.8	6.5
Cr. Masterpiece	65	7.0	22	28	29	7.4	8.0	8.0
Cherokee Beauty	56	3.7	19	27	37	2.7	3.5	4.0
Donald Vandemark	73	4.6	27	27	33	5.2	5.8	5.6
Lacemaker	63	5.2	29	38	43	6.2	6.8	6.0
Fancy free	66	5.2	23	29	45	7.5	4.5	6.2
Avalanche	73	7.2	32	48	49	7.5	8.2	7.4
Black Out	69	3.6	21	29	44	2.6	5.0	4.6
Vigour	77	4.0	24	28	35	4.5	5.0	4.0
Kalyani	61	3.9	15	28	43	3.5	5.0	3.5

* Bhattacharjee *et al.* (1971).

Whipker and Hammer (1997) drenched the substrate of potted (1.5 l pot) dahlia cvs. Red Pigmy and Golden Emblem with paclobutrazol (1-16 mg/pot), ancymidol (0.5-8 mg/pot) and uniconazole (0.125-2 mg/pot) to obtain marketable plants. Marketable potted plants of cultivar Red Pigmy were produced with doses of 2-4 mg paclobutrazol, 0.25-0.5 mg uniconazole or 0.5 mg ancymidol/pot. All treatments reduced total plant height of the more vigorous cv. Golden Emblem by >11 per cent compared with the untreated control, the latter having a height of 82.1 cm. Marketable potted plants of this cultivar were produced with 4-8 mg paclobutrazol, 0.5-1 mg uniconazole or 2 mg ancymidol/pot. Shravan-Dasoju *et al.* (1998) drenched paclobutrazol (0, 2 and 4 mg a.i./pot) to *cv.* Red Pigmy potted dahlias, grown in substrates containing 50 per cent, 60 per cent, 70 per cent or 80 per cent (by volume) sphagnum peat or coir, with the remainder being perlite. Growth was significantly reduced as paclobutrazol concentration increased. Coir-based substrates did not reduce the activity of paclobutrazol drenches compared with peat-based substrates, although to compensate for the greater amount of plant growth in coir-based substrates, higher concentrations of paclobutrazol was suggested to reduce plant growth.

Dormant tubers of *D. variabilis* cvs. Golden Emblem and Red Pigmy were potted in 1.5 l pots. Eleven days later, ancymidol at 1.5 and 2.0 mg *a.i.*/pot; paclobutrazol at 0.24, 0.47, 0.95 and 1.9 mg *a.i.*/pot; uniconazole at 0.24 and 0.47 mg *a.i.*/pot were applied as soil drenches (Whipker *et al.*, 1995). The ancymidol drenching reduced inflorescence height rather than leaf canopy height and took more days to flower. Uniconazole caused height reduction of 20 per cent and produced marketable pot plants of *cv.* Red Pigmey. Paclobutrazol reduced height of Golden Emblem.

Khan and Tewari (2003) reported that GA (90 ppm) increased plant height and leaf area, while, chloremquat (6000 ppm) resulted in maximum reduction. Similarly, chloremquat (4000 ppm) increased stem diameter, number of branches and leaves. Significant reduction in number of days to full bloom was observed in chloremquat (6000, 4000 and 2000 ppm). Flower diameter, shelf life and number of flowers were observed maximum with chloremquat (4000 ppm). Whipker *et al.* (2011) revealed that dahlia tubers presoaked with fluprimidol at 11.6 and 20 mg/l caused 26 and 22 per cent shorter canopy height, respectively. However, tuber soaks with 20 and 40 mg/l delayed flowering by 11 to 16 days, respectively.

Akbar *et al.* (2011) revealed that the application of chloremquat (5500 ppm and 4500 ppm) reduced plant height and leaf area but, increased stem diameter and number of branches per plant. Singh *et al.* (2017) obtained better plant spread, number of leaves and number of branches with GA_3 200 ppm at 30, 45 and 60 days after transplanting in *cv.* 'Suryodya'.

Potted dahlias can be obtained with dwarf varieties grown from seeds or vegetatively-propagated tubers after reducing its size by means of growth retardants (Anónimo, 2005). Foliar application of phytoregulator Nettle (*Urtica dioica*) at 20 g fresh weight per liter of water showed greater height, with a value of 6 cm over 30 days which surpassed TF Biozyme growth of 5 cm and Agromyl V growth of 4 cm. Chemical phytoregulators increased leaf area and presented a larger bud number (Jiménez, 2001).

Both foliar and substrate applications of Unicozole-p (Sumagic) were efficient in reducing the height of dahlia (*Dalhia variabilis*). It was observed that a single application of Unicozole-p (Sumagic) is sufficient. A height reduction of 40-44 per cent (3 ppm to substrate or 10 ppm to foliage) was achieved without affecting number of flower buds and head diameter. When increasing growth regulator dose, the accumulation of total dry matter decreased in the stem and increased root/shoot ratio (Laguna *et al.*, 2003).

The concentrations of 0, 50, 100, 200 and 400 ml/l paclobutrazol to *Dahlia pinnata* Cav plants were applied every 14 days (Pichardo-Ruiz *et al.*, 2003). After seven dosages, paclobutrazol 400 ml/l recorded 45, 30 cm high plants with short internodes, higher chlorophyll content as well as increased weight and volume of tuberous roots, compared to the control.

6.2.2 Effect on Flowering

Mittal (1967) found that 100 and 200 ppm GA_3, applied twice at vegetative stage advanced flowering. While Mathur and Sharma (1969) observed that GA_3 accelerated flowering of dahlia plants in inductive day length. Significant increase in flower production of dahlia by ethrel spray at 500 and 1000 ppm, and a trend of increase in flower yield by CCC and MH was also recorded by Pappiah and Muthuswamy

(1974). Increased flower number was reported with the treatment of GA_3 in cv. Cr. Masterpiece (Bhattacharjee *et al.*, 1976).

In an experiment on different cultivars of dahlia, Bhattacharjee *et al.* (1971) also obtained large size flowers with SADH spray. In another experiment, SADH (1000-5000 ppm), TIBA (500-2000 ppm) and ethrel (2000 ppm) delayed flower bud appearance by 6-15 days, whereas MH (500-1000 ppm), GA_3 (10-100 ppm) and NAA (10-100 ppm) advanced it by 4-5 days (Bhattacharjee, 1984). The effectiveness of chemicals also varied with cultivars. Maiko and Yashchenko (1980) reported that treatment with ethrel and SADH at 10 and 20 mg/l, respectively, stimulated flowering in dwarf *cvs.* Eternal Fire and Park Princess, but were less effective for tall cultivars Torch and Gerrie Hoek.

Potted dahlia was sprayed with 1000 ppm kamposan (ethephon), or 5000 ppm dextrel (both ethylene producing substances) or with 40 ppm paclobutrazol (as cultar) on 3 occasions at weekly intervals. Dextrel and kamposan delayed flower initiation by up to 30-40 days. Cultar had a negligible effect on flowering date (Rounkova, 1989). The application of CCC at 1.5 kg *a.i.*/ha at stem elongation stage and paclobutrazol at 1.0 kg *a.i.*/ha at the appearance of the first visible flower bud significantly increased the seed yield of field grown hybrid dahlia cv. Unwins Mixed. The former through the production of more seed heads/plant and the latter through a significant increase in seeds/seed head and seed heads/plant (Phetpradop *et al.*, 1994). Chloremquat 3500 ppm recorded early flowering. Increase in number of flowers and flower diameter were observed with 5500 ppm and 4500 ppm, respectively.

Enhanced plant height, more number of leaves, per cent of mineral elements (NPK), chlorophyll content, increased number of inflorescence, prolonged flowering period, keeping quality but delayed appearance of first inflorescence with the application of K_2SO_4 5 g/l and ascorbic acid 100 mg/l in Dahlia *cv.* Arizona (Anonymous, 2016).

Singh *et al.* (2017) obtained better flower diameter, early flower emergence, number of flowers per plant and maximum shelf life with GA_3 (200 ppm) at 30, 45 and 60 days after transplanting in *cv.* 'Suryodya'.

Sajid *et al.* (2017) found that MH 1000 ppm was very effective and recorded minimum plant height and leaf number, stem diameter, primary and secondary branch number. Maximum days to flower bud appearance and colour break, maximum flower diameter, flower fresh weight and minimum peduncle length was observed with ethephon 1000 ppm. Highest flower number was recorded with MH 500 ppm while maximum flower bud diameter with MH 1000 ppm. Among all growth regulators Maleic hydrazide was found more effective followed by Ethephon and Alar in modifying the plant architecture of dahlia *cv.* Charmit.

6.2.3 Effect on Tuberous Root Formation

Pappiah and Muthuswamy (1974) reported significant increase in the tuber production with the application of CCC, ethrel and MH, and the effect was more pronounced with CCC. Read *et al.* (1972) observed that foliar spray of SADH at

2500 ppm and CCC at 1500 ppm increased the number and size of the tuberous roots of *Dahlia pinnata* under short-day condition (Table 4). It has also been noted by Moser and Hess (1968) that SADH promoted tuberization in dahlia under long photoperiods and simultaneous application of GA_3 inhibited it. Pot-grown plants of cv. Kelvin Rose treated with CCC (2500-5000 ppm), ethephon (1000-2000 ppm) and SADH (2500 ppm) significantly increased the number and weight of tuberous roots (Bhattacharjee, 1984). Preplant soaking of dahlia cuttings in ethrel at 10 mg/l improved early tuber formation and soaking in SADH at 20 mg/l improved subsequent tuber growth. Spraying young plants with ethrel at 100 mg/l or SADH at 200 mg/l was a more effective method of stimulating tuber development than soaking the cuttings (Maiko and Yashchenko, 1980).

Table 4: Effect of Growth Retardants on the Gormation of Tuberous Roots under Non-inductive (Long-day) Conditions (cv. Arthur Godfrey)*

Treatments	*Application (Days after potting)*	*Mean No. of Tuberous Roots*	*Mean Diam. of Tuberous Roots (mm)*	*Mean Crown Weight (g)*
Untreated	–	4.6	4.48	5.74
SADH 2500 ppm	9	4.3	8.52	8.60
SADH 2500 ppm	9 and 23	5.2	9.42	13.90
SADH 5000 ppm	9	4.2	5.62	6.30
Chlormequat 1000 ppm	9	4.6	8.02	11.76
Chlormequat 2500 ppm	9	5.0	4.06	6.40

* Read *et al.* (1972).

Aung and Purohit (1990) reported the importance of cytokinins and cytokinin-like substances in the inductive (cell division) processes of formation of storage organs of dahlia tuber and significance of carbohydrates in the developmental stages.

Beura and Maharana (1990) applied GA_3, MH and Alar as foliar sprays to rooted dahlia cuttings 25 days after transplanting. The highest shoot : root ratio (2.12) and lowest number of tubers/plant (6.25) and tuber yield (110 g) were obtained with 200 ppm GA_3. The lowest shoot : root ratio (0.88) and highest number of tubers/plant (10.75), tuber yield (233.75 g), tuber length (10.88 cm) and tuber diameter (2.95 cm) were obtained with 2500 ppm alar. Dry matter content of tubers was highest (25.62 per cent) with 500 ppm MH. Suma *et al.* (1994) obtained greatest tuber production of pot grown dahlia *cv.* Formal Decorative with 4000 ppm alar spraying.

Pichardo-Ruiz *et al.* (2003) found that after seven dosages of paclobutrazol (400 ml/l) to *Dahlia pinnata* Cav plants recorded increased weight and volume of tuberous roots, compared to the control.

7.0 Propagation

Dahlias are propagated from seeds, tuberous roots and cuttings. Multiplication by grafting is also possible. Propagation through meristem culture may be used for producing virus-free plants.

Propagation of Dahlia

Pod

Seeds

Seedlings

Tubers

Planting of Tubers

Growth of Seedling

Stem Cutting

Planting with Rooting Hormone

Rooting of Cuttings

7.1.0 Seed Propagation

Seed propagation is generally adopted for raising dwarf bedding singles and by breeders to develop hybrids. Plants of large flowering types raised from seeds shown a heterogenous mixture. Sowing of seeds should be done thinly in shallow boxes or seed-pans containing porous soil, covered with a layer of screened leaf mould and watered through a fine rose. Germination of seeds takes about ten days and the favourable temperature is from 18°C to 28°C. Seedlings are ready for transplanting in three to four weeks. Safronova (1969) obtained fully viable seeds from open pollination of double dahlias, and good germination was observed when sown directly in the ground.

Samant and Das (1990) observed that seeds of *D. variabilis* germinated in 3-4 days between 15 July and 15 September and in 6-12 days between 15 November and 15 December sowing. Although some seeds germinated in 18 days during January none of the seedlings survived. No seeds germinated in the period between February and June due to high temperature and low humidity. Seeds sown during

August and September at 15-days intervals germinated in 4-5 days and showed significant differences in growth, flowering and seed production. Seed yield was highest (0.52 g/plant) from sowing on 14 September.

Dahlia seeds sown in a medium containing equal parts of peat moss and vermiculite and treated with 1 per cent solution of roots (a commercial growth stimulant containing *Ascophyllum nodosum* extracts) at the time of sowing, resulted in earlier germination (Poincelot, 1993).

7.2.0 Vegetative Propagation

7.2.1 Division

Propagation by division can be done by separating the tubers each with a piece of stem. As the buds are not on the tubers, but on the crown to which tubers are attached, care should be taken that each division has at least one bud, otherwise the shoot will not develop. It is, therefore, suggested to allow bud development by placing the roots in a warm, moist place for a short time before dividing the roots. Lloyd (1968) reported that dahlia plants grown in clay pots under short-day condition produced more healthy tubers.

7.2.2 Cutting

Dahlias are commonly propagated from cuttings. Terminal stem cuttings of 6 to 8 cm length with two to four leaves, are prepared with the help of a sharp knife. The cut end is treated with IBA powder and planted in washed coarse sand for rooting and is kept in humid and shady place. A mixture of perlite and sand may also be used for satisfactory rooting (Mustafajeva, 1985). Miller (1982b) described three methods of taking dahlia cuttings out of which two methods are worth trying : (i) when the shoots from tuber have two sets of leaves, they can be removed from the base by giving a cut below the first node just above the root. The cuttings are then planted in vermiculite, treating with rooting powder. This is a rather simple method and used by many dahlia growers ; (ii) the second method is done by taking leaf cutting. The shoots are raised from tuber as in first method up to three full sets of leaves. The stem is then cut 0.5 cm above and below each pair of leaves and outer half of each leaf is removed. The stem is split in half and each leaf cutting is planted in pot for rooting after treatment with rooting hormone. Bhattacharyya *et al.* (1976) reported the rooting performance of 65 cultivars of dahlia, where percentage of survival of rooted cuttings ranged from 40 to 85 depending upon the cultivar. In a trial under glass, Winter (1979) recorded that potting soil fertilized with nitrochalk in a temperature range of 20°C – 25°C produced more number of cuttings from mother tubers. Biran and Halevy (1973) found an inverse relationship between the rooting response of the cuttings and the growth rate of the buds. Reproductive buds suppressed rooting more than vegetative ones. Removing the growing points together with a few terminal leaves increased the rooting percentage of cuttings. Read and Hoysler (1969) observed that herbaceous cuttings dipped momentarily in solutions of SADH at 1000 and 5000 ppm produced significantly greater weight and number of adventitious roots than the untreated cuttings. Conversely, similar treatments with CCC caused a marked depression of adventitious root production.

Cuttings of *D. variabilis* were treated with Seradix B-1 (containing IBA) and planted in coarse sand each month from February to November. Successful rooting and survival became less likely as the year progressed from February to May but more again from August to November. Values were highest in November (94 per cent). Cuttings planted in June and July failed to root (Borah *et al.*, 1995).

7.2.3 Grafting

Though not very common and practical mode of propagation, grafting is performed for preservation of rare and weak growing types. The top of the tubers are cut slantingly upward and the scion downward placed together and tied with raffia or polythene strip. They are then planted in a pot. Miller (1982a) tried two methods of grafting, *viz.*, English and French. English method consisted of removing the bud from one tuberous root and replacing it with the bud from that of another cultivar. In French method, the buds were removed with wedge-shaped cuts from the rootstock and scion material, *i.e.*, cutting was inserted into the space where the buds were removed from the aborted root. The cut end of the cutting was wedge-shaped so that it fits perfectly. He found that English method was better.

7.3.0 Micropropagation

Virus was eliminated from infected *D. pinnata* cv. Gillian by culture of 0.2-1 mm meristem tips. Of 102 cultures, 19 grew into plantlets and 13 survived transplanting. Initially all 13 survivors were dahlia mosaic caulimovirus (Da MV)-negative on *Verbesinaencelioides*, but after 10 months 3 of them developed symptoms and were Da MV-positive (Mullin and Schlegel, 1978). Wang *et al.* (1988) reported that meristem propagation is effective tool for eradicating Da MV or the Da MV-cauliflower mosaic caulimovirus (CMV). The Elisa and dot-blot hybridization techniques were effective to detect the presence of CMV and Da MV in leaves, stems and tubers.

Gavinlertvatana *et al.* (1979) studied the influence of stock plant pre-treatment with photoperiod and daminozide on ethylene and CO_2 levels and callus formation

Micropropagation of Dahlia

Micropropagation of Dahlia Hybrid Plants [*Dahlia variabilis* (Wild.) Desf.] via Direct and Indirect Organogenesis Techniques.

A and B – Direct adventitious shoot regeneration from shoot tip cultured on MS medium supplemented with different combinations of NAA and BA (2.0 + 2.0, 2.5 + 2.5, 3.0 + 3.0 mg/l, respectively (Left to right). C, D, E, F – Callus formation, growth and development from shoot tips cultured on MS medium supplemented with 3.0 mg/l NAA and 3.0 mg/l BA concentrations. G, H, I, J, K – Indirect shoots regeneration and growth stages from developing callus of shoot tip cultured on MS medium supplemented with 2.0 mg/l NAA and 2.0 mg/l-1 BA concentrations. L – Indirect shoots regeneration from callus cultured on MS medium supplemented with different combinations of NAA and BA (2.0 + 2.0, 2.5 + 2.5 and 3.0+3.0 mg/l, respectively (Left to right) after six weeks. M - Indirect shoots regeneration from callus cultured on MS medium supplemented with 2.0 mg/l NAA and 2.0 mg/l BA concentrations after eight weeks. N - Rooting shoot cultured on half strength of MS supplemented with 0.6 mg/l concentration of IBA after eight weeks. O - Plant of dahlia after six weeks from the acclimatization (Source: Ibrahim and Daraj (2015).

from dahlia leaf segment cultures. Larger callus and higher levels of ethylene and CO_2 were detected when tissues from stock plants under SD were incubated under SD (SD–SD) than when incubated under LD. Tissue under LD–LD or LD–SD formed similar amount of callus to SD–LD cultures but ethylene and CO_2 levels were higher compared with water-sprayed controls, 2500 mg/l daminozide sprayed on SD stock plants one day before explant removal promoted callus formation and ethylene and CO_2 production in tissue incubated under either SD or LD. Daminozide sprayed on LD stock plants had no influence on callus or ethylene, but CO_2 was increased under SD or LD incubation. Rashid and Power (1983) reported that cell suspension protoplasts of *D. variabilis* did not divide.

Shoot-tip explants of *D. variabilis* were cultured on MS medium supplemented with IAA at 0, 0.05 or 0.1 mg/l and BA at 0, 0.05, 0.1 or 0.2 mg/l and produced shoots with almost all combinations of IAA and BA. Shoot tips cultured on media supplemented with BA tended to produce red-pigmented shoots. Rooting varied between treatments. Shoots were obtained from flower-bud explants on MS medium supplemented with 0.1 mg IAA + 0.25 mg BA/l (Hernandez-Perez and Mejia-Munoz, 1994).

Organogenesis has been reported in dahlia by Shatilo *et al.* (1991). The isolated meristems of the cultivar Triomphe de Paris were cultured on solid medium with GA_3 at 4 mg/l, with subsequent transfer of the regenerants to liquid medium containing 0.1 mg IAA/l. Rapid propagation of dahlia by organogenesis is reported by Wei *et al.* (1990). The results indicated that GA_3 is necessary to increase the shoot proliferation of stem slice explants. The best results were obtained by culture on MS medium supplemented with 3 mg BA + 0.2 mg NAA + 10 mg GA_3 + 50 mg adenine/l. Root induction was most effective on half strength MS medium supplemented with 0.2 mg NAA + 20 g sucrose/l. After three weeks' culture, all shoots formed roots. *In vitro* plantlets were successfully transplanted and they show normal growth, flowering and reproduction. No somaclonal variation was observed. Shatilo (1991) studied the patterns of rhizogenesis in shoots of dahlia *in vitro*. A consistently favourable effect in inducing root formation by IAA and its synthetic counterpart IBA at concentrations of 0.03 and 0.1 mg/l was reported.

Al-Ahmer *et al.* (2010) indicated 100 per cent response of apical shoots and single nodal explants when cultured on free MS medium. Highest number of shoots per explants (4.0) was achieved when shoots were cultured on MS medium supplemented with 2 µM of both BA and IAA. Addition of 60 mg/l adenine sulphate to the medium increased the shoot number to 6.8 with 7.27 cm length. Moreover, addition of 300 ml/l alcoholic extract from immature seeds of Johnson grass increased the shoot number to 2.5 per explant while 200 ml/l aqueous extract of mature seeds of Johnson grass produced taller shoots of 7.76 cm. Further, highest root number per explants (5.0) of 5 cm length were achieved in MS + 3 µM IBA. Shoots cultured on ½ MS + 45 g/l sucrose + 3 µM IBA produced more number (8.0) and longer roots (10.40 cm). Acclimatization of plantlets on 1:1 peatmoss and loam under controlled humidity resulted in 100 per cent success.

Ibrahim and Daraj (2015) reported that MS medium fortified with 2 mg/l NAA + 2 mg/l BA produced more number of direct shoots per shoot tip (10) with longer shoots (6.67 cm). Whereas, the shoot tip cultured on MS medium supplemented with 1.5 mg/l NAA and BA each did not respond to adventitious shoot formation after 8 weeks of culture. Moreover, the shoot tips and nodal segments grown on 3 mg/l of NAA and BA each produced callus in the range of 23.33-75.0 per cent after 8 weeks of culture. The treatment 2 mg/l as well as 2.5 mg/l NAA and BA each were superior over 3 mg/l in terms of number and length of regenerated shoots from developing callus of shoot tip. The direct and indirect shoots were rooted when cultured on half strength MS medium was supplemented with 0.6 mg/l IBA. The dahlia plants thus produced by micropropagation have been acclimatized at high success of 100 per cent. Liudmila *et al.* (2019) cultured nodal explants from *in vitro* seedlings in modified MS medium in a semi-solid state supplemented with 0, 0.1, 0.5 and 1 mg/l of IAA. Highest number of roots (3.9), length (3.76 cm) and rooted explants (87.5 per cent) after 30 days were obtained with MS medium + 0.5 mg/l IAA.

Sediva *et al.* (2006) observed highest regeneration rate of primary explants when cultured on MS medium + 0.5 mg/l Zeatin. To eradicate Dahlia mosaic virus, apical meristems excised from *in vitro* culture of Dahlia were grown on the above media. Efficiency of meristem culture in elimination of DMV was 50 and 9 per cent for plantlets developed from meristems in size 0.4-0.5 mm and 0.9-1.0 mm, respectively.

8.0 Soil and Climate

Dahlias grow well in any type of rich and porous soil. They generally prefer well-drained, deep, fertile and moist soil with pH 6.5 (Swami Vinayananda, 1985). An open and sunny place but sheltered from exposure of strong wind is ideal. A cool atmosphere free from frost is also necessary. Effects of environment on growth and flowering in dahlia have already been discussed.

9.0 Cultivation

Mariña (2015) reviewed the recent work done on various cultivation aspects of Dahlia.

9.1.0 Selection of Cultivars

To obtain more flowers/plant, stress should be given on the number of branches/plant during selection of cultivars and strains. Plant height and number of leaves also showed a slight positive effect on the number of flowers (Suman *et al.*, 1991). The cultivars Kenya and Vigour were superior with respect to plant height and flower size. Flower longevity on the plant was best in Kenya (13.3-15.5 days), followed by Kelvin Rose, Black Out and Vigour (Mishra *et al.*, 1990). In a mild winter with a temperature ranging between 15 and 27°C, Beura and Maharana (1992) reported that cultivars Croydon Apricot and Tenzing Norkey showed the most rapid flower initiation and development, and were therefore best suited for places having a short winter or for late planting. Swami Lokeswarananda and Yallen were best for cut flower production, having the longest flower stalks. White Kenya produced the largest flowers, followed by Yellow Kenya. Arthur Hambly had the longest flowering span followed by Avalanche.

Ahmed and Gul (2002) found that varieties Siedler Stolz, Hajley Jane, Mystery day and Vuurvogel were superior for number of branches per plant, number of leaves, number of buds, number of flowers, vase life and tuber weight per plant. The varieties Procyon and Hajley Jane perform better for plant height, days to sprouting, number of plants per tuber and number of tubers per plant cultivar. The cultivars Vuurvogel, Siedler Stolz, Mystery Day, Hajley Jane and Procyon found successful and recommended for general cultivation.

Mahawer *et al.* (2010) found that on the basis of floral characters, yield and economic parameters cv. NT Pompon was found best under Udaipur conditions. Kumar *et al.* (2010) recommended 'NT Pompon' and 'Jyotsna' for cut flower and exhibition/standard purposes, respectively under Udaipur condition.

Gupta *et al.* (2015) under Allahabad agro-climatic conditions found that Eternity Sports was better for plant spread, number of tubers per plant and flowering duration, however, for number of primary branches per plant and number of flowers per plant S.P. Glory India was better. The variety Kenya White performed better for single flower weight and flower yield per plant, while, S.P. Sri Radha for tuber yield per plant. Baburao *et al.* (2017) found that Sourave performed better for number of primary branches per plant, number of tubers per plant, tuber yield per plant, single tuber weight and tuber yield/ha with good benefit cost ratio of (3.57).

Manjula *et al.* (2017) reported that 'Joyal Singh' recorded early first flowering, 50 per cent flowering, complete flowering and duration of flowering; Sourav for number of flowers per plant and number of tubers per plant with respect to yield parameter and with respect to quality parameters genotype Kamala showed highest weight of flower, whereas, tuber weight, flower diameter, number of petals per flower were recorded highest in the Gagi, Pue Sona, Bharati.

Verma and Kulkarni (2017) found that maximum plant height, number of branches, stem girth, plant spread, intermodal length, number of flowers and tuber per plant were recorded in 'Hiranmoyee'.

Shukla *et al.* (2018) found that maximum weight of single flower, flowering duration, flower yield per plant and per hectare, weight of single tuber, number of tuber per plant, yield of tuber per plant and per hectare was found in hybrid S.P. Romia.

Bajaraya *et al.* (2019) recommended 'Prime Minister' for number of flowers and tubers and 'Korean Orange' for flower diameter, bloom life of flowers and tuber weight under Gwalior agro-climatic conditions. Mounika and Saravanan (2019) reported that 'Duet' found suitable for early flower bud initiation, 'Yellow Ball' for flower diameter, 'Ice Cube' for single flower weight and 'Scarlet Striped Pink' for number of flowers per plant.

9.2.0 Soil Preparation

Dahlias are grown both in pots and in ground in well drained soil, rich in organic matter. The ground should be prepared thoroughly and manured a few weeks before planting. Decomposed organic matter in soil is necessary for a better root growth and subsequent effect on flowering. Half of the recommended rate or nitrogenous fertilizer and full amount of P and K are applied before planting of cuttings. Effective drainage of excess water is considered essential.

Several studies related to the effect of different growth media on the development of Dahlia in pots have been carried out (Kiram *et al.*, 2007). The best medium was sand+silt+leaf mould for improving plant height, stem diameter, number of branches per plant, minimum days to flowering, number of flowers per plant, number of petals per flower, flower diameter and vase life. The medium composed of leaf mould alone also did well.

Younis *et al.* (2008) found that leaf manure as substrate media improved plant height, length of side branches, number of flowers, blooming period, size of flower and number of side branches in *Dahlia coccinia*. Similarly Chemical analysis of potting media showed that pH was in required range for best growth of *Dahlia*. Nitrogen, Phosphorous and Potassium were best taken from leaf manure. Leaf manure proved superior growing medium for *Dahlia coccinia* plant.

Bergi *et al.* (2014) reported that potted dahlia grown on cocopeat + FYM (1:1) was found better for leaf area, specific leaf weight, leaf chlorophyll content and dry matter accumulation with greater net assimilation rate (NAR) and relative growth rate (RGR). The same potting media also maintained higher leaf and petal relative water content, sugar with greater membrane stability index with prolonged flowering duration, flower diameter and yield.

Kareem *et al.* (2014) conducted a research to assess the growth performance of dahlia on various crop residues as potting media. Pot research was conducted by transplanting 20 days old seedlings on different crop residues, including, silt as control, FYM+Silt (1:3), FYM+Silt (1:3) cockscomb and FYM+Silt (1:3) maize crop residues as treatment. Completely Randomized Designs (RCBD) was applied to test the significance and compared means of treatments. Data was recorded for different parameters including plant height (cm), total number of leaf, total number of branches, total number of flower, fresh weight(g), dry weight (g), total bulbs, bulb diameter, flower diameter. All the results were statistically significant and the results related with plant growth indices indicated that the maximum values for plant height (39.33cm), total number leaves (53), total flowers (4.66), fresh weight(g) (76.03), dry weight (g) (14.00), bulb diameter (8.33), flower diameter (11.66) were found in slit. The combination of Slit + FYM and crop residues has little effect on all the parameters studies.

9.3.0 Planting

Dahlias are generally planted in September-October in the plains, and in April in the hills. Hong and Jeong (1988) sowed dahlia seeds at monthly intervals from February to July and reported that late sowing caused early flowering but of short duration. Afzal *et al.* (2000) found highly significant result when they tasted a local dahlia variety on four sowing dates at fortnight interval, starting from 10th of September. Its performance was judged for the traits *viz.*, seed germination time, plant height, days to flower, blooming period, flower size and seed yield/ replication. Seed sowing of 25th of September took minimum time to germinate. Maximum plant height, flower size, seed yield per plot and longest blooming period

Cultivation of Dahlia

Planting of Tubers

Growth of Plants from Tubers

Row System Cultivation

Flowering in Open

Disbudding

Pinching

Staking with Mulching

Flowering in Dahlia

was also observed when the seeds were planted on 25th of September whereas the seeds sown on 25th October took minimum days to flowering. The planting distance varies according to the type, *i.e.*, about 75 cm in case of tall, large-flowered types, 60 cm for medium-sized plants and 30 to 45 cm in case of dwarf cultivars. In an experiment on the planting density, Quagliotti and Gullino (1967) reported that plant growth and flower quality were not affected in *cv.* Graziella by any of the spacings compared *viz.*, 30 x 50 cm, 45 x 65 cm, 55 x 50 cm and 65 x 95 cm. The number of flower heads per plant increased with greater spacing between plants, whereas the number per unit area decreased.

9.4.0 Irrigation

Dahlias require abundance of water not only superficially but also in depth. A judicious application of water during a severe dry spell is very beneficial. Over-watering causes a rapid but soft growth, as a consequence, roots become feeble, few buds are formed and they generally blast before developing into flowers. Plants in beds should be watered thoroughly only when they need it. The amount of water in ground should be such that it moistens the full depth of 40 cm and the whole bed remains moist for few days (Swami Vinayananda, 1985). Chandraju *et al.* (2013) sown dahlia seeds in different pots and irrigated with raw water (RW), 1:1, 1:2 and 1:3 spent wash (SW). They investigated the nature of germination, growth and yield of seeds. It was found that, the germination, growth and yield was very good (100 per cent) in 1:3 SW irrigation, while very poor (25 per cent) in 1:1 SW, moderate (80 per cent) in 1:2 SW and 95 per cent in RW irrigation. This concludes that the diluted spent wash can be conveniently used for irrigation purpose without adverse effect on soil.

9.5.0 Manuring and Fertilization

Proper nutrition is essential for successful cultivation of dahlia. Nitrogen, phosphorus and potassium greatly influence the growth and flowering of this plant. El-Gamassy and Moustafa (1963) recorded improvement in growth, quality of flowers and yield of tubers when nitrogen and phosphorus were applied in optimum doses. Improvement in vegetative growth and flowering of dahlia with medium to high dose of nitrogen was also reported by Rahman and Mitra (1974). In a sand culture experiment, Jana *et al.* (1974) found that high doses of nitrogen and phosphorus proved very effective in improving the vegetative growth and flowering (Table 5). From a field trial on *cv.* Black Out, Bhattacharjee and Mukherjee (1981) concluded that 40 kg N, 50 kg P_2O_5 and 40 kg K_2O per acre were optimum for flower yield, while omission of nitrogen markedly reduced the plant growth and flowering. It is suggested to apply full dose of phosphorus and potash and half dose of nitrogen before planting and the remaining half of nitrogen, forty days after planting.

Liquid organic manure is also useful and should be applied biweekly after the plants are established.

Table 5: Effect of Nitrogen, Phosphorus and Potassium on Growth and Flowering of Dahlia*

Treatments	*Height of Plant (cm)*	*Mean No. of Leaves on Main Stem*	*Mean No. of Branches*	*Mean No. of Flowers*	*Size of Terminal Flower (cm)*
NPK (control)	74.1	26.2	8.5	9.6	17.0
(3N) PK	97.3	27.7	13.0	16.5	16.0
N (3P) K	93.5	27.0	8.7	10.6	15.8
NP (3K)	96.2	26.5	7.3	10.1	16.4
(1/10N) PK	17.4	16.6	nil	nil	–
N (1/10P) K	42.9	19.4	nil	1.5	4.3
NP (1/10 K)	72.5	25.0	1.1	4.8	14.7
C.D. at 5 per cent	10.7	4.6	–	3.8	–

* Jana *et al.* (1974).

Foliar spray of urea one table spoonful in a gallon of water may be given once a week till the initiation of bud. In order to get large size bloom, a fertilizer mixture (3 parts superphosphate, 2 parts muriate of potash and 1 part urea) at the rate of one tea spoonful per pot (20 cm) should be applied at the time of bud formation. Fertilization should be discontinued when the buds begin to show colour (Niyogi, 1973). Efficacy of different organic fertilizers was tested by Miller (1983) and it was found that blood and bone meal produced the tallest plant with largest average bloom and improved the tuberous root production qualitatively and quantitatively over other treatments. Winter (1979) reported that application of the granular organic NPK fertilizer at 2 kg/14.20 m^2 reduced the loss of cuttings after planting and doubled the yield of tubers compared with plants receiving no fertilizer. Incorporation of composted municipal sludge (Techna Gro) as a soil additive to a depth of 1.5 inches and a mulch 2 inch deep soil or municipal sludge mulch significantly increased the growth of dahlia *cv.* Figaro White (Smith and Treaster, 1991, 1992).

Singh and Gupta (1995a,b) studied the effect of N, P and K on the number, size and weight of tubers/plant in *cv.* Powder Puff. Nitrogen at 75 mg/kg potting medium produced the highest number (6.70) and yield of tubers/plant (104.34 g) while P at 75 mg/kg significantly improved tuber formation. Availability of K increased with increasing levels of N and potash. Singh *et al.* (1996) also reported that plant N content increased significantly with increasing levels of N, P and K application. Plant P and K contents decreased with increasing N application but increased with increasing P and K application. High rates of N, though promoted vegetative growth and increased the size of flowers in dahlia, the quality deteriorated and they wilted faster than those treated with optimum rates and balanced fertilizers.

Ahmed *et al.* (2004) found that urea, P_2O_5 and FYM in different combinations showed significant results on plant height, number of branches/plant, early flowering and number of flowers/plant, while, number of leaves were affected by the combination of urea + P_2O_5 + FYM. Size of flowers were significantly increased by application of urea with combination of potash and FYM and also by using single P_2O_5, while single FYM and urea did not affect size of flowers.

Julio *et al.* (2011) reported that application of P at 120 kg/ha and planting 6.25 plants/m^2 increased the production of root dry matter Also, removal of leaf from the middle and lower strata satisfied the 'sink' from the root. Gupta *et al.* (2016) envisaged that application of 30 g N and 24 g K per m^2 improved plant height, flowering duration, flower size, stem length, number of cut stems/plant, weight of tuberous roots/plant and total number of tuberous roots/plant.

Sheergojri *et al.* (2013) found that N at 50 kg/ha recorded early tuber emergence, first flower bud appearance, full opening and longer flowering duration; 75 kg/ha N resulted in maximum plant height, number of primary branches per plant, length of primary branches, leaf area, flower diameter, fresh flower weight and longer vase life; 100 kg/ha P resulted in early tuber sprouting, minimum days to floral bud initiation, taller plants, bigger and longer flowers, extended flowering period, more number of flowers per plant and prolonged vase life; Azotobacter improved plant height, length of primary branches, flower diameter, peduncle length and yield. The interaction effect of N 75 + P100, N 75 + Azotobacter and P 100 + Azotobacter showed significant improvement for all parameters. Among all treatment combinations, 75 kg/ha N + 100 kg/ha P + Azotobacter obtained maximum plant height, number of primary branches per plant, length of primary branches, leaf area, flower diameter, flowering duration, flower fresh weight and longer vase life in 'Pink Attraction'.

Trivedi *et al.* (2015) reported that vermicompost 2.5 t + *Azotobacter* 2 kg + PSB 2 kg/ha found better for plant height, plant spread, number of leaves, number of primary branches, days to first bud appearance, single flower weight, flower diameter and flowering duration cv. Blackout. Saravanan and Barik (2016) observed that dose of 50 per cent NPK + 50 per cent vermicompost per ha recorded delayed first flower bud initiation, flowering duration, maximum flower stalk length, flower diameter, flower longevity, vase life, number of flowers per plant and per hectare, flower yield per plant, and tuber yield per ha.

Younis *et al.* (2014) reported that 7.5 ml/l micronutrient and 6 ml/l macronutrient proved best with maximum plant height, number of leaves, side shoots, leaf area, number of tubers/plant, flower diameter and earliest flower emergence in dahlia.

Shivhare and Sharma (2012) conducted a greenhouse experiment to determine the phytotoxic effect of heavy metals, Ni and Pb on the growth of Dahlia *cv.* Gerogina wild and found that the overall length of the plant reduced with increase in the concentration of heavy metals in soil.

Kumar *et al.* (2019) concluded that the application of N_{80} + P_{100}+ K_{100} + Azotobacter @ 2.0 kg/ha resulted in maximum flower diameter, flower weight, flower yield per plant, flower yield per plot, flower yield per hectare. The application of N_{100} + P_{85} + K_{100} + PSB @ 2.25 kg/ha resulted in maximum weight of single tuber, number of tuber per plant, number of tuber per plot, tuber yield per plant, tuber yield per plot, tuber yield per hectare.

Alzahrani *et al.* (2020) investigated the influence of inorganic amendments: SDS and EDTA in improving plant metal tolerance. Dahlia plants were grown in pots containing sterilized sand and Hoagland's medium with different levels of Cd (0, 10, 25 or 100 mg/kg) and Pb (0, 100, 500 and 5000 mg/kg) with or without inorganic amendments. They revealed that both the Cd and Pb significantly reduced plant growth, pigment content, relative water content, superoxide dismutase, peroxidase, ascorbate peroxidase and catalast along with protein and total soluble sugar. The heavy metal stress also increased the production rate of reactive oxygen species (ROS) by increasing malondialdehyde (MDA) and hydrogen peroxide (H_2O_2) production in leaf. Inorganic amendments alone or in combination considerably alleviated the heavy metal induced toxicity with maximum effect under EDTA + SDS.

9.6.0 Staking

Support should be given to each plant immediately after they have started growth, as new growth is somewhat soft and liable to be affected by strong wind. Stakes may be painted green and the base tarred. The tying should be secured. For prize flowers separate stake for each side branch in addition to the main support is also necessary. Supporting stake helps the vertical growth of the plant. Bamboo or polythene ring can be fixed in case of pompon or other small flowering types.

9.7.0 Mulching and Weed Control

Mulching is an important operation in dahlia cultivation. This can be done with grass clipping, old hay and sawdust showing several beneficial effects (Leroux, 1982). It helps to maintain an even soil temperature and causes earlier and better bloom. Mulch slows down water evaporation and reduces the irrigation requirement. It also keeps the weeds under control. Using black plastic mulch for growing dahlia is also beneficial as it improves growth, flowering, tuberous root formation and suppresses weed growth.

Clay and Davison (1968) tested the efficacy of different types of herbicides in a field experiment and observed that lenacil, linuron and simazine gave good control of weeds. Pre-emergent herbicides like pennant (metolachlor) at 4.6 and 9.1 kg/ha, ronstar 2G (oxadiazon) at 4.5 and 9.0 kg, rout 3G (oxyfluorfen + oryzalin) at 3.4, 6.8 and 13.6 kg, surflan AS (oryzalin) at 2.3 and 4.6 kg and treflan 5G (trifluralin) at 4.5 kg successfully controlled *Setaria glauca* (*S. pumilal*), *Echinochloa crus-galli*, *Poa annua*, *Capsella bursa-pastoris*, *Senecio vulgaris* and *Amaranthus retroflexus* in pot grown dahlia (Staats and Klett, 1993). Southward *et al.* (1998) recommended that alachlor, chlorpropham, chlorthal-dimethyl, pendimethalin, trifluralin, methabenzthiazuron, and haloxyfop as the safest weedicides in dahlia cultivation.

They also concluded that herbicides like EPTC, oryzalin, and oxyfluorfen tended to cause significant injury when applied in the initial phase of plant growth.

9.8.0 Pinching

In order to promote branching in dahlia in beds for garden display or cut flower, pinching is to be done after two or three pairs of new leaves appear. Apart from increasing the number, this operation also improves the quality of flower, though the size will be smaller. In pot plants with single flower for exhibition at flower shows, tip is not removed, while the axillary branches are cut to improve the size and quality of the flower from the apical bud. Flowering in dahlia may be delayed by stopping. When the plants are about 15 cm tall, the tip of the plant is pinched off. If stopped at this time, the first two branches would flower about 15 days later than from the centre bud. For timing blooms, stopping is practised at some chosen time and at various stages of growth (Swami Vinayananda, 1985). Pinching of forced tuberous-rooted cultivars Park Princess and Miramar was evaluated as a method for increasing flower production and plant quality. Pinched plants produced more flowers, flowered later, had smaller flowers and were taller than non-pinched controls. On an individual plant basis, pinching at node 4 generally gave the best result while pinching at node 2 resulted in the maximum delay and lowest number of flowers. The more distal the pinch, the greater the number of laterals formed on both cultivars, and the higher the percentage of laterals flowering on Park Princess. On a population basis, pinching of plants with single strong shoot, at node 3 or 4 resulted in the best compromise between increased flower production and delayed flowering and increased plant height (Barrett and Hertogh, 1978).

The effects of pinching above node 4 on the main stem on seed production of field grown hybrid dahlia cv. Unwins Mixed (*D. Pinnata* × *D. coccinea*) were investigated by Phetpradap *et al.* (1994). Pinched plants reduced the spread of flowering and resulted in seed heads at approximately the same height above ground level, but did not increase seed yield.

When dahlia plants are grown from tubers, except the strongest one, all shoots appearing from the tuber are removed.

9.9.0 Pruning and Disbudding

If the branches threaten to become a thick mass, the shoots are thinned to keep the centre of the bushes open. For large decorative cultivars four to five main branches may be retained, while small flowering types may carry eight to ten branches. The number of flowers should be restricted by disbudding, especially on the large bloom type. To get the largest bloom, all buds except the centre bud and all lateral branches should be removed. Mostly in all types of dahlias, the accompanying buds should be removed in the very early stage. Disbudding and removal of lateral branches should not be done at a time. This can be staggered over a few days.

10.0 Greenhouse Cultivation

Many gardeners love dahlias for their wide range of colours and large, showy size. But the drawback to dahlias in northern climates is that they cannot tolerate severe cold, so the tubers must be dug out each fall and planted outdoors each spring when all danger of a frost has passed. This means that it can be well into summer before northern dahlia growers see blooms. However, greenhouse owners can get flowers much earlier than this by starting their dahlias in the greenhouse in February or early March.

One or more dahlia potting trays, about 6″ deep can be set up in the greenhouse. The soil should be slightly acidic, with a pH of about 6.5. Add a little peat moss if the soil is more alkaline. Add a little compost and bone meal to the mix to help promote growth. Most gardener store their dahlia tubers for the winter in sawdust or sand, but it can simply placed in trays under a greenhouse bench. At planting time, carefully break up the clusters of tubers, making sure that each separate tuber has at least one eye to produce a new plant stalk. Then plant each tuber in a potting tray or pot to a depth of 10 – 12 cm and water it in. Use just enough water to moisten the soil, not soak it. Too much water can rot the tubers. Keep the trays at a temperature of about 18-21°C during day time and 17-18°C during night. The trays will also require about eight hours of sunlight daily, which is impossible in winter, so to provide additional lighting. This can be done with 100-watt-equivalent LED lights. When stalks begin to show, start fertilizing the plants. At this early stage, apply fertilizer diluted to about a quarter of full strength. Add a bit more bone meal to the garden soil when planting (about a teaspoon in each hole), and stake the stalks to reduce wind damage. Good crop growth is obtained with constant 100 – 150 ppm N with additional micronutrients added from a complete N-P-K water-soluble fertilizer such as 15-5-15 or 17-5-17; the majority of N should be in nitrate form. Avoid high-phosphorous fertilizers, which can reduce plant quality by encouraging soft, stretched growth.

11.0 Diseases and Pests

11.1.0 Diseases

Important diseases of dahlia and their control measures are given below :

Fungal Diseases

11.1.1 Powdery Mildew

Powdery mildew (*Erysiphe polygoni* and *E. cichoracearum*) is one of the most common fungal diseases of dahlia. Sanabria-de-Albarracin and Albarracin (1988) reported that *Erysiphe cichoracearum* caused powdery mildew on *D. pinnata*. Perwezb *et al.* (1989) observed the occurrence of powdery mildew on *D. variabilis* during September-April in Aligarh and its adjoining areas, India. A white to grey powdery mass appears on the upper surface of the leaves. The disease is prevalent at the early stage and again at the end of the flowering season. This can be controlled by spraying 0.1 per cent bavistin or karathane (Swami Vinayananda, 1983). Spraying of karathane (dinocap), carbendazim and morestan (quinomethionate) at 0.2 per cent gave 92.9 per cent, 88.2 per cent and 87.56 per cent control of powdery mildew on dahlia, respectively (Perwez*et al.*, 1994). Punlerdmatee *et al.* (2017) found that yeast strain CMY044 showed efficacy in reducing the disease incidence of *E. cichoracearum* on dahlia leaves to as low as 0.00 per cent compared to control plants. Sharma *et al.* (2017) reported the first occurrence of powdery mildew on Dahlia by *Golovinomyces cichoracearum* in Himachal Pradesh which caused withering and death of plants with advancement of disease.

11.1.2 Leaf Blight

Rhizoctonia solani caused leaf blight disease on dahlia in the high rainfall zone of West Bengal. Applications of bavistin (carbendazim) and dithane M-45 (mancozeb), separately or together, were effective in reducing the severity of the disease (Jana *et al.*, 1989).

11.1.3 Botrytis Blight

This disease is caused by the fungus *Botrytis cinerea*. During dull, cloudy weather, and in propagating houses where ventilation is not adequate, the young shoots and the flower buds are infected with the grey mould, which causes fading and browning of petals and destruction and death of other infected parts. In moist weather the buds may be attacked by a soft rot and all infected parts may be covered with the greyish powdery mould. Spraying with zineb, captan or ferbam will control the disease (Pirone *et al.*, 1960).

11.1.4 Stem Rot

The stem rot fungus (*Sclerotinia sclerotiorum*) attacks the main stem and branches near the base of the plant; a white mould may completely encircle the stem. Plants growing in heavy wet soil infected with this fungus ultimately wilt and die. The same fungus may attack young plants in propagating houses, where it causes them to wilt and rot. To control the disease heavy soil should be made porous with a mixture of sand and good drainage be provided. Plants should not be crowded. Soil sterilization with steam or chemicals is also very effective to control the disease.

11.1.5 Wilt

This disease is caused by *Verticillium alboatrum* and *Fusarium* sp. These soil fungi enter the roots, in which their presence is made evident by brown or black streaks along the conducting tissue. The plants wilt and die because of the blockage or destruction of the water-conducting vessels, or death of the living tissues due to the toxins produced by the fungus. Since these wilt fungi may attack roots in storage, discoloured or decayed parts should be cut away before planting. Wilted plants should be destroyed and only healthy tubers should be used in propagation.

11.1.6 Smut

The disease, caused by *Entyloma dahliae*, is characterized by presence of more or less circular spots, at first yellowish green, later becoming brownish. The infected parts dry out and turn brown. In a trial on the control of smut, several fungicides were applied of 15-day intervals and dithane Z-78 (zineb) and dithane M-45 (mancozeb), each at 0.2 per cent gave the best result (Misra, 1978). The use of dithane Z-78, blitox and benlate (benomyl) was also suggested by Upadhyay and Bhandari (1985).

Phytophthora and *Pythium* are other fungi reported in dahlia. Turban (etridiazole) 5 per cent granular and subdue, 2E (metalaxyl) 25 per cent E.C. were evaluated to control *Phytophthora* spp. and *Pythium* spp. and both the fungicides were effective (Powell, 1981). Of the 21 fungi associated with dahlia seeds, *Alternaria alternata*, *Curvularia pallescens*, *Phoma putaminum*, *Aspergillus flavus*, *A. niger*, *A. repens*, *Drechsler ahawaiiensis* and 3 *Fusarium* spp. were found to cause varying reduction in seed germination in pathogenicity tests (Shrotri *et al.*, 1985).

Bacterial Diseases

11.1.7 Crown Gall

This disease is caused by *Agrobacterium tumefaciens*. Large tumors of abnormal tissue develop at the base of the plant and on the roots. Infected plants become stunted and the shoots spindle. The disease can be prevented by dipping roots at planting in streptomycin solution. Roots and crowns of plant with tumors should be destroyed and changing the location of planting will also help to curb the disease (Pirone *et al.*, 1960).

11.1.8 Bacterial wilt

Plants infected by the bacterium *Pseudomonas solanacearum*, usually droop and wilt suddenly. Stem near the soil rots and, if cut yellowish masses of bacteria ooze out. The control measures are similar to those of crown gall disease.

11.1.9 Bacteriosis

This disease, caused by *Erwinia cytolytice*, was first observed on dahlias grown at the New York Botanic Garden. It appears as browning and softening of the stem ultimately resulting in rots. Tubers are also affected by this disease, and it can also be controlled by following the measures suggested for crown gall disease.

Viral Diseases

11.1.10 Diagnosis of Mosaic Virus

Shatilo (1994) reported that the presence of viroplasms in dahlia cytoplasms was the only specific diagnostic criterion for dahlia mosaic caulimovirus (Da MV). Albouy *et al.* (1994) studied the efficiency of double antibody sandwich (DAS) Elisa tests, antigen coated indirect Elisa and dot immunobinding (Dot blot) Elisa in detecting viral diseases of dahlia. All 3 techniques were adequately sensitive, however, Dot blot Elisa was faster and cheaper than the other assays. Iftikhar *et al.* (2020) developed a loop-mediated isothermal amplification (LAMP) assay for Dahlia Mosaic Virus (DMV) detection which targets the reverse transcriptase region of DMV genome. Amplification was detected through fluorescence detection and agarose gel electrophoresis. The assay detected DMV in DNA concentration is low as 10^{-3} mg which will be helpful in rapid and sensitive detection of DMV and in reducing the virus incidence through production of virus-free planting material.

11.1.11 Mosaic

Mosaic, caused by *Marmor dahliae*, is the most serious viral disease of dahlia. The leaves become mottled and develop pale green bands along the mid-ribs and larger secondary veins. They are dwarfed and show a general mosaic or pale yellow spotting. To prevent the disease, the roots of infected plants should not be used. Ruthless rouging of infected plants will often prevent the spread of the disease to healthy plants. Control of the peach aphid (*Myzus persicae*), the vector of mosaic

Diseases of Dahlia

Botrytis Blight

Dahlia Smut

Wilt

Powdery Mildew

Crown Gall

Leaf Curl Virus

Ring Spot Virus

Dahlia Mosaic

virus, will also help to prevent the disease. As mentioned earlier, culture of meristem was found to be a very effective method to eliminate the disease from the infected plant (Mullin and Schlegel, 1978).

The virus infected dahlia plants showed symptoms in late June or, in certain cultivars, in August. Potato X potexvirus, potato Y potyvirus and cucumber mosaic cucumo-virus showed the highest level of infection in *cvs.* Zvaizgnite, Alva's Doris, Vuurvogel, Prefait Demange, Honeker and Doris Day. The major vectors were *Myzus persicae* and *Aphis fabae*. The most effective methods to prevent virus dissemination were elimination of infected plants, chemical control of winged aphids and use of virus-free planting material (Prokoshina *et al.,* 1990). A potyvirus infected dahlia plant showed chlorotic spots and slight distortion in cvs. Hangzhon, Zhejiang in the autumn. A spherical virus, identified as cucumber mosaic cucumovirus, was isolated from the same plant (Chen and Li, 1990). Albouy *et al.* (1992) identified dahlia mosaic caulimovirus, tomato spotted wilt tospovirus, cucumber mosaic cucumovirus and tobacco streak ilarvirus in 30 commercial dahlia cultivars. For the first time, Kuluev and Chemeris (2007) amplified and cloned Dahlia mosaic virus promoter, and reported the homology between this promoter and the promoters of other caulimoviruses.

Pahalwatta *et al.* (2008) reported that members of the genus Caulimoviridae were shown to integrate their genomic sequences into their host genomes and exist as endogenous pararetroviral sequences (EPRV). However, members of the genus Caulimovirus remained as exception and exist only as episomal elements in the infected cell. It was also evidenced that DNA genome of a new and distinct Caulimovirus species, associated with dahlia mosaic, is integrated into its host genome (dahlia). Southern blot hybridization of total plant DNA from dahlia seedlings showed the presence of viral DNA into the host DNa using cloned viral genes. Fluorescent *in situ* hybridization using labeled DNA probes from the D10 genome localized that viral sequence in dahlia chromosomes. The natural integration of Caulimovirus genome into its host and its existence as EPRV suggested the co-evolution of this plant-virus pathosystem.

Eid *et al.* (2011) also found that four wild Dahlia species in their natural habitats from west-central Mexico harboured Dahlia mosaic virus (DMV-D10), an endogenous plant pararetro virus which was associated with cultivated species. This suggested a possible emergence, co-existence and co-evolution of DMV-D10 and its host dahlia plant. Abdel-Salam *et al.* (2010) confirmed DMV-D10 by amplification of ORF1 (encoding for the movement protein) using species-specific primers (D10F1/R1). The expected size (900 bp) was amplified from 36 samples without any infection of DMV or DCMV. Sequence comparisons of the Egyptian DMV-D10 ORF1 amplified from the samples revealed that the amplicon had highest sequence identity (96 per cent) with that of DMV-D10 (isolated from US dahlia cultivars). Cluster dendrogram based on the amino acid sequences of movement protein of all known caulimoviruses placed D10-US, D10-NZ, D10-DC (*D. coccinea*) and D10-Egypt in one clade.

11.1.12 Spotted Wilt

This disease, caused by *Lethum australiense*, is also known as ring spot and appears as a characteristic ring pattern on the leaves. Spraying of 0.1 per cent malathion or metacid to control the vector insects is recommended. The disease can be checked by using virus-free tubers or cutting. Gofflot and Verhoyen (1990) reported the occurrence of tomato spotted wilt virus (TSWV) in Belgium. TSWV occurrence is related to a significant increase in *Frankliniella occidentalis* populations. Other possible vectors are *Thrips tabaci, F. schultzei* and *F. fusca*. The strain of TSWV was also found in naturally growing *Dahlia* spp. in Himachal Pradesh, India. It appeared from the natural reservoirs of the virus and/or vector (Khurana *et al.*, 1990). Sherwood *et al.* (1991) reported the tomato spotted wilt virus in greenhouse dahlia, in Oklahoma and Cleveland in 1990.

11.1.13 Chrysanthemum Stunt Viroid (CSVd)

Chrysanthemum Stunt Viroid (CSVd) is known to infect a wide range of *Dahlia spp.* Shunsuke *et al.* (2020) detected CSVd in 94 of 214 dahlia samples collected from across 8 different Japanese Prefectures. Six sequence variants of CSVd were detected from 32 CSVd-infected dahlia samples including 3 new variants and 3 previously isolated variants from dahlias or chrysanthemums. Five CSVd variants were infected and replicated in both dahlia and chrysanthemum plants by sap and RNA inoculation three months post inoculation. The sequence variation among these variants did not significantly affect the infection efficacy. However, CSVd reduced fresh plant weight, plant height, size of compound leaves, flower diameter and fresh flower weight of dahlias. In addition, it altered the colour ray florets and increased the rate at which flowers exposed their discs and disc florets. It was concluded that CSVd infects a wide range of dahlias in the field and is an important pathogen to manage for their cultivation as it negatively affects the productivity.

11.1.14 Dahlia Latent Viroid (DLVd)

Two viroids belonging to the family Pospiviroidae, Potato Spindle Tuber Viroid (PSTVd) and Dahlia Latent Viroid (DLVd) were reported separately from dahlia plants in Japan and the Netherlands, respectively. A total of four dahlia samples infected with PSTVd-dahlia produced two specific bands, one corresponding to PSTVd infection and other being unknown which were corresponding with DLVd reported from the Netherlands. Further, out of 78 dahlia leaf samples collected randomly, 38 were detected as DLVd insisting that DLVd was widespread in Japan. To evaluate the pathogenicity of DLVd and PSTVd, these viroids were co-inoculated into wild hop and dahlia seedlings. DLVd alone was inoculated into *Chrysanthemum, Tithonia, Helianthus* and *Glebionis*. Inoculated dahlia seedlings were positive for both DLVd and PSTVd, but wild hop and other plants were negative. Dahlia seedlings that were positive showed mild leaf curling in the early stage of growth. Moreover, a possible negative synergistic interaction of DLVd and PSTVd during dahlia cultivation must be considered (Tsushima *et al.*, 2015).

Pests of Dahlia

Leaf Hopper

Thrips Infestation

Red Spider Mite Infestation

Aphids Infestation

Dahlia Borer

Broad Mite Injury

Harvesting of Dahlia

Flower Harvesting

Harvested Tubers

11.2.0 Pests

11.2.1 Aphids

Aphids can not only transmit virus diseases, such as mosaic, spotted wilt but they also suck sap from plant tissues which affects plant growth. Several species of aphids are found on dahlia, the most common being *Myzus persicae*, *Brachycaudus helichrysi* and *Aphis fabae*. Spraying with 1 per cent metacid, malathion or basudin controls these pests.

11.2.2 European Corn-borer (*Pyrausta nubilalis*)

The young larvae feed on the tender bud ends, flower parts and leaves, causing them to become distorted and turn brown. The affected tips usually die completely after which the borers bore down through the stem causing a wilting of all parts above the lowest region of invasion. The pest can easily be controlled by spraying monocrotophos every week.

11.2.3 Leaf Hopper (*Empoasca fabae*)

This insect caused a discolouration of the leaf which appears first along the margin and spreads toward the mid-vein. The affected area is at first pale yellowish, later it becomes brown and brittle. The infested plants often become stunted unless the leaf hoppers are controlled. Spraying with 1 per cent malathion or metacid is effective to control leaf hoppers.

11.2.4 Thrips

Three species of thrips, *Heliothrips haemorrhoidalis*, *Frankliniella tritici* and *Thrips tabaci* are apt to infect the flowers. Singh and Arora (1989) reported the incidence and damage of *Heliothis armigera* on *D. variabilis*. They rasp the surface and feed on the exuding juice and the lower surface of the petals turns whitish and withers. The measures suggested to control leaf hoppers may be followed for these pests also.

11.2.5 Red Spider Mite

Red spider mites (*Tetranychus urticae*) are serious pests on dahlia. They damage foliage by feeding on the undersides of leaves, causing a white speckling, but as numbers increase webbing may be spun round leaves and buds. Metasystox, wettable sulphur, or karathane may be used to control the pests. In greenhouse, control of mites on dahlia by treatment with plictran was reported by Wickizer and Meisch (1977).

12.0 Harvesting and Postharvest Management

12.1.0 Harvesting

The best time to cut dahlia flowers is in the cool of the morning. Snip horizontally with pruners or scissors by taking stems long enough. Choose flowers that are open or nearly open because the buds will not open once they're cut. It should be cut as much length as it requires for the bouquet, but try to make the cut just above a set of leaf nodes and side buds. New shoots will grow from those nodes. After harvesting dahlia flowers, make a fresh horizontal cut at the bottom of the stem and place the cut ends in about 2-3 inches of very hot (not quite boiling) water. Let the stems stay in the water for at least one hour. This hot-water treatment conditions the stems so the blooms will last four to six days. Once the cut stems have been conditioned, strip off all leaves that would be below the water line in flower vase. When leaves stay underwater, they decay and release bacteria that shorten the vase life of the flowers. Change the water in the vase every two or three days, adding a floral preservative to help the blooms last longer.

Blooms are cut in the early morning, the sun being avoided by all means, and kept immediately in a container half filled with water. The flowers should be cut with as long stem as possible. The container with cut flowers is to be placed in a cool and dark place for conditioning before packing.

12.2.0 Postharvest Management

12.2.1 Vase Life

Dole *et al.* (2004) obtained longest vase life when floral buds cut at the breaking stage (one petal open). The buds required 3.1-4.5 days to fully open and lasted 12.1-12.8 days when placed in 2 or 4 per cent sucrose or commercial holding solutions. Many of the buds held in water (control) either did not fully open or did not attain the flower size and color of buds held in sucrose or commercial holding solutions. Therefore, flowers harvested when fully open are best used for local sales, while those cut as buds should be used for wholesale marketing. Pre-treatments with commercial hydration solutions had no effect on vase life. Dahlia had a vase life of 6-9 days that could be increased to 10-11 days using commercial holding solutions such as Chrysal Professional 2 Processing Solution or Floralife Professional. Floral foam had no effect on vase life. They suggested that the optimum handling procedures for cut dahlias are: 1. cut in bud stage, 2. place in commercial holding solutions, 3. cold store at 34°F for less than one week.

Dipping the ends of the stems in boiling water for a second, or sealing them with candle flame will make the blossoms last longer. As an alternative, flowers may be cut in the early morning and immediately placed in 5 cm deep hot water (100°C) allowing them to set for four to five hours before arranging (Crockett *et al.*, 1971). According to Kaicker (1984), blooms can be held for 3-5 days after pre-treatment of the base of the stem in boiling water for 30 seconds to prevent blocking of water channels. Semi-open flowers of the cultivar Purple Gem, on 30 cm leafless stalks, were placed for 24 hours in glucose or sucrose solution containing preservatives and then transferred to containers with distilled water or they were held in these solutions. Vase life was longest (8.7 days) with 10 per cent glucose + 0.2 mM $AgNO_3$ or with 10 per cent glucose + 0.2 mM $AgNO_3$ + 8 HQS at 200 mg/l (8.2 days) compared with 4.3 days in the control (Lukaszewska, 1983). In another experiment, ethephon (50 ppm) decreased and a solution of 10 per cent glucose + $AgNO_3$ (50 ppm) increased the longevity of cut flowers. The use of ethephon and the preservative solution gave intermediate results (Lukaszewska, 1986).

Crude extracts containing an invertase inhibitor prepared from wilting petals of *dahlia pinnata* cv. Sniezna Kula, reduced invertase activity (20-30 per cent) of fresh petals. It indicates that invertase activity during petal senescence is controlled by the synthesis of an inhibitor (Halaba and Rudnicks, 1989).

Mahawer *et al.* (2008) observed maximum water uptake, vase life, flower diameter and gain in flower diameter without bent neck and fully opened flower with sucrose 2 per cent and 25 ppm NaOCl.

Dole *et al.* (2009) reported that Cut dahlia flowers could be held in unamended tap or deionized (DI) water with no effect on vase life. However, commercial holding solutions FLP (Floralife Professional) or CP2 (Processing solution) improved consumer vase life of dahlia compared to DI water. Floral foam or 2 per cent or 4 per cent sucrose had no effect on the consumer vase life of dahlia. Cut dahlia flowers were unaffected by exogenous ethylene.

Peck *et al.* (2016) found that Configure (benzyladenine) applied by postharvest dip did not affect the vase life of dahlia 'Park Princess' or 'Karma Yin Yang' flowers. Vase life for both cultivars was 7 days with all stems showing petal wilt at the end of vase life.

Azuma *et al.* (2019) revealed that CMIT/MIT antibacterial treatment extended the vase life of Dahlia 'Kamakura', 'Magic Pink' and 'Purple Stone', whereas, the GLC + CMIT/MIT increased the fresh weight of flowers along with vase life. A comparison of two cultivars with relatively long and short vase lives ('Moon Waltz' and 'Port Light Pair Beauty') showed that a longer vase life was related to higher carbohydrate content in the petals.

Bergmann *et al.* (2019a) expressed that 24-h pulse of cut dahlias in a solution containing both benzyladenine and gibberellic acid at 10-20 mg/l improved flower quality after 4 day in the vase and prolonged vase life regardless of handling method (dry packed or held in water from harvest through delivery).

Bergmann *et al.* (2019b) found that combination of 20 ppm each of BA and GA resulted in the same or better flower quality after 4 days and the same or longer vase life than either PGR used alone at 20 ppm. The maximum benefit of the combined PGRs was achieved at a lower concentration for dry-handled flowers than for wet-handled flowers. Flowers pulsed with 2 ml/l Chrysal BVB did not perform as well as the better BA/GA combinations.

Alkaç *et al.* (2020) studied the effect of deionized water (control), sucrose + deionized water, thyme oil, lavender oil, carvacrol, thymol, sodium hypochlorite, and gibberellic acid were used as vase solutions to extend the vase life of 'Le Castel' *Dahlia* (star flower). The most extended vase life was measured in the gibberellic acid treatment (8.22 days), the highest proportional fresh weight was measured in the carvacrol in the 6th day (139.78 per cent), the highest daily solution uptake was measured in the thyme oil in 0-2 days (11.7 g/day FW) and the highest total solution uptake was measured in the thyme oil (27.5 g/stem). It was concluded that the vase life of earlier harvested flowers was longer than of late harvested.

Azumaa *et al.* (2020) investigated the role of ethylene in dahlia flower senescence by comparing the responses to ethylene, ethylene inhibitors, and ethylene production among cultivars. Exogenous ethylene significantly accelerated petal abscission in seven cultivars and petal wilting in other five cultivars out of 12 cultivars. Whole

Storage of Dahlia Bulbs

Dahlia Tubers after Harvesting

Storage of Tubers

Packaging and Marketing of Dahlia

In Card Board Box

Polythene Packing

Dahlia in Local Market

Dahlia in International Markets

florets and detached receptacles (with bracts) produced different amounts of ethylene; ethylene production was higher in 'Carnelian' and 'Port Light Pair Beauty' than 'Heavenly Peace' and 'Purple Stone'. Onset of senescence was delayed in detached petals compared with attached petals, suggesting that petal abscission was induced by ethylene produced by ovary and receptacles. The ethylene action inhibitor 1-methylcyclopropene inhibited petal abscission and delayed petal wilting in eight cultivars. Moreover, the silver thiosulfate complex delayed petal wilting of 'Carnelian'. They suggested that ethylene plays a role in senescence of cut dahlia flowers and ethylene inhibitors can extend their vase life.

12.2.1.1 Storage

Dole *et al.* (2004) reported that one week of cold storage at 1°C reduced vase life up to 2 days. Ethylene at 0.1 to 1.0 ppm had no effect on dahlia stems, indicating that they are not ethylene sensitive. STS and 1-MCP had no effect on vase life. Stems could be cold stored at 1°C for up to one week, but reduced the vase life to 6.6 days as compared to 8.4 days for unstored flowers. Storage of cut stems in commercial holding solutions such as Chrysal Professional 2 Processing Solution or Floralife Professional increased vase life compared to storage in water only or dry storage.

12.3.0 Lifting and Storage of Tubers

After flowering when the plants are almost dried and colour of the stem turns yellow, the plants are cut leaving only 15 cm stem from the ground. The tuberous roots are taken out with a forked hoe and are allowed to dry for 3 or 4 days in a shady place. Before storing, the tubers should be treated with 0.2 per cent captan for 30 minutes. They can be stored on sand floor in a cool place for several months. The ideal storage temperature should be between 4°C and 7°C. Amongst 3 storage systems (sand, cold storage and ambient conditions), cold storage (5-6°C) was the best for dahlia tubers (Beura and Maharana, 1992). If the air is dry in the storage, the tuberous roots should be covered with peat moss, sand, ashes or vermiculite or may be wrapped with newspaper to prevent shrivelling. Ciobanu *et al.* (2015) found that substrate composed of peat + sawdust (70 + 30 per cent) was best substrate for storage and stratification of tuberous roots in five dahlia varieties.

Tubers can also be stored in perforated plastic bags or plastic-lined boxes filled with slightly dampened vermiculite or styrofoam. Broken tuberous roots must be cut sharply across and cut-surface be dusted with fine sulphur and dried. Tuberous roots in storage should be examined occasionally. Alternatively where there is no chance of frost damage, after drying the tuberous roots for 5 – 7 days in a shady place, these may be planted in well-drained sandy soil and the pots may be kept in semi-shade during the hot months. The stock plant for propagation either in pot or ground should be protected from excess moisture and water stagnation to prevent rotting of tubers.

12.4.0 Late Cutting

This is the most satisfactory method of dahlia preservation. Tender basal shoots from the plants in flower are taken as cuttings and the resultant plants (stock plants) are kept in growing condition. Cuttings from these stock plants are made during the rainy season in order to make additional stock plants. Pinching off the growing tips of the stock plants should be done regularly as this helps to produce better cuttings in a much larger number. During summer months the stock plants should be kept in such a place where they get only the morning sun for 2 hours. Late cutting method of dahlia preservation is particularly suitable for the plains (Swami Vinayananda, 1985, 1986).

13.0 Biotechnology

13.1.0 Molecular Markers

Behr and Debener (2004) studied the pollination behavior of dahlia using molecular markers. From two dahlia genotypes, seeds were harvested from open pollination flowers and from flowers covered with polyethylene bags. Although protected from pollinating insects, the later produced viable seed. Analyses with RAPD markers demonstrated that the plants from seed obtained under protection were most probably the result of self-pollination, whereas, 94.0 and 87.5 per cent of the open-pollinated seed resulted from outcrossing. These results indicated that some genotypes could be easily self-pollinated and therefore did not possess any self-incompatibility. Moreover, the outcrossing rate of open-pollinated progeny exceeded 90 per cent.

Wegner and Debener (2008) analysed genetic distances among 19 Dahlia cultivars, 3 genotypes of wild species and a hybrid using 1432 AFLP markers generated by 10 Hind/Mse primer combinations. Dendrogram was constructed by UPGMA method in which the cultivars were clustered together in a group whereas, wild species and hybrid showed a larger genetic distance. The level of genetic similarity of the cultivars ranged between 0.68 and 0.77. Only two varieties expressed a similarity of 0.92. It was concluded that relatively larger genetic distance among cultivars assures its vast potential for variety development.

Schie *et al.* (2014) generated SSR and AFLP marker data from two segregating populations of garden dahlias for determining the level and type of ploidy. The results revealed that dahlias are segmental allo-octaploids which originated from autotetraploid ancestral genomes.

13.2.0 Sequencing

Dahlia variabilis shows an exceptionally high diversity of floral forms and colours. Due to lack of comprehensive transcriptomic resources for this species, studies on genetic controls of pigment patterns have been limited. Lehnert and Walbot (2014) reported the sequencing, assembly and annotation of the transcriptome of developing leaves, stems and floral buds. This revealed 35,638 contigs to contain complete coding sequence including 20,881 annotated by similarity to UniProt. Moreover, the expression pattern of contigs was found consistent with tissue-specificity. These results would accelerate further research into genetic controls of pigmentation and floral form of *Dahlia variabilis.*

13.3.0 Germplasm Conservation

The diversity available with respect to floral characters is sufficient but *in situ* field conservation is costly and labour intensive. However, *in vitro* conservation

Exhibition of Dahlia

Outdoor Exhibition

Indoor Exhibition

technique offers the best alternative methods for germplasm conservation. Shivayogeppa *et al.* (2010a) standardized the growth and storability protocol for *in vitro* conservation of dahlia microshoots using different concentrations of mannitol, alar and maleic hydrazide (MH). Microshoots cultured on MS medium supplemented with MH at 40 mg/l recorded significantly higher number of days for bud sprouting (27.5), lease shoot length (4.25 cm), maximum storage period (134.00 days) and good recovery (83 per cent) followed by alar at 40 mg/l with 22.7 days for sprouting, shoot length of 3.40 cm, storage period of 129 days and 91 per cent recovery. Thus, the best treatment was mannitol at 20 g/l which registered 29.20 days for sprouting, 4.80 cm shoot length, storage period of 125 days and 89 per cent recovery. Cultures grown on MS medium without sucrose recorded maximum number of days for sprouting (21.60), least shoot length (4.90 cm) and longest storage period of 131 days followed by MS medium + 0.5 per cent sucrose. Maximum number of days for sprouting (21.5) with shoot length of 4.20 cm and culture storage period of 126 days was recorded with ¼ MS medium, whereas minimum days for sprouting (7.80), highest shoot length (10.50 cm) and lowest culture storage period (91.30 days) was observed in MS medium + 2 mg/l BAP. Highest recovery percentage of 95 per cent was recorded in polypropylene cap as bottle enclosure, while a lowest recovery of 41 per cent with cotton plug (Shivayogeppa *et al.*, 2010b).

14.0 References

Abdel-Salam, A.M., Al-Khazindar, M.M., Eid, S.G. and Pappu, H.R. (2010) *African J. Biotech.*, **9**(41) : 6835-6839.

Afzal, M., Ahmed, M.M., Shah, R.A. and Awan, B.M. (2000) *Pak. J. Biol. Sci.*, **3**(1) : 150-152.

Ahmed, M. J. and Gul, S. (2002) *Asian J. Plant Sci.*, **1**(5) : 565-566.

Ahmed, M., Khan, M.F. and Hussain, A.H.A. (2004) *Int. J. Agric. Biol.*, **6**(2) : 393–395.

Akbar, P.I., Mohan, B. and Kumar, V. (2011) *Asian J. Hort.*, **6**(1): 247.

Al-Ahmer, S.M., Salman, M.A. and Hamad, M.S. (2010) *Anbar J. Agril. Sci.*, **8**(1) : 148-161.

Albouy, J., Kusiak, C., Louanchi, M., Wang, W. and Zaayen, A. van (1994) *Acta Hortic.*, **377** : 189-195.

Albouy, J., Wang, W., Mevel, A., Morvan, C. le and Le Morvan, C. (1992) *PHM Revue Horticole*, **329** : 21-26.

Alkaç, O.S., Öcalan, O.N. and Güneo, M. (2020) *Ornam Hotic.*, **26**(4) : 607-613.

Alzahrani, Y., Alharby, H.F., Hakeem, K.R. and Alsamadany, H. (2020) *Notulae Botanicae Horti Agrobotanici Cluj-Napoca*, **48**(2).

Anónimo (2005) Boletín de flores. Métodos de propagación en el cultivo de flores y species ornamentales. FIA, 2005, no. 19, enero. ISSN 0718-0349.

Anonymous (2016) *Diyala Agril. Sci. J.*, **8**(1).

Aung, L.H. and Purohit, S.S. (1990) *Hormonal regulation of plant growth and development*, Vol. V, pp. 91-105.

Azuma, M., Onozaki, T. and Ichimura, K. (2019) *Hort. J.*, **88**(1) : 106-115.

Azumaa, M., Onozakia, T., Ichimuraa, K. (2020) *Sci. Hortic.*, **273**. https://doi.org/10.1016/j.scienta.2020.109635.

Baburao, D.S., Kullur, L.R., Manavi, G.H. and Prasad, V.M. (2017) *Int. J. Curr. Microbiol. and Appl. Sci.*, **6**(12) : 2013-2016.

Bajaraya, B., Kanawija, A., Jaysawal, N., Dubey, A., Parveen, S. and Pawaiya, S. (2019) *J. Pharmacognosy and Phytochem.*, **8**(1) : 98-102.

Balsevièius, A., Narijauskas, R. and Pauþaitë, G. (2016) *Agric. and Food*, **4** : 36-46.

Barrett, J.E. and Hertogh, A.A.De (1978) *J Amer. Soc. Hort. Sci.*, **103** : 775-778.

Behr, H. and Debener, T. (2004) *Eur. J. Hortic. Sci.*, **69**(5) : 177-183.

Bergi, U., Khan, G.U., Neelofar, N., Khan, F.A., Nazki, I.T. and Dar, M.A. (2014) *J. Plant Pest Sci.*, **2**(1): 33-40.

Bergmann, B.A., Ahmad, I. and Dole, J.M. (2019a) *Can. J. Plant Sci.*, **99** : 97-101.

Bergmann, B.A., Ahmad, I., Dole, J.M. (2019b) *The Cut Flowers Quarterly*, **31**(1). NC State Extension, NC State University.

Beura, S. and Maharana, T. (1990) *Orissa J. Hort.*, **18** : 48-51.

Beura, S. and Maharana, T. (1992) *Orissa J. Agric. Res.*, **5** : 174-178.

Bhattacharjee, S.K. (1984) *Punjab Hort. J.*, **27** : 138-144.

Bhattacharjee, S.K. and Mukherjee, T.P. (1981) *Punjab Hort. J.*, **21** : 226-230.

Bhattacharjee, S.K. and Wahi, S.D. (1982) *Hortic. J.*, **1** : 75-80

Bhattacharjee, S.K., Gupta, D.K. and Bose, T.K. (1971) *Phyton, (Argentina)*, **28** : 61-65.

Bhattacharjee, S.K., Mukhopadhyay, T.P. and Bose, T.K. (1976) *Indian Agric.*, **20** : 193-199.

Bhattacharjee, S.K., Mukhopadhyay, T.P. and Bose, T.K. (1978) *Indian Agric.*, **22** : 41-48.

Bhattacharyya, A.P., Pandey, H.S. and Yadav, L.P. (1976) *Prog. Hort.*, **8** : 51-56.

Biran, I. and Halevy, A.H. (1973) *Physiol. Plant.*, **28** : 244-247.

Borah, S.K., Nath, J.C. and Mohan, N.K. (1995) *South Indian Hort.*, **43** : 156-157.

Brondum, J.J. and Heins, R.D. (1993) *J. Amer. Soc. Hort. Sci.*, **118** : 36-42.

Bye, R. and Linares, E. (2008) *Biodiversitas*, **76** : 13-15.

Canham, A.E. (1969) *ShinfieldProgr.*, **14** : 38-39.

Chandraju, S., Thejovathi, C. and Chidan Kumar, C.S. (2013) *Chem. Sci. Trans.*, **2**(2) : 635-641.

Chen, J.H. and Li, D.B. (1990) *Acta Agriculturae Universitatis Zhejiangensis*, 16 : 78-82.

Ciobanu, I., Buta, E., Husti, A. and Cantor, M. (2015) *Agricultura, Agricultural Practica and Science Journal*, **95**(3-4).

Ciobanu, I., Cantor, M., Buta, E., Gocan, T. and Crisan, I. (2016) *Bulletin of Univ. Agril. Sci and Veter. Med. Cluj-Napoca: Hort.*, **73**(2): 135-140.

Clay, D.V. and Davison, J.G. (1968) *Proc. 9th Brit. Weed Control*, pp. 965-970.

Crockett, J.U. and editors (1971) *Time Life Books*, New York.

Cruz-Alvarez, O., Mejia-Munoz, J.M., Martinez-Damian, M.T., Rivera-Espejel, E., Garcia-Mateos, M.R. and Colinas-Leon, M,T.B. (2019) *Notulae Botanicae Horti Agrobotanici Cluj-Napoca*, **47**(3).

Darlington, C.D. (1973) *Chromosome Botany and the Origin of Cultivated Plants*, George Allen and Unwin, London.

Das, P.K., Dube, S. and Dey, A.K. (1978) *Indian Hortic.*, **23** : 19, 21.

Deguchi, A., Tatsuzawa, F., Hosokawa, M., Doi, M. and Ohno, S. (2016) *Hort. J.*, **85**(4): 340-350.

Devi, M.S., Seetharamu, G.K., Patil, B.C., Hanchinamani, C.N., Kukanoor, L., Satish, D. and Nishani, S. (2020) *Int. J. Curr. Microbiol. and Applied Sci.*, **9**(8): 2944-2950.

Dohare, S.R., Gill, H.S. and Arora, R.S. (1974) *Indian Hortic.*, **19** : 21.

Dole, J.M., Fanelli, F.L., Fonteno, W.C., Harden, B.T. and Blankenship, S.M. (2004) *Special Research Report* #422: Postproduction.

Dole, J.M., Viloria, Z., Fanelli, F.L. and Fonteno, W. (2009) HortTechnology, **19**(3) : 593-600.

Durso, M. and Hertogh, A.A.De (1977) *Proc. Amer. Soc. Hort. Sci.*, **102** : 314-317.

Eid, S., Saar, D.E., Druffel, K.L. and Pappu, H. R. (2011) *Plant Pathol.*, **60**(2) : 378-383.

El-Gamassy, A.M. and Moustafa, M.B. (1963) *Ann. Agric. Sci. Cairo*, **8** : 357-381.

Fedorov (1979) *Chromosome Number of Flowering Plants*. Reprinted by Otto Koelt, Science Publishers, West Germany, pp. 97.

Fischer, D., Stich, K., Britsch, L. and Grisebach, H. (1988) *Arch. Biochem. Biophys.*, **264** : 40-47.

Forkmann, G. and Stotz, G. (1984) *Planta*, **161** : 261-265.

Gardner, R. (1963) *Expt. Hort.* No. 9, pp. 90-97.

Gatt, M., Hammett, K. and Murray, B. (2000) *Plant Syst. Evol.*, **221** : 25-33.

Gavinlertvatana, P., Read, P.E., Wilkins, H.F. and Heins, R. (1979) *J. Amer. Soc. Hort. Sci.*, **104** : 849-852.

Giannasi, D.F. (1975) *The Flavanoid, Systematics of the Genus Dahlia* (*Compositae*), Memoirs of the New York Botanical Garden, **26**, No. 2.

Gofflot, A. and Verhoyen, M. (1990) *Parasitica*, **46** : 85-88.

Gupta, A.K., Jaiswal, N.K. and Saravanan, S. (2015) *Int. J. Agril. Sci. and Res.*, **5**(1): 55-58.

Gupta, Y.C., Dinesh, R.V., Kashyap, B., Bhatia, S. and Sharma, P. (2016) *Curr. Hortic.*, **4**(2): 48-53.

Halaba, J. and Rudnicki, R.M. (1989) *Acta Hortic.*, **251** : 119-120.

Hall, O.G. (1969) *ShinfieldProgr.*, **14** : 36-37.

Harborne, J.B., Greenham, J. and Eagles, J. (1990) *Phytochem.*, **29** : 2899-2900.

Hasegawa, A., Kobayashi, N., Muraki, C., Takagi, T. and Koutaka, T. (2009). *Acta Hortic.*, **813:** 587 - 592.

Heins, R.D., Kaczperski, M.P., Wallace, T.F.Jr, Lange, N.E., Carlson, W.H. and Flore, J.A. (1995) *Acta Hortic.*, **396** : 285-296.

Hernandez-Perez, F. and Mejia-Munoz, J.M. (1994) *Revista Chapingo Serie Horticulturae*, **1** : 63-66.

Hong, Y.P. and Jeong, J.H. (1988) *Res. Reports Rural Dev. Admn., Horticuture*, **30** : 104-110.

Hossain, M.M., Mukherjee, D. and Mohanty, B.K. (1999) *Adv. Plant Sci.*, **12** : 355-359.

Ibrahim, M. and Daraj, I.A. (2015) *Advances in Agriculture and Botanics*, **7**(1): 28-35.

Iftikhar, R., Moyo, L. and Pappu, H.R. (2020) *Annals of Applied Biol.*, **176**(2).

Jana, B.K., Laha, S.K. and Khatua, D.C. (1989) *Pesticides*, (Bombay), **23** : 33-34, 40.

Jana, B.K., Roy, S. and Bose, T.K. (1974) *Indian J. Hortic.*, **32** : 182-185.

Jiménez, V. (2001) *Aplicación de los fitorreguladores del crecimiento (Agromyl V, Biozyme TF) y un mejorador natural de Ortiga (Urtica dioica) en Dhalia.* [Tesis de Licenciatura]. Universidad Popular Autónoma, San Luis Tlaxiatemalco, Xochimilco, D.F, México. pp.70.

Julio, A.Y.R., Delgado-Martinez, R., Morales-Rosales, E.J., Laguna-Cerda, A., Franco-Mora, O. and Sanchez, U.E. (2011) *Phyton: Int. J. Exp. Bot.*, **80**(1) : 107-112.

Kaeser, W. (1989) *J. Microscopy*, **154** : 273-278.

Kaeser, W., Koyro, H.W. and Moor, H. (1989) *J. Microscopy*, **154** : 279-288.

Kannangara, T. and Booth, A. (1978) *Zeitscherift fur Pflanzenphysiologie*, **88** : 333-339.

Kareem, A., Shafqutsaeed, Rehman, S. and Aslamkhan, M. (2014) *Global Journal of Science Frontier Research.*, **14**(7) : 41-44.

Khan, F.U. and Tewari, G.N. (2003) *Indian J. Hort.*, **60**(2) : 192-194.

Khurana, S.M.P., Garg, I.D., Behl, M.K. and Singh, M.N. (1990) *Nat. Academy Sci. Letters*, **13** : 297-299.

Kiram, M., Baloch, J., and Waseen, K. (2007) *Park J. Biol. Sci.*, **10**(22) : 4140-4143.

Krismiyanto, L., Suthama, N. and Wahyuni, H.I. (2014) *J. Indonesian Trop. Animal Agric.*, **39**(4) : 217-223.

Kuluev, B.R. and Chemeris, A.V. (2007) *Russian J. Genet.*, **43**(12) : 1413-1414.

Kumar, L., Mahawer, L.N., Shukla, A.K., Kaushik, R.A. and Upadhyay, B. (2010) *Prog. Hortic.*, **42**(2) : 157-161.

Kumar, N., Prasad, V.M. and Yadav, N.P. (2019) *J. Pharmacogn. Phytochem.*, **8**(4) : 2265-2267.

Kumar, S., Sharma, R., Datta, K.S. and Nanda, K.K. (1978) *Indian J. Plant Physiol.*, **21** : 261-264.

Laguna, A., Valeriano, J. and María, G. (2003) *Ciencia Ergo Sum.*, **11**(001) : 59-64.

Legorreta, S.S., Villanueva-Carvajal, A., Ej, M., Laguna-Cerda, A. and Dominguez-Lopez, A. (2016) *Phyton: Int. J. Exp. Bot.*, **85**(1) : 63-70.

Lehnert, E.M. and Walbot, V. (2014) *Frontier in Plant Science*, 10.3389/fpls.2014.00340.

Leroux, G. (1982) *Dahlias of Today, Puget Sound Dahlia Association*, pp. 26-27.

Liudmila, J., Milvia, F., Aida, G., Sucel, I. and Joel, V. (2019) *Cultivos Tropicales*, **40**(1).

Liudmila, J.M. (2015) *Cultivos Tropicales*, **36**(1): 107-115.

Lloyd, F.G. (1968) *9th Orig, Rep. Exp. Husb. Fms Exp. Hort. Sta.*, pp. 66-67.

Lukaszewska, A.J. (1983) *PraceInstytutuSadownictwa w Skierniewicach*, **8** : 199-205.

Lukaszewska, A.J. (1986) *Acta Hortic.*, **177** : 663-666.

Mahawer, L.N., Kumar, L., Shukla, A.K. and Bairwa, H.L. (2010) *Indian J. Hortic.*, **67**(2) : 234-237.

Mahawer, L.N., Shukla, A.K., Bairwa, H.L., Sarolia, D.K. and Kumar, L. (2008). *J. Ornam. Hort.*, **11**(2): 118-120.

Maiko, T.K. and Yashchenko, N.P. (1980) *Byulleten' Glavnogo Botanicheskogo Sada*, No. 118, pp. 61-64.

Malik, S.A., Rather, Z.A., Wani, M.A., Din, A. and Nazki, I.T. (2017) *J. Exp. Agric. Int.*, **15**(3) : 1-7.

Manjula, B.S., Nataraj, S.K., Hegde, P.P., Anitha, G. and Ayesha, N. (2017) *Environ. Ecol.*, **35**(4C) : 3158-3161.

Mariña, L. J. (2015) *Cultivos Tropicales*, **36**(1) : 103-110.

Mastalerz, J.W. (1976) *Sci. Agric.*, **24** : 11.

Mathur, S.N. and Sharma, R.A. (1969) *Abstr. 11th Int. Bot. Congr., Seattle*, p. 142.

Mehlis, E. (1976a)*Stsche Gartenbau*, **30** : 36.

Mehlis, E. (1976b) *Dtsche Gartnerborse*, **76** : 1071.

Mehlis, E. (1978) *Dtsche Gartenbau*, **32** : 1833.

Miller, H. (1982a) *Dahlia of Today, Puget Sound Dahlia Assoc.*, pp. 28.

Miller, H. (1982b) *Dahlia of Today, Puget Sound Dahlia Assoc.*, pp. 29-31.

Miller, H. (1983) *Dahlia of Today, Puget Sound Dahlia Assoc.*, pp. 38.

Mishra, H.P., Singh, K.P., Mishra, G.M. and Prasad, B. (1990) *Haryana J. Hort. Sci.*, **19** : 284-290.

Misra, R.L. (1978) *Prog. Hortic.*, **10** : 34-46.

Misra, R.L. (1990) *Prog. Hortic.*, **22** : 36-39.

Misra, R.L., Saini, H.C., Dhyani, D., Verma, T.S., Thakur, P.C., Singh, A. and Kumar, R. (1990) *Indian J. Gen. Plant Breeding*, **50** : 51-55.

Mittal, S.P. (1967) *Madras Agri. J.*, **54** : 103-107.

Moe, R. (1978) *Gartneryrket*, **68** : 62-64.

Moser, B.C. and Hess, C.E. (1968) *Proc. Amer. Soc. Hort. Sci.*, **94** : 595-603.

Mounika, T. and Saravanan, S.S. (2019) *Int. J. Curr. Microbiol. and Applied Sci.*, **8**(8): 2389-2397.

Mukhopadhyay, R., Bhattacharya, S. and Moulishabiwas (2013) *J. Adv. Pharm. Edu.*, **3**(4).

Mullin, R.H. and Schlegel, D.E. (1978) *Plant Dis. Reptr.*, **62** : 565-567.

Mustafajeva, A.K. (1985) *Referativny Zhurnal*, **55** : 9, 55, 835.

Nigam, S.K. and Misra, G. (1992) *J. Oil Technol. Asso.*, India, **24** : 17-18.

Nimbalker, C.A., Dhane, A.V. and Bajaj, V.H. (2006) *J. Ornam. Hortic.*, **9**(2): 122-125.

Nitsch, E. (1992) *German Federal Ruplic Patent*, DE 4101910, p. 4.

Niyogi, S.G. (1973) *Annual Flower Show Souvenir, Agri-Hort. Soc. India*, pp. 23-28.

Ohno, S., Hori, W., Hosokawa, M., Tatsuzawa, F. and Doi, M. (2016) *Hortic. J.*, **85**(2): 177-186.

Ohno, S., Hori, W., Hosokawa, M., Tatsuzawa, F. and Doi, M. (2018) *Planta*, **247**(2): 413-428.

Okada, H., Tauchi, Y., Karasawa, Y., Kikumura, T. and Kitamura, Y. (2020) *Shokubutsu Kankyo Kogaku*, **32**(1).

Onozaki, T. and Azuma, M. (2019) *Hort. J.*, **88**(4) : 521–534

Otani, Y., Chin, D.P. and Mii, M. (2013) *Plant Biotech.*, **30**(2): 135-139.

Pahalawatta, V., Druffel, K. and Pappu, H. (2008) *Virology*, **376**(2): 253-257.

Pappiah, C.M. and Muthuswamy, S. (1974) *South Indian Hort.*, **22** : 77-80.

Paul Hale (1983) *Dahlia of Today, Puget Sound Dahlia Association*, pp. 36-37.

Peck, L., Latimer, J. and Scoggins, H. (2016) *Grow Research Update*, #2016.01.

Perwez, M.S., Husain, S.T. and Akram, M. (1994) *Ann. Plant Protection Sci.*, **2** : 80-81.

Perwez, M.S., Mohd-Akram and Akram, M. (1989) *Acta Botanica Indica*, **17** : 267.

Phetpradap, S., Hampton, J.G. and Hill, M.J. (1994) *New Zealand J. Crop Hort. Sci.*, **22** : 313-320.

Pichardo-Ruiz, F. D., Villegas-Monter, Á., Hernández-Livera, A. and Colinas-León, M. T. (2003) *Revista Chapingo Serie Horticultura*, **9**(1) : 151-161.

Pirone, P.P., Dodge, B.O. and Rickett, H.W (1960) *Diseases and Pests of Ornamental Plants*, Constable and Co. Ltd., London.

Pizzetti, I. and Cocker, H. (1975) *Dahlia in Flowers - A Guide to Your Garden*, Harry N. Abrams Inc., New York.

Poincelot, R.P. (1993) *J. Sustainable Agric.*, **3** : 99-110.

Powell, C.C. (1981) *Ohio Florists' Assoc. Bull.*, No. 624, pp. 1-4.

Pradeep, M.K., Pratibha, S., Kamal, G.K., Harish, T. and Pradeep, S. (2012) *Indian J. Fibre and Textile Res.*, **37**(1): 83-86.

Prokashina, I.I., Shatilo, V.I. and Keldysh, M.A. (1990) *Byulleten Glavnogo Botanicheskogo Sada*, **157** : 69-72.

Punlerdmatee, B., Cheewangkoon, R. and To-anun, C. (2017) *International Journal of Agricultural Technology.*, **13**(1) : 1-17.

Pupykina, S., Nikolaevna, M., Ljudmila, D. and Aleksandrovna, G.K. (2015) *Chem. Plant Raw Material*, **2**: 77-83.

Quagliotti, L. and Gullino, G. (1967) *Reprint from atti Conv. Naz., Foric. Trieste*, p. 18.

Rahman, N.M. and Mitra, S.N. (1974) *Bangladesh Hort.*, **2** : 15-18.

Rashid, A. and Power, J.B. (1983) *In Protoplasts, Poster Proc.*, BaselsSuitzerland, Birhauser Verlag (1983), pp. 30-31.

Read, P.E. and Hoysler, V.C. (1969) *J. Amer. Soc. Hort. Sci.*, **94** : 314-316.

Read, P.E., Duham, C.W. and Field House, D.J. (1972) *HortScience*, : 62-63.

Robertson, C. (1981) *Dahlias of Today, Puget South Dahlia Association*, p. 12.

Robertson, C. (1982) *Dahlias of Today, Puget South Dahlia Association*, pp. 22-23.

Rounkova, L.V. (1989) *Acta Hortic.*, **251** : 281-288.

Runkova, L.V. (1985) *Acta Hortic.*, **167** : 69-77.

Saar, D. E., Sorensen, P. D. and Hjerting, J. P. (2003) *Acta Botánica Mexicana*, **64** : 19-24.

Safronova, A.I. (1969) *Bjull. Gos. Nikitsk Bot. Sada*, **1** : 14-17.

Samant, P.K.S. and Das, P. (1990) *Orissa J. Agric. Res.*, **3** : 192-196.

Sanabria-de-Albarracin, N. and Albarracin, N.S. de (1988) *Fitopatologia Venezolana*, **1** : 59-64.

Saravanan, S. and Barik, I. (2016) *Int. J. Multidisciplinary Res. and Develop.*, **3**(8): 260-263.

Saryono, Hendris, S., Fitriyah, D., Jose, C., Nugroho, T.T. and Ardhi, A. (2015) *J. Chem. and Pharm. Res.*, **7**(9): 201-208.

Saryono, Octarina, N., Yuharmen, Pratiwi, N.W. and Ardhi, A. (2017) *Nusantara Biosci.*, **9**(2): 146-151.

Schie, S., Chaudhary, R. and Debener, T. (2014) *Plant Genome*, **7**(3). https://doi.org/10.3835/plantgenome2014.01.0002

Sediva, J., Novak, P., Kanka, J. and Laxa, J. (2006) *Acta Hortic.*, **725.**

Sekerci, Dalda, A. and Güloen, O. (2016) Overview of dahlia breeding. Scientific Papers. Series B, Horticulture. Vol. LX.

Sharga, A.N. (1975) *Prog. Hort.*, **7** : 29-38.

Sharma, K., Verma, S., Gautam, H.R. and Handa, A. (2017) *Int. J. Curr. Microbiol. and Applied Sci.*, **6**(6): 1595-1598.

Shatilo, V.I. (1991) *Byulleten' Glavnogo Botanicheskogo Sada*, No. 159, pp. 72-75.

Shatilo, V.I. (1994) *Byulleten Glavnogo Botanicheskogo Sada*, No. 170, pp. 120-124.

Shatilo, V.I., Keldysh, M.A. and Morozova, S.E. (1991) *Byulleten' Glavnogo Botanicheskogo Sada*, No. 159, pp. 68-72.

Sheergojri, G.A., Neelofar, Z.A., Rather, Z.A., Khan, F.U., Nazki, I.T. and Qadri, Z.A. (2013) *Applied Biol. Res.*, **15**(2): 121-129.

Sherwood, J.L., Broembsen, S.L. von and Von-Broembsen, S.L. (1991) *Plant Dis.*, **75** : 750.

Shivayogeppa, G., Adiga, J.D., Prabhuling, G., Reddy, B.S. and Sathyanarayana, B.N. (2010b) *Acta Hortic.*, **865.**

Shivhare, L. and Sharma, S. (2012) *Int. J. Curr. Res. and Review*, **4**(19): 25.

Shravan-Dasoju, Evans, M.R., Whipker, B.E. and Dasoju, S. (1998) *HortTechnol.*, **8** : 595-598.

Shrotri, S.C., Gupta, J. and Srivastava, R.N. (1985) *Indian J. Mycol. Plant Pathol.*, **14** : 266-269.

Shukla, P., Prasad, V.M., Burondkar, S.S. and Ainarkar, A.A. (2018) *J. Pharmacognosy Phytochem.*, **7**(5) : 1109-1113.

Shunsuke, A., Kandai, Y., Terufumi, N., Yoshihiko, H. and Yosuke, M. (2020) *European J. Plant Pathol.*, **156**(1).

Singh Zile, Z., Gupta, A.K. and Singh, Z. (1996) *Environ. and Ecol.*, **14** : 940-943.

Singh, D. and Arora, J.S. (1989) *J. Insect Sci.*, **2** : 168-169.

Singh, J.N., Singh, D.K. and Sharma, K.K. (1994) *Orissa J. Hortic.*, **22** : 10-12.

Singh, V.K., Prasad, V. M. and Saravanan, S. (2017) The Allahabad Farmer, **LXXIII**(3) : 43-45.

Singh, Z. and Gupta, A.K. (1995a) *Crop Res.* (Hissar),**10** : 174-178.

Singh, Z. and Gupta, A.K. (1995b) *Crop Res.* (Hissar), **10** : 327-333.

Smith, A.W. (1963) *A Gardeners Book of Plant Names*, Harper and Row Publishers, New York.

Smith, E.M. and Treaster, S.A. (1991) *OARDC Special Circular*, Ohio Agricultural Research and Development Centre, No. 137, pp. 19-21.

Smith, E.M. and Treaster, S.A. (1992) *Special Circular Ohio Agricultural Research and Development Centre*, No. 140, pp. 1-4.

Sorensen, P.D. (1969) *Rhodora*, **71** : 309-416.

Southward, R.C., Harrington, K.C., Hampton, J.G. and Han, H. (1998) *Proceedings Agronomy Society of N.Z.* **28** : 11-15.

Staats, D. and Klett, J.E. (1993) *J. Environ. Hort.*, **11** : 78-80.

Stich, K., Halbwirth, H., Wurst, F. and Forkmann, G. (1994) *Zeitschrift fur Naturforschung, Section C, Biosciences*, **49** : 11-12.

Suma, B., Philipose, J. and Joshua, P. (1994) *South Indian Hortic.*, **42** : 340-341.

Suman, C.L., Wahi, S.D. and Bhattacharjee, S.K. (1984) *Indian J. Hortic.*, **41** : 299-303.

Suman, C.L., Wahi, S.D. and Bhattacharjee, S.K. (1991) *Agri. Sci. Digest Karnal*, **11** : 190-194.

Suraya, N., Piska, F., Pratiwi, N.W., Ardhi, A. and Saryono (2019). *IOP Conference Series: Materials Science and Engineering*, **532**(1).

Suzuki, H., Nakayama, T., Yonekura-Sakakibara, K., Fukui, Y., Nakamura, N., Yamaguchi, M., Tanaka, Y., Kusumi, T. and Nishino, T. (2002) *Plant Physiol.*, **130**(4): 2142-2151.

Swami Vinayananda (1983) *Dahlias of Today, Puget South Dahlia Association*, pp. 24-25.

Swami Vinayananda (1985) *Dahlia Growing, Associated Publishing Company*, New Delhi.

Swami Vinayananda (1986) In : *Orna. Hort. India*, ICAR, New Delhi.

Thevissen, K., Cammue, B.P.A., Lemaire, K., Winderickx, J., Dickson, R.C., Lester, R.L., Ferket, K.K.A., Even, F.V., Parret, A.H.A and Broekaert, W.F. (2000) *Proceedings of the National Academy of Sciences of the United States of America*, **97**(17): 9531-9536.

Tiyagi, S.A. and Wani, A.H. (1991) *Curr. Nematol.*, **3** : 119-122.

Trivedi, H., Punetha, P., Shekhar, C. and Singh, N. (2015) *Ecol. Environ. and Conserv.*, **21**: 247-250.

Tsushima, T., Matsushita, Y., Fuji, S. and Sano, T. (2015). *New Dis. Rep.*, **31**: 11.

Upadhyaya, J. and Bhandari, T.P. (1985) *Indian Phytopath.*, **38** : 338-339.

Verma, J. and Kulkarni, B.S. (2017) *Int. J. Curr. Microbiol. App. Sci.*, **6**(11) : 402-409.

Vikas, H.M., Patil, V.S., Agasimani, A.D. and Praveenkumar, D.A. (2011) *Int. J. Sci. and Nature*, **2**(2): 372-375.

Wang, W.C., Tronchet, M., Larroque, N., Dorion, N. and Albouy, J. (1988) *Acta Hortic.*, **234** : 421-428.

Wani, A.H. and Ansari, A.P. (1993) *Curr. Nematol.*, **4** : 81-84.

Wegner, H. and Debener, T. (2008) *European J. Hortic. Sci.*, **73**(3).

Wei, S.L., Li, P.M., He, Z.P. and Han, B.W. (1990) *Acta Hortic. Universitatis Pekinensis*, **16** :381-385.

Whipker, B.E. and Hammer, P.A. (1997) *HortTechnol.*, **7** : 269-273.

Whipker, B.E., Eddy, R.T. and Hammer, P.A. (1995) *HortScience*, **30** : 1007-1008.

Whipker, B.E., McCall, I., Buhler, W. and Krug, B. (2011) *Acta Hortic.*, **886.**

Whitley, G.R. (1985) *J. Ethnopharmacol.*, **14** : 75-82.

Wickizer, S.L. and Meisch, M.V. (1977) *Arkansas Farm Research*, **26** : 6.

Willis, J.C. (1966) *Flowering Plants and Ferns*, Cambridge Univ. Press, U.K.

Winter, J.A.T.De (1979) *Bloembollencultuur*, **89** : 1046.

Younis, A., Ahmad, M., Riaz, A. and Khan, M.A. (2008) *Acta Hortic.*, **804** : 191-196.

Younis, A., Anjum, S., Riaz, A., Hameed, M., Tariq, U. and Ahsan, M. (2014) *American-Eurasian J. Agril. and Environ. Sci.*, **14**(2): 137-142.

Zimmer, K. (1989) *Deutscher Gartenbau*, **43** : 884-887.

8

ALSTROEMERIA

C. Aswath, V.A. Parthasarathy and T.K. Bose

1.0 Introduction

Alstroemeria is popularly known as Inca or Peruvian lily. It is grown in garden for their showy multicoloured flowers. Now the flowers of *Alstroemeria* hybrids are one of the ten important cut flowers in the world market. The spotted flowers in various shades of pink, rose, red, white, yellow, orange and bronze are popular as cut flowers. In India, it is being grown in the Nilgiris, Bangalore, Himachal Pradesh and Uttarkhand. Considering its attractive flowers, many commercial cultivars have been successfully introduced in the market, using this species as cut flower and more recently as a potted plant and as a garden flower (Bridgen, 2018). Nowadays, alstroemeria is one of the most important cut flowers in the world market (Lim *et al.*, 2012) and many new varieties are released every year.

2.0 Importance and Uses

Alstroemeria is a hardy perennial, easy to grow in bed, border and pot, and is used for garden display. In a dry border, it may be allowed to remain undisturbed for years. Large clump of any of the varieties will produce brilliant flowers. Straight and erect flower stems of various colour are used as cut flowers. Cut flowers stay fresh in vase for several days. Cultivation of alstroemeria hybrids is economic if the first flower crop is finished in a mobile glasshouse in time for it is to be used for another crop (*e.g.*, chrysanthemums) and again in the autumn for a second alstroemeria crop (Spier and Milde, 1977).

3.0 Origin, History and Distribution

Alstroemeria is a genus native to South America with about 60 described species, 31 of them native to Chile and Brazil which are the main diversity centres of this genus (Reeves, 1992; Aros *et al.*, 2019). Alstroemeria is herbaceous, perennial and rhizomatous plants with large and attractive flowers, growing in a wide range of habitats across the Andes Mountains and the rainforests in Brazil (Aros *et al.*, 2019).

Uphof (1952) reported that it has approximately 62 species, 30 species were recorded in Chile and 23 species in Brazil (Bayer, 1989). Tombolato and Mathes (1998) suggested that the number of Brazilian species is around 50 to 100.

The genus was established by Linnaeus in 1762, by the description of Feuille of three Chilean species, in 1714, under the epithet *Hemerocallis.* The type species is the *A. pelegrina* by Feuille, which was grown in the Incas' gardens. By 1754, this species was found growing in Spain by Alstroemer (Uphof, 1940, 1952; Stinson, 1942).

Since the time of Linnaeus, morphological variability was noted among different populations. At that time there had been confusion on the separation and naming of the *Alstroemeria* species. Leaves of the vegetative stalks, hairs and the flower shape, size and colour were utilized for identification (Stinson, 1942, 1945, 1952; Traub, 1973; Bayer, 1989).

The *Alstroemeria* species were found in different environmental conditions-swamps, under-woods, thorn shrub land or woodland, desert areas or rocky hillsides

Uses of Alstroemeria

In Bed

In Border

In Pot

In Flower Vase

As Cut Flowers

In Ikebana

In Greenhouse

up to an altitude of 3700 m rainforest and thickets along the seashore and in rupi colours meadows (Uphof, 1952).

There are at least two Brazilian scented species, one *A. caryophyllaea*, was not collected for about 150 years, after which it was rediscovered in the state of Espirito Santo and taken to the United States (Foster, 1945). The other scented plant is a strain of *A. sellowiana*.

In a survey of the Index Kewensis, by 1895 there were 35 epithets for Brazilian species. Several names were written on herbarium species which could not be confirmed and their validity was ignored (Tombolato and Mathes, 1998). Hannabal (1941) stated that *A. inodora* and *A. brasiliensis* are synonyms of *A. psittacina*.

4.0 Morphology, Species and Cultivars

4.1.0 Morphology

Alstroemeria are bulbous, perennial, about 1.2 m and erect; leaves resupinate; flowers zygomorphic, borne on umbel-like inflorescences, rarely solitary, 10-30 flowers per inflorescence; stamens six, slightly curved, usually unequal, bigger or shorter than the tepals; ovary inversely conical; protandry common in the genera (Brydon, 1940; Uphof, 1952; Miyake, 1989; Bayer, 1989).

4.2.0 Species

Morphological characters of some important species and hybrid cultivars are given below:

Alstroemeria aurea syn. *A. aurantiaca*

Plants tuberous perennial, 45 to 100 cm high; leaves lanceolate, scarcely stalked, up to 10 cm long; flowering stem branched, up to over 90 cm height with 40 – 50 flowers; flowers about 3.8 cm long, orange, 2 upper segments acute, streaked red, outer broadly ovate, blunt, tinged red, point green. June to August. Cultivars of this species are *concolor* 6 flower without stripes; *lutea* 6 bright yellow flowers.

A. brasiliensis

Plants tuberous, hardy; leaves linear, 7-10 cm long; those of sterile stem lanceolate, 7-10 cm long, mainly in upper half; flowering stem 90 – 120 cm height, flowers reddish-yellow, 3.8 cm long; inner segment spotted brown, in an umbel of five, 1 – 3 flowered rays.

A. campaniflora

Plants rhizomatous; stem erect, 90 cm height; leaves dense in upper part, narrow oblong-lanceolate, 6.5 – 15 cm long, glabrous, leathery, glaucous beneath; flowers 10 – 23 in a compound umbel, tubular, green; segments about 3 cm long, inner spotted dark within.

A. caryophyllea

Leaves spatulate, oblong; flowering stem 20 – 30 cm height, flowers scarlet or scarlet striped, very fragrant; perianth 2-lipped, peduncles long. Suitable for warm house.

Species of Alstroemeria

Alstroemeria aurea

Alstroemeria ligtu hybrids

Alstroemeria hookeri

Alstroemeria psittacina

Alstroemeria paupercula

Alstroemeria pelegrina

Alstroemeria violacea

Alstroemeria versicolor

Alstroemeria spathulata

A. hookeri

Plants dwarf, up to 60 cm, tuberous perennial; leaves narrow linear, 2.5 cm long; stems 15 cm long, bears up to 6-rayed umbels; flowers pink, inner tepals yellow flashes and streaked with purple.

A. ligtu

Known as St. Martin's Flower. Plants tuberous perennial; leaves thin, narrow linear, up to 7.5 cm long; flowering stem 45 – 60 cm high, with 20 – 30 flowers; inflorescence umbel with 3 – 8 rays, branched, flowers pale lilac to reddish or whitish, 3.8 cm long, inner upper pair narrower, usually yellow in upper half with purple stripes. Var. *angusfolia* : leaves narrow, flowers pale pink; var. *pulchra* : flowers larger, perianth segments acute, all with purple spots in upper half.

A. pelegrina

Leaves lanceolate, up t– 5 cm long; flowering stem about 30 cm high; inflorescence sometimes many-flowered in garden forms; flowers 5 cm long, lilac or bright rosy lilac, striped outside, inner segments with a yellow base, purple spots, outer segments with large purplish blotches. Many colour forms occur and in garden forms the flowers are more numerous than in the wild. Var. *alba* : well marked white form, known as Lily of the Incas; var. *rosea* : flower deeper rose, rather tender.

A. pulchella

Leaves oblong to oblong-spatulate, scarcely stalked, those of barren stems long stalked; flowering stem about 90 cm or more high, leafy; inflorescence simple, 4- to 6-flowered, pedicels 3.8 cm long; flowers about 3.8 cm long; segments somewhat unequal, pale red below, green above, the inner brown with reddish-brown spots. Hardy *A. pulchella* has crossed with other species, *A.* x *Ereemboulfii* with white flowers spotted purple being one of the hybrids.

A. spathulata

Leaves in upper part of stem rather dense, oblong spatulate, up to 5 cm long, obtuse, blunt, crisped and scarious at margin; flowering stem 30 cm high; inflorescence simple, up to 10-flowered; flowers reddish, 3.8 cm long, outer segments broad, obtuse, green pointed, rose, inner narrower, acute, with yellow blotches and red spots, toothed. Var. *nellii* : robust form.

A. versicolor

Leaves thin, narrow linear; flowering stem about 30 cm high; inflorescence with few branches and linear bracts; flowers about 2.5 cm long, yellow with purple flecks, the inner perianth segments narrow. Rather tender.

A. violacea

Leaves spreading, scattered, up to 2.5 cm long, on sterile shoots larger, ovate-oblong, 5-nerved; flowering stem 30 – 60 cm high; flowers bright lilac, 3.8 to 5 cm long, in a 5-rays umbel on forked pedicels; outer segments obovate, blunt, shortly cuspidate, inner oblong-acute, spotted.

Some other hybrids are *A.* 'Parigo Charm', *A.* 'Rosy Wings', *A.* 'Sonata' and *A.* 'Sweetheart'.

4.3.0 Cultivars

Other important cultivars are :

White - Amanda, Mona Lisa, White Wings.

Red - Carmen, King Cardinal, Red Sunset, Valiant.

Pink - Capitol, Fiona, Olympic, Rito, Trident, Veronica, Pink Triumph.

Orange - Harlequin, Sunrise.

Yellow - Canaria, Eleaner, Friendship, Orchid, Zebra, Yellow King.

Red-purple - Marina, Purple Joy, Sunstar.

Bronze - Butterscotch.

Lavender - Barbara, Butterfly, Jubilee, Jupiter.

Salmon - Atlas.

Hybrid Varieties (Reid, 2006)

The 'Dr Fleming' series, or 'Orchid' types, are infertile hybrids of *A. aurantiaca*. 'Walter Fleming' has yellow and white flowers. However, new Dutch and English varieties have superseded these hybrids. For example, 'Canaria' is a mutant of 'Walter Fleming' and 'Rosita' is a mutant from the triploid variety 'Regina'. Several European companies such as van Zanten, van Staaveren, Konst and Hilverda offer new *Alstroemeria* hybrids. These are protected by Plant Breeders Rights legislation and plants are relatively expensive.

5.0 Genetics and Breeding

5.1.0 Genetics

The chromosome numbers of alstroemeria indicate that all the species are diploid, 2n=16. Of 25 cultivars examined, four were also 2n=16, 12 had 2n=24 (3X), six had 2n=32 (4X) and the remaining had either an extra or missing chromosome from the 3X or 4X situation (Rees, 1992).

Five colour types from one seed source were examined cytogenetically (Rustanius *et al.*, 1991). The somatic chromosome numbers were all 2n = 2x = 16. Karyotype analysis revealed that all 5 plants had the same chromosome constitution. Chromosome pairs 2, 3, 5 and 8 had satellites. Chromosome complements of the *A. ligtu* hybrid were unique in that they contained 2 pairs of satellited metacentric chromosomes that were not found in any alstroemeria cultivars. Alstroemeria cultivars comprise diploid, triploid and tetraploid hybrids derived from inter-crossing of either diploid Chilean species or Brazilian and Chilean species whose genomes are clearly differentiated. On the basis of a study of 9 diploid interspecific hybrids (4 between Chilean species and the rest involving Chilean and Brazilian ones), Ramanna and Mariani (1992) suggested that the polyploid forms originated through a gradual process of sexual polyploidization involving mainly 2n gametes,

Cultivars of Alstroemeria

Casablanca

Friendship

Red Coat

Inca exotica

Evening Song

Marina

Orange Glory

Rebecca

Sussex Gold

Jupiter

Mona Lisa

Yellow King

as well as other types of functional gametes produced by sterile diploid interspecific hybrids and triploids.

5.2.0 Breeding

5.2.1 Interspecific Hybridization

The ornamental value of new hybrids of alstroemeria has been usually assessed using morphological descriptors considering flower size and color, stem length (Aros *et al.*, 2015) and vase life (Leverentz *et al.*, 2002), which are the main characters chosen by ornamental plant breeders. Although an important character, well appreciated by consumers (Aros *et al.*, 2015), floral scent has been poorly studied in this species. So far *Alstroemeria caryophyalleae*, a species endemic to Brazil, is the only scented species described within the genus (de Assis, 2004) but little is known about the heredity of this character. Aros *et al.* (2019), quoting Porat *et al.* (1993), opined that apparently floral scent has been unintentionally selected by breeders considering a putative negative association between vase life length and fragrance.

Today's alstroemerias are hybrids. The commercial hybrids are derived from species *A. pelegrina, A. violacea* and *A. aurantica* and were bred by two companies, Parigo in England and Van Staaveren in the Netherlands in the 1960s (Rees, 1992).

The alstroemeria cultivars produced as cut flowers are interspecific hybrids of relatively few species. Methods for production of a large number of different hybrids were investigated by Kristiansen and Vainstein (1995). Viable hybrids between *Alstroemeria* species were obtained using ovule culture. Most of the hybrids were male sterile, but hybrids from some combination were highly fertile. The percentage of germinating ovules increased with increasing age of the ovules at the time of incubation, and it was possible to make cross throughout the year. Pollen from most species could be stored for at least 3 months at 5°C and low humidity. Actively growing young rhizomes of several cultivars, mostly sterile hybrids, were X-irradiated at 50-150 rad/min. The optimum dose was about 400 rad for diploid and 500-600 rad for triploids. No mutant showed any sign of chimaerism despite the fact that the buds on the rhizomes possess multicellular apices. Several of the mutants were of considerable ornamental value and 5 were introduced into commerce (Broertjes and Verboom, 1974).

Aros *et al.* (2019), quoting the work of Vonk Noordegraaf (1981), indicated that the first alstroemeria hybrids were obtained through the interspecific hybridization between two Chilean species (*Alstroemeria aurea* × *Alstroemeria violacea*). Interspecific hybridization followed by embryo rescue has been one of the most common breeding techniques to obtain new varieties. Aros *et al.* (2019) initiated studies to develop alstroemeria hybrids through interspecific crosses using novel scented parental lines and assessing these hybrids. A total of 3669 embryos were rescued, from which 18 flowering plants were morphologically evaluated. Flower stem length ranged from 25 (14E07) to 83 cm (14A11) and most of the flowers showed pink/white colors with stripes over the inner tepals. Only 3 hybrids were perceived as scented and one of them (14E08) was evaluated through GC-MS analysis, detecting 9 Volatile Organic Compounds (VOC), all of them monoterpenes. Molecular analysis using RAPD markers clustered two main groups: (A) hybrids with *A. pelegrina* as parental line and (B) hybrids coming from the crossing UC05 x C3 and its reciprocal. Moreover through this analysis was possible to sub-cluster hybrids coming from different seasons. A protocol for obtaining alstroemeria hybrids was successfully developed for this species using molecular and phenotypic evaluation to provide information about their pedigree and to spot those traits which are attractive for the ornamental plant trade (Aros *et al.*, 2019).

5.2.2 Mutation Breeding

Gamma irradiation has been widely used as a breeding technique to obtain new cultivars in ornamental species such as Alstroemeria, where several cultivars have been obtained through rhizome radiation (Aros *et al.*, 2012). The optimum dosage for an appropriate induction of mutation must be considered for breeding purposes and it depends mainly on plant susceptibility. Thus, *in vitro* cultured rhizomes of *Alstroemeria aurea* were irradiated with a gamma source using different dosages to evaluate the direct effect produced. Damage and number of rhizome sprouting were observed and recorded during 61 days after irradiation. At the end of this period, rhizomes were weighted and mortality was evaluated. Both mortality and weight increased depending on dosage. All irradiated rhizomes showed early sprouting in comparison with control (0 Gy) and no significant difference in final number of shoots after 61 days among irradiated treatments was observed. Bleaching and necrosis was observed in all irradiated rhizomes and was more evident at higher doses. LD_{50} was established at about 40 Gy and the optimum dosage to induce mutation was suggested between 2.5 and 5 Gy, when the growth was reduced in 50 per cent, and probably this dosage could be used for breeding purposes (Aros *et al.*, 2012).

6.0 Growth and Flowering

The growth and flowering of this crop is primarily regulated by genetic and environmental factors. Light, temperature, CO_2, soil moisture and nutrients, use of growth regulating chemicals, size of bulb, shoot thinning *etc.*, also influence the growth and flower production.

6.1.0 Environmental Factors

6.1.1 Light

The day length requirement for flower induction (11-14 h) varies with cultivars, 11 h for *cv.* Beauty and 13-14 h for Regina and Starosa. Light deficiency leads to drying of floral buds, and high temperature (25°C) to blind shoots (Milde, 1983). Plomacher (1978) also reported that alstroemeria produces blind shoots with insufficient light. Increasing day length to 13 h advanced flowering of cv. Orchid and made the stems shorter. Too long a day, however, could reduce the number of flowers per inflorescence and the number of shoots. A night break resulted in higher yields in the second cutting and an earlier second cutting, but lower yields in the first cutting than the day length extension (Meij, 1976). Verboom (1979) recorded that a photoperiod of 14 h or more, particularly in combination with a temperature above 18°C, reduced growth and decreased the number of shoots produced and the number of flowers in the inflorescence. Increasing the day length in winter accelerated bud initiation, resulting in earlier flowering and shorter flower stems. In almost all cultivars a day length of 13 h was optimal in winter and spring and enhanced plant growth after the first flowers were cut in June.

Plant height was reduced when they were grown under 13 h photoperiods. The number of generative shoots was not affected by photoperiod. The number of vegetative shoots was related inversely to the length of the photoperiod. The number of days to flower decreased as photoperiod increased (Healy *et al.*, 1993).

Krogt (1985) in an investigation on cultivars Red Sunset, King Cardinal, Jacqueline and Rosario, reported that supplementary lighting for 0-9 h from 2 August, the numbers and percentage of flowering shoots of all cultivars during the winter and spring were increased by supplementary lighting for 6 or 9 h. Flowering was also advanced.

In order to optimize the daily investment of light energy from artificial sources with respect to photoperiod and photosynthetic fluence rates and to elucidate possible links between reactions to lighting regimes and root-zone temperature, Bakken and Baevre (1999) carried out experiment on 2 cultivars of alstroemeria. The more photons (photosynthetically active radiation, PAR) that were supplied to the plants per day (8, 11 and 13 mol/m^2), the higher was the total production of flowering stem. The total yield from regimes with 13 mol/m^2/day was higher when the light was spread over 20 and 16 h compared to 2 h. In treatments with soil cooling, the plants flowered continuously under all combinations of photo periods and photosynthetic fluence rates. It was concluded that it might be more cost-effective to spread the daily investment of light over 20 rather than 16 or 12 h when the total energy budget and CO_2 costs were taken into consideration.

After spring flowering, alstroemerias often undergo a rest period induced by long days and high temperatures, and during the summer few shoots or flowers are formed. In an attempt to stop the onset of dormancy by short day (SD) light treatment for 9 h was given from 1 June to 30 June, 15 or 30 July, or from 1 July to 30 July. Controls were grown in natural days. In the first few weeks after SD treatment started, large numbers of new shoots were formed on the rhizomes, up to 5 times the numbers formed under natural day length conditions. However, over 90 per cent of these shoots remained vegetative, associated partly with day length and partly with high temperatures which suppressed flower induction. Rhizome formation and growth were also suppressed by SD treatment; after SD treatment ceased, however, the number of flowering plants increased and there was an increase in rhizome numbers and weight (Darnbre, 1987).

6.1.2 Temperature

Effects of raising the temperature to 13-14°C for 1 or 2 months from 15 November, 1 or 15 December, vs. a constant minimum temperature of 5°C were studied by Blomme and Dambre (1984). Raising the temperature early for 1 month markedly increased the numbers of young shoots and weight of the rhizomes, and advanced flowering by 4-6 weeks. Healy and Wilkins (1982) determined that the number of days to flowering (DTF) was inversely related to the number of weeks (0-8) that *A. hybrida*, cv. Regina, plants remained at 5°C (a vernalizing temperature). One-year-old plants maintained at 21°C (a non-inductive temperature) and not divided before the 5°C treatment, showed an increase in total shoot production and delayed DTF, compared with plants which were divided. Total shoot and flowering shoot production were not affected by increasing the time at 5°C when the plants were maintained at 21°C and divided before this treatment. In another experiment on 3 alstroemeria cultivars (Valiant, Parade and Butterfly), Bond and Alderson (1993) observed that the number of lateral rhizomes, aerial shoots and tubers produced by rhizome segments (splits) were not significantly affected by increasing the temperature from 8 to 13 or 18°C. Increasing the temperature significantly decreased the number of plants establishing from rhizome 'splits' obtained from plants grown in greenhouse soil. For maximum rhizome production a temperature between 13° and 18°C was required. In an experiment employing rhizome splits from container grown plants, temperature had no effect on plant establishment. The presence or absence of the root system on the splits at planting was important in plant establishment, with removal of the root system producing a significant decrease in the number of plants establishing. It is suggested that the quality of the root system, and damage sustained by it prior to planting, are more important than temperature for the successful establishment of plants. Flowering responses of alstroemeria cultivars Regina, Carmen (syn. Cana) and Wilhelmina to air and soil temperatures for producing cut flowers in a soil cooling system were studied by Doi *et al.* (1998). Chilling requirements for flowering shoot production were 3-6 weeks at 2°C for Wilhelmina, 6 weeks for Carmen and more than 10 weeks for Regina. Regina plants stopped producing flowering shoots when rhizomes were kept at 25 or 18°C but continued to produce flowers at 14°C, irrespective of air temperature. Rhizomes of Regina plants, subjected to 15°C or lower for at least 6 h per day, continued to produce flowering shoots, although the number of vegetative shoots increased compared with rhizomes' exposed to a continuous chilling treatment. Chilled Regina and Carmen plants ceased flowering subsequently when they were exposed to 28/22°C (12 h day/12 h night) for 4 weeks, respectively. From the results of an extensive investigation on the effect of temperature in alstroemeria, Doi *et al.* (1999) concluded that plants of Regina and Carmen, when kept at a minimum of 10°C in winter, continually produced flowering shoots if they were planted by 21 May and 20 June, respectively, provided that the soil temperature was maintained at 14°C at night. After these critical dates it became almost impossible to keep Regina and Carmen below 15°C and 17°C, which are the critical and threshold temperatures for flowering. Cut flower yield and quality obtained by this method were poor. Potted Regina plants which were previously exposed to chilling at 2°C for 10 weeks and then transplanted to a chilled soil bed on 10 June produced flowering shoots in autumn. When plants were raised at a minimum of 20°C prior to chilling, they produced more vegetative shoots after transplanting than plants raised at 10°C. The former plants also produced more flowering shoots and high quality cut flowers in early autumn. Carmen plants grown in a chilled soil bed, kept below 17°C for only 6 h at night, produced more vegetative shoots and higher quality cut flowers in autumn than plants grown in a bed which was cooled continuously.

Dias-Tagliacozzo *et al.* (2005) studied environmental factors that could influence the germination, growth and flowering of a Brazilian cultivar, an unnamed *Alstroemeria caryophyllaea* hybrid that flourishes in tropical conditions. Seed germination of this cultivar is indifferent to light or dark. The rate of germination was different in seeds maintained at 25°C and 17°C, however, the percentage of germination after 45 days was 88 and 82 per cent, respectively. A temperature of 30°C partially inhibited this process; only 50 per cent of seeds germinated at this temperature. Rhizomes were maintained for 2, 4 and 8 weeks at 5°C and 25°C to evaluate the influence of vernalization. It was observed that although flowering occurred independently from the temperature, the percentage of flowering plants was higher at 25°C (70 per cent) comparing to 5°C (20 per cent). It appears that cold has a negative effect on flowering of this cultivar. Effect of photoperiod is qualitative in this cultivar. All plants flowered at both photoperiods, plants maintained with short days (8h) bloomed before plants maintained with long days (18h).

Greenhouse conditions influence the growth and flowering and cultivars also show variation in their response to these conditions (Beltran *et al.*, 1998). Alstroemeria cultivars Bloemfontein, Fury and Yellow were grown in greenhouses clad with glass or low density polyethylene. In the plastic greenhouses, the temperature was kept at a minimum of 7°C or was uncontrolled and the day length was 12 h or of natural length. In the glasshouses, the temperature was kept at a minimum of 7°C but natural day lengths and 12 h were compared. Bloemfontein and Fury gave 54 per cent lower cut flower yields under glass than under plastic, compared with 27 per cent lower in Yellow. In the plastic greenhouses, flowering was 1 month earlier if heated. Day length had no effect on flowering except in the unheated plastic greenhouse where it advanced flowering by up to 1 month. Cut flower yields were highest from Bloemfontein followed by Fury. In a greenhouse trial with cvs. Atlas and Monika, day time air temperatures of 14-25°C, 23-22°C and 17-23°C were compared from February to July 1986 (Keil, 1986). All treatments were similar until outside temperatures began to rise. From the end of May, the two cooler treatments gave higher yields than the two warmer ones. Plants in cooler treatments had shorter cymes and stems. Plants in the warmest treatment produced tall, weak stems with long cymes. Spieser (1987) used water-to-water heat pumps to maintain soil temperature in the 12-15°C range for alstroemeria production in two 16 × 61 m wooden greenhouses with double polyethylene, air-inflated cladding. The system worked successfully in spring and summer.

6.1.3 Carbondioxide (CO_2)

Milde (1983) reported that CO_2 supply increased yields of *cvs.* Canaria and Orchid. Verboom and Staaveren (1978) found that raising the glasshouse atmospheric CO_2 concentration to 0.12 per cent increased cut flower yields of alstroemerias by an average of 10 per cent with 2 cultivars (Canaria and Orchid). With *cv.* Regina yields were not increased but all 3 cultivars responded with greater earliness. The beneficial effect of CO_2 supplementation was greater with higher temperature. When CO_2 content was raised from 0.03 per cent to 0.12 per cent a slight increase in production occurred but when CO_2 content was raised to 0.11 per cent at night plus a rise in temperature, production was increased by 10 per cent in cvs. Canaria and Orchid although Regina showed no increase (Plomacher, 1978).

6.1.4 Light and Temperature

Healy *et al.* (1982) in an investigation on the role of temperature and photoperiod, reported that alstroemeria plants grown at 22°C did not flower regardless of photoperiod treatments. If grown at 13°C the plants flowered sooner under long photoperiod than under natural days (ND). Incandescent or red light treatments applied as a night interruption (NI) promoted earlier flowering than NI-far-red, ND or short days (SD). Plants grown under SD produced the fewest flower shoots. When plants were rotated every 20, 30 or 40 days between SD and NI-Inc light treatments, the days to flower were delayed compared with plants grown continuously (non-rotated) under NI-Inc. Days to flower were reduced when the plants were transferred monthly (December to June) from SD to either ND, 20 h Inc or 10 or 20 h of high-intensity discharge (HID) lights. Flowering was hastened by 20 h of HID lighting. Maintaining plants Under SD past January delayed the start of flowering regardless of subsequent light treatments.

Cervelli *et al.* (2000) studied the effect of temperature and light intensities on rooting, growth and flowering of alstroemeria. After 2 or 7 weeks from start of rooting plants of cultivars King Cardinal and Yellow King were treated for 5 weeks at 10°C at different light intensities (2 or 110 mmol/m^2/sec) or maintained for the same period in a greenhouse (untreated plants). A total of 90-100 per cent of 7 week plants rooted. Among the 2 week plants the percentage of rooting was highest in untreated plants (87.9100 per cent), lowest on cold-treated plants lighted at low intensity (25.8-29.1 per cent). Growth of shoots and roots was higher in 7 -week than in 2-week plants. Cold-treated high-lighted plants showed a growth similar to or higher than low-lighted ones. Growth of untreated plants was always high. Cut flower production was not affected by different treatments in King Cardinal, while all the cold-treated Yellow King plants showed better results than untreated ones. The stem grade in both cultivars was positively influenced by the cold treatment, irrespective of the different modalities. Flowering was late in untreated plants of both cultivars. High intensity lighting advanced (King Cardinal) or delayed (Yellow King) the flowering of cold-treated plants.

6.1.5 Light and CO_2

The effects of CO_2 enrichment and supplementary lighting on the production and flower stem quality of five alstroemeria cultivars (Barbara, Fiona, Helios, Mona Lisa and Tiara) were investigated by Labeke and Dambre (1998). CO_2 enrichment upto 900 µl/l increased the number of flower stems and the flower stem quality, regardless of the use of supplementary lighting. Supplementary lighting also increased flower stem production and quality, although to a lesser extent than did CO_2 enrichment. The combination of supplementary lighting and CO_2 enrichment resulted in increased flower stem production for Fiona, Helios and Mona Lisa, and increased flower stem quality for Barbara, Fiona and Mona Lisa (Tables 1 and 2).

Table 1: Main Effect of CO_2 Enrichment on Production and Flower Stem Quality of *Alstroemeria* Cultivars. The data are means for the two CO_2 levels and the two growing seasons for the period November-April*

Cultivars	*CO_2 Enrichment*	*Total Shoot Production (m^2)*	*Total No. of Flower Stems (m^2)*	*Flowering Percentage*	*Average Stem Weight (g1100 em)*	*No. of Cymes per Infloresc-ence*
Barbara	–	120.3 a	89.2 a	74.3 a	67.4 b	5.3 b
	+	120.2 a	85.0 a	79.8a	70.6 a	5.4 a
Fiona	–	254.2 b	125.9 b	50.0 a	49.5 a	4.3 a
	+	299.0 a	148.5 a	50.5 a	50.6 a	4.3 a
Helios	–	171.7 b	137.1 b	79.9 a	57.2 a	3.7 a
	+	216.8 a	167.2 a	77.3 a	56.2 a	3.8 a
Mona Lisa	–	205.8 a	75.1 b	36.5 a	62.2 a	4.9b
	+	235.3 a	89.0 a	38.2 a	63.7 a	5.0 a
Tiara	–	134.3 b	73.6 b	56.8 a	61.5 a	4.7 b
	+	189.5 a	107.5 a	60.0 a	62.3 a	4.8 a

Mean separation for each cultivar within column by Toke's test, P= 0.05.

* Labeke and Dambre (1998).

Table 2: Main Effect of Supplementary Lighting on Production and Flower Stem Quality of Alstroemeria Cultivars. The data are means for the two CO_2 levels and the two growing seasons for the period November-April*

Cultivars	*Supple-mentary Lighting*	*Total Shoot Production (m^2)*	*Total No. of Flower Stems (m^2)*	*Flowering Percentage*	*Average Stem Weight (g/100 cm)*	*No. of Cymes per Inflores-cence*
Barbara	–	109.6 b	78.4 b	72.8 a	66.3 b	5.2 b
	+	130.9 a	95.9 a	73.2 a	71.7 a	5.5 a
Fiona	–	262.8 b	131.7 b	50.6 a	48.4 b	4.2 b
	+	290.4 a	142.7 a	49.9 a	51.7 a	4.3 a
Helios	–	178.2 b	140.3 b	78.7 a	54.9 b	3.7 b
	+	210.2 a	164.0 a	78.5 a	58.4 a	3.8 a
Mona Lisa	–	201.5 b	78.1 b	38.8 a	59.3 b	4.8 b
	+	239.6 a	85.9 a	35.9 a	66.6 a	5.0 a
Tiara	–	139.0 b	91.5 a	66.0 a	57.9 b	4.7 b
	+	175.8 a	89.7 a	50.8 b	65.9 a	4.8 a

Mean separation for each cultivar within column by Tuke's test, P=0.05.

* Labeke and Dambre (1998).

6.1.6 Soil Temperature and Light

Lin (1984) studied the effects of soil temperature and supplementary lighting. Soil cooling by circulating water at 10°C resulted in an increased number of flowering shoots in both spring-summer (21 March to 20 September) and autumn-winter (21 September to 20 March) during the first year of a 2-year production scheme. Daily 16 h supplementary lighting with high pressure sodium (HPS) lamps reduced flowering shoots in spring-summer. In winter, soil cooling + HPS lighting produced flowers. Lin (1985) also reported that soil cooling to less than 14°C overcame the typical non-flowering of Regina plants during the November-May period. High intensity supplementary lighting with HPS lamps had little effect on or reduced the number of flowering shoots produced in non-cooled benches. HPS lighting combined with soil cooling increased the number of flowering shoots, shoot length and the total number of flower buds per shoot, and decreased the number of aborted flower buds per shoot.

Bridgen and Bartok (1990) recorded that cooling of medium (10-15°C) resulted longer flowering at the end of the summer and was reinitiated 1 month earlier than in the noncooled control. Total flower production increased when the growing medium was cooled. Flower production was greatest in the summer for plants with tubes 10 cm below the surface and least for plants in non-cooled medium. This trend continued when flowering was reinitiated; however, by April of the following year, all treatments resulted in similar levels of flower production.

Labeke and Dambre (1993) examined the flowering response to soil cooling alone or with supplementary lighting (40 $\mu mol/m^2/s$) in five cultivars in computer controlled greenhouse. Long days (14 h) were provided in all treatment combinations from September until December. Plant response to the different treatments was cultivar dependent. Soil cooling (13-15°C) reduced total shoot production in cultivars Annabel, Mona Lisa and Yellow King, but increased the percentage of flowering shoots in Annabel, Mona Lisa and Red Sunset and Yellow King. The total number of flowering shoots increased for all cultivars while the number of cymes per inflorescence increased only in Mona Lisa, Red Sunset and Yellow King.

The effect of a constant 16°C and a non-controlled soil temperature on the flowering of 4 cultivars grown in a greenhouse was studied by Blom and Piott (1990). Soil temperature regime did not influence either the start or the cessation of flowering. During spring/summer, production was 15 per cent lower under the constant soil temperature, irrespective of cultivar. During autumn/winter, the effect of the constant soil temperature was cultivar dependent; the yield of cv. Red Sunset was increased by 150 per cent while that for Rio fell by 22 per cent relative to the non-controlled conditions. Annual production was not affected but the ratio between the spring/summer and autumn/winter yields was 3.1 and 2.2 for non-controlled and constant soil temperatures, respectively.

6.1.7 Growth Regulators

Spray treatments of Arest (A-Rest, ancymidol), cycocel (chlormequat) or B-Nine (daminozide) were only moderately effective in controlling plant height in potted alstroemeria *cv.* Majoinnies while paclobutrazol was the most effective.

The number of days to flower was increased by up to 5 days by paclobutrazol application (10 or 20 ppm). The number of generative shoots was not affected by the growth regulator treatments (Healy *et al.*, 1993). When Alar (daminozide) at 0-3 g/l was applied to 10month-old plants in early September, flower stalk lengths were reduced by 19-23 cm in all treatments. Flower stalk numbers, however, were reduced by concentrations of 2 g/l and above (Blomme and Dambre, 1982).

7.0 Propagation

Alstroemerias are generally propagated by rhizome. Seed propagation though uncommon, is necessary for crop improvement. Multiplication by division of established clump, division of rhizome and micropropagation has also been found successful.

7.1.0 Seed Propagation

Seeds should be sown in containers as soon as ripe. The seedlings are planted out without damaging the tubers.

7.2.0 Vegetative Propagation

Alstroemeria can easily be propagated by rhizome. Milde (1983) reported that vegetative propagation was possible by division of rhizomes with single buds. In an experiment on propagation, Szczepaniak (1997) suggested that sections of 2-year-old rhizomes planted in May gave the best result. The propagation coefficient was highest and rooting was comparable with that of rhizomes planted in August and better than that of those planted in February. The longer rhizome sections produced better quality plants but increased less in size than the shorter sections. Differences were noted between the rhizome structure of the different cultivars which may have an influence on the propagation coefficient.

7.3.0 Micropropagation

Alstroemeria is commonly been propagated vegetative by rhizome division, with low efficiency, high time consumption and a high risk of virus dissemination. *In vitro* propagation has several advantages, particularly in terms of efficiency and has been applied to the micropropagation of alstroemeria (Aros *et al.*, 2017). Various explants have been used for microporpagation. Rhizome sections have been the most common explant used for the *in vitro* propagation of alstroemeria (Pumisutapon *et al.*, 2012) although some experiments have also considered aerial explants such as stem (Kim *et al.*, 2006; Guzmán *et al.*, 2018), leaf (Nasri *et al.*, 2013) and flower buds (Pedraza-Santos *et al.*, 2006). Nodal leaf explants taken from *in vitro* grown seedlings are good starting material for micropropagation (Lin *et al.*, 1997). Lin *et al.* (1998) reported direct shoot regeneration from leaf explants of a tetraploid breeding line.

Bridgen *et al.* (1992) studied in *in vitro* nutrition to develop an optimum growing medium for alstroemeria. Photosynthetic photon flux densities (PPFD) and media composition were found to affect both *in* and *ex vitro* performance of *Aistroemeria* hybrids (Kristiansen, 1999). The number of aerial shoots increased with increasing PPFD at 3 per cent sucrose while at 5 or 7 per cent sucrose PPFD did not affect the number of shoots.

Propagation of Alstroemeria

Pods Containing Seeds

Seedlings

Planting in Field

Rhizomes

Planting in Pot

Growth of Seedling

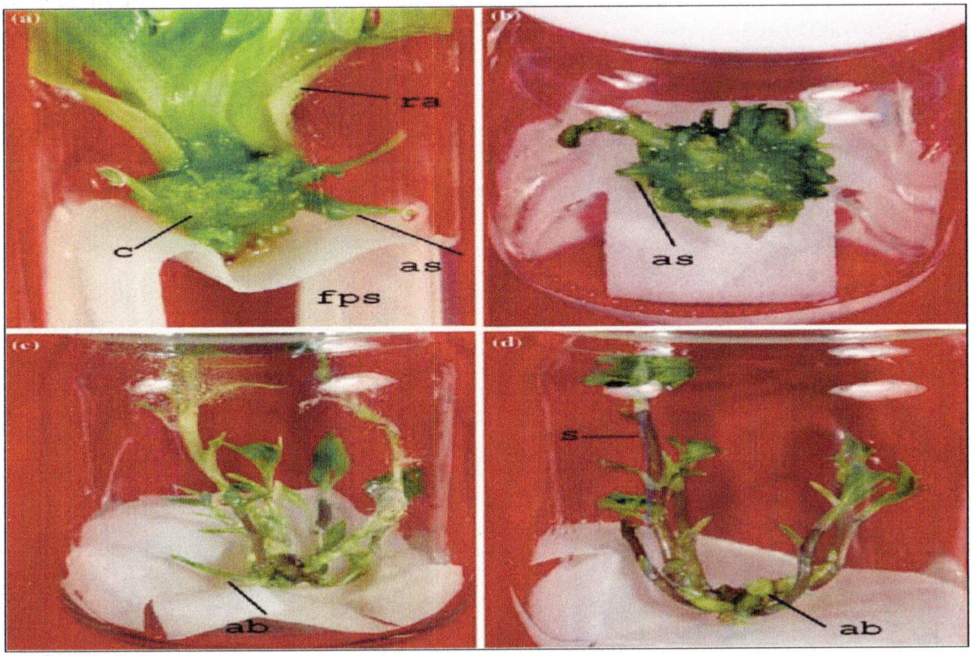

Micropropagation (a) Shoot induction on a liquid MS medium, from explants supported by filter paper. (b) mass of new shoots derived from an initial explant. (c) shoot multiplication on a liquid MS medium. (d) plantlet ready for acclimation (ab, axillary bud; as, adventitious shoot; c, callus; fps, filter paper used as explant support; ra, immature floral apex; s, stem) (Source: Pedraza-Santos *et al.* (2005)).

Monette (1987) found that BA at 0.5-4.0 mg/l or 2iP at 0.5-8.0 mg/l, especially the former, stimulated the growth of axillary rhizome initials and formation of erect shoots in *cv.* Zebra. Lin and Monette (1987) regenerated plantlets from rhizome-tips on solid and liquid media based on MS salt formulation. The quality of the cultures was superior when intact rather than longitudinally sliced rhizome-tips were used as explants, and when a temperature of 8°C rather than 22°C was used at the initiation stage. More roots were produced on rhizome-tips containing a rhizome apical meristem than on rhizome sections lacking such a meristem. The rooted plantlets acclimatized successfully and developed into true-to-type flowering plants. Gabryszewska (1995) found the regenerative ability of the explants from various organs like apical and axillary rhizome-tips to be better in MS with BA at 2 mg/l and NAA at 0.5 mg/l.

Ziv *et al.* (1973) standardized protocol for micropropagation from young actively growing tissue explants in 2 media from inflorescence stem taken at distance of 1-2 mm below the apex. Plantlets formed from regenerating buds. Root and bud regeneration took place directly from the stem tissue and not from the callus. White's medium and MS medium supported growth and regeneration equally well. King and Bridgen (1987) reported *in vitro* induction of axillary and adventitious shoots of alstroemeria inflorescence stem using half-strength MS basal medium which was supplemented with BA and Kinetin both at 1.0-16.0 mg/l separately and in combination.

Aros *et al.* (2017) described an efficient method for the *in vitro* propagation of *Alstroemeria pallida* Graham, a Chilean native species of high ornamental value. The highest explant weight was observed in rhizomes grown in culture medium supplemented with 3.5 g/l agar (3.79g), and using 2.0 g/l BAP showed the highest weight increase (3.33 g) after 8 wk. The proliferation rate rose with increasing concentrations of BAP, whereas low concentrations of BAP promoted longer shoots. Guzmán *et al.* (2018) developed *in vitro* propagation protocols using aerial explants of *Alstroemeria caryophyllaea* and *A.* 'Sweet Laura', two scented genotypes. Aerial explants of alstroemeria were regenerated by testing three types of aerial explants and four concentrations of cytokinins. Both *A. caryophyllaea* and *A.* 'Sweet Laura' showed the highest induction rate from shoots, with values of 35 and 63.3 per cent, respectively. For both genotypes, the optimum BAP concentration for direct organogenesis from buds was 1.0 mg/l. Finally, a successful protocol for *in vitro* propagation of *A. caryophyllaea* and *A.* 'Sweet Laura' through direct organogenesis from aerial explants was generated.

8.0 Soil and Climate

8.1.0 Soil and Substrates

In open cultivation, alstroemerias require moist but well drained soil with rich in humus. In greenhouse they are grown in a mix of loam, leaf mould and sharp sand (Brickell, 1996). Plants are tolerant of various substrates at pH 6.0-6.5. They have a high nutrient requirement and are salt-sensitive (Milde, 1983). In 3-year trial on 8 substrates, in *cv.* Regina, the number of inflorescences was highest (23/plant) in plants grown on a 1:1 mixture of hotbed muck soil and high peat (Lisiecka, 1981;

Lisiecka and Szczepaniak, 1983). Bridgen and Soriano (1986) compared two growing media and found soil, peat moss and perlite (1: 1: 1) medium better for stem and flower production. Of the 5 substrates tested, the flower yield was highest in plants grown on the conifer needle substrate (Dinova *et al.*, 1988). Lisiecka *et al.* (1992) found that the best substrate used was a 1 : 1 mixture of hotbed muck soil and peat.

Healy and Lang (1984) reported that planting in soil benches produced consistently more flowers than gravel benches in all but one cultivar (Maxima), probably due to better soil temperature control. The highest yielding cultivars in soil benches were Orchid, Canaria and Red Surprise (63.5, 55.0 and 45.1 flower stems/ft^2, respectively, from 1 February to 31 August). The cultivar Maxima in gravel benches yielded 46.3 flower stems/ft^2 over the same period.

8.2.0 Climate

Alstroemerias grow well in both sunny situation and partial shade. Depending on the temperature and light intensity, Spier and Milde (1977) described the growing condition and cultivation of alstroemeria and suggested favourable temperature of 10–12°C in winter and 14-16°C later and light 15 W/m^2 in winter to 60 W/m^2 during vigorous growth.

9.0 Cultivation

9.1.0 Selection of Cultivars

The success of alstroemeria cultivation depends on proper choice of the cultivars. There are many cultivars and new cultivars are being added every year.

Waithaka and Chepkairor (1988) investigated the growth and flowering responses of 3 alstroemeria cultivars to the outdoor climatic conditions in Kenya and reported that *cv.* Pink Perfection produced significantly more total shoots (vegetative plus reproductive) than Marina and Cannen. Cannen, Pink Perfection and Marina started flowering 19, 30 and 35 weeks after planting, respectively. The number and percentage of flowering' shoots were highest in Carmen, followed by Pink Perfection and Marina. The average length of inflorescences produced by Pink Perfection, Marina and Carmen was 94.3, 82.9 and 47.0 cm, respectively. Peak production of marketable cut flower occurred 31, 40 and 44 weeks after planting for Carmen, Pink Perfection and Marina, respectively.

Alstroemeria cultivars Canaria, Orchid, Regina and Harmony were evaluated by Kwiatkowska and Brzozowska (1980) in greenhouse. In the 1st year Harmony gave the highest flower yield, followed by Regina. In the 2nd and 3rd year no appreciable differences in flower yield between cultivars were noted. In the 1st and 2nd year inflorescence quality was best in Canaria and Orchid. In the 3rd year inflorescence quality was poor in all cultivars. The highest return/m^2 was from Canaria, followed by Orchid. Productivity appreciably declined in the 3rd year of cultivation.

Ten alstroemeria cultivars grown in the greenhouse for cut flowers produced flowers from the end of February or mid-May until the beginning of July and again starting about 4 weeks later from the end of July until the end of November or beginning of December. Cultivars Orchid, Canaria, Zebra and Red Sunset produced the earliest blooms in February and March. The highest yields were with Orchid and King Cardinal (431.5 and 420.2 stems/m^2, respectively). Zebra and Canaria produced the highest percentage of best quality (extra) blooms (Milde, 1981).

Nijssen and Hoogeveen (1990) reported cut flower yields, flower quality and vase life for the new cultivars Apollo, Bianca, Bingo, Cavalier, Tiara and Westland. The crop was planted on 8 September, 1987 and the trial ended in April, 1989. In general, winter yield (7 November, 1988 to 26 February, 1989) was not high; Bingo (322 stems/m^2) and Cavalier (313 stems/m^2) gave the heaviest total crops.

In Poland, cultivars Appelblossom, King Cardinal and Pink Triumph gave particularly good results, and Cana and Red Sunset showed a tendency to flower in winter (Lisiecka *et al.*, 1992). Ballerina, Bonanza, Carmen, Fidelio, Orchid Flower, Regina, Maxima and Staroza are the most popular cultivars in Ukraine (D-yachenko, 1991).

The short cultivars with high quality cut flower production throughout the year are cvs. Patricia, Pink Triumph, Rosario and Yellow King (Milde, 1983). Of four greenhouse cultivars Samora was the highest yielding with its first flowers being harvested in November while Jubilee also had many Class I blooms but was the most susceptible to cold. These provide flowers from late June to October (Anon., 1986).

Szczepaniak (1998) reported that cultivars which flowered abundantly in untreated plastic tunnel included Harmony and Mona Lisa. Cut flower yield in the first year was 3 times lower than in the second and third year. Mona Lisa, Harmony, Rosita and Pink Triumph produced high quality flower shoots. The yield dynamics were similar in the majority of cultivars, and inflorescence production peaked in July and August. In South Ukraine it was possible to grow alstroemeria in the open but the rhizomes had to be covered with 30-40 cm leaf layer for winter protection (D-Yachenko, 1991).

9.2.0 Planting

In a series of greenhouse trials over many years, several cultivars were grown in containers (for substrate, fertilizer and irrigation trials) or in beds to determine which factors affect flower yield and quality. Lisiecka *et al.* (1992) reported that planting in February rather than in November delayed flowering by 1-3 weeks and the flowering peak by one month (although yields during the first year of cultivation were lower with February planting than November planting).

Lisiecka and Szczepaniak (1993) observed that planting in November produced the most inflorescences/year and planting in May the fewest. Yield and flowering date differences were marked only in the first year. Planting after November (the traditional time) delayed flowering by 3 (February planting) and 12 (May planting) weeks. Delay in planting did not greatly improve autumn flowering. February planting could be recommended as it does offer the opportunity to reduce winter fuel bills. Healy *et al.* (1993) in the experiment on regulating plant growth reported that potted alstroemeria cultivars available for containerized production were 50 per cent shorter than cut flower cultivars, which were still excessive for most markets. When plants of the cultivars Mojoinnier and Bird of Paradise were planted on a

Cultivation of Alstroemeria

Seedling Stage

Growth of Seedlings

Cultivation in Polyhouse

Flowering in Open Field

monthly basis, plant height showed a cyclic pattern with the tallest plants developing from spring and summer plantings and the shortest plants from winter planting. Flowering behaviour of cultivars Yellow King, Mona Lisa, King Cardinal and Rosita was studied by Cocozza *et al.* (1996) by planting rhizomes in early October, early November or early December. Staggering the planting dates did not extend the flowering period sufficiently. Rhizome quality had no effect on cut flower yield or quality. King cardinal was generally the highest yielding cultivar but was the earliest only in 1992. Yellow King was the lowest yielding, the latest and had the shortest flowering period.

9.3.0 Planting Density

In a plant spacing trial with 4-10 plants/m^2 flower quality was not adversely affected by close planting, but the average yield/plant declined as plant density rose (Blomme and Dambre, 1979). In greenhouse, *cv.* Regina produced 2 crops (spring and autumn) per year and a density of 2-3 plants/m^2 gave the best flowering stem (Cireasa, 1983). Rhizomes of vigorous *cv.* Regina planted in raised gravel at densities of 3.5, 1.75 and 1.2 sq ft/plant resulted in April to August flower stem yields of 35.3, 48.4 and 51.9/sq. ft, respectively, and 123.0, 84.4 and 60.5/plant, respectively. The data suggested that a density of 2.0 sq ft/plant would provide the most economic yields. For less vigorous cultivars a density of 1.75 sq ft/plant was recommended (Healy and Lang, 1985).

9.4.0 Irrigation

Investigation on salt tolerance of alstroemeria, using irrigation water with EC ranging from 0.2 – 4.0 dS/m at 25°C revealed that the salinity threshold value was below 0.8 dS/m and the yield (weight of flowers) decreased by 15 per cent per dS/m (Sonneveld, 1988). There was no difference in salt sensitivity between the 2 cultivars. The reduction in yield was caused by reduction in the number and weight of peduncles. The number of flowers per peduncle was affected slightly. For better growth, flowering and yield of cut flowers, the alstroemeria plants be irrigated with the 10 l/m^2 (October- February) and 12.5 l/m^2 (March-September) under Himachal Pradesh conditions (Kaushal *et al.*, 2018).

9.5.0 Manuring and Fertilization

In a fertilizer trial comparing the effects of N and K_2O at 5-15 and 20-30 g/m^2, respectively, with P_2O_5 at 10 g/m^2, added between the end of November and mid-March, the highest numbers of flowering shoots and flowers were obtained with Nand K_20 at 10 and 25 g/m^2, respectively (Blomme and Dambre, 1982). Responses to 4 rates of N and 3 of K_2O (weekly applications up to 300 mg/12.51 pot) were assessed in 2 consecutive crops of cvs. Orchid and Carmen. In both crops N and K had favourable effects on flowering stem production, total fresh shoot weight and flower numbers/stem. Best results were obtained when N : K_2O ranged from 1 : 1 to 1 : 1.4. The best treatment in the second crop for cv. Orchid resulted in foliar levels of 4.9 per cent N and 3.9 per cent K, and for Carmen 5.1 per cent N and 3.7 per cent K (Bik and Berg, 1982). Meij (1976) reported that following a normal base dressing, the highest yields of glasshouse alstroemerias, cv. Orchid, were achieved with K top dressing, both in the second year and in the 2-year total. The effects of Ca and N on cut flower production of cv. Parigo Pink were determined in separate greenhouse experiments (Smith *et al.*, 1998). Ca was supplied as $Ca(NO_3)_2$ and CaC12 at 0, 1, 2, 4, 8 and 12 mmol Ca/l (in tap water containing 0.2 mmol Ca/l). N was supplied as KNO_3 and $Ca(NO_3)_2$, providing 0,3.5, 7, 14,28.5 and 57 mmol total N/l (in tap water containing <0.2 mmol N/l). Nutrient solutions were applied at 7- or 10-day intervals to plants growing in a soilless medium in 2.6 or 5.5 1 containers. Flower production was not affected by Ca supply, but increased with N supply to a maximum of 4 sterns/plant on a weekly basis at 28.5 mmol N/l, then decreased to <3 stems/plant at 57 mmol N/l. On a dry weight basis, leaf N was maintained at 45 ± 3 and 52 ± 5 g N/kg in plants supplied with 28.5 and 57 mmol N/1, respectively, but was <40 g/kg when N was supplied at 14 mmol N/l. It is suggested that N fertilizers should be managed to maintain leaf tissue N at 45 g N/kg in alstroemeria. In a series of experiments on nutrition in container grown plants of alstroemeria using peat substrates enriched with basal fertilizers 150 mg N + 100 mg P + 300 mg K + 3 g Ca + MIS-4 (a compound fertilizer containing Mg + micronutrients at 200 mg/dm^3) and weekly top dressing of N at 0, 175, 350, 700 or 1400 mg/plant, Szczepaniak (1991a) recorded the best results with regard to various aspects of flowering by applying N at 175 mg/plant. Flowering duration depended more on ambient temperature and solarization in a given year than on N nutrition. Plants showed high tolerance to a wide range of N nutrition, and a significant crop reduction occurred only with N at 1400 mg/plant. In a similar experiment, the plants receiving the optimal N rate of 175 mg/plant, the N levels in the apical leaves was 4.37-4.92 per cent in early April and 4.93-5.64 per cent during intensive vegetative growth. Leaf P, K, Ca and Mg contents depended mainly on the stage of plant development and, to a lesser degree, on different N rates (Szczepaniak, 199Ib). Szczepaniak (1991c) also reported that plants receiving the optimal N rate (determined earlier) of 175 mg/plant, the substrate content of NH_4 + NO_3 was 170-270 mg N/dm^3 during intensive vegetative growth. Substrate P, K, Ca and Mg contents depended more on the plant development stage than on the applied N rate. K supply at flowering is important for vase life. Fe deficiency showed easily in cvs. Starosa and Regina (Milde, 1983).

A substrate of pH <6.5, manured with farmyard manure or peat, followed by liquid feeding is required. Two slow release fertilizer mixtures nitrophoska perfekt (14N : 5P : 20 K : 2Ca) were applied at 30 or 40 g N/m^2. The cultivar Vasa Canaria produced the highest cut flower yield (731 stems/m^2 over 2 years) with the higher rate of nitrophoska perfekt, which was double the yield obtained with the lower N rate. The cut flower yield of cv. Vasa Rosita was similar (510-531 stems/m^2) at both fertilizer levels (Hermann, 1986). Lisiecka and Szczepaniak (1983) reported that yield of flower on plant grown singly in 10 litre containers on a peat substrate were highest at NPK levels of 30 : 150-300 : 600 mg/l applied in the 1st year and at 300-450 : 300 : 900 mg/l in the 2nd year. In an experiment on treatment with seven resin-coated fertilizer (RCF) to post production flowering and growth of potted alstroemeria, Chiari *et al.* (1999) concluded that potted plants did not need fertilizer for continued flowering, and fertilizer application favoured production of vegetative stems relative to flowering stem. However, application of RCF to marketable plants prevented chlorosis, increased fresh weight, and, if low to moderate rates of formulations with N: P ratios of at least 6: 1 were applied, did not inhibit flowering.

Studies on N accumulation in the vegetative mass in *cv.* Regina showed that increasing N level in plant tissues, as a result of N fertilization, resulted in a decrease in flower yield (Dinova *et al.*, 1988).

According to Reid (2006), *Alstroemeria* is salt sensitive. Trials with some varieties showed that growth slows as soon as the EC rises above 0.5 mS/m. For every unit of EC above that, growth decreases by 15 per cent, yield by 10 per cent and stem length by 5 per cent. Suitable fertiliser for *Alstroemeria* is a carnation feed given below (Reid 2006).

Ammonium nitrate - 5.3 kg

Potassium nitrate - 10.0 kg

Magnesium sulfate - 2 kg

Librel BMX® - 50 to 100 g

Iron Chelate - 50 to 100 g

Copper chelate - 50 to 100 g

This makes up 100 L of stock solution which should be diluted at 1:200 or an EC not exceeding 1.8 S/cm. Successful growers in Western Australia are using N: K ratios ranging from 0.8:1 to 1.3:1 (Reid, 2006).

Foliar Analysis

The recommended nutrients levels for normal healthy growth are given in Table 3. Leaves to be sampled are recently matured leaves from the top of the plant.

Table 3: Recommended Nutrient Levels for Alstroemeria*

Element	*Concentration*
Nitrogen (per cent dry weight)	3.8-5.6
Phosphorus (per cent dry weight)	0.3-0.7
Potassium (per cent dry weight)	3.7-4.8
Calcium (per cent dry weight)	0.6-1.8
Magnesium (per cent dry weight)	0.2-0.4
Iron (ppm)	175-275
Zinc (ppm)	35-68
Manganese (ppm)	13-50
Copper (ppm)	2.7-15
Boron (ppm)	13-50
Molybdenum (ppm)	59-87

*Source: www.hilagric.ac.in/edu/coa/vegetables/lectures/VSF_233_HSC/VSF_233_HSC_Lect_9.pdf

9.6.0 Shoot Thinning

In a trial comparing several methods of enhancing shoot formation for a second crop in autumn after spring flowering, Blomme and Dambre (1982) reported the numbers of new shoots formed by the autumn were markedly increased by removing the remaining shoots + the non-flowering shoots, but these new shoots did not flower.

Cutting down was detrimental. In a trial in which either all remaining shoots + all flowering shoots or only the thin flowering shoots + the remaining shoots were removed after autumn cropping, at different times between the first week of January and the second week of February, the numbers of flowers subsequently harvested were greatest after removing only the thin flowering shoots + the remaining shoots. Thinning the shoots of *cv.* Regina did not increase productivity in the autumn and is not recommended (Meij, 1976).

Alstroerneria cultivars King Cardinal and Diamond were grown by Bakken (2000) with soil cooling for 50 weeks under high pressure sodium lamps at a PPFD of 150 μmol/m^2/s (photosynthetically active radiation, PAR) for 20 h per day. The total shoot production and the morphology and yield of flowering stems from plants subjected to different thinning practices were recorded. For both cultivars, the total shoot production was higher when stems were thinned and harvested by cutting rather than pulling from the rhizome. The yield of flowering stems was not influenced by technique for thinning and harvest. For Diamond the highest yields of stems of marketable quality occurred when 30 per cent of the shoots that remained after harvest were removed every third or every fifth week. For King Cardinal the optimal thinning seemed to be a removal of 30 or 60 per cent of the shoots every fifth week.

9.7.0 Mulching

The effect of mulching from June to October and/or lowering the soil temperature (by 2.1-2.8°C) from June to October were tested on alstroerneria cultivars Flamengo and Jubilee (Wiel, 1989). Soil cooling plus mulching increased Flamengo flower production by about 30 per cent mainly by extending the harvest later into the year. Mulching had no effect on Flamengo flower production in non-cooled soil. Mulching, with or without soil cooling, increased Jubilee flower production by about 16 per cent but, as Jubilee has extended winter flowering, had no effect on the harvest period.

10.0 Diseases and Pests

10.1.0 Diseases

Among various diseases affecting *Alstroemeria, Fusarium, Phythium, Rhizoctonia* and virus are important.

Fungal Diseases

Kaminska and Korbin (1999) reported that experimental infection of alstroemeria seedlings with naturally infected lily (*Lilium*) Casablanca with stunting and flower bud deficiency phytoplasma resulted 3-4 weeks after top grafting in chlorotic and/ or necrotic stripes, whitening of the leaves, shoot necrosis and die back. Chang and Mirza (1993) surveyed the incidence and severity of alstroemeria root rot disease complex for 15 cultivars and the highest disease incidence (DI) occurred on Orange

Monarch (67.9 per cent) while the highest disease severity (DS) rating was observed on Saxony (47 per cent). The lowest DI and DS values occurred on Paloma (16.6 per cent and 11 per cent respectively). The pathogens most frequently isolated from cultivars in the greenhouses examined were *Fusarium* sp. and to a lesser extent *Pythium* sp. and *Rhizoctonia solani.*

Chang *et al.* (1993) isolated pathogens from the basal stems of diseased alstroemeria plants showing symptoms including dark brown stripes along leaf margins, leaf chlorosis, plant wilting, and browning or rotting of the basal stem, rhizome and fibrous roots. The pathogens isolated most frequently were *Fusarium* sp. (40.5 per cent of plants examined); *Pythium* sp. and *Rhizoctonia solani* were less frequent (5.5 and 6.8 per cent, respectively), *F. oxysporum and P. ultimum* were the most frequent. Chang and Mirza (1994) isolated *Sclerotium rolfsii* (*Corticium*) from diseased rhizomes of greenhouse-grown alstroemeria in Alberta, Canada. The underground portions of the plants were attacked resulting in leaf chlorosis which started on the lower leaves and gradually moved upwards. The characteristics of the isolates from diseased plants were compared with those from other crops. This was the first report of *S. rolfsii* on alstroemeria in Canada.

The effect of *Fusarium oxysporum* (F), *Pythium irregular* (P) and *Rhizoctonia solani* (R), singly and in combination on root and stem rot of cv. Jacqueline showed that the average number of stems and stem height per plant varied with the pathogen treatment. P + R caused more severe disease than P alone, but the combination of the 3 pathogens induced the greatest plant mortality (Chang *et al.*, 1994).

Viral Diseases

Although several studies have attempted to produce virus-free Alstroemeria plants by using a meristem culture (Chiari and Bridgen, 2002), viruses continue to be a major problem in Alstroemeria species since meristem-derived Alstroemeria plants can be re-infected during harvest or transport. Numerous viruses are currently considered critical for Alstroemeria. Examples include Alstroemeria carla virus (AlCV), Alstroemeria mosaic virus (AlMV), cucumber mosaic virus (CMV), tomato spotted wilt virus (TSWV), Alstroemeria streak virus (AlSV), and Impatiens necrotic spot virus (INSV) (Van Zaayen 1995). Of these viruses, AlMV is considered the most endemic. Moreover, it is difficult to detect or prevent AlMV contaminations in Alstroemeria cultures.

Van Hoff *et al.* (1970) reported that a virus isolated from alstroemeria cv. Orchid with mottled leaves resembled tomato spotted wilt virus but inoculation tests on tomato were negative. Brunt and Phillips (1981) reported that alstroemeria mosaic virus, occurred in plants with leaf chlorosis and broken flowers, and had filamentous particles c. 12 × 750 nm, another had filamentous particles c. 650 nm long and was serologically related to carnation latent and chrysanthemum B viruses. Tobacco rattle and tobacco streak viruses were also present. Studies on alstroemeria mosaic virus, the most widespread of 4 viruses indicated that it was a distinct potyvirus. A carlavirus, tobacco rattle virus and a possible strain of tobacco streak virus were also reported (phillips *et al.*, 1983). A putative potyvirus was isolated by Chiko and Godkin (1984) from mottled leaves of alstroemeria. A virus with spherical particles

Diseases and Pests of Alstroemeria

Pythium Rot

Rhizoctonia Wilt

Rhizome Rot

Yellow Spot Virus

Leaf Spots

Alstroemeria trialeurodes

Nematode Infection

Leaf Minor Infestation

c. 29 nm diameter was isolated from apparently symptomless cultivar Orange Beauty and did not react serologically with antisera to 12 nepoviruses. A technique was developed by Hakkaart and Versluijs (1988) for elimination of alstroemeria mosaic virus from infected cultivars. Meristems were excised from rhizome tips and placed on a medium containing indole-3-butyric acid at a cultivar dependent concentration. After 6-8 weeks shoots had formed and root formation was better on filter paper bridges in a liquid medium than on a solid medium. Virus elimination was cultivar dependent, but plants reacting negatively in serological tests were obtained with most cultivars.

Bouwen *et al.* (1996) reported several virus infecting alstroemeria. The 2 most important are alstroemeria mosaic potyvirus (AIMV) and alstroemeria streak potyvirus (AlSV). Four viruses were identified by Bakardzhieva *et al.* (1995) including alstroemeria mosaic potyvirus (AIMV), arabis mosaic nepovirus (ArMV), cucumber mosaic cucumovirus (CMV) and tobacco rattle tobravirus (TRV). Of them, AIMV was the most widespread. Shiota *et al.* (1997) detected three viruses, alstroemeria mosaic potyvirus, an unidentified potyvirus and cucumber mosaic cucumovirus, from alstroemeria hybrid plants showing mosaic and white streak symptoms in China and Japan. The distribution of alstroemeria mosaic potyvirus (AIMV) was investigated by Blank *et al.* (1994) and found erratic. A survey of viruses affecting alstroemeria plants, cultivated in Italy and/or imported from the Netherlands, was conducted (Table 4).

Table 4: Viruses Found in European *Alstroemeria* sp. Plants**

Virus	*Taxonomic Group*	*Vectors*	*Distribution*
AIMV	Potyvirus	Aphids	World-wise
U.V. (*)	Potyvirus	Aphids	Holland
FMV	Potyvirus	Aphids	Italy
U V. (*)	Potexvirus	Aphids	Denmark
U V. (*)	Closterovirus	Aphids	Italy
ACa V	Carlavirus	Aphids	Holland, Italy
LSV	Carlavirus	Aphids	Holland, UK. Italy
TSV	Ilarvirus	Thrips	Holland
TRV	Tobravirus	Nematodes	Holland, U. K.
TSWV	T.M. (°)	Thrips	Holland, Italy
ArMV	Nepovirus	Nematodes	Denmark, Italy

(*) U. V. = Unidentified virus; (°) T.M. = Type member.

** Bellardi *et al.* (1994)

The viruses found were alstroemeria mosaic potyvirus, freesia mosaic potyvirus, alstroemeria virus (lily symptomless carlavirus), tomato spotted wilt tospovirus and arabis mosaic nepovirus (Bellardi *et al.*, 1994). Zaayen *et al.* (1994) detected simultaneous occurrence of a second potyvirus which was frequently found in a study to improve detection of alstroemeria mosaic potyvirus (AIMV) in *Alstroemeria* sp. by Elisa. This second virus, tentatively named alstroemeria streak potyvirus (ASV), was multiplied in indicator plants. Bellardi *et al.* (1992b) found that alstroemeria plants of Italian and Dutch origin were infected by alstroemeria mosaic potyvirus, freesia mosaic potyvirus, alstroemeria virus, tomato spotted wilt tospovirus and arabis mosaic nepovirus. During a survey in Italy on viruses infecting *Alstroemeria* sp., Bellardi *et al.* (1992a) recorded that of antisera to 9 potyviruses reacted with the particles, only freesia mosaic potyvirus (FMV) and alstroemeria mosaic potyvirus (AIMV) gave a strong decoration of the virus particles in immuno EM studies.

Bellardi and Bertaccini (1992) reported tomato spotted wilt tospovirus (TSWV) for the first time from Italy on *Alstroemeria* sp. The affected plants showed chlorotic and yellow rings on the leaves. TSWV was identified on the basis of electron microscopy studies and serological tests. Occurrence of tomato spotted wilt tospovirus (TSWV) was reported also by Latham and Jones (1997).

Virus symptoms found to alstroemeria cultivars Campfire, Canarie, King Cardinal and Orchid were reported by Navalinskiene *et al.* (1993). Isometric virus particles were detected in orchid and alstroemeria mosaic potyvirus (AIMV) was found in Campfire, Canarie, King Cardinal and Rosario. A technique was developed for obtaining a purified preparation of AIMV with concentration of c. 9 µg/100 g. Habenaria mosaic potyvirus (HaMV) was isolated and identified from *H radiata* showing severe leaf mosaic. It showed distant serological relationships to alstroemeria mosaic potyvirus (Inouye *et al.*, 1998).

A virus isolated from *Alstroemeria* sp. showing chlorotic mosaic along veins was characterized and identified by Yasuda *et al.* (1998). The isolate AlMV-Y showed a narrow host range of 4 species belonging to 3 families among 34 species of 12 families tested. It consisted of filamentous virus particles, 700-800 nm long and 12 nm wide.

Nested-PCR with group specific primers and RFLP analyses indicated that the phytoplasmas infecting alstroemeria belonged to the aster yellows group (16Srl) subgroups 16Srl-B, I-C and I-G and to the peach × disease group m (16SrIII) (Bertaccini *et al.*, 1996). In *A. brasiliensis* polymerase chain reaction (PCR) amplification indicated a 92 per cent homology with lily mottle virus, LMoV (potyvirus) (Bouwen *et al.*, 2000).

10.2.0 Pests

10.2.1 Leaf Miner

Liriomyza trifoii was recorded in Yugoslavia damaging alstroemeria (Kornnenovic and Pagliarini, 1980).

10.2.2 Trialeurodes

An overall rate of parasitism of >70 per cent was observed by *Encarsia formosa* to control *Trialeurodes vaporiorum* on *Alstroemeria* sp. in the greenhouse (Arnparo-Espanol and Corredor, 1998).

10.2.3 Nematodes

Cotten *et al*. (1990) found *Pratylenchus bolivianus* in 10 soil and 8 root samples from 35 alstroemeria crops in England and Wales. In a 40-week pot experiment, population development and reproduction of *Pratylenchus bolivianus* on alstroemeria cv. Jubilee were investigated by Amsing (1996). Thirty-two weeks after planting in soil initially infested with 24 *P. bolivianus*/100 cm^3 soil, nematode populations in subterranean plant parts and soil reached maxima of 3,350/10 g and 266/100 cm^3, respectively, giving an overall reproduction factor of × 59.

11.0 Harvesting and Postharvest Management

11.1.0 Harvesting

Flowers are harvested when colour appears on first flower. The flowers are sensitive to ethylene and the leaves are fragile and turn yellow after harvest.

11.2.0 Yield

Milde (1985) evaluated the flower production of cultivars Orchid, Canaria and King Cardinal. All cultivars flowered twice each year, between April and mid-June and again between August and November. The yields over 4 harvests for Orchid, Canaria and King Cardinal were 272, 286 and 369 stems per m^2, respectively. Orchid and Canaria produced 60 per cent Extra Class (>0.8 m long), 27 per cent Class I (>0.6 m) and 12 per cent Class II (< 0.6 m) stems, while, King Cardinal produced 44 per cent Extra Class, 39 per cent Class I and 17 per cent Class II stems. Cultivar Orchid produced the most inflorescences (346/m^2), compared with 336 for cv. Stavifa Canaria, and the average for all cultivars was 241/m^2 (Spier and Milde, 1977).

11.3.0 Postharvest Management

The main problem with alstroemeria is the development of early chlorosis in the leaves and the subsequent abscission of the petals (Halevy and Mayak, 1981). This is induced by the imbalance of hormones after harvest, and the use of hormones delays chlorophyll depletion and postpones flower senescence (Van Doorn and Stead, 1997). Ethylene plays a significant role in regulating the aging process of most plant organs such as flowers. Aging in the flower stalk is divided into two categories: ethylene-sensitive stalk, and ethylene-insensitive stalk. Extracellular synthesis of ethylene in the first group is carried out through a specific biochemical pathway that results from flower development and pollination. The vase life of Alstroemeria (cv. Rebecca) flowers is terminated when the tepals abscise (Chanasut *et al*. (2003). Abscission was accelerated by both chloroethylphosphonic acid (CEPA) and 1-aminocyclopropane -1-carboxylic acid (ACC). Petals abscised 24 h earlier compared with controls, when isolated cymes were placed in 340 nM CEPA, and earlier still when higher concentrations were used. This suggests that flowers of this cultivar are very ethylene sensitive. Treatment with silver thiosulphate (STS) overcame the effects of exposure to CEPA and delayed perianth abscission of untreated isolated flowers by 3- 4 days. The inclusion of 1 per cent sucrose in the vase solution also extended longevity but not by as much as STS treatment; combined STS and sucrose treatments did not increase longevity beyond that of

Harvesting and Marketing of Alstroemeria

Full Bloom Stage

Harvesting in Greenhouse

Packing at Tight Bud Stage

Packing at Full Bloom Stage

Flowers in Market

Cut Flowers

either treatment alone. However, removal of the young buds from the axil of the first flower was the most effective treatment to extend vase-life and encouraged the growth and development of the remaining flower. Flowers on cut inflorescences from which young axillary buds were trimmed more than doubled in fresh weight 6 days after flower opening compared with an increase of only 70/80 per cent in those untreated or treated with STS and/or sucrose. Growth was less in isolated cymes but followed a similar pattern. The effect of STS and/or sucrose treatment was synergistic with the trimming treatment and thus the vase life of trimmed, STS and sucrose-treated flowers was over 7 days longer than that for untreated controls (Chanasut *et al.*, 2003).

11.3.1 Vase Life

Cut flowers are stored at 2-4°C in water for 2-3 days, but dry storage is not recommended. Flower stems were immersed to a depth of 15 cm in various solutions (deionized) water (control), 4 per cent sucrose, silver thiosulphate complex (STS), 2.5 per cent chrysal, STS plus 4 per cent sucrose, or 2.5 per cent chrysal plus 4 per cent sucrose for 1 h (pulsing) and then transferred into fresh deionized water containing a biocide. Chepkairor and Waithaka (1988) reported that the vase life of Carmen, Pink Perfection and Marina was increased by 8, 7 and 6 days, respectively, when flower stems were pulsed in STS solution. Carmen had the longest vase life in all treatments followed by Pink Perfection and Marina. Use of 150 mg/l AOA prolonged the vase life of Alstroemeria (Naghiloo *et al.*, 2019).

Some studies have indicated the positive herbal essential oils in preventing floral senescence and increasing vase life of cut flowers (Razi, 2017). It has also been reported that in addition to prolonging shelf-life, use of these treatments leads to an increase in the activities of (SOD) and (CAT). At the same time, some sprays with putrescine and spermine decreases the content of (MDA), showing a significant relationship between the content of these enzymes and the vase life of cut alstroemeria flower (Alborz *et al.*, 2015).

The effect of dill and cumin essential oils on physiological and microbiological traits of alstroemeria cut flowers was reported by Langroudi *et al.* (2020). Hydro-distillation method and GC isolated the essential oils of seed and GC/MS examined the chemical compositions of the samples. The results showed that the dill essential oil treatment (100 ppm) in pre-harvest was the best treatment for all known traits. 11 colonies of bacteria were identified in stem end of cut alstroemeria flowers; which are *E. coli*, *Enterobacter*, *Klebsiella*, *Proteus*, *Serratia*, *Citrobacter*, *Pseudomonas*, *Staphylococcus aureus*, *Streptococcus*, *Bacillus cereus* and *Actinomycetes*. Due to the positive impact of dill essential oil (100 ppm) in the pre-harvest method in improving traits associated with vase life of cut alstroemeria flowers, these treatments are recommended.

12.0 Biotechnology

12.1.0 Tissue Culture

12.1.1 Micropropagation

Germinated embryos showing one or more shoots were cultured in glass culture tubes containing 15 ml of MS medium supplemented with 30 g sucrose/l, 7 g agar/l and a pH adjusted to 5.7, and was transferred to a growth room under 21 ± 1°C and a 16 h photoperiod. After formation of rhizomes, plants were sub-cultured into 200 ml flasks containing 45 ml of MS medium supplemented with 30 g sucrose/l, 7 g agar/l and 2 mg 6-Benzylaminopurine l-1 (BAP) to promote shoot growth. Plants were subcultured every 3–4 weeks and transferred to fresh media, keeping the same conditions described above, until a minimum of three clones were obtained. One of the clones was transferred to 300 ml magenta containers with 20 ml liquid MS medium supplemented with 30 g sucrose/l, and 1 mg Indole- 3-butyric acid l-1 (IBA) to promote rooting.

12.1.2 Acclimatization

Rooted plants were placed into plastic pots (400 ml) containing a substrate mixture of 60 per cent peat moss and 40 per cent perlite. Another clear plastic container was placed and initially sealed over the plastic pot, to start opening gradually every 3–4 days until the containers with plants were completely discovered. After 3 weeks plants were transferred to larger pots (3 l), containing the same mixture of substrate mentioned before, and moved to a greenhouse to complete acclimation at growing conditions of 15 to 25 °C and < 40 per cent RH.

12.1.3 Embryo and Ovule Culture

Buitendijk *et al.* (1995) reported that interspecific hybridization in the genus *Alstroemeria* is hindered by post fertilization barriers. To create interspecific hybrids between *Alstroemeria, A. pelegrina, A. magnifica, A. inodora* and *A. psittacina* in diallelic combinations, an ovule culture technique was developed. Harvesting ovules before the onset of endosperm degeneration *i.e.*, at 14 days after pollination, cutting them into halves and culturing the micropylar halves in a rotating liquid culture medium containing 6 per cent sucrose at 21°C in the dark, led to successful embryo rescue. Hybrids were obtained in 100 per cent of the interspecific combinations. All full grown plants showed interspecific morphological traits and analysis of chromosome complement confirmed their hybrid nature. Diploid hybrid plants were obtained in all the 20 interspecific 3x-2x combinations. Half-ovule culture of 3x-4x and 4x-2x crosses resulted in 43 triploid hybrid plants.

An early post fertilization ovule culture technique in *Alstroemeria* was described by Jeu *et al.* (1995) in which young ovules was cultured *in vitro* two days after pollination. This resulted in the direct production of seedlings in normally incongruous cross combinations. Culture of ovules in crosses between *Alstroemeria ligtu* hybrids and *A. pelegrina* var. *rosea* recorded by Ishikawa *et al.* (1997) revealed that hybrid embryos aborted around 28 days after pollination in the cross *A. ligtu* hybrid × *A. pelegrina* var. *rosea*. When ovules with placenta were transferred to 2 g/l gelrite solidified MS containing 3 per cent sucrose before embryo degeneration, plantlets were formed.

Shinoda and Murata (2003), quoted by Hoshino (2008), carried out great number of cross hybridizations through ovule cultures using *A. angustifolia, A. aurea, A. diluta, A. garaventae, A. hookeri, A. ligtu, A. magnifica, A. magenta, A. pelegrina, A. presliana, A. pulchra, A. versicolor, A. zoellneri, A. inodora* and *A. psittacina*. The ovule cultures after diallel crossing showed interspecific incompatibility in many combinations; hybrid embryos germinated in 196 out of 210 crossing combinations. In particular, *A. angustifolia*, *A. ligtu*, *A. magenta* and *A. pelegrina* showed high cross compatibility as ovule parents. Therefore, *Alstroemeria* breeding through interspecific hybridization has been successful resulting in progress of breeding novel cultivars (Hoshino, 2008).

Consequently, the most reliable embryo rescue techniques for the production of interspecific hybridizations in *Alstroemeria* are summarized as follows:

1. Ovaries are harvested 2-14 days after pollination.
2. Ovules are cultured on 2 g/l gellan gum-solidified MS medium or 1/2 strength MS medium supplemented with 30-90 g/l sucrose.
3. The ovules are kept under 12-16-h photoperiod at 20°C.

Hoshino *et al.* (2006) indicated that the ovules of *A. aurea* can be used to develop a reliable protocol for isolating living egg cells and zygotes involved. They outlined the protocol as follows. Fresh ovule sections were prepared with a surgical blade and treated with an enzyme solution. Subsequently, these ovule sections were dissected using a hand-made glass needle with fine tip under an inverted microscope. The egg cells successfully isolated by this procedure were collected using microcapillaries connected to a computer-controlled micropump. For zygote isolation, ovules were excised from the ovaries 24 h after self-pollination. By treating the excised ovules with an enzyme solution and subsequently dissecting them by using a glass needle, the zygotes were successfully isolated from the ovules and collected with a microcapillary. Egg cells and zygotes were viable for up to 2 h following isolation, as determined by fluoresce in diacetate staining. These procedures may be utilized for *in vitro* fertilization, enabling artificial fusion between a single isolated egg cell and sperm cell. This concept can be applied for obtaining crosses of sexually incompatible *Alstroemeria* species through *in vitro* fertilization (Hoshino, 2008).

A Protocol for embryo culture for interspecific hybridisation was developed by Aros *et al.* (2019) is given here:

Hybridization was performed through manual pollinations between the genotypes. Pollinations were performed using fresh dehiscent anthers; considering the protandric condition exhibited by these species, eventually pollen was collected from mature anthers and stored (4°C) to perform pollination during the following 1 or 2 days. After 12 – 14 days post pollination, fruits were harvested and disinfected at the Tissue Culture Laboratory by direct flaming them with ethanol (95 per cent) for 5 s. Fruits were dissected and cut into halves and the embryos were rescued and sown in Petri dishes (15 x 9 x 60 mm) containing 20 ml of ½ strength Murashige and Skoog (1962) medium, supplemented with 30 g sucrose/l,7 g agar/l and a pH adjusted to 5.7. The Petri dishes were maintained under complete darkness in a growth room at 24 ± 1°C until germination (8 to 12 weeks).

12.1.4 Somatic Embryogenesis

It was interesting to note that kinetin stimulated the production of embryogenic callus in alstroemeria in combination with NAA. Gonzalez and Aldetson (l992a) obtained callus from mature embryos of cv. Butterfly on MS medium supplemented with either 2 or 4 mg/l picloram or 4 mg 2,4-D/1 combined with BA or kinetin (0-4 mg/l). Shoots regenerated from torpedo-like structures (somatic embryos) formed when callus was transferred to regeneration media containing BA plus picloram or 2,4-D. Gonzalez *et al.* (1992a,b) excised somatic embryos of alstroemeria (*cv.* Butterfly) and cultured on solid MS medium alone or on MS medium supplemented with BA + coconut water or GA_3. After 4 weeks of culture, the highest percentage of cultures with single shoot plantlets (25 per cent) was obtained from the medium supplemented with GA_3 and cultured at 15°C, but the maximum percentage of cultures with callus (75 per cent) was obtained from the same medium at 25°C.

A simple procedure to induce callus growth and plantlet regeneration was developed by Hutchinson *et al.* (1994). The callus obtained from mature zygotic embryo cultured on basal medium (MS salts + B5 vitamins) with 20 μM Kin and 10 or 20 μM NAA could be maintained for one year without losing potential. When cultured on medium without growth regulators, plantlets regenerated via somatic embryogenesis. Gonzalez and Alderson (1995) also attempted callus from roots, stem segments, shoot-tips and rhizome buds of cv. Carmen and mature embryos of cv. Butterfly. Explants were cultured on MS basal medium supplemented with various concentration of 2,4-D, picloram, NAA, Kin, BA and/or GA_3. The best results were obtained with mature embryos. Lin *et al.* (2000) cultured stem segments of seedlings from two *Aistroemeria* breeding lines on MS medium supplemented with 4 mg/l 2,4-0, 0.5-1.0 mg/l BAP. Soft callus was initiated which became compact after subculture on a medium with only 0.5 mg/l B.A. Proembryos developed from friable embryogenic calli via embryos into plants after subculture on medium supplemented with 0.1 mg/l BA.

Kim *et al.* (2006) reported on somatic embryogenesis from nodes with axil tissue. Organogenesis from leaf and stem has not been considered possible in monocotyledons. The development of culture system from leaf and stem in *Alstroemeria* is valuable, although successful organogenesis from leaf and stem is strongly genotype-dependent at the present. These results could expand the possibilities of use of culture systems for *Alstroemeria*. By utilizing the embryogenic callus culture in *Alstroemeria*, isolation of protoplasts, subsequent protoplast culture and regeneration of plants from protoplasts were achieved (Kim *et al.*, 2005). Such techniques will promote research on somatic hybridization between cross-incompatible species as well as on direct gene transfer with electroporation (Hoshino, 2008).

12.2.0 Biochemical and Molecular Markers

12.2.1 Phylogenetic Relationship and Identification of Genotypes

Few reports are available on molecular markers and isozyme. Anastassopoulos and Keil (1996) used RAPO markers to study genetic variation of *Alstroemeria*. They used RAPO markers in different genotypes obtained by traditional breeding. All genotypes examined including commercial alstroemeria varieties could be distinguished on the basis of their RAPO profiles. Progeny plants could be distinguished from their parents.

Kamstra *et al.* (1997) used FISH to localize two species-specific repetitive DNA sequences, A001-I and D32-13 and two highly conserved 25S and 5S rDNA sequences on the metaphase chromosomes of two species of *Alstroemeria, A. aurea* and *A. inodora.* There were notable differences in number and distribution of rDNA sites between the two species.

Dubouzet *et al.* (1998) extracted nucleic acid from frozen leaves of 7 *Alstroemeria* species (*A. aurea, A. brasiliensis, A. inodora, A. ligtu, A. paupercula, A. pelegrina* and *A. psittacina*), five orchid types, three butterfly types and five hybrid types. Sub-group analysis between the parental species and the cultivar types indicated close relationships between (i) the orchid types, *A. aurea* and *A. paupercula,* (ii) the carmen types, *A. aurea* and *A. paupercula,* (iii) the butterfly types, *A. psittacina,* and (iv) the hybrid types, *A. paupercula* and *A. aurea.*

In order to estimate the genetic diversity within the *Alstroemeriaceae* family, Aros *et al.* (2006) evaluated nine wild alstroemeria accessions, 10 commercial varieties and the monotype *Leontochir ovallei* using two different methods, RAPD analysis and UPOVs morphological descriptors. DNA from leaves, roots and tepals were analyzed by RAPDs with eight primers that generated 236 RAPD bands. Dendrograms obtained allowed identification of five main clusters: *A. garaventae* alone, wild alstroemerias, commercial varieties, *A. exserens* and *A. spathulata* together, and *L. ovallei.* Twenty-five morphological descriptors related to stem, leaf and inflorescence characteristics were evaluated and a resulting dendrogram was analyzed containing two main clusters: one grouping all commercial varieties plus *A. magnifica* ssp. *magnifica* and the other one with the rest of the wild alstroemerias. With both methods, enough informative information data was obtained to place the wild alstroemerias and commercial varieties hierarchically in different clusters. In this respect, morphological analysis grouped all commercial varieties closer to each other. Also, morphological descriptors grouped all wild alstroemerias except *A. magnifica* ssp. *magnifica,* whereas, RAPD markers grouped seven out of the nine wild alstroemerias, leaving in another cluster *A. garaventae, A. exserens* and *A. spathulata.* These results suggest that also in alstroemeria, RAPD markers are a useful tool for the protection of new releases from a breeding program.

12.2.2 Detection of Diseases and Genes of Resistance

During a survey of viruses infecting *Alstroemeria,* plants showing virescent and malformed flowers were found by Bertaccini *et al.* (1996) under greenhouse conditions. After DNA extraction, 17 samples were tested with direct and nested-PCR using several phytoplasma 16S ribosomal primer pairs. In 5 samples the presence of a 1.2 kb DNA product identical to those observed in phytoplasma strain was detected.

12.2.3 Markers for Vase Life Gene

Naghiloo *et al.* (2019) conducted studies to explore the effects of amino oxy acetic acid (AOA) on the expression of ethylene biosynthesis pathway genes (1-aminocyclopropane-1-carboxylate synthase (*ACS*) and 1-aminocyclopropane-1-carboxylate oxidase (*ACO*) and vase life of the Alstroemeria cut flowers 'Mayfair' and 'Frosty' varieties. Also, the effect of AOA on at the rates of 0, 50, 100 and 150 mg/l was studied on the morphological and physiological traits. The comparison of the vase life of Mayfair and Frosty showed that Frosty had a longer vase life than Mayfair. Frosty exhibited lower levels of cell degradation, higher relative water content, more chlorophyll content, more total soluble solids, and higher peroxidase level than Mayfair. Higher AOA levels were significantly associated with higher activity of catalase and peroxidase, so that the 150 mg/l AOA had the greatest effect on the activity of these two enzymes. Also, the increase in AOA rate to 150 mg/l resulted in the retention of leaf chlorophyll and the reduction of leaf chlorosis percentage in both studied varieties.The results indicated that the highest *ACO* and *ACS* gene transcripts were related to the 7th flower opening stage. The use of AOA was effective in reducing the gene expression in Frosty. Also, the application of 150 mg/l AOA had a significant reducing effect on the expression of the *ACS* gene in Frosty (Naghiloo *et al.*, 2019).

12.3.0 Genetic Transformation

van Schaik *et al.* (2000) reported the first genetic transformation work in Alstroemeria. They confirmed transient gene expression in embryogenic callus after particle bombardment. Foreign gene expression examined with a β-glucuronidase (GUS) assay indicated the reducible transformation conditions in which the *GUS* gene was driven by the maize ubiquitin promoter; suitable parameters for the assay equipment have been reported. Although stable transformed calluses were obtained after visual selection based on luciferase activity, no plant regeneration was observed during the 2 years of maintenance.

Hoshino *et al.* (2000), quoted by Hoshino (2008) reported on the genetic transformation of *Alstroemeria* via co-cultivation of embryogenic suspension cells with *Agrobacterium tumefaciens*. Embryogenic cell suspension cultures were established via ovule culture of an interspecific cross between *A. pelegrina* and *A. magenta*. Cell suspension cultures could be initiated from the callus using liquid MS media containing picloram. An *Agrobacterium* mediated transformation system was developed by using the embryogenic suspension cells. The suspension cells were co-cultivated with *A. tumefaciens* strain EHA101/pIG121Hm or LBA4404/pTOK233, both of which harboured the binary vector carrying the *neomycin phosphotransferase II (NPTII)*, *hygromycin phosphotransferase* (*HPT*) and the introncontaining β-*glucuronidase* (*intron-GUS*) genes in the T-DNA region. Transgenic plants were obtained from co-cultivated cells after selection on a hygromycin-containing medium.

Lin *et al.* (2000) indicated success in producing transgenic plants by particle bombardment. They used a *luciferase* gene from a firefly as the reporter gene under the control of the maize ubiquitin promoter (*Ubi1*) and attempted transferring this gene into friable embryogenic calluses and proembryos. Following visual selection, transgenic plantlets were regenerated through embryogenesis. Moreover, another selection by phosphinothricin (PPT) helped in obtaining *bar* and *GUS* gene-incorporated plants after particle bombardment. The successful transformation obtained by Lin *et al.* (2000c) appears to be due to the selection of elite embryogenic callus lines with a high regeneration potential and maintaining these cells till abiolistic procedure was performed (Lin *et al.*, 2000).

Transgenic alstroemeria plants resistant to Alstroemeria mosaic virus (AlMV) were generated through RNA mediated resistance (Kim, 2020). To this end, the friable embryogenic callus (FEC) of alstroemeria was induced from the leaf axil tissue and transformed with a DNA fragment containing the coat protein gene and 32 non-translated region of AlMV through an improved particle bombardment system. The bar gene was used as a selection marker. More than 300 independent transgenic FEC lines were obtained. Among these, 155 lines resistant to phosphinothricin (PPT) were selected under low stringent conditions. After increasing the stringency of PPT selection, 44 transgenic lines remained, and 710 somatic embryos from these lines germinated and developed into shoots. These transgenic shoots were then transferred to the greenhouse and challenged with AlMV. In total, 25 of the 44 lines showed some degree of resistance. PCR analysis confirmed the presence of the viral sequence. Virus resistance was observed at various levels. Establishment of an efficient transformation system for Alstroemeria will allow inserting transgenes into this plant to confer resistance to viral and fungal pathogens. Accordingly, this is the first report on the production of a transgenic virus-resistant Alstroemeria and lays the foundation for alternative management of viral diseases in this plant (Kim, 2020).

13.0 References

Alborz, Z., Habibi, F. and Mortazavi, S.N. (2015) *J. Crops Improv.,* **17** : 241-255.

Amparo-Espanol, J. and Corredor, P.D. (1998) *Agronomia Colombiana,* **5** : 97-102.

Amsing, J.J. (1996) *Nematologica,* **42** : 71-79.

Anastassopoulos, E. and Keil, E. (1996) *Euphytica,* **90** : 235-244.

Anonymous (1986) *Grower,* **106** : 22-25.

Aros, D., Spadafora, N., Venturi, M., Nu´n˜ez-Lillo, G., Meneses, C., Methven, L., Rogers, H. (2015) *Sci. Hortic.,* **185** : 183–192.

Aros, D., Suazo, M., Rivas, C., Zapata, P., Ubeda, C. and Bridgen, M. (2019) *Euphytica,* **215** : 93.

Aros, D., Meneses, C. and Infante, R. (2006) *Sci. Hortic.,* **108** : 86-89.

Aros, D., Vásquez, M., Rivas, C. and Prat, M.L. (2017) *Chilean J. Agric. Res.,* **77** (1) : 95-99.

Aros, D., Valdés, S., Olate, E. and Infante, R. (2012) *Rev. FCA UNCUYO,* **44**(1) : 191-197.

B1omme, R. and Dambre, P. (1984) *Verbondsnieuws voor de Belgische Sierteelt,* **28** : 221-227.

Bakardzhieva, N., Svrakov, K. and Denkova, S. (1995) *Inter. confer. plant virol.* Apriltsi, Bulgaria, 19-23 September, 1994, Rasteniev" dni-Nauki, 1995, **32** : 121-122.

Bakken, A.K. (2000) *Sci. Hortic.,* **85** : 285-293.

Bakken, A.K. and Baevre, O.A. (1999) *Sci. Hortic.,* **80** : 225-233.

Bayer, E. (1989) *Herbertia,* **45** : 63.

Bellardi, M.G. and Bertaccini, A. (1992) *Colture Protette,* **21** : 81-83.

Bellardi, M.G., Bertaccini, A and Betti, L. (1992a) *Informatore Fitopatologico,* **42** : 18-22.

Bellardi, M.G., Bertaccini, A and Betti, L. (1994) *Acta Hortic.,* **377** : 73-80.

Bellardi, M.G., Vibio, M. and Bertaccini, A (1992b) *Plant Dis.,* **76** : 643.

Beltran, E., Delgado, M., Porcel, M.A., Miralles de Imperial, R., Bellido, N. and Bigeriego, M. (1998) *Colture Protette,* **27** : 103-105.

Bertaccini, A, Vibio, M., Bellardi, M.G., Danielli, A and Loebenstein, G. (1996) *Acta Hortic.* **432** : 312-317.

Bik, RA. and Berg, T.J.M. van den (1982) *Acta Hortic.,* **126** : 287-292.

Blank, C.M. de, Zaayen, A. van, Bouwen, I., De Blank, C.M. and Van Zaayen, A (1994) *Acta Hortic.,* **377** : 199-208.

Blom, T.J. and Piott, B.D. (1990) *HortScience,* **25** : 189-191.

Blomme, R. and Dambre, P. (1979) *Verbondsnieuws voor de Belgische Sierteelt,* **23** : 561-565.

Blomme, R. and Dambre, P. (1982) *Verbondsnieuws voor de Belgische Sierteelt,* **26** : 643-646.

Bond, S. and Alderson, P.G. (1993) *J. Hortic. Sci.,* **68** : 847-853.

Bouwen, I., V1ugt, RAA. van der and van der Vlugt, R.A.A (2000) *Plant Dis.,* **84** : 103.

Bouwen, I., Vlugt, R.A.A. van der, van der Vlugt, R.A.A. and Zaayen, A van (1996) *Acta Hortic.,* **432** : 72-74.

Brickell, C. (1996) *A-Z Encyclopedia of Garden Plant,* Dorling Kindersley Limited, London.

Bridgen, M.P. (2018). Breeding Alstroemeria. *In: Van Huylenbroeck Johan* (ed) *Handbook of plant breeding: ornamental crops,* vol 11. Springer, Berlin, pp 231–236.

Bridgen, M.P. and Bartok, J. (1990) *HortScience,* **25** : 1592-1594.

Bridgen, M.P. and Soriano, C. (1986) *Cannecticut Greenhouse Newsl.,* **132** : 5-6.

Bridgen, M.P., Elliott, G.C. and Smith, M.A. (1992) *In vitro,* **28** : 3.

Broertjes, C. and Verboom, H. (1974) *Euphytica,* **23** : 39-44.

Brunt, A.A. and Phillips, S. (1981) *Annual Report of the Glasshouse Crops Research Institute,* pp. 151.

Brydon, P.H. (1940) *Herbertia,* **7** : 204-205.

Buitendijk, J.H., Pinsonneau, N., Donk, AC. var, Ramanna, M.S. and Lammeren, AA.M. van (1995). *Sci. Hortic.,* **64** : 65-75.

Cervelli, C., Mascarello, C., Farina, E., Proft, M.P. de and Verhony, M.N.J. (2000) *Acta Hortic.,* **516** : 107-113.

Chanasut, U., Rogers, H.J., Leverentz, M.K., Griffiths, G., Thomas, B., Wagstaff, C. and Stead, A.D. (2003) *Postharvest Biol. Technol.,* **29** : 324-332.

Chang, K.F. and Mirza, M. (1993) *Canadian Plant Diseases Survey*, **73** : 3-8.

Chang, K.F. and Mirza, M. (1994) *Plant Pathol.*, **43** : 406-409.

Chang, K.F., BlnT, D.J.S., Hwang, S.F. and Mirza, M. (1994) *Zeitschrift fur Pflanzenkrankheiten and Pflanzenschutz*, **101** : 460-466.

Chang, K.F., Mirza, M. and Hwang, S.F. (1993) *J. Phytopathol.*, **137** : 207-213.

Chepkairor, M.J. and Waithaka, K. (1988) *East Afr. Agric. Fores. J.*, **53** : 229·231.

Chiari A, Bridgen M.P. (2002) *Plant Cell, Tiss and Organ Cult.*, **68** : 49-55.

Chiari, A, Elliott, G.C. and Bridgen, M.P. (1999) *HortScience*, **34** : 657-659.

Chiko, AW. and Godkin, S.E. (1984) *Research Highlights. Saanichton Research and Plant Quarantine Station*, pp. 4-5.

Cireasa, R. (1983-84) *Lucrari Stiintiftce.Institutul Agronomic "Ion lonescu de la Brad", Horticultura*, **27-28** : 69-70.

Cocozza-Talia, M.A., Pacucci, G., Pczzuto, A, Lucia, B. de and De Lucia, B. (1996) *Colture Protette*, **25** : 63-68.

Cotten, J., Bartlett, P.W. and Webb, R.M. (1990) *Indian Phytopath.*, **43** : 311-312.

D'-yachenko, A.D. (1991) *Introduktsiya* i *Akklimatizatsiya Rastenii*, **13** : 71-73.

Dambre, P. (1987) *Verbondsnieuws voor de Belgische Sierteelt*, **31** : 911-921.

Dias-Tagliacozzo, G.M., Tombolato, A.F.C. and Válio, I.F.M. (2005) *Acta Hortic.*, **683** : 287-291.

Dinova, M., Boboshevska, D. and Racheva, Ya. (1988) *Acta Hortic.*, **226** : 563-566.

Doi, M., Cheng, Z.Y., Saito, K., Sumitomo, E., Inamoto, K., Imanishi, H. and Cheng, Z.Y. (1999) *J. Japanese Soc. Hortic. Sci.*, **68** : 160-167.

Doi, M., Cheng, Z.Y., Tsuji, M., Imanishi, H. and Cheng, Z.Y. (1998) *Jap. Soc. Hortic. Sci.*, **67** : 965-969.

Dubouzet, J.G., Murata, N. and Shinoda, K. (1998) *Sci. Hortic.*, **73** : 37-44.

Foster, M.B. (1945) *Herbertia*, **12** : 44-48.

Gabryszewska, E. (1995) *Acta Agrobotanica*, **48** : 95-104.

Gonzalez, B.M.E. and Alderson, P.G. (1992a) *J. Exp. Bot.*, **43** : 205-211.

Gonzalez, B.M.E. and Alderson, P.G. (1992b) *Phyton.*, **53** : 95-101.

Gonzalez, B.M.E. and Alderson, P.G. (1995) *Phyton*, **57** : 67-75.

Guzmán, C., Prat, L., Rivas, C. and Aros, D. (2018) *Cien. Inv. Agr.*, **45**(2) : 158-168.

Hakkaart, F.A. and Versluijs, J.M.A. (1988) *Netherlands J. Plant Pathol.*, **94** : 49-56.

Halevy, A.H. and Mayak, S. (1981). *Hort. Rev.*, **3** : 59-143.

Hannabal, L.S. (1941) *Herbertia*, **8** : 161-164.

Healy, W., Klick, S. and Roh, M.S. (1993) *Acta Hortic.*, **377** : 25-29.

Healy, W.E. and Lang, D. (1984) *Res. Bull., Colorado Greenhouse Growers' Association,* **414** : 1-2.

Healy, W.E. and Lang, D. (1985) *Res. Bull Colorado Greenhouse Growers Association,* **415** : 2-3.

Healy, W.E. and Wilkins, H.F. (1982) *Sci. Hori.,* **17** : 383-390.

Healy, W.E., Wilkins, H.F. and Celusta, M. (1982) *J. Amer. Soc. Hort. Sci.,* **107** : 1046-1049.

Hermann, P. (1986) *Zierpflanzenbau,* **26** : 228-229.

Hoshino, Y., Murata, N., Shinoda, K. (2006) *Annals of Bot.,* **97** :1139-1144.

Hoshino, Y., Sato, H., Murata, N. and Shinoda, K. (2000) *The Asia Pacific Conference on Plant Tissue Culture and Agribiotechnology,* pp. 93.

Hoshino, Y. (2008) Advances in *Alstroemeria* Biotechnology In: Floriculture, Ornamental and Plant Biotechnology Volume V ©2008 Global Science Books, UK. pp. 540-549.

Hutchinson, M.J., Tsudita, J.M. and Saxena, P.K. (1994) *Plant Cell Rep.,* **14** : 184-187.

Index Kewensis (1895) *Plantarum Phanerogamarum,* Oxford University Press, New York Tamus 1, Supplementum la 18.

Inouye, N., Maeda, T., Mitsuhata, K. and Garn, I.W (1998) *Bulletin Res. Inst. Bioresources,* Okayama University, **5** : 155-168.

Ishikawa, T., Takayama, T., Ishizaka, H., Ishikawa, K. and Mii, M. (1997) *Breed. Sci.,* **47** : 15-20.

Jeu, M.J.de, Laschuit, J., Chevalier, F., Visser, R.G.F. and Weiss, D. (1995) *Acta Hortic.,* **420** : 62-64.

Kaminska, M and Korbin, M. (1999) *Acta Physiologiae Plantarum,* **21** : 21-26.

Kamstra, S.A., Kuipers, A.G.J., Jeu, M.J. de, ramanna, M.S. and Jacobsen, E. (1997) *Genome,* **40** : 652-658.

Kaushal, S., Dilta, B.S., Gupta, Y.C., Kumar, P., Spehia, R.S and Gupta, R.K. (2018) *J. Pharmacog. Phytochem.,* **7**(3) : 2950-2955.

Keil, L. (1986) *Res. Bull. Colorado Greenhouse Growers' Association,* **436** : 1-2.

Kim, J.B, Raemakers, C.J.J.M., Jacobsen, E. and Visser, R.G.F. (2006) *Plant Cell, Tiss. and Organ Cult.,* **86** : 233-238.

Kim, J.B. (2020). *J. Plant Biotechnol.,* **47** : 164–171.

Komnenovic, R. and Pagliarini, N. (1980) *Acta Entomologica Jugoslavica,* **16** : 127-135.

Kristiansen, K. and Vainstein, A. (1995) *Acta Hortic.,* **420** : 85-88.

Kristiansen, K., Ornstroup, H. and Brandt, K. (1999) *Plant Cell Tissue and Organ Cult.,* **56** : 145-153.

Krogt, T.M. van der (1985) *Vakblad voor de Bloemisterij,* **40** : 50-57.

Kwiatkowska, B. and Brzozowska, J. (1980) *Prace Instytutu Sadownictwa i Kwiaciarstwa w Skierniewicach B.,* **5** : 31-40.

Labeke, M.C. van and Dambre, P. (1993) *Sci. Hortic.,* **56** : 135-145.

Labeke, M.C. van and Dambre, P. (1998) *Sci. Hort.,* **74** : 269-278.

Langroudi, M.E., Hashemabadi, D., Kalatejari, S. and Asadpour, L. (2020) *J. Ornam. Plants,* **10**(1) : 15-25.

Latham, L.J. and Jones, R.A.C. (1997) *Australian J. Agric. Res.,* **48** : 359-369.

Leverentz, M.K., Wagstaff, C., Rogers, H.J., Stead, A.D., Chanasut, U., Silkowski, H., Thomas, B., Weichert, H., Feussner, I. and Griffiths, G. (2002) *Plant Physiol.,* **130**(1) : 273–283.

Lim, S.S., Lee, S.I., Kang, S.C. and Kim, J.B. (2012) *J. Plant Biotechnol.,* **39**(4) : 219–224.

Lin, H.S, van der Toorn, C, Raemakers, K.J.J.M, Visser, R.G.F, de Jeu M.J, Jacobsen, E. (2000) *Molecular Breed.,* **6** : 369-377.

Lin, H.S., Jeu, M.J. de and Jacobsen, E. (1997) *Plant Cell Rep.,* **16** : 770-774.

Lin, H.S., Jeu, M.J. de and Jacobsen, E. (1998) *Plant Cell Tiss. and Organ Cult.,* **52** : 165-169.

Lin, W.C. (1984) *HortScience,* **19** : 515-516.

Lin, W.C. (1985) *HortScience,* **20** : 378-380.

Lin, W.C. and Monette, R.L. (1987) *Plant Cell Tiss. Organ Cult.,* **9** : 29-35.

Lisiecka, A. (1981) *Prace Instytutu Sadownictwa i K wiaciarstwa w Skiemiewicach, B, Rosling Ozdobne,* **6** : 5-12.

Lisiecka, A. and Szczepaniak, S. (1983) *Prace Instytutu Sadownictwa i Kwiaciarstwa w Skiemiewicach B Rosliny Ozdobne,* **8**: 93-99.

Lisiecka, A. and Szczepaniak, S. (1993) *Prace z Zakresu Nauk Rolniczyeh,* **75** : 107-113.

Lisiecka, A., Szczepaniak, S. and Saniewski, M. (1992) *Acta Hortic.,* **325** : 379-385.

Meij, *J.* (1976) *Vakblad voor de Bloemisterij,* **31** : 14-15.

Milde, H. (1981) *Gb + Gw,* **82** : 778-780.

Milde, H. (1983) *Gb + Gw,* **83**: 81-89.

Milde, H. (1985) *Gb + Gw,* **85**: 1330-1331.

Miyake, I. (1989) *Herbertia,* **45** : 40.44.

Monette, P.L. (1987) *Biotech. Agric.,* **20** : 1-18.

Naghiloo, S., Rabiei, V., Soleimani, A., Khalighi, A. and Arkinezhad, M. T. (2019) *Genetika,* **51** : 861-876.

Nasri, F., Mortaza, S.N., Ghaderi, N., and Javadi, T. (2013) *J. Hortic. Res.,* **21**(2) : 23-30.

Navalinskiene, M., Samuitiene, M. and Jackeviciene, E. (1993) *Eksperimentine Biologija,* **2** : 57-63.

Nijssen, H.M.C. and Hoogeveen, M.G. (1990) *Vakblad voor de Bloemisterij*, **45** : 26-28.

Pedraza-Santos, M.E., López-Peralta, M.C., González-Hernández, V.A., Engleman-Clark, E.M. and Sánchez-García, P. (2006) *Plant Cell, Tiss. Organ Cult.*, **84** : 189-198.

Phillips, S., Brunt, A.A., Stone, O.M. and Pawley, R.R. (1983) *Annual Report Glasshouse Crops Research Institute*, pp. 142-143.

Plomacher, H. (1978) *Zierpjlanzenbau*, **18** : 578-579.

Porat, R., Borochov, A. and Halevy, A.H. (1993) *Plant Growth Regul.*, **13**(3) : 297-301.

Pumisutapon, P., Visser, R.G.F., and de Klerk, G.J. (2012). *Plant Cell, Tissue and Organ Cult.*, **110**(3) : 395-400.

Ramanna, M.S. and Mariani, A. (1992) *Proc. workshop: Gametes with somatic chromosome number in the evolution and breeding of polyploid polysomic species: achievements and perspectives*, Perugia, Italy, 9-10 April, 1992, pp. 83-89.

Razi, M. (2017) *J. Ornam. Plants*, **7** : 81-91.

Rees, R.A. (1992) *Ornamental bulbs, corms and tuber*, pp. 38.

Reid, A. (2006) Alstroemeria. Department of Agril., and Food, Western Australia, Perth. Bulletin 4688.

Rustanius, P., Hang, A., Hughes, H.G. and Tsuchiya, T. (1991) *HortScience*, **26** : 902-904.

Shiota, A., Uematsu, S., Kobayashi, Y., Mikoshiba, Y. and Honda, Y. (1997) *Proc. Kanto Tosan Plant Skierniewicach B*, **5** : 31-40.

Smith, M.A., Elliott, G.C. and Bridgen, B.P. (1998) *HortScience*, **33** : 55-59.

Sonneveld, C. (1988) *Acta Hortic.*, **228** : 317-326.

Spier, P. and Milde, H. (1977) *Gb + Gw*, **77** : 774-776.

Spieser, H. (1987) *American Society Agricultural Engineers*, **87** : 4026.

Stinson, H.L. (1942) *Herbertia*, **9** : 41-52.

Stinson, H.L. (1945) *Herbertia*, **12** : 48-51.

Stinson, H.L. (1952) *Plant life*, **8** : 99-105.

Szczepaniak, S. (1991a) *Roczniki Akademii Rolniezej W Poznaniu*, Ogrodnicto, **19** : 117-127.

Szczepaniak, S. (1991b) *Roczniki Akademii Rolniezej W Poznaniu*, Ogrodnicto, **19** : 139-148.

Szczepaniak, S. (1991c) *Roczniki Akademii Rolniezej W Poznaniu*, Ogrodnicto, **19** : 129-138.

Szczepaniak, S. (1997) *Zeszyty Problemowe Postepow Nauk Rolniczyeh*, **449** : 183-192.

Szczepaniak, S. (1998) *Folia Universitatis Agriculturae Stetinensis*, Agricultura, **70** : 129-136.

Tombolato, A.F.C. and Matthes, LAF. (1998) *Acta Hortic.*, **454** :

Traub, H.P. (1973) *Plant Life*, **29** : 100.

Uphof, T.C.Th. (1940) *Herbertia*, **47** : 194-199.

Uphof, T.C.Th. (1952) *Plant life*, **8** : 37-53.

Van Doorn, W.G. and A.D. Stead, (1997). *J. Exp. Bot.*, **48** : 821–837.

van Schaik, C.E., van der Toorn, C., de Jeu, M.J., Raemakers, C.J.J.M., Visser, R.G.F. (2000) *Euphytica*, **115** : 17-26.

Van Zaayen (1995). Alstroemeria. In: Loebenstein G, Lawson RH and Brunt AA (eds) Virus and Virus-like Diseases of Bulb and Flower Crops, John Wiley and Sons, Chichester, UK. pp 237-249.

Van, H. Hoff, Hakkaart, FA., Scholten, G., Bos, L. and Beuzenberg, M.P. (1970) *Proefstation Voor de bloemisterij in Nederland te Aalsmeer*, pp. 140.

Verboom, H. (1979) *Vakblad voor de Bloemisterij*, **34** : 44-45.

Verboom, H. and Staaveren, M.C. van (1978) *Vakblad voor de Bloemisterij*, **33** : 23,61.

Waithaka, K. and Chepkairor, M.J. (1988) *East Africa Agric. Fares. J.*, **53** : 213-220.

Wiel, A. van de (1989) *Vakblad voor de Bloemisterij*, **44** : 32-33.

Yasuda, S., Saka, K. and Natsuaki, K.T. (1998) *Japanese* 1. *Trap. Agrie.*, **42** : 85-93.

Zaayen, A. van, Blank, C.M. de, Bouwen, I., Van Zaayen, A. and De Blank, C.M. (1994) *European J. Plant Pathal.*, **100** : 85-90.

Ziv, M., Kanterovitz, R. and Halevy, AH. (1973) *Sci. Hortic.*, **1** : 271-277.

www.hilagric.ac.in

9

AMARYLLIS AND HIPPEASTRUM

S.R. Dohare, C. Aswath and M. Mitra and V.A. Parthasarathy

1.0 Introduction

Amaryllis is closely related to *Hippeastrum* and is commonly known as belladonna lily. *Hippeastrum,* known as Royal Dutch Amaryllis is also called Trumpet lily or Night Star lily. Plants of both the genera are successfully grown in plains as well as in the hilly areas. Native to Peru and South Africa, the genus *Amaryllis* comes from the Greek word *amarysso,* which means "to sparkle". Bulbs were brought to Europe in the 1700s and have been known to bloom for up to 75 years. Today, most amaryllis are hybrids but are classified in the genus *Hippeastrum* (Meyer and Weisenhorn, 2018). *Hippeastrum* sp., commonly known as Amaryllis, is commercially used as a potted plant and for cut flowers. Hippeastrums are often erroneously described as Amaryllis (*Amaryllis belladona*) although these two plants have distinct difference between them (Sultana *et al.,* 2010). It is a geophyte native to South America, whereas the true genus Amaryllis is native to Africa. The genus Hippeastrum has about 90 species and hundreds of cultivars. For many years there was confusion among botanists over the generic names Amaryllis and *Hippeastrum,* resulting in the common name amaryllis used for cultivars of this genus, while the generic name Amaryllis was applied to bulbs from South Africa, usually grown outdoors (https://plants.ces.ncsu.edu/plants/hippeastrum/).

2.0 Importance and Uses

2.1.0 Aesthetic Uses

Hippeastrums are important bulbous flowering plants. The spectacular flowers have various shades of bright colour, and wide range of variation exists in cultivars developed by hybridization. They are suitable for planting in pot, greenhouse, window garden, bed, rockery, shrubbery and also in landscaping. They are also popular as cut flowers because of their large size, attractive colour and good keeping quality. This plant is highly valued in international markets, especially in winter time. Brazil produces about 17 million bulbs per year for marketing, of which 60 per cent is exported worldwide (Inkham *et al.,* 2019). Amaryllis gradually became a very popular flower in the West with greater than 300 cultivars. As the cultivation of amaryllis and its hybrids is now observed in many countries around the world, including the Netherlands, South Africa, Japan, Brazil, and the USA, the original founder species has lost genetic relationships amongst many cultivars and hybrids due to long-term artificial cultivation and breeding (Wang *et al.,* 2018).

2.2.0 Therapeutic and other Uses

Even though they are valued for their beauty, they also contain alkaloid which has medicinal value. De Andrade *et al.* (2012) presented an exhaustive chemical and biological review of the alkaloids found in the genus *Hippeastrum*. Over the last 50 years, the bulbous genus *Hippeastrum* has yielded 64 different alkaloids, together with others whose structures remain undefined. Further studies on the isolation of these compounds are called for, especially after recent biological studies showing their significant antiparasitic, sychopharmacological and AChE-inhibitory

Uses of Amaryllis and Hippeastrum

In Bed

In Pathway of Garden

In Pot

In Home Decoration

In Flower Vase

As Cut Flowers

In Flower Arrangement

activities. Notably, some *Hippeastrum* species are able to produce a high level of galanthamine, comparable with species of other genera currently being used for the commercial production of this alkaloid. The lack of biological activity shown by most of the alkaloids found in the *Hippeastrum* genus may be due to the small amounts isolated. Consequently, their synthesis or *in silico* studies will facilitate further bioactivity assessment.

3.0 Origin, History and Distribution

All the species of wild amaryllis are natives of tropical and subtropical America, mostly to Mexico and West Indies, southward to Chile and Argentina. Only the species *Amaryllis reginae* crosses the ocean to another continent.

The history of cultivated amaryllis was grouped into five periods by Traub (1958). The first period (1689-1799) began with the first published record of an *Amaryllis* species later named as *Amaryllis belladonna* (Linnaeus, 1737, 1751). The descriptions of five species, namely *Amaryllis belladonna, A. reginae, A. Striata, A. reticulate* and *A. Vittata* were noted during the period. The second period was between 1799 and 1865 when first amaryllis hybrid, *Amaryllis* × *johnsonii* was developed in the year 1799 (Herbert, 1837; Baker, 1888). During that period 11 additional *Amaryllis* species were included. The species were *Amaryllis minata, A. elegans, A. psittacina, A. aulica* and *A. calyptrate, A. stylosa, A. correiensis* and *A. harrisonii, A. breviflora, A. cybister* and *A. moreliana*.

The third period (1866-1918) began with the introduction into cultivation of *A. leopoldii* and *A. pardina*. In addition to these, fourteen species were also described during this period, *viz., A. bluemenavia, A. scopulorum, A. andreana, A. mandonii, A. angustiflia, A. petiolata, A. canterai, A. flammingera, A. tucumana, A. damaziana, A. fusca, A. crociflora, A. viridiflora* and *A. forgetii*. The firm of James Veitch and Sun made crosses between *A. pardina* and *A. leopoldii*, which resulted as Reginae hybrids.

The fourth period (1919-1949) was ushered in to impose the plant quarantine and adoption of vegetative propagation. During the World War I the amaryllis trade in Germany and France was seriously affected and there was a temporary setback in the Netherlands and Denmark. In England, the growing of amaryllis declined after 1918. The introduction of plant quarantine in the United States in 1919 restricted the import of amaryllis to only a limited number of disease-free bulbs for use as propagation stock. During that period only three new species were described, *viz., A. immaculata, A. kromeri*, and *A. aglaiae*. In Northern United States, twelve hybrids were imported from England in 1909 to produce with breeding work. These were Diana, Venus, Seraphis, Progress, Achilles, Crimson King, Virgin Queen, Monarch, Vesta, Bacchus, Adonis and Scarlet King. Similarly, in California, both Reginae and Leopoldii hybrids were developed. The breeding work was undertaken in southeast of the United States and European countries during that period. In addition, the breeding work was also done in Australia. An excellent hybrid was bred from bulbs imported from the Netherlands and England (Strout, 1954). In South Africa, A. C. Buller and in Brazil, Joao Dierberger developed new strains of amaryllis. A National organization, American Amaryllis Society was established in 1933 and became affiliated with American Plant Life Society in 1944. During that period advancement in cultivation of hybrid amaryllis was spectacular. Important achievements were the standardization and application of rapid methods for vegetative propagation of amaryllis. The Christmas forcing was another achievement. A number of local, national and regional amaryllis shows were held in Orlando, Florida, Montebello and California by the American Amaryllis Society.

The fifth period (from 1950 to the present time) is noted for development of hybrid amaryllis. This period led to the origin of intermediate and short stature hybrid clones. A large number of plant explorers including Roman Ferreyra of Peru, Ernesto Miranda de Neto of Brazil, Mulford B. Foster of Orlando. Florida, Harlod N. Moldenke of Yonkers, New York and Ira S. Nelson of Louisiana brought back new plant material. Studies of these materials revealed several new species, such as *A. ferreyrae*, *A. fosteri*, *A. espiritensis*, *A. oconequensis*, *A. maracasa*, *A. traubii*, *A. barreirasa*, *A. apertispatha*, etc. The Herbert Medal in 1951 was received by Foster from the American Amaryllis Society for his outstanding work in bringing back new species (Foster, 1951). This period is also known for Henry Miniature hybrids (Traub, 1951). The floriferous hybrids of Belladona division were named as *Amaryllis* × *gladwynensis* (Henry, 1951).

Even after the controversy on the name of two genera, *Hippeastrum* and *Amaryllis* both are synonymous for Trumpet lily or Night Star lily. There had been the different views regarding the generic name of *Hippeastrum*. Linnaeus (1753) recognized its generic name as *Amaryllis* L., later on Herbert (1819) typified it as *Hippeastrum* Herb.

During the International Botanical Congress held in Paris in 1959 it was concluded that *Amaryllis belladona* and *Amaryllis* in botanical sense applied to the South African species while *Hippeastrum* spp. to the American species and hybrids. According to Bailey (1929), most of the species of *Hippeastrum* were first described in the genus *Amaryllis,* but that genus differs in its solid scape and absence of scales between the filaments.

To the conflict of opinion, however, it was discussed at length by Moore (1963). Before and after Moore, several authors recognized the generic identity of *Hippeastrum* (Wilkinson,1943; Everett, 1954; Brants and Wanduvan, 1965; Gorer, 1970; Rees, 1972; Willis, 1973). Six species of *Hippeastrum* have been mentioned in the Firminger's Manual of Gardening in India. The taxonomy of the genus has been controversial. In 1753, Carl Linnaeus created the name *Amaryllis belladonna*, the type species of the genus *Amaryllis*. At the time both South African and South American plants were placed in the same genus; subsequently they were separated into two different genera. The key question is whether Linnaeus's type was a South African plant or a South American plant. If the latter, *Amaryllis* would be the correct name for the genus *Hippeastrum*, and a different name would have to be used for the genus discussed here. Alan W. Meerow *et al.*, have briefly summarized the debate, which took place from 1938 onwards and involved botanists on both sides of the Atlantic. The outcome was a decision by the 14th International Botanical Congress in 1987 that *Amaryllis* L. should be a conserved name (*i.e.* correct regardless of priority) and ultimately based on a specimen of the South African *Amaryllis belladonna* from the Clifford Herbarium at the Natural History Museum in London (https://en.wikipedia.org/wiki/Amaryllis). Forty-two species transfers are made from Amaryllis to Hippeastrum in accord with the decision, at the Fourteenth International Botanical Congress in 1987, on the application of the former name to a South African genus. This completes the validation of combinations required in Hippeastrum, and will thus aid nomenclatural clarity and enable botanists and horticulturists to refer to these species by their correct names (Meerow *et al.*, 1997).

Finally, it was accepted that *Amaryllis* has one species only, *A. belladona* and *Hippeastrum* with 75 species. Some of the important species of *Hippeastrum* and their place of origin are listed below:

Hippeastrum advenum	:	Chile
H. andreanum	:	Colombia
H. aulicum	:	Brazil and Paraguay
H. bagnoldi	:	Argentina and Chile
H. bicolour	:	Chile
H. bifidum	:	Argentina and Uruguay
H. blumenavia	:	Southeast Brazil
H. breviflorum	:	Argentina
H. calyptratum	:	Brazil
H. candidum	:	Argentina
H. cybister	:	Bolivia
H. elwesii	:	Argentina
H. equestre	:	South America, Guiana to Peru and West Indies
H. forgetii	:	Peru
H. leopoldii	:	Peru
H. miniatum	:	Peru
H. organense	:	Brazil
H. pardinum	:	Peru
H. phycelloides	:	Chile
H. pratense	:	Chile
H. procerum	:	Brazil
H. psittacinum	:	Brazil
H. reginae	:	Brazil
H. reticulatum	:	Brazil
H. roseum	:	Chile
H. rutilum	:	Venezuela
H solandriflorum	:	South America
H stylosum	:	Guiana and Brazil
H vittatum	:	Peru

4.0 Morphology and Classification

Amaryllis is a genus of a single species of South African bulbous plant with deciduous foliage. Leaves after flowers, numerous, strap-shaped; flowers large, sweet-scented, stalked; spathe 2-1eaved; umbels few-flowered; perianth tube short, funnel-shaped, 6-parted, somewhat irregular; segments many-nerved, broad, undulate, spreading at tips; stamens attached at top of tube, unequal, declinate; anthers versatile; style declinate; stigma somewhat 3-1obed; scape tall, solid, compressed; capsule obovate; seeds globose, fleshy (Chittenden and Synge, 1956).

Hippeastrum is a South American bulbous plant, closely related to Amaryllis, Habranthus, Pyrolirion and Zephyranthes. Bulbs tunicated; leaves linear or strap-shaped; flowers declinate, funnel-shaped, segments of 4 different sizes; stamens clustered, of 4 different lengths; spathes of 2 equal and opposite simple valves free of one another to the base; peduncle 2- to several flowered, rarely only 1, each with a bract below it; fruit a 3-celled capsule (Chittenden and Synge, 1956). *Hippeastrum* consists of 50-60 entirely New World species, though one species, *H. reginae* Herbert, appears to have been introduced to Africa. The species are concentrated in two main areas of diversity, one in eastern Brazil, and the other in the central southern Andes of Peru, Bolivia, and Argentina, on the eastern slopes and adjacent foothills (Meerow, 2009).

According to Garcia *et al.* (2019) quoting the work of Meerow *et al.* (1999, 2000, 2010), the molecular era of plant systematics has altered our understanding of relationships within Hippeastreae and brought significant changes to the classification of Amaryllidaceae sensu stricto (s.str.), including the recognition of : (1) an American clade of Amaryllidaceae that is divided into the hippeastroid clade and the Andean tetraploid clade (including tribes Eustephieae, Hymenocallideae, Clinantheae, and Eucharideae) (Meerow *et al.*, 1999, 2000); (2) tribe Griffinieae within the hippeastroids, including Griffinia Ker Gawl. and Worsleya (W. Watson ex Traub) Traub, which is sister to Hippeastreae (Meerow *et al.*, 2000); and (3) Pyrolirion Herb. as the first branch in Eustephieae rather than being closely related to Habranthus Herb. and Zephyranthes Herb. (Meerow, 2010). Traditionally, Pyrolirion had been considered close to the latter genera due to morphological similarities (*i.e.*, tubular spathe valve, single-flowered scape). A robust generic classification for Amaryllidaceae has remained elusive mainly due to the lack of unequivocal diagnostic characters, a consequence of highly canalized variation and a deeply reticulated evolutionary history (Garcia *et al.*, 2019). A consensus classification is proposed here, based on recent molecular phylogenetic studies, morphological and cytogenetic variation, and accounting for secondary criteria of classification, such as nomenclatural stability. Using the latest sub tribal classification of Hippeastreae (Hippeastrinae and Traubiinae) as a foundation, we propose the recognition of six genera, namely Eremolirion gen. nov., *Hippeastrum, Phycella, Rhodolirium, Traubia* and *Zephyranthes.* A subgeneric classification is suggested for *Hippeastrum* and *Zephyranthes* to denote putative subclades. In *Hippeastrum,* they recognize *H.* subg. *Hippeastrum* and *H.* subg. *Tocantinia.* In *Zephyranthes, Z.* subg. *Eithea, Z.* subg. *Habranthus, Z.* subg. *Myostemma* (= core Rhodophiala clade), *Z.* subg. *Neorhodophiala* subg. *nov.* and *Z.* subg. *Zephyranthes* are recognized. Descriptions, synonymy, taxonomic keys, and new combinations are provided for each genus and subgenus (Garcia *et al.*, 2019).

5.0 Species and Cultivars

5.1.0 Species

5.1.1 Amaryllis

A. belladonna (Belladonna Lily)

Leaves strap-shaped, channelled, acute, produced in winter or early spring; flowers fragrant, rose-red in type but variable; scale 45 – 76 cm long, solid; umbel few flowered. August to October. Very variable in colour and size of flower but always beautiful, bulb large. Var. *blanda:* flowers large, white fading to blush; var. *maxima*: robust, flowers many, large, rose; var. *pallida:* flowers pale rose-pink; var. *purpurea:* flowers purple on segments; var. *rosea:* flowers rose, segments striped white, pointed, late; var. *spectabilis:* umbels large, flowers rose, white within.

5.1.2 Hippeastrum

H. aulicum

Leaves broad strap-shaped, not glaucous, closely striate, rather obtuse; flowers large, 12 to 15 cm long, rich crimson, green at base with a dark blotch of deep purple above the green; segments slender-pointed, striate; scape 30 – 45 cm long, rounded, glabrous, usually 2-flowered. Var. *platypetalum:* more robust with broader perianth segments.

H. bagnoldi

Leaves linear, glaucous, 30 cm long; flowers yellow, tinged red, erect or nearly so, 3.5 – 5.0 cm long; tube very short; segments 1.25 cm wide; umbel 4 – 6-flowered; scape 30 cm long; bulb 5 cm thick, tunic nearly black. Var. *punctatum:* drooping flowers, tube green, segments milky-white with numerous small red dots, revolute at tips.

H. bicolor

Leaves about 4, with flowers, linear-obtuse, 45 – 60 cm long; flowers bright red, yellowish green towards base, ascending, narrow-funnel-shaped; umbel 4 to 9-flowered; scape slender, 30 – 45 cm long; bulb globose, 5 cm across. October. Var. *magnifica:* flower over 7.6 cm long.

H. equestre

Leaves lorate, about 5 cm wide, developing fully after flower, then about 45 cm long; flowers bright green, red at base, 10 – 12.5 cm across, tube 2.5 cm long, green; segments broadened from base, then pointed; scape 2 to 4-flowered, 30 – 60 cm long, rounded, glaucous; bulb globose, 5 cm thick, tunic brown, neck short. A variable species which crosses freely. Var. *fulgidum*: flowers bright orange, margined white; var. *ignescens*: deep scarlet, barred white, throat white; var. *major:* very large bright orange, green at base; var. *pyrrochroum*: flowers deep red, greenish yellow in throat; var. *splendens*: larger in all parts, red.

H. leopoldii

Leaves lorate, 45 – 60 cm long; flowers 15 – 17.5 cm across, bright red, greenish white in throat; segments white towards tip, 5 cm wide, keel 2-flowered, white, in lower half; tube short; scape stout, usually 2-flowered; bulb 5 or 7.5 cm thick.

H. pardinum

Leaves 6, with flowers, 60 cm long, 3.80 cm wide, light green; flowers over 15 cm across, spreading, rich cream dotted all over with crimson; tube very short; scape 45 cm long, 2-flowered, glaucous; bulb small, rather thick neck, 7.5 cm long. Var. *tricolor*: whitish and greenish chequered red without, dull crimson chequered white within, lower part of flower white with a few red spots.

H. pratense

Leaves dark green, linear, about 25 cm long; flowers bright scarlet sometimes feathered yellow at base, about 6.3 – 7.5 cm across; scape 3 or 4-flowered, about 30 cm long.

H. procerum

Leaves about 12, falcate, ensiform, 60 to 90 cm long, 3.2 – 5 cm wide; flowers lilac, tube very short; segments 13 – 15.2 cm long, less than 2.5 cm wide; scape 4- to 12-flowered, 30 – 45 cm long, 2-edged, stout; bulb large, neck 25 – 30 cm long. Winter.

H. reginae

Mexian Lily. Leaves 60 – 90 cm long, 3.8 cm wide; flowers bright red, with large greenish-white star in throat; segments 10 – 12.5 cm long, tube about 1.25 cm long, green; scape 2 to 4-flowered, 30 cm long; bulb 5 – 7.5 cm thick.

H. reticulatum

Leaves dark green with ivory-white midrib, 30 cm long, 5 cm wide, with flower; flowers soft pink and white, veins darker, about 7.2 cm across; scape 5 or 6-flowered; bulb globose, neck short.

H. rutilum

Leaves 6 – 8, lorate, 30 cm long; flowers bright crimson, keeled green half-way, tube green; filaments red; segments 7.2 – 10 cm long; tube 1.9 cm long; scape 2- to 4flowered, 30 cm long; bulb 5 to 7.5 thick. Var. *acuminatum*: flowers red. Leaves with ashy bloom; Var. *citrinum*: flowers bright yellow; var. *crocatum*: flowers saffron, segments undulate; var. *fulgidum*: large than type, deep crimson.

H. stylosum

Leaves 4 – 6, lorate, more glossy, purplish at base; flowers 10 cm across, pale brownish-pink, with specks and veins deeper; anthers straw-coloured, striped

Species of Amaryllis and Hippeastrums

Amaryllis belladonna

Hippeastrum aulicum

Hippeastrum equestre

Hippeastrum leopoldii

Hippeastrum pardinum

Hippeastrum puniceum

Hippeastrum reginae

Hippeastrum reticulatum

Hippeastrum rutilum

Hippeastrum striatum

Hippeastrum stylosum

Hippeastrum vittatum

red; tube 1.3 cm long; segments under 2.5 cm wide; style long exserted; scape 3- to 8-flowered, 45 – 60 cm long; bulb globose, 7.5 cm thick, tunic pale, neck short.

H. vittatum

Leaves 6 – 8, after flower, 60 cm long, bright green; flowers whitish with double red stripes, keel white, 10 – 12.5 cm across; segments obovate-oblong, about 3.8 cm across; tube about 2.5 cm long; scape 3 to 6-flowered, up to 90 cm high; bulb globose, 7.5 cm thick.

Some of the species are synonyms of *Amaryllis* spp. (Moore, 1963).

5.2.0 Cultivars

The cultivars grown in gardens are hybrids of *Amaryllis* or their derivatives. Continuous breeding and careful selection over a long period have resulted in production of the present-day cultivars with numerous shades of the finest and bright colour and also possessing other desirable characters like longer flowering duration and increased flower size, *etc*.

Barberton : Dark rose-pink flowers (Brickell, 1996)

Cape Town : Deep rose-red flowers

Hathor : White flowers, bud pink in colour

Johannesburg : Free flowering, flowers pale-pink

Kimberley : Deep carmine-pink flowers with white centres

Some important cultivars of *Hippeastrum hybridum* are as follows:

Apple Blossom

Flowers white with dawn-pink. The lower petals are lighter in colour, almost white with dash of dawn-pink. The stamens are white and the throat is faintly greenish with a dark red ring.

Bahia

Plants moderately tall, scapes 60-65 cm tall bearing 4 flowers per scape, pedicels are 3.5-4.5 cm long. The flowers with luminous, crystalline character, 12 cm long, laterally and dorsoventrally they are 16.5 cm and 18.5 cm wide, respectively. The flowers are red and white with a red picotee, red occurring as dense striations on the tepals. Stamens and styles red.

Beautiful Lady

Light mandarin-red flowers. The petals have azalea-pink in the middle. The stamens have some colour shade, whereas the throat is somewhat darker.

Bouquet

A unique salmon colour, with begonia-pink and a dark throat. Stamens of the same colour shade.

Cultivars of Amaryllis

Capetown

Glory Red

Green Dragon

Johannesburg

Pink Glory

Red Lion

Green Valley

Benito

Velvet Nymph

Cultivars of Hippeastrums

Apple Blossom

Christmas Star

Flamingo

Liberty

Picotee

Star of Holland

Red Pearl

Red Peacock

Pink Rival

Ruby Star

Opal Star

United Nation

Daintiness

Flowers are porcelain-rose in colour with a crimson-rose glow towards the centre. The throat is veined with a soft green, the stamens being again porcelain-rose.

Fire Dance

An attractive Dutch vermilion with dark violet-red in the throat. All the broad and flat petals are of self colour and recurve symmetrically. The blooms are nearly 25 cm in diameter.

Front Page

Flowers are solid scarlet with scarlet stamens and a darker blood-red throat.

Ludwig's Scarlet

Flowers magnificent dark red strongly accentuated in the throat. The four blooms stand proudly on robust scapes. The inner petals are slightly bearded.

Pink Favourite

Colour is the camellia-rose. The tips of the petals are lighter rose, imparting a special freshness to the flowers. Towards the centre of the flowers the colour is darker rose. The stamens are rose as well. Flowers red-margined white, 12 cm across.

Salmon Tower

Flower colour light mandarin-red with a suggestion of azalea-pink towards the middle of the petals with stamens of the same colour shade and darker in the throat.

Scarletto

A sparkling bright scarlet with a very dark red throat of a solid blood-red and with stamens of the same colour shade.

United Nations

Ideal colour combination of pure white petals with vermilion stripes on each of the six petals, though least on the lowest petal, and slightly green in the throat.

Rio

Moderately tall, scape 44-58 cm tall; 4 flowers with 3.5-4.5 cm long pedicels, flowers funnel-form, 15-16 cm long, 17 cm wide laterally, 19 cm wide dorsoventrally. Tepals deep pink and overlain silvery white stripe. Stamens and styles deep pink.

Sampa

Semi dwarf plants, 38-52 cm tall scapes bearing 5-8 flower/scape, pedicels 5-6 cm long; flowers broadly funnel-form, 13-14 cm long, 13 cm wide laterally 15 cm wide dorsoventrally. Three upper tepals white striated red-purple with a white median keel. The lower, lateral, outer tepals white striated with red-purple in the upper half and have a lower white zone. Stamens and styles white.

T.S.S. No.1-Pink Pearl

It is a double-flowered and fragrant Amaryllis cultivar (Liu and Yeh, 2015). It was a cross between 'San Remo' (Single-flowered, female) x 'Blossom Peacock' (Double-flowered, male). Female parent has showy white and pastel-colored flowers, linked with fragrance (Meerow, 2009, 2000). It is selected for cut flower or landscape use in subtropical and tropical areas.

Boca (Meerow, 2014)

A new and distinct triploid (2n=33) Hippeastrum interspecific hybrid. The flowers are densely striated purple pink (red purple RHS 57A) on their inner surfaces, with an irregular marginal white RHS 155D picotee and a faint white RHS 155D keel on the outer lateral segments. It was selected for its "butterfly" floral shape and unique color pattern. It shows excellent heat resistance, grown under 50 per cent shade, and resistance to Red Scorch fungus (*Staganospora curtisii*).

Miami (Meerow, 2014)

It was developed from the cross Hippeastrum Mount Blanc x (*H. ambiguum* x *H. papilio*) x *H. Bouquet*. It is new amaryllis with novel floral form and coloration and heat tolerance. It was selected for intense orange red color contrasted by a broad median white stripe on the tepals. It has excellent heat resistance, grown under ambient conditions and 50 per cent shade, and resistance to Red Scorch fungus (*Staganospora curtisii*).

Orlando (Meerow, 2014)

A new and distinct aneuploid (2n=43) moderately tall Hippeastrum hybrid from the cross [(*Hippeastrum ambiguum* x *H. papilio*) x *H. Brasilianum*] x Hippeastrum Wonderland. It was selected for large flowers with deep pink-purple coloration contrasted by a broad median white stripe on the tepals. It has excellent heat resistance, grown under ambient conditions, under 50 per cent shade, and resistance to Red Scorch fungus (*Staganospora curtisii*).

Tampa (Meerow, 2014)

Its bulbs are 6.0 – 9.7 cm in diameter, producing three to five offsets annually. Leaves are located, six to eight per bulb, 45 – 55 cm long, 4.2 – 4.7 cm wide at the base, 6.5 – 7.1 cm wide at the midpoint, and 4.0 – 4.3 cm wide 5 cm from the acute apex.

Jax (Meerow, 2014)

Its leaves are locating with obtuse apices and entire margins, four to six per bulb. Flowers are trumpet-shaped, slightly nodding. The six tepals are undulate along their margins, of medium texture, appear luminous and crystalline, the adaxial surface lightly striated amethyst on a white background, most prominently on the lateral inner tepals, with a picotee of the same color and a green throat.

Opal Star (Ziv et al., 2012a)

The new Hippeastrum originated from a cross WT30 (wild-type *Hippeastrum papillio*) x 97/4 (interspecific Hippeastrum hybrid). It is characterized by star shaped flower, measuring 16 – 18 cm in length and about 12 – 16 cm in diameter, and with a unique color pattern of red (RHS 42 B) with yellow-greenish (RHS 146 D) stripes, and dark red (RHS 46A) veining.

Ruby Star (Ziv et al., 2012b)

A new Hippeastrum originated from a cross WT30 (wild-type *Hippeastrum papillio*) x 97/4 (interspecific Hippeastrum hybrid). It was selected for single flowering It is characterized by star shaped flower, 14 – 17 cm in diameter, and with a unique, color pattern of primarily dark red (RHS 53B) with a yellow-greenish (RHS 145B) stripes and greyed-purple (RHS 183B) veining.

6.0 Genetics and Breeding

6.1.0 Genetics

A number of *Amaryllis* species was studied cytologically by Inariyama (1937), Sato (1938), Baldwin and Speese (1947), Neto (1948), Ficker (1951), Schmidhauser (1954), Sharma and Jash (1958), Traub (1958), Narain and Khosoo (1968) and Burnham *et al.* (1971). The plants studied by these workers had, in every case, a basic chromosome number of eleven (x= 11) which forms a series from 2x through 3x, 4x, 6x or 7x. These investigations are likely to include *Hippeastrum* species also.

Flory and Robert (1981) found karyotype in *A. striata* (tetraploid) with exceptions, in the proportion of one different type of chromosome. Most of the *Hippeastrum* species were diploid (2n=22) but *H. blumenavia* was found to contain 2n=20 chromosomes which supported its segregation from the genus as *Griffivia blumenavia* (Arroyo, 1982). The basic chromosome number of Hippaestrum species is *x*=11 and all species show a constant bimodal basic karyotype of four short [metacentric (m) or submetacentric (sm)] and seven long [four sm and three subterminal (st)] chromosomes. Polyploidy is the only conspicuous change in the karyotypic constitution, with triploids, tetraploids, and pentaploids being known (Naranjo and Poggio, 1988; Poggio *et al.*, 2007). In another paper, Poggio *et al.* (2007) presented the karyotype and DNA content of 12 diploid species of *Hippeastrum* from South America. The variation in genome size is compared with the karyotype and DNA content of *Amaryllis belladonna* from South Africa. The *Hippeastrum* species present a uniform and bimodal basic karyotype formula, but significant differences are found in the total chromosome volume (TCV) and nuclear DNA content. A positive correlation between the DNA content and TCV is also observed. The karyotype's constancy is a product of changes in DNA content occurring in the whole chromosome complement. The DNA addition to the long and short sets of chromosomes varies independently. In species with higher DNA contents, the short chromosomes add equal DNA amounts to both arms, maintaining their metacentric morphology, whereas the long chromosomes add DNA only to the short arm, increasing the chromosome symmetry. These data show that the evolutionary changes in DNA amount are proportional to chromosome length, maintaining the karyotypic uniformity *A. belladonna* has a larger DNA content and possesses a karyotype different from that of *Hippeastrum* spp., supporting the distinction between the two genera and upholding the name *Amaryllis* for the South African entity against *Hippeastrum* for the South American genus.

Hybridization has been one of the important single factors involved in the origin of garden amaryllis. The basic point here is that hybridization has been between taxa which was compatible and where good deal of gene exchange was possible. Reginae and Leopoldii group hybrids were evolved in this manner.

There is marked self-sterility in diploid *Amaryllis* species (2n = 22), but the polyploids tested are self-fertile (Traub, 1958). According to Cage (1975), sterility can successfully be eliminated by resorting to either siblings' crosses or back-crosses on good ancestors, rather than continuing selfing. Use of certain chemicals like gibberellins has also been found useful.

The *Amaryllis* species offer excellent opportunities for the study of inheritance patterns. Although there is some variability in species material, plants from wild sources are usually uniform in genetic characters and some predictions can be made of traits expected in the primary crosses. With a record of the pedigree of species hybrid, predictions can be extended to the following generation with some degree of accuracy.

Most floral colours in amaryllis have been reported to be determined by the presence or absence of 2 anthocyanin pigments (Harbome, 1967; Love *et al.*, 1971). In diploid species and their hybrids, cyanidin (pink to rose) appears to be dominant over pelargonidin (salmon to scarlet). Expected classic ratios were obtained in crossing diploid amaryllis with these floral colours. Using pollen of a back-cross, *H. reticulata* × 'Mrs. Garfield' progeny produced from a salmon-flowered seed parent, was pink or salmon in a 1: 1 ratio. A sibling of this pollinator could yield all pink flowers or the same ratio in this cross. Since 'Mrs. Garfield' is a hybrid of *H. reticulata,* it appears that one of the other species in the background of this hybrid had the genes for salmon flower. Yellow or green is the result of a different pigment, plastids. Four species have been of interest in breeding for yellow flowers, *H. evansiae, H. aglaiae, H. parodii* and *H. anzaldoi.* Progeny from crosses among the above had flowers of pale-yellow colour. The yellow was marked in crosses of these species with others containing anthocyanins in the epidermis. Preliminary evidence from crosses made by several breeders suggested that yellowness vs. greenness is not a simple dominant recessive inheritance pattern. Hybrids of *H. evansiae* × *H. cybister* generally had an overall colour effect between the parent species. Among species to be considered would be *H. fosteri* and the others with bright yellow throat (Bell, 1977).

The white midrib leaf stripe of *H. reticulata* var. *striatijolia* is also inherited as a single dominant gene. One of the few species to produce seeds by self pollination, a selfed plant yielded progeny in the ratio of 3 striped: 1 plain green (Bell, 1978).

Higher floret number appears to be dominant, but not inherited as a simple trait in the classic ratios. Useful here as breeding materials for high floret number are such species as *H. cybister, H. fosteri, H. angustifolia* and *H. reticulata.* Even the second-generation hybrids of *H. cybister* tend to retain the high floret number. A trait considered undesirable for pot plants is the long scape length found in species, such as *H. cybister, H. angustifolia* and *H. fosteri.* The hybrid *H. evansiae* × *H. cybister* had tall scapes of *H. cybister* but succeeding generations of these varied in scape length. Again, no clear pattern of inheritance was noted.

Flowering season also appears to be controlled by multiple factors. But, with pollen storage and the occasional off-season bloom, hybrids can be made with the few which are autumn flowering with the vast majority which flower during the spring months. Hybrids of *H. reticulata* are useful for autumn flowering (Traub, 1958), but seem to inherit the lack of hardiness of this species.

Transgressive segregation seems to be one of the important ways by which flower size was increased considerably. Fragrance is slight or entirely lacking in most amaryllis. This is particularly true of the Dutch hybrids. Some species like *H. brasiliana* and *H. fragrantissima* and hybrids do have fragrance (Cothram, 1979). Narain and Khoshoo (1977) reported that fragrance was associated with white colour in many-flowered *H. tucumann* and 3-flowered *H. viridiflora*. Doubleness also occurs very seldom, but one of the *H. belladonna* variations, *H. belladonna florepleno* became completely double (Cothram, 1979).

6.2.0 Breeding

The improvement of amaryllis and hippeastrum depends on various characters from breeders' point of view. To qualify for acceptance a new hybrid clone has to confirm to or surpass some very stringent criteria. According to Barnhoorn (1976), following points to be taken into account are number of blooms per stem, number of stems per bulb (size of bulb taken into account), length of stems, time taken from potting till flowering (time of potting taken into account, *viz.*, early or late in season), clarity/pureness of colour as sought, shape and general appearance of blooms, size of individual blooms, public acceptance of colour, sturdiness of stems, aspect of blooms in relation to stem and to other blooms on the same stem, ease of and dependability in forcing (important for commercial pot plant growers), size of bulb needed to produce 2 good stems, time-lag between appearance of first and second flower stem, lasting time of blooms, whether plant produces foliage together with or after flowers, ability of bulbs to reproduce fast by vegetative means naturally (*i.e.*, number and size of off-shoots formed per year), growth rate of bulb per year, resistance to diseases, shape and general appearance of the bulb, condition of root system, ability of bulb and roots to stand up to long periods of storage under controlled temperatures.

According to Meerow (2009), quoting Traub (1934), Hippeastrum 'Johnsonii', generally acknowledged as the first amaryllis hybrid, was a primary hybrid of *H. vittatum* and *H. reginae* made in England in 1799. The historical concentration of commercial breeding efforts among the various populations of tetraploids may exist for several reasons: (1) desirable characteristics of flower size, scape number, and plant vigor are already stabilized in the hybrid races; (2) sterile triploid progeny result when diploid species are crossed with tetraploid hybrids (3) many of the diploid species are not readily available; and (4) self-incompatibility, which occurs in most diploid species and diploid hybrids, generally breaks down in the tetraploid hybrids, thereby allowing breeders to obtain a segregating F_2 generation.

Fragrance has not been a major target of amaryllis breeders who traditionally select for color characteristics (Meerow, 2009). Introducing fragrance into non scented varieties has the potential to increase demand for amaryllis plants and add to their value as cut flowers as well. Later Merrow *et al.* (2017) reported the analysis of floral fragrance volatiles from two scented amaryllis species, *Hippeastrum brasilianum* and *Hippeastrum parodii*. Whereas the headspace of *H. brasilianum* is dominated by a large peak of (Z)-b-ocimene (88 per cent of total peak area), *H. parodii* has two large peaks comprising 78 per cent, identified as eucalyptol (1, 8-cineole, 30 per cent) and (Z)-b-ocimene (48 per cent). The two species also differ in other constituent compounds. Overall, *H. parodii*, though producing less total volatile organic compounds (VOCs), has a more diverse bouquet. This species also exhibits an inverse pattern of emission between eucalyptol and b-ocimene, both daily and across the 4-day period from anthesis to senescence of the flower. They compare a complex amaryllis hybrid and a bat-pollinated species, *Hippeastrum calyptratum*. The hybrid had a very different complement of emissions. The bat-pollinated species shared some of the same constituent volatiles as *H. brasilianum* and *H. parodii*, but at lower percentages, and emitted a more diverse assortment of compounds. Meerow *et al.* (2017) concluded that both *H. brasilianum and H. parodii* attract lepidopteran pollinators but suggest that *H. parodii* may also attract other insects.

To conserve the various characteristics, germplasm collection is a major part for improvement.

6.2.1 Hybridization

Selection of parents having desirable characters and process of selfing and crossing is necessary for raising the hybrids. There are two methods of production of F_1 hybrids: (i) line hybrid method, and (ii) group hybrid method.

6.2.1.1 Line Hybrid Method

This method is ideally suited for obtaining superior and uniform seedlings which possess hybrid vigour. Inbred lines are increased and maintained, and the crosses of superior lines are made on large scale. The heterosis as applied to the line breeding should not only include more vigorous growth but result in large blooming size of bulbs along with flower size, colour intensity, duration of bloom, *etc.*

6.2.1.2 Group Hybrid Method

In this method different hybrids and species are collected and assessed in observation rows. The superior ones are selected from germplasm reservoir. The crosses are made among species and hybrids, and promising ones are selected for garden use. As a result of this method, two outstanding hybrid groups namely, Reginae and Leopoldii have been evolved (Traub, 1958; Bell, 1977; Dix, 1977; Prakash, 1981). Pollens mature a number of days before the stigma of the same flower is ready to receive it. Storing of pollen is essential to make crosses. When the pollens mature, the two lobes of each anther burst open. The anther shrinks and the pollen grains are fully exposed on the surface. These may be collected in test-tube or petridish. The anthers with exposed pollen are very carefully removed from the filaments by means of metal forceps and are placed in a petridish. The pollens from one flower thus collected should carefully be preserved in the closed petridish in a cool dry place.

In controlled pollination when it is desired to exclude unwanted pollens it is necessary to remove the anthers of the flower (emasculation) at the time of flower opening. A thin cloth-bag or perforated polythene-bag may be placed over the emasculated flower and tied with a string around the base to prevent the entrance of insects that might carry unwanted pollen to the stigma. When the stigma becomes receptive in 3 days from flower opening the bags should be removed from the emasculated flower and the selected pollen applied to the stigmatic surface. This may be done with the help of a fine camel-hair brush. A little quantity of the stored pollen may be collected at the tip of the brush and the pollens gently brushed on the surface of the stigma. A label indicating the names of the parent should be used and the bag replaced over the pollinated flower and tied as before. As soon as the flower starts withering, the bag should be removed permanently. Success in crossing indicates the swelling of ovary and formation of fruit (a 3-celled capsule) which ultimately ripes within 20-25 days after pollination. When the fruits ripe they become yellowish in colour. Care should be taken to collect the mature seeds, otherwise seeds shatter after the fruits burst open. The flat, winged seeds should be air dried. Seedlings are raised by sowing the seeds in beds or pot which usually flower within 2 – 3 years. Meerow (1998) developed few F_3 progenies through reciprocal crosses of *H. lapacense* and *H. cardinasianum*. *H. sryelossum* cv. Surya Kiran was developed from a cross between red and pure white form (Singh *et al.*, 1994; Singh, 1994). The red colour flowers with white strips arising from the throat.

6.2.2 Mutation

The hereditary make up in the cells may change due to crossing over, translocations, and deletions in chromosome number and due to change of gene in quality or kind in chromosomes. Thus, the new variants are created due to mutation.

6.2.2.1 Natural or Spontaneous Mutation

Traub (1953) reported natural hexaploid (6x = 66) mutants in *A. belladonna* and septaploid in *H. blumenavia* (7x = 77).

6.2.2.2 Induced Mutations

Mutation has been derived by use of physical and chemical mutagens.

Physical mutagen: Russel (1969) irradiated seeds of amaryllis Dutch hybrid with 15000 R. The seedlings were divided in two groups. The first lot was sown after irradiation and the second lot of irradiated seeds was further treated with 0.2 per cent colchicine. The seedlings on treatment with 15000 Rand 0.2 per cent colchicine showed curled leaves.

Chemical mutagen: The seeds of *H. evansiae* were placed in a sterile culture, and after germination the seedlings were transferred to aqueuous solution of 0.2 per cent colchicine with 2 per cent sucrose for 24 hours. They were washed with sterile water and transferred to mineral medium solidified with 1 per cent agar. The root tips of seedlings were studied and found to be tetraploid (Bell and Stiles, 1974).

6.2.3 Polyploid Breeding

Polyploidy has been an important means of improvement and some of the species of hippeastrum evolved are tetraploid. Of particular interest are the triploid hybrids which were outstanding and surpassed both diploid and tetraploids in all floral characters (Narain and Khoshoo, 1977). Meerow (2000) obtained three triploids (Rhio, Bahia and Sampa) of *Hippeastrum* through breeding programme. *H. papilio* was crossed with *H. ambiguum*. *H. lapacense* and *H. pardinum* and diploid phenotypes were obtained. Selected diploids were then crossed with commercial tetraploid varieties Apple Blossom, Dutch Belle and White Christmas as pollen parents.

7.0 Growth and Flowering

Growth and flowering of amaryllis and hippeastrum are influenced by several factors. Among the various external factors, light and temperature play important role in the developmental processes of the plants. The March lily (*Amaryllis belladonna*) is a late summer flowering bulbous plant of South Africa. The inflorescence bears 2-12 soft pink fragrant funnel-shaped flowers on 'naked' (leafless) stems. It makes an excellent potted plant and fresh cut flower lasting up to a week in a bouquet. The factors influencing the morphological changes to promote flower-bud initiation which include physical traits of bulb size, age and dormancy; planting establishment with relevance to methods and bulb disturbance and the environmental factors of temperature, watering, day length and season (Wilmot and Laubscher, 2019).

7.1.0 Environmental Factors

7.1.1 Light

Treatment with long day was found to improve the growth and flowering. Bose *et al.* (1981) recorded that the number and length of leaves were maximum in the long day (16 h) treated plants while least number of leaves were produced under short day (9 h) condition. Increased day length promoted the number, weight and diameter of bulblets, while plants under short day failed to produce bulblet and flowers (Table 1). Similarly, Hayashi (1974) observed slight reduction of growth under short day.

Intensity of light also affected growth and flowering. It is commonly observed that when plants are grown in shady places, the number and size of flowers are considerably reduced, while flowering is improved in higher light intensity. Huh *et al.* (1998) observed that high light intensity (full sunlight) enhanced the quality of flowers, but shortened the length of the flower stalk and reduced flower longevity in hydroponically grown bulbs of *H. hybridum* cv. Red Lion. In an experiment on hippeastrum, high light intensity between 20000 and 25000 lux markedly increased the number of leaves and number and weight of bulb lets per plant compared with lower light intensities. Higher percentage of plants produced flowers under this treatment. Reduction of light intensity adversely affected flowering and bulblet formation and up to 2000 lux suppressed flowering (Jana and Bose, 1987).

Table 1: Effect of Day Length on Flowering of *Hippeastrum hybridum**

Treatment	*Days to Flower Appea-rance*	*Number of Flower Stalk per Plant*	*Length of Flower Stalk (cm)*	*Number of Flower per Plant*	*Diameter of Flower (cm)*	*Length of Flower (cm)*	*Life of Flower (Day)*
Long day	309	1.5	58.0	4.6	16.6	15.9	5
(16 h light)							
Natural day (light period varied between 10 h 32 min and 13 h 28 min.)	312	1.6	47.2	5.0	15.2	13.3	6
Short day (9 h light)	-	-	-	-	-	-	-
CD. at 0.05 per cent		N.S.		N.S.	1.14	1.30	
CD. at 0.01 per cent		-		-	N.S.	1.79	

* Bose *et al.* (1981).

7.1.2 Temperature

For the best results, amaryllis should be grown under conditions where the night temperature does not fall below 60°F (15.6°C). The minimum day temperature should range somewhat higher. The flowering of *Hippeastrum* is regulated by bulb size, and the plant does not require low temperatures for flower initiation. However, since floral initiation alternates with leaf formation, flowering may become poor when the plant is grown at high temperatures (23-27 °C) (Inkham *et al.*, 2019). The effects of storage and growth temperatures on growth, flower quality, and bulb quality of *Hippeastrum* sp. were studied (Inkham *et al.*, 2019). Bulbs of *Hippeastrum* sp. 'Apple Blossom' with circumferences in the range of 22-24 cm were selected and stored for 12 weeks at one of three different temperatures (*i.e.*, room temperature [RT: 30±2 °C], 13±2 °C, and 15±2 °C). After storage, the bulbs were grown in a growth chamber at one of three growth temperatures-that is, RT (30±2 °C), 15±2 °C, and 25±2 °C. The results showed that storing bulbs at RT for 12 weeks reduced their quality in terms of fresh weight, diameter, and firmness, whereas storing bulbs at 13 °C or 15 °C retained their quality in terms of fresh weight and firmness. Electron scanning imaging of the starch granules in the bulb scales showed that prior to storage, the leaf base cells were filled with large starch granules of either lenticular shape or spherical shape. After storing for 12 weeks at RT, the starch granules in the bulb scales were sharply reduced in number and were predominantly of the small, spherical form. A growing temperature of 25 °C could stimulate better growth with respect to plant height, flower diameter, bulb diameter, bulb fresh weight, and bulb dry weight, compared with RT (30 °C). A significant interaction between storage and growing temperatures was found for all parameters except for stalk length. A bulb storage temperature of 15±2 °C for 12 weeks combined with a growing temperature of 25±2°C produced the best results for both flower and bulb production (Inkham *et al.*, 2019).

Traub (1958) suggested the temperature range from 68° – 80°F (20-26.6°C), under greenhouse culture for both bottom heat and air temperature. Rapid temperature changes should be avoided. When flowers begin to open, the plants are transferred to the living room with floor temperature of 72 – 75° F (20-24.8°C) and air temperature of 68 – 72°F (20-22.2°C). In India, amaryllis grows best in outdoor conditions at the temperature of about 25°C and relative humidity ranging from 50 – 90 per cent. However, it can be grown up to 35°C. Kurki (1958) found that the flower buds initiated at 15-17°C. Hayashi and Suzuki (1970) observed that the best growth of amaryllis was at 23°C day and 18°C night temperature. The highest day and night temperature of 28/23°C promoted leaf growth but inhibited the bud development and flower bud initiation. One year old, 2-leaved seedling of hippeastrum cv. Star of Holland when exposed to 24°C day and 17°C night temperature produced intermediate vegetative growth and enlargement of mother and daughter bulbs and was optimum for flower stalk elongation and flowering (Ijiro and Ogata, 1997). Huh *et al.* (1998) suggested that for flower induction of *H. hybridum* cv. Red Lion, grown in hydroponic vessel culture, the bulbs with diameter less than 7 cm and those above 7 cm be exposed to 5°C for 60 and 45 days, respectively. Hong (1970) reported that the bulbs of hippeastrum stored at 30°C for 10 days resulted in early flowering while 40°C was less effective. Huh *et al.* (1998) reported that only the bulbs of *H. hybridium* cv. Red Lion grown in hydroponic vessel culture with diameter above 10 cm produced flowers after dry storage at 30°C for 30 days. It was recorded by Hayashi (1972) that the growth of leaves and roots and thickening of bulbs in seedlings were best at 20°C compared with lower temperature. Bose *et al.* (1979) reported that bulbs stored at 5, 10, 17 and 30°C for 30, 60 and 90 days showed interesting variation in subsequent growth and flowering. Irrespective of the duration of storage, bulbs stored at 30°C developed maximum number of leaves and bulblets. The number of flower stalk on each plant, length of the stalk and number of flowers were markedly stimulated under higher storage temperature. It also increased the size and life of flower. Adverse effect of storage temperature on growth and flowering was observed with the decrease in temperature. Reduction in growth and flowering was also recorded with longer duration of storage. The adverse effect of low temperature storage was very much pronounced when the bulbs were stored for 60 and 90 days at 5°C (Table 2).

7.2.0 Growth Regulators

Soaking of hippeastrum bulbs in 3 concentrations each of IAA, GA_3, CCC and ethrel showed various responses to growth and flowering (Table 3). IAA increased the weight and number of bulblets while GA_3 enhanced the flower diameter and bulb weight. CCC (1000 ppm) increased the number of flowers per plant. IAA at 100 ppm and GA_3 at 10, 100 or 1000 ppm twice as foliar spray at an interval of 15 days promoted the number of bulblets on the treated plants while high concentration of CCC and ethrel (1000 ppm) increased the weight of bulb lets. All concentrations of IAA, GA_3 and 1000 ppm CCC increased the number and size of flowers (Bose *et al.*, 1980a).

Table 2: Effect of Temperature and Duration of Storage of Bulbs on Growth and Flowering of *Hippeastrum hybridum**

Treatments	*Height of Plant (cm)*	*Number of Leaves*	*Number of Bulblets*	*Weight of Bulblets (g)*	*Diameter of Bulblets (cm)*	*No. of Flower per Plant*	*Diameter of Flower (cm)*
Temperature							
30°C	47.6	4.8	1.3	14.3	2.3	4.5	13.0
17°C	42.0	4.2	0.9	11.3	2.1	2.8	12.4
10°C	40.8	3.8	0.8	9.3	1.9	2.7	12.4
5°C	39.8	3.6	0.7	6.3	1.4	1.5	11.4
C.D. at 5 per cent	4.67	N.S.	N.S.	2.04	0.54	0.47	0.74
Duration of storage							
30 days	44.7	3.8	1.5	15.8	3.6	3.4	12.9
60 days	42.8	4.0	0.7	9.5	3.4	3.2	12.8
90 days	40.1	4.4	0.3	5.8	3.1	2.0	11.3
C.D. at 5 per cent	N.S.	N.S.	0.39	1.76	N.S.	N.S.	0.64

* Bose *et al.* (1979).

Table 3: Effects of Soaking Bulbs of *Hippeastrum hybridum* with Regulators on Growth and Flowering (Non-significantly differing parameters omitted)*

Treatments	*Regulator in ppm*	*Number of Leaves*	*Weight of Bulb (g)*	*Weight of Bulblet (g)*	*Number of Flowers per Plant*	*Diameter of Flower (cm)*
Control		3.6	32.0	10.1	2.0	9.2
IAA	1	3.3	42.0	13.0	2.7	10.4
IAA	10	3.3	43.4	16.0	2.8	11.8
IAA	100	3.3	48.3	17.5	3.3	11.8
GA_3	10	4.2	49.0	14.4	3.0	10.6
GA_3	100	4.2	50.3	18.4	3.0	10.9
GA_3	1000	4.7	51.6	24.0	3.3	12.5
CCC	10	3.7	36.5	15.5	2.5	10.4
CCC	100	3.8	37.0	16.5	2.5	10.4
CCC	1000	4.0	45.0	20.0	3.8	10.9
Ethrel	10	3.5	32.0	10.1	2.0	9.9
Ethrel	100	4.0	40.0	11.5	2.8	10.0
Ethrel	1000	4.0	54.5	15.0	2.8	11.0
C.D. at P=0.05		0.80	3.10	1.79	1.32	1.43
C.D. at P=0.01		1.06	4.10	2.37	1.75	1.89

* Bose *et al.* (1980a).

Bulbs treated with DCPTA (2,3,4-dichlorophenoxytriethylamine) at 30 µM reduced the number of emergent inflorescences and bulb biomass increase, with no significant change in days to first flowering or number of flowers produced per scape as compared to those untreated in hippeastrum F_1 hybrid H-8-14 (*H. cardenasianum* × *H. papilio*) (Meerow *et al.*, 1994).

It was also reported that spraying *Hippeastrum vittatum* plants with kinetin at 60 ppm and mineral fertilization with N:P:K at the ratio of 2:1:2, respectively induced prospective effects on vegetative growth and flowering parameters with higher bulbs productivity (Youssef and Abd El-Aal, 2014).

8.0 Propagation

Plants of both the genera, *Amaryllis* and *Hippeastrom* can be propagated from seeds and by vegetative methods. Seed propagation is used by breeders for development of hybrids while to maintain the purity of a cultivar or sterile hybrid clones, separation of offset is usually practised.

8.1.0 Seed Propagation

Amaryllis is cross pollinated in nature and after pollination and fertilization the ovules gradually develop into mature seeds in the ripe capsule. When mature, the capsule bursts and exposes the seeds. The seeds are air dried for 24 – 48 hours at room temperature if they are to be stored. However, the best germination is obtained when the seeds are sown in bed immediately after harvesting. The seeds in unsealed envelopes or open containers retain viability for a short period in humid climate (Traub, 1953) and for longer period in dry atmosphere (Brown, 1952). Seeds in sealed containers may be stored at a temperature of 4.4° – 7.2°C (40° – 45°F). The seeds may be sown in pots, wooden boxes, raised beds in half shade and in glasshouses. Seeds usually germinate within 7 – 10 days after sowing and seedlings are ready for transplanting in pots or beds in 2- to 3-leaf stage. Sometimes parthenogenetic seeds develop due to spontaneous growth of the female germ cell (Hannibal, 1955).

Amico *et al.* (1994) reported that stratification of seeds of *A. belladona* for 30 or 60 days at 4°C increased the germination speed (9.1 and 15 days, respectively) and emergence (8.3 and 20.4 days, respectively) but showed a decreasing trend in germination and emergence percentage. They found a positive corelation between seed size and germination percentage.

Successful attempts to culture ovules and immature seeds were also made by Bell (1972, 1973). Usually, tetraploid × diploid crosses are made for these purposes.

8.2.0 Vegetative Propagation

Multiplication of plant from seed will show wide variation in flower colour whereas separation of bulb lets is a very slow process for obtaining a large number of plants in a short time. Normally, a plant produces 2 – 3 bulblets in a year of growth. A number of experiments have been conducted in different parts of the world to increase the number of plantlets vegetatively. Plants may be raised from scale section, scale-stem section, and stem sections of bulbs. Andrade-Rodriguez *et al.* (2015) lists four methods of propagation of *Hippaestrum*. The first method

Propagation of Amaryllis and Hippeastrum

Seed Pod of Amaryllis

Germination of Seeds

Bulbs of Amaryllis

Division of Bulbs

Separation of Bulb

Pot Planting

Planted in Pot

Growth of Bulbs

Micropropagation

Plant regeneration in *Amaryllis belladonna* L.: A. Cultured bulb scale after 3-5 d; B. Proliferation of single shoot after 7-10 d; C. Formation of double shoots after 10-14 d; D-E. Formation of multiple shoots after 10-14 d; F. Callus formation after 10-14 d from bulb scale explant; G. Shoot initiation from callus of bulb scale; H. Multiple shoot(s) formation from cultured bulb scale I. Callus formation from mature embryo explant; J. formation of globular stage somatic embryo from cultured mature embryo; K. Callus formation from leaf disc explant; L. Formation of shoot(s) from cultured leaf disc; M-N. In vitro rooting; O. Regnerants transferred in Environmental Growth Cabinet after 40-45 d and P. Regenerants transferred in Net House for hardening (Source: Veeraballi et al. (2017)).

is using seeds, the most efficient propagation method of bulbous plants whose vegetative propagation is slow is usually used to develop new varieties due to the large variation in the characteristics of flowers, plant form and flowering time; the second method is dividing bulbils and this method is used for cultivars that produce at least three bulbils; the third method is twin cataphylls used in most propagation cases propagation *in vitro* is the fourth method that has been used successfully for bulbous and in particular for *Hippeastrum*, however, to use this method fancy facilities and equipment are required. Amaryllis (*Hippeastrum* × *johnsonii*) is commonly propagated by three methods: seeds, offset bulblets, and twin scaling. Sectioning the bulb into 8 segments was the best method for producing an acceptable number of vigorous bulblets (Kharrazi *et al.*, 2017b). Twin scales prepared from outermost layers of mother bulb had greater areas and so had more stored nutrition. Therefore, their capacity for producing plantlet was higher in comparison to innermost layers. For obtaining the best quality of produced bulblet, use of the outermost twin scales is recommended (Kharrazi *et al.*, 2017a).

8.2.1 Bulb Scale Sections

This is a method of separating and sprouting of bulb scales which have the power of regenerating bulblets in a region from the base to about 0.5 cm above it. The scales are placed in slightly slanting position in moist fine sand and at a constant temperature of about 30°C. After three months the plantlets from scales are lifted for planting (Luyten, 1926; Alkema *et al.*, 1977).

In order to evaluate different cutting methods of the mother bulb to induce the formation of bulbils and determine better method for the propagation of *Hippeastrum*, Andrade-Rodriguez *et al.* (2015) studied nine cutting methods of bulbs (8 cm) of *Hippeastrum hybridum* Hort. as 1) the whole bulb, 2) basal section, 3) cut in basal section, 4) basal section without central bud, 5) cutting basal section in half, 6) basal section into quarters, 7) longitudinal cut into quarters, 8) longitudinal cut into eighths and 9) basal section into quarters, separating each quarter into five groups of twin cataphylls). When the main bud was preserved, no bulbils were formed; when using the basal section: eliminating the central bud, cutting in half, and cutting into quarters, 5-6 bulbils were obtained; when the entire bulb was used and it was cut longitudinally into four and eight sections, 10 and 13 bulbils were obtained, respectively. In contrast, when the basal section was cut into four parts, separating each quarter into five twin-cataphyll groups, 25 bulbils were obtained. It was concluded that *Hippeastrum* propagation was greater when the basal section of the mother bulb was cut into quarters and each quarter was separated into 5 twin-cataphyll groups.

The effects of different media, temperatures and plant growth regulators on the *ex-vitro* propagation of twin-scale cuttings as well as on the relationship between starch degradation and bulblet development of *Hippeastrum vittatum* cv. 'Red lion' were studied (Zhang *et al.*, 2013). Sawdust was the most optimal medium for the propagation of twin-scale cuttings resulting in the lowest percentage of rotting scales and the best quality bulblets. Furthermore, 25°C was a more suitable temperature than 30 or 20°C. Gibberellic acid (GA_3), indole-3-butyric acid (IBA) or

1-naphthyleneacetic acid (NAA) did not result in scale propagation and bulblet development. In particular, NAA treatment significantly increased the frequency of rotten scales. During bulblet formation and development, starch content in the mother scales declined earlier and more rapidly, especially in the outer scale of twin scales. The culture of scales at 30°C resulted in a distinct increase in the consumption of starch in the scale. The starch content in bulblets increased during development. Starch accumulation in bulblets at 25°C was most remarkable at the last stage of development. The dominant soluble sugars in the bulbs of *H. vittatum* were fructose, sucrose and glucose, fructose content being highest in the mother scales. However, the key soluble sugar in the bulblets was sucrose and its content increased linearly over the first 50 days (Zhang *et al.*, 2013).

8.2.2 Fractional Scale Stem Sections

Development of plantlets from small sections of scale and stem has been reported by a number of workers (Traub, 1934; Heaton, 1934; Fosler, 1953; Bonde, 1955). Selected bulbs of amaryllis or hippeastrum are cleaned by removing the roots, leaves and dry scales. The knife used in the cuttage operation is sterilized to avoid spread of diseases. The bulbs are cut first horizontally keeping about 1.5 cm scale portion and basal stem portion of similar thickness and then longitudinally into small pieces, each containing scale and stem portions. Thus, 40 pieces of scale-stem sections can easily be obtained from each bulb of about 6.0 cm diameter. Cut-pieces are treated with fungicide and planted in vermiculite or sand media for rooting. The cut-pieces become reddish in colour, gradually turn greenish and finally shoot arises from the junction of scale-stem section within 35 – 40 days after planting. Roots come out from the stem portion during or before the appearance of shoot. These plantlets are transferred in pot containing well drained compost. Bose *et al.* (1980b) reported higher percentage of regeneration from smaller sections (36 pieces/ bulb) in vermiculite and it also promoted the growth of the plantlets compared with sand, leaf mould and sand + leaf mould mixture (1:1). Percentage of regeneration up to 24 pieces was more in sand in comparison with sand + leaf mould (1:1) or leaf mould alone. In laboratory and glasshouse trials the bulbs were cut vertically into 8 or 16 parts and set in 1:1 perlite and peat mixture. Bulblets formed more readily in glasshouse than in laboratory (Gyenes, 1975). Treatment with fungicides is very useful in preventing rotting of cut sections. They also noted that treatment with 0.5 per cent solution of ziride proved very effective in preventing rotting of smaller scale-stem sections and also stimulated regeneration of plantlets (Table 4).

8.2.3 Individual Scale-Stem Sections

In this method, the bulbs are cut into scale-stem sections as described earlier. Each section containing about 8 to 10 rows of scales, and a portion of stem is further separated into individual scale section having a small piece of stem. The sections are also treated with fungicide and planted in vermiculite or sand for regeneration. High humidity is maintained to prevent drying of scales. Shoot arises from the junction of the scale and stem within 40 – 60 days after planting depending on the atmospheric humidity as well as the size of sections. Generally, the roots come out from the lower portion of the young shoot. By this way, 120 to 125 plantlets are obtained from a single bulb of about 6.0 cm diameter. Bose *et al.* (1980b) also observed that 78 per cent sections from outer rows of scale-stem produced plantlets, while 61 per cent of the sections from the inner rows regenerated. Earlier appearance of shoot and root was recorded in sections from outer 3 rows of scales than those from the inner rows. Sakanishii (1962) also reported that better generation from outer scales was due to higher sugar content.

Table 4: Effect of Fungicides on the Regeneration of Plantlets from Sections of Bulbs of *Hippeastrum hybridum**

Number of Sections from each Bulb	Treatments					
	Control		Ziride		Unizeb	
	1	2	1	2	1	2
8	100	2.0	100	2.6	100	2.2
12	100	1.5	100	2.2	100	1.6
16	100	1.0	100	1.6	100	1.0
20	100	1.0	100	1.0	100	1.0
24	100	1.0	100	1.0	100	1.0
28	82	1.0	100	1.0	100	1.0
32	21	1.0	100	1.0	37	1.0
36	-	-	58	1.0	-	-
40	-	-	24	1.0	-	-
C.D. at 5 per cent	10.32	N.S.	11.45	N.S.	11.86	N.S

1. Percentage of section producing plantlets; 2. Number of plantlets from each section.

* Bose *et al.* (1980b).

8.2.4 Stem Sections

This method was reported by Klotzbach (1957). In this method, the basal stem of the bulb was sliced off just below the scale and exposed surface was dusted with rooting hormone. After 8-9 weeks from planting, when a ring of tiny bulblets appeared around the base, the mother bulb was set in a gravel hydroponic bed and fed regularly with a standard nutrient solution. This method could not be used for commercial production of bulbs.

8.2.5 Micropropagation

Mii *et al.* (1974) reported organ formation from bulb scale of hippeastrum, *in vitro*, and found NAA and kinetin to be effective in bud and root development. Callus cultures were successfully established from explants of bulbs, floral peduncles of hybrid amaryllis. Callus induction was dependent upon the presence of a synergistic combination of coconut milk and 2, 4-0 in the medium (Bapat and Narayanaswamy, 1976). Successful micropropagation from scale-stem explants was reported by Bose and Jana (1977) using modified Knudson's basal medium and NAA which promoted the growth of root and shoot. GA_3, on the other hand, delayed regeneration and inhibited the growth of plantlets. 2, 4-D at 0.2 and 0.5 mg/l proved to be toxic.

Bruyn *et al.* (1992) reported that twin scale and immature scapes explants could only generate plantlets in *A. belladona* and obtained highest multiplication rate of twin-scales explants in MS medium supplemented with 22.2 µM BA and 0.54 µM NAA and for plantlet initiation 2-3 per cent sucrose concentration produced best results. Prasad and Chaturvedi (1993) induced mature bulb-scale segments (basal) of an *Amaryllis* hybrid to regenerate multiple plantlets in MS medium supplemented with 5 mg BA, 0.5 mg NAA, 15 mg adenine sulphate and 5 mg ascorbic acid/l. The rate of multiplication progressively increased with successive subcultures. After 4 years 10-fold multiplication occurred in 8 weeks as compared to 4-fold initially.

Saker *et al.* (1998) standardized a reliable protocol for *in vitro* propagation of *H. vittatum.* Shoot proliferation was greater from shoot-tips than from bulb-scales when cultured on MS medium supplemented with 4 mg/l NAA and 8 mg/l BA. Multiplication rate varied between multiplication media. Adding adenine sulphate at 60 mg/l to medium containing 4 mg NAA + 8 mg/l BA increased shoot development.

Haung *et al.* (1990) regenerated protocorm like bodies from single-scales of *H. hybridum* cultured *in vitro* which ultimately formed bulblets. Okubo *et al.* (1999) reported that auxin (NAA) had a negative role in bulblet production on single-scales, whereas the basal plate had a significant role as have the outer scales in twin-scale propagation of new plant. The use of antiauxin (naphthylphthalamic acid, TIBA, PCB or morphactin) and scales attached to a portion of basal plate are recommended for *in vitro* bulblet propagation of *Hippeastrum* × *hybridum* cv. Apple Blossom. Twinscale explants of *cv.* San Antonio Rose bulbs responded on MS salt medium supplemented with NAA (2 mg/l) and was successfully transferred to soil (Smith *et al.*, 1999).

9.0 Soil and Climate

9.1.0 Soil

Amaryllis and hippeastrum prefer rich sandy loam or loamy soil with proper drainage and plenty of moisture. The soil pH should range between 6.1 and 7.0 (Traub, 1958). They may also be grown in soils having pH from 6.1 to 7.8 (Bear, 1953; Small, 1954).

Plants of *H. hybridum cv.* Red Lion when grown either hydroponically or in vessel culture filled with peat moss, vermiculite or bark, top growth was better when grown in later than when grown hydroponically, but such media delayed flowering (Huh *et al.*, 1998).

9.2.0 Climate

Both amaryllis and hippeastrurn can be grown under wide environmental conditions ranging from tropical to subtropical or temperate climate. Under extreme conditions, however, the bulbs are lifted from soil and stored. Plants require sufficient light for proper growth and flowering (Doran, 1974). In temperate regions, they are grown under the protection of glass, while in subtropical and tropical regions they produce fine blooms in the open.

10.0 Cultivation

10.1.0 Planting

Before planting of bulbs, soil should be pulvarized and properly manured. Planting should be done from December to January in the plains and in February-March in the hills (Jindal, 1968). Size of bulb, depth of planting and spacing between plants has been found to influence growth and flowering. In order to determine the optimum bulb size, depth of planting and spacing for the production of bulb lets and flowers of *H. hybridum* an experiment was carried out. Among the three sizes, large bulbs (5.1 – 5.5 cm diameter) produced larger number of bulbs lets and flowers than planting at a depth of 5.0 cm below the soil (Table 5). Wider spacing (25 cm apart) improved the production of bulblets and size of flowers. Flower production was improved when large bulbs were planted at a shallow depth and 20 cm apart (Jana and Bose, 1981a).

Table 5: Effect of Bulb Size and Depth of Planting on Flowering of Plants*

Treatments		*Number of Flowers per Plant*	*Diameter of Flower (cm)*	*Length of Flower (cm)*
Bulb size	S_1 (4.0-4.5 cm)	1.4	11.2	12.5
	S_2 (4.6-5.0 cm)	1.7	11.8	12.5
	S_3 (5.1-5.5 cm)	2.6	12.0	12.8
C.D. at P=0.05		0.69	0.57	0.19
Depth	D_1 bulb neck above soil surface	2.1	11.9	12.7
	D_2 bulb neck 5 cm below the soil	1.6	11.5	12.5
C.D. at P= 0.05		0.38	N.S.	N.S

* Jana and Bose (1981a).

10.2.0 Irrigation

Bulbs are usually planted in moist soil. If the soil is dry, watering is done few days before planting in an optimum moisture level. Generally, irrigation is done at an interval of 15-20 days, if the weather is dry. Watering should be done during the period of vegetative growth and it is withheld after flowering till new growth starts. Doran (1974) reported that the plants should go to 40 centibars or more before watering.

10.3.0 Manuring and Fertilization

Proper nutrition is very important for normal growth and development of plants. Nitrogen, phosphorus and potassium markedly influence these processes. In a sand culture experiment on hippeastrum, treatment with high level of nitrogen (3N) caused maximum plant growth and number of flower stalks and flowers. High dose of phosphorus and potassium also proved beneficial. Deficiency of these nutrients resulted in low yield of bulb and reduced flower production. The adverse effects of low levels of nitrogen (1/10 N) and phosphorus (1/10 P) appeared at the early stage of growth and production of flower was severely affected (Table 6). The effect of potassium deficiency, on growth and flowering, however, was less pronounced (Bose *et al.*, 1980c).

Table 6: Effect of Nitrogen, Phosphorus and Potassium on Flowering of *Hippeastrum**

Treatments	*Days to Flower Bud Appearance*	*Number of Flower Stalk per Plant*	*Length of Flower Stalk (cm)*	*Number of Flower per Plant*	*Diameter of Flower (cm)*	*Length of Flower (cm)*
NPK (control)	339	1.2	41.1	2.6	11.9	12.8
(3N) PK	328	1.6	39.8	5.2	12.0	12.8
N (3P) K	335	1.4	47.4	4.0	12.5	13.4
NP (3K)	334	1.4	43.4	3.4	12.0	12.9
(1/10 N) PK	340	0.6	33.9	1.4	9.6	11.1
N (1/10 P) K	348	0.8	27.7	1.4	10.3	12.0
NP (1/10 K)	342	0.8	35.0	1.8	11.5	12.5
C.D. at P=0.05	-	0.69	-	1.49	0.84	1.02
C.D. at P=0.01	-	0.92	-	2.01	1.14	1.38

* Bose *et al.* (1980c).

In *Hippeastrum*, bulbs grew best when fertilized with 25 mM N (as NH_4NO_3) and enriched with CO_2 (carbon dioxide) (Silberbush *et al.*, 2003). *Hippeastrum* bulbs are a strong sink for nutrient accumulation during the initial stages of bulbil development. Once leaves begin to supply assimilates for plant growth, bulbs become a source of nutrients for the developing plant (Stancatoa *et al.*, 1995). When the N concentration was lower than that supplied to the control, the order of N content was leaves> roots> bulbs, but when the N concentration was higher than the control, the order was bulbs>leaves>roots. However, the correct balance of NH_4^+ to NO_3^- nitrogen in a fertilizer mix for *Hippeastrum* plants is largely unknown (Vazquez *et al.*, 2015).

The correct balance of NH_4^+ to NO_3^- nitrogen in a fertilizer mix for *Hippeastrum* plants is largely unknown. Hence, Vazquez *et al.* (2015) applied nitrogen 2 times weekly following irrigation at either 0.6 g (high), 0.3 g (medium) or 0.15 g (low) total N every four months. Nitrogen was supplied in different combinations of NO_3 and/or NH_4. Nitrate: NH_4-N ratios were either 100 per cent NO_3:0 per cent NH_4 (100NO_3), 70 per cent NO_3:30 per cent NH_4 (70NO_3), 50 per cent NO_3:50 per cent NH_4 (50NO_3) (second group only), 30 per cent NO_3:70 per cent NH_4 (30NO_3), or 0 per cent NO_3/100 per cent NH_4 (100NH_4). Growth in bulb diameter after one year of fertilizer treatments not only increased from 0.15 to 0.6 g N (low to high level), but also differed with the form of N supplied to the plant. The largest diameter bulbs were produced in the 70NO_3 and 50NO_3 high N treatments. Within any NO_3/NH_4-N ratio grouping, fertilization at the high N rate resulted in larger diameter bulbs. No significant differences existed between treatments in the number of bulbs produced. Bulb growth was greater with a portion of N supplied as NO_3 than with

Cultivation of Amaryllis and Hippeastrum

Seedlings in Greenhouse

Flower Bud Stage in Greenhouse

Cultivation in Greenhouse

Flowering of Hippeastrums in Greenhouse

Flowering of Hippeastrum in Open Field

NH_4-N alone. These results indicate that application of N as a mixture of NH_4 and NO_3 at 0.6 g per 4 months produces the largest increase in bulb diameter (Vazquez *et al.*, 2015).

Springer John (1937) obtained better size of amaryllis bulbs using lower dose of fertilizer mixture, *i.e.*, 100 lbs of NPK (4:5:5) compared with 400 lbs/acre. In California, application of 25 tonnes of dairy manure before planting amaryllis and a commercial fertilizer dose of 12:16:12 (NPK) at the rate of 2 tonnes per acre after starting a new growth improved flowering, and produced 4–6 flowers on each stalk (Rice, 1955). In Japan, Tsukamoto and Fujioka (1956) applied N, P and K solution to the bulblets of hippeastrum to determine the best fertilizer rates for plant growth. Treatment with 300 ppm each of NPK or 225 ppm N, 150 ppm P and 300 ppm K was found most effective in this respect. Nautiyal and Bajpai (1979) recorded maximum flower production in *A. belladonna* using N, P and K at the rate of 60, 30 and 30 g/m^2, respectively.

In an experiment on the effect of nutrition on amaryllis under field condition, Joiner *et al.* (1966) observed increased number of large bulbs with higher levels of nitrogen and phosphorus. Response of hippeastrum to nitrogen and phosphorus was also investigated by Jana and Bose (1980). Application of 200 kg nitrogen per hectare showed significant increase in weight of bulbs and number of bulblets and flowers. Treatment with phosphorus at the rate of 400 kg significantly increased the number of flower stalks and flowers, but it did not improve the production of bulblets. Low dose of nitrogen and phosphorus inhibited the growth and flowering (Table 7). The maximum yield of bulbs, bulblets and flowers were obtained with a combined application of 200 kg N, 400 kg P and 200 kg K per hectare.

Table 7: Effect of Nitrogen and Phosphorus on Flowering of Hippeastrum*

Treatments (kg/ha)	*Number of Flowers per Plant*	*Diameter of Flower (cm)*	*Length of Flower (cm)*
Nitrogen			
0	2.1	12.0	12.5
50	2.6	12.6	12.6.
100	2.9	12.7	12.1
150	3.0	12.9	13.0
200	3.5	13.1	13.0
C.D. at P = 0.05	0.30	0.40	0.23
C.D. at P = 0.01	0.41	0.56	0.02
Phosphorus			
0	2.5	12.6	12.7
200	2.8	12.6	12.8
400	3.1	12.8	12.9
C.D. at P = 0.05	0.23	N.S.	N.S.
C.D. at P = 0.01	0.32		

* Jana and Bose (1980).

In a similar study Bhattacharjee *et al.* (1982) reported that application of N, P and K appreciably improved the production of flowers, and good result with weekly foliar feeding of nutrients was reported by Tebban (1955).

A fertilizer combination of 4.37 to 8.75 g/m^2 of N and 9.38 to 12.5 g/m^2 of K, applied with the 2 parallel lines of double deep irrigation method, were found to be optimum for good bulb diameter, plant leaf area and dry weight (Alekperov *et al.*, 1995). El-Naggar *et al.* (2009) studied the effect of different growing media (clay, composted leaves and sand + composted leaves (1: 1 v/v) and fertilizer rates of mineral NPK (19: 19:19) at 0.0, 2.5, 5 g/plant and their combined effect on the vegetative growth, flowering qualities, bulbs production and chemical constituents of leaves of Amaryllis (*Hippeastrum vittatum*, Herb.) in Egypt. The highest level of complete fertilizer (5 g/plant) was the best for growth analysis, bulb fresh and dry weight and chemical constituents of leaves. Applying the complete fertilizer of N, P and K (19: 19: 19) at 5 g/lant grown in composted leaves medium or its mixture with sand (1:1 v/v) gave the maximum beneficial effect on the vegetative growth characteristics, flowering and bulbs and bulblets production. The highest increase in total chlorophylls content, N, P and K contents in leaves and total carbohydrates content in bulbs were obtained by using the highest rate of complete fertilizer (5g) with the growing medium of sand + composted leaves (1: 1 v/v). It is recommended that to produce high quality of *Hippeastrum vittatum*, plants for different decorative purposes in landscaping by cultivated them in the mixture of composted leaves medium or its mixture with sand (1:1 v/v) and applying 5 g/plant complete fertilizer NPK (19:19:19) monthly during the growing season.

The greenhouse horticulture industry applies more water, fertilizers, and pesticides per unit of production area than any other agricultural system. Minimizing water and fertilizers requirements for greenhouse production represents a major sustainability challenge, since many growers are faced with higher fertilizers costs, decreasing availability of good quality water, as well as restrictions and regulations imposed by many governments (Rouphael *et al.*, 2016). Implementing nutrient management strategies in soilless culture, which improve water use efficiency (WUE) and limit the loss of eutrophying elements without affecting crop performance, is a priority for the floriculture industry. Hence, Rouphael *et al.* (2016) assessed the effect of two nutrient management strategies, based on electrical conductivity (EC) or nitrate-nitrogen ($N\text{-}NO_3^-$) concentration control on plant growth, ornamental quality, plant–water relations, mineral composition, and WUE of greenhouse Hippeastrum grown in semi closed soilless system. The recirculating nutrient solution was discharged whenever a threshold EC value of 3.0 dS/ml was reached (EC based strategy), or when $N\text{-}NO_3$ concentration decreased below the limit of 1.0 mol/ml (nitrate-based strategy). There were no significant differences in terms of plant growth parameters, stomatal resistance, leaf water relations, and macronutrient composition in plant tissues between the two nutrient management strategies. In the EC and the nitrate-based strategies, the recirculating nutrient solution was flushed 10 and 5 times, respectively. The water loss (WL) and the total water use (W_{use}) in the EC-based strategy were significantly higher by 261.1 per cent and

61.5 per cent, respectively, compared with the N-NO_3 based strategy. In contrast with the EC-based strategy, the adoption of the N-NO_3 based strategy significantly minimized the nitrate, phosphate, and potassium emissions to the environment. The effective WUE of the system (WUES) recorded in the N-NO_3 based strategy was higher by 55.9 per cent compared with the one recorded with the EC based strategy. In contrast with the EC based strategy, the adoption of the N-NO_3 based strategy significantly minimized the nitrate, phosphate, and potassium emissions to the environment. The effective WUE of the system (WUE_S) recorded in the N-NO_3 based strategy was higher by 55.9 per cent compared with the one recorded with the EC-based strategy (Rouphael *et al.*, 2016).

10.4.0 Weeding

Weeding is an important operation in the commercial cultivation of hippeastrum. Hand weeding is laborious and more expensive than chemical weed control. Pre-emergence application of weedone (2,4,5-T) at 10 litres per hectare proved very effective in controlling the weeds (Jana and Bose, 1981b).

11.0 Diseases and Pests

Amaryllis and hippeastrum are attacked by several diseases and pests which cause extensive damage to the plants and bulbs during storage.

11.1.0 Fungal Diseases

Among the different fungal diseases reported from different places, red leaf spot caused by *Stagonospora curtisii* is very common in the USA, Europe and India. The disease appears in the form of spots on the leaves which ultimately turn yellow and fall off (Munjal and Kapoor, 1962; Pag, 1964; Tapio, 1966; Leeuwen, 1974). The disease may be controlled by spraying with 0.75 per cent copper oxychloride or by warm water treatment of bulbs at 43.5°C for 2 hours (Schenk, 1954). Gill (1959) reported that in amaryllis, application of zineb and ferbam at 2 lb/100 gal is more effective in controlling *Stagonospora curtisii* than captan and copper fungicides. Leeuwen (1974) suggested the combination of hot water treatment + 0.5 per cent benlate dip to control the disease. In field trials, a dip containing 1 per cent aglitorn and 0.2 per cent benlate was also effective. *H. papilio* cvs. Rio, Sampa and Bahia are reported to be resistant to *Stagonospora curtisii* (Meerow, 2000).

Fungal diseases of amaryllis other than red leaf spot, reported by Weiss (1947) are soft rot of bulbs caused by *Rhizopus stolonifer,* bulb decay caused by *Sclerotium rolfsii,* greymould caused by *Botrytis cinerae,* root rot caused by *Pythium debaryanum,* black mildew caused by *Asterinella hippeastri,* leaf spot caused by *Cercospora amaryllidis,* brown leaf spot caused by *Aecidium amaryllidis* and anthracnose caused by *Steptogloeum amaryllidis.* These diseases can also be controlled by treatments suggested to control the red leaf spot of amaryllis. Maiti *et al.* (1979) reported *Phoma glomereta* on *Hippeastrum hybridum. Helminthosporium rostratum* (Lal and Tandon, 1970) and *Fusarium* sp. (Pag, 1964; Leeuwen, 1974; Tapio, 1966) also adversely affect the growth, bulb formation and flowering in hippeastrum.

11.2.0 Viral Diseases

Cucumber Mosaic Virus (*Amaryllis mosaic*)

This disease is widely spread in the USA, India, Europe and other parts of the world where amaryllis is grown. Large yellowish green patches appear at random over the leaf surface. The vector can be controlled by spraying 0.2 per cent Malathion regularly at 15-day interval. Sometimes, this disease also spreads through weeds and these should be eradicated from the amaryllis plots. Stouffer (1963) suggested that CMV can be controlled by treating bulbs at temperature of 25°C and light intensity of 1000 ft-candles. Aseptic shoot-apex culture was employed in an attempt to recover virus-free plant material from the infected bulbs. Bulb cuttage was a source of shoot- apices which subsequently was cultured on liquid nutrient medium at a temperature of 25°C and light intensity of 1000 ft-candles (Nowicki and Eno-Rourke, 1974).

Tomato Spotted Wilt

This disease has been found in England and India. Brierley (1948) reported that the leaves of the affected plants showed numerous yellow spots and the plants ultimately died. The symptoms are similar to amaryllis mosaic virus but are coarser. This disease is spread by thrips and may affect many other plants of Amaryllidaceae. The thrips can be controlled by spraying 0.2 per cent Malathion. According to Pramer (1955), the antibiotics streptothriein and terramycin suppress two of the virus diseases, tobacco ring spot and tobacco mosaic. All mosaic-infected plants should be burnt as soon as identified. The control of possible insect or mite carriers is particularly important under outdoor amaryllis culture, nearby weeds that might harbour virus diseases should also be eradicated.

11.3.0 Pests

11.3.1 Caterpillar

During June-August, plants are attacked by the most common caterpillar, *Polytela gloriosae.* They eat all the leaves within 4-5 days. Sachan (1969) reported that application of parathion at 0.02 per cent controlled the caterpillars within 48 hours.

11.3.2 Thrips

In both greenhouse and field, several species have been reported in the. USA. But *Taeniothrips simplex* is commonly known in India, Thrips suck the juice from leaves, flower stalks and flowers. It can be controlled by weekly application of phosphomidon at 0.05 per cent and endosulfan at 0.1 per cent.

11.3.3 Aphids, Mealy Bugs and Scale Insects

These insects puncture the tissues and suck the sap of the surface. The aphids and of mealy bugs and scale insects at younger stages may be controlled with a spray containing 15 g of nicotine sulphate and 25 g of soap chips in one gallon of water or by 0.2 per cent of malathion at 10-day interval.

Diseases and Pests of Amaryllis and Hippeastrum

Red Leaf Spots of Amaryllis

Brown Leaf Spots

Bulb Rot

Hippeastrums Mosaic Virus

Amaryllis Caterpillars

Amaryllis Thrips

Aphids Infestation

11.3.4 Mites

Mites also damage the tissue and suck the sap from beneath the surface. There are several mites, such as spider mites (*Tetranychus bimaculatus*), bulb scale mites (*Tarsonemas laticeps*) and the bulb mites (*Rhizoglyphus echinopsus* and *R solani*).

The spider mites may be controlled by spraying of aramite 0.2 lb per gallon of water. For controlling bub scale mites and bulb mites, dipping of bulb in water at 43°C for 60 minutes is suggested. After treatment the bulbs should be allowed to cool gradually before planting.

11.2.5 Other Insects

In addition to the above insects and pests, grasshoppers (*Romalea microptera*), blister bettle (*Epsicanla smgosa*) and narcissus bulb fly (*Lampetia equesms*) cause damage to plant. Reports from Bulgaria for the first time indicated the infestation of Amaryllis by *Plusia chalcites,* was maximum during April-June (Lecheva and Loginova, 1988).

12.0 Harvesting and Postharvesting Management

12.1.0 Harvesting of Flowers

In Europe, amaryllis is extensively used in the florist trade as cut flowers. To meet the demand during Christmas, amaryllis is grown under glass. The scapes are cut when the buds are fully elongated. Harvesting should be done early in the morning and the stalk placed in water. Cut flowers may be stored in refrigerated showcases or storerooms to minimize the spoilage.

Flower scapes should be packed in paper cartons with adequate protection so that cut flowers reach the customer in good condition. Scapes packed in moist cloth remain fresh for a longer period.

12.2.0 Harvesting of Bulbs

In both amaryllis and hippeastrum, after the bulbs mature and the leaves become yellow and dry, the bulbs are dug out and tops cut away. The best time to harvest amaryllis bulbs under Florida conditions is during the semi-dormant period at the end of the summer growing season, ranging from August 1 to September 15 before the fall flush of growth appears. If the bulbs cannot be dug at this time, the next harvesting period follows after the maturing of the fall flush of growth, from December 1 to January 1 (Traub, 1958).

12.3.0 Postharvest Management

12.3.1 Vase Life of Flowers

Hippeastrum is a glamorous flower, a flower of perfection and excellence for floral decoration. The average life of a flower stalk is about 15 to 20 days *in situ,* while that of an individual flower varies from 5-6 days. Vase life of a flower varies from 10-15 days depending on the number of flowers per stalk, environment and change of water. The flower remains fresh for a longer period if it is kept in 3 per cent sugar solution or 25 ppm maleic hydrazide (Jana, 1979). Dijkherizen (1980) found equal duration of blooming in buds of attached and detached flowering stalk.

Hippeastrum (Hippeastrum *hybridum* Herb. *Amaryllidaceae*) is used for either fresh cut flowers or flowering potted plant. The flowers of *Hippeastrum* 'Minerva' are susceptible to mechanical damage during handling and occupy more space during shipping and storage (Abdel Kader, 2012). Studies were conducted on the earliest possible flower bud stage to cut the inflorescences and the effect of flower preservative solutions on flower opening, vase life and quality of *Hippeastrum* 'Minerva' cut flowers (Abdel Kader, 2012). Inflorescences were harvested at three different stages of flower bud development and placed in three different holding solutions. Stems cut at stage III and placed in deionized water opened more satisfactorily (4 flowers/stem) than flowers of stems cut at stage I or stage II (less than 4 flowers/stem), but wilted earlier than those two stages. Within each stage of bud development, stems placed in 8-HQS plus sucrose solution maintained their positive water balance for a longer period, had higher average solution uptake, higher increase in fresh weight and longer vase life than those placed in 8-HQS only, while those placed in DI water (control) were the least. In each comparable solution, stems cut at stage I maintained their positive water balance for a longer period, had higher average solution uptake, higher increase in fresh weight and longer vase life than stems of stage II, while those of stage III were the least. At the time of harvest, buds cut at stage III had higher glucose and fructose concentrations than those cut at stage I or stage II. However, eight days after starting the experiment, highest fructose and glucose concentrations were found in flowers of stems cut at stage I and placed in 8-HQS plus sucrose (Anbdel Kader, 2012; Hassan, 2009). Vase life of the Hippeastrum flower varies from 10-15 days depending on the number of flowers per scape, environment and change of water. The combination of 4 per cent sucrose and 0.75 mM aluminium sulphate (0.48) with the longest vase life of 10.33 days of cut Hippeastrum flower cv. 'Apple Blossom' (Jamil *et al.*, 2016).

A study involving flower senescence (tepal senescence) and evaluation of the efficacy of certain chemical formulations on vase life (of its cut scapes) was conducted in *Amaryllis belladonna* (*Gul et al.*, 2020). For studies related to senescence, flower development has been divided into easily recognizable six morphological stages (stages I to VI) and changes in some important morphological and biochemical parameters including fresh and dry weight, MSI, soluble protein content and soluble solid content were recorded. SDS-PAGE of crude protein extract was also performed. Moreover, the cut freshly cut healthy scapes were subjected to different chemical formulations (provided as vase solutions) including sugar source (sucrose) and antimicrobial compounds [$Al_2(SO_4)_3$ and $CoCl_2$] for assessing the efficacy of these chemicals in improving their postharvest performance. The results indicate that the loss of floral fresh and dry weight, decline in MSI, decline in the content of soluble proteins and changes in the soluble solid content (sugar status and phenolics) may be considered as important parameters of tepal senescence in *Amaryllis* and that postharvest performance in cut scapes can be improved to a great extent by the combined application of antimicrobial compounds and sugar source at given concentrations in vase solutions (Gul *et al.*, 2020).

12.3.2 Grading of Bulbs

The American Amaryllis Society has established a set of improved standards for trade of bulbs. Similarly, standards have been provided by the American Association of Nurserymen and by the Dutch growers.

12.3.2.1 American Amaryllis Society's Standards

American Amaryllis Society presented an outline of improved standards for grading bulbs. The standards for flower size, bulb size, and number of bulbs packed per carton and flower shape and form apply to all the groups of amaryllis.

The Flower Size Standard

The bulbs are graded in three grades. The bulbs in first grade should be capable of producing well shaped, clear coloured flowers, 20 cm. in diameter across the face of flower, a second-grade flowers are 15-20 cm in diameter and in third grade there is no set standard. It may contain any kind of Reginae and Leopoldii strain seedlings.

Bulb Standard

The bulbs are also graded in three grades. The bulbs in first and second grades may be of any size, greater than 6.5 cm in diameter when cured. They must be sound, free from diseases, insects or mechanical injury, well cured with roots trimmed to not less than 7.5 cm from root base and with leaves trimmed to the neck of bulbs. In grade three, the bulbs must be free from insects, pests, diseases and mechanical injuries. The bulbs in this grade are further graded after curing in three sizes 6.25-7.5 cm (medium) 7.5-9.0 cm (large) and more than 9.0 cm in diameter (top).

Curing Standard

All the bulbs are cured in trays or racks for atleast 30 days before sizing and packing.

Packing Standard

The bulbs in all grades are to be packed in paper cartons of 30 × 25 × 38 cm, 100-6 cm, 50-7.5 cm and 35-9 cm bulbs in each, with adequate protection by paper to ensure that the bulbs reach the customer in good condition during cold weather.

12.3.2.2 American Association of Nurserymen Standard

The following standards are adopted by the American Association of Nurserymen in cooperation with American Wholesale Dealers Association. The bulbs are sold under grade names, such as forcing size to size, *etc*. The standards for amaryllis bulbs are designated in centimetre in diameter, (i) fancy 9 cm up, (ii) top size 8.0-8.5 cm, (iii) large 7.5-8.0 cm, (iv) medium 6.25-7.50 cm, (v) small 5.75-6.00 cm, and (vi) under 5.75 cm, not acceptable.

12.3.2.3 Dutch Amaryllis Bulb Standards

These are similar to those of American Amaryllis Society in bulb sizes, but the bulb size is expressed as bulb circumference in centimetres. The standard size of named clones is 26 cm.

Harvest and Marketing of Amaryllis and Hippeastrum

Hippeastrums in Full Bloom in Open

Cut Flowers

Harvesting of Bulbs

Harvesting Stage Greenhouse

Bulbs in Market

The bulbs are packed in two methods to avoid losses during transit. The first method used by the Netherlands exporters who ship the bulbs to USA, consists of packing in heavy wooden crates made of half-inch tongue-and-groove lumber glued at the seams. This forms a warm light container which is then lined with several layers of felt paper. The bulbs are packed in sawdust or crimpled paper or wood shavings. Before packing bulbs should be completely cured. In the second method, bulbs are shipped in similar containers but they are packed in coarse vermiculite. This increases aeration and its highly absorptive quality lessens danger from sweating (Traub, 1958).

Hertogh *et al.* (1998) suggested packing bulbs of Hippeastrum cv. Apple Blossom and Red Lion with wood wool in perforated polyethylene bags and placing in perforated cardboard boxes to obtain best results in terms of rooting system, basal roots, low disease incidence and early flowering. Packing the bulbs loose in polyethylene bags also produced early flowering, packing systems had no significant effect on floral stalk length, number of flower/stalk or overall flower quality.

12.4.0 Storage

Bulbs are best stored in cold storage. They may also be stored in trays or boxes where temperature and moisture conditions can be controlled at least partially. According to Dijkherizen and Leeuwen (1975) the bulbs of Hippeastrum cv. Scarlet Glow should be kept in cold storage at 13° or 17°C for 10-12 weeks. Bulbs of hippeastrurn cv. Apple Blossom and Red Lion packed with wood wool in perforated polythylene bags and placed in perforated cardboard boxes were reported to be stored as such for 84 days at 9°C (Hertogh *et al.*, 1998).

13.0 Biotechnology

13.1.0 Molecular Markers

Phuon (2014) conducted to clarify the distribution and genetic diversity of *Hippeastrum* accessions cultivated in Vietnam. Twenty-five *Hippeastrum* accessions collected from 17 provinces in the northern, central, and southern regions of Vietnam were included. Results of floral observation and RAPD analysis indicated that there are four distinct groups of *Hippeastrum* being grown in Vietnam. These include *cam dai* (*H. puniceum* (Lam.) Kuntze syn. *H. equestre* Herbert), *do dai* (*H.* x 'Johnsonii'), *do cam dai* (*H. striatum* Lamarck), and the hybrid cultivars. In the RAPD analysis result, 25 primers produced a total of 230 distinct bands, of which 167 were polymorphic (about 72.6 per cent). The number of polymorphisms per primer ranged from 1 to 13, and one primer did not reveal any polymorphisms. Their phylogeny indicated four separate groups, with dissimilarity values ranging from 0.2 – 0.3 according to the phylogeny based on RAPD results. Some variation was observed between northern, central, and southern *H. puniceum* populations. Extensive variation in flower color, shape, and size was detected among accessions of the hybrid group. While *H. puniceum* is the most widely distributed species from the northern to the southern regions, *H.* x 'Johnsonii' has been observed only in the northern region, but not in the central or southern regions, so the distribution of *Hippeastrum* within Vietnam might depend on adaptation to environmental conditions. The results have distinguished the *do dai* and the *cam dai* groups, although a single scientific name has been applied to both extant groups in Vietnam. In view of the Plant protection rights, markers will be useful to identify the parentage of the hybrid. Hybrids with common parents (one or more) clustered together indicating their level of genetic similarity. From this study, clear genetic associations of various Hippeastrum cultivars and hybrids with parents were found. Cluster analysis of RAPD profiles of Hippeastrum hybrids and/or varieties where the pedigree is not known can now be used as a tool to study ancestral relationships (Chakrabarty *et al.*, 2007).

The genetic diversity of 62 cultivars (including 60 cultivars introduced from Netherlands and 2 local cultivars in Suzhou) of *Hippeastrum* spp. was analyzed by ISSR marker method, and the cluster analysis on 62 cultivars was carried out by UPGMA method (Lin *et al.*, 2012). On the basis, by means of comparison and selection of specific bands, the ISSR fingerprint of tested cultivars was constructed by ways of schematic diagram with black and white square lattices. The amplification result shows that total 118 bands are amplified from genomic DNA of 62 cultivars by 11 primers, in which, there are 109 polymorphic bands with the polymorphic band percentage (*PPB*) of 92.4 per cent, and *PPB* of bands amplified by five primers reaches to 100.0 per cent. The change range of genetic similarity coefficient among 62 cultivars is larger with a value from 0.3714 – 0.8429, showing that there are rich genetic variation and genetic diversity among different cultivars. The cluster analysis result shows that 62 cultivars can be divided into seven groups at the genetic similarity coefficient of 0.63. Most cultivars with similar morphology are clustered together, in which the genetic similarity coefficient among some similar morphology cultivars with white and single petal is higher (about 0.8) and most of these cultivars are clustered together, indicating that their homology is higher. While, two local cultivars in Suzhou have the highest genetic similarity coefficient (0.8429), meaning that the two cultivars may possess a same origin. Cultivar 'Xiaohongxing' has a specific band at 450 bp in amplification pattern of primer UBC873 and cultivar 'Elvas' has a specific missing band at 3000 bp in amplification pattern of primer UBC835, therefore, these two specific bands can be used for identification of cultivars 'Xiaohongxing' and 'Elvas', respectively. According to ISSR amplification bands by primers UBC835 and UBC873, the ISSR fingerprint of 62 cultivars is combined and constructed, and all of cultivars tested can be identified with this fingerprint (Lin *et al.*, 2012).

According to Wang *et al.* (2018), there are two main explanations for the complex population structure of amaryllis. The first explanation involves the abundance, wide distributional range, and long history of species breeding within this plant group. Amaryllis is native to the mountainous areas of South America, including Brazil, Peru, Argentina, and Bolivia, and this ornamental plant has been bred in the UK since the 1690s. Although the first recorded documentation of *Hippeastrum* hybrids was in 1799 in the UK as 'Johnsonii' (*H. vittatum* x *H. regime*), the introduction of these central and South American flower crops into Europe occurred later during the 19th century. Thus, after numerous breeding experiments, *Hippeastrum* was divided into four key hybrid groups: Reginae Leopoldii, Vittatum and Reticulatum (Callaway and Callaway, 2000). An additional reason for this confusion is the existence of a polyploid variety of amaryllis, *H. hybridum*, a bulbous perennial with ploidy level that ranges between diploid and octoploid. Examples include *H. aulicum*, *H. machupichense*, *H. psittacinum*, and *H. solandrifoliu* (2n = 22); *H. forgeti* (2n = 2x + 1 = 23) and *H. argentinum* (2n = 33); *H. reginae* (2n = 44); and *H. rutilum* (n = 55) (Arroyo, 1982). As a direct result of this diversity, amaryllis has gradually become an increasingly popular flower globally. This polyploidy has resulted in a large floral organ and a high yield, which have both attracted researchers. Thus, a big flower has always been a clear goal of breeding experiments. These characteristics also indicate that the complex structure of amaryllis relationships have proved difficult to untangle (Wang *et al.*, 2018).

The genomic structure of amaryllis (*Hippeastrum hybridum*) and the extent of their genetic resources remain limited. The development of molecular markers is therefore of great importance to accelerate genetic improvements in *Hippeastrum* species. Wang *et al.*, (2018) defined a total of 269 unique genes that might regulate the flower spathe development of amaryllis. In addition, 2000 simple sequence repeats (SSRs) were detected based on 171,462 *de novo* assembled unigenes from transcriptome data, and 66,4091 single nucleotide polymorphisms (SNPs) were also detected as putative molecular markers. Twenty-one SSR markers were screened to evaluate the genetic diversity and population structure of 104 amaryllis accessions. A total of 98 SSR loci were amplified for all accessions. The results reveal that Nei's gene diversity (H) values of these markers ranged between 0.055 and 0.394, whereas the average values of Shannon's Information index (I) ranged between 0.172 and 0.567. Genetic tree analysis further demonstrates that all accessions can be grouped into three main clusters, which can be further divided into two subgroups. STRUCTURE-based analysis revealed that the highest K values were observed when K=5, K=6, K=7 and K=8. The results of this study enable large-scale transcriptomics and classification of *Hippeastrum* genetic polymorphisms and will be useful in the future for resource conservation and production.

13.2.0 Tissue Culture

An effective micro-propagation via callus-mediated (indirect) somatic embryogenesis was established, using *in vitro* bulblets as explants in *Hippeastrum hybridum* (Gong *et al.*, 2012). 2,4-D (0.5 mg/l) and 6-BA (0.5 mg/l) had important role in inducing embryogenic callus from *in vitro* bulblets and in multiplication of callus during sub-cultures. The percentage of friable callus formation was 83.3 per cent in primary culture. The 6-BA (2.5-5.0 mg/l) had a notable effect on development of somatic embryos. The rate of callus with more than twenty somatic embryos was about 50 per cent for 50 days. Development of somatic embryos had no difference on the same 6-BA media supplemented with different concentrations of NAA (Gong *et al.*, 2012).

Rooting is a crucial step for successful micropropagation. For rooting of bulblets of *Hippeastrum johnsonii* different concentrations of 2-iP (0, 14, 16 and 18 mg/l) and NAA (0, 14, 16 and 18 mg/l) were tried (Zakizadeh *et al.*, 2013). A well-developed root system was achieved in MS medium containing 16 mg/l 2-iP and 4 mg/l NAA. The highest root length (2.71 cm), longest root (3.53 cm) and root number (2.77) were obtained in the MS medium supplemented with 4 mg/l NAA. The plantlets containing roots were acclimatized (90 per cent survival percentage) by transferring them to plastic pots containing coco peat, compost and sand.

Since the natural multiplication rate of Hippeastrum is slow, twin scaling might be developed to overcome this deficiency. *In vitro* plantlets production has already been established through twin scaling. But *in vitro* bulb production is more advantageous than raising plantlets regarding time, storage, handing, establishment of planting materials *etc*. (Sultana *et al.,* 2010). Sucrose level at 90 g/l produced the maximum average weight as well as the highest regeneration percentage. The increasing rate of CCC increased the number and average weight of bulb. The maximum bulb formation observed in media supplement with 6.0 mg/l BAP and 500 mg/L CCC fortified with 90 g/l sucrose (Sultana *et al.,* 2010).

Different concentrations of different elicitors and additives like methyl jasmonate, spermine, casein hydrolysate, or progesterone combined with 16 mg/l 2 iP+4 mg/l naphthalene acetic acid (NAA) were investigated in order to obtain higher multiplication rate and growth of *Hippeastrum vittatum* bulbs in vitro on MS basal medium. The highest multiplication rate (8.2 bulbs/explant) was attained with 80 mg/l spermine, while the highest bulb fresh weight (1.23 g/bulblet) was obtained with 4 mg/l methyl jasmonate. Progesterone at 20 mg/l or casein hydrolysate at 2.0 g/l gave the highest leaf length (14.1 and 13.2 cm, respectively). It can be advised to use 80 mg/l spermine combined with 16 mg/l 2 iP+4 mg/l NAA to obtain the highest number of bulbs per explants with moderate leaf length and bulb fresh weight. Chemical analysis showed alternations in the alkaloid type ratio and number of compounds in the bulbs treated with 4mg/l of methyl jasmonate (Zayed *et al.,* 2011). The *in vitro* propagation of the new hybrid *Hippeastrum* x *chmielii* was analysed by Ilczuk *et al.* (2005). Propagation on solid medium, permanent immersion (flasks) and in a temporary immersion system (TIS) using quarters of *in vitro* grown bulblets as explants was compared. The TIS was constructed as twin-flasks system, the components of which were inexpensive. Within 4 weeks of culture on solid medium 3.9 new bulbs were formed per initial bulb and by using liquid culture systems this number was increased to 6.3–6.5. The number of daily immersions in TIS only slightly influenced the propagation rates and no significant differences to the permanently immersed treatments were observed. By adding flurprimidol (0.1–1.0 mg/l) to the medium the propagation rate nearly doubled in liquid as well as in solid medium. Moreover, the size of the newly formed bulbs was increased by this inhibitor of gibberellic acid biosynthesis. Flurprimidol strongly reduced the elongation of the leaves while root formation was enhanced.

13.2.1 Acclimation

The substrate is also a critical issue for the transplanting of micropropagated seedlings. It should have a good capacity of humidity retention without compromising the aeration of the root system, and should be chemically inert to allow nutritional manipulation according to species (Heintze *et al.,* 2018). In order to characterize chemically and physically and to compare five substrates in acclimatization, micropropagated seedlings of *Hippeastrum reticulatum* var. *striatifolium* (Herb),

Exhibition of Amaryllis and Hippeastrum

seedlings with an average of four roots, two leaves and bulbs with a 4.7 mm diameter, were transplanted to 50-cell polyethylene trays each containing the following substrates, vermiculite, carbonized rice husk (CRH), commercial (peat + vermiculite + CRH), coconut fiber, and composted Pinus bark. After transplantation, the trays were kept for 17 days at a greenhouse with intermittent irrigation by nebulization, and 80 days at a greenhouse on a bench with intermittent irrigation by capillarity. The experimental design was randomized blocks with seven replications and ten seedlings per plot. The acclimatization of seedlings was efficient for the substrates tested despite the differences verified by chemical and physical characterization. The mean survival rate was 96 per cent. However, seedlings acclimatized in composted Pinus bark substrate showed a qualitative improvement due to a higher formation of roots, leaves and fresh matter (Heintze *et al.*, 2018).

14.0 References

Abdel Kader, H.H. (2012) *World Appl. Sci. J.*, **19**(7) : 943-950.

Alekperov, C., Barukhin, E., Ben-Asher, J., Ephrath, J. and Lamm, E.R. (1995) *Proc. of the fifth Intemational Microirrigation. Congress.* Orlando, Florida, USA, pp. 206.

Alkema, H.Y, Lccnwcn, C. and Van, G.M. (1977) *Vakblad voor de Bloemisterij*, **32** : 24-25.

Amico-Roxus, U., Roxas, U.A., Zizzo, G.V., Agnello, S., Sciortino, A. and Iapichino, G. (1994) *Acta Hartic.*, **362** : 281-288.

Andrade-Rodriguez, M., Guillen-Sanchez, D., Villegas-Torres, O.G., Ayala-Hernandez, J., Lopez-Martinez, V. and Vargas-Araujo, J. (2015) *Revista Chapingo Serie Horticultura*, **21**(1) : 57-69.

Arroyo, S. (1982) *Kew Bull.*, **37** : 211-216.

Bailey, L.H. (1929) *The Standard Cyclopedia of Horticulture*, The Macmillan Company, London, New York.

Baker, J.G. (1888) *Hand Book of the Amaryllidaceae including Alstroemerieae and Agareae.* London.

Baldwin, J.T. and Spccse, R.M. (1947) *Bull. Torrey Bot. Club.*, **74** : 250-254.

Bapat, V.A and Narayanaswamy, S. (1976) *Bull. Torrey Bot. Club.*, **103** : 53-56.

Barnhoorn, E. (1976) *Plant Life*, **32** : 59-63.

Bear, E.E. (1953) *Soils and Fertilizers*, John Wdey and Sons, New York.

Bell, W.D. (1972) *Plant Life*, **28**: 72-76.

Bell, W.D. (1973) *Plant Life*, **29** : 66-68.

Bell, W.D. (1977) *Plant Life*, **33** : 65-68.

Bell, W.D. (1978) *Plant Life*, **34** : 95-98.

Bell, W.D. and Stiles, H.D. (1974) *Plant Life*, **30** : 89-90.

Bhattacharjee, S.K., Yadav, L.P. and Mukhopadhyay, T. (1982) *Indian Agric.*, **26** : 193-197.

Bonde, R. (1955) *Marine Farm Res.*, **3** : 7-9.

Bose, T.K. and Jana, B.K. (1977) *Indian J. Hort.*, **34** : 446-447.

Bose, T.K., Jana, B.K. and Mukhopadhyay, T.P. (1979) *Punjab Hort. J.*, **19** : 205-207.

Bose, T.K., Jana, B.K. and Mukhopadhyay, T.P. (1980a) *Sci. Hortic.*, **12** : 195-200.

Bose, T.K., Jana, B.K. and Mukhopadhyay, T.P. (1980b) *Indian J. Hortic.*, **37** : 415-417.

Bose, T.K., Jana, B.K. and Mukhopadhyay, T.P. (1980c) *Punjab Hortic. J.*, **20** : 222-226.

Bose, T.K., Jana, B.K. and Mukhopadhyay, T.P. (1981) *Indian J. Hortic.*, **38** : 110-112.

Brants, H.D. and Wanduvan, H.J. (1965) *Netherland J. Pl. Pathol.*, **71** : 145-151.

Brickell, C. (1996) *A-Z Encyclopedia of Garden Plants*, Dorling Kindersley, London.

Brierley, P. (1948) *Herbertia*, **15** : 113-119.

Brown, H. (1952) *Plant Life*, **8** : 112.

Bruyn, M.H. de, Ferreira, D.l., Slabbcrt, M.M., Prcforius, J. and Dc Bruyn, M.H. (1992) *Plant Cell Tiss. Organ Cult.*, **31** : 179-184.

Burnham, M., O'Rourke, E.N., Love, J.E. and Schmit, EB. (1971) *Plant Life*, **27** : 91-95.

Cage, J.M. (1975) *Plant Life*, **31** : 65-72.

Callaway, D.J. and Callaway, M.B. (2000) *Breeding Ornamental Plants.*, pp. 174-195.

Chakrabarty, D., Gupta, V.N. and Datta, S.K. (2007) *Plant Biotechnol. Rep.*, **1** : 211-217.

Chittenden, F.J. and Synge, P.M. (1956) *Dictionary of Gardening*. The Royal Horticultural Society, Oxford University Press, London, pp. 95-998.

Cothran, C.D. (1979) *Plant Life*, **35** : 61-75.

de Andrade, J.P., Pigni, N.B., Torras-Claveria, L., *et al.* (2012) *Rev. Latinoamer. Quím.* **40/2** : 83-98.

Dijkherizen, T. and Leeuwen, A.J.M. Van (1975) *Vakblad voor de Bloemisterij*, **30** : 12-13.

Dijkherizcn, T. (1980) *Bloembollencultuur*, **80** : 890-891.

Dix, J.F.C. (1977) *Bloembollencultuur*, **76** : 391.

Doran, J.L. (1974) *Plant Life*, **30**: 97-103.

El-Naggar, A.H. and El-Nasharty, A.B. (2009) *American-Eurasian J. Agric. and Environ. Sci.*, **6**(3) : 360-371.

Everett, T R. (1954) *The American Gardener's Book of Bulbs*, Random House, New York.

Ficker, T. (1951) *Plant Life*, **7** : 68-71.

Flory, W.S. and Robert, F.C. Jr. (1981) *Plant Life*, **27** : 43-56.

Fosler, G.M. (1953) *Fm. J.*, December, pp. 90-91.

Foster, M.B. (1951) *Plant Life*, **7** : 12-18.

García, N., A.W. Meerow, Leuenberger, S.A., Oliveira, R.S., Dutilh, J.H., Soltis, P.S. and Judd, W.S. (2019) *Taxon.*, **68**(3) : 481-498.

Gill, D.L. (1959) *Plant Dis. Reptr.*, **43** : 1272-1273.

Gong, X.Q., CuiRong, C.R.Y., Qu F.N., Chen, L.X., Yu, W.S. and Xie, X.X. (2012) *Acta Hortic. Sinica.*, **39** : 381-386.

Gorer, R. (1970) *The Development of Garden Flowers,* Hippeastrum, Eyro and Spottis Woodc Ltd., pp. 202-204.

Gul, F., Tahir, I. and Shahri, W. (2020) *Plant Physiol. Rep.*, **25** : 315–324.

Gyenes, L. (1975) *Kerte'szet es Szoleszet.*, **24** : 7.

Hannibal, L.S. (1955) *J. Royal Hort. Soc.*, **80** : 130-131.

Harbome, J.B. (1967) *Comparative Biochemistry of the Flavanoids.* Academic Press, London and New York.

Hassan, F.A.S. (2009) *Acta Agronomica Hungarica*, **57**(2), doi: 10.1556/AAgr.57.2009.2.

Haung, V.W., Okubu, H. and Vemoti, S. (1990) *Sci. Hort.*, **42** : 151-160.

Hayashi, I. (1972) *Bull. Kanagawa Hort. Expt. Sta.*, **20** : 95-102.

Hayashi, I. (1974) *Bull. Kanagawa Hort. Expt. Sta.*, **22** : 116-119.

Hayashi, I. and Suzuki, M. (1970) *Bull. Kanagawa Hort. Expt. Sta.*, **18** : 171-188.

Heaton, I.W. (1934) *Yearb. American Amaryllis Soc.*, **1** : 78-80.

Heintze, W., Milani, M., Pradella, E.M., Fior, C.S. and Schafer, G. (2018). *Ciência Rural, Santa Maria*, 48:12, e2018. http://dx.doi.org/10.1590/0103-8478cr20180205.

Henry, M.G. (1951) *Plant Life*, **7** : 118-121.

Herbert, W. (1819) *Bot Mag.*, London.

Herbert, W. (1837) *Amaryllidaeae*, London.

Hertogh, A.A. de, Gallitano, L.B. and De Hertogh, A.A. (1998) *Hort. Technol.*, **8** : 175-179.

Hong, Y.P. (1970) *Rept. of Hort. Stn. Suwan*, S. Korea, pp. 57-63.

Huh, M.R., Ko, Y.S., Kang, H. and So, I.S. (1998) *J. Korean Soc. Hort. Sci.*, **39** : 629-634.

Ijiro, Y. and Ogata, R. (1997) *J. Japanese Soc. Hort. Sci.*, **66** : 575-579.

Ilczuk, A., Winkelmann, T., Richartz, S., Witomska, M. and Serek, M. (2005) *Plant Cell, Tiss. and Organ Cult.*, **83** : 339-346.

Inariyama, S. (1937) *Sci. Rep. Tokyo Bountike Digaku Sect B*, **3** : 95-113.

Inkham, C., Piriyapongpitak, P. and Ruamrungsri, S. (2019) *Hortic. Environ. Biotechnol.*, **60** : 357–362.

Jamil, M.K., Rahman, M.M., Hossain, M.M., Hossain, M.T. and Karim, A.J.M.S. (2016) *Bangladesh J. Agril. Res.*, **41**(2) : 221-234.

Jana, B.K. (1979) *The Lal-baugh*, **24** : 17-20.

Jana, B.K. and Bose, T.K. (1980) *Indian Agric.*, **24** : 23-30.

Jana, B.K. and Bose, T.K. (1981a) *Indian Agric.*, **25** : 39-48.

Jana, B.K. and Bose, T.K. (1981b) *South Indian Hort.*, **29** : 215-219.

Jana, B.K. and Bose, T.K. (1987) *Indian Agric.*, **31** : 87-92.

Jindal, S.L. (1968) *Ornamental Bulbous Plants*, ICAR, New Delhi.

Joiner, J.N., Sheehan, T.J. and Merecith, W.C. (1966) *A.R. Fla. Agric. Exp. Sta.*, pp. 152-153.

Kharrazi, M., Tehranifar, A., Nemati H., and Bagheri, A. (2017b) *Hortic. Sci. Technol.*, **35**(3) : 373-380.

Kharrazi, M., Tehranifar, A., Nemati, H. and Bagheri, A.R. (2017a) *J. Hortic. Sci.*, **31**(1) : 33.

Klotzbach, C.L. (1957) *Plant Life*, **13** : 107-109.

Kurki, L. (1958) *Puutarha*, **61** : 414.

Lal, B. and Tandon, R.N. (1970) *Mycopath. Mycol Appl*, **42** : 289-291.

Lecheva, I. and Loginova, T. (1988) *Rasteniev dni-Nauki*, **25** : 87-94.

Leeuwen, A.J. Van (1974) *Vakblad voor de Bloemisterij*, **29** : 15.

Lin, Z., Chun, X.Y., Zhong, C.H., Zhen, Z.Y., Ming, L.X. and Tao, L.W. (2012) *J. Plant Resour. Environ.*, **21** : 48-54.

Linnaeus, C. (1737) *Origin of Name, Amaryllis, Critica Botanica*, pp. 30-118.

Linnaeus, C. (1751) *Derivation of Name Amaryllis, Philosophia Botanica*, pp. 140-170.

Linnaeus, C. (1753) *Species Plantarum*, **1** : 292-293.

Liu, M.C. and D.M. Yeh. (2015) *HortScience.*, **50**(10) : 1588-1590.

Love, J.E., O'Rurke, E.N., Burnham, M., Carson, J. and Schmitz, F.B. (1971) *Plant Life*, **27** : 95-97.

Luyten, (1926) *Herbertia*, **2** : 115-122.

Maiti, S., Sahambi, H.S. and Jana, B.K. (1979) *Indian Phytopathol.*, **32** : 120-121.

Meerow, A.W. (1998) *Proc. Florida State Hort. Soc.*, **101** : 285-288.

Meerow, A.W., Svenson, S.E. and Kane, M.E. (1994) *HortScience*, **29** : 1149-1150.

Meerow, A.W. (2009) Isr. J. Plant Sci., ***57** : 303-313.*

Meerow, A.W. (2014) *HortScience.*, **49**(8) : 1102–1107.

Meerow, A.W. (2010) Convergence or reticulation Mosaic evolution in the canalized American Amaryllidaceae, pp. 145-168.

Meerow, A.W., Fay, M.F., Guy, C.L., Li, Q.B., Zaman, F.Q. and Chase, M.W. (1999) *Amer. J. Bot.*, **86**(9) : 1325-1345.

Meerow, A.W., Guy, C.L., Li, Q.B. and Yang, S.L. (2000) *Syst. Bot.*, **25**(4) : 708-726.

Meerow, A.W., Reed, S.T., Dunn, C. and Schnell, E. (2017) *HortScience*, **52**(12) : 1853-1860.

Meerow, Alan W., Van Scheepen, Johan and Dutilh, Julie H.A. (1997) Transfers from *Amaryllis* to *Hippeastrum* (*Amaryllidaceae*). *Taxon*, **46**(1): 15-19.

Merrow, A.W. (2000) *HortScience.*, **35** : 147-149.

Meyer, M. and Weisenhorn, J. (2018) https://extension.umn.edu/houseplants/amaryllis.

Mii, M., Mari, T. and Iwsae, N. (1974) *J. Hort. Sci.*, **49** : 241-244.

Moore, H.E. Jr. (1963) *Baileya*, **11** : 15-16.

Munjal, R.I. and Kapoor, J.N. (1962) *Indian Fungi-VII*, **15** : 259-263.

Narain, P. and Khoshoo, T.N. (1968) *J. Cytol Genet.*, **3** : 41-45.

Naranin, P. and Khoshoo, T.N. (1977) *Indian J. Hort.*, **34** : 80-85.

Naranjo, C.A. and Poggio, L. (1988) *Kew Bulletin.*, **42** : 317-325.

Nautiyal, M.C. and Bajpai, P.N. (1979) *Plant Sci.*, **11** : 75-83.

Neto, T.D.M. (1948) *Herbertia*, **15** : 25-29.

Nowicki, M.F. and Eno-Rourki, J.R. (1974) *Plant Life*, **30** : 108.

Okubo, H., Huang, C.W. and Kishimoto, F. (1999) *J. Japanese Soc. Hort Sci.*, **68** : 513-518.

Pag, H. (1964) *Phytopath.*, **51** : 317-323.

Phuon, P.T.M., Isshiki, S. and Miyajima, I. (2014) *J. Fac. Agr., Kyushu Univ.*, **59**(2) : 235-241.

Poggio, L., González, G. and Naranjo, C.A. (2007) *Bot. J. Linn.*, **155** : 171-178.

Prakash, S. (1981) *Plant Life*, **37** : 114-118.

Pramer, D. (1955) *Scientific Amer.*, **192** : 82-90.

Prasad, R.N. and Chaturvedi, H.C. (1993) *Indian J. Exp. Biol.*, **31** : 242-246.

Rees, A.B. (1972) *The Growth of Bulbs, Academic Press,* London and New York, p. 311.

Rice, W.E. (1955) *Plant Life*, **11** : 83-84.

Rouphael, Y., Raimondi, G., Caputo, R. and De Pascale, S. (2016) *HortScience*, **51**(6) : 684-689.

Russel, H. (1969) *Plant Life*, **25** : 82-88.

Sachan, I.N. (1969) *Indian J. Hort.*, **26** : 94-96.

Sakanishii, Y. (1962) *J. Jap. Soc. Hort Sci.*, **31** : 174-184.

Saker, M., Rady, M. and EI-Babr, M. (1998) *Egyptian J. Hort.*, **25** : 113-128.

Sato, D. (1938) *Cytologia*, **9** : 203-242.

Schenk, P.J. (1954) *Cult Hand.*, **20** : 279-280.

Schrnidhauser, T.F. (1954) *Dissertation in Univ. Virginia.*

Seberg, O., Petersen, G., Barfod, A.S. and Davis, J.I. (eds.), *Diversity, phylogeny and evolution in the monocotyledons*. Aarhus: Aarhus University Press.

Sharma, A.K. and Jash, M. (1958) *Phyton.*, **11** : 103-110.

Silberbush, M., Ephrath, J.E., Alekperov, C. and Ben-Asher, J. (2003) *Sci. Hortic.*, **98** : 85-90.

Singh, H.P., Dohare, S.R. and Singh, B. (1994) *Indian Hort.*, **38** : 3-4.

Small, J. (1954) *Modern Aspects of pH with Special Reference to Soils*, D. VanNostrad Co. N.Y.

Smith, R.H., Burrows, J. and Kurten, K. (1999) *In vitro Cell. Dev. Biol. Plant.*, **35** : 281-282.

Springer John, R. (1937) *Herbertia*, **4** : 243-244.

Stancatoa, G.C., Mazzafera, P. and Magalhães, A.C. (1995) *Sci. Hortic.*, **62** : 81-87.

Stouffer, R.F. (1963) *Proc. Fla. St. Hort. Soc.*, **76** : 462-466.

Strout, E.B. (1954) *Plant Life*, **10** : 53-61.

Sultana, J., Sultana, N., Siddique, M.N.A., Islam, A.K.M.A., Hossain, M.M. and Hossain, T. (2010) *J. Cent. Eur. Agric.*, **11**(4) : 469-474.

Tapio, E. (1966) *Ann. Agric. Fenn.*, **5** : 26-27.

Tebban, F. (1955) *Plant Life*, **11** : 93-96.

Traub, H.P. (1934) *Yearb. American Amaryllis Soc.*, **1** : 72-74.

Traub, H.P. (1951) *Plant Life*, **7** : 29-31.

Traub, H.P. (1953) *Euclides*, **13** : 289-298.

Traub, H.P. (1958) *The Amaryllis Mannual*, The Macmillan Co., New York.

Tsukamoto, Y. and Fujioka, S. (1956) *J. Hort. Ass. Japan*, **25** : 208-212.

Vazquez, C., Reed, S.T. and Dunn, C. (2015). *Agric. Sci.*, **6** : 1547-1554.

Veeraballi, T., Tripathi, M.K., Vidhya, S.M. and Patel, R.P. (2017) *Medicinal Plants*, **9**(2:) 114-128.

Wang, Y., Chen, D., He, X. *et al.* (2018) *Sci. Rep.*, **8** : 10645, https://doi.org/10.1038/s41598-018-28809-9.

Weiss, F. (1947) *Plant Dis. Reptr.*, **31** : 459-465.

Wilkinson, A.E. (1943) *The Flower Ebcyclopedia and Gardeners Guide*, The Neo Home Library, New York.

Willis, J.C. (1973) *Dictionary of Flowering Plant and Ferns*, Cambridge Univ. Press.

Wilmot, C.M. and Laubscher, C.P. (2019) *Acta Hortic.*, **1237** : 137-144.

Youssef, A.S.M. and Abd El-Aal, M.M.M. (2014) *J. Plant Prod., Mansoura Univ.*, **5**(3) : 357-381.

Zakizadeh, S., Kaviani, B. and Onsinejad, R. (2013) *Ann. Biol. Res.*, **4**(2) : 69-71.

Zayed, R., El-Shamy, H., Berkov, S., Bastida, J. and Codina, C. (2011) *In vitro Cell. Dev. Biol. Plant*, **47** : 695-701.

Zhang, W., Song, L., Teixeira da Silva, J.A. and Sun, H. (2013) *Sci. Hortic.*, **160** : 230-237.

Ziv, D.S., Ziona, N., Cohen, A., Ziona, N., Ion, A., Fintea, C., Baizerman, M. and Daous, A. (2012a) US PP22,805 P3 dated Jun. 19, 2012.

Ziv, D.S., Ziona, N., Cohen, A., Ziona, N., Ion, A., Fintea, C., Baizerman, M. and Daous, A. (2012b) US 2011/019 1926 P1(43) Pub. Date: Aug. 4, 2011.

10

PEONY

T. Yatung and V. Bhargav

1.0 Introduction

Peonies are hardy herbaceous or shrubby perennials, belong to family *Paeoniaceae*. The plant when brought under cultivation, can survive up to 100 years. The majority of peony plants in gardens are hybrids of original true varieties. In ancient days, the Chinese considered peony as 'the flower phase', and thus it had gained the reputation in the ornamental garden as 'the first beauty'. Duc to their varying forms of flowers, rich palette of horticultural varieties and deep ethnobotanical history in Chinese culture, tree peonies have been referred to as 'the king of flowers' (Cheng, 2007; Li *et al.*, 2011). Earlier, peonies were popular amidst the rich class of the society but due to its availability and affordability in the present time, it is offered by all section of the society as a token of good luck since it is symbolic of 'financial prosperity'.

In the language of flowers, peonies signify romance, elegance, honour, wealth and abundance. Peonies have become a very popular decorative plant in Europe, it is known by the name 'the rose without thorns' and are frequently grown in gardens. Peony is the state flower of Indiana, a state of United States. In China, it is known as *fuguihua* or the flower of 'riches and honour', 'prime minister of flowers' and is regarded as China's unofficial national flower. There are about 1,800 cultivated species of peony in the world and more than 1,000 come from China. Today, cut peony flower is produced in more than 25 countries; Europe, the USA and Canada being the primary market.

Though the peony flowers are highly valued in the international market and are available for a short period from late spring to early summer (May-June), depending on the location in Northern Hemisphere. To meet out the early market demand (January- August), in the recent days the greenhouse cultivation of peony has been taken up in Israel, Italy and southeast France while cultivation has also extended in Southern Hemisphere (Chile, New Zealand, Australia) to supply cut flowers during late season (October-January), thus making the peony cut flower available throughout the year. Italy is the largest grower of peony while the Netherlands is the world's largest producer and China being the No.1 exporter in the world (Evans *et al.*, 1990; Halevy *et al.*, 1995).

2.0 Importance and Uses

2.1.0 Aesthetic Uses

Varieties of cut flowers are in demand in the global market, which is met by the local and big flower growers. Amidst all other traditional cut flowers the herbaceous peony is gaining importance recently due to large, scented flowers having wide variation in flower colour ranging from red, pink, yellow and white to unique colours such as purple, blue, green that are used in wide range of flower arrangement and as ornamental plant in the garden where it has a long-standing reputation as the 'queen of the garden'. It is one of the most popular flowers used for the preparation of wedding bouquets and is the official flower for 12th wedding anniversary in

Uses of Peony

Border Plant

Mass Planting

Pot Plant

As Specimen

Cut Flowers

In Flower Vase

Peony Flower Tea

Peony Jam

Peony Multi-purpose Oil

China. Peony is associated with and grown traditionally for sale during Mother's Day (Stevens *et al.*, 1993). In China, Japan, the USA, Canada and Europe tree peonies are popularly grown as a garden plant. Chinese peony (*Paeonia lactiflora* Pall.) is the most popular peony species grown for cut flower industry; occupying the top list of 20 most important cut flowers. The use of cut peony is limited owing to the availability only in local markets and for a short harvest time of 3-4 weeks. The Netherlands, New Zealand, Israel, USA, Chile are the largest suppliers of cut peonies, followed by Germany, France and Poland (Skutnik *et al.*, 2020).

2.2.0 Therapeutic and Other Uses

Besides grown as garden ornamental, peonies hold a long history of more than 2000 years for use in traditional medicine in China, Japan, Korea and Europe. The entire plant part, the roots, bark, seed and flowers were thought to have medicinal uses. Peony was called as *ebisugusuri* meaning 'foreign medicine' since the roots was used to treat convulsions. Peonies were considered a symbol for healing by ancient Greeks and Christians from Middle Ages and were used to treat stomach pains, bladder issues, jaundice and even nightmares. The roots are used to treat wounds, rheumatism, gout, osteoarthritis, asthma, nervous disorders like epilepsy, hysteria, convulsions as well as to hemorrhoids and are known to have sedative and antispasmodic action due to volatile oil and peoniflorine present in the root. Root and seed were frequently used in the past for treatment of headache and asthma and to relieve pain associated with childbirth. Peonies have many other therapeutic uses.

3.0 Origin and History

3.1.0 Origin

There are about 35 species of peony that are native to Asia, Southern Europe and Western North America (Halda *et al.*, 2004). Of 35 species, 15 species and 5 subspecies are widely distributed in China, which is considered as the centre of origin (Yan *et al.*, 2020). The Chinese peony (*Paeonia lactiflora* Pall.) is native to central and eastern Asia from eastern Tibet across northern China to eastern Siberia. Common garden peony (*Paeonia officinalis*) is native to mountainous areas of Southern Europe, which was introduced to Central and Western Europe and North America. There are six known subspecies: *P. officinalis* subsp. *arietina*: native to Italy, Albania and Northern Caucasus, *P. officinalis* subsp. *banatica*: native to Bosnia and Herzegovina, Hungary, Romania and Siberia, *P. officinalis* subsp. *huthii*: native to South-East France and North-West Italy, *P. officinalis* subsp. *italica*: native to Italy. *P. officinalis* subsp. *microcarpa*: native to northern Portugal, Spain and southern France, *P. officinalis* subsp. *officinalis*: native to Austria, Switzerland, Hungary, Romania, Slovakia, France, Italy, Slovenia, Croatia, Albania and introduced to the United States in Kentucky, Massachusetts, Vermont, West Virginia and Canada in Ontario. *P. tenuifolia* is native to the Caucasus Mountains of Russia and the Black Sea coast of Ukraine, spreading to Bulgaria, Romani, Siberia and Kazakhstan. *P. anomala* is native of northern Europe, Russia and northern Mongolia and south to the Tien Shan Mountains. *P. clusii* is endemic to the island of Crete in Greece and have small population (Tzanoudakis, 1983). *P. obovata* is widely distributed in China (Hong *et al.*, 1988; Hong *et al.*, 2001).

3.2.0 History

In Europe, herbaceous peonies are a part of ancient Greek mythology and are highly regarded for their medicinal properties. The name peony is named after 'Paeon', the Greek God of medicine. It is believed that Paeon, student of Asclespiusa (the god of healing and medicine) had used roots of peony to heal the wounds of God Pluto after he had fight with Hercules. Out of jealousy and anger Asclespiusa threatened to kill Paeon. To save him from wart of Asclespiusa; Zeus blessed Paeon with a gift of immortality by transforming him into a Peony plant. Herbaceous peony, a perennial herb is a traditional Chinese flower having a cultivation history of more than 3900 years. The flower is called 'Sho Yo' in Chinese meaning most beautiful, and was honoured as the main flower in the Garden of Imperial Palace. Fascinated by the beauty and fragrance of the peony flower, poets during the reign of Tang Dynasty (AD 618-907), proclaimed peony as the 'Beauty of the Empire'. During Sung dynasty (10th century AD) peony cultivation had spread throughout China and Louyang, the capital city became the centre for peony culture. The city of Caozhou, presently known as HeZe of the Ch'ing dynasty (1644-1911) became the second centre for peony culture. In both the cities, in the present day, an International Peony Fair is held each year. By the eighth century, peony had spread to Japan where it was cultivated for beauty and further hybridization resulted in development of more than 300 cultivars. Japanese breeder Toichi Itoh developed intersectional hybrids in 1948 known as 'Itoh Hybrids'.

First botanical garden in England was established in 1621 which had a collection of two types of peony and later by mid-19th century, 250 types of *Paeonia* spp. were recorded (Abbey, 2015). Chinese peony seeds were first introduced to Europe about 1,200 years ago by Italians and first tree peony was brought from China in 1789 by Dr. Duncan and planted at Kew Garden. European breeders J. Calot, F. Crousse, A. Dessert, V. Lemoine, N. Lemon and E. Mechin in late 1800's and early 1900's developed varieties of *P. lactiflora*: Le Printemps' and Avant Garde. In 19th century, nearly 300 peonies were introduced into England by Kelway and son nursery. 'Peony valley' a perfume made of peony petal was made by Kelways in 1930's. The popularity of peonies spread across Europe and eventually to North America. *P. officinalis* was the first species to be grown in the gardens of the United States. Numerous *P. lactiflora* cultivars were introduced in the mid 1800's, making it a dominant type in the USA.

In 1904, the American Peony Society (APS) was formed and along with the Cornell University, it had taken up the first project to study the commercial value and nomenclature of the peonies. In Russia, the first garden varieties were introduced in the 17th century, during the monarchy of 'Peter the Great', where peony was cultivated for medical and botanical studies. A botanical study of peony in Russia is attributed to 1733, when Peter Simon Pallas had described more than 20 species. To fulfil the high demand for early market niches, cultivation of peony was started in Israel in the early 1990s (Kamenetsky and Dole, 2012). The growers in the West developed better cultivation techniques as well improved the ornamental qualities of flower by crossbreeding imported cultivars.

4.0 Taxonomy, Morphology and Classification

4.1.0 Taxonomy

Peony belongs to genus *Paeonia* of the order Saxifragales (Jian *et al.*, 2008) under family *Paeoniaceae.* Friedrich K.L. Rudolphi for the first time used the family name 'Paeoniaceae' in 1830 (Halda *et al.*, 2004). Earlier, the genus was included under family Ranunculaceae. Genetic analysis shows that peonies are most closely related to Crassulaceae which includes *Sedum* and *Crassula* (McGeorge, 2006). The genus *Paeonia* has been classified by many taxonomist, the most popular being developed by Sir Frederick J. Stern (1946), the English peony expert. The genus has been divided into 3 sections, 4 subsections 16 groups, 33 species and 13 botanical varieties. The genus is assigned under three sections: *Moutan*, *Onaepia* and *Paeoniae*. The section *Moutan* is divided into two subsection : *Delavayanae,* consisting of *P. delavayi* and *P. ludlowii*; *Vaginatae,* consisting of *P. catayana*, *P. decomposita*, *P. jishanensis*, *P. osti*, *P. qiui* and *P. rockii*. *P. brownii* and *P. californicum* are included under *Onaepia* section. Rest of the species are included under section *Paeonia*. *P. suffruticosa* is a hybrid of species exclusively belonging to *Vaginatae* subsection; consisting most of the cultivars of tree peony (Ji *et al.*, 2012). 25 species, including 40 subspecies and varieties were documented by Halda *et al.* in 2004. Classification of Paeonia by Stern, (1946) is given in Figure 1.

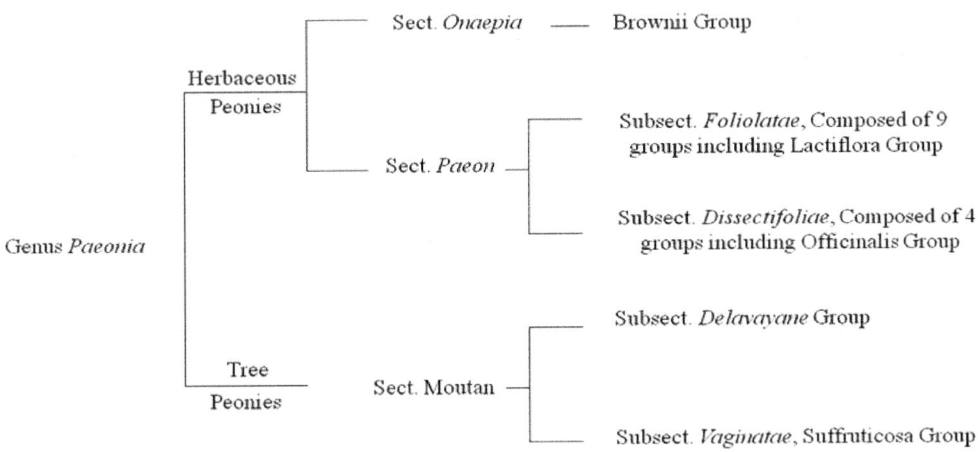

Figure 1: Composition and Classification of *Paeonia* (Based on Stern, 1946).

4.2.0 Morphology

The herbaceous peonies are bushy and grow upto a height of 50-100 cm. Peonies grow from underground tuberous stem (crown) comprising of two types of roots; the bulky storage roots and thin fibrous roots (Kapinos and Dubrov, 1993). In *Paeonia officinalis*, the roots put forth adventitious buds while in *Paeonia tenuifolia*, large clumps are formed in the underground stolon. The stem colour varies from green, pink to red with broad leaf to grass like leaf shape, which is oppositely arranged biternately. The foliage is a compound leaf comprising of a primary terminal leaflet and two lateral leaflets. At the base of the stem, the eyes/underground bud develops on the crown after commencement of flowering. The eyes are protected by several sheaths and emerge after fall in the spring. The new shoots grow rapidly and reach final height in 60-70 days.

Flowering begins in early spring from April and continues till the end of June depending on the location and cultivar grown. The flowers are large with five persistent sepals, five to ten petals, one to eight pistils, stamens upto 140 depending upon the species. The pods are fleshy and become leathery or woody with maturity, the seeds are smooth, large and brown to black colour. The plant remains green till September to October and then starts senescence, followed by overwintering for a period of 3 – 4 months.

4.3.0 Classification

On the basis of peonies introduced from China to countries such as England, USA and France in the 18th and 19th centuries, most of the research on the classification, description and identification of peony species has been conducted by scientists from Western countries. Peony species and cultivars are classified based on morphological characters, patterns of flavones/flavonol compounds in the petals or DNA analysis (Hosoki *et al.*, 1997). Based on flower colour, the cultivars are classified into 8 groups, consisting of white, yellow, light pink, pink, red, purple, dark red and orange; these being an important basis for classification. Based on the growth habit, peonies are classified mainly into two groups: Herbaceous peony and Tree peony; the third type includes the Itoh hybrids, an intermediate of two.

4.3.1 Herbaceous

These types of peonies are short with annual stems arising from the underground stem (crown). During summer, buds of various sizes develop on the current season growth. The smaller buds remain dormant while the larger buds grow into stem during the growing season. Differentiation of flower bud starts in October and comes to flowering in late spring and early summer after the annual shoot completes its growth and development and dies back in December (Kamenetsky *et al.*, 2012).

4.3.2 Tree

Tree peonies are shrub growing approximately up to 2 m height having perennial woody stem. Large buds develop at the tip of the annual growth during summer. It shed leaves during early fall and bears flower in early spring.

4.3.2 Itoh (or 'Intersectional')

It is an intersectional hybrid developed by Toichi Itoh from Tokyo in 1948 by crossing herbaceous *P. lactiflora* cultivar 'Katoden' with pollen from the yellow tree peony Alice Harding. These are herbaceous, having leaves like tree peonies and are wilt resistance. The flowers are large, comes to bloom from late spring to early autumn. Similar to herbaceous peony, Itoh after completion of annual growth in early summer the stem dies back in December (Pauwels and Vervoort 2009).

5.0 Species and Cultivars

The genus *Paeonia* consists of more than 35 woody and herbaceous species.

5.1.1 Herbaceous Species

Herbaceous peony contains nearly 30 species, namely *Paeonia lactiflora* (Chinese or common garden peony), *Paeonia officinalis* (European or common peony), *Paeonia kesrouanensis* (Keserwan peony), *Paeonia mascula* (Balkan, wild or male peony), *Paeonia parnassica* (Greek peony), *Paeonia tenuifolia* (Steppe peony), *Paeonia veitchii* (Veitch's peony), *Paeonia californica* (California peony or wild peony), *Paeonia cambessedesii* (Majorcan peony), *Paeonia brownii* (Brown's peony), *Paeonia algeriensis, Paeonia anomala, Paeonia arietina, Paeonia broteri, Paeonia clusii* ssp. *clusii, Paeonia clusii* ssp. *rhodia, Paeonia coriacea, Paeonia daurica, Paeonia daurica* ssp. *coriifolia, Paeonia daurica* ssp. *macrophylla, Paeonia dauricass*p. *mlokosewitschii, Paeonia dauricass*p. *tomentosa, Paeonia dauricass*p. *velebitensis, Paeonia daurica* ssp. *wittmanniana, Paeonia corsica, Paeonia emodi, Paeonia intermedia, Paeonia mairei, Paeonia obovata, Paeoniawillmottiae, Paeonia peregrina, Paeonia sterniana.*

5.1.2 Woody Species (about 8 species)

Rock's peony or tree peony: *Paeonia rockii*; synonym *Paeonia suffruticosa* subsp. *rockii* (Chinese tree peony/moutan peony), *Paeonia delavayi* (Delavay's tree peony), *Paeonia jishanensis* (Jishan peony), *Paeonia ludlowii* (Ludlow's tree peony), *Paeonia decomposita, Paeonia ostii* (Osti's peony), *Paeonia qiui* (Qiu's peony).

Description of important herbaceous peony species:

Paeonia lactiflora (syn *P. Albiflora*) (Chinese/Common Garden Peony)

The plant grows to a height of 60-100cm. Leaves are broad and lobed having pinnate section with lanceolate leaflets. The flowers are large, fragrant, available in different shades (white, to pink, rose and red) and forms (single, semi-double, double, bomb) with yellow stamen. The plant comes to bloom from May to June. These are prolific bloomer and are suitable for cut flower production.

Paeonia officinalis (Common Peony)

The plant is a perennial herbaceous shrub growing to a height of 60-70 cm. The leaves are pinnately arranged with lanceolate leaflets. The flowers are large having 10-13 cm diameter, with shades of pink to deep red colour in single or double form. It blooms in late spring.

Paeonia tenuifolia (Thin-leaved peony, Narrow-leaved peony, Steppe peony, Fern leaf peony)

It is an herbaceous perennial species growing to a height of 30-60 cm. The stems are hairless with alternately arranged compound leaf, which are finely divided to thread like segment resembling the fern leaf. The flowers are single type, scented, red coloured with numerous yellow stamens. The species comes to flowering earlier than other species of peony in May and dies earlier as well. It produces two-three hairy carpels which develop to fruit called follicles.

Paeonia anomala **(Anomalous peony, Pink peony, Xinjiang peony)**

It is a non-woody species growing to a height of 1 m having irregular fleshy taproot of 50 cm length and 2 cm width. The leaves are biternate compound leaf which is arranged alternately on the stem. The leaves turn vivid red at the end of the growing season. The plant comes to flowering from April to July, depending on the altitude; one to two flowers are borne per stem. The flower is a single type consisting of two to five leaf like bracts, three to five leathery sepals, six to nine oblong corolla which range in shed from pink to white and yellow stamen. The carpels are initially pale yellow with reddish stigma which later turns green and forms fruit which may be hairy or hairless.

5.2.0 Cultivars

The flower comes in wide range of shape and colours ranging from white and cream to a rare lemon colour to coral, pink and red. The blooms may be either single, semi-double, double or bomb in form based on type of flowers (Eason *et al.*, 2002).

Single: Broad single guard petals (5-10) are arranged in a flower encircling the functional stamens and carpels. This type of arrangement is generally found in wild forms. *eg.* Imperial Red.

Japanese: A single or double row of guard petals encircle the thickened staminodes containing viable pollen, carpels are visible, *eg.* Gertrude Allen.

Anemone: It resembles Japanese type in which a single or double row of guard petals encircle the staminodes that are narrow incurved petal-like but contain non-functional stamens and visible carpels.

Semi-double: In this type, some of the stamens are transformed into petals which are surrounded by a single or double row of guard petals and carpels are arranged in the centre. *eg.* Coral Charm, Coral Sunset.

Double: The flower have multiple rows of petals. The stamens and sometime the stigmas (non- functional) are transformed into petals, making the main body of flower. The guard petals are shorter than the pataloids, thereby giving it a globular shape. *eg.* Sarah Bernhardt.

Bomb: It is similar to the double type but the guard petals are longer than the petaloids, giving it a ball shaped centre; it has non-functional stamens and carpels. *eg.* Raspberry Sundae.

Most of the cut flower peony cultivars belong to *P. lactiflora*. Varieties having semi-double, double and double bomb flower forms are suitable for cut flowers. Few of the cultivars dominating the world market for cut flowers are given below.

Sarah Bernhardt: It is a late season, double type; producing deep pink fragrant flowers exclusively for cut flower, high-yielding and resistant to adverse environmental condition. The plant requires full sunlight for growth. Approximately 50 per cent production of peony for cut flower in the world belongs to this cultivar.

Species of Peony

Paeonia lactiflora

Paeonia officinalis

Paeonia tenuifolia

Paeonia anomala

Flower Forms

Single

Japanese

Anemone

Semi-double

Double

Bomb

Cultivars of Peony

Canary Brilliants

Dr. Alexander Fleming

Pink Hawaiian Coral

Yellow Crown

Red Charm

Coral Sunset

Sarah Bernhardt

Festiva Maxima

Kansas

Duchess de Nemours

Propagation of Peony

Pod with Seeds

Germination of Seeds

Division of Crown

Renewal Buds with
Crown Segment

Shoot Cutting

Root Cutting

Micropropagation of Peony-Shoot Tip Culture

Cultivation of Peony

Open Field Cultivation

Cultivation Under Protected Structure

Disbudding

Red Sarah Bernhardt: It is a late season cultivar requiring full sunlight for its growth. The flowers are scented, red coloured, double type, exclusively meant for cut flower.

Festiva Maxima: Flower is large, double type, scented and white coloured with red tint in the centre, mid season variety (May-June). It grows well under full sunlight.

Dr. Alexander Fleming: It is a cross between Sarah Bernhardt and Bunker Hill which was bred in 1950 by Dutch breeder Blonk. The flower is a double type, deep pink colour, mid season variety (May-June) and highly fragrant. Produces many side buds, thereby extending the bloom time. It can grow both in full sunlight and partial shade.

Duchess de Nemours: It is a mid season cultivar having double type, white coloured flower with creamish centre. The plant grows well under full sunlight. It is also suitable for border planting, mass planting.

Kansas: It is an early, double type, having red colour flower, which comes to bloom in mid to late spring and prefers full sunlight for its growth though it can tolerate partial shade.

Red Charm: It is a cross between *P. lactiflora* and *P. officinalis*. The flowers are deep red in colour, bomb type, an early season cultivar which prefers full sunlight.

Karl Rosenfeld: The flowers are cherry red, fragrant and double type. It requires full exposure to sunlight for its growth, and is good for mass planting. The plant comes to bloom from late spring to mid- summer.

Monsieur Jules Elie: The hybrid was developed in 1888 by Crousse in France. The flowers are light pink, fragrant and double-bomb type. It comes to bloom from early to mid spring. The foliage is purple when immature and later turns to dark green colour at maturity. The plant is susceptible to sooty mould but tolerant to acidic soil and grows well under both full sunlight and partial shed.

Edulis Superb: It is a pink colour, double flowered cultivar developed by Lemon in 1824 in France. The plant comes to flowering very early in the season and is highly fragrant. It requires full sunlight for its growth.

Crinkled White: The cultivar is a mid-season producing single white- coloured scented flower from late spring to early summer. It prefers full sunlight for growth but can be grown under partial shed as well. Suitable for border planting, mass planting as well.

Shirley Temple: The flowers are double light pink which fades to white at maturity, and have fragrance which is long lasting. It is a mid-season cultivar which can grow both under full sunlight and partial shade.

Coral Sunset: It is a semi-double type, fragrant flowers having intense rose coral colour which gradually fades to yellow. Mid-season cultivar (late spring to early summer), good for cut flower and mass planting. It was first bred by Samuel Wising in 1965.

Maestro: The plant is a mid-season cultivar, producing semi-double flower with deep red to burgundy colour.

Nick Shaylor: It is a late season cultivar producing double type of rose shape, light pink colour flower which later turns white; grows well under full sunlight. It was first bred by Allison in 1931 and later became famous in the United States.

Pink Hawaiian Coral: It is a hybrid of *P. lactiflora* 'Charlie's White' and *P. peregrina* 'Otto Froebel'. The flowers are large, fragrant, and double with pink colour petal and require full sunlight which comes to blooming from late spring to early summer. The plant requires staking, and is cold hardy, tolerant to heat and humidity, pest and disease resistant. It is suitable for garden, cut flower and border planting.

Itoh Cultivars

Yellow Crown: It is a mid-season hybrid. The plant produces semi double, fragrant, yellow flower with a hint of red at the base. The leaves turn red in spring and autumn, floriferous producer. The plant is tolerant to frost, grows both under full sunlight and partial shed. It is ideal for growing in pots and container and cut flower as well.

Yellow Dream: It is a cross between Kakoden (semi-double herbaceous peony) and Alice Harding (tree peony). The plant produces semi-double, fragrant, yellow flower. It is a vigorous plant reaching up to a height of 60 – 90 cm and blooms from late spring to early summer. The plant is tolerant to frost, grows both under full sunlight and partial shed. It is ideal for growing in pots and container and cut flower as well.

Scarlet Heaven: The plant comes to flowering in late spring to early summer. It produces single, deep scarlet red colour fragrant flower. It is cold hardy, resistant to diseases, grows well under full sunlight and does not require staking. Suitable for cut flower, border and mass planting, and in pot and container.

Canary Brilliants: The plant bears yellow to creamy apricot, semi-double to double type flower with light spicy fragrance. It can produce 30-50 blooms per season and can be grown in both full sun and partial shade in pots, container. The plant comes to bloom in late spring.

Scrumdiddlyumptious: It is a mid-season cultivar, producing flowers from late spring to early summer. The flowers are large, fragrant, double type, creamy white with soft pink hue. The plant grows well under full sun and partial shed and is suitable for cut flowers, border plant, mass planting and container garden.

6.0 Genetics and Breeding

6.1.0 Genetics

Most of the herbaceous peony species belong to section *Paeonia* that have evolved through reticulate evolution. Most of the *Paeonia* species are compatible, thus making interspecies hybridization possible. The ploidy level in species is basically correlated with the geographical distribution and presence of pubescence on the dorsal side of the leaf (Hong *et al.*, 2001). All the species under section *Moutan* and *Onaepia* are diploid (Hong *et al.*, 1988). *Paeonia* section comprises of both diploid (2n=10) and tetrapolid (2n=20) species, while few species have both the ploidy level. This is the only section of three taxonomic sections which comprises of tetraploid species constituting more than one-third of the 25 species in the section. Polyploids in peony have probably originated due to genomic divergence between diploid progenitors. Majority of the tetraploid species are allopolyploids derived from crosses between diploid species (Sang *et al.*, 2004) and are found in the Mediterranean region. It is believed that during Pleistocene era, due to climate change the tetraploid species diverged from diploid species. The tetraploid species are more hardy and adaptive to harsh environment than the diploid species. Therefore, it is due to this reason that the tetraploid species are gaining popularity (Rodrigo-Lopez, 2010). The species under section *Paeonia* having diploid chromosome number includes *P. anomala* (2n=10; hybrid of *P. lactiflora* and *P. veitchii*), *P. lactiflora, P. veitchii, P. tenuifolia, P. emodi* (hybrid of *P. lactiflora* and *P. veitchii*), *P. broteri* (hybrids of *P. lactiflora* and *P. obovata*), *P. cambedessedesii* (diploid hybrid of *P. lactiflora* likely *P. mairei*, but possibly also *P. obovata*), *P. sterniana* (diploid hybrids of *P. lactiflora* and *P. veitchii*), *P. clusii* (hybrids of *P. lactiflora* and *P. obovata*), *P. rhodia* (2n=10, hybrids of *P. lactiflora* and *P. obovata*), *P. daurica* ssp. *coriifolia, daurica, macrophylla* and *mlokosewitschii*. The tetraploid species includes *P. arietina, P. officinalis, P. parnassica, P. banatica, P. russi* (2n=10, hybrid of *P. lactiflora* and *P. mairei*), *P. peregrina* (hybrid of *P. anomala* and either *P. arietina, P. humilis, P. officinalis, P. parnassica or P. tenuifolia*), *P. coriacea* (hybrids of *P. lactiflora* and *P. obovata*), *P. mascula* ssp. *hellenica* (hybrids of *P. lactiflora* and *P. obovata*), *P. mascula* ssp. *mascula* (2n=20, hybrids of *P. lactiflora* and *P. obovata*), *P. daurica* ssp. *tomentosa* and *P. daurica ssp. wittmanniana* (hybrids of *P. lactiflora* and *P. obovata*).The autotetraploid species having both the diploid and tetraploid cytotype includes *P. obovata, P. mairei* and *P. clusii* ssp *clusii* (*2n=10, 20*).The most recognizable of these species *P. officinalis* or common garden peony, native to Europe which has been hybridized with others in the genus to produce bigger and better flowers, is an allotetraploid with two sets of chromosome from different parents (2n=20), is a nothospecies. *P. peregrina* is thought to be the one of the parent while the other parent might be the one related to *P. parnassica, P. arietina* and *P. humilis*.

6.1.1 Genes Responsible for Flower and Leaf Colour

The demand for peony in market is determined by the flower colour. Different composition and concentration of pigments such as flavonoids, carotenoids and betacyanins within the epidermal cell of petal is responsible for various shades of flower colour. Colouration is dependent on the expression of various genes responsible in flavonoid biosynthesis. Flavonoids comprises of chalcones, anthoxanthins (flavone and flavonol) and anthocyanins. Anthocyanin and its derivatives are the main pigments responsible for colouration with anthoxanthins acting as co-pigments. The major anthocyanin gene involved in colouration includes peonidin 3,5-di-o-glucoside (*Pn3G5G*), pelargonidin 3,5-di-o-glucoside (*Pg3G5G*), cyanidin 3,5-di-o-glucoside (*Cy3G5G*), peonidin 3-o-glucoside (*Pn3G*) and cyanidin 3-o-glucoside (*Cy3G*), peonidin 3-o-glucoside-5-o-arabinoside (*Pn3G5Ara*), cyanidin

3-o-glucoside-5-o-galactoside (*Cy3G5Gal*) and pelargonidin 3-o-glucoside-5-o-galactoside (*Pg3G5Gal*) (Jia *et al.*, 2008); the composition and content of these genes determine the flower colour in peony. According to Zhao *et al.*, 2012, eighth genes encoding phenylalanine ammonia-lyase (*PAL*), chalcone synthase (*CHS*), chalcone isomerise (*CHI*), flavanone 3-hydroxylase (*F3H*), flavonoid 3'-hydroxylase (*F3'H*), dihydroflavanol 4-reductase (*DFR*), anthocyanidin synthase (*ANS*), UDP- glucose: flavonoid 3-o-glucosyltransferase (*UF3GT*) are responsible for flower colour in peony. The pink colour is governed by DFR, red by ANS, and higher expression of anthocyanin o-methyltransferase (*AOMT*) is responsible for purple colour. The gene expression is higher in flower when compared to other plant parts and expression changes according to developmental stage of flower. The expression of *PAL* is high during bud and withering stage, the expression of *CHS* and *CHI* gene remains high throughout the developmental stage of flower while the expression of *F3H*, *F3'H*, *DFR*, *ANS* and *UF5GT* (Flavonoid 5-O-glucosyltransferase) gene diminishes with the developmental stage. The expression of PlUF3GT gene gradually increases with the developmental stage of flower. The blotch at the base of petal and carpel coat covering the tomentum is controlled by dominant gene.

Flower colour is also regulated by the transcription factors besides structural genes. The concentration of anthocyanin in red and white petal is controlled by the transcription factor *PsMYB* and *PsWD40*; different level of expression of both the transcription factor is responsible for multicolour flowering *P.* × *suffruticosa*. The petal blotch is regulated by *Ps MYB12* factor which activates the *PsCHS* gene in tree peony that is responsible for darker colour in the base of petal than other part of the petals. Besides *Ps MYB12*, the purple blotch in tree peony may also be regulated by *MYB-1* and *MYB-5* factor. The expression of transcription factor *bHLH3* and *MYB22* is high in white and purple petals respectively; the activity of both the transcription factor thus regulates the shades of petal in tree peony.

The foliage colour during the initiation stage is red, as the leaves expand, the red colour gradually fades and turns green. The accumulation of anthocyanin in red leaves is due to high expression of *CHS*, *DFR*, and *ANS* and low expression of leucoanthocyanidin reductase (*LAR*) and anthocyanidin reductase (*ANR*) genes. In springs, *MYB2* a repressor; inhibiting the expression of anthocyanin is activated by replacing *MYB1*. The expression level of *CHS*, *DFR*, and *ANS* gene is regulated by the transcription factor complex *MYB2* + *bHLH1* + *WD40-1*, leading to change in leaf colour to green from red.

6.1.2 Genes Governing Flower Type

The wild peony species produces single flower type; only after domestication and rigorous selection, now flowers of different forms ranging from semi-double, double and double bomb are available due to centripetal petalization of stamens. Development of floral organ in peony is controlled by gene belonging to MADS-box family; *APETALA2* (*AP2*) gene being one of this family regulates the formation of petals and sepals in herbaceous peony (*P. lactiflora*). *AP2* is an A-class gene in the ABCDE model responsible for organ development in flower as well as in various physiological and biochemical reaction (Wu *et al.*, 2018). The gene also encodes the transcription factors of family *AP2/EREBP* that plays an important role in formation of floral meristem, as well sepal and petals. The *AP2* gene is correlated with the formation of petal and the expression of this gene is higher in the outer petal than in the inner petal.

The expression of *PlAP1, PlAP2,* and *PlSEP3* gene increases with increase in degree of petal formation in *P. lactiflora* while *PlAP3-1, PlAP3-2,* and *PlPI* gene expression decreases. The expression of gene *PlAP2* also determines the formation of carpel besides determining the formation of sepals and petals. *PlAP1* and *PlSEP3* genes determine the formation of sepals and petals, expression of *PlAP3-1, PlAP3-2,* and *PlPI* genes mainly determine the formation of stamens and petals in flower. The genes *MADS5, MADS7* and *MADS12* are the key genes responsible for petalization of stamen.

6.1.3 Gene Governing Flowering Time

The transcription factor *PsCOL4* regulates the flowering time in *P.* × *suffruticosa;* the expression is highest in the stem and leaves during early stage of bud initiation and later decreases as the bud develops, besides inducing the expression of downstream gene Flowering locus T (*FT*) and Suppressor of Overexpression of Co1(*SOC1*) to ensure flowering. GA_3 regulates the expression of *PsSVP* gene during vegetative stage in *P.* × *suffruticosa,* mostly in leaves and stems. The suppressive expression of *PsSVP* (Short Vegetative Phase) gene in aborted flower bud is much higher than the normal flower bud, thus inhibiting flower formation. The genes responsible for secondary flowering in *P.* × *suffruticosa* include *PsAP1, PsCOL1, PsCRY1, PsCRY2, PsFT, PsLFY, PsLHY, PsGI, PsSOC1,* and *PsVIN3.* The gene *PsFUL1* is an important transcription factor regulating the floral transition and flowering in tree peony, thus responsible for occurrence of early blossoms.

The expression of few genes is regulated by external factors such as photoperiod, temperature, exogenous application of plant hormone *etc. PdFT* is the key gene involved in photoperiodic pathway that determines the flower regulation in *P. delavayi.* The expression of *PdFT* in flower bud is enhanced by the application of GA_3 @ 1000mg/l and defoliation; while, the expression decreases under low temperature (4°C) and short day (8 hrs photoperiod) condition. The exogenous application of GA_3 induces and inhibits the expression of *PsCPS* and *PsGA2ox,* gibberellin related gene in tree peony. Defoliation in addition to an exogenous application of GA_3, inhibit the expression of *PsNCED* and *PSbZIP* genes that coincides with ethylene production. The gene *PsGA2ox* inhibits flowering in tree peony.

6.1.4 Genes Governing Resistance to Abiotic Stress

The activity of antioxidant enzymes *viz.,* superoxide dismutase, peroxidase, and catalase and expression of Heat Shock Protein (HSP) is enhanced at high day (35°C) and night temperature (27°C) in heat resistant *P. lactiflora* cultivar. The genes regulating these enzymes and HSP remove the extra oxygen atom thus preventing plants from dying. The expression of *PlHSP70* gene increases with the raising temperature and thus provides tolerance against heat. Micro RNAs play an important role in posttranscriptional level in *P. lactiflora.* The expression of miR172c-3p, miR395a, miR397a, miR408-5p, and miR827 is enhanced under high temperature thus indicating the post-transcriptional regulation under heat stress. The glycerol-3-phosphate acyltransferase gene (*GPAT*) in *P. lactiflora* (*PlGPAT*) plays an important role in regulation against cold (Li *et al.*, 2018).Tryptophan decarboxylase (TDC) is the first enzyme gene engaged in melatonin biosynthesis pathway which converts tryptophan into tryptamine, thus protecting plants from drought and salt stress by increasing the expression of gene involved in oxidation-reduction process that protects the cell membrane and delays plant senescence, also increasing the potential of photosynthesis. Despite the presence of *PlTDC* gene in *P. lactiflora* that provides protection against water and salt stress, under extreme drought condition, the growth and flowering is affected (Zhao *et al.*, 2019). *PlWRKY13* gene belonging to family *WRKY* is responsible for resistance against abiotic (low and high temperatures, water logging, and salt), and biotic (red spot disease caused by *Alternaria tenuissima*) stress in *P. lactiflora*.

6.1.5 Genes Governing the Postharvest Life of Flower

The longevity of cut flower after harvest depends upon the rate of ethylene production, sugar accumulation and water metabolism. Concentration of ethylene is an important factor that determines the quality of cut flower during preservation. The genes involved in biosynthetic pathway of ethylene include *PsACS* and *PsACO*. In tree peony, *PsACS1* encoding the synthesis of enzyme 1-aminocyclopropane-1-carboxylic acid (*ACC*) is the key gene involved in ethylene biosynthesis of cut flower while *PsACO1* encoding the enzyme 1-aminocyclopropane-1-carboxylic acid oxidase (*ACO*) regulates the post-transcriptional level during flower opening. The transcription of *PsACS1* is inhibited by glucose of hexokinase signal transduction pathway, thus delaying the cut flower senescence. Aquaporins are the channel proteins that form pores to facilitate water transport between the cells. Aquaporin genes (*AQPs*): *PlPIP1;2*, *PlPIP2;1* and *PlNIP*, help water regulation in *P. lactiflora* thereby improving the keeping quality of cut flower in vase. The sucrose transporter gene (SUT):*PlSUT2* and *PlSUT4* and invertase gene (*INV*): *PlCWIN1*, *PlVIN1*, *PlCIN1*, *PlCIN2* and *PlCIN3* present in herbaceous peony are responsible for maintaining the vase quality of cut flowers when sucrose was applied exogenously (Xue *et al.*, 2018).

6.2.0 Breeding

In the ancient times, peony was cultivated and bred in order to harvest roots for medicinal purpose and until 11th century not for ornamental purpose (Abbey, 2015). Record of scientific hybridization coincides with the emergence of plant collectors in the 17th century who brought specimen from China through trade between the East and the West. The number of peony cultivar increased in West due to introduction of *P. lactiflora* which was crossed with *P. officinalis*, the species from Europe. The hybrids had better bloom with new shades like yellow, which was unseen in the plant before.

Besides, application in traditional medicine, peony is also popular for landscape architecture, potted plants and cut flowers. In the present day, though available for a very short period, peony is gaining importance in cut flower industry because of different flower form and shades. Peonies for cut flower must have long, strong stem to hold the large flower head. Despite having large flower of marketable size, there are very limited cultivars having good stem length and girth; majority being double flower in pink and red shade. Therefore, effort is being made to breed cut flower variety having greater shades of colour and shapes such as bomb type, good fragrance, year round producer, longer shelf life, resistance to biotic and abiotic stress *etc.*, despite of many problems in breeding programme, such as partial knowledge on the extent of germplasm resources, rudimentary breeding techniques, fragmentary breeding information, unclear breeding objectives and low breeding efficiency (Yang *et al.*, 2020).

Though the number of cultivars currently exceeds more than 8000, but the use of species from the wild in breeding programme is very limited, confining to *P. decomposita*, *P. rotundiloba*, *P. rockii* subsp. *rockii* and *P. ludlowii* in Moutan sect. which are rarely used, under *Paeonia* section, only few species *viz. P. lactiflora*, *P. officinalis*, *P. peregrina*, *P. daurica* sub sp. *macrophylla*, *P. tenuifolia*, *P. emodi* and *P. anomala* are used, while no species from *Onaepia* section are used. For the development of new cultivars, utilization of wild genes is very much essential *eg.* Lutea hybrid peony developed from cross between *P. lutea* and *P. delavayi*. The limited use of species thus gives opportunities to the breeders to exploit the wild resources. *P. lactiflora* is commonly used as one of the parents in all the breeding programme owing to its wide ecological adaptation, resistance to environmental stress, and variability in floral form and colour in the hybrids. Hybrid cultivars produced by crossing two or more sections of *Paeonia* are called herbaceous hybrid peonies; it comes to flowering much earlier than the *Lactiflora* peonies; having a prolonged flowering period of new shades of flower colour such as yellow, coral, bright red *etc.* In the present day, more than 1000 herbaceous peony hybrids are registered under American Peony Society.

The most common breeding method used in peony breeding is the selection and hybridization. During selection, the best progeny from the naturally cross pollinated population is selected and propagated, and further multiplied by vegetative means to maintain the true to type nature of the new cultivar.

6.2.1 Selection

Breeding by selection was started way back 1500 years ago in China for selecting fruit trees and found practical in tree peony breeding (Chen, 1953), when grafting was incorporated into propagation as of the Tang dynasty (Li, 1984). Successive repeated selection of peonies of higher ornamental value played a significant role in breeding of tree peonies. Natural bud mutations enhance the chances of selection in peony. The typical example of mutation selection is 'Shima-nishiki' a Japanese variegated cultivar from a bud mutation in 'Taiyao', in 1974. The former resembles the later almost in every aspect, except for petal variegation (Hashida, 1993). 'Shou An Hong' is a triploid cultivar evolved through natural mutation.

In Sichuan province, 'Shao Xin Chun' was a chance seedling from 'Xiang Yun'. Several open pollinated seedlings were selected and carried out on a greater scale at the time of Ming Dynasty (1368-1644) in Anhui Province and Quin Dynasty (1616-1911) in Caozhou and Shangdong province of which some are still be seen growing today. Similarly, in Japan during the Edo period (1600-1867) several classic cultivars termed Botan, or Japanese tree peonies (*P.* × *suffruticosa* Japan Group) which are most common in European gardens because of more frequent business from Japan with Europe (Hashida, 1996; Cheng and Li, 1998). In the last few decades, a number of cultivars were selected in *P. rockii* 'Gansu Group' from the open pollinated seedlings. The relationship between cultivars originated in China and species of tree peonies is given in Figure 2.

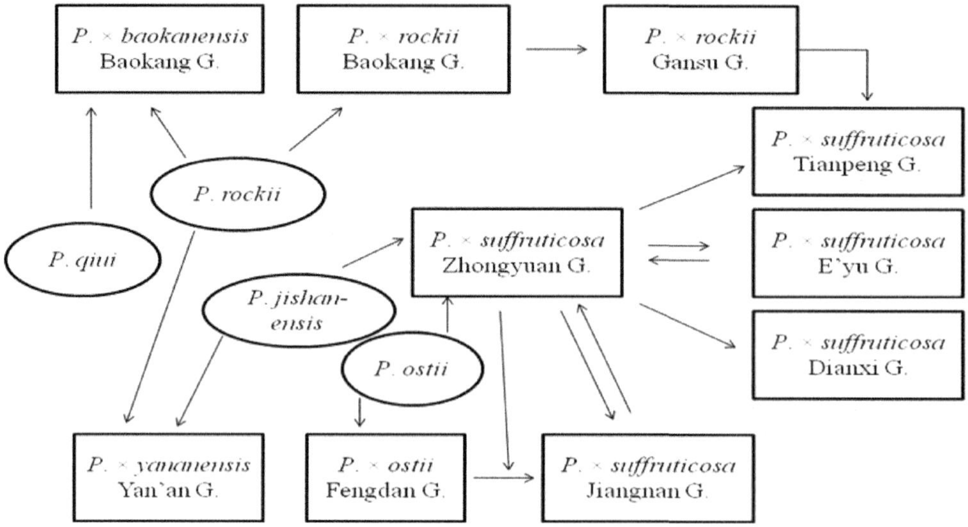

Figure 2: Relationship between Cultivar Groups Originated in China and Species of Tree Peonies (Cheng, 2007).

6.2.2 Hybridization

Hybridization in herbaceous peony is mainly done to improve the cut flower quality. For cut flower production, generally intra-species crossing in *P. lactiflora* is done besides interspecies crossing, which is mostly involved for generating diversity in herbaceous peony. Shirley Temple, Mother's Choice, and Shirayuki-hime are few of the cut flower cultivars developed through intra-species hybridization. Intra-species hybrids of *P. officinalis* and *P. tenuifolia* have also been developed for use in landscaping. The first interspecies hybrid in herbaceous peony was developed by crossing *P. lactiflora* (♀) with *P. wittmanniana* (♂).

Many peony species have been bred through hybridization and first intersectional hybrid (Itoh hybrid) cultivar was developed in Japan using herbaceous peony and tree peony. Of the three sections of peony, only the two sections, *Paeonia* and *Moutan* have been successfully used in hybridization programme till date using *Lactiflora* peonies as the female parent (♀) and Lutea hybrid peonies as the male parent, (♂). The leaf and flower types in these hybrids are mostly similar to the tree peony while the ecological habitat and flowering time is similar to the herbaceous peony. Generally, intersectional hybrids have wider adaptability, less susceptible to diseases and with more abundant flowers of white, yellow, pink, orange, purple, and scarlet shed. More than 150 cultivars are registered. Gene responsible for yellow or orange colour pigment (chalconaringenin 22-O-glucoside) in Lutea and Itoh hybrid is present only in the petals of wild peony species which is transferred through hybridization as well the gene responsible for fragrance in sub-section. *Delavayanae* flower is also transferred to offspring by this method.

Hybridization in peony involves both intra-sectional and inter-sectional crosses. The most successful intersectional hybrids were bred using *P. lactiflora* (♀) × Lutea hybrid peony (♂); though some successful cross combination also involve *P. lactiflora* (♀) × (*P.* × *suffruticosa*) (♂), *P. lactiflora* (♀) × *P. delavayi* (♂), (*P.* × *suffruticosa*) or Lutea hybrid peony (♀) × *P. lactiflora* (B♂), and *P. delavayi* (♀) × *P. lactiflora* (B♂). Intersectional crossing involving species of section Onaepia was carried out but only few selected combinations produced very few seeds and no cultivars were developed from such crosses.

In tree peony, intra-subsectional and inter-subsectional crossing is possible. Under section *Moutan*, all the species in subsections are cross compatible; the offspring are fertile and produces large number of viable seeds. The ornamental traits and resistance in tree peony are enhanced by crossing species of subsection. *Vaginatae* with *P.* × *suffruticosa*, crossing different cultivars of *P.* × *suffruticosa* with each other, and crossing wild species of same subsection; eg., Angel Emily, Hei Fu Ren, Wen Hai, Zi Yu Zui Xue, and Fei Hua Si Meng in *P.* × *suffruticosa*. Subsect. *Vaginatae* and subsect. *Delavayanae* when crossed, the offspring does not yield viable seed. The cross between *P. delavayi/P. lutea* (♀) and *P.* × *suffruticosa* (B♂) yield plants with heavy flowers of single or semi-double type and of different colours such as crimson, scarlet, apricot yellow and amber to golden yellow and lemon yellow but the F_1 hybrids are sterile. The sterility in hybrids of *P. lutea* can be overcome by back crossing for several generations.

Irrespective of the fact that many hybrid cultivars have been bred, breeding in peony is difficult due to limitation, such as poor self-compatibility, high chromosomal heterozygosity and long juvenile period. The limitation of conventional breeding can be overcome by molecular breeding which shortens the breeding cycle.

6.2.3 Mutation

Besides selection and hybridization, mutation breeding is also used to develop a new cultivar but the application and effectiveness is limited due to extremely large seed and thick testa of peony, which limit the penetration of chemical mutagen and radiation. Spontaneous mutation in the nature also gives rise to new mutant having variation. The cultivar developed through bud mutation is maintained through asexual propagation. Few cultivars that have occurred through bud mutation in Itoh hybrid and Lactiflora peony cultivars include 'White Emperor' and 'Alpine Aire, and Ann Styer', respectively and 'Shima-nishiki, Hua Er Qiao and Ice Age' in *P.* × *suffruticosa*.

7.0 Growth and Flowering

The peony produces seeds which are brown to black, smooth and relatively large. Seeds are prerequisite for propagation of the plant as well development of new cultivar by hybridization. During development of seeds, two types of embryo form successively; the sexual embryo is formed by syngamy, while the somatic embryo is formed by formation of embryoid cell from a somatic cell in the protoderm of sexual embryo (Batygina and Vasilyeva, 2003). The seeds have deep morpho-physiological dormancy to survive the adverse environment condition. Seed dormancy can be either epicotyl in all the species or both hypocotyl and epicotyl like in tree peonies. The dormancy is broken by exposure of seeds to low temperature. The seed germination in most of the herbaceous species is hypogeal, while epigeal germination is seen in *P. tenuifolia*. Due to underdeveloped embryo, seeds of many species take more than 8–9 months to germinate under natural conditions and sometimes may take up to 5-7 years. The survival percentage of the seedlings is very low (Halda *et al.*, 2004; Ren *et al.*, 2019).

In herbaceous peony, the monocarpic annual stems develop from underground perennial storage organ (crown) having large number of perennial roots. Peony has a sympodial growth habit. At the base of the annual stems, renewal bud of different sizes forms from crown surface which emerge as a stem after falls in the growing season. Renewal buds are protected by 4 – 8 layers of specialized leaf which firmly envelops the bud inside. The large buds produce new stems the following spring from March depending on the cultivar and climatic condition, while the smaller buds remain dormant till next season, or produces only small sprouts. Flowering begins in April and continues until the end of May. Flowers arise from both terminal and axillary floral bud in the current season stem. The terminal flowers are larger in size as compared to the axillary flowers. The plant undergoes resting period for 3 – 4 months during winter (Figure 3).

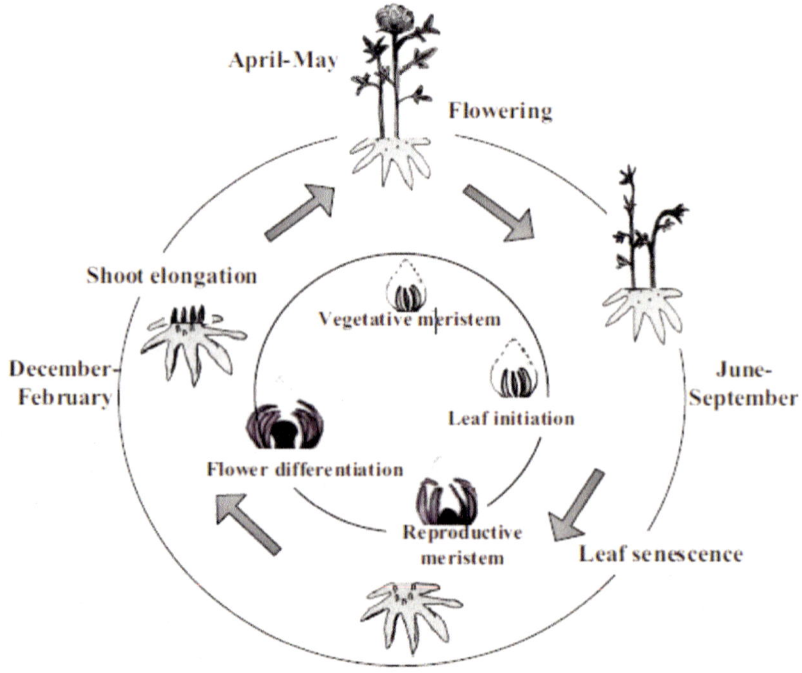

Figure 3: Annual Cycle and Floral Development of Herbaceous Peony 'Sarah Bernhardt' (Barzilay *et al.*, 2002).

7.1.0 Monocarpic Shoot Morphogenesis and Flowering

After completion of flowering in the annual stem, at the end of June/July, monocarpic shoots begin to develop in the renewal bud by formation of leaf primordia in the apical meristem. After initiation of leaf primordial in all the renewal buds by end of August, 4 leaf-like bracts arise symmetrically on the periphery of apical meristem revealing the end of vegetative stage. By the end of September or early October the meristem of all the buds become generative which is marked by visible change in shape of the apical meristem from flat to round and then to dome shape but remains undifferentiated. By late October the apical meristems expand and begin flower differentiation. Differentiation of floral parts occur centripetally, beginning from the periphery of the generative meristem. The differentiation of the floral bud ceases after the formation of floral parts- bracts, calyx, corolla, androecium and gynoceium by the beginning of December. After the termination of differentiation, the androceium and gynoecium are not formed. Each renewal bud contains 3–6 axillary shoots besides the central flowering shoot. The apical meristems become generative simultaneously with the central floral bud in larger axillary shoots along with differentiation of floral parts occurring concurrently in both the central and axillary buds. The smaller axillary shoots however, remains vegetative. The fully differentiated monocarpic shoot slowly elongate within the renewal bud. After prolong chilling period, with the rise in temperature during spring, the dormancy is broken and flower stem elongates. The ability of the bud to come out of dormancy and swell up is the determining factor for the number of days to anthesis which varies with the cultivar. Morphological structure of the herbaceous peony is given in Figure 4. for reference.

Larger plant produces more flowering stem owing to more accumulation of resources for reproduction. Peony is a hermaphrodite plant but gender modification is dependent on the size of plant. Larger plant produces more female flowers than the smaller plants (Kamenetsky and Dole, 2012). Peonies are deciduous in nature which undergoes dormancy during winter. Carbon required for breaking the dormancy and initiating the growth during subsequent spring is supplied from the reserve stored in the roots (Walton *et al.*, 2007). The plant is rich in all the sugar forms comprising of fructose, glucose, sucrose and myo-inositol, starch being the main carbohydrate. In the spring, with the resumption of growth, concentration of starch declines in the roots while begin to increase after shoot emergence and continue throughout the flowering period till mid-summer, thereby giving advantage to the plant to survive under adverse climatic condition specially under short growing period. The flower bud accumulates starch prior to flowering; initiation of starch hydrolysis coincides with the loosening of the tightly furled bud which later hydrolyses rapidly with the complete opening of the flower. Stem elongation and flower development are influenced by change in temperature regime. Unfavourable temperature regime can cause either too slow or too fast elongation of stem or flower abortion resulting in no flowering or flower of low quality. Cheng *et al.*, (2009) have divided the growth and development of peony in seven stages. Stage 1 comprising of onset of re-growth of plant after dormancy which is evident by swelling of bud from the crown. 2nd stage involving bud sprouting and shoot emergence followed by shoot elongation and appearance of flower bud in the 3rd stage. Stem length is about 2/3rd to 4/5th of its final height in the end of third stage; petiole elongation occurs with the opening of the leaflets which remain curved, flower bud becomes visible. Early flower abortion can be noticed at this stage. In the 4th stage, extension of leaf occurs which become flat and droopy and flower buds are visible above the upper leaves. Enlargement of flower bud and secretion of exudates occur in the fifth stage following extension of flower stem above upper leaves. Sixth stage indicates the bud maturity; the flower buds show change in petal colour which is followed by opening of flower in the final stage.

Figure 4: Morphological Structure of Herbaceous Peony.

(A) Entire plant prior to flowering (B) Schematic presentation of the monocarpic shoot (adapted from Cheng *et al.*, 2009). (C) Renewal buds (eyes), developing on underground crown. (D) Underground organs: crown, storage roots and developing shoots (Kamenetsky and Dole, 2012).

8.0 Propagation

Peonies are propagated by both sexual and vegetative means.

8.1.0 Seed Propagation

Commercial propagation of peonies through seeds is not recommended due to complex nature of seed germination, long juvenile period and variation in seedling trait. The mature peony seeds have hard seed coat and most of the species have epicotyl dormancy. The seeds require warm stratification for embryo growth and radical protrusion followed by cold stratification for epicotyl growth. By artificial means, the plant comes to first flowering 27 months earlier than the seed sown in the soil, where it takes 2 seasons to break the dormancy and come to germination.

The seedling plant has long juvenile period of about 4 – 6 years and the commercial flower harvesting stage is attained after 7 years. Seeds should be placed in moist soil-less media, for a month in dark at warm temperature (20°C). The seeds after emergence of radical are transferred to pots containing growing media and are kept for one month under greenhouse condition. After a month, the pots are placed back in cool (4°C) and dark room for 15 – 30 days to stimulate aerial growth. The pots are placed back in green house (15-25°C) after development of first leaf (Jacob *et al.*, 2006).

8.2.0 Vegetative Propagation

Commercial propagation in peony is done through vegetative means. Several vegetative methods are involved in propagation of peony.

8.2.1 Crown Division

It is the most popular technique used for propagation of herbaceous and Itoh peonies. For successful propagation, defoliation is done in late summer or early autumn after digging out the plants with crown and roots intact. Division of crowns having at least one bud is used, however, the crowns with three to five buds per division produces strong and healthier plants. In Dutch, the division of crown is done from October to November after harvest and stored till January to March prior to shipment of planting material (Kamenetsky and Dole, 2012).

8.2.2 Renewal Buds with Small Crown Segment

The technique is similar to crown division. The crowns after harvest in late summer are divided into many small fragments, each containing a small bud. The propagules are planted 3 – 4 cm deep in beds and after one year new and well developed plants are replanted in the main field. This method is applicable for mass multiplication of rare varieties and rejuvenation of old plants (Kamenetsky and Dole, 2012).

8.2.3 Shoot Cuttings

The success rate through green shoot cutting is relatively low which accounts for 15-20 per cent. This method is not suitable for many hybrids but is preferred for propagating very rare varieties and when plant is to be propagated without injuring the crown of mother plant. Cuttings are made two weeks prior to flowering. Only one-third of the shoots are removed from an adult plant. Two to three cuttings containing leaves (2) and internodes are made from each shoot. For shoot tip cuttings, in order to reduce the water loss by transpiration, the lower leaves and two third of the upper leaves are removed. The cuttings taken from the lower part of the stem which are located close to the crowns give best rooting. Rooting is increased by treating cuttings with indole-3-acetic acid (IAA) or naphthalene acetic acid (NAA) @ 5-10 mg/l for 12 – 24 h prior to planting. The cuttings treated with 2000 mg/l indole-3-butyric acid (IBA) by quick dip method gives rooting percentage up to 86.7 per cent (Guo *et al.*, 2009). The cuttings are planted in well drained substrate with the lower node inserted at a depth of 3 – 4 cm and 10 – 20 cm spacing to avoid rotting and disease infection. The optimum temperature for rooting is at 20 – 25°C with 90-95 per cent relative humidity. In the initial first two weeks, shading should be provided to the cuttings and prophylactic spray of fungicide should be given to prevent rotting. Callusing takes place 3 weeks after planting while roots are visible 1.5 – 2 months later. Irrespective of rooting, the plant does not survive until the formation of renewal bud in the lower node, formation of bud being the critical stage for development of cutting. Rooted cuttings are to be protected during the first winter. The cuttings commence to flowering three to four years after planting.

8.2.4 Root Cuttings

The roots are divided into 1 – 3 cm long sections during autumn and planted in sand under protected structure. To protect the cuttings from frost injury, the temperature is maintained at 3-8°C. Renewal bud begins to form during February-March and with raising temperature during April-May, new shoots start to sprout. After 3 years, the plantlet is ready for transplanting in the main field (Kamenetsky and Dole, 2012). This method is successful for propagation of *P. officinalis, P. peregrina* and *P. tenuifolia* hybrids, that produces renewal bud on storage roots. Few hybrids *viz.*, Carol, Henry Bockstoce, Ellen Coweley are successfully propagated by this method. It is not feasible for *P. lactiflora* hybrids, as adventitious roots do not produce buds (Kamenetsky and Dole, 2012).

8.2.5 Layering

This method is otherwise also called as 'Chinese technique' and was initially used for propagation of tree peony, later it was also adopted for herbaceous peony (Rogers, 1995; Uspenskaya, 2002). In this technique, a portion of stem is forced to produce roots while still attached to the mother plant either in early spring before the new growth occurs or in the late summer or early fall. When the bud begins to sprout in lower branches during spring, these are exposed and covered with a growing mixture of sand and compost incorporated with 300 – 400 g of bone meal and 150 g super phosphate per plant. The branch is covered with a frame of 50 × 50 cm dimension and 40 cm height. After the shoots have sprouted, the frame is filled with sand/compost up to 10 cm with elongating stem, the substrate is filled up till 25 – 30 cm height followed by regular watering for successful rooting. In the late summer, the rooted shoots are separated from the mother plant and planted in the nursery. The same mother plant is used only once in 3 – 4 years (Kamenetsky and Dole, 2012).

8.3.0 Micropropagation

Conventional propagation method by division of rhizome containing 3 – 5 eyes in peony is very slow. The problem of long juvenile period and multiplication of virus free planting material or new cultivars can be overcome by micropropagation technique, thus tissue culture is employed to improve the efficiency of propagation in peony. The mass production of peony cultivar is possible through rescue of seed embryo of the progeny from distantly hybridised germplasm. Despite the efficiency of propagation, tissue culture encounters the problem of explants browning, difficulty in rooting, induction of shoot from callus and lack of protocol for somatic embryogenesis. Thus, overcoming these limitations will help realise the full potential of micropropagation. According to Albers and Kunneman (1991), the most important factor for getting aseptic culture is the year of initiation, and the cultivars. The growth rate of explants in the initial period is very slow. The percentage of contamination in initiation medium of herbaceous peony (*P. lactiflora* and *P. officinalis*) ranged from 22-100 per cent because the buds grow on sub-soil while in tree peony (*P. suffruticosa*) it is negligible as the axillairy buds are not sub-soil. The multiplication rate of explant in Lepoivre based medium ranges from 1.3 – 2.9 at 15°C, 35mmol/s/m^2 light intensity and a day length of 16 hrs. Low temperature of 1-4°C has influence over multiplication rate. The rooting percentage is higher in IAA or IBA (0.1 mg/l) than in NAA at 12.5°C under dark condition rather than in white or red light after 12 weeks of culture. About 50 per cent of the plantlets are lost during hardening due to microbial infection. Gabryszewska (2010) reported the possibility of leaves as an important factor stimulating shooting and rooting and inhibiting formation of new adventitious buds. In the culture media devoid of plant growth regulators, the number of roots in explants containing leaves enhanced with the increasing concentration of glucose. No rooting was observed in explants devoid of leaves in the culture media containing glucose (30 g/l), IBA (1 mg/l), GA_3 (1 mg/l) or kinetin (1 mg/l).

Hosoki *et al.* (1989) reported formation and growth of axillary buds in the shoot tip of *Paeonia lactiflora* cvs. 'Takinoyosooi' and 'Sarah Bernhardt' when cultured in MS media supplemented with 6-benzylaminopurine (0.5 mg/l) and gibberellic acid (1 mg/l). Continuous shoot multiplication was done every 36th day by vertically splitting the shoot axis and subsequent division of elongated axillary shoots. Rooting percentage was high (57–100 per cent) on paper-bridge liquid medium when supplemented with 1 mg/l indole-3-butyric acid. From a single bud in a year, 700 and 300 plants of cv. 'Takinoyosooi' and 'Sarah Bernhardt' could be obtained, respectively.

Callusing of flower buds and pedicels are induced by culturing in MS media supplemented with 2-isopentenyladenine (2iP) and PIC (Radtke, 1983). Filaments and petals serve as a suitable explant to obtain callus in tree peony as these parts do not produce phenolic compounds unlike other parts which prevent the growth of tissue and are capable of regenerating shoots after 8 weeks in media supplemented with thidiazuron (Beruto *et al.*, 2004). The developmental stage of bud and the capability to micropropagate are linearly correlated. The success rate of culture is higher in the buds with expanded leaves which give rise to higher number of axillary buds, avoiding callusing. The multiplication rate of the plant decreases with increasing number of subcultures and are dependent on the genotype of the cultivar, which varies between 2 – 5 per cycle. Contamination and browning of the media adversely affect the growth of plantlets. Herbaceous peonies can be propagated by somatic embrogenesis of zygotic embryo obtained from callus, microspore or anther culture. The anthers of *P. albiflora* give rise to embroid in growing media (MS culture) containing 0.2 mg/l NAA and 0.4 mg/l TDZ. Rooting and formation of adventitious bud occur in media containing 0.5 mg/l BA and 2μg/l TDZ, while secondary embryogenesis takes place in MS media containing 0.002 mg/l TDZ (Lee *et al.*, 1992). Low temperature (5°C) treatment of *P. albiflora* cotyledon 10 days prior, improve the somatic embryogenesis in MS medium containing 2,4-D (0.2

mg/l) and BA (0.2 mg/l) while in media containing only 2,4-D (0.5 mg/l) non-embryogenic callus formation occurs followed by high rooting percentage when supplemented with glutamic acid (Kim and Lee,1995; Kim and Lee, 1996). High temperature (20-25°C) induces dormancy in herbaceous peony shoots, reduces the growth and multiplication rate, causing senescence and death of the leaves. Exposure of shoots to low temperature (4°C) for 2 months prior to culture reactivates the growth and development of shoots. The fresh weight of shoots is very high in first subculture of non-dormant shoots which produces more number of shoots and leaves when propagated after cooling at 15°C. The dormant (non-cooled) shoots produce lesser number of shoots and leaves than the non-dormant (cooled) shoots. As the temperature increases, the shoot and leaf growth decreases in dormant and non-dormant shoots. The multiplication rate of the shoots is independent of temperature and decreases with subsequent subculture (Gabryszewska, 2004).The dormancy in zygotic embryo of *P. lactiflora* is broken by GA_3 treatment @ 3mg/l along with application of 1g/l lactoalbunin hydrolysate (LH), 0.1g/l vitamin C and 3g/l AC (Active Charcoal) in MS media (Lan, 2003). Glucose (17.13g/l) enhances the axillary bud induction in embryo culture followed by GA_3 (1.84mg/l) and sucrose (11.02g/l) (Yu *et. al.*, 2008)

The tree peony has both hypocotyl and epicotyl dormancy, germination is induced by breaking both the dormancy. After ripening, the seeds are given cold stratification to break the dormancy. Hypocotyl dormancy of embryo is first broken under dark using the MS medium, and subsequently epicotyl dormancy is broken under blue light by supplementing the MS medium with 1.0 mg/l GA_3 and 0.5 mg/l BA (Ren *et al.*, 2019).

9.0 Soil and Climate

9.1.0 Soil

Peonies are well adapted to all types of soil but well drained clay loam soil rich in organic matter gives good flower yield. The plant can survive in acidic, saline soil but optimum pH for the growth is 6.5. Good drainage is important to avoid root rot and fungal diseases. Soil temperature greatly influences the growth and development in peony because the renewal bud of peony is located underground, thus apical meristem and root growth is affected by it. High soil temperature causes slow release of dormancy; poor stem elongation and often causes flower abortion thus having negative effect on plant phenology. The release of dormancy is measured by chill-heat unit (Stevens *et al.*, 1993).

9.2.0 Climate

Peonies are often found to be growing at high altitudes on cliffs in rocky and inhospitable terrain in its natural habitat. It grows successfully in areas having mild to cold winter but recently it is also under cultivation in the areas having warmer climatic condition. Herbaceous peonies require winter chilling to set flowers and therefore are grown for cut flower production in climates that have cold winters (Eason *et al.*, 2002). The plant is independent of temperature and photoperiod for the vegetative growth, floral initiation and differentiation while dormancy breaking of both seed and bud, and floral stalk elongation are dependent on temperature. Peonies need an extended period of cold, at least 900 hours around or below freezing in order to undergo dormancy and bloom in the following springs. Generally, transition from vegetative to the reproductive phase in peony occurs from September to October, thus temperature regime during the summer period may accelerate or delay intrabud florogenesis. In absence of low temperature, the plant fails to put flower (Jacob *et al.*, 2006). In warmer areas like Iserial, prolonged chilling prior of 8-10°C night temperature and 16-26°C day temperature with 17°C average temperature from November to February promotes dormancy release and flower stem elongation. A temperature regime of 2°C for 60 days or 1-7°C for 3-12 weeks were beneficial for breaking dormancy in Coral Sunset, Monsieur Jules Elie, and Sarah Bernhardt cultivars (Din *et. al*, 2014). The sprouting is delayed if duration of chilling is prolonged after completion of chilling treatment. The stem height, flower size, rate of flowering in peony cultivars is positively correlated to chilling. The crop prefers full sunlight for growth but few species can also be grown under partial shed.

10.0 Cultivation

10.1.0 Open Field Cultivation

10.1.1 Land Preparation

Land is prepared a month prior to planting by giving deep ploughing to a depth of 30-90 cm. Big clods are broken and soil made into fine tilth, deep ploughing facilitates root growth owing to tap root system, thereby increasing the plant size and amount of bloom. Farm Yard Manure (FYM) is added during land preparation. Peonies being a perennial crop with a productive year of 25 years or more require good supply of organic manure prior to planting.

10.1.2 Planting

The plant grows well under full sunlight, but also performs well under partial shed. Peonies do not respond well to transplanting therefore, the location subjected for planting should be permanent as the plant is perennial in nature. Healthy root cuttings having 2-3 buds from commercial varieties are collected and planted the following winter. Planting of injured root cuttings and cuttings having broken bud is not preferable as these are prone to rotting or give rise to weak plant which is not desirable. The root cuttings must be kept moist prior to planting. In case of dried roots, these are soaked for several hours in water before planting. When plants are to be kept longer for a year, long trenches are dug and root cuttings are planted 15 cm apart which is then covered with soil. Plant enters dormancy by last week of September thus planting is done during this time before the snow, preventing the root cuttings from damage. The plant is less vigorous during the first year due to under developed root system. The cuttings which are stored in cold storage may also be used for planting in the spring.

The planting depth depends on the soil type. For clay soil, planting of root cuttings is done at a depth of 2.5 – 5.0 cm while in light soil, it is planted at a depth of 5.2 – 7.6 cm. Planting too deep should be avoided as it hampers the growth of emerging buds by developing many branching stems between the roots and the shoots. Planting too shallow will kill the roots due to winter frost or may be damaged by the injury during cultivation which causes rotting. Extra care is required for the smaller divisions, which should not be planted deeper than 2.5 cm depth. While planting, care should be taken not to plant the roots upside down. Soil is compactly filled after planting which is followed by watering and mounding with soil few centimetres above to protect the bud from winter frost and roots from heaving. The following spring, the mounds are levelled. As the plant grows old, crown is pushed outward which is covered with soil (Stevens *et al.*, 1993).

10.1.3 Spacing

Spacing mainly depends on the method of planting. Generally, a spacing of 60-90 cm is maintained between the plants and 1.2 m between the rows. Closer spacing less than 60 × 60 cm accommodates more number of plants but reduces the longevity of the plant from predicted 10-25 years to the lower end. In a double row system, spacing of 45 cm between the rows and 25 cm between the plants are maintained (Stevens *et al.*, 1993; Klingman, 2002).

10.1.4 Nutrient Management

FYM @ of 8-10 t/ha should be applied at the time of field preparation. Application of FYM just prior to planting should be avoided to prevent rotting of eyes. Fertilizer should be applied according to the need of the plant. Band application of NPK @ 35:35:35 kg/acre/year gives good plant growth and flower production. Fertilizer application is done twice a year; once, during shoot emerging stage in the month of February- March when half the amount is applied and the other half during fall (September- October) when plant undergoes dormancy. During fertilizer application, care should be taken to apply fertilizer in the drip circle (15-45 cm) away from the plant and mixed properly with the soil. Excess application of nitrogen should be avoided (Stevens *et al.*, 1993).

10.1.5 Irrigation

For good production of cut flowers, optimum amount of irrigation is to be supplied. The plant does not give good yield under rainfed condition; therefore assured irrigation for good cut flower production is given through drip irrigation. Overhead irrigation is not recommended as the quality of the flower gets deteriorated by spotting of petals, splashing of soil on the foliage and spread of the diseases. The amount of water required and the frequency of irrigation depend on the weather and crop maturity. Irrigation is scheduled, based on the soil moisture content in the root zone. Excess or less irrigation should be avoided which leads to root rot and reduction in crop production, and quality, respectively. The plant can withstand drought upto some extent but optimum amount of water should be supplied for proper growth and development. Water deficiency during flowering reduces the yield as well as the flower quality.

10.1.6 Weeding

Weeds in the plot reduce the flower production in peony. Common weeds that grow in the peony field include *Abutilon theophrasti, Amaranthus blitoides, Amaranthus spicatus, Andropyron repens, Capsella bursapastoris, Chenopodium album, Cirsium arvense, Crepis biennis, Digitaria sanguinalis, Echinochloa crusgalli, Galinsoga parviflora, Matricaria matricarioides, Panicum capillare, Polygonum aviculare, Populus deltoids, Portulaca oleracea, Senecio vulgaris, Solanum nigrum, Setaria lutescens, Setaria verticillata* and *Taraxacum officinale.*

Pre-emergence application of weedicide is least effective and reduces the weed population by 29 per cent while post-emergence application controls the weeds by 76 per cent (Calkins *et al.*, 1996). *Senecio vulgaris* (Common groundsel) is the most prevalent weed during cool and wet condition, and is found to grow throughout the growing season as long as moisture is available. Pre-emergence application of Balan 2.5G (benefin) @ 3.36 kg a.i./ha, Stomp 60WDG (Pendimethalin) @ 4.48 kg ai/ha or Snapshot 80DF (isoxabenjoryzalin) @ 4.48 kg a.i./ha helps control grasses and broad leaf weeds. The phytotoxicity of herbicide depends on the species, age and time of application. Generally, young plants are susceptible to herbicide injury than the well-established plant. The herbicides are less phytotoxic when applied in dormant stage during fall (September- October) than during the active growing stage in spring (February-April).

Herbicide application is followed by two-three hoeing and hand weeding to control the weeds effectively. During hoeing care should be taken to avoid injury to the roots as feeder roots grow on the surface and thus hoeing should be done not more than 5 cm deep near the crown (Calkins *et al.*, 1996).

10.1.7 Mulching

Mulching also effectively controls weed besides maintaining soil temperature, prevents soil splashing during rain, prevents damage to plant from alternate freezing and thawing and gives good quality plant. After winter, in spring the mulches are removed to avoid disease build up in the emerging shoots. Mulching materials such as, dry leaves, wheat straw, wood chip, newspaper *etc.*, are used. Generally, mulching is done in the initial one year of planting while for smaller divisions and young seedlings, it is done for 2-3 years. Avoid using manures and dead leaves and stems of peony as mulching materials. The mulches are removed immediately in springs when the shoot breaks out.

10.1.8 Disbudding

Depending on the market demand disbudding is practiced. Lateral buds growing in the leaf axils are removed when axillary buds start to form. For production of big size single flower per stem, the lateral flower buds are removed keeping only the terminal bud intact, to divert the flow of all the plant resources to a single bud. To promote the development of lateral buds in spray type cultivars, the terminal bud is removed. Lateral buds are allowed to grow for flower having short stem length, and more number of smaller size flower. Staking is not required for the plant which is grown for cut flower production as the flowers are harvested in the bud stage (Stevens *et al.*, 1993).

10.2.0 Protected Cultivation

Peonies are drought tolerant, cold hardy and generally resistant to diseases. It is popular in temperate region for different flower forms and beautiful shades of colour. Though in great demand, the availability of peony flower is only for a short period from late spring to early summer, thus emphasis is made for availability throughout the year by growing in non-traditional areas of southern hemisphere and forcing it under protected cultivation. Site selection is an important operation for production of peony. The crop grows well in well-drained soil rich in organic matter, supplemented with optimum fertilization during early spring and immediately after dormancy. The plant produces flowers successfully in areas having 2-3 months of low temperature, followed by warm temperature in summer. For growing peony under protected cultivation, plants are either planted in pots under controlled environment condition or in field which is covered with mulch. In pots under protected structure, early crop can be obtained from mid-January. In field condition, planting is done under uncovered greenhouse/high poly tunnel, the plants are initially exposed to low temperature until they receive a certain number of predetermined chill units, then they are covered with plastic sheets, to raise the temperatures. The plant commences flowering one month earlier than the one planted in open field condition (Kamenetski, 2007). Peony being a day neutral plant, flower initiation is independent of day length but is dependent on the temperature, thus providing supplemental lighting or shading is insignificant for flower production. Under greenhouse, the temperature is regulated for crop production. Shortening of dormancy period is advantageous for the cultivation of peony under protected cultivation. Growing of root cuttings or crown division stored at 2°C for six weeks reduces the interval for dormancy breaking and helps produce best quality bloom. The root cuttings/divisions are either planted in containers or in ground, the growing media is drenched with 250 ml GA_3 @ 100mg/l. Shoot and bud emergence occurs in about 8 weeks at 22°C day and 10°C night temperature. In the ground, the plants are spaced at 60 cm apart between the row and within the row. Overhead irrigation is avoided to prevent the build up of diseases; water and fertilizer is applied through drip system which helps save water as well makes nutrient available in the root zone. Fertilization is done once during shoot emergence and the other when the plant has entered dormancy. New shoots and flower buds form on crown in 12 weeks interval at 28°C and 10°C day and night temperature, respectively. Protected cultivation provides the scope of harvesting flowers earlier by 8 weeks from 12 weeks of flower bud initiation by shortening the time thus making the flowers available in offseason. Depending on the cultivar, harvesting for cut flower is done when the buds are of 3.1 – 4.4 cm in diameter at marshmallow stage; the buds are soft yet remain closed with at least 30 cm stem height (Stevens *et al.*, 1993). During harvest when temperature is above 26°C, harvesting may be done once a day, since heat speeds up the blooming process. Gray mould is a serious problem in peony under protected cultivation with 16-24 °C optimum temperature for growth and sporulation (Shining, 2011). To prevent disease build up, proper air circulation should be allowed between the plants and prophylactic spraying of fungicide is done.

11.0 Diseases and Pests

11.1.0 Diseases

Peony is infected by many fungal, bacterial and viral diseases; however most economically important being gray mould and red leaf spot.

11.1.1 Botrytis Blight or Gray Mould (*Botrytis cinerea* or *Botrytis paeoniae*)

Brown to black rot is observed below ground at the base of stems. Grayish fungal spore growth is visible on stems just above soil line. Large irregular, brown spot develop on leaves, the infected flowers turn brown and later gray fuzzy fungal spores appear both on infected stem and flower. To control the disease, avoid overhead irrigations; maintain low humidity under protected cultivation. Remove the infected and dead plant parts, clean the basin area during rainy season. Soil application and foliar spray of mancozeb or ferbam @ 2 g/l during early spring in the month of March as a prophylactic measure and repeat at 7 – 14 day intervals when necessary.

11.1.2 Phytophthora Blight (*Pytophtora cactorum*)

The growth of *Pytophtora* is favoured by cool and wet weather during spring. The disease is severe in ill drained soil and the pathogens spread through splashing water, contaminated tools, soil or planting material. The pathogen inhabits the soil and with favourable condition, it initially infects the roots and later spreads to the collar portion. Use healthy planting material, clean equipment and uproot the plant with infected root along with the adjacent soil and destroy the plant. Spray Mancozeb or Maneb @ 2g/l on infected foliage and shoots; soil drench with these fungicides along with 1g/l carbendazim during rainy season at an interval of 7-10 days.

11.1.3 Leaf Blotch, Red Spot or Measles (*Cladosporium paeoniae*)

The fungus survives on dead plant debris and in soil. The spores spread by rain or wind. The disease is noticed in the field after completion of flowering. Symptoms are visible by presence of small, circular dark red spots on the leaves, making it look like a measles. The spots later coalesce to give an irregularly shaped purplish blotchy appearance to the upper leaf surface. Under surface of infected leaf turns brown. The infected young leaves become distorted. Reddish streaks are noticeable on the infected stem. Remove and destroy the infected plant parts. Drenching of soil around the basin with Mancozeb, Dithane M-45 or copper based fungicide @ 2g/l before the shoot emergence in spring controls the disease.

Other foliar diseases occurring in peonies include- bacterial blight (*Xanthomonas hortorum*), southern blight (*Sclerotium rolfsii*) and powdery mildew (*Erysiphe paeoniae*).

11.1.4 Watery Soft Rot or White Mold (*Sclerotinia sclerotiorum*)

The disease survives in soil as sclerotia, it spread from plant to plant and through air borne spore. Initially water soaked pale or dark brown lesions develop on infected stem. In advance stage, shoots become stringy and wither. White mycelial growth occurs on the infected crown surface. Large black sclerotial mass forms in the pith and on surface of infected tissue. Brown lesions are also noticed on the

Diseases and Pests of Peony

Botrytis Blight in Leaf

Botrytis Blight in Bud

Leaf Blotch

Tobacco Rattles Virus

Thrips

Rose Chafer

Foliar Nematode

Root Knot Nematode

foliage and flower which are covered with white mycelial growth during humid weather. Remove and destroy the infected plant. Replanting in the infected area should be avoided. Avoid overhead irrigation. Soil treatment with *Coniothyrium minitans* reduces the severity of the disease. To check further spread of the disease, soil drenching with 20 l/m^2 formaldehyde @ 20ml/lit water is recommended around the infected plant. Drenching the soil with Mancozeb or Carbendazim @ 1 g/l is effective to control the disease.

11.1.5 Verticillium Wilt (*Verticillium albo-atrum, V. dahliae*)

It is a soil borne disease and spreads through water, contaminated implements and infected plant debris. The disease occurs during cool weather but the symptoms are not visible until temperature raises. Initially, the symptoms can be noticed by leaf yellowing at the margin and inter-venial areas, which begins at the tip of the leaf and later progress downward. Uproot the infected plant and destroy. Avoid replanting in the infected area. Use healthy planting material for planting. Soil solarisation reduces the pathogen build up in the soil.

11.1.6 Peony Ring Spot or Peony Mosaic or Tobacco Rattles (Tobacco Rattle Virus (TRV) earlier known as Peony Ring Spot Virus or Peony Mosaic Virus)

The virus is transmitted by nematodes of genera *Trichodorus* and *Paratrichodorus*. It is an important plant virus of vegetables, ornamentals and weed hosts; it was first reported in Chinese herbaceous peony plants in Yangzhou City, Jiangsu Province in 2016. The disease is identified by the presence of alternate green and yellow concentric rings or yellow-green mosaic or mottle on the leaves (He and Chen, 2016). Removal of infected plants also prevents spread of the virus from adjacent plants via vector or mechanical transmission. Remove other host species, and weeds from the vicinity. Soil fumigation with the chemicals such as Basamid @ 20g/l prior to planting controls the nematodes, thus reduces the disease infection.

11.2.0 Pests

Peony is a hardy plant; thus, is not subjected to major pest infestation.

11.2.1 Thrips (*Rhipiphorothrips cruentatus/Scirtothrips dorsalis*)

It is the most common insect pest of peony. In spring, when new leaves mature and flower bud begins to form, thrips attack the plant. Damage can be seen by silvery flecks on the flower and foliage, shrivelled leaf and deformed buds when large population is available, otherwise the insect goes unnoticed because of small size. Thrips are managed by releasing predatory insects like parasitic wasp or by spraying with neem oil or insecticidal soap. Pesticides such as Acephate @ 0.75g/l, or Fipronil @ 1.5 ml/l is sprayed when infestation is noticed.

11.2.2 Rose Chafers (*Cetonia auranta*)

The beetles occasionally attack the plant and are readily visible on plant. They have tan-colour body, reddish brown head and thorax. Adult emerge from the soil in June and early July and feed primarily on flower making big holes and also skeletonise the leaves. The female lays eggs in the soil, grub emerge from egg and feed on roots of weeds and grasses. The larvae do not cause economic damage and undergo hibernation during winter. The beetles are handpicked and killed when in small number. Physical barrier such as cheese cloth can be used to protect the plants. Insecticides such as Indoxacarb @ 1.0g/l or Chlorantraniliprole 0.4 ml/l is sprayed in case of severe infestation.

11.2.3 Root Knot Nematodes (*Meloidogyne* spp.)

Numerous small galls are formed on the feeder root. The plants become stunted, chlorotic or water stressed. The plants fail to produce flower. Hot water treatment of infected roots at 47°C for 30 minutes can kill the nematode. Use clean and healthy planting material for propagation as well as for planting. Prior to planting, soil fumigation with nematicide is effective to control nematodes. Apply carbofuran 3G granules @ 1 kg a.i./ha.

11.2.4 Foliar Nematodes (*Aphelenchoides* spp.)

Foliar nematodes spread through water and infected plant debris. Nematodes feed within the leaf tissue, thus causing damage to the plant. Infestation in plant can be detected by the presence of angular purplish leaf spot in the intervenial region. In severe case, the foliage becomes completely necrotic. Maintain clean basin since nematode survives in the plant debris. Avoid overhead irrigation to prevent spread of the nematode. Foliar nematode can be controlled up to some extent by spraying Dichlorvos or Dimethoate @ 2ml/l.

12.0 Harvesting and Postharvest Management

12.1.0 Harvesting

12.1.1 Stages for Harvesting

Flowers are mostly harvested at the 'marshmallow' stage between 3.1-4.4 cm diameter which refers to the firmness of the flower bud at harvest. This stage can be checked by holding the stem at the base of the bud between the index and middle fingers and pressing down with the thumb, on top of the bud. The firmness of the bud is similar to that of a stale marshmallow (Auer and Holloway, 2008). The vase life of the flower is significantly influenced by the cutting stage but is not significant for commercial purpose as the difference in vase life is less than a day. Early cutting of flower is commercially significant due to longer time between cutting and full bloom. Cutting of the flowers at the latest possible stage shows uniformity in opening time of flower and subsequent vase life of flowers (Eason *et al.*, 2002). Flowers are more susceptible to bruising or damage during shipping when cut at the late stage due to more open buds. Flowers are cut at later stage when meant for selling as loose flower for local market. It is always preferable to cut flowers at early stage for long distance shipment rather than at late stage as at later stage, the buds are more susceptible to damage.

Table 1: Bud Maturity Scale for *Paeonia lactiflora* (Eason *et al.*, 2002)

Stage	*Description*
1.0	Hard tight bud with little petal colour
1.5	Hard tight bud with slight colour
2.0	Firm tight bud, petals showing colour
2.5	Slightly softer bud,"Squeezable" (Marshmallow stage)
3.0	Loose or soft bud, outer petal loosening, petals still firm
4.0	Very loose bud, hollow feeling
5.0	Almost out, petals curved inward
6.0	Open (optimum), petals reflexed
7.0	Petals reflexed and wilted or dropping

Figure 5: Flower Index of Herbaceous Peony (Yu *et al.*, 2011).

Stage 1: The bud shows less than 1/3rd colour of the whole bud. Sepals slightly open but the buds remain close. Stage 2: The bud shows more than 2/3rd colour and the bud remains firm. Stage 3: The outermost layer of petal opens slightly while the inner whorl is still tight. The bud is slightly softer than stage 2. Stage 4: The outer whorls of petal open, the bud is loose but the petals are still firm. Stage 5: The petals of inner whorl open but, the petals remain curved inward. Stage 6: The petals are completely open and reflexed. Stage 7: The petals are reflexed, flowers become senescent and the petals drop off.

Harvesting of flowers at the proper developmental stage is crucial for successful storage and maximum vase life (Heuser and Evensen, 1986). The flowers are ready for harvest three to five years after planting. The time and rate of blooming depend on the weather condition during spring months. In warmer season, the harvesting period is earlier and the interval is shorter between the harvesting. In warm spring, the harvesting period is short and may last only for three weeks (Holloway *et al.*, 2005). The harvesting may be extended up to six weeks by planting early, mid and late season cultivars. After each harvest, disbudding of lateral buds may be done in many cultivars to ensure the large terminal flowers (Rogers 1995). Buds are removed by pinching before the stems have elongated. Disbudding is repeated regularly to remove all the developing buds.

The flowers are harvested with one flower bud and multiple compound leaves per stem. Harvesting may be done at least twice a day to ensure collection of all the flowers at optimal harvest stage (Gast, 1997). Under long day condition, if the temperature is high (>26°C), three or more cuttings is recommended as heat speeds up the blooming process. About 50 – 75 per cent of stems can be harvested per plant leaving some over matured flowers on the plant (Rogers, 1995). Harvesting is advised early in the morning after the dews have dried as during this hour, the metabolic processes in plant are low and stems remain turgid (Rogers, 1995). Harvesting is also recommended in the evening when carbohydrate levels are highest to promote longer vase life (Byczynski, 1997). Harvesting is not recommended when the temperatures are high as this will put the flowers under stress (Armitage *et al.*, 2003). Flowers harvested in the heat require more energy to cool and contain lesser nutrients for prolonging shelf-life in storage (Rogers, 1995).

12.2.0 Postharvest Management

The cut stem is kept in the clean water 20 minutes after harvest at 4°C for 1-2 hours followed by drying prior to grading (Stevens *et al.*, 1993). Peonies are graded according to the bud diameter and length of the stem. The flowers having crooked or weak stem if harvested should not be sold as these cannot support the bud. According to USDA grade, buds should not be less than 25mm in diameter, the stems relatively straight and at least 60 cm long for top quality (USDA, 1938). In New Zealand, bud diameter scale followed for harvesting different grades are: AAA grade- 45mm, AA grade- 40mm, and A grade-35mm. The minimum required cutting length for the lowest grade of cut peonies is 60 cm (Holloway *et al.*, 2008).

In general, the stem length during the first year of planting is longer due to availability of reserved carbohydrates in the roots while in the subsequent years it becomes shorter due to influence of environmental factor of the region (Auer and Holloway, 2008). For longer stem length, fewer stems per plant can be cut, retaining sufficient number of leaves for plant health. To facilitate photosynthesis and accumulation of reserve for growth and flowering in the next season, at least three sets of leaves at the base of each cut stem is retained after every harvest (Rogers1995).

Postharvest flower longevity in *Paeonia* spp. is influenced by the bud maturity at harvest (Heuser and Evensen, 1986). After harvest, the flowers are cooled at desired storage temperature to remove the field heat. The flowers are immediately cooled for 20 minutes at 4° C followed by pulsing for 15-30 minutes in a solution containing 4 per cent sucrose + 1 per cent STS (Rogers, 1995; Rehman *et al.*, 2013). Pulsing is done to prolong the storage by treating the flowers with floral preservative. Many cultivars do not show significant response to pulsing in bloom size, vase life and time to opening following cold storage (Gast, 1997), and thus should be avoided unless the benefits are significant. Apart from postharvest treatments and storage conditions several other factors contribute to the longevity of cut peonies, one among these being the genetical factors which are specific to the cultivars. The vase life is also affected by the pre-harvest conditions such as rainfall and temperature during flower bud formation and its growth (Skutnik *et al.*, 2020). The buds after harvest should not be placed in the water for longer time (Wet storage); this may lead to premature opening of bud also known as 'blast' (Stevens *et al.*, 1993).

12.2.1 Sorting, Bunching and Packaging

The flowers are sorted and bunched in number of five or ten stems (Stevens, 1998) and then stored horizontally wrapped in florist wrap at 1-2°C. With proper care buds can last up to four weeks. The bud opens when stem is placed in water (Finnegan, 2014). Each bunch in a box contains flowers of like cultivar, stem length, bud quality and size. Making of bunch with the longest stems first minimises the amount of cut off to make flowers with shorter stems. Flowers in bunches are secured with a rubber band two or three inches above the stem; ensuring the flowers are tightly held together but loose enough to prevent breaking of stem.

The USDA has strict grading standards specifically for *Paeonia* spp. in effect since 1938. These standards cover proper stem length 45-60 cm) and overall general health of the bloom, the grades should consists fresh flower with strong, well trimmed unbroken stems which are fairly straight; the buds should be of proper shape, firm with expanded calyx abut not over matured, free from puff balls, bull heads and wood heads. The stem and bud should be free from damage caused by pests and diseases or mechanical injury, or freezing injury. Bunches are packed in boxes of 90 x 40 cm, with 20-40 bunches per box with heads facing either end. Shredded plain papers are placed between the heads to prevent bruising and prevent movement of flowers during transit. Multiple small refrigerant gel packs are placed inside the box for thermo regulation.

12.2.2 Storage

Selection of best cultivars and good storage management prolongs the shelf life of the cut flowers from seven days to several weeks. Pre-cooling is the first step to be practise, to improve and extend the shelf life of flower in storage. Pre-cooling suppresses the enzymatic degradation and respiratory activity, slows down the rate of transpiration, minimizes ethylene production and inhibits the growth of organism causing decay (Wilson *et al.*, 1999b). Flowers prior to long long-term storage are checked for signs of infection or decay by examining the samples for any cuts, severe bruises, or rotting tissue (Wilson *et al.*, 1999a). Optimal condition for storage is at 0–1°C air temperatures and 75-80 per cent relative humidity (Armitage *et al.*, 2003). The flowers should be stored at high humidity but to reduce the development and spread of diseases (Byczynski, 1997) especially *Botrytis*, the flowers should remain dry. Under proper conditions peonies can be stored up to four weeks before vase life is reduced (Gast, 2000; Pertwee, 2000) while some of the cultivars can be stored up to 73 days (Ohlander and Watson,1951).

12.2.3 Vase Life

Vase life and days to flower opening varies depending on the cultivar, length of storage and their interaction. Addition of floral preservatives *viz.*, Floralife,

Hinsdale, Illinois in the vase solution increases fresh weight of the flower, improves quality by increasing the flower diameter and colour intensity, delays senescence and prolong the vase life (Heuser and Evensen, 1986). The preservatives contain sugar, mostly sucrose and biocides such as 8-hydroxyquinoline-sulphate (8-HQS) or citrate (8-HQC), silver nitrate ($AgNO_3$) *etc.* The sugar serves as a respiratory substrate and maintains the hydrological balance by increasing the concentration of osmolyte, also exogenous application of sugar enhances the flower colour by stimulating the synthesis of pigments, and limits ethylene biosynthesis. The biocide prolongs the vase life of a flower by preventing microbial build up in the holding solution. Application of Nanosilver @ 30mg/l vase solution helps in prolonging the vase life and quality of cut flower by preventing the bacterial growth (Zhao *et al.*, 2018). Peony is an ethylene sensitive crop, when treated with β-cyclodextrin-based nanosponge-1-methylcyclopropene (1-MCP) complex (β-CD-NS complex), it enhances the efficacy of 1-MCP thereby delaying the bud opening and senescence (Seglie *et al.*, 2013).

13.0 Yield

Harvesting of flower is avoided in the initial three years of planting. Harvesting begins from fourth year onward; single plant yields about 20-30 flowers in a year. Only one-third of the flowers are harvested and remaining are pinched off to maximise leaf surface and allow development of plant. A healthy and vigorous plant yields about 36-50 flowers, of which, half are left unharvested to maintain the plant.

14.0 Biotechnology

14.1.0 Molecular Marker-Assisted Breeding

The origin of peony cultivar is very complicated; the morphological markers alone do not serve the purpose of identifying the cultivar due to several limitations. Therefore, molecular markers having several added advantages, such as they are highly abundant, store house of information, insensitive to environment factor *etc.*, help in conventional breeding program. In molecular-assisted breeding of peony, mainly two processes are involved at present: identification of the authenticity of seedlings at early stage and quantitative trait locus (QTL) mapping of phenotypic traits. Though the species are cross compatible, certain amount of self-compatibility is noticed which accounts for seed setting (<5 per cent) due to self pollination. During hybridization, improper handling of crossing materials may lead to self pollination or faulty hybrids and peony having a long juvenile period of three years, the true hybrid cannot be detected. Thus, the molecular markers help identify the authenticity of hybrid seedling at early stage, thereby reducing the maintenance cost.

Several molecular markers, including amplified fragment length polymorphism (AFLP), inter-simple sequence repeat (ISSR), sequence-related amplified polymorphism (SRAP), and simple sequence repeats (SSRs) markers, are used for identifying the hybrids. SSR markers are most suitable for identifying the authenticity of Peony hybrid due to its advantage of simple operation, high accuracy, and good stability, hyper-variability, less use of DNA and co-dominance nature.

Genetic linkage maps, permitting the elucidation of genome structure, are one of most powerful genomic tools to accelerate marker-assisted breeding. Cai *et al.*, (2015) developed genetic linkage map for tree peonies (*Paeonia* Sect. *Moutan*) by using Specific- locus amplified fragment sequencing (SLAF-seq) in an F_1 population, derived from the cross *P. ostti* 'FenDanBai' × *P.* × *suffruticosa* 'Hong Qiao', to identify sufficient high-quality markers for the construction of high-density genetic linkage map in tree peonies. 309,198 high-quality SLAFs were detected from a total of 78 GB sequenced data, of which 85,124 (27.5 per cent) were polymorphic. Subsequently, 3518 of the polymorphic markers, which were successfully encoded in Mendelian segregation types and were in conformity with the criteria of high-quality markers, which were defined as effective markers and used for genetic linkage mapping. An integrated genetic map was constructed which comprised of 1189 markers on the five linkage groups, and spanned 920.699 centiMorgans (cM) with an average inter-marker distance of 0.774 cM. This provides a solid foundation not only for marker-assisted breeding, but also for genome sequence assembly of tree peony.

14.2.0 Genetic Engineering

Peony as a cut flower is gaining popularity among the consumers due to its flower colour and form. Novel cultivars for flower colour, form and enhanced vase life can be developed through genetic engineering. Chalcone isomerase (*CHI*) is a key enzyme that converts yellow chalcone to colourless naringenin, playing an important regulatory role in colour formation of ornamental flowers. Coding sequence of *CHI* in herbaceous peony was determined using rapid-amplification of cDNA ends (RACE) technology, and subsequently detected the expression pattern of *CHI* in the inner and outer petals at different developmental stages using qRT-PCR. The upstream promoter sequence of *CHI* was cloned using genome walking technology and predicted the location of CpG islands and 5′ truncation. The result showed that the full-length cDNA sequence of *CHI* was 898 bp, and the 5′-upstream core promoter was located at –1 651 to –2 050 bp region, which contained one CpG island (–1 897 to –2 010 bp) and several important binding sites of transcription factor, such as Sp1, serum response factor (SRF), activating protein (AP)-2alpha and CCAAT/enhancer binding protein (C/EBP) alpha. Expression of *CHI* at different developmental stages was generally higher in inner petals than those in outer petals and the maximum at the bud stage (Wu *et al.*, 2018).

A comprehensive reference gene set was constructed by transcriptome sequencing to obtain *R2R3 MYB* genes. Based on the assembly, 57 *R2R3 MYB* genes were identified containing full-length open reading frames, which were clustered into 35 clades by phylogenetic analysis. *PsMYB57* clustered with anthocyanin regulation genes in *Arabidopsis* show transcription in the buds and young leaves. The overexpression of *PsMYB57* and four anthocyanin structural genes, including *NtCHS, NtF3′H, NtDFR,* and *NtANS,* were upregulated. The two endogenous *bHLH* genes *NtAn1a* and *NtAn1b* were also upregulated and may work in combination with *PsMYB57* to regulate the anthocyanin structural genes. Functional prediction of the *R2R3 MYBs* verified that *PsMYB57* regulates anthocyanin biosynthesis in the buds and young leaves in tree peony (Zhang *et al.*, 2020).

Dehydrins (DHNs) proteins play a protective role in plants under different abiotic stress conditions. A stress-responsive dehydrin gene *PlDHN1* (GenBank accession number: KX524516) was cloned from the buds of the *Paeonia lactiflora* 'Dafugui'. Sequence alignment and phylogenetic analysis showed that the *PlDHN1* sequence possessed an 8-serine residue stretch followed by three conserved repeats of a lysine-rich segment, characteristic of the typical SK_3-type dehydrin. Subcellular localisation analysis revealed that the *PlDHN1* protein is mainly present in the cytomembrane and nucleus. qRT-PCR analysis indicated that the *PlDHN1* genes are highly expressed more in buds than in other tissues. Furthermore, the *PlDHN1* transcript in buds accumulated during the winter and peaked in harsh December with gradual decrease prior to spring. The expression of *PlDHN1* was up-regulated in response to cold, heat, waterlogging and abscisic acid (ABA) treatment, most significantly for heat and waterlogging (Chen *et al.*, 2018).

15.0 References

Abbey, M., (2015) *Paeonia spp. production and future developments*. University of Minnesota

Albers, M.R.J. and Kunneman, B.P.A.M. (1991) *II International Symposium on Propagation of Ornamental Plants*, **314**: 85-92.

Armitage, A.M. and Laushman, J.M. (2003) *The Production of Annuals, Perennials, Bulbs, and Woody Plants for Fresh and Dried Cut Flowers.* Portland, Oregon, Timber Press.

Auer, J.D. and Holloway, P.S. (2008) *University of Alaska Fairbanks.*

Barzilay, A., Zemah, H., Kamenetsky, R. and Ran, I. (2002) *HortScience*, **37**: 300-303.

Batygina, T.B. and Vasilyeva, V.E. (2003) *Acta Biologica. Cracoviensia. Sereries Botanica*, **45**: 27-36.

Beruto, M., Lanteri, L. and Portogallo, C. (2004) *Plant Cell Tiss.Org.*, **79**: 249-255.

Byczynski, L. (1997) *The Flower farmer: An Organic Grower's Guide to Raising and Selling Cut Flowers.* White River Junction, Vermont, Chelsea Green Publishing Company.

Byrne, T. and Halevy, A.H. (1986) *J. Amer. Soc. Hort. Sci.*, **111**: 379– 383.

Cai, C., Cheng, F. Y., Wu, J., Zhong, Y. and Liu, G. (2015) *PloS one*, **10**: e0128584.

Calkins, J.B., Swanson, B.T. and Newman, D.L. (1996) *J. Envi. Hort.*, **14**: 221-227.

Chen, J.Y. (1953) *New Sci.*, **3**: 16-38.

Chen, Y., Ma, Y., Guo, J. and Guo, X. (2018) *J. Hortic. Sci. Biotechnol.*, **93**: 557-565.

Cheng, F. (2007) *Int. J. Plant Breed.*, **1**(2): 89-104.

Cheng, F., Zhong, Y., Long, F., Yu, X. and Kamenetsky, R. (2009) *Isr. J. Plant Sci.*, **57**: 357-367.

Cheng, F.Y. and Li, J.Y. (1998) *Nat. Sci.*, **34**: 109-116.

Cheng, Y., Kim, C.H., Shin, D.I., Kim, S.M., Koo, H.M. and Park, Y.J. (2011) *J. Medi. Plants Res.*, **5**(31): 6744-6751.

De-Yuan, H., Zhi-Xian, Z. and Xiang-Yun, Z. (1988) *J. Syst. Evol.*, **26**(1): 33-43.

Dijk, B.V. (1988) *Bloembollencultuur*, **99**(11): 14-15.

Eason, J., Pinkney, T., Heyes, J., Brash, D. and Bycroft, B. (2002) *New Zealand J. Crop Hort. Sci.*, **30**(1): 61-67.

Evans, M.R., Anderson, N.O. and Wilkin, H.F. (1990) *HortScience*, **25**: 923-924.

Fangyun, C. and Dezhong, C. (1998) *J. Beijing Forestry University*, **20**(2): 27-32.

Finnegan, M. (2014) *CAFRE*. December: 1-5

Fitzgerald, D. (2004) *Peony-A Future Crop for Alaska.*

Gabryszewska, E. (2004) *V International Symposium on In Vitro Culture and Horticultural Breeding*, **725**: 471-476.

Gabryszewska, E. and Kawa-Miszczak, L. (2010) *Biotechnol.*, **2**: 172-179.

Gast, K.L.B. (1997) *Report of Progress*, Agricultural Experiment Station, Kansas State University.

Gast, K.L.B. (2000) *Report of Progress*, Agricultural Experiment Station, Kansas State University.

Guo, X., Fu, X., Zang, D. and Ma, Y. (2009) *Sci. Hortic.*, **119**(2): 177-181.

Halda, J.J. and Waddick, J. W. (2004) *The genus Paeonia*. Oregon, USA: Timber Press.

Halevy, A.H., Weiss, D., Shlomo, A., Levy, M. and Cohen, M. (1995) *Bul. Israeli Grower Asn.*, **5**: 58-63.

Hashida, R. (1993) *Jap. Bot. Soc.*, pp. 10-148.

Hashida, R. (1996) *Amer. Peony Soc. Bull.*, **297**: 26-29.

He, Z. and Chen, C. (2016) *Plant Disease*, **100**(12): 2543-2543.

Heuser, C.W. and Evensen, K.B. (1986) *J. Amer. Soc. Hort. Sci.*, **111**(6): 896-899.

Holloway, P. S., Matheke, G. E., Hanscom, J., Gardiner, E. and Weber, J. (2008) *Annual Flowering Plant Trials* 2007.

Holloway, P.S., Hanscom, J.T. and Matheke, G.E. (2005) *Research Progress Report no. 44*, University of Alaska Fairbanks Agricultural and Forestry Experiment Station.

Holloway. P. (2007) *Georgeson Botanical Garden*, Fairbanks, Alaska.

Hong, D.Y., Pan, K.Y. and Rao, G.Y. (2001) *Plant Syst. Evol.*, **227**(3-4): 123-136.

Hosoki, T., Ando, M., Kubara, T., Hamada, M. and Itami, M. (1989) *Plant Cell Rep.*, **8**(4): 243-246.

Hosoki, T., Kimura, D., Hasegawa, R., Nagasako, T., Nishimoto, K., Ohta, K. and Haruki, K. (1997) *Sci. Hortic.*, **70**(1): 67-72.

Jacob, Y., Mastrantuono, S. and Ferrero, F. (2006) *XXII International Eucarpia Symposium, Section Ornamentals, Breeding for Beauty*, **714**: 59-66.

Ji, L., Wang, Q., da Silva, J.A.T. and Yu, X.N. (2012) *Sci. Hortic.*, **143**: 62-74.

Jia, N., Shu, Q.Y., Wang, L.S., Du, H., Xu, Y.J. and Liu, Z.A. (2008) *Sci. Hortic.*, **117**(2): 167-173.

Jian, S., Soltis, P.S., Gitzendanner, M.A., Moore, M.J., Li, R., Hendry, T.A., Qiu, Y.L., Dhingra, A., Bell, C.D. and Soltis, D.E. (2008) *Syste. Biol.*, **57**(1): 38-57.

Kamenetski, R., Barzilay, A. and Cohen, M. (2007) *International Conference on Quality Management in Supply Chains of Ornamentals*, **755**: 121-126.

Kamenetsky, R. and Dole, J. (2012) *Floric. Ornam. Biotech.*, **6**: 62-77.

Kapinos, D.B. and Dubrov, B.M. (1993) *Peonies in the Garden*, Minioin, Tumen, Russia, pp. 190.

Kim, Y.S. and Lee, B.K. (1995) *J. Korean Soc. Hort. Sci. (Korea Republic)*, **36**(2): 255–262.

Kim, Y.S. and Lee, B.K. (1996) *J. Korean Soc. Hort. Sci.*, **37**(6): 827–830.

Klingman, M.A. (2002) *Production and transportation considerations in the export of peonies from Fairbanks, Alaska* (Doctoral dissertation).

Lan, H.Y. (2003) *Studies on embryo culture of Paeonia lactiflora Pall. and paeoniflorin content in the roots* (Doctoral dissertation, M. Sc. Thesis. Nei Meng Gu University (in Chinese)).

Lee, B.K., Ko, J.A. and Kim, Y.S. (1992) *J. Korean Soc. Hort. Sci. (Korea Republic)*, **33**(5): 384–395.

Li, J.J., Zhang, X.F. and Zhao, X.Q. (2011) *Beijing: Encyclopaedia of China Publishing House*, p. 206.

Li, X., Liu, P., Yang, P., Fan, C. and Sun, X. (2018) *PloS one*, **13**(8): 1-13

Li, Y.L., Wu, D.Y., Zhang, S.L. and Li, X. (1984) *Sci. Bull.*, **8**: 500-502.

Matheke. G. (2007) *Georgeson Botanical Garden*, Fairbanks, Alaska.

McGeorge, P. (2006) *Peonies First*. Albany: Firefly books.

Ohlander, M.C. and Watson, D.P. (1951) *Proc. Amer. Soc. Hort. Sci.*, **58**: 371-376.

Pauwels,I., Vervoort, G. (2009) *Pioenen*. Uitgever Lannoo Uitgeverij, pp. 143.

Pertwee, J. (2000) *International Cut Flower Manual*. Doetinchem, Netherlands, Frank Paulter Berg.

Radtke G.M. (1983) *Am. Peony Soc. Bull.*, **246**: 19-23.

Rehman, S. I., Qadri, Z. A., Sheikh, M. Q. and Ahmed, Z. (2013) *Afr. J. Agri. Res.*, **8**(20): 2360-2365.

Ren, X., Liu, Y. and Jeong, B.R. (2019) *Plants*, **8**(10): 356.

Rodrigo-Lopez MJ. (2010) *Paeonies Literature Review*. Belfast.

Rogers, A. (1995) *Peonies*. Portland, Oregon, Timber Press Inc.

Sang, T., Pan, J., Zhang, D., Ferguson, D., Wang, C., Pan, K.Y. and Hong, D.Y. (2004) *Biol. J. Linnean Soc.*, **82**(4): 561-571.

Seglie, L., Devecchi, M., Trotta, F. and Scariot, V. (2013) *Sci. Hortic.*, **159**: 162-165.

Shen, M., Wang, Q., Yu, X. and da Silva, J.A.T. (2012) *Sci. Hortic.*, **148**: 30-38.

Shining, C., Yan, L. and Zengpu, L. (2011) *Chinese Agricultural Science Bulletin*, **6**: 18.

Skutnik, E., Rabiza-Œwider, J., Jêdrzejuk, A. and £ukaszewska, A. (2020) *Agron.*, **10**(11): 1631.

Song, C., Wang, Q., da Silva, J.A.T. and Yu, X. (2018) *J. Amer. Soc. Hortic. Sci.*, **143**(4): 248-258.

Stern, F.C. (1946) *A study of the genus Paeonia*, Royal Horticultural Society, London, pp. 1-14.

Stevens, A.B. (1998) *Field grown cut flowers: A practical guide and sourcebook* (No. 635.9 St471f Ej. 1 019072).

Stevens, S., Stevens, A.B., Gast, K.L.B., O'Mara, J.A., Tisserat, N.A. and Bauernfiend, R. (1993) *Peonies. Kansas State University. Cooperative Extension Service, Manhattan, Kansas. Pub. MF-1083.*

Toma, F., Selaru, E., Georgescu, M.I., Petra, S., Pãdure, I.M., Zamfir-Vasca, D. and Venat, O. (2012) *AgroLife Scient. J.*, **2**: 103-112.

Tzanoudakis, D. (1983) *Nordic J. Bot.*, **3**(3): 307-318.

United States Department of Agriculture. (1938) *United States Standards for Grades of Cut Peonies in the Bud*. Agricultural Research Service.

Uspenskaya.M. (2002) *Peonies*, Phyton, Mpscow, pp. 208.

Walton, E.F., McLaren, G.F. and Boldingh, H.L. (2007) *J. Hort. Sci. Biotech.*, **82**(3): 365-370.

Wilson, L.G., Boyette, M.D and Estes, E.S. (1999b) *Part II: Cooling. North Carolina State University.*

Wilson, L.G., Boyette, M.D. and Estes, E.S. (1999a) *Part I: Quality Maintenance. North Carolina State University.*

Wu, Y. Q., Zhu, M. Y., Jiang, Y., Zhao, D. Q. and Jun, T. A. O. (2018) *J. Integr. Agric.*, **17**(1): 122-129.

Wu, Y., Tang, Y., Jiang, Y., Zhao, D., Shang, J. and Tao, J. (2018) *Biosci. Rep.*, **38**(6).

Xue, J., Tang, Y., Wang, S., Yang, R., Xue, Y., Wu, C. and Zhang, X. (2018) *Postharvest Biol. Technol.*, **143**: 92-101.

Yang, Y., Sun, M., Li, S., Chen, Q., da Silva, J.A.T., Wang, A., Yu, X. and Wang, L. (2020) *Hort. Res.*, **7**(1):1-19.

Yom Din, G., Cohen, M. and Kamenetsky, R. (2014) *J. Hort.*, **2**: 2376-0354.

Yu, H.X., Zhou, M.Y., Wang, M., Liu, Q.Q., Huang, Z.L., Gong, Z.Y., Liang, G.H. and Gu, M.H. (2008) *Agri. Life Sci. Ed.*, **29**(4): 84-89.

Yu, X.N., Guo, P.P., Lu, G.P. and Zhang, Q.X. (2011) *J. Forestry Res.*, **22**: 137-140.

Zhang, Y., Xu, S., Cheng, Y., Wang, J., Wang, X., Liu, R. and Han, J. (2020) *BMC Genetics*, **21**: 1-11.

Zhao, D., Cheng, M., Tang, W., Liu, D., Zhou, S., Meng, J. and Tao, J. (2018) *Protoplasma*, **255**:1001-1013.

Zhao, D., Tao, J., Han, C. and Ge, J. (2012) *Mol. Boil. Rep.*, **39**: 11263-11275.

Zhao, D., Zhang, X., Wang, R., Liu, D., Sun, J. and Tao, J. (2019) *Environ. Exp. Bot.*, **162**: 345-356.

https:/www.appleyardflowers.com › flowerdiaries › history-p., (2020).

https:/*www.proflowers.com* › blog › peony-meaning, (2016)

http://hyg.ipm.illinois.edu/pastpest/199814d.html

www.petalrepublic.com › floristry-and-floriculture-stati.

11

ANEMONE

A. Sumalatha, V. Bhargav, T. Yatung and K. Dutta

1.0 Introduction

Anemone, is a genus of flowering plants, belonging to family *Ranunculaceae* is perennial in nature. These gorgeously coloured poppy like flowers with wide array of petal colours varying from purple-blue, red, pink and white; having single and double type long lasting blooms with tall and sturdy stems, anemones are widely used as decorative flowers.

2.0 Importance and Uses

2.1.0 Aesthetic Uses

Anemones are excellent cut flowers and are used as dry flower, in container planting, bouquets, table decoration and potted plants and in vase. Geophytes like anemone can give an attractive aesthetic look when used as a bush in groups or with other herbaceous species or as a solo plant. When used as a border plant along the entire length of a grassed area, gives a vibrant look to it. *Anemone blanda* have light coloured flowers with frail petals and are suitable for planting in the lawns. *Anemone blanda* and *Anemone ursinum* can be grown around or under the deciduous trees as it does not abstain the root growth of the tree, instead helps in better development of trees. Group planting of anemone on rock garden gives an aesthetic look. Anemone, having different shades of flower colour and fragrance and long lasting flowers are suitable for flower beds which gives it a feeling of continuity is no match by other summer annuals. Spring blooming Anemone flowers are either grown alone or in combination with other flowers such as *Petunia, Tagetes, etc.*, to extend the flowering period and give a complementary or contrasting effect to the garden. *Anemone nemrosa* can be used along with ground covers to create beautiful effect in the garden.

2.2.0 Therapeutic and Other Uses

It has wide medicinal properties. Chau *et al.* (2012) reported a new saponin (anemonerivulariside A), three known saponins (huzhangoside A, Huzhangoside C and prosapogenin CP_6), two steroids and three phenolic compounds from whole plants of *A. rivularis.* In traditional Chinese medicine (TCM) and folk medicine, *Anemone* is used in heat-clearing and detoxification, warming and orifice-opening, as pesticide, in dysentery, malaria, tinea, ulcers and sores, arthritis, traumatic injury, pharyngolaryngitis, parasitic diseases and hepatitis. A broad spectrum of pharmacological activities, including antitumor, antimicrobial, anti-inflammatory, sedative and analgesic activities, and anticonvulsant and antihistamine effects have been observed (Sun *et al.*, 2011). Some anemone compounds and extracts display immunomodulatory effects, cures traumatic injury, rheumatic arthralgia, enteritis, dysentery, stomatch worms, sharp pain, snakebite, cold, tumour and gonorrhoea. They also exhibit therapeutic effects like relieving oppression, pus drainage, dissipating phlegm stasis, relieving dyspepsia, detoxification, vermifuge, eparsalgia cough, malnutrition and indigestion in children, malaria, dysentery, sore furuncle carbuncle, dermatitis, ascariasis *etc.* (Hao *et al.*, 2017).

Uses of Anemone

Along Garden Pathway

As Specimen

Along Side of Waterbody

In Pot

In Garden Bed

In Rock Garden

In Flower Vase

As Cut Flowers

In Flower Arrangement

3.0 Origin and History

3.1.0 Origin

The genus *Anemone* includes about 85 species, most of which are native to the temperate and mountainous regions of the Northern Hemisphere. The species are equally distributed across Europe, North America and Asia, but only few are endemic to South America or Southern Africa. The Mediterranean basin is the centre of diversity for many species, from which several cultivated varieties are derrived; these include *A. coronaria*, (poppy anemone), *A. hortensis*, A. *pavonina* and *A.* × *fulgens*. Wild populations of *A. coronaria* are frequently found throughout the Eastern

Mediterranean littoral, from Greece, through Southern Turkey and Syria, to Israel. Sporadic populations extend eastwards towards Northern Iraq and westwards along the Mediterranean shores of Italy, Southern France and Northern Africa.

3.2.0 History

Pliny recorded the presence of both wild and cultivated forms of anemone during the Roman era. At the end of the Middle Ages, Ferrari listed several anemone varieties in his *Florum cultura* (1631). *A. coronaria* is believed to have been introduced to England before 1596, as it was described and faithfully portrayed in the floricultural volume Johnson's Gerarde and was very popular during the reign of Queen Elizabeth I (1558– 1603). By 1629, Parkinson listed 60 species and varieties of *Anemone.*

Anemone coronaria is a winter-flowering species, both in wild as well as under cultivation (Yonash *et al.*, 2004). This species is considered as the progenitor for majority of varieties currently grown for cut-flower production and garden plants. *A. hortensis*, *A. pavonina* and *A.* × *fulgens* (the interspecific hybrid of *A. hortensis* and *A. pavonina*) are popularly grown as garden plants.

Anemone has several different meanings depending on the culture and context in which the flower is being used. It is commonly called as the 'poppy anemone', 'Spanish marigold' or 'windflower'. Several of the Western meanings of anemone flowers pertain to the Greek mythology on the origin of anemone flower featuring Adonis and Aphrodite. The goddess Aphrodite kept the mortal man Adonis as a lover; when Adonis was gored by a wild boar, at his death, Aphrodite's tears mixed with his blood and gave rise to the anemone. Ovid had described the etymology of anemone, in reference to the frailty of the petals which could be easily blown away by the wind. Hence name windflower is used for the whole genus as well as the woody anemone *A. nemorosa*.

In other cultures, the meaning differs. In Chinese and Egyptian cultures, the anemone flower is considered as a symbol of illness due to its colouring. According to the Eastern culture, anemone is considered as a symbol of bad luck. The Japanese anemone is associated with ill tidings.

4.0 Morphology and Classification

4.1.0 Morphology

Plant habit is typically herbaceous, with a perennial underground modified stem (tuber, rhizome or corm); leaves are with long petioles and one or many erect flower stalks, which have an involucres bract near or remote from the flower. Flower is characterized by five or more petaloid sepals, numerous free stamens and one ovulated carpel.

4.2.0 Classification

According to horticultural classification, there are mainly three groups of Anemone:

1. *Spring-flowering species*: These are found in woodland and alpine meadows, often tuberous or rhizomatous *e.g. A. nemorosa, A. blanda.*
2. *Spring- and summer-flowering*: These are species from hot dry areas, with tuberous roots *e.g. A. coronaria.*
3. *Summer- and autumn-flowering*: These species have fibrous roots and thrive in moist and partial shade *e.g. A. hupehensis.*

5.0 Species and Cultivars

5.1.0 Species

Anemone blanda

It is a fall-planted, spring-flowering ephemeral species which is grown commercially in large-scale. It is most commonly available in bluish violet colour (usually erroneously called 'Blue Shades' despite its flower being more purple than blue) that varies from intense to pale, depending upon the environmental factors. A white-flowered form is the second most common type and the rare cultivated form is a pale pink. The violet and especially pink forms sometime possess petals that fade to white near the flower centre.

Anemone coronaria

Among the most well-known anemones, it is often called as poppy anemone. It is a tuberous-rooted plant with parsley like divided leaves and large poppy like blossoms on stalks of 15–20 cm high. It can be planted in the fall without extra protection or in spring in cooler areas. If planted in fall it will flower in the spring and if planted in the spring it will flower in late summer. The flowers are typically scarlet, crimson, bluish purple, reddish purple or white. There are also double-flowered varieties, in which the stamens in the centre are replaced by a tuft of narrow petals. The hybrids are used as garden plant. Hybrids of the 'de Caen' and 'St. Brigid' groups are the most prevalent in the market. In certain areas of Israel, large numbers of red-flowering non-hybrid *A. coronaria* are grown.

Anemone hupehensis

Anemone hupehensis and its white cultivar 'Honorine Joubert', the latter

Species of Anemone

Anemone blanda

Anemone coronaria

Anemone hupehensis

Anemone nemorosa

Anemone hortensis

Anemone vitifolia

Anemone fulgens

Anemone parviflora

Anemone virginiana

Cultivars of Anemone

Hollandia

Mr. Fokker

Sylphide

The Bride

His Excellency

The Admiral

The Governor

Mount Everest

Lord Lieutenant

especially, is well-known autumn-flowering selection. They perform better in well-drained but moisture-retentive soil and reach 60–100 cm in height, blooming continually for several weeks.

Anemone hupehensis, *A. vitifolia* and their hybrids are particularly attractive to honeybees. A number of low-growing species are native to UK, such as *A. nemorosa* and *A. apennina*, which prefer woodlands and other shady places for their growth. *A. hortensis* and the hybrid *A. fulgens* are spring flowering, having less-divided leaves than others and have rose-purple or scarlet flowers.

5.2.0 Cultivars

Anemone coronaria cultivars are classified into two major groups: 'De Caen' and 'St. Brigid'.

De Caen: It represents a collective name for single flower cultivars, such as 'Hollandia' (purple), 'Mr. Fokker' (blue), 'Sylphide' (mauve), 'The Bride' (white), and 'His Excellency' (scarlet with a white eye).

St. Brigid: This group includes cultivars with semi-double and double flowers, such as 'The Admiral' (dark pink), 'The Governor' (carmine red), 'Lord Lieutenant' (blue-violet), and 'Mount Everest' (white). Two minor groups are also recognised: 'Rissoana', consisting of very rustic types which flower precociously (in November) and 'Grassensis', typified by their large double flowers, blooming in spring.

Most modern cultivars have been derived either by selection from old 'De Caen' types or from crosses between 'De Caen' cultivars and selections from wild populations. 'Wicabri', 'Mona Lisa', 'Cristina', 'Jerusalem', 'Tetranemone', 'Mistral' and 'San Piran' are cultivated throughout Europe, the USA and Israel include a number of sub-cultivars distinguished by their flower colour.

6.0 Genetics and Breeding

6.1.0 Genetics

The karyological and nucleolar studies on *Anemone coronaria* revealed that, $x = 8$ and all the species are diploid in nature with a chromosome number of $2n= 2x= 16$, but tetraploid cultivars have also been developed having the somatic chromosome number ($2n=4x=32$).

6.2.0 Breeding

6.2.1 Morphology and Pollination

Flowers with 4–27 sepals are produced singly, in cymes of 2–9 flowers or in umbels, above a cluster of leaf or sepal like bracts. The sepals vary in shades of colours. The gynoecium have one ovule. The flowers have nectaries, but petals are missing in the majority of species. The fruits are ovoid to obovoid shaped achenes, the achenes are beaked and some species have feathery hairs attached to these achenes.

Anemone coronaria is mostly a cross-pollinated species. Its stigma is receptive ten days before the anther dehiscence. Enforced self-pollination generates inbreeding depression. The mode of pollination in *A. coronaria* is mainly through bees, flies and beetles which are attracted to the flowers to feed on nectar (Dafni *et al.*, 2020). Each flower produces 200–300 seeds, covered with a woolly or glabrous coat.

6.2.2 Breeding Objectives

a) To breed hybrids/varieties with novel flower colour like yellow which is naturally not found in anemone.
b) Uniformity in flowering.
c) Development of varieties which are resistant to diseases.
d) To develop varieties which are resistant to high temperatures.
e) To develop varieties that promotes flowering under longer photoperiods.
f) To improve in longevity of cut flowers.
g) To develop new cut flower cultivars which are early flowering, having long and robust stems with large multi-sepaled flower.
h) To have a compact growth habit, erect leaves with a short petiole, profuse and coordinated flowering for garden and pot plants.

6.2.3 Breeding Methods

In recent years, new tools that may support breeding activities were established: *in vitro* propagation of plant, androgenetic plant production, characterization of populations and cultivars by molecular markers and detection of Single Nucleotide Polymorphism in cDNA. In consideration to the potentials of such tools, two major strategies have been proposed for breeding new cultivars (Laura and Allavena, 2007).

a) Crosses between selected heterozygous parents and
b) Crosses between inbreed homozygous or androgenetic parents in order to obtain true F_1 hybrids.

Over the last three decades, *A. coronaria* breeding methods have remained largely unchanged. However, recent advances in the fields of micropropagation, androgenesis and molecular markers present a number of new opportunities. The means to micropropagate selected plants are now well established, which can be employed to propagate and maintain the parental material required for seed production. Androgenetic plants have been derived from elite cultivars suitable for cut flower production. Doubled Haploid (DH) technology provides a rapid means to achieve homozygosity, thereby accelerating breeding programs over the conventional selfing which takes nearly six years to breed pure lines. Development of new cultivars using Doubled Haploid technology reduces the time period to lower end by one year besides guarantee of uniformity and improved quality of the F_1 hybrid seeds.

Application of horizon technologies such as, androgenesis that facilitates wide hybridisation, the induction of genetic variation via mutagenesis; the establishment of Marker Assisted Selection and the creation of new traits by genetic engineering in anemone. AFLP provides an effective means for the evaluation of genetic diversity, and to know about the origin of the cultivars. Molecular markers are also informative in assigning seed lots, rhizomes, cut flowers or pot plants of a particular cultivar or even sub-cultivar and represent a way forward for the application of MAS in *A. coronaria* breeding and protection of breeder rights (Laura and Allavena, 2007).

Prior to choosing strategies for appropriate genetic improvement in anemone, breeders have to consider the following features of *A. coronaria* into account: allogamy, protogyny, inbreeding depression, slow generation turnover (one per year), sufficient seed production, technical difficulties associated with vegetative cloning needed for phytosanitary reasons, low genetic variation for economically relevant traits (*e.g.* resistance to diseases, specific colours) (Laura and Allavena, 2007).

6.2.4 Sub-cultivars

Current cultivars cluster into a series of sub-cultivars, which are defined on the basis of flower colour; within each cluster, other traits, such as flower yield, precocity and flower quality are genetically variable. This non-uniformity reflects the dual aims of breeding: first to maintain a high level of vigour and second to provide a uniform produce. Seed lots of sub-cultivars are prepared by deliberate hand crossing selected plants of distantly related families, which are maintained by sib crossing. Thus, individuals within each family are typically heterogeneous and partially heterozygous, so that progeny are inevitably non-uniform. In a final step, a commercial seed lot of each sub-cultivar is generated by bulking the seed lot of several crosses. Sub-cultivars are therefore, plant populations and their genetic composition can vary through the years as a consequence of seed lot assembling with material of diverse origin (Laura and Allavena, 2007).

6.2.5 Autopolyploidisation

Autopolyploidisation is a quite different breeding strategy to breed new varieties, using synthetic autopolyploidisation. The cultivar 'Tetranemone', bred from 'Wicabri' using this technology, is characterised with greater flower stalk length, sepal thickness and flower size. However, flowering time is delayed in autumn. Recently, in tetraploids 'Mona Lisa' was induced and compared to 'Tetranemone', tetraploid 'Mona Lisa' and the 'Tetranemone' × 'Mona Lisa' hybrids, with their originating diploid cultivars, hybrids maintained. The advantage of tetraploid, was early flowering as 'Mona Lisa' and produced a comparable number of stems. Triploid genotypes combined the productivity of the diploid progenitor cultivars and some of the advantageous traits of the tetraploids. The introgression of new genetic variation (*e.g.* yellow flower colour, erect leaves) was attempted by crossing *A. coronaria* with some of its related species. However, interspecific sterility was found to hinder the transfer of useful genes. Similarly, hybrids between 'Tetranemone' and *A.*× *fulgens* were sterile, though both parents are tetraploid, sharing the same somatic chromosome number (2n=4x=32).

In recent years, a number of techniques to support the conventional breeding of *A. coronaria* have been developed. These include the *in vitro* propagation of seedlings, the generation of androgenetic plants, the *in vitro* regeneration from somatic tissue, the genetic characterization of populations and cultivars using molecular markers and the discovery of single nucleotide polymorphisms (SNPs) in expressed sequences.

6.2.6 Anther Culture

The genome of each androgenetic plant is unique, as it reflects the outcome of a single meiotic event in the parent plant. Thus, a population of androgenetic plants, derived from a single hybrid plant, can be expected to include a large number of alternative allelic combinations. Furthermore, there is evidence that certain allele recombinations are expressed in androgenetic plants, otherwise which remain hidden in the parent plant or in other hybrid derivatives. This hidden variation may prove to be a valuable resource for future plant improvement. Furthermore, the androgenesis route may avoid any problems related to poor male transmission that can arise during conventional breeding programmes due to pollen competition or to pollen-stigma interactions. This phenomenon, which can lead to the loss of certain allele combinations, is less likely to be an issue in androgenesis. There are some evidences that, androgenesis generates variation spontaneously in ploidy and so allows for the recovery of haploid, doubled haploid, triploid and tetraploid plants (Laura *et al.* 2006).

Laura *et al.* (2006b) regenerated somatic embryos and plantlets from anthers of *A. coronaria* elite cultivars: 'Cristina' (white, blue, pink, red, fuchsia, purple), 'Mona Lisa' (white, blue, red, purple, bicolour red and bicolour blue), 'Tetranemone' (white, blue), 'Wicabri' (blue, red, fuchsia, white, bicolour red) and 'Tetra Elite' (white), and up to 16.9 regenerants per 100 cultured anthers were obtained. The androgenetic origin of regenerated plants was ascertained by both cytological and molecular tests. Root apex chromosome counting showed that 2 of 19 regenerants were haploids and that 11 regenerants had either a 2x karyotype, or were x and 2x mixoploids. Two plants had 2x and 4x cells. RAPD-based DNA fingerprinting showed that all the regenerants tested differed genetically from their anther donor, confirming their androgenetic origin. Visual observation and molecular marker analysis proved that anther somatic tissue did not participate either in embryo formation or in plant regeneration. Androgenetic plants were acclimatized in the greenhouse and produced flowers. A high regeneration capacity in term of recurrent somatic embryogenesis or adventitious bud formation from androgenetic embryos or from roots of regenerated androgenetic plants respectively was observed on medium used to rescue anther-culture derived embryos (Laura *et al.* 2006b). The high regeneration ability is thought to either have resulted from selection during the androgenetic process (genetic basis), and/or reflect an epigenetic effect associated with the juvenile state of explants.

Embryo formation was induced in anther cultures of *Anemone virginiana* and some other species of the same genus. The culture medium which included activated charcoal gave higher percentage of embryos. Embryo formation was also induced when cytokinin 2ip ($10^{-6}M$) was included, but at a lower percentage. This indicates that abscisic acid (ABA) can be involved in the normal inhibition of embryo formation in pollen (Johansson and Eriksson, 1977).

6.2.7 Embryo Culture

Whole ovules or ovaries have been cultured in *A. coronaria* where the embryo is too small to manipulate without risk of damage. Embryo culture must be initiated before embryo abortion, but the rate of success improves as the embryo matures. Young embryos require a complex medium with high sucrose concentrations, while more mature embryos can usually develop on a simpler medium with low levels of sucrose (Laura and Allavena, 2007).

6.2.8 Interspecific Hybridization

Transfer of the foliar habit of *A.× fulgens* (erect leaves with short petiole) into *A. coronaria* would facilitate the breeding of cultivars suitable for pot plant culture. At the same time *A. × fulgens* is monomorphic with respect to flower colour (red only), and could be greatly improved by the introgression of other colours from *A. coronaria*. Interspecific hybrids between tetraploid *A. coronaria* and *A. × fulgens* are possible, but sterility largely blocks their utilisation. Where hybrid plants do yield viable pollen, androgenesis can be attempted (Laura *et al.* 2006b); either to reduce chromosome number of the hybrid or to isolate individuals carrying a recombinant chromosome set (Kopecky *et al.* 2005).

6.2.9 Three Way Hybrids

The breeding of true F_1 hybrids will provide a high level of product uniformity, but needs substantial time and/or cost associated with the process of producing pure lines and later by the poor fertility of the parental, Double Haploid (DH) or inbred lines. On the other hand, hybrids made between selected heterozygous parents could speed up the breeding cycle, at the expense of some loss of crop uniformity. A trade-off between uniformity and productivity could be found by the synthesis of three way hybrids, made by crossing a highly fertile partially heterozygous line as female to an androgenetic line as male. Whatever the approach adopted, the preservation and multiplication of parent materials by tissue culture is essential for sustained success. The variable ploidy among androgenetic plants may facilitate the selection of the tetraploid and triploid lines needed to create polyploid varieties (Laura and Allavena, 2007).

6.2.10 Mutation Breeding

Mutation breeding has been largely neglected in anemone, possibly because of the large size of the progenies that need to be screened to identify recessive mutations. The use of anther culture derived haploids, combined with *in vitro* mutagenesis, may speed up the recognition of induced mutants, since mutations are not masked at the haploid level. Since mutagenesis frequently abolishes function, it may be particularly suitable to modify aesthetic characters such as flower colour, and plant architecture. In *A. coronaria*, some potential targets for mutagenesis are the gibberellin-insensitive (GAI) and brassinosteroid (BR) biosynthetic or signaling genes, since these are the major genes that determine the plant height (Wang and Li, 2006). *A. coronaria* lacks yellow sepal colour, but the flavonoid pathway is present, like in *Dianthus* and *Callistephus*, disruption of the chalcone isomerase, dihydroflavonol 4-reductase or flavanone 3-hydroxylase gene leads to the accumulation of yellow pigment (Forkmann and Martens 2001). Hence mutagenesis may be a practical approach to create yellow flowers in *A. coronaria*.

6.2.11 Disease Resistance

The introduction and naturalization of novel pathogens and pests pose one of the biggest challenges in contemporary plant pathology. Chromosome doubling can be an effective method to obtain plants that are tolerant to pests and diseases in both artificial and natural environments. The susceptibility of the induced tetraploid of *Anemone sylvestris* to artificial *Phytophthora plurivora* root infection was compared with the susceptibility of the diploid. The tetaploids were vital, uniform, with compact growth and there was no mortality observed. However, diploids were in poor health and showed unbalanced development with decreased survival percentage. Induced tolerance to *P. plurivora* is unknown till date in *A. sylvestris*. (Sediva *et al.*, 2019)

7.0 Growth and Flowering

In natural habitat anemone starts it annual growth cycle with active growth seen after the first autumn rains and flowers during early winter and a period of dormancy is observed during the dry and warm summer. Under Indian conditions anemones can be planted in spring season in hills; flowering starts in early June. Winter season is suitable for planting in North Indian plains and other cooler parts of India and flowering is seen in early summer.

To bring forward the flowering by an additional four weeks, tubers are vernalised before planting. The process includes of four major steps: soaking in water for 24–48 h; disinfection with fungicide; four to five week storage at <5 °C and planting of sprouted tubers in the field or greenhouse. In the Mediterranean temperate coastal area forced tubers, planted in the open field in the second fortnight of August, start to flower in the beginning of October and continue throughout winter and spring, providing a source of cut flowers for a period of six to seven months. Removal of the dried flowers can encourage continuous flowering and prevents seed development.

7.1.0 Photoperiod

In *Anemone coronaria* summer dormancy can be induced by long photoperiods even in the middle of the winter. Day length does not affect the flowering time, but dormancy induced by long day causes early cessation of flowering. Long day treatment also results in longer scapes (Ben-Hod *et al.*, 1989).

8.0 Propagation

Most cut flower cultivars belong to *A. coronaria* and are propagated through seeds. It can also be vegetatively propagated through one year old bulbs/corms/tubers.

8.1.0 Seed Propagation

A. coronaria is grown over the winter for cut flower production and in late winter to spring as a pot and garden plant. Seeds harvested in spring usually germinate in late summer or early autumn. The seeds have low germination rate, but seed stratification for 1-2 months, enhances the germination percentage. To achieve this, seeds should be mixed with moist coarse sand or peat (1:3) and water should be sprinkled over it to keep it moist. As soon as seeds get swelled, some substratum should be mixed and keep it in a ventilated room at 5°C. After few days, germination of seeds takes place and during the spring, the seedlings are ready for transplanting. Seedlings transplanted in pots, start flowering the following spring.

8.2.0 Vegetative Propagation

Corms are commercially used for propagation. Prior to planting the corms are given hot water treatment for few hours to break the dormancy which facilitate proper sprouting. After hot water treatment, the corms are planted at a depth of 5 cm in a pot containing planting mixture, consisting of moist peat and sand. Regular watering should be done to maintain the soil moisture. Excess watering should be avoided. Soaking anemone bulbs by wrapping them in cloth for about six hours in a plastic bag can also induce sprouting. Later, the sprouted bulbs of anemones can be directly planted in the ground.

Tubers (corms) are frequently used by growers in order to shorten the time from planting to harvest. These can be derived from seed within one growth cycle in the nursery. Intact tubers, rather than fragmented materials are usually preferred for phytosanitary reasons.

8.3.0 Micropropagation

In order to rationally protect and utilize the rare medicinal plant resources, plant tissue culture on *Anemone flaccida* was studied. The optimal explant was screened and effects of different plant growth regulators on callus induction and bud differentiation were tested. The combination of 2.0 mg/l 6-BA and 1.0 mg/l NAA added in the basic MS medium was propitious for callus induction and bud differentiation. The rate of callus induction attained was 85.7 per cent, and the adventitious bud or corm-like structure was easy to differentiate when healthy bud germinating from rhizome was chosen as explant. Rooting of the regenerated adventitious bud or corm-like structure (CLS) was induced well on the 1/2MS+0.2 mg/l NAA medium under 5°C (Zhang *et al.*, 2010).

Ruffoni *et al.* (2005) established clones of selected plants under *in vitro* conditions from the portion of the floral stalk, overlapping the bracts. This approach could overcome the problem of contamination encountered with tuber-derived explants. Up to 47 per cent of the genotypes tested generated bud primordial when cultured on a medium containing zeatin, and a histological analysis revealed that *de novo* meristematic patches were formed, which later develop into adventitious buds. Under natural conditions, shoots and new flowers arise sporadically at the base of the floral bracts; thus, it is possible to generate shoots from pre-existing meristems. These *in vitro* formed shoots were micropropagated and successfully acclimatized after induction of rooting in the presence of indole-3-butyric-acid and when the basal tuber attains a threshold weight of 5 g. The mature plant developed from these explants showed uniformity and no signs of somaclonal variation. Clones were also derivable using seeds as explants.

Propagation of Anemone

Seeds

Sowing of Seeds

Emergence of Seedlings

Growth of Seedlings

Planting in Field

Flowering in Field

Corms

Planting of Corms

Growth of Seedlings

Cultivation of Anemone

Corms of Anemone

Sprouting of Corms

Planting Under Polytunnel

Mulching

Cultivation in Greenhouse

Flowering in Greenhouse

Flowering in Open

9.0 Soil and Climate

Garden cultivated anemones generally grow best in well drained fertile loamy soils. The ephemeral *A. blanda* does not require much moisture during the summer, when it is dormant. *A. coronaria* can grow in a wide range of soils from acidic to alkaline. Majority of the anemone species grow well in partial shade, or in full sun when protected from the scorching sun in southern direction. Under controlled conditions, the highest number and the best quality *A. coronaria* flowers are produced under regime of 5–10 °C night temperature and 14–18 °C day temperature. In cooler regions, for cut flower production, anemone is grown under greenhouse condition and supplementary heating should be provided.

10.0 Cultivation

10.1.0 Time of Planting and Corm Size

The best time for planting of anemone is second fortnight of February to last week of March in the hills. However, in North Indian plains, first week of October is ideal. Planting large sized corms of 3-4 cm in circumference, results in flowering during the month of January to February. Planting of hardy species such as *A. blanda*, in the month of October commences flowering from mid spring (March).

In anemone, one-year corms and two- year old corms are known as pips and buttons respectively. Irrespective of the age, larger size corms (3-5 cm circumference) comes to flowering early and gives higher yield (Haegeman and Van Onsem 1970).

10.2.0 Planting

Corms or pre-sprouted corms can be planted directly in the main field. Pre-sprouted corms come to flowering earlier than the unsprouted corm. Planting is done at a depth of 5-7.5 cm with the pointed end facing downward. Anemones normally start to flower about three months after planting. Fall-planted (October) corms bloom in early spring and continue 8-10. Late winter-planted corms flower from mid spring and continue for about six weeks. Under Indian conditions anemones can be planted in spring (February- March) season in hills and winter season (October- November) in North Indian plains and other cooler areas.

10.3.0 Manuring and Fertilization

Application of compost at the rate of 50 t/ha provided the maximum number of flower stems, stem length and stem diameter (Aydinsakir *et al.*, 2009). Since, anemone is a tuberous plant, it has to supplied with high amount of phosphorous and potassium. Application of NPK along with well decomposed FYM gives good yield per hectare.

10.4.0 After Care

Care should be taken to maintain optimum soil moisture for proper growth and development of anemone. Over irrigation of the field, enhances the incidence of the root rot disease, and lack of soil moisture, especially during the formation of buds, is not conducive for growth and flowering of anemones. To maintain the optimum moisture in the field, crop should be cultivated in well drained soil. After planting it is highly desirable to mulch the field with 5 cm layer of dried leaves or turf. During the spring and moderate summers, the field is irrigated at weekly interval, during severe summers, plants are irrigated daily, especially during flowering time.

10.4.1 Anemone Care after Flowering

After the flowering is completed corms/bulbs should be dug out from the soil and prepared for winter storage: the corms/bulbs should be dried, foliage trimmed and kept in cool and dark place. To prevent the moisture loss, store the bulbs are stored in moist sand or peat. In areas having warmer winters, the bulbs are stored in the field by covering with dried leaves to prevent the damage of corms/bulbs from sudden frost.

11.0 Harvesting and Postharvest Management

The most ideal time of harvesting anemone flowers for the extended vase life is, when the flower has grown about 1 cm above the little "collar" leaf on the stem. Early harvesting results in smaller flowers having longer vase life. While, delayed harvesting results in decreased vase life, but the flowers are larger in size, which is suitable for use in flower arrangements for weddings and other events. The stem length is short at the beginning of the season, but later attains up to 30- 45 cm.

The flowers can be stored with optimum vase life, leaf quality, petal quality and flower marketability for 8 days at 16±2°C and 50 per cent RH under 300 Lux light intensity. Pre-harvest or postharvest spray of GA_3 @ 50 ppm enhances the flower quality of cut flowers when placed in vase solution containing silver nitrate (1mM) plus 30 ppm sucrose (Sharifani *et al.*, 2004).

12.0 Diseases, Pests and Disorders

12.1.0 Diseases

Anemone being a hardy crop, is susceptible to very few diseases and pests.

12.1.1 Anemone Mould

It is caused by *Botrytis cinerea*, during the moist humid conditions a grayish thin mass of mycelium grow on the infected plant surfaces. The infected plants when disturbed, large dusts like clouds of fungal spores are released. Moist and high relative humidity also trigger the release of spores. Poor air circulation adds to the survival and growth of the pathogen. The disease spread stops under warm to hot weather conditions. Spray Chlorothalonil or Mancozeb or Dithane M-45 @ 0.2 per cent for good control of the disease.

12.1.2 Anemone Crinkle

Colletotrichum acutatum is the causal organism. High temperatures (25-30° C) and humidity are conducive for growth and development of the pathogen. The incidence of the disease is most severe in the early planting crops. The disease can be controlled by soaking anemone corms in a solution of captan at 3 g/l, and harvesting of flowers in May/June before the hot weather helps in escaping the disease attack. The disease can be prevented by avoiding closer planting, which increases the humidity around the plant canopy, avoid excess irrigation, prune and destroy the infected plant parts and foliar sprays with Mancozeb or Dithane M-45 @ 2g/l.

12.1.3 Downy Mildew (*Personospora ficariae*)

A fine white mildew growth is observed on the under surface of the leaf surface. The infected leaves gets discoloured and distorted; under severe infection the stems topple down. Use of broad spectrum protectant fungicides such as Copper oxychloride or Chlorothalonil or Mancozeb and Dithane M-45 @ 2g/l effectively controls the infection.

12.1.4 Anemone Mosaic

The disease is caused by Anemone mosaic virus. The infected plants have mottled leaves and broken and distorted flowers. This virus can cause winter browning and can contribute to crinkle in anemones. Weeds acts as natural host for AMV and the virus is transmitted through aphids. To control the virus spread, remove the infected plants and destroy. To control the vector effectively, spray with neem oil/neem soap @ 5ml/l or any systemic insecticide such as imidachloprid @ 0.4 ml/l or acephate @ 0.75g/l.

12.1.5 CMV and Tobacco Necrosis Virus

Cucumber Mosaic Virus and Tobacco Necrosis Virus also affect *Anemone coronaria*. Affected plants show the mosaic pattern of chlorosis on leaves and eventually leaves become necrotic. Control of the respective vectors (aphids and *Olpidium brassicae*) reduces incidence of both the viruses. It is advisable to use virus free healthy planting material for planting.

12.2.0 Pests

12.2.1 Leaf Eating Caterpillar (*Agrotis segetum*)

Larvae feed on the leaves of anemone. During the initial stages larvae feed on the leaves by scraping the chlorophyllus portion, in the later stage, the larvae voraciously feed and skeletonise the plant completely. Spraying of neem based products such as neem seed kernel extract or neem oil @ 5 ml/l is effective during the initial stages of infestation. In case of severe infestation, spraying of chloranthoniliprole @ 0.4 ml/l or indoxicarb @ 1g/l is effective.

12.2.2 Aphids (*Myzus persicae*)

Nymphs and adults suck the sap from the lower portion of the leaves resulting in yellowing of leaves. In severe cases, the whole plant will become devitalised and drying of leaves is observed. To control the aphids, spraying of water with high pressure is helpful. Under severe infestation, spray of acephate @ 0.75 g/l or imidacloprid @ 0.4 ml/l is effective.

12.2.3 Thrips (*Rhipiphorothrips cruentatus*/*Scirtothrips dorsalis*)

Thrips are tiny insects which are black or straw in colour. They suck the plant sap by scrapping from the flower petals, leaves and stems. This results in discoloring,

Diseases and Pests of Anemone

Botrytis Mould

Downy Mildew

Mosaic

Leaf Eating Caterpillar

White Fly

Aphids

Thrips

Nematode Affected Leaf

Harvesting and Marketing of Anemone

Full Bloom Stage

Harvesting of Flowers

Harvesting stage in Greenhouse

Freshly Harvested Flowers

Cut Flower Vase

Loose Flowers

Pot Plants for Marketing

Cut Blooms for Market

Anemone Flowers in Market

flecking or silvering of the infested surface. Proper cultural and sanitation measures help in avoiding the thrips infestation. Spraying of fipronil @ 1.5 ml/l or spinosad @ 0.3 ml/l will be effective to control the thrips infestation.

12.2.4 White Flies (*Bemisia tabaci*)

These are the tiny insects, coated with white waxy bloom. Nymphs and adults suck sap from the plant, results in yellowing of leaves and reduces the vigour of plant. Under severe infestation spraying of imidacloprid @ 0.4 ml/l or acetamiprid @ 0.3 g/l is effective.

12.2.5 The Root Knot Nematode (*Meloidogyne incognita*)

Meloidogyne incognita, may cause severe problem when the plant is raised on the nematode infested soils. Nematodes suck the sap from the roots of the plants. As a result plants become devitalised and stunted in growth, thus reducing in flower size and yield. Soil sterilization with methyl bromide and fumigation with organic halogenates before planting are recommended for nematode control. Biologically root knot nematode can be controlled by the application of entomo-pathogenic fungi such as *Purpureocillium lilacinum, Pochonia chlamydosporia* in the infested soil.

12.2.6 Slugs and Snails

Anemones are resistant to pest attack, but slugs and snails can cause severe damage during the monsoon. These pests can be easily destroyed by gathering them during the night hours. Under the severe infestation, dusting with metaldehyde @ 25kg/ha can effectively controls the slugs and snails.

12.3.0 Disorders

12.3.1 Flower Abortion

Calcium deficiency induces a range of physiological disorders in plants. The disorders typically appear in young growing tissues that are characterized by high demand for Ca and restricted Ca supply due to low transpiration. Leaf damage and flower abortion occurs due to calcium deficiency in anemone. Application of $Ca(NO_3)_2$ @ 110 ppm through fertigation prevents flower abortion and improves the yield and quality of flowers (Bernstein *et al.*, 2018).

12.3.2 Bending

Anemone is prone to stem bending when shipped horizontally, keeping in upright position in cold storage and while transporting, to prevent stem bending.

13.0 Biotechnology

13.1.0 Molecular Markers

Molecular markers can complement breeding programmes in a number of ways. Chief among these is marker-assisted selection (MAS), an approach which effects selection for a desirable trait by means of selection for a linked marker. Linkage is established by genetic mapping, and DH populations are widely used for this purpose. Because of their homozygosity, DH lines are genetically stable and infinitely replicable and are thus ideal material for the genetic analysis of quantitatively inherited traits, since repeated trials are needed to isolate the environmental component of the overall phenotypic variance. Molecular markers have been used in combination with DH-populations for QTL (quantitative trait locus) mapping in the major crop species. Anther culturability appears to be quantitatively inherited, under the control of several nuclear-encoded genes. Earlier, it was to possible to determine the heritable response of anther culture between the cultivars in traits such as callus induction and plant regeneration but now, development of MAS even allows the pattern of inheritance to be defined. Mapping genes for anther culturability simplifies the selection of individuals with a good response to anther culture and thus helps in the establishment of the DH population that is required for the mapping of economically relevant traits (Laura *et al.* 2006).

13.2.0 Genetic Engineering

The major limitation to perform genetic engineering in *A. coronaria* is the lack of a reliable protocol for plant regeneration from somatic tissue. Recent observations made it clear that certain androgenetic plants undergo recurrent somatic embryogenesis or adventitious bud formation; and the availability of wide host range of *Agrobacterium tumefaciens* strains, have opened new perspectives for *A. coronaria* genetic engineering. Mono or oligogenic characters, such as resistance to diseases, tolerance to high temperature, absence of vernalisation requirement, increased vase life and flower colour traits which are absent either in *A. coronaria* or any of its interfertile species, could be readily transferred by genetic engineering.

14.0 References

Aydinsakir, K., Unlu, A., Yilmaz, S., and Ari, N. (2009) *Acta Hort.*, **807**: 359-364.

Ben-Hod, G., Kigel, J., and Steinitz, B. (1989) *Sci. Hortic.*, **40**: 247-258.

Bernstein, N., Sacks, M., Snir, P. and Rosenberg, R. (2018) *Isr. J. Plant Sci.*, **65**: 97-108.

Chau, T. A. M., Nguyen, M. K., Phuong, T. T., Hwang, I., Kim, D. and Na, M. (2012) *Biochem. Syst. Ecol*, **44**: 270-274.

Dafni, A., Tzohari, H., Ben-Shlomo, R., Vereecken, N. J. and Neeman, G. (2020) *Plants*, **9**: 397.

Forkmann, G. and Martens, S. (2001) *Curr. Opin. Biotech.*, **12**: 155–160.

Haegeman, J., and Van Onsem, J. G. (1970) *Meded. Rijksstat. SierplTeelt Melle*, 20.

Hao, D. C., Gu, X., and Xiao, P. (2017) *Acta Pharm. Sin. B*, **7**: 146-158.

Jacob, Y., Mastrantuono, S. and Ferrero, F. (2006) *XXII Eucarpia Symposium. Section Ornamentals:* Breeding for beauty (Eds. Mercuri, A. and T. Schiva Sanremo), Italy, pp. 11– 15.

Kopecky, D., Lukaszewski, A. J. and Gibeault, V. (2005) *Crop Sci.*, **45:** 274–281.

Laura, M., Allavena, A., Magurno, F., Lanteri, S. and Portis, E. (2006a) *J. Hort. Sci. Biotech.*, **81**: 621–626.

Laura, M., Safaverdi, G. and Allavena, A. (2006b) *Plant Breed.*, **125**: 629–634.

Ruffoni, B., Semeria, L., Laura, M., Savona, M. and Bisio, A. (2005) *Propg. Ornam. Plant,* **5**(2): 74–77.

Sediva, J., Mrazkova, M., Zahumenicka, P., Cusimamani, E. F. and Zahradnik, D. (2019) *Sci. Hortic.*, 256: 108579.

Sharifani, M., Parizadeh, M. and Mashayekhi, K. (2005) *Acta Hortic.*, **682**: 701–707.

Sun, Y.X., Liu, J.C. and Liu D.Y. (2011) *Pharmazie*, **66**:813–821.

Wang, Y. and Li, J. (2006) *Curr. Opini. Biotech.*, **17**: 1–7.

Yonash, Y., Jinggui,F., Shamay, A., Paz, N., Lavi, U. and Cohen, A. (2004) *Euphytica*, **136**: 51–62.

Zhang, Z., Geng, M., Yang, T., Liu, W., Wang, J., Xin, L. and Zhu, D. (2010) *Zhongcaoyao Chinese Traditional and Herbal Drugs*, **41**: 1168-1172.

12

CANNA

Shreya, Dhawani A. Patel and Reeta Bhatia

1.0 Introduction

Canna, is a well-known ornamental perennial herb, grown throughout the tropics and subtropics (Tanaka 2001a). The generic name 'Canna' is derived from the Latin word *'kanna'* means 'a reed' referring to its herbaceous stem. Canna is a free flowering and easy to grow crop highly valued for its attractive foliage and magnificent flowers, widely used in landscaping. Canna represents tremendous diversity with respect to its colour and is available in various shades of red, orange, yellow and pink. Multi-coloured and bi-coloured forms are available in this crop and some varieties even have variegated foliage with attractive pattern that are in great demand (Bihari *et al.*, 2009). Due to its attractive foliage and availability of spectacular shades of flower colour, it is drawing tremendous attention of gardeners.

2.0 Importance and Uses

2.1.0 Aesthetic Uses

Canna is one of the most spectacular garden plants, best known for their marvellous foliage and vibrant blooms. The important species of canna that are being cultivated in India include, *C. indica*, *C. glauca* and, *C. iridiflora* (Roy, 2009) along with the hybrid species, *Canna generalis*. Cannas are being considered as impressive summer bedding plant and are ideal as herbaceous borders, shrubbery border, background plantings, and as a patio or decking plant. Both tall flowering and dwarf cannas can create a spectacular display and can easily complement the colour scheme of any garden. They can be planted as large border or in small corner bed. The tall varieties are ideal for planting at the back of the border and can act as an attractive screen. The dwarf varieties can be used for planting in containers or at the front of the border. Besides, cannas can be used to create a living hedge to define an entrance, hide an unwanted view or divide one garden area from another. They are ideal for planting along a fence or near a stone wall. Cannas grow well even in soggy soil where other plants may struggle. They are being used as traditional gift on the occasion of Father's day in Vietnam. Cannas are believed to have high tolerance to contaminants and hence they can be utilised to extract many undesirable pollutants in a wetland environment. The recent study by Gunarathna *et al.* (2016) have revealed that cannas can be used as a phytoremediation agent to mitigate the higher pollution concentration.

2.2.0 Therapeuticand other Uses

Apart from the ornamental significance of canna, all parts of canna have commercial importance. Its rhizomes are rich source of starch and have great economic significance as edible or laundry starch. The edible canna, *Canna discolour* Lindl. is especially being cultivated for its rhizomes which have high nutritive value and starch content. The edible rhizomes of canna can be eaten raw (Hair and Maynard, 2003) or after cooking or boiling. They are also used as feed for livestock, have pharmaceutical and chemical utility (Imai 2008; Koyama 1984). Its decoction is used as remedy in various women complaints and it is also believed to be diuretic, demulcent and diaphoretic (Duke *et al.*, 1985; Imani *et al.*, 2014). The young shoots of canna can be utilized as green vegetable. Leaves of *Canna indica* are used for treating malaria and fumigated leaves have insecticidal properties (De Kamer and Maas, 2008). The red flowers of *Canna indica* contain anthocyanin pigments (Al-Snafi, 2015) and *Canna generalis* acts as water purifier (Tran *et al.*, 2019). The seeds are used as beads for making different kind of jewellery or made into rosaries.

3.0 Origin and History

The genus is considered to be indigenous to central-south America and West Indies and have widely naturalized in the tropical and sub-tropical parts of the world (Roy and Banerji, 2006; Tanaka, 2001a). Until 1500, cannas were not being known to the European botanist in spite of their wide utility as food crop for thousands of years. It was first documented in the book 'The Vienna Code' during 1536-1566. Canna may have arrived in Europe from the Americas as early as 1492 during Columbus's travels. Canna has now become naturalized in many regions, including South-East Asia and the Pacific islands. It is widespread in its distribution and has become invasive in Queensland (Batianoff and Butler, 2002) and South Africa. Presently, cannas are being grown in almost every tropical countries of the world. It is also likely to be more distributed in sub-tropical and Mediterranean regions.

4.0 Taxonomy

Canna is the only genus in the family *Cannaceae*, belonging to the order Zingiberales. Genus Canna has 19 species which is classified by Nobuyuki Tanaka, a prominent botanist from Japan in 2001. The species included under Tanaka's classification are: *Canna amabilis, Canna bangii* Kraenzl, *Canna coccinea* Mill., *Canna compacta* Rosc, *Canna discolor* Lindl., *Canna flaccid* Salisb, *Canna glauca, Canna indica* L., *Canna iridiflora* Ruiz. and Pav, *Canna jaegeriana* Urb., *Canna tuerckheimii* Kraenzl., *C. liliiflora* Warsz. ex Planch,*Canna paniculata* Ruiz. and Pav., *Canna patens* (Roscoe) Tb. Tanaka, *Canna pedunculata* Sims., *Canna plurituberosa* T. Koyama and Nob. Tanaka, Canna *speciosa* (Roscoe) Hook, *Canna stenantha* Nb. Tanaka, *Canna jacobiniflora* T. Koyama and Nb. Tanaka (Mishra *et al.*, 2013).

The plants under genus Canna are perennial, rhizomatous with simple, erect and leafy stem. Alternate or sub opposite, simple leaves with sheathing petioles. The average height of the plant is 2 m, the stem is non-branched and bears green or purplish leaves or foliage. The Inflorescence is a terminal raceme or few-branched panicle. Although canna flowers are very attractive they have very poor vase life (Segeren and Maas, 1971). Flower of canna is showy, bisexual, and asymmetrical. It contains three sepals, three petals, six stamens and one pistil. The sepals are persistent and all the true petals united in a tube at the base. Apart from having true petals, five false petals are also found which are actually modified stamens. Out of the six stamens, five are modified into staminodes (sterile stamen) and petaloid (sterile stamen imitating petal).

5.0 Species and Cultivars

5.1.0 Species

The five elemental species of genus *Canna* includes, *C. indica* Linn. (*C. nepalensis*), *C. glauca* Linn., *C. iridiflora* Ruiz and Pav., *C. warscwiczii* A. Dietr. and *C. flaccida* Salisb (Khosoo and Mukherjee, 1970).

Uses of Canna

Canna in Trophy Formation

In Pot

In Mixed Border

As Ground Cover

In Water Garden

As Focal Point

Canna Hedge

In Miniature Water Garden

Dwarf Canna along Walkways

Canna for Deck Planting

For Patio Planting

As Bedding Plant

Uses of Canna

Canna Border along the Fence

Grown for Garden Display

Planted along the Entrance

Species of Canna

Canna flaccida

Canna indica

Canna irridiflora

Canna glauca

***C. tulianensis* Nb. Tanaka**: It is one of the recently discovered species from eastern Honduras which is characterized by its multi-branched inflorescences and more height (Tanaka, 2008).

***C. flaccida* Salisb**: It is regarded as one of the parents of the Italian group of cannas which grows well as water canna and possesses yellow flowers (flower lip is wavy) in cluster at the top of the stalk.

C. glauca: It bears lemon yellow or bright yellow delicate flower. *Canna glauca* var. *glauca* and *Canna glauca* var. *siamensis* (widespread in South and Southeast Asia) are the two varieties of *Canna glauca* which differentiated within the past few centuries (Tanaka, 2001b).

***C. indica* Ruiz and Pav (Syn. *Canna edulis* Ker Gawl):** *Canna indica* L. is one of the most popular species of *Canna*, known as Indian shot, African arrowroot, edible canna, purple arrowroot, Sierra Leone arrowroot. It bears red, yellow or pink coloured small flowers. Different sub species of *Canna indica* based on the flower colour includes, *Canna indica* var. *indica* L. (red colour flowers), *Canna indica* var. *flava* (yellow colour flowers), *Canna indica* var. *maculate* (yellow flowers with red spot), *Canna indica* var. *sanctae rosea* (pink flower colour) and *Canna indica* var. *warszewiczii* (red coloured flowers) (Tanaka, 2001b).

C. iridiflora: Possesses carmine-red to purple flower having drooping nature, foliage is green (pale inner edge and dark outer edge) spreading stems and gently spreading leaves creating goblet shaped clump.

***C. generalis* (**Canna lily, Kaana Vazha, Common garden canna**):** It is a hybrid resulted from cross between *Canna indica* and *Canna flaccida*. The flower colours vary from crimson-red to yellow, orange and pink. Bicoloured flowers with streak and stripes in contrasting colours are also available (Bailey, 1923; Thekkayam, 2009).

C. discolor Lindl.**:** It is the edible canna and also known as 'Achira' in Colombia. *Canna discolor* is a cross between *C. indica* var. *indica* and *C. plurituberosa* and is triploid in nature. Sub species of *Canna discolor* are; *Canna discolor* var. *discolor* (a triploid, sterile), *Canna discolor* var. *rubripunctata* (yellow flowers, many red spots and sterile) and *Canna discolor* var. *viridifolia.*

5.2.0 Cultivars

Cannas are available in wide range of shades, ranging from creamy white, yellow, orange and pink to an intense maroon red. Some cultivars possess dark reddish purple leaves and some of the cultivars even have striped yellow leaves. Generally, the cultivated cannas are classified in two groups based on their height; tall cultivars having height ranges from 80-160 cm and dwarf cultivars whose height ranges from 50-80 cm.

Classification of Canna Based on Flower Morphology

Alipore Hybrids	They are the improved crozy type which bears markedly large sized flower with wider range of colours.
Bouquet	The ideal variety in this group is 'Cupid' of canna lily or *Canna generalis* which bears flowers on closely branched spikes.
Crozy and gladiolus flower	These groups of hybrid were developed by Pierre Antoine Marie Crozy in about 1850 and Vilmorin in 1880, which resembles the gladiolus flower.
Candleabra	This group possesses branched flower stalk and contained 8-12 spikes instead of two or three.
Dreadnaught	This group advances over the gladiolus or crozy type canna both with respect to the individual flower and its bunch.
Miniature	This group is having small blooms obtained by crossing Society's (The Agri-Horticultural Society of India) dwarf hybrid with *Canna indica* and having compact and neat spikes.
Giant or Orchid flower	This group has originated from Italy and have large blooms which resemble with Iris Flag due to its silky appearance and softness.

(http: / / www. indiaagronet. Com / horticulture / CONTENTS / cannas. htm)

Classification of Canna

Self coloured	Bharat, Indiana, Black Knight, Aga Khan, Rose queen, Perfection, Professor, Golden Standard, Bridal Veil, Plume, Pink Satin, Gloria, Diana, After Glow, Aida, City of Portland, Trinarcia Variegate, Wyoming
Bicoloured	Eureka, King Alfred, Anarkali, Prince Philip, Olive, Dream, Isobel, Bardora, Julia, Imperator, Sweet Heart, Morning Glow, Angle's Robe, New Red, Yellow King Humbert
Spotted	Queen of Italy, Gladiator, Louise, Percy Lancaster, Goliath, Arjun, Queen Elizabeth, Fair Maid
Margined	Orange King, Aristocrat, Rosemond Coles, Yellow Gal, Carmine King, Star of India, Masterpiece
Blotched	Dragon's Tongue, Daphne, Rajaji, Stripped Queen, Bengal Tiger

(http: / / www.indiaagronet.com / horticulture / CONTENTS / cannas.htm; Roy and Banerji, 2006)

6.0 Genetics and Breeding

The in-depth studies on evolutionary history, breeding and genetics of canna were conducted during 1960-1970 at National Botanic Gardens, Lucknow in India by T. N. Khoshoo. He has published his conclusive research findings in well-known book, 'The Origin and Evolution of Cultivated Canna'. The chromosome number of genus *Canna* is considered to be remarkably stable within the family Cannaceae. The karyotypic analysis of 22 taxa of the genus *Canna* have revealed that most of the species under this genus are diploid (2n=18) with a basic chromosome number of n=9. The triploid somatic chromosome number (2n=27) was observed only in one species *i.e. C. discolour*. The basic chromosome number in the genus *Canna* has been established as n= 9 (Tanaka, 2009).

The flower colour in different species is controlled by a number of genes and their complex gene interactions such as intensifiers, inhibitors, lethals, *etc.* The recombination and segregations of these complex genes during interspecific hybridisation have led to the development of diverse genotypes with new colours and colour combinations. The repeated cycles of hybridization have played a vital role in the evolution of cannas. The continuous breeding efforts in ornamental cannas over the last 150 years have led to reduction in the height of the plant, increased the flower size and staminode width, increased the length of the flowering period, improved the flower placement (higher above the leaves and more erect), flower durability and cold hardiness, and produced self-cleaning plants (the spent flowers fall off automatically and do not need to be pruned). The efficient vegetative propagation in this crop has further allowed the fixing of the useful genotypes, although they may contain a high degree of heterozygosity and sexual sterility (Khoshoo and Mukherjee, 1970).

6.1.0 Introduction

It is one of the basic means for creating variability in any crop species. During 1885-1904 around 51 cultivars of canna were introduced at the Agri-Horticulture Society of India from Italy and USA (Percy Lancaster, 1976). Some of the important introductions are 'Austria', 'Italia', 'Professor Traub', 'Alemannia', Indiana', 'King Herbert', 'Mrs. Kate Grey', 'Pennsylvania', 'Philadelphia', 'Queen of Italy', 'Winter's Colossal', 'Wyoming' 'America', 'Burgundia', 'Africa' 'Asia', 'Attika', 'Bavaria', 'Borussia', 'Britannia', 'Crown Prince of Italy' (Roy and Banerji, 2006).

6.2.0 Mutation Breeding

Natural or spontaneous mutations have played a pivotal role in the evolution of modern day garden canna. The event of gene mutation and re-patterning of chromosomes and recombination during inter-specific hybridization are the main force behind the evolution of garden canna (Khoshoo, 1970). Along with triploidy, somatic mutations have also speeded up the evolution in this crop (Khoshoo and Guha, 1975). The spontaneous mutations have led to the modification in flower colour in the form of spots and stripes, which are generally unstable. The colour mutations in canna usually involve a change from red to yellow. A number of bud mutations have been reported in canna, a pure red cultivar, 'King Humbert', sported into the yellow flower cultivar 'Queen of Italy'. *Canna generalis* cv. 'Trinacria Variegata', a bud sport of cv. 'Trinacria; has variegated leaf (yellow stripe on green leaf) along with small yellow flower with white median stripes (Roy and Banerji, 2006).

High level of heterozygosity, sterility and long duration between sexual generations, are the major hurdles in improvement of canna through hybridization and selection. Thus, natural and induce somatic mutations may offer valuable tool of producing new variation in canna. Physical mutagen like gamma radiation have been successfully used to induced mutants with leaf variegation and novel flower colour in 'Rosemond Coles' cultivar (Khoshoo and Guha, 1975). It has been observed that with the increase in the mutagen dose, the plant development decreased and plants showed stunted growth, sterility, twisted peduncle, chlorophyll variegation and other morphological abnormalities (Khalabardin, 1962). The triploids are considered more radiation resistant than the diploids either due to the buffering effect of the presence of duplicate loci or reduced inter-phase chromosome volume in triploids (Khoshoo and Guha, 1975). Chuantang and Yazhi (1989) compared the mutagenic effects of the two mutagens *viz.*, gamma radiation and fast neutrons in canna cv. 'Crimson Canna'. It was observed that gamma-rays were better than fast neutrons in induction of novel mutants in canna.

The Gamma Irradiation Service and Nuclear Technology Research Centre (GISC), Thailand have developed and released twenty two mutants in canna. Some of the promising mutants includes: Pink Peeranuch, Cream Prapunpong, Yellow Arunee, Orange Siranut, Pink Pannee, Orange Rungsee, Pim Ngern, Wanvisa and Red Ridthee.

6.3.0 Polyploidy

Generally canna species are diploid with few exceptions with the existence of triploids. Autotriploids and segmental allotriploids are reported in canna (Belling, 1925; Mukherjee and Khoshoo, 1967). Polyploid cannas tend to be larger, stronger, more substantial, and more persistent in every respect. Tetraploids have been induced in *Canna generalis* and *Canna indica* using colchicine at the rate of 0.5 to 0.7 per cent for six hours. The induced tetraploids had thicker and broader, deep green leaves and exhibited stunted growth as compared to diploid. With the production of tetraploids, the number of flowers per inflorescence were reduced but the size of individual flower was enhanced (Oomer, 1978; Mukherjee, 1970; Khoshoo and Guha, 1975).

6.4.0 Hybridization

Hybridization has played a pivotal role in the evolution of garden cannas. Transgressive segregation during the process of hybridization have led to the improvement in several ornamental traits in canna including plant height, flower size and length and breadth of staminodes. The technique of hybridization has been widely used to create genetic variability in canna. Various group of canna have been created by crossing different species. The hybridization work in canna was initiated in France by M. Annee. The first hybrid was developed in 1848 by crossing two parents *Canna glauca* and *Canna indica*. The Annee Canna were bred primarily for foliage attributes and their flowers were salmon-yellow or orange-yellow or tinged with rose-red. Annee released nearly 20 hybrid lines by the 1870s with names such as Canna Annei-rosea, Canna Annei-rubra, and Canna Annei-marginata *etc.* The another canna hybrid group, 'Ehemann Canna' was developed by crossing

Canna iridiflora with *Canna warscewicsii*. This group of canna hybrid bears flowers of magenta red colour.

From the 1860s to 1903, Pierre Antoine Marie Crozy, bred another group of canna hybrids also known as French Dwarf Canna/Gladiolus flowered Canna/ Crozy group of canna by crossing *Canna nepalensis* (syn. *C. indica*) with *Canna warscewiczii* and *Canna warscewiczii* with *Canna aureopicta*. The Crozy group hybrids are considered to be the most cold-hardy of the garden canna. A German botanist, Carl Ludwig Sprenger created the Sprenger's group of canna/Italian Canna/ Orchid-flowered Canna by crossing existing cultivar of Italy with *Canna flaccida*, with the aim to introduce bright yellow flower colour to the canna gene pool. The change in morphological traits in cultivated form from wild form has been achieved in terms of changed floral color, form of leaves, reduced plant height, spikes well above foliage, erect flowers, free flowering, increased in the color spectrum and flowers size, increased thickness of floral parts, intensified color and durability of flowers and self-shedding flowers.

7.0 Growth and Flowering

The growth and flowering pattern in canna differ according to the area of its cultivation. In tropical and sub-tropical regions, it is cultivated as perennial crop whereas, in temperate region it is cultivated as annual crop. In the tropical and sub-tropical condition, canna is planted during the month of March-April and it begins to sprout nearly after one month of planting. The adult phase of canna bears 7 – 11 leaves and at that stage it attain the height of 1-2 m. Thus, the vegetative growth of canna is achieved from beginning of June-July depending on the climatic conditions. The flowering in canna begins from the month of September and continues at the end of May.

Bruner *et al.* (2001) determined effects of growth retardants namely, Cutless (Flurprimidol) and B-Nine/Cycocel on vegetative growth and flowering of *Canna* x *generalis* 'Florence Vaughan'. Cutless was found to be very effective in suppressing plant height in canna. The reduction in height was up to 34 per cent when used at the lower rate (25 ppm) and up to 41 per cent at the higher rate (50 ppm). Height suppression from B-Nine/Cycocel treatments was less than from Cutless.

8.0 Propagation

Canna can be propagated both by asexual (division of rhizomes) and sexual methods (seeds). Since canna is highly heterozygous crop and seeds take very long time to germinate after sowing, vegetative propagation through division of rhizomes is commonly practiced.

8.1.0 Seed Propagation

Propagation of canna using seeds become essential for breeders when it is required to raise new varieties after hybridizing two parents. The seed of canna is generally scarified before sowing as its seed coat is very hard which delays the germination process. The presence of water repellent substances like silica, callose and lignin in the hydrophobic part of exotestal layer are mainly responsible for the impermeability of the seed (Graven *et al.*, 1997). The presence of this sort of water impermeable layer in the seed coat is a rare character of family *Cannaceae* which is not found in the other families of the Zingiberales (Kress, 1990; Grootjen and Bouman, 1988; Graven *et al.*, 1997). Moreover, there is striking difference in the ovules structure *i.e.*, pachychalazal development and seed coat anatomy in *Cannaceae* and other Zingiberales families. Canna often take 40 days or even more days for germination (Grootjen and Bouman, 1988).

To speed up the germination process, the seeds are scarified using sandpaper. Sometimes, integument is ruptured manually or a small portion of it is cut off with a sharp razor blade without injuring the embryo. Treatment of the seed with hot water for 24 hours can also enhance germination in canna. Acid scarification by treating seeds with Sulphuric acid for 4 hours was also found effective (Imani *et al.*, 2014). The scarified seeds of canna can germinate under a wider temperature range from 10-40°C but the optimum temperature for germination is 15-35°C (Souza Filho *et al.*, 2011).

8.2.0 Vegetative Propagation

Propagation of canna can be easily accomplished through rhizomes. Clumps of rhizomes are dug out from a healthy mother plant, lifted with the help of forked hoe and divided into many pieces keeping the eye intact. Strong splits are selected from the divided pieces and stored in shady place for a week before planting. The plants raised through this method are true-to-the-type and exhibit uniform flowering (Roy and Banerji, 2006). The rhizomes start sprouting once the temperature rises above 16°C. However, the optimum temperature range for sprouting of canna rhizome is between 20 - 25°C. Generally, it takes 20-25 days to sprout after planting of rhizome (Flora of North America Editorial Committee, 2000). The rhizomes of canna are highly perishable and hence they should always be stored in cool, dry and well-ventilated place after harvesting (Gade, 1966). In the area where temperature falls below -10°C during winters, the rhizomes are uplifted from soil and over wintered by keeping above 10°C to prevent freezing injury and later on replanted in the spring (Cooke, 2001).

8.3.0 Micropropagation

It is a rapid means of *in-vitro* propagation where millions of identical plantlets can be produced within a short span of time. Plant growth regulators like auxins, cytokinins *etc.*, play an important role *in-vitro* multiplication. Cannas have a reputation of being difficult-to- micropropagated candidates (Cook, 2001). Apical dome culture or meristem culture coupled with 4 weeks thermotherapy treatment at 31.6°C produces virus free plants in canna (Prestowitz, *et al.*, 2007). Wafa *et al.* (2016) developed an efficient protocol for micropropagation of *Canna indica* L., using rhizome explants (excised from two-month-old aseptic seedlings). Complete plant regeneration was obtained on MS medium supplemented with 3.0 mg/l BAP and 1.5 mg/l NAA.

9.0 Soil and Climate

9.1 Soil

Generally canna can be grown in any soil but they do not prefer clayey soil. Sandy loam soil rich in organic matter is best for canna cultivation. The land chosen for its cultivation should not be too acidic or alkaline. They prefer well-drained soil that is either neutral or slightly acidic (5.5-7.5 pH). The soil must not be water logged and have proper drainage facility as excess of water is damaging to the rhizomes.

9.2 Climate

The best performance of canna can be seen when it is grown in full sunlight with moderate water supply. These plants prefer full sun for luxuriant growth of vibrant leaves and flowers, but they can survive in a little shade. It can be cultivated in warm, temperate, tropical or subtropical regions which are located at an elevation of 900-2,000 meters from mean sea level. The optimum temperature for its growth is from 18 – 32 °C and requires minimum annual rainfall from 300 mm – 1,200 mm. As cannas are native to tropical zones, so they grow well under humid conditions.

Even though it is of perennial nature, it can be grown as annuals in temperate zones for a tropical or exotic look in the garden. It is highly sensitive to frost and top growth is often killed by even light frosts, but the rootstock/rhizome can survive under several degrees of frost. In cooler zones, they can be grown as annuals or the rhizomes can be dug up in the fall and stored indoors for replanting during the next spring. While in arid regions it is cultivated in water by planting them within a pot submerged into the water. As canna plants have large leaves, windy locations must be avoided for its cultivation as wind can tear off the leaves.

10.0 Cultivation

10.1.0 Land Preparation

Soil preparation before planting is an important aspect for successful cultivation of this crop. Before planting of canna, the soil should be dug up to the depth of 40-50 cm for drying. Weeding, manuring, fertilization should be done as per the requirement before planting (Roy and Banerji, 2006).

10.2.0 Planting

Cannas are highly sensitive to cold, hence they should be planted in spring season once the danger of frost is over and temperature has risen above 18°C. Under northern plains of India, planting is generally done in February-March with the rise in temperature to get the full benefit of monsoon and to obtain a strong clump before the winter. The rhizomes (with 3-5 eyes) are planted horizontally with the eye/growing points facing up. They are planted at a depth of 10-15 cm and are often covered with 7 – 15 thick layer of soil/mulch. It can also be grown in pot using 2 parts of stable manure along with 1 part of garden soil as potting media. The potted plants should be repotted at 6-9 months interval as the roots become pot bound.

Leaf Variegations in Canna

Tropicana

Australia

Pretoria

White Tiger

Bengal Tiger

Cleopatra

Source: https://www.plantdelights.com/blogs/articles/canna-lily-bulbs-plant-canna-lilies

Methods of Propagation in Canna

Canna Seed

Removal of Seed Coat

Germination of Canna Seeds

Digging of Rhizomes

Cutting of Rhizomes

Rhizomes with Eyes

Methods of Propagation in Canna

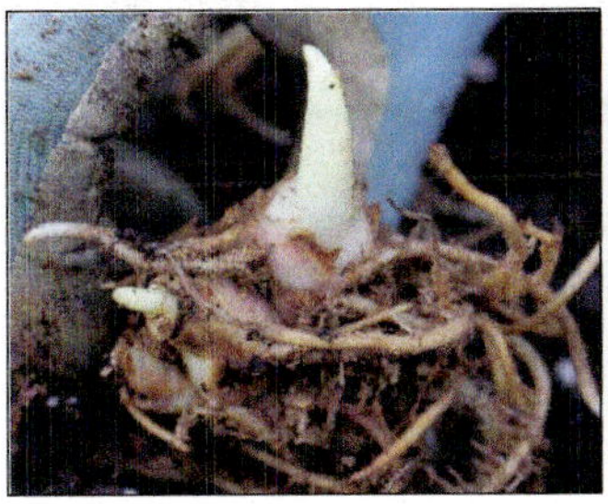

Sprouted Canna Rhizome

Planting of Rhizomes

Shoot Emergence from Rhizomes

Source: https://www.dirtonmyhands.com/canna_winter

Micropropagation of Canna

Direct Organogenesis from Rhizome Explants in ***Canna × generalis*** 'Raktima'.

Source: Singh *et al.* (2019).

Cultivation of Canna

Canna Seedling for Planting

Cultivation in Open

10.3.0 Spacing and Planting Density

The spacing of canna during planting is based upon the type of cultivar chosen. The plant to plant distance should be maintained at 45-50 cm for the dwarf type and 60-70 cm for tall type. The tall and vigorous varieties can be planted at a distance of 90 cm. The row to row distance should be maintained at 75 cm for dwarf cultivars and 1 m for tall cultivars (Roy and Banerji, 2006). Approximately 3000-4500 kg of rhizomes are required for planting one hectare land and the planting density should not exceed 22,500 plants/ha.

10.4.0 Planting and Care for Water Canna

Water canna are basically hybrids of *C. glauca*. They are ideal for growing in wet lands or bog soils. Planting of water canna is often done in baskets with up to 15 cm water above the rootstock/rhizome. The baskets (approx. 30 cm in diameter) are filled with loam based compost and after planting, these baskets are kept in shallow water to acclimatize the canna plants. The water cannas are generally fed using slow releasing fertilizers.

10.5.0 Mulching

A thick layer of mulch about 7-10 cm deep is often applied to newly planted canna in order to retain moisture during hot, summer weather. Mulching during the growing season helps to prevent weeds growth and provides a way to add fresh compost to give plants extra nutrients. While, mulching during the winter season can be practiced as a way of overwintering the canna plants. Kwack *et al.* (1990) studied the effects of polythene mulch during early spring on growth and flowering of Canna hybrida 'Amercan Red Cross'. Pre-emergence mulching using vinyl film enhanced the sprouting, improved the vegetative growth and advanced flowering by at least 25 days. Sometimes, mulching during winter season is practiced as a mechanism to overwinter the canna plants.

10.6.0 Intercultural Operations

Canna is a moisture loving plant, hence frequent watering should be done all through the growing season to maintain sufficient soil moisture. The frequency of watering is often reduced once flowering is over in canna. Frequent weeding, watering and digging of soil is recommended for better growth and flowering of canna. The periodical removal of dried leaves, dead flowers and shoots is necessary to promote new growth. Generally canna bears 2-4 spikes per stem and spent spikes should be removed from the base once the flowering is over.

10.7.0 Manuring and Fertilization

Canna is a heavy feeder and requires good amount of manures and fertilizers for growth of their foliage and flowering. Organic manure like compost, cow dung manure, neem cake and bone meal should be incorporated to the soil during land preparation. For better foliage growth and flowering, application of well decomposed FYM at the rate of 3.5 kg along with nitrogen, phosphorous and potash at the rate of 20 g per square meter area is recommended (Roy and Banerji, 2006;

Raychoudhary and Misra, 2001). After the first flush of flowering is over, leaf-mould or cow dung manure should be applied to the beds to ensure good flowering in the second flush.

Konnerup and Brix (2010) studied the effects of inorganic nitrogen (N) source (NH_4^+, NO_3 or both) on growth, biomass allocation, photosynthesis, N uptake rate, nitrate reductase activity and mineral composition of *Canna indica* in hydroponic culture. The relative growth rates, biomass allocation and plant morphology of *C. indica* were indifferent to N nutrition. It was revealed that NH_4^+ fed plants had higher concentrations of N in the tissues, lower concentrations of mineral cations and higher contents of chlorophylls in the leaves compared to NO_3 fed plants suggesting a slight advantage of NH_4^+ nutrition.

11.0 Diseases and Pests

Canna leaves have a waxy coating that helps the plant to resist fungal diseases. They are also generally resistant to pest problems and only few incidences of pest and diseases have been reported in this crop.

11.1.0 Fungal Disease

Canna rust (*Puccinia thaliae*), collar rot and fungal tip burn are some of the prevalent diseases of this crop. It appears as rusty-orange colored pustules spread by splashed water on the back of the leaves which eventually turn black and die. It may cause systemic damage to canna plant by affecting the leaves, stems, and flowers. Canna rust is more prevalent in areas of high relative humidity, extended periods of foliage wetting further increases the incidence of this disease. It can be controlled by spraying mancozeb (2 g per litre). The problem of fungal tip burn of the leaves and collar rot are controlled by the application of bavistin (2 g per litre) at fortnight interval (Brickell, 1996). Proper sanitation should be maintained to avoid incidences of diseases in canna. The diseased plants, its parts and its debris should be removed to reduce the risk of the spread of diseases

11.2.0 Viral Disease

Mild to strong symptoms of streaked leaves, stunted growth and distorted blooms are often seen in canna due to the attack of viruses, like Canna yellow streak virus, Canna yellow mottle virus and Hippeastrum mosaic virus (Reddy, 2015). The canna virus often causes distortion and discolouration of the leaves. Maintaining proper sanitation around canna plants and controlling the aphids can prevent incidence of viral diseases in canna. The virus infected plants should be removed as and when seen so that the disease does not spread to other plants.

11.3.0 Pest Attack

During the monsoon, due to high relative humidity it is infested by several pests like aphids, beetles, caterpillars, mites, slugs and snails. Most of these pests attack the leaves and flowers of canna and may cause substantial damage to the plant (Roy and Banerji, 2006). Aphids are a particular problem when growing canna under tropical climate. They can be controlled by spraying of insecticides like monocrotophos (1 ml/L) or fenazaquin 10 EC (1 ml/l).

Diseases of Canna

Symptoms of Canna Rust

Healthy Plant **Virus Infected Plant**

Symptoms of Viral Disease on different Parts of Canna.

Source: http://www.canna-virus.org.uk

Pests of Canna

Leaf Roller

Lesser Canna Leaf Roller

Leaves Damaged by Leaf Roller

Aphids Feeding on Leaves

Japanese Beetle Damaging Canna

Japanese Beetle Feeding on Flowers

Source: https://aggie-horticulture.tamu.edu/galveston/weekly_Q and A/f and b_11. Htmhttps://garden.org/ideas/view/LarryR/882/Beating-the-Beetles/

Harvesting and Storage of Canna

Harvesting Stage

Harvesting of Rhizomes

Storage of Rhizomes

Canna Flowers in Market

Canna leaf rollers: Incidences of two species of leaf roller namely, canna leaf roller (*Calpodes ethlius*) and lesser canna leaf roller (*Geshna cannalis*) have been reported in canna. They roll up the leaf and eat them, can leave huge holes in the leaves, thus ruin the aesthetics of the plant. There are several ways to manage infestations of the leaf roller caterpillars. Cutting dead canna plants to the ground in the late winter is a good way to reduce populations of leaf rollers. The application *Bacillus thuringiensis* (BT; Dipel or Bio-Worm-Killer) as foliar spray is effective against this pest. It is highly effective against the larvae and adult stages of the moth. Application of systemic insecticides (Carbaryl) has been found effective against these pests

12.0 Harvesting and Overwintering of Canna

Canna blooms make excellent short term cut flowers for just one day and have very poor vase life. However their foliage, especially the variegated types, are often used with other long lasting blooms to make spectacular floral arrangements. The studies conducted by Van-Doorn (1999) revealed that the poor vase life of canna is attributed to inhibition of water uptake by the cut stems. The blockage of water uptake in canna is mainly due to exudation of mucilage at the cut surface of the flowers (Roy and Banerji, 2006).

In warm climates, canna rhizomes can be left in the ground during winter, and the plants will behave as perennials, coming back year after year. However, in colder climates, at the end of the growing season (in the fall season), when leaves have completely died up, the canna plants are cut back to the ground level, and rhizome clumps are carefully dig up. During winters, the rhizome clumps can be stored in peat or vermiculite at a spot where temperatures does not dip below 4° C. During storage, the humidity of the mix should be maintained around 50 per cent and the temperature should be maintained at 6 to 8 °C. The container grown cannas can be stored in their containers without digging them up.

13.0 Biotechnology

Molecular markers are considered valuable tools for the assessment of genetic diversity. Patra *et al.* (2008) used the PCR based molecular markers (RAPDs and ISSRs) to characterize the genetic diversity in ten cultivars of Canna lilies (*Canna* Linn.). The dendrogram constructed based on RAPD as well as the ISSR markers grouped the ten cultivars of canna into two clusters. *Canna* x *generalis* 'Lucifer' was grouped separately from rest of the varieties. All the other nine varieties were again sub-divided to two categories. *Canna* x *generalis* 'Scarlet' was separated out and the varieties *Canna indica* 'Tropicanna' and *Canna indica* were grouped under a single sub-cluster sharing a common node at 56 per cent similarity level. Though RAPD and ISSR method cannot delimit the species and hybrids but can provide useful information for breeding program. Molecular markers are also highly reliable tools for detection of somaclonal variations. Mishra *et al.* (2015) assessed genetic fidelity of *in vitro* regenerated plants of *C. Indica* using RAPD and ISSR markers. No evidence of polymorphism was detected between parent plant and the *in vitro* regenerated plants using RAPD and ISSR analysis suggesting true-to-the-type nature of the regenerants.

Protoplast fusion has been attempted between *Canna indica* and *Canna edulis* to generate interspecific hybrid varieties of Canna (Mishra *et al.*, 2015). *In vitro* generated leaves and shoots of *Canna indica* and *Canna edulis* were used as the source for isolation of protoplasts. The protoplasts from both the species were isolated in the enzymatic mixture consisting of cellulase (1 per cent) and pectinase (0.5 per cent) with an incubation temperature of 24±3°C for 16-18 hrs in dark. The isolated protoplasts of both the species were successfully fused using Polyethylene glycol method. Singh *et al.* (2019) successfully optimised the protocol for adventitious shoot regeneration and *Agrobacterium*- mediated transformation in *Canna × generalis*. The transformation was successful, when the rhizome explants of canna were co-cultivated in *Agrobacterium* suspension for 60 minutes. The explants were then washed with 500 mg/l cefotaxime solution and subjected to 100 mg/l kanamycin selection. The treated shoots were excised and cultured on IBA-supplemented media for root development. Transgene integration in the putative transformants was confirmed by PCR assay and copy number by Southern blot hybridisation analysis.

14.0 References

Al-Snafi, A.E. (2015) *Int. J. Pharmacol. Toxicol.*, **5**(2): 71-75.

Bailey, L.H. (1923) *Gentes. Herbarum.*, **1**(3): 120.

Batianoff, G.N. and Butler, D.W. (2002) *Plant Protection Quarterly*, **17**(1): 27-34.

Belling, J. (1925) *J. Hered.*, **16**:465-466.

Bihari, M., Narayan, S., Singh, L., *et al.* (2009) *J. Ornam. Hort.* **12** (3):217-221.

Bruner, L.L., Keever, G.J., Kessler Jr, J.R. and Gilliam, C.H. (2001) *J. Environ Hort.*, **19** (4): 180–183.

Chuantang, N., and Yazhi, L. (1989) *AGRIS*, 2: 33-39.

Cooke, I. (2001) *The Gardener's Guide to Growing Canna*. Timber Press, Portland, Oregon.

De Kamer, H.M. and Maas, P.J. (2008) *Blumea.*, **53**: 247318.

Duke, J.A. and Ayensu, E.S. (1985) *Medicinal Plants of China.*, **1**: 232.

Gade, D. W. (1966) *Econ. Bot.*, **20**(4): 407-415.

Graven, P., de Koster, C.G., Boon, J. J. and Bouman, F. (1997) *Plant Syst. and Evol.* **205** : 223–240.

Grootjen, C. J. and Bouman, F. (1988) *Ann. Bot.* (New Series), **61**: 363– 371.

Gunarathna, M.H.J.P., Ranasinghe, A.I., Rathnayake, S.C. and De Costa, T.K. (2016). *Intl. J. Adv. Agric. Environ. Engg.*, **3**: 2349-1531

Hair, S.K.O and Maynard, D.N. (2003) *Encyclopedia Food Sci. Nutr.*, pp. 5962-5965.

http://www.indiaagronet.com/horticulture/CONTENTS/cannas.htm

Imai, K. (2008) *Japanese J. Pl. Sci.*, **2**: 46–53.

Imani, A.F., Sardoei, A.S. and Shahdadneghad, M. (2014) *Int. J. Adv. Biol. Biomed. Res.*, **2**(1): 223-229.

Khoshoo T.N. and Guha I. (1975) *In:* Nair P.K. (Ed.) *Glimpses in Plant Research.*, Vikas Publishing House. New Delhi, India, **3**: 1-81.

Khoshoo, T.N. (1970) *La cellule*, **68**: 109-134.

Khoshoo, T.N. and Mukherjee, I.(1970) *Theor. Appl. Genet.*, 40(5): 204-217.

Konnerup, D. and Brix, H. (2010) *Aquatic Botany.*, **92**: 142–148.

Koyama, T. (1984) *Plant Resources Studies*. pp.152-155.

Kress, W. J. (1990) *Ann. Missouri. Bot. Gard.*, **77**: 692– 721.

Mishra, S., Yadav, A. and Singh, S.K. (2013) *J. Harm. Res. Pharm.*, **2**(2): 131-144.

Mishra, T., Goyal, A.K. and Sen, A. (2015) *Horticulturae*, **1**: **3**-13.

Mishra, T., Goyal, A.K., Bhattacharya, M. *et al.* (2015) *Res. Plant. Biol.*, **5**: 20–24.

Mukherjee, I. and Khoshoo, T.N. (1967) *J. Cytol. Genet.*, 2: 1-5.

Patra, B., Acharya, L. and Mukherjee, A.K. (2008) *Int. J. Integr. Biol.*, **2** (2): 129-137.

Prestowitz, C., Sherry Kitto, S., Harbage, J. and Lyons, R. (2007) *Hort. Sci.*, **42**(4): 929.

Reddy, P. P. (2015) *Plant Protection in Tropical Root and Tuber Crops*, 281-291.

Roy, R. K. and Banerji B. K. (2006) *Adv. Ornam. Hort.*, **3**: 256-269.

Roy, R.K. (2009) *Indian J. Plant Genet. Resour.*, **22**:3.

Segeren, W. and Maas, P.J.M. (1971) *Acta Botanica Neerlandica.*, **20**: 663-680.

Singh, R., Dubey, A.K. and Sanyal, I. (2019) *Hortic. Plant J.*, **5** (1): 39-46,

Souza Filho, P.R.M., Tozzi, H.H. and Takaki, M. (2011) *Seed Sci. and Technol.*, **39:** 243-247.

Tanaka, N. (2001a) *Makinoa New Seri.*, **1**: 1–74.

Tanaka, N. (2001b) *Makinoa New Seri.*, **2** (1):34–43.

Tanaka, N.(2008) *J. Jpn. Bot.*, **83**: 7–10.

Tanaka, N., Uchiyama, H., Matoba, H. and Koyama, T. (2009) *Plant Syst. Evol.*, 280: 45–51.

Thekkayam, S.G. (2009) *Ornamental Plants*, 204-205.

Tran, H.D., Vi, H.M.T, Dang, H.T.T. and Narbaitz, R.M. (2019) *Global J. Environ. Sci. Manage.* 5(3): 331-344.

Van-Doorn, W.G. (1999) *Acta Hortic.*, **482**: 65-69.

Wafa, S.N., Taha, R.M., Mohajer, S. *et al.* (2016) *Biomed. Res. Int.*, pp. 282.

13

ORNITHOGALUM

Zahoor Ahmed, R.K. Dubey and I.T. Nazki

1.0 Introduction

Ornithogalum is a genus of flowering perennials belonging to the family *Asparagaceae*. This genus is represented by bulbous monocots with highly heterogeneous natural habitats. The botanical name of the genus is derived from a combination of two Greek words '*ornithos*' meaning 'bird' and '*gala*' meaning 'milk'. This name is believed to refer to the white color of its flowers. This name is also thought to have its origin from *Ornithogalum umbellatum,* the species referred to in the Bible as 'Dove's Dung' as the scattered white flowers when seen from a distance resemble bird droppings. Another term used for ornithogalum is 'Chincherinchee' which is the English translation of the South African name of these flowers, 'tienkenrientjee'. It is also known by several other common names including, 'Star of Bethlehem flower', 'Sun Star', 'Grass Lily', 'Eleven O'clock Lady'.

2.0 Importance and Uses

2.1 Aesthetic Uses

Ornithogalum is ideal for cut flower production as it possess all the characteristics of a good cut flower including long lasting vase life, long stems and diverse flower colours. The exceptionally long lasting cut flower of 'Chincherinchee' are highly suitable for making bouquets. Dyes can also be added to the vase water which can change the white colorations of its petals. *Ornithogalum thyrsoides* was the first fresh cut flower to be exported from South Africa (Lighton, 1960). Ornithogalum is ideal for planting in beds, sometimes in borders and perennial gardens for landscape purpose. They can also be grown in a rockery, border, or containers. These plants can also be naturalized in grasses, meadows, shrubbery borders, woodland margins, and other wild areas where they do not become a problem. These bulbous perennial plants produce offsets and are able to spread very quickly, with some species becoming invasive if not controlled properly.

2.2 Therapeutic and Other Uses

Biochemical analysis of *Ornithogalum* species have revealed the presence of steroidal glycosides (Komissarenko, 1971), monoterpene-lactone (Tang *et al.*, 2002) and homoisoflavone (Kuroda *et al.*, 2002), some of which possess antimicrobial, cytotoxic, cytostatic, anticancer, antioxidant, mould inhibiting and insect deterrent properties (Ebrahimzadeh *et al*, 2010). The bulbs of *O. saundersiae* offer valuable raw materials for pharmacological studies, because they contain biologically active compounds like 3β, 16β, 17α-trihydroxycholest-5-en 22-one 16-O, (2-O-4 methoxybenzoyl-β-D-xylopyranosyl)-(2-O-acetyl-α-L-arabinopyranoside) including OSW-1 saponins having anticancer activity (Salachna and Zawadziñska, 2013). The diluted extracts of *O. umbellatum* are useful in the treatment of stomach diseases such as gastric ulcers, peptic ulcers, duodenal ulcers, acidity *etc.* (Bhattacharya *et al.*, 2016).

The species *O. thyrsoides* and *O. conicum* from the Cape are known to be extremely poisonous and may also poses severe health risk if consumed by human beings. All parts of the plants are poisonous and poisoning of grazing livestock is

Uses of Ornithogalum

As Cut Flower

In Wedding Bouquet

Table Decoration

Planted as Edge

Herbacious Border

As Pot Plant

Garden Display

Indoor Decoration

In Japanese Floral Arrangement

well known. The typical symptoms of ornithogalum poisoning include blindness, which may become permanent in severe cases. In severe cases, there may be sudden heart failure. Other symptoms are a loss of appetite, diarrhea, weakness of the hindquarters causing a swaying gait and rapid deterioration of health condition (Obermeyer, 1978).

3.0 Origin and History

Ornithogalum is mostly native to Southern Mediterranean Europe and Southern Africa but some species are native to the Caucasus regions. Based on the location from where they have been brought, ornithogalums are grouped into three distinct classes: Tender species from (1) Southern Africa and (2) Tropical Africa; and Hardy species from (3) the Mediterranean and Caucasus regions. Nodding Star of Bethlehem has been reported to be invasive in Maryland and Pennsylvania. The *Ornithogalum* genus comprise of about 150 species of bulbous plants mainly found in the Mediterranean (and also in the temperate) regions of Africa, Asia and Europe (Bryan, 1989). The genus is widely distributed across Europe, Asia (reaching Afghanistan to the East), Africa (excepting the Tropic of Cancer band) and Madagascar (Zahariadi, 1965).

4.0 Taxonomy

Ornithogalum was originally described by Linnaeus in 1753, with 12 species being placed in the *Hexandria monogynia* (six stamens, one carpel). When Adanson (1763) formed the family *Liliaceae* in 1763, he initially placed ornithogalum in it, where it remained till this family was dismembered towards the end of the 20th century. Specifically, he included the genus with the onions (now Allioideae). Baker (1871) placed *Ornithogalum* in the tribe Scilleae, one of eight tribes into which he divided the *Liliaceae* (Baker, 1873). *Ornithogalum* is one of four genera in the tribe Ornithogaleae, the largest tribe within the subfamily Scilloideae of the *Asparagaceae*. The taxonomy of genus *Ornithogalum* is under constant revision and numerous taxonomical studies based on morphology, anatomy, cytology, seed micro-morphology and molecular phylogenies have been published till date (Rat, 2017). On the basis of floral and reproductive characters, many sections and subgenera have been accepted within ornithogalum. However, Manning *et al*. (2004) reduced all these genera to *Ornithogalum*, which resulted in over 300 species and was thus coincident with subfamily Ornithogaloideae. Manning *et al*. (2009) reconsidered the taxonomic identity of taxa within this subfamily and recognized only four genera: *Ornithogalum*, *Albuca*, *Dipcadi* and *Pseudogaltonia*.

The plants belonging to *Ornithogalum* genus are perennial bulbous geophytes. The bulbs that grow in winter rainfall regions are mostly small in relation to the size of the plant, whereas those from the summer rainfall areas are large, firm and globose. *O. paludosum* and *O. esterhuysenae*, grow well in permanently wet habitats and do not form a normal swollen bulb as there is no need for reserve food storage. The underground part in these species is more or less rhizomatous or there is a slight bulge (Obermeyer, 1978). Leaves vary from one to numerous and are basal, narrow, and long linear to oblong lanceolate (lance shaped), sometimes with a white longitudinal band on the adaxial (upper) side. In nature, the leaves are usually deciduous, in response to predominantly dry conditions but, when growing in a mild climate all the year round, these become larger, firmer and more numerous. The flowers are showy, diurnal or rarely nocturnal, bisexual and occur in racemes. The inflorescence is corymbose or pseudo corymbose. Flower is a perianth consisting of 6 petal like structures called tepals with a longitudinal green band only visible on the abaxial (lower) side. The perianth segments are 6 in number and each flower has 6 stamens arranged in 2 whorls. Fruits are 3 angled capsules that are obovate or oblong having black and shiny seeds. Seeds are usually globose with a prominently reticulate (net like pattern) testa having firm endosperm.

5.0 Species and Cultivars

The genus *Ornithogalum* is one of the largest genera in the family *Asparagaceae*. It comprises of about 160 species (Stevens, 2016). Muller-Doblies and Muller-Doblies (1996) recognized 123 species only for South Africa, forty of which were described as new taxa. Only a few out of 120 species of *Ornithogalum* genus have been extensively researched and introduced into cultivation as ornamental plants (Duncan, 2013). According to another estimate by Bryan (1993), there are about 150 different species available in *Ornithogalum*. Some of the important species are discussed as:

5.1.0 Species

Ornithogalum arabicum

The common names for this species include 'Arab's eye', 'Lesser cape-lily' and 'Arabian starflower'. It is native to Europe and Africa. The flowers are grouped in large, white and strongly perfumed raceme. The flower petals resemble stars, have yellow anthers and a large black ovary situated in the center. The plant usually blooms from the end of spring until the beginning of summer, and it also endures lower temperatures, but requires sunny expositions (Raluca *et al.*, 2017)

O. umbellatum

It is also known as 'Garden Star of Bethlehem', 'Splendid Ornithogalum', 'Nap at Noon', 'Sleepydick' and 'Grass lily'. Flowers are star shaped with white colour on the inner side and green on the outer side. It is herbaceous plant that grows from oval bulbs renewed every year. The leaves are narrow and linear and flowers are white star shaped arranged on the scape which can grow up to a height of 30 cm and each inflorescence contains 3-10 flowers. The flower is green outside with white margins. The flowers bloom from May up to June (Krichfalishi *et al*, 1989; Moret, 1992).

O. dubium

The common names of this species include, 'Sun Star', 'Yellow Chincherinchee' or 'Orange Star Flower'. It is a frost tender bulbous plant native to South Africa. It is mainly grown for cut flower and flowering pot plant production. It grows up to a height of 10-50 cm, with 3-8 linear erect leaves. The flowers are yellow up to orange and rarely white, usually with a brown or green center. Leaves are narrow,

Species of Ornithogalum

Ornithogalum umbellatum

Ornithogalum dubium

Ornithogalum saundersiae

Ornithogalum nutans

Ornithogalum thyrsoides

Ornithogalum arabicum

Source: https://www.gardenershq.com/Ornithogalum-Star-Bethlehem.php., https://commons.wikimedia.org/wiki/File:Ornithogalum_arabicum_JPG1b.jpg., https://en.wikipedia.org/wiki/Ornithogalum_saundersiae

lance shaped, and yellow-green. The capsule is rounded, deep in 3 corners and with prominent nerves. The flowers appear from August up to December. The plant begins to grow during autumn, when winter rains begin and dies after the spring bloom, when the hot summer weather begins.

O. nutans

It is also known as the 'Nodding Star of Bethlehem' or the 'Drooping Star of Bethlehem'. It can reach to a height of 15-45 cm with an erect, cylindrical and green scape. The bulb measures 20-25 mm, while the leaves are 2-3 linear, 15-35 cm in length and 0.2-0.4 cm in width, with a full margin (Raluca *et al.*, 2017) and are strapped with a central vein of silver. The flowers are just like bell shaped stars, striped green and white, and are fragrant. The raceme is cylindrical with 7-12 flowers. The fruit measures 10 mm in length and is ovoid, while the seeds measure 1.5-2 mm (Raluca *et al.*, 2017). These are usually grown beneath shrubs or rose bushes, in rock gardens, bed or border and are considered to be invasive in many areas.

O. thyrsoides

It is commonly referred to as 'Chincherinchee' or the 'Wonder Flower'. Chincherinchee is the English translation of the South African name, 'tienkenrientjee' which refers to the noise making properties of its seed pods when dried. It is a fleshy, succulent, dark green, hairless plant with 3-6 flat or concave, strap like, basal leaves about 15 cm long and 3 cm wide that die back after flowering. The flowering head is held well above the leaves by a simple, cylindrical stem about 30 cm tall. The flowers are white with darker center giving one of its common names, 'Black eyed Susan'. It has a conical spike and flower stalks are leafless and produce either a tight cluster of 30-50 flowers in a raceme or a loose corymb. It blooms in summer if planted in spring, or in spring if planted in the autumn or appear from late spring to midsummer (October to February). Flowers are long lasting and show phototropic movement.

O. saundersiae

The common name of *Ornithogalum saundersiae* Bak. syn. *Galtonia saundersiae* is 'Giant Chincherinchee' having its origin from Southeast Africa. It has white flowers with a characteristic ovary similar to a black olive gathered in large inflorescences at the end of long leafless stems (De Hertogh and Le Nard, 1993). It is a highly valuable species, recommended for cultivation in gardens and green areas, as well as cut and pot plant (Salachna and Zawadzinska, 2015).

5.2.0 Cultivars

Several horticultural hybrids have been produced for the flower market including the white flower varieties 'Chesapeake Snowflake' and 'Chesapeake Starlight'. These were produced from backcrosses of the South African cultivar of *Ornithogalum thyrsoides* 'Bok Bay'. Orange and yellow flowered varieties include 'Chesapeake Blaze', 'Chesapeake Sunburst' and 'Chesapeake Sunse'. 'Chesapeake Starlight' is another variety of *Ornithogalum thyrsoides*.

6.0 Genetics and Breeding

6.1 Genetics

Cytological studies of this plant species have shown a great variation in somatic chromosome number ranging from 2n=6 being normal and 2n=12 of frequent occurrence in vivo system (Bhattacharya *et al.*, 2016). Chromosome analysis of primary explants of *O. virens* revealed that all the cells were diploid with somatic chromosome number, 2n=6 (Nayak *et al.*, 1997). According to Obermeyer (1978), South African species have range of somatic chromosomes numbers having 2n=6, 10, 12, 14, 16, 18, 24, 54. Measurements of nuclear DNA content, meiotic and mitotic studies of karyotypes of 8 chromosome races (2n = 4, 6, 8, 10, 12, 16, 26) revealed that the African 2n = 12 race is the most primitive and has given rise to the other chromosome races. The 2n=6 race had a significantly higher fitness level

Morphology of Ornithogalum

Ornithogalum broteroi

a: Flower; b: Bract; c: Inner stamen (left) and outer (right); d: Gynoecium; e: Sections of the mature capsule; f: Capsules before dehiscence; g: Inflorescence and leaf; h Bulb.

***Source*: Martinez-Azorin *et al.* (2007).**

Ornithogalum baeticum

a: Flower, frontal view; b: Flower, dorsal view; c: Whole plant; d: Bract; e: Inner stamen (left) and outer (right); f: Gynoecium; g: Ccapsule, lateral view; h: Capsule, apical view; i: Capsule, transverse section; j: Portion of one leave; k: Bulb.

***Source*: Martýnez-Azorin *et al.* (2010).**

than the 2n=12 race (Stedje, 1989). Some species having different ploidy levels like *Ornithogalum umbratile* is usually hexaploid (2n=54+0-4 accessory chromosomes), but occasionally heptaploid with chromosome number, 2n=63+1B (Tornadore, 2003). The basic chromosome number of ornithogalum in sub-Sahara members was found to be x = 6 but the Eurasian members have the base number of x = 7, 8, 16 with *O. umbellatum* polyploidy series showing a base number x = 9 (Goldblatt and Manning, 2011).

6.2.0 Breeding

6.2.1 Polyploidy

Only a few South African species of *Ornithogalum* are used in breeding programs due to incompatibility reactions. Polyploidization can play an important role in solving incompatibility problems that existed in breeding the two common species *O. thyrsoides* and *O. dubium* which in turn created more variation. Since *Ornithogalum* (2n=12) is a bulbous plant, application of 0.2 per cent colchicine for 48 h on explants was most effective in creating maximum tetraploids (Blomerus, 2002). 2,4-D played an important role in increasing the frequency of polyploids particularly hyper tetraploids in callus cultures of *O. thyrsoides* after 120 days in culture whereas, NAA on the other hand helps to maintain a high percentage of diploid cells (Nayak and Sen, 1993).

6.2.2 Hybridization

Griesbach (1993), successfully developed interspecific hybrids between *O. dubium* (short inflorescence with orange flowers) and *O. thyrsoides* (tall inflorescences with white flowers) using the *in- vitro* embryo rescue approach. The developed interspecific F_1 hybrids were found to be completely intermediate in flower color and inflorescence habit but exhibited severe meiotic abnormalities. The backcross progenies segregated for flower color and inflorescence traits, and introgressed seedlings with orange pigmented flowers on tall inflorescences were obtained in the interspecific population. Van-Raamsdonk (1985), studied the fertility relationships of *O. umbellatum*, *O. angustifolium* and 9 other related taxa. *O. umbellatum* and *O. algeriense* are polyploid, whereas *O. monticolum* is diploid in nature. The intraspecific crossing of the later two species resulted in good seed set and germination, but viable seeds were not produced so far in crosses between either of these species. Bulbil producing species have a lower seed set after self pollination than species which do not produce bulbils.

7.0 Growth and Flowering

Ornithogalum is herbaceous perennial with annual tops that die off with the onset of summer drought. Seeds germinate from autumn to spring and form a small bulb in the first year and the bulb continue to grow into a flowering size bulb to develop inflorescence. Several factors like temperature, light, irrigation, growth regulators *etc.* are involved in influencing the growth, flowering and development of this crop.

7.1.0 Effect of Temperature and Light

In *O. arabicum,* the primordia of the outer and inner tepals of the first floret appear in late October and development of flower buds continued slowly until mid April and the plants flower in mid-May. Cut flowers with long scapes can be obtained in March by exposing plants to natural low temperature until late January (Shimada *et al.,* 1995). To control growth and development, the bulbs need to be prepared well in advance before planting. The flower induction process takes place at a cooler temperature with a relative humidity of 60 percent under controlled environment, very similar to that of dry vernalization of tulips. The bulbs are kept in this environment for three weeks or until roots start to develop. The bulbs harvested at immature physiological stage followed by storing them at 30 °C for 6 weeks can be used for forcing. Postharvest high temperature treatment can thus be used to force immature bulbs that were harvested before the senescence of the leaves (Roh *et al.,* 2007).

Lee and Miller (2015) determined the effects of pre-plant storage temperature and duration and greenhouse growing temperature on the growth and flowering in *O. dubium* (three cultivars) and *O. thyrsoides* (one cultivar). Bulbs were stored at five different temperatures for 1 to 4 weeks before planting. Within the range of 9 to 27 °C, lower preplant storage temperature resulted in early flowering and taller plants. Pre-plant storage of bulbs at 9 °C for 3 weeks, hastened flowering at least by 12 days as compared with control (27 °C). At 9 °C, as preplant bulb storage duration increased from 0 to 4 weeks, plants flowered more quickly and grew taller. Within the range of 13 to 21 °C, 17 to 18 °C, forcing temperatures gave the best combination of forcing time and plant quality.

De Hertogh and Gallitano (1997) concluded that the optimum circumference or size of *O. dubium* bulbs to be used for forcing is 4/5 and 5+ cm. It was suggested that bulbs should be stored at 25°C after harvesting. Before planting, the bulbs should be exposed to four weeks at 17°C for flowering potted plants and 9°C for cut flower forcing. Bulbs should be forced in a greenhouse with minimum 22°C day and 18°C night temperatures. They require at least 96 days for attaining flowering stage. Either a 9 hour short day or long day (3 hour night break) can be used for forcing. The latter is probably needed under very short winter days. A three week pre-planting temperature treatment of *O. dubium* bulbs at 13°C resulted in significantly longer flowering stems than treatments of 2, 9 or 25°C. Warm day/ night temperatures of 27/22°C and long day conditions induced early anthesis. Stem length was recorded maximum at moderately low production temperatures, and long days further increased it. Under the two lower temperature regimes, high number of flowers developed per inflorescence and, again, long days enhanced this number. However under the highest temperature regime tried, this effect was reversed (Luria *et al.,* 2002).

The root primordia of *O. thyrsoides,* 'Chesapeake Starlight' bulbs become clearly visible at the end of the 20°C treatment and at the same time the scape of the first inflorescence was clearly formed. Treatment of *bulbs* at 28 °C during the first 6 weeks (week 0–6) accelerated flowering. Sizes of bulb did not have any effect on flowering (Roh and Hong, 2007). Chilling of bulbs for 3 weeks at 10°C before planting accelerated flowering of the first inflorescence and shorten leaf length as compared to non-chilled bulbs (Wang and Walter, 2006).

Table 1: Effect of Controlled Growth Temperature and Day Length on Flowering of *O. dubium*

Temp (°C)	Days to Flowering		Stem Length (cm)		No. of Flowers/Stem	
	Long Day	Short Day	Long Day	Short Day	Long Day	Short Day
17/12	121±1	127±3	37.9±1.8	33.4±1.3	23.1±1.9	20.0±6.2
22/17	90±3	102±2	29.9±1.2	25.0±0.8	26.7±4.2	16.6±2.5
27/22	75±2	89±2	26.3±1.9	19.7±0.9	19.1±4.0	11.0±2.2
32/27	90±4	100±2	19.6±1.1	16.1±0.8	9.8±1.2	16.2±4.0

Source: Luria *et al*., (2002).

7.2.0 Growth Regulators

Plant growth regulators are being used as a part of cultural practices because of quicker impact on vegetative growth as well as flower yield. The physiological activities of flowering crops can be regulated by the application of growth regulators (Pal, 2019). Soaking of Chincherinchee (*O. thyrsoides*) bulbs in GA_3 (150 ppm) for 24 hours produced best results in terms of plant growth and longer flowering duration (Anu *et al.*, 2005). Fascination (formulation of gibberellin GA_{4+7} and cytokinin benzyladenine) and Pro-Gibb applied in Ornithogalum 'Chesapeake Snowflake' accelerated flowering of the first inflorescence but higher rates of these growth regulators were found to cause abortion of inflorescence (Wang and Walter, 2006). Jeung Keun *et al*. (2000) observed that flowering of *O. dubium* was accelerated when bulbs were dipped in 100 ppm GA_3 and Promalin (benzyladenine + gibberellins) solution but number of flowers was reduced by Promalin treatment. Flowering of *O. arabicum* was not influenced by Promalin. When Promalin at 50 ppm and GA_3 at 200 and 400 ppm were sprayed, flowering of *O. dubium* was accelerated.

Benzyl adenine (100 ppm) in *O. thyrsoides* improved growth, flowering and bulb production (Slathia *et al.*, 2018). On the other hand, growth retardant Topflor 015 (30 ppm) applied as a foliar spray on *O. saundersiae* inhibited plant growth (by 52.5 per cent), caused delayed flowering and reduced flower traits. The use of flurprimidol resulted in increased relative chlorophyll content (Salachna and Zawadzinska, 2015).

8.0 Propagation

8.1.0 Seed Propagation

Seeds should be first sown in propagation flats or trays or cold frame as soon as they are ripe during autumn. Place the seeds on the surface and cover lightly, leaving them undisturbed for the first dormancy period. Place the flats on the ground in an area that offers shade, preferably close to a wall that faces north. Cover it with a glass/plastic material and keep the medium moist and keep them at a temperature of 16°C. Alternately, stratified seeds kept under refrigeration for a period of three weeks can also be sown in spring in a moist growing medium.

Propagation Techniques in Ornithogalum

Bulbs of Different Species of Ornithogalum

Ornithogalum umbellatum

Ornithogalum dubium

Ornithogalum caudatum

In vitro **Propagation of** ***Ornithogalum thyrsoides*** **from Establishment of Cultures to Flowering (*Source*: Joung *et al.,* 2020).**

Germination takes place in one to six months. In *O. longibracteatum*, both low (10°C) and high (40°C) temperatures result in poor germination. The optimum temperature for seed germination and seedling survival rate is 22.9°C and 10 to 25°C, respectively (Kulkarni *et al.*, 2005). It takes about four years for *ornithogalum* to produce first bloom from seed germination.

8.2.0 Bulbs

Various methods of propagation through bulbs like scaling and use of cuttings, scooping, scoring or coring of bulbs can be employed for producing bulblets. It was found that the average number of bulblets per bulb produced by scooped bulbs, scales, cuttings, scored or cored bulbs were 40.7, 21.2, 19.8, 10.1 and 7.4, respectively. In scaling, scale pieces without a piece of the basal plate formed the largest and the highest number of bulblets compared to the scale pieces with disc and twin scales. Flowered sized bulbs produced more bulblets than non-flowered bulbs (Kariuki, 2008). *O. thyrsoides* bulbs having 3-4 cm size produced best vegetative and floral growth and bulb production (Slathia *et al.*, 2018)

In *O. saundersiae,* twin scale cuttings weighing 2.1-4.1 g, encapsulated in gellan and chitosan or in iota-carrageenan and chitosan produced bulblets having greater fresh weight, circumference and number of adventitious roots (Piotr, 2014). Bulb dormancy is induced as temperatures rise above 32°C and can be broken by high temperature treatments rather than low in commercial operations.

8.3.0 Leaf Cuttings

The mature vegetative leaves of non-flowering ornithogalum treated with 1000 ppm IBA and planting in a medium with 24.5°C bottom heat produced best results (Blomerus and Schreuder, 2002).

8.4.0 Micropropagation

During 1970s, slow propagation rate and susceptibility to mosaic virus were the two major constraints to successful commercialization of ornithogalum. The slow natural propagation of most the commercial cultivars and their susceptibility to *Ornithogalum mosaic virus* (OMV), make *in-vitro* propagation of 'virus free' plants imperative to sustain production (Niederwieser, 2004). A variety of ex-plants can be utilized for establishment *in-vitro* culture in ornithogalum. In *O. arabicum,* no difference was observed in the numbers of bulblets regenerated from the explants taken from innermost (youngest) and outermost (oldest) scales, but for explants taken from 10 mm above the basal plate, the innermost scales showed a higher rate of bulblets regeneration (Yanagawa and Ito, 1988). In ornithogalum cv. 'Rollow', rapid propagation was achieved by using small explants (0.5 × 0.5 cm) taken from the basal end of leaves and cultured in a vertical polar position without wounding on modified MS medium containing 0.1-1.0 mg NAA and 1-2 mg BA/l (Landby and Neiderwieser, 1992). The best regeneration from twin scales and *in-vitro* regenerated bulblets in *O. ulophyllum* was obtained on MS medium containing BAP (2 mg/l), NAA (0.5 mg/l) and BAP (2 mg/l), NAA (1 mg/l), respectively (Ozel *et al.*, 2008). In *O. thyrsoides*, leaf explants taken from proximal position produced the highest mean number of shoots (JeungKeun *et al.*, 2005). The explants taken from inflorescence stalk harvested just before floral sprouting showed efficient propagation rate than explants taken from leaves and bulbs (Marin *et al.*, 2009).

Various naturally occurring and synthetic hormones are available to induce the processes of dedifferentiation and redifferentiation (Petti, 2020). Among different species, *O. umbellatum* showed the highest bulblet regeneration (Karaguzel *et al*, 2012). In *O. thyrsoides*, the highest mean number of shoots were induced on MS basal medium, supplemented with 30 g/l sucrose, 0.8 mg/l ancymidol, 0.5 mg/l NAA and 1.5 mg/l BA (JeungKeun *et al.*, 2005). Thidiazuron at a concentration of 5uM promoted direct adventitious shoot proliferation and at lower concentrations (0.1-1.0 µM), its effect was similar to the combined effect of NAA+BAP in promoting morphogenic responses from shoot tip explants of *O. saundersiae* (Hutchinson, 2004). MS basal medium supplemented with 60 µM abscisic acid resulted in maturation of oval and globular shaped embryos and complete plantlets were produced when the mature embryos were separated and cultured on half-strength MS basal medium without growth regulators (Malabadi and Staden, 2004)

9.0 Soil and Climate

9.1.0 Soil

Ornithogalum grows best in loam or sandy loam soils rich in humus and organic matter. *O. thyrsoides* showed greater stem and inflorescence length, flower stem diameter and bulb weight when grown in un-amended soil as compared to the soil amended with different substrates like degraded granite or perlite (Bonomelli *et al.*, 2017). These plants prefer medium moisture and well drained soil mixtures and grow preferably well in the soil having a pH range of 5.5 – 6.0. *O. saundersiae* has been found to be sensitive to salinity stress as the plants exposed to salinity stress exhibited lower fresh weight of leaves, inflorescences and bulbs and delayed flowering (Salachna *et al.*, 2016).

9.2.0 Climate

Ornithogalum requires full to partial sun for good vegetative growth and abundant flower production. Good air circulation around plants must be maintained. For early autumn planting in warm areas, use of at least 50 per cent shading is recommended to increase plant height. These grow well in areas receiving medium to high rainfall and a temperature range of 14-24°C. Warm day/night temperatures of 27/22°C and long day conditions induce early anthesis in this crop. Stem length was highest at moderately low production temperatures and long days while as the effect was reversed under high temperature regime (Luria, 2002).

Growing of ornithogalum at low temperatures is also possible, but bench time is significantly increased. Depending on growing temperature, the production cycle for the crop is between 100 and 120 days, even longer under low temperature growing conditions. Supplemental light under greenhouses especially in temperate regions has shown excellent crop development. High light in combination with high temperatures, however, is detrimental to the crop. Shimada *et al.* (1995) reported that

O. arabicum plants subjected to 5-13°C for 8 weeks from the tepal to outer stamen formation stages, showed accelerated flowering and scape elongation.

10.0 Cultivation

10.1.0 Planting

Ornithogalum is an easy to grow plant, however it needs careful attention when it comes to maintenance. The normal size of bulbs for producing flowers is 12-14 cm in *O. arabicum* and 14 cm or more in *O. saundersiae*. In case of *O. saundersiae,* medium to large sized bulbs (5-11 cm) are recommended for production of high quality cut flowers (Kariuki and Kako, 1999). Parent bulbs of *O. saundersiae* having a circumference of 22-24 cm produced the highest number of bulblets (Piotr, 2014). The rate of growth, time to flowering and total yield of daughter bulbs was found to be proportional to size of the planted bulb (Luria *et al.*, 2002).

Table 2: Effect of Parent Bulb Size on the Yield of Twin Scale Cuttings and Bulblets of *Ornithogalum saundersiae*

Circumference of Parent Bulb (cm)	*Number of Cuttings per Parent Bulb*	*Fresh Weight of Cutting (g)*	*Number of Bulblets per Parent Bulb*	*Fresh Weight of Bulblet (g)*
16-18	28.5 ± 1.09 d	1.75 ± 0.09 d	8.10 ± 0.07 d	0.82 ± 0.03 c
18-20	32.0 ± 0.10 c	2.00 ± 0.15 c	9.80 ± 1.02 c	1.18 ± 0.08 b
20-22	39.0 ± 3.07 b	2.95 ± 0.44 b	22.1 ± 2.57 b	2.41 ± 0.21 a
22-24	52.8 ± 4.60 a	3.70 ± 0.20 a	35.6 ± 4.60 a	2.97 ± 0.02 a

Source: Piotr, (2014).

10.2.0 Time of Planting

Planting time varies depending on geographic location, however the prefer method of production is to divide offsets or bulbs in September or October. The larger bulbs are planted directly in their permanent position, while the smaller bulbs are first planted in pots or propagation block and allowed to establish for a year. They are finally transplanted in late or early summer. The studies conducted under Punjab (India) conditions revealed that planting *O. thyrsoides* bulbs in October produced maximum plant spread and height, bulb weight and number of bulblets per bulb than planting in January. October planting also resulted in the earliest flowering with the longest scapes, while as January planting resulted in late flowering with the shortest scapes (Bhardwaj and Kumar, 2001).

10.3.0 Planting Density and Depth

Planting density depends on the type of species and size of bulbs used. The planting density of 80 bulbs per m^2 for *O. arabicum*, 60 bulbs per m^2 for *O. saundersiae* and approximately 110-80 bulbs per m^2 for *O. thyrsoides* is recommended. In this species, wider spacing (20 × 30 cm or 20 × 40 cm) resulted in the more number of bulblets per plant, bulb weight and scape number. But leaf area decreased significantly with increased plant density (Bhardwaj and Kumar, 2001). In *O.*

Commercial Crop Ornithogalum Grown under Greenhouse

***Source*: http://sunvalleyfloralfarms.blogspot.com**

arabicum, bulblets planted at the density of 25 bulblets per m^2 produced heavier bulbs (25 g) after 10 months and planting density of 50 per m^2 produced flowering sized bulbs (Cervelli, 1994).

Ornithogalum bulbs should be planted at a depth of 7 to 15 cm. An investigation carried out by Kariuki (2003) in *O. saundersiae* revealed that the best growth and flowering characters were observed with bulbs planted at shallow depth (5 cm). While, the deep planting (20 cm) adversely affected the growth and flowering parameters.

10.4.0 Fertilization

Ornithogalums are light feeders and over feeding may results in browning of foliage tips. It is a long season crop and hence requires a good nutrient supply spread over a longer period for better vegetative and bulb growth. The P and K requirements are not very high but adequate potassium levels were found to be essential at the early growing stage due to the low potassium content of the bulbs. At the same time, the nitrogen content of bulbs was found to be high and thus a low nitrogen supply at the early growing stages had little effect on yield (Claassens and Beusichem, 1990).

The plants should be fertilized in April by supplying manure into the soil and light fertilizing until the end of the summer. Use of fertilizer N:P:K in the ratio of 5:2:4 in a well drained potting mixture is recommended, with application of trace elements during flower stem elongation in *O. dubium*. Growing conditions should be mild with high light intensity (Littlejohn *et al.*, 2000).

In case of pot culture, a general purpose house plant liquid fertilizer will do well for the plants. When the plants are new and germinating, apply fertilizer every two weeks. For pot culture under greenhouses, a complete 17-10-27 liquid fertilizer mixture is recommended. About 75 – 100 ppm nitrogen fertilizer solution is sufficient. For mature plants, withhold fertilizing after the growing season ends.

10.5.0 Irrigation

The plants have average water requirement and they perform well with regular watering throughout the growing season, *i.e.*, from early spring through early summer. During the growing season, the planting medium should always be kept moist. The pH of irrigation water should be maintained at 6.0 to prevent bacterial wilt outbreak.

The potting media needs to be well drained, allowing excellent aeration and avoiding both water logging and under watering conditions. Intermittent drying out of the media causes desiccation of delicate roots and plants usually do not recover from a damaged root system. Watering should be limited to the early morning. For most of the production cycle, watering two to three times per week is sufficient. Kulkarni *et al.* (2005) observed that watering had a pronounced positive effect on the seedling survival percentage and seedling growth of *O. longibracteatum*. Watering three times a week resulted in rapid seedling growth compared to watering once or twice a week. A study conducted by Karaguzel *et al.* (2011) on 12 species of ornithogalum revealed that irrigation levels at 1 liter/plant week and 2 liter/plant week had significant effects on growth, flowering and bulb production. Once the flowering is over, and the leaves have started to become yellow, irrigation should be withheld until new growth appears.

11.0 Diseases and Pests

The plants are mostly free of serious insect and pest problems. Bulb rot and root rot may be a problem in plants where the soil does not have good drainage and gets too soggy. Improving the soil drainage may improve the plant health. Some of the important diseases of this crop have been listed as under:

Bacterial soft rot is one of the most destructive diseases affecting ornithogalum. The commercial potential of ornithogalum is severely hampered by its susceptibility to bacterial soft rot. The disease is caused by the bacterium *Pectobacterium carotovorum* subsp. *carotovorum*. The disease may infect plants in the greenhouse, and eventually spread among bulbs during storage or during shipment of cut flowers, causing severe economic losses (Cohen *et al.*, 2011). Ornithogalum mosaic virus (OrMV) causes severe leaf mosaic symptoms as well as flower deformation in ornithogalum.

12.0 Harvesting and Postharvest Management

When first floret is fully coloured and open, it is ready for harvesting. Flowers can be stored at 2°C for more than a week. The bulbs need to be lifted from the soil when they are in the first two growth years. Once the blooms have faded, clip off the spent blooms to prevent self seeding and over spreading.

13.0 Bulb Storage

The bulbs upon arrival for storage should be plump and free from mold and other signs of disease. The bulbs should be stored in well ventilated area and need to be protected from direct sunlight, relative humidity should be moderate (50 to 60 percent) as excessive relative humidity may promote bacterial soft rot (*Erwinia carotovora*) and blue mold. The shoot apex of the bulb is vegetative at harvest stage and bulbs should be stored dry at 28-30°C for 6 weeks for rapid inflorescence development (Roh and Hong, 2007). Storage of *O. dubium* hybrid bulbs at 10°C for 6 weeks after harvest delayed the initiation of floral organs indicating that these were dormant and dormancy was not broken, whereas, bulbs stored at 22 and 28°C for 6 weeks formed the primary inflorescence indicating that dormancy was broken within this temperature range (Roh and Joung, 2004). When bulbs of *O. arabicum* were stored for 16 weeks after harvest at various combinations of 30 and 20°C, the percentage of bulbs producing flowers increased as the length of time at 30°C increased (Shimada *et al.*, 1995). The bulbs of *O. saundersiae* either solarized or stored at 20 °C for 2 months showed a good sprouting rate, flowering rate, number of flower stalks per bulb and flower stalk length. Storage of bulbs for 4 months at 3 °C resulted in upto 100 per cent decay of the bulbs while storage at 8 °C for 4 months exhibited delayed flowering (MeiLing *et al.*, 2006). Flowering of *O. arabicum* and *O. dubium* were accelerated when bulbs were stored at 10, 15 and 20°C for 4 – 6 weeks as compared to control (25°C). Flowering of *O. dubium* was accelerated when plants were forced at 19/13°C as compared to those forced at 13/10°C. The

Harvesting and Marketing of Ornithogalum

Harvested Flowers

Harvested Bulbs

Packing in Polythene

Cut Flowers in Market and Display.

number of flowers in *O. dubium* was decreased when storage duration of bulbs was increased from 2 – 6 weeks, regardless of temperature (JeungKeun *et al.*, 2000).

14.0 Biotechnology

14.1.0 Genetic Transformation

More advanced technology like gene transfer and marker assisted hybrid selection, may also be applied to develop cultivars resistant to virus and to increase efficiency of hybrid selection (Niederwieser, 2004). Hybrid embryos developed from incompatible crosses of *O. dubium* were rescued by culturing them on basal medium containing 70 g sucrose/l with no added growth regulators. Later on the ovules were transferred after 14 days to another medium containing 10 g sucrose/l for germination to occur (Niederwieser *et al.*, 1990).

One of the effective and sustainable ways to manage the plant pathogens is to use genetic modification and genome editing to expand the genetic tools available to breeders. At the same time these solutions must be effective with no negative effects on plant agronomy or introducing a potential allergen or toxin. Resistance to virus infection would be the ultimate solution and will reduce the production cost considerably (Niederwieser, 2004).

Genetic transformation mediated by bombardment with microscopic metal particles carrying target genes is the preferred method for the introduction of foreign genes into monocotyledonous plants like ornithogalum. The cells were bombarded with a pCAMBIA2301 vector, carrying nptII gene conferring kanamycin resistance and GUS reporter gene. The plants thus developed retained their stable transgenic state when grown in the greenhouse for two seasons (Cohen *et al.*, 2004). Genetic transformation of ornithogalum for Ornithogalum mosaic virus (OMV) resistance was performed by using the *Agrobacterium tumefaciens* strain AGL1. The gene constructs used contained either the viral coat protein or replicase genes under control of either the Delta SVB or UBQ3 promoters. Leaves from *in vitro* plants of an *O. dubium* × *O. thyrsoides* hybrid (line A_2) were used as explants which were wounded to promote infection in the *Agrobacterium* suspension. Co-cultivation continued for 4 days in the dark, pre-selection was performed on 250 mg/l cefotaxime (in the dark) and selection was performed on 80 mg/l kanamycin and 250 mg/l cefotaxime. Eighteen putative transgenic lines were selected, multiplied *in- vitro* and transferred to a glasshouse for hardening. PCR and real-time RT-PCR were performed and the integration of the transgenes into these plants was confirmed (van Emmenes *et al.*, 2008).

14.2 Molecular Markers

RAPD and AFLP markers have been used to identify the parentage of putative ornithogalum hybrids wherein about 169 random amplified polymorphic DNA (RAPD) bands and 298 amplified fragment length polymorphism (AFLP) fragments were developed (Young Hee and Roh, 2004).

15.0 References

Adanson, M. (1763) "VIII Liliaceae IV Cepae". *Familles des plantes Part II.* Paris: Vincent. *p*.50.

Anu, Kumar R and Dhatt, K. K. (2005) J. Ornam. Hort., **8**(2): 131-133.

Baker, J. G. (1871) *J. Linnean Soc. Bot.,* **11** (54–55): 349–436.

Baker, J. G. (1873) *J. Linnean Soc. Bot.,* **13** (68): 209–292.

Bhardwaj, B.M. and Kumar, R. (2001) *J. Ornam. Hort.,* **4**(2): 98-101.

Bhattacharya, S., Ghosh, B. and Choudhury, M. (2016) *Plant Tissue Culture and Biotech.,* **26**(1): 1-14.

Blomerus, L.M. (2002) *Acta Hortic.,* **570**: 251-254.

Blomerus, L.M. and Schreuder, H.A. (2002) *Acta Hortic.,* **570**: 293-296.

Bryan, J.E. (1989) *Bulbs.* Christopher Helm., London. **2**:297-299.

Cervelli, C. (1994) *Colture Protette,* **23**(3):75-77.

Claassens, A. S. and Beusichem, M. L. (1990) *Plant nutrition- physiology and applications*: **41** Springer, Dordrecht, *pp*. 649-652

Cohen, A., Lipsky, A., Arazi, T. *et al.* (2004) *Acta Hortic.,* **651**(2): 131-138.

Cohen, A., Lipsky, A., Arazi, T. *et al.* (2011) *Acta Hortic.,* **886**: 49-58.

De Hertogh, A. A. and Le Nard, M. (1993) *Elsevier, Amsterdam,* 761–764.

De Hertogh, A.A. and Gallitano, L. (1997) *Acta Hortic.,* **430**: 227-232.

Ebrahimzadeh, M. A., Nabavi, S. M., Nabavi, S. F. and Eslami. B. (2010) *Trop. J. Pharm. Res.,* **9**(2): 141-148.

Goldblatt and Manning. (2011) *Bothalia,* **41**: 83-134

Griesbach. R. J. (1993) *J. Am. Soc. Hortic Sci.,* **118**(3):409-414. 1993.

Hutchinson, M.J., Onamu, R. and Obukosia, S.D. (2004), *J. Agric. Sci and Tech.,* **6**(1): 60-68.

JeungKeun, S., Lee, A. K., Lee, J. S. and Maloupa, E. (2000) *Acta Hortic.,* **541**: 335-341.

Joung,Y. H., Wu X., Roh, M. S. (2020) *Sci. Hortic.,* **269**: 1093-95.

Karaguzel, O., Aydinsakir, K. and Kaya, A. S. (2011) *Ziraat Fakultesi Dergisi, Akdeniz Universitesi,* **24**(2): 125-130.

Karaguzel, O., Kaya, A., Biner, B. and Aydinsakir, K. (2012) *Afr. J. Agricl. Res.,* **7**(17): 2669-2673.

Kariuki, W (2003) *Acta Hortic.,* **624**: 217-221.

Kariuki, W (2008) *Acta Hortic.,* **766**: 135-141.

Kariuki,W. and Kako, S. (1999) *Sci. Hortic.,* **81**(1): 57-70.

Komissarenko, N. F. (1971) *Khimiya Prirodnykh Soedinenii,* **7**(1): 33-35.

Krichfalishi M, G.N., Krichfalushi, V.V., and Komendar, V. I. (1989) *Tiscia*, **24**: 3-10.

Kulkarni, M. G., Sparg, S.G. and van Staden, J. (2005) *Sci. Hortic.*, **107**(1): 103-109.

Kuroda, M., Mimaki, Y., Yokosuka, A. *et al*. (2002) *J. Nat. Prod.*, **65**(10): 1417-1423.

Landby, P.A. and Neiderwieser, J.G. (1992) *J. S. Afr. Soc. Hortic. Sci.*, **2**(1): 50-54.

Lee, J. and Miller, W, B. (2015) *HortSci.*, ***50**(2): 1784-1790.*

Lighton, C. (1960) *Cape Floral Kingdom*. Juta and Company Ltd, Cape Town.

Linnaeus, C. (1753) *Hexandria Monogynia* **1**: 306-308.

Littlejohn, G. M., Blomerus, L. M. and Maloupa, E. (2000) *Acta Hortic.*, **541**: 253-256.

Luria, G., Watad, A.A., Cohen-Zhedek, Y. and Borochov, A. (2002) *Acta Hortic.*, **570**: 113-119.

Malabadi, R. B., van-Staden, J. (2004) *S. Afr. J. Bot.*, **70**(4): 618-621.

Manning J. C., Forest, F., Devey, D. S., *et al*. (2009) *Taxon.*, **58**(1):77–107.

Manning, J. C., Goldblatt, P. and Fay, M. F. (2004). *Edinb. J. Bot.*, **60**(3):533–568.

Marin, L.J., Gonzalez, A., Cos, J. (2009) *Acta Hortic.*, **812**: 161-164.

Martinez-Azorin, M., Crespo, M.B. and Juan, A. (2007) *Analues del Jardin Botanico de Madrid*, **64** (1): 7-25.

Martinez-Azorin, M., Crespo, M.B. and Juan, A. (2010) *Plant Systema and Evol.*, **289**: 181-211.

MeiLing, L., ChaurChuang, H., Chen-I J. U., TsuLiang, C. (2006). *J. Taiwan S. Hort. Sci.*, **52**(1): 53-60.

Moret, J. (1992). *Nord. J. Bot.*, **12**(2): 183-195.

Nayak, S. and Sen, S. (1993). *Cytobios.*, **76**: 306-307, 209-216.

Nayak, S., Sen, S., Nayak, S. and Sem, S. (1997). *Cytobios.*, **91**: 366-367, 135-142.

Niederwieser, J.G. (2004). *S. Afr. J. Bot.* **70**(1): 47-51.

Niederwieser, J.G., Van-de Venter, H.A. and Robbertse, P. J. (1990) *HortSci.*, **25** (5): 565-566.

Obermeyer, A. A. (1978). *Bothalia*,**12**(3): 323-376.

Ozel, C. A., Khawar, K.M., Karaman, S., *et al*., (2008). *Sci. Hortic.*,**116**(1): 109-112.

Pal, S. L.(2019). *J. Pharmacog. Phytochem.*, **8**(3): 789-796.

Petti, C. **(**2020) *Plants*, **9**(8): 929.

Piotr, S. and Agnieszka, Z. (2014) J Basic Appl. **10**: 514-518

Raluca, E., Dinca L. and Cantar I. C. (2017) *J. Hortic. For. and Biotechnol.*, **21**(3): 89- 95.

Rat, M., Andric, A., Anackov, G. (2017). *Plant Syst. Evol.*, **303**:573-586

Roh, M. S. and Hong, D. K. (2007) *Sci. Hortic.*, **113**(1): 60-69

Roh, M. S. and Joung, Y. H. (2004) *J. Hortic. Sci. Biotechnol.*, **79**(4): 576-581.

Roh, M. S., AeKyung, L. and Keun S. J. (2007) *Sci Hortic.*, **114**(2): 138-141.

Salachna, P and Zawadzinska, A. (2013) *Afr. J. Agric Res.*, **8**(49): 6625-6628.

Salachna, P. and Zawadzinska, A. (2015). *Folia Pomeranae Universitatis Technologiae Stetinensis. Agricultura, Alimentaria, Piscaria et Zootechnica*, **318**(34), 57–64.

Salachna, P., Zawadzinska, A. and Podsiadlo, C. (2016). *Acta Sci. Pol-Hortoru.*, **15**(1):123-134.

Shimada, Y., Mori, G., Imanishi, H. (1995). *J. Japanese Soc. Hort. Sci.*, **64**(3): 617-623.

Slathia, D., Pandey, R. K, Hussain S., *et al.* (2018). *J. Pharmacog. Phytochem.*, **7**(5): 2500-2503.

Speta, F. (1998). In: Kubitzki, K. (ed.) The families and genera of vascular plants, vol **3** Springer, Berlin *pp*, 261–285.

Speta, F. (2000). *Beitra¨ge zur Naturkunde Obero¨sterreichs* **9**:743–792

Stedje, B. (1989). *Plant Syst. Evol.* **166 (**1-2): 79-89.

Tang, Y., Yu, B., Hu, J., Wu, T. and Hui, H. (2002) *J Nat. Prod.*, **65**(2): 218-220.

Tornadore, N., Marcucci, R. and Garbari, F. (2003) *Taxon.*, **52**(3): 577-582.

Van Raamsdonk, L.W. D. (1985) *Plant Syst. Evol.* **50**: (179–190).

van-Emmenes, L., Veale, A., Cohen, A., and Arazi, T. (2008) *Acta Hortic.*, **766**: 477-484.

Wang, J. and Walter, V.R. (2006) *Agric. Sci. and Technol.*, **7**(1): 23-28.

Wang, J., Walter, V. R. (2004) HortSci. 39(4):823E-823 DOI: 10.21273/HORTSCI.39.4.823E

Yanagawa, T. and Ito, I. (1988) *Journal of the Japanese Society for Horticultural Science*, **57**(3): 454-461.

YoungHee, J. and Roh, M. S. (2004) *J. Hortic. Sci. and Biotechnol.*, **79**(2): 316-321.

Zahariadi, C. (1965) *Revue roumaine de biologie. serie botanique* **10**(4):271–291.

https://en.wikipedia.org/wiki/Ornithogalum.

http://www.herbiguide.com.au/Descriptions/hgChincherinchee.htm.

http://www.learn2grow.com/plants/ornithogalum-thyrsoides/

https://en.wikipedia.org/wiki/Ornithogalum_arabicum

https://www.gardenershq.com/Ornithogalum-Star-Bethlehem.php.

https://www.greenhousegrower.com/crops/grow-ornithogalum-in-the-greenhouse/

https://www.asa.co.il/how-to-grow-ornithogalum-dubium-as-cut-flower/

https://www.flowerbulbs.com/1038/ornithogalum

14

ACHIMENES

K. Dutta

1.0 Introduction

Achimenes is a perennial rhizomatous flowering plant of family Gesneriaceae (Bailey, 1963) consisting about 25 species, distributed in tropical and subtropical regions of the world. It is commonly known as 'Magic Flowers', 'Widow's Tears', 'Star of India', 'Orchid Pansy', 'Cupid's Bower' or 'Hot Water Plant' (Ramirez Roa, 1987). It is well known for its floral diversity among closely related species (Roalson *et al.*, 2003). The name achimenes derived from the Greek word 'cheimanos' meaning 'tender' or 'sensitive to cold' (Weber and Skog, 2007).

2.0 Importance and Uses

Achimenes is an important bulbous plants mostly grown in pot, beds, gardens and as flowering house plants with attractive blossoms of white, pink, blue or purple colour which begin blooming from the month of June. These are also excellent for hanging basket displays under trees or in shady areas. Plants produce blooms profusely in indoors and used for interior decoration and when planted outside they flower from early summer in outdoors and can be grown in beds, rockery, in pots and hanging baskets.

3.0 Origin and History

The genus *Achimenes* is found throughout Mexico and Central America, with the center of diversity in central and southern Mexico (Ramírez, 1987) with one species (*Achimenes erecta*) occurring naturally in the West Indies. Large number of species are found in Mexico. Several species and hybrids are widely cultivated and naturalized outside their native range. Achimenes were first cultivated in the late 1700s, after being introduced to England, popular as potted flowering plants during the Victorian period. In the 1940s, more breeding and cultivar selection was undertaken in Germany, the Netherlands and the United States. It is now widely grown and distributed in Europe and United States (Miller, 2011).

4.0 Morphology

Most Achimenes have 'scaly rhizomes', a underground structures adapted from stems and leaves of the plant. Each scaly rhizome (1 – 3 cm long) is made up of a central stem with modified leaves (the 'scales') packed around the stem. The upper stems and leaves of the plant are dried during drought and sprouts from the rhizomes in the rainy season. The trumpet-like flowers can be singles or double, generally 2-4 cm in diameter. Plant growth habits for different cultivars may vary, ranging from spreading or pendulous flowering stems to compact, multi-branching flowering stems. The flowers are available in a wide spectrum of colors including white, scarlet, salmon, pink, blue, lavender, purple and even yellow. Their pubescent foliage ranges from bright to dark green in color, with some varieties with bronze leaves.

Uses of Achimenes

In Garden

In Pot

In Home Decoration

In Garden Bed

In Container

In Window Box

In Hanging Basket

5.0 Species and Cultivars

5.1.0 Species

The species is extensively hybridized with many of the hybrids involving the large-flowered species *A. grandiflora and A. longiflora.* Many of the species and their hybrids have large, brightly colored flowers and are cultivated as ornamental flowering plants in greenhouse and beds.

Achimenes erecta (syn. *A. coccinea*)

It is also known as 'cupid's bower', 'nut orchid', and 'magic flower'. Despite its name, *A. erecta,* is a plant trailing in nature. It produces a display of bright red flowers and dark green foliage. The reddish green stem carries groups of hairy, heart-shaped leaves. The flowers last only a few days, but are produced over a long period from June to October.

A. grandiflora

It is native to Mexico and Central America, growing in mountainous regions from Chihuahua to Nicaragua. It is cultivated as an ornamental plants in other places because of its showy purple flowers (DeCandolle, 1839).

A. longiflora

It can grow up to 60 cm, arising from small rhizomes. The hairy leaves have saw toothed edges and can be up to 8 cm long and 3 cm wide. The flowers are produced from June to October and are usually blue with a white throat. They can be quite large, up to 5 cm long and 7 cm across.

A. cettoana

It is native to Chiapas State, Mexico. It is a compact-growing species, possibly most compact of all the species in cultivation. The small plants can produce a profusion of lilac flowers.

A. mexicana

It is characterized by blue or purple corolla, paired by a pure white throat with a broad entrance. It is likely pollinated by bees, as distinct from species like *A. longiflora* which have narrow throat openings suitable for the proboscis of a pollinating moth. This species is restricted to several states in Mexico, unlike other species which are widely distributed through Central America.

A. admirabilis

It is native to Mexico and has an unusual bright orange to red colour flower and a yellow throat. The pattern of spots in the throat varies considerably.

A. antirrhina

It is a very attractive and variable species, lower colors range from red to spotted yellow. It is a native of south-western Mexico, and is well suited for indoor culture.

Species of Achimenes

Achimenes admirabilis

Achimenes antirrhina

Achimenes candida

Achimenes cettoana

Achimenes erecta

Achimenes longiflora

Achimenes misera

Achimenes grandiflora

Achimenes pedunculata

Achimenes mexicana

Achimenes glabrata

Achimenes heterophylla

A. heterophylla

It has variable orange flowers with yellow throat.

A. pedunculata

It is a tall-growing species, native to southern Mexico.

A. candida

It is widely distributed in western Central America, has very uncommon white flowers. There is significant variation in flower form, size and the dot pattern in the throat.

A. misera (syn. *A. warszewicziana*)

It is native throughout most of Central America from Mexico to Honduras, produces very small flowers.

A. glabrata

It appears in two forms; both have white flowers with a golden yellow throat, but one version has a pattern of fine purple spots over the central part of the corolla.

5.2.0 Cultivars

There are hundreds of cultivars available with single or double flowers having different colours. Popular cultivars are as follows

'Purple King', deep purple flowers, 5 cm across, flower from July to September.

'Charm', flowers 5 cm in diameter, charming coral pink flowers tinged crimson.

'Glory', orange-red flowers with a small touch of yellow in the throat.

'Desire', double red flowers, medium green leaves.

'Camille Brozzoni', purple medium flowers, leaves are large.

'Amie Saliba' is a hybrid made by Serge Saliba and named for his wife. It is a vigorous hybrid, gives yellow flowers.

'Blue Sparks', flower color is blue, white spotted.

'George Houchw', pink flower, dark green leaves.

'Rosea', flower colour is deep rose,

'Elegens', large size deep rosy-lilac flowers

The other popular cultivars are 'Harry Williams', 'Jewel Pink', 'Judith', 'Longiflora Alba', 'Pink cloud', 'Prima Donna', 'Snow Princess' etc.

6.0 Genetics and Breeding

The chromosome number of achimenes is 2n = 22. The mutant variety of achimenes 'Early Arnold' was officially approved in 1971. It was developed by irradiation of freshly detached leaves using 20-40 Gy of X-rays or 10-20 Gy of fast neutrons. It is an early maturing variety, gives 1-2 weeks early flowering than other varieties (Sigurbjornsson and Micke, 1974). 'Tetra Linda' is discovered as a

Cultivars of Achimenes

Purple King

Charm

White Glory

Sugarland

Odessa Blue

Amelia Vitorino

'Camille Brozzoni'

'Amie Saliba'

'Blue Sparks'

Serge Saliba

Elegance

Jwele Pink

mutation tetraploid cultivar in India by colchicine treatment. It has a distinct white flower including the throat and a flower size larger than 'Rosa Charm' and 'Allrote Charm'. The shape and growth habits of 'Tetra Linda' are also not as compact as 'Rosa Charm' (Michelssen, 1989).

7.0 Growth and Flowering

Achimenes is a day-neutral plant. The flowers are initiated when the third or fourth leaf node is developed. In achimenes, species are divided into different pollination classes, those with humming bird-, butterfly-, bee-, and female euglossine bee-pollination (Roalson *et al.*, 2003). The flowers within each of these groups share similar characters. For example, flowers pollinated with humming bird tend to have red, tubular flowers with ample nectar provided as a reward, while those with bee-pollination are small, white, funnelform flowers. The genetic basis for transitions in floral form, including traits such as flower color, thought to be closely associated with pollinators, has been studied extensively in many plant systems (Hoballah, 2007). The high level of floral divergence among closely related species in achemenes is driven by mechanisms such as pollination, hybridization, or time of flowering (Wiehler, 1983).

Six cultivars of achimenes, 'Blau Import', 'Blauer Planet', 'Linda', 'Prima Donna', 'Schneewittchen', and 'Tetraelfe', were grown for 16 weeks under short (SD, 8 hours) or long days (LD, 16 hours) at 21°C in a greenhouse. Long day increased plant height, shoot mass, number of flowers, and accelerated time to anthesis and depending on the cultivar, increased number of rhizomes. It was found that long day promotes shoot growth and flowering and several other characteristics of growth and development (Vlahos, 1990).

The effects of supplementary irradiance (20 μmol/m^2/s for 6 hours) with incandescent light (I) or fluorescent compact gas-discharge lamps (CF) vs. a basic irradiance (96 μmol/m^2/s for 12 h) with fluorescent (F) light at 17 or 25°C were studied for *Achimenes* hybrids 'Flamenco', 'Hilda', and 'Rosenelfe'. All three cultivars produced a longer stem and more flowers at the higher temperature. The growth responses were influenced by an interaction between temperature and cultivar (Vlahos *et al.*, 1992).

8.0 Propagation

Achimenes is propagated by seeds, division, tip cuttings and pinecone- looking rhizomes. The plant is generally propagated by rhizomes which give flowering within two to four months after planting, depending on the cultivar. Stem tip and leaf cuttings are also used for clonal propagation, produce flowering plants in three months.

The plants enter into dormancy after flowering. Rhizomes can be harvested and stored in dry conditions, where they can remain dormant for a substantial period of time (one to five months) depending on the cultivar. After the dormancy period is over, rhizomes can be directly planted into pots or beds. Rhizomes stored longer than 8 weeks resulted in decreased shooting and rooting responses (Miller and Bridgen, 2010). Minimum dormancy periods may vary with cultivars, but

Propagation of Achimenes

Separation of Rhizomes

Rhizomes Ready for Planting

Sprouting from Rhizomes

Commercial Propagation of Achimenes

Achimenes Tip Cutting and Planting

Cultivation of Achimenes

Cultivation in Pots

Cultivation in Open Field

generally four weeks is a sufficient rest period. Rhizomes are available to plant from late January to April. Plant the rhizomes (1-3 cm long) in pots in 2-3 cm depth, laid flat on the surface.

Achimenes is also propagated from tip cuttings or division of scaly rhizomes. A single scale from a rhizome can produce a new plant (Kim, 2007). Achimenes can be grown from seeds, but there can be variability in the resulting plants. Seed-grown plants may flower in five to six months. Commercial propagation from softwood cuttings in early spring under mist using bottom heat technique and leaf cuttings have also been successfully done by Miller, 2005. The bulblets formation in *A. longiflora* cv. 'Major' was investigated under controlled conditions in growth chambers. The bulblet formation in *Achimenes longiflora* cuttings was found to be inhibited by long day treatment such as continuous illumination. It was also observed that the degree of inhibition depends on the quality of the light. High intensity far red and especially infrared emission was needed to effectively suppress the bulbing. When red light (660 nm) was substituted for 20 per cent of the white light, the bulbing was promoted over 400 per cent. Among the growth regulators GA_3 and NAA reflected inhibitory effect whereas BA strongly stimulated bulblet formation (Deutch, 1974). Scaly rhizome formation in *Achimenes* cv. 'Primadonna' was investigated under controlled conditions in growth chambers. When plants from seeds were grown at 30-25°C (day/night), scaly rhizome formation was delayed. Plants from cuttings grown at 30-25°C showed a delay in scaly rhizome formation. The number of scaly rhizomes and fresh weight of rhizomes were lowest at 30-25°C. These results indicated that high temperature at 30-25°C retards the scaly rhizome formation in *Achimenes* (Koike and Suzuki, 2005).

9.0 Soil and Climate

9.1.0 Soil

Achemenes is generally grown in soil rich in organic matter, well-drained soil, bright indirect light, and constant supply of moisture. It can grow in peat moss-based potting medium with perlite or vermiculite.

9.2.0 Climate

Achimenes species and hybrids are commonly grown as greenhouse plants or outdoors as bedding plants in tropical, subtropical and temperate regions. It requires moderate to high humidity for vegetative growth. It has winter dormancy; the scaly rhizomes are kept dry until they sprout again in the spring. Some of the species and their hybrids are moderately hardy and can be grown outdoors year-round. A plenty of bright, indirect sunlight is required for achemenes. In direct sunlight the tender leaves may scorch or dry. Filtered light from a south or west-facing window is best for achemenes. Set plant about 20-25 cm under a fluorescent light for 12 hours a day in indoor condition. The average room temperature should be 18-24°C. Achimenes can tolerate temperature as low as 12°C, but anything above 25°C will cause flower buds to shrivel and dry up.

10.0 Cultivation

Rhizomes are available to plant from late January through to April. Plant the rhizomes, 1-3 cm in length in pots, laid flat on the surface. The best compost for achimenes is vermicompost or any organic compost. Plant the rhizomes 1-2 cm deep and with up to 5 rhizomes in a 13cm pot, smaller pots can of course be used when planting 1-3 rhizomes. The soil should be kept moist during the growing season, soggy or waterlogging condition may quickly cause rotting of rhizomes. If the soil medium is allowed to dry out, the plant will go dormant. Generally 40-50 per cent relative humidity should be maintained around the plants by applying proper irrigation. Misting of water on leaves can accelerate dark spots on leaves due to fungus attack. Achimenes is most suitable plants for container growing. Containers with good quality, well-drained soil with adequate drainage holes should be selected as achimene rhizomes cannot withstand waterlogged condition. The container is placed in light to moderate shady areas, full sun may burn the plants.

11.0 Harvesting and Postharvest Management

Achimenes flowers can be harvested within 6-8 weeks after planting depending on whether plants are propagated from rhizomes or stem cuttings. Little research has been done regarding the postharvest care of achimenes. It is reported that achimenes are highly sensitive to ethylene. Flowers and flower buds begin to abscise within 24 hours when exposed to less than 3 ppm ethylene. Spraying of silver thiosulfate at 0.3 to 0.5 μM decreased negative effects of ethylene and increased plant longevity. To avoid chilling injury, plants and rhizomes of achimenes should not be exposed to temperatures below 5°C, as chilling injury can occur below this temperature (Miller and Bridgen, 2011).

12.0 Diseases and Pests

There are few diseases affecting achimenes, among them botrytis is very prominent; particularly in high humid conditions. Waterlogging results wilting of foliage in achimenes. The achimenes is commonly affected by thrips, aphids, whiteflies and mites. The long tubular flower shape of achimenes provides a perfect condition for thrips to feed, which can cause serious foliar and flower damage and reduce marketability. Dry condition may encourage red spider mites to attack the plant.

13.0 Biotechnology

The phylogenetic relationships, proper supraspecific classification, evolution of floral forms and pollination syndromes, the origins of floral spurs, chromosome evolution, and biogeographic patterns using nrDNA ITS and cpDNA trnL-F spacer sequences were analyzed using maximum parsimony and maximum likelihood methodologies. Phylogenetic hypotheses support the non-monophyly of most of the supraspecific classification units currently recognized in achimenes, multiple origins of each of the pollination syndromes and the major morphological characteristics used to define these syndromes, multiple origins of floral spurs, multiple tetraploid

Diseases and Pest of Achimenes

Rhizome Rots

Botrytis Blight

Scale in Achimenes

Harvesting and Marketing of Achimenes

Harvesting Stages of Flowers

Harvesting and Storage of Rhizomes

Achimenes in Market

events, and sympatric distributions of many closely related species (Roalson *et al.*, 2003).

To explore the comparative transcriptome analyses of flower development in four species of achimenes RNA-seq were used and investigate gene expression and sequence evolution. The assembled transcriptomes contain between 29,000 and 42,000 genes expressed during development. In the experiment the sequence orthology and co-expression clustering were combined with analyses of protein evolution to identify candidate genes for roles in floral form evolution. Over 25 per cent of transcripts captured were distinctive to achimenes and overrepresented by genes involved in transcription factor activity. Using model-based clustering approach dynamic, temporal patterns of gene expression was found among species. Selection tests provide evidence of positive selection in several genes with roles in pigment production, flowering time, and morphology (Roberts and Roalson, 2017).

14.0 References

Bailey, L.H. and MacMillan Co. (1963) *Standard Cyclopedia of Horticulture.*

Broertjes, C. (1976) *Euphytica*, **25**: 297–304.

DeCandolle, A. (1839) *Prodromus Systematis Naturalis Regni Vegetabilis*, **7** : 536.

Deutch, B. (1974) *Physiologia Plantarum*, **30**: 113-118.

Henry M. Cathey (1968) *Home and garden* bulletin no. 151: 1-15.

Hoballah, M. E., Gübitz T., Stuurman J., Broger L., Barone M., Mandel T., Dell'Olivo A. (2007) *The Plant Cell*, **19**: 779–790.

Kim, (2007) *The Gesneriad Society*, Inc., pp, 10.

Koike, Y. and Suzuki, S. (2005) *Acta Hortic.*, **673**: 569-573.

Michelssen, K. (1989) *United States Patent* (19), Michelssen pp.1-4.

Miller, C. and Bridgen, M. (2011) *Crop cultivation*, pp. 30-33.

Miller, C. T. (2005) *Acta Hortic.*, **673**: 349–54.

Miller, C.T. and Bridgen, M.P. (2010) *Achimenes*, **20**: 377-380.

Ramírez R.M.A. (1987) *Ph.D. dissertation*, Universidad NacionalAutonoma de Mexico, Mexico City, Mexico.

Roalson, E. H., Skog, L. E. and Zimmer E. A (2003) *Systematic Botany*, **28**: 593-608.

Roberts, W.R. and Roalson, E.H. (2017) *BMC Genomics*, **18**: pp, 240.

Sigurbjornsson, B. and Micke A. (1974) *IAEA, Vienna*, pp. 303-343.

Vlahos, J.C. (1990) *HortScience*, **25** :1595-1596.

Vlahos, J.C., Martakis, G.F.P. and Heuvelink, E. (1992) *HortScience*, **27** :1269-1271.

Weber, A. and Skog L.E. (2007) The genera of Gesneriaceae.

Wiehler, H. (1976) *Selbyana*, **1**: 374-404.

Wiehler, H. (1983) *Selbyana*, **6**: 1–219.

15
TORCH GINGER

Vanlalruati, Prativa Anand and Reeta Bhatia

1.0 Introduction

Torch ginger [*Etlingera elatior* (Jack) R. M. Smith], is a perennial herb which belongs to the family *Zingiberaceae*. It is amongst the most cultivated species of the genus *Etlingera*. It is commonly known as 'Torch Ginger' or 'Wax Flower' due to striking resemblance of its inflorescence to a flaming torch (Wong, 2008). The other common names for this plant include, 'Ginger Flower', 'Red Ginger Lily', 'Torch Lily' and 'Wild Ginger'. Its exotic characteristics and elegant flowers make it as one of the potential tropical ornamental plants for commercialization.

2.0 Importance and Uses

2.1 Aesthetic Uses

Torch ginger is a popular ornamental due to its high durability, versatility, low cost of cultivation and spectacular floral display. It is one of the most sought after tropical ornamental plants, extensively used for landscaping and in large decorative flower arrangements. The bright showy inflorescence of this plant is commonly used as cut flower. It is commercially cultivated as cut-flower in Hawaii and Australia (Wagner *et al.*, 1999; Ibrahim and Setyowati, 1999). In Australia, Brazil, Hong Kong, Thailand and United States, torch ginger has been widely used in different floral arrangements. In the temperate climates, it is mostly planted as a backdrop for lawns and smaller flowering ornamentals.

2.2.0 Therapeutic and Other Uses

Essential oil extracted from different parts of this plant including rhizome, leaf and inflorescence is reported to have the phytochemical and pharmacological properties (Abdelmageed *et al.*, 2011; Chaon and Ding, 2016). It is used as a therapeutic agent for treating sore throats, sea and travel sickness, relieve cramps and rheumatic pains. The extracts from torch ginger leaf and inflorescence have exhibited antibacterial activity against gram-positive bacteria (Wijekoon *et al.*, 2011). Inflorescence bud extract has anti-fungal properties to control anthracnose diseases and inhibit *Colletotrichum gloeosporioides* mycelial growth (Punnawich *et al.*, 2009). The rhizome, leaf and inflorescence are also used as a condiment and as a natural ingredient in cosmetics such as soap, shampoo and perfume (Voon *et al.*, 2012). The daily intake of its raw inflorescence is believed to reduce diabetes and hypertension (Wijekoon *et al.*, 2011; Lim, 2017). Its stems can also be utilized as a potential raw material for the manufacture of paper (Ravindran, 2017). The rhizome of torch ginger is used for the extraction of a yellow dye (Ibrahim and Setyowati, 1999).

3.0 Origin and History

Etlingera elatior is native to Indonesia, Malaysia, and southern Thailand. It is distributed throughout South-east Asia, New Guinea, the Bismarck Archipelago, the Philippines, Polynesia and Australia (Queensland) (Ravindran, 2017). It was described in 1792 by Paul Dietrich Giseke and named after the German botanist Andreas Ernst Etlinger. It has been introduced as an ornamental in gardens and yards

Uses of Torch Ginger

In Bed **In Pot** **In Home Decoration**

In Flower Arrangements

throughout the tropics (Ibrahim and Setyowati, 1999). It has been commercialized as a cut-flower and hedge plant in tropical America. In America, its introduction was probably during the last century from Malaysia and Indonesia. In Panama and Costa Rica, torch ginger is reported in herbarium collections in 1970s and 1980s respectively (Missouri Botanical Garden, 2014).

4.0 Taxonomy

Etlingera is one of the largest genus of the family *Zingiberaceae*. It was described in 1792 by Paul Dietrich Giseke and named after the German botanist Andreas Ernst Etlinger. The genus *Etlingera* includes approximately 110 species, widely distributed in tropical and sub-tropical Indo- Malaysia and tropical Australia (Stevens, 2012). Several forms of torch ginger have been observed in wild and in cultivation. The cultivated ones have flowers with white-edged labellum, while the wild type have yellow-edged labellum. The red- bracted are most popular among the three forms (*viz.*, Red, Pink and White) and are popularly used as ornamentals and cut flowers, whereas, the pink involucral bracts are normally consumed as spice. White blooms have been reported to be rare compared with other colour blooms.

Torch ginger is a coarse herb often found growing in large colonies. Pseudostems (shoots) 3–6 m tall. Leaves numerous; ligule to 1.5 cm long, shortly 2-lobed, glabrous; petiole 1.5–4 cm long; leaf blades lanceolate, 38–85 x 8–18 cm, glabrous. The inflorescence a terminal, ovoid head of spirally imbricated flowers, surrounded at base by showy involucral bracts; peduncle 60–150 × 0.8–1.5 cm, clothed with green, glabrous sheaths; involucral bracts 3–6 × 2–3 cm, spreading, the upper obtuse to emarginate, the lower abruptly narrowed to a caudate tip, crimson-pink, glabrous; floral bracts similar to involucral bracts but pinkish, smaller; bracteoles tubular, approximately 2 cm long, unilaterally split; calyx 3–4cm long, unilaterally split, the apex 3-toothed; corolla pink to red, sometimes white; labellum deep crimson with white or yellow margin; filament short, flat, whitish pubescent; anther red, longer than filament. Fruiting head greenish or reddish, globose, 2–2.5 cm in diameter, short-pubescent; seeds many, black (Acevedo-Rodríguez and Strong, 2012).

Flowering in torch ginger occurs all through the year round. During the peak flowering period (January-April), 18 – 24 flowers bloom on each inflorescence. The life span of one inflorescence of torch ginger is of 3 – 4 months. Peak anthesis occurs between 5.30 - 6.30 a.m. (Aswani *et al.*, 2013). Morphological characters of Zingiberaceous flowers are thought to have coevolved along with their pollinators. The flowering phenology or the morphology of reproductive organs of the plants is influenced by the interactions of certain pollinators with plants and their behaviour.

5.0 Species

The important species under the genus *Etlingera* includes: *Etlingera elatior, Etlingera corneri, Etlingera punicea, Etlingera fimbriobracteata, Etlingera venusta, Etlingera fenzlii, Etlingera giseke, Etlingera linguiformis, Etlingera orbiculata, Etlingera coccinea etc.*

6.0 Genetics and Breeding

The somatic chromosome number of *E. elatior* varies from 2n = 48 to 2n = 50 (Eksomtramage *et al.*, 2002). To sustain in the ornamental and cut flower industry,

Colour Forms of Torch Ginger

White Torch Ginger (*Etlingera elatior*)

Pink Torch Ginger (*Etlingera elatior*)

Red Torch Ginger (*Etlingera elatior*)

Species of Etlingera

Etlingera corneri

Etlingera punicea

Etlingera fenzlii

Etlingera fimbriobracteata

Etlingera fimbriobracteata

Etlingera venusta

Etlingera fulgens

Etlingera maingayi

Etlingera giseke

Etlingera linguiformis

Etlingera orbiculata

Etlingera coccinea

Source: Helen, 2013; http://tropical.theferns.info/viewtropical.php?id=Etlingera+punicea; https://www.flowersofindia.net/catalog/slides/Elongated per cent 20Torch per cent 20Ginger.html

torch ginger requires continuous improvement in certain traits such as flower color, morphology, longevity, size, odour, and decreasing the time to flower formation (Yunus *et al.*, 2013). However, the conventional breeding in this crop is hindered due to cross incompatibility, poor fruit set, and low seed production. The *in vitro* mutagenesis can serve as a valuable tool for improvement of *E. elatior*. Yunus *et al.* (2013) developed the protocol for *in- vitro* shoot regeneration and mutant induction in torch ginger. They irradiated the *in- vitro* buds using gamma irradiations and optimised 10 Gy as the LD_{50} dose. Further screening of the regenerated mutants using molecular markers revealed that the gamma irradiated regenerants were distinct from the untreated control.

Fereol *et al.* (2010) attempted intergeneric hybridization between *Alpinia purpurata* 'Eilen McDonald' and *Etlingera elatior*. Both species have an equal chromosome number 2n=4x=48. Intergeneric hybrids were successfully generated among these two genera and about 20 per cent of the pollinations yielded viable hybrid seeds. The intergeneric hybrid seedlings were fertile and produced more flowers compared to *A. purpurata*. Fruits were formed from these flowers without manual pollination.

7.0 Growth and Flowering

7.1.0 Temperature

Environmental factors play an important role in growth, flowering and overall plant metabolism of torch ginger. It prefers partially shaded condition and is extremely sensitive to cold temperatures. In general, *Etlingera* requires high temperatures, light and humidity for its robust growth. They are highly sensitive to frost, and they may grow poorly in areas where night temperatures fall below 10°C. In temperate climate, it should be grown in greenhouse and regular feeding and repotting should be done. It shows profuse growth when grown at 18-25°C temperature with relative humidity of more than 80 percent. They need to be shaded during late spring and summer but low light condition during winter period causes them to grow taller and weaker.

7.2.0 Growth Regulators

Torch ginger often grows up to 3 meter in height and have limited utility for use as pot plant. Muangkaewngam and Te-chato (2018) studied the effect of Paclobutrazol (PBZ) on growth, morphological and physiological characteristics of torch ginger. PBZ was found effective in reducing the plant height, and lowest height was reported when PBZ was applied @ 400 mg/l. PBZ also affected the guard cell size. The highest width and length (29.81 and 54.95 µm, respectively) was recorded at 400 mg/l. For pot plant production of torch ginger, PBZ @ 279.19 mg/l is recommended.

8.0 Propagation

Torch ginger can be propagated sexually (seeds) and asexually (rhizomes and cuttings). It takes a year from planting to flowering in case of asexual propagation.

Various Techniques of Propagation in Torch Ginger

The Rhizomes to be Used as the Planting Materials

Three-Week-Old Seedlings of Etlingera

Source: Helen, 2013

Source: https://globinmed.com/index.php?option=com_content and view=article and id=83473:kantan and catid=798:k

Micropropagation of Etlingera Elatior Using Sucker Explant

Source: Yunus *et al.*, 2012

8.1.0 Seed Propagation

Fresh *Etlingera* seeds give better germination and viability of seeds greatly decreases with the age. The fresh seeds take around two months to germinate, while dried seed might take a year for germination. Germination of dried seeds can be improved by soaking them overnight in warm water. The pre-soaked seeds can be sown in 5–10 cm pots and covered lightly with compost. In cold places, pots may be sealed with plastic bag to retain moisture and placed in a warm chamber with a temperature of 25–30°C. Sealing should be opened to prevent condensation after the first leaves have appeared. The seedlings should be transferred into individual pots after acclimatisation. Care should be taken not to over-pot the seedlings as too much moisture in larger pots will cause rotting of roots (Helen, 2013).

Soilless media consisting of medium to large-sized composted bark are often used as a growing medium. Perlite, vermiculite, charcoal *etc.* can be used to provide aeration and to retain moisture.

8.2.0 Vegetative Propagation

Torch ginger can be propagated by using matured rhizomes. The rhizomes segments selected for planting should contain at least two growing buds. The freshly cut rhizomes can be directly planted in the field. The mature rhizomes often become tough and fibrous over time and hence a sharp implement is required to divide them. While planting, the roots should be shortened by one-third to a half and in older rhizomes dead roots should be pruned out. To prevent excessive transpiration, the pseudostems should be reduced by about one-half to two-thirds. Separated clumps should be potted with compost and lightly covered. Deep planting should be avoided to prevent rotting of rhizomes.

Cuttings are usually taken from younger growth at the outer edges of the clump or in pot. The cutting of torch ginger comprises of a section of rhizome (15–25 cm long) along with at least one developed leafy shoot and several growth buds. The cuttings are often planted into a shallow seed tray. The cuttings should be shifted to propagation chamber with bottom heat and high humidity until rooting. Cuttings taken in winter take longer time to root due to low light levels (Helen, 2013).

8.3.0 Micropropagation

E. elatior is commercially propagated through rhizome, however, proliferation rate is very low, and there is high risk of spreading systemic infections using this technique. Thus, micropropagation is a viable alternative for the production of disease-free plants with higher multiplication rate. Successful shoot induction, multiplication and complete plant regeneration from suckers of *Etlingera elatior* was obtained on MS media supplemented with 13.32µM 6-benzylaminopurine (BAP). MS medium without plant growth regulators has been recommended for profuse and strong rooting. Chaidir *et al.*, 2015 also reported that combination of 40 mg/l sucrose and 1mg/l BAP was the best for induction of healthy and green shoot.

9.0 Soil and Climate

9.1.0 Soil

Torch ginger can be grown in wide range of soils. It prefers humus-rich and well-drained soils with a pH range between 5.6 and 7.5 for optimum growth and flowering.

9.2.0 Climate

Torch ginger prefers wet and moist climate and hence mostly grown in tropical and subtropical regions. It is tolerant to acid soils and shaded conditions and has a medium drought tolerance (Ibrahim and Setyowati, 1999). It can be successfully grown in areas where the mean annual temperatures range from 10°C – 35°C, and grows well in areas where temperatures do not fall below 10°C at night and relative humidity is more than 80 per cent. It shows robust growth when planted in moist and relatively shaded locations. When planted in tropical climate, it requires some shade to protect from extreme heat.

10.0 Cultivation

10.1.0 Preparation of Soil

Proper land preparation is very important for good plant growth. It helps to reduce weed problems and improve soil structure. Before planting, the soil should be prepared well up to a depth of 30 cm. As torch ginger are sensitive to potassium deficiency, good quality fertilizer with high levels of potassium should be incorporated in the soil before planting to about 30 cm deep. They should be fertilized monthly throughout the growing season.

10.2.0 Spacing

Planting distance can significantly influence the floral attributes of torch ginger. It should be planted at a spacing of 5.0 m x 3.0 m with a population density of about 660 clumps/ha. The rhizomes should be planted at a depth of 5-8 cm deep into the soil. Organic mulching using rice straw should be applied around the planting point to conserve moisture and control the weeds. Under subtropical climate, a spacing of 3.0 x 2.0 m is recommended for production of quality flowers.

10.3.0 Irrigation

It has medium water requirements and cannot tolerate standing water. It is not drought-tolerant and soil should be kept moist consistently. Organic mulching can also be applied to conserve the moisture content.

10.4.0 Manuring Fertilization

Both organic manure and inorganic fertilizers are required for optimum growth of torch ginger. Torch ginger prefers potassium rich fertilizers. For optimum growth of torch ginger, the recommended dose of organic manure is 2,000 kg/ha/year. The inorganic fertilizer (N:P:K:Mg=12:12:17:2) should be applied at 500 kg/ha/year. The fertilizers doses should be divided into 3 equal portions and applied at 4 months

Various Developmental Stages of Torch Ginger

Full Grown Plants of *Etlingera elatior*

Flower Starting to Open

Flowering Stem

Freshly Harvested Flowers of Torch Ginger

***Source*: http://tropical.theferns.info/viewtropical.php?id=Etlingera+elatior**

intervals. Dolomitic limestone can be used as magnesium source, and slow-release granular fertiliser is added to the mix to provide essential nutrients. Fertilizer with N:P:K ratio of 3:1:1 can be used during spring and early summer while a balanced 1:1:1 feed is used in summer. A fertilizer ratio of 3:1:6 (N: P: K) can be used in winter to promote flowering in the following season.

11.0 Harvesting and Postharvest Management

Torch ginger generally takes about 12 months to start flowering when propagated through rhizomes, but a longer time is needed when seeds are used. Flowering occurs throughout the year and continuous harvesting is possible (Ibrahim and Setyowati, 1999). The flowers buds are ready for harvesting at about 4-5 weeks after its emergence form the soil surface. The flower buds should be harvested when they are fully-grown but before they start to bloom. While harvesting, the flower stalks should be cut at about 10 cm above the soil surface. Each matured plant will produce about 4 flower buds per year.

The floral stems harvested at the semi-open stage showed better visual quality and superior physiological parameters than floral stems harvested at the open stage. Different harvest stages, starting from 'closed' (stage in which the basal bracts begin to expand) to the full expansion define its commercialisation potential (Mattos *et al.*, 2017). The longevity, commercialization period and vase life is affected by the floral opening stage. The high mass of inflorescence (more than 1 kg) limits it cultivation as it complicates not only the harvesting and packaging process, but also the transportation.

Choon *et al.* (2018) observed longer vase life in tight bud (closed and pigmented bud) stage inflorescence as compared with the inflorescences harvested at torch show (all outer bracts unfolded) stage. Addition of sucrose to the vase solution promoted opening of true flowers whereas true flowers fail to open with the use of hydroxy quinoline sulphate (8-HQS) and distilled water. Unemoto (2011) assessed the influence of 1-methylcyclopropene (1-MCP) on ethylene action and its association with preservative solutions on torch ginger. The association was found to promote greater longevity and better quality in the postharvest conservation of torch ginger for additional 3 days. 1-MCP was most efficient at the concentration of 1.5 g/m^3.

12.0 Diseases and Pests

12.2.1 Diseases

The first incidence of *Curvularia lunata* var. *aeria* leaf blight on *Etlingera* was reported from Nagaland, India (Kithan and Daiho, 2014). The initial symptoms of this disease appear as small brown water soaked flecks on the upper leaf surface, which later coalesce to form dark brown lesions with a well-defined border. Lesions often merged to form large necrotic areas, covering more than 90 per cent of the leaf surface, which causes plant death. Treatment of rhizome and foliar spray of metalaxyl and mancozeb have been found to be very effective in management of this pathogen (Kithan and Daiho, 2014).

12.2.0 Pests

In greenhouse, common pests include mealy-bugs, root aphids, ants, cockroaches, spider-mites, scale, slugs, worms *etc.* Adoption of integrated pest management strategy is necessary to limit the damage. Pyrethroid insecticide based on deltamethrin 25g/l (2.8 per cent w/w) can be sprayed in severe cases. Spider-mite can be controlled with its natural predator *Phytoseiulus persimilis.* Cockroach and slug can be controlled with slug pellets, cockroach traps and by practising good hygiene. Biological control of slugs and sciarid flies can be done with Nemaslug and *Hypoaspis mites* (a predatory mite), respectively.

13.0 Biotechnology

13.1.0 Isolation of Protoplast

Da Silva (2012) standardised the protocol for protoplast isolation in *E. elatior*. They isolated the protoplasts using different tissues including, *in-vitro* leaves, *in-vitro* pseudostems, and leaves from the plants cultivated hydroponically, on six different enzymatic combinations. The effects of four incubation time periods, the rotary system (40 rpm), and three concentrations of mannitol (0.5, 0.6 and 0.7 M) on protoplast isolation was also studied. It was observed that the *in vitro* leaves were the best source of explants for isolation of protoplast, exhibited a greater viability and yielded larger protoplasts *in E. elatior*. The optimal incubation period was found to be 15 hours. Greater yields were obtained using a rotatory system with protoplasts incubated in the dark. They optimised 3 per cent Cellulase 'Onozuca' R-10 + 2 per cent Meicelase + 1 per cent Driselase + 1 per cent Dextran + 5 mM MES and 0.6 M mannitol as the best enzymatic combination for protoplast isolation.

13.2.0 Genetic Diversity

Ismail *et al.*, (2019) evaluated 57 accessions of torch ginger from Malaysia and assessed the pattern of genetic diversity and association among the characteristics using ISSR and SSR marker. Considerable degree of genetic variation was recorded among accessions of torch ginger. SSR marker was found to be more effective in evaluating diversity levels, genetic structure, relationships and determining gene diversity, polymorphic information content (PIC), and heterozygosity in this population. RAPD analysis is a highly effective method for detection of DNA polymorphism and early screening of potential mutant lines of torch ginger (Yunus, 2013).

14.0 References

Abdelmageed, A.H.A., Faridah, Q.Z., Nur Amalina *et al.* (2011) *J. Med Plants Res.*, **5**(15): 3432-3443.

Acevedo-Rodriguez, P. and Strong, M. T. (2012) *Catalogue of the Seed Plants of the West Indies*. Washington DC, USA: Smithsonian Institution. pp. 1192.

Aswani, K., Sabu, M. and Smisha, K.P. (2013) *Int. J. Plant, Animal and Envtl. Sci.*, **3** (4): 75-80.

Chaidir, L., Hasani, L., Diana, A. *et al.* (2015) *IOP Conf. Series: Earth and Environmental Science*, **334**:1-3.

Chan, E.W.C, Lim, Y.Y. and Omar, M. (2007) *Food Chem.*, **104**: 1586-1593

Choon, S.Y., Ding, P., Mahmud, T.M.M. and Shaari, K. (2018) *Acta Hortic.*, 1213.

Da Silva Junior, J. M., Paiva, R., Campos, A.C.A.L. *et al.* (2012) Protoplast production and isolation from *Etlingera elatior*. *Acta Sci. Agron.*, **34**: 45-50.

Eksomtramage, L., Sirirugsa, P., Jivanit, P., Maknoi, C. (2002) *J. Sci. Tech.*, **24**:311-319.

Fereol, L., Luc-Cayol, F. and Guitteaud, M. (2010) *Acta Hort.*, **855:** 131-136.

Helen, Y. (2013) SIBBALDIA: *The Journal of Botanic Garden Horticulture*, **11**: 71-85.

Ibrahim, H. and Setyowati, F.M. (1999) *Plant Resources of South-East Asia No. 13: Spices*. Leiden, The Netherlands: Backhuys Publisher, *pp*. 123-126.

Ismail, N.A., Rafii, M.Y., Mahmud T. M. M. *et al.* (2019) *Bio. Med. Res. Int.*, 1-14.

Kithan, C. and Daiho, L. (2014) *J. Exp. Biol. Agric. Sci.*, **2**(3): 308-315.

Mattos, D.G., Paiva, P.D.O., Nery, F.C. *et al.* (2017) *Postharvest Biol. Tec.*, **127**: 35-43.

Missouri Botanical Garden, (2014) Tropicos database. St. Louis, Missouri, USA: Missouri Botanical Garden.

Muangkaewngam, A. and Te-chato, S. (2018) *Int. J. Agril. Tech.*, **14**(4): 559-570

Punnawich, Y., Montree, I., Warin, I. and Kan, C. (2009) Philipp. Agric Sci., **92**(3): 265-270.

Ravindran, P. N., (2017) *In*: Ravindran P.N (ed), *The encyclopedia of herbs and spices* Vol 1 and 2, Wallingford, UK., p 1128.

Unemoto, L.K., de Faria, R.T., Takahashi, L.S.A. *et al.* (2011) *Acta Sci.*, **33**(4) : 649-653.

Voon, H.C., Bhat, R. and Rusul, G. (2012) *Compr. Rev. Food. Sci. F.*, **11**: 34-55.

Wagner, W. L., Herbst, D. R., Sohmer, S. H.(1999) *Manual of the flowering plants of Hawai*, Vols. *1 and 2*. Honolulu, USA: University of Hawai Press/Bishop Museum Press.

Wijekoon, J.O. Karim, A.A. and Bhat, R. (2011) *Int. Food Res. J.*, **18**: 1415-1420.

Wong, W. (2008). www.greenculturesg.com/articles/(accessed 02.02.10).

Yunus, M. F., Aziz, M.A. Kadir, M. A. *et al.* (2012) *Sci. Hort.*, **135**:145–150.

Yunus, M. F., Aziz, M.A. Kadir, M. A. *et al.* (2013) *Turk J Biol.*, **37**: *716-725.*

16

VICTORIA LILY

Ritu Jain, Varun M. Hiremath, Babita Singh and A.K. Tiwari

1.0 Introduction

Victoria amazonica (Poepp.) J. C. Sowerby is an aquatic, tropical bulbous flower that belongs to the family *Nymphaeaceae* under order *Nymphaeales*. It is commonly called as Giant waterlily, Amazon waterlily, Royal waterlily or Victoria lily. Victoria lily is well known for its huge circular leaves. It is a short-lived perennial with a life span of more than two years in its natural habitat. It is national flower of Guyana.

2.0 Importance and Uses

2.1.0 Aesthetic Uses

These plants are found in the aquatic section of nearly every botanic garden because of their highly valued ornamental features. The lovely circular floating leaves are the unique characteristics of Victoria lilies that could be the object of interest to many visitors in the gardens.

2.2.0 Therapeutic and other Uses

In Brazil, the riverine inhabitants of the Amazon consume its seeds and rhizomes as food (Piedade and Junk, 2000). The root, stem and seed are edible and have been used in traditional medicine (Castner *et al.*, 1998). Its seeds have been used as a food source in Central America, and once known as 'water corn' in Paraguay, because of its large seeds that were used to make flour. Native people of Brazil use seeds to make flour (Les, 2003). The flower seeds when ground, bolted, and baked form a delicious food item, considered a luxury in native country (Seaman, 1891).

3.0 Origin and History

It is an aquatic, herbaceous, short lived perennial with rhizomatous roots inhabiting still rivers in the North of South America, and East of the Andes. Victoria was documented in extant versions of the Indian legend *i.e.*, the *Tupi* legend wherein it is mentioned that the *Tupi* Indians regarded *Victoria* flowers as objects of great beauty (Prance and Arius, 1975). The Bohemian botanist and naturalist Tadeas Haenke first discovered *Victoria amazonica* in 1801 during his exploration on Bolivian flora. The plant was first sighted on the banks of river Momore, one of the great tributaries of the River Amazon. Eduard Poeppig was first to publish an account of Victoria under the name *Euryale amazonica* in 1832. Schomburgk made an initial attempt to cultivate it by transplanting from lakes and streams to Georgetown, British Guiana. Gardeners of Victoria were in huge competition to establish this plant in Europe. It is grown as an annual in Royal Botanic Garden, Kew, England under specially constructed small, square glasshouse popularly known as Waterlily house. It was designed specifically to display the giant Amazon waterlily, a natural wonder of the age (Petko and Holovko, 2020).

It was first successfully brought to bloom during November, 1849 in a specially built greenhouse at the Duke of Devonshire's estate at Chatsworth. It was later grown at Royal Botanic Gardens, Kew and from there its seeds were distributed to Europe, Asia and America. Thomas Meehan succeeded to bloom the first plant of Victoria in the United States in 1851 in the garden of Caleb Cope, Springbrook, Pennsylvania. Edward Rand sent seeds of this plant from Para, Brazil, to Edmund Sturtevant in Bordentown, New Jersey. In India, Victoria species are maintained at Acharya Jagadish Chandra Bose Indian Botanic Garden, Calcutta and Lalbagh Botanic Garden, Bengaluru. Several ornamental nurseries at Kadiyam, Rajahmundry, Andhra Pradesh have been maintaining these amazing waterlilies.

4.0 Taxonomy

Magnificent plant structure of Victoria has always attracted many gardeners and botanists to understand the curious floral biology of this plant. Most of the significant studies have been carried out in greenhouses in Europe (Prance and Arius, 1975). Prance (1974) has described the nomenclatural and taxonomic history of the genus wherein the exact botanical name for Amazonian species is *Victoria amazonica*. The most unique feature of plant is its leaves, which are borne on a long petiole (approx. 6m) radiating from the submerged rhizome. The undersides of the leaves are considered as a marvel of structural engineering, support for the highly buoyant pad surfaces. Leaves are huge, orbicular, peltate, floating, entire and full green adaxially with the edge turn upwards forming a rim of around 5–10 cm height and the basal surface is a coppery colour with a prominent network of flattened ribs radiating from the petiole and cross ribs, all bearing sharp, stout prickles (Allen, 1854). Buoyancy of the plant is achieved by its symmetrically arranged large air canals in the ribs and stems of both leaves and flowers (Seaman, 1891).

The plants are armed, everywhere except the top surface of the pads with thorns. It is possibly a defense against herbivorous fishes and manatees. Each plant has 10–15 floating leaves and the total leaf area per plant differed considerably (minimum value 12.1–17.2 m^2, maximum value 27.6–31.2 m^2). Leaves have an average life span of 35.5 days. Flower is borne solitary on an aculeate, terete pedicel arising from the rhizome, longer than leaf petiole, pear-shaped in bud. It is epigynous, fragrant, 30–50 cm across, hemicyclic, dichlamydeous, bisexual with vespertine anthesis (Rosa-Osman *et al.*, 2011). The calyx, consisting of four large and four small spiny sepals is coriaceous, fleshy and green becoming reddish purple. Petals are numerous, inserted on the throat and torus of the calyx; outer ones are completely reflexed. The petals are erect and white in colour during the first night as flower opens the next morning. During the evening of the second day, the petals will turn variegated white pink or red. Stamens are numerous, inserted with the petals on the torus in about three rows; the stamens, including the outer staminodia (infertile stamens) petaloid, middle fertile stamens and inner staminodia (syn. paracarpels), are erect and form an approx. 12 mm tunnel into the floral chamber (Seymour and Matthews, 2006). Pollen of *Victoria* is generally united into tetrads; each grain is zonasulculate (Schneider and Williamson, 1993). The filaments are subulate, erect, outer filaments petaloid, connective long appendiculate, filaments of the sterile stamens of the innermost whorl with fleshy appendages; anthers introrse. Carpels are numerous, crescent or horseshoe shaped up to the time of anthesis, and then postgenital fusion occurs forming the ovary. The ovary is globose below and concave campanulate at the top, with many locules containing numerous anatropous ovule; style is wanting, and stigmas are forming radiating lines on the top of the ovary (Schneider, 1976).

Victoria Lily

Victoria amazonica at the Pantanal Matogrossense National Park, Mato Grosso, Brazil

Waterlily house at Royal Botanic Garden, Kew, England

Victoria lily at the Sir Seewoosagur Ramgoolam Botanic Garden, Port Louis, Mauritius

Victoria amazonica Lily Pads

Victoria amazonica in the South Australian Botanic Gardens

Source: https://www.gardenclinic.com.au/how-to-grow-article/travels-with-victoria?pid=44216

Leaf and Floral Morphology of Victoria Lily

Lower Side of Pads Showing Network of Flattened Ribs

Flower Bud of Victoria Lily

Victoria amazonica Flower

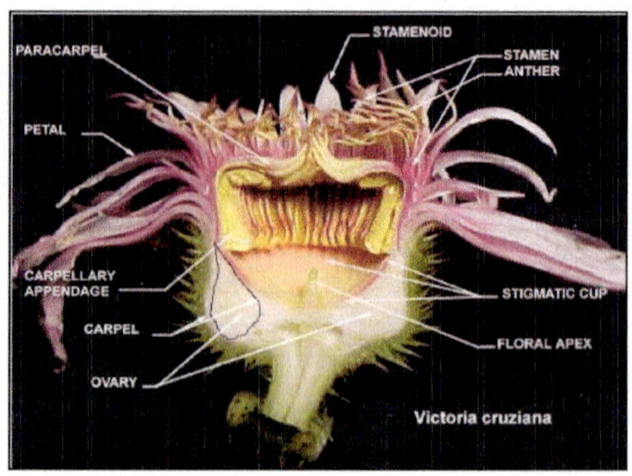

Flower Anatomy of *Victoria cruziana* (Schneider, 1976)

Victoria Longwood at 'Longwood' Gardens, Pennsylvania, USA

Victoria Seeds

The fruit is fleshy, campanulate and irregular dehiscent, pseudo-syncarpy, prickly, 11 cm across (Rosa-Osman *et al.*, 2011). Fruits rupture along the dorsal carpellary walls due to the enlargement of numerous seeds and swelling of mucilage contained in the locules. Arils surround the seeds of *Victoria* (Schneider and Williamson, 1993). Seeds are many and ellipsoid globose with horny testa and aril that acts in the water dispersion. Fruits ripen under water having several cavities containing a number of seeds of shape and size of small peas. They have a thick, black skin and are white from inside.

5.0 Species and Cultivars

The genus *Victoria* is comprised of two known species viz, *Victoria amazonica* and *Victoria cruziana*. English botanist and horticulturist John Lindley established a separate genus under the name *Victoria* during 1837 and described the species *'regia'* in honor of Queen Victoria. *Victoria amazonica* is the most accepted nomenclature for Tropical Amazonian species whereas, less tropical species discovered at Corrientes was named *Victoria cruziana* by Alcide d'Orbigny in 1840. *Victoria amazonica,* commonly called as Royal Waterlily, Amazon Waterlily which grows in quiet pools and inlets of the mighty Amazon River. *Victoria cruziana* (formerly called *Victoria trickeri*) is commonly called as Santa Cruz Waterlily. It got its name by d'Orbigny in honour of the Spanish general Santa Cruz who joined and helped him during his expeditions (Knotts, 2020). In Argentina and Paraguay, it is called 'Yrupe' that means 'water platter'. *Victoria cruziana* is reported to be more tolerant of cooler temperatures than the former species. Schneider (1976) studied flower anatomy and distinguished *Victoria amazonica, Victoria cruziana* and their hybrid *Victoria 'longwood'* with respect to flower attributes. All Victoria lilies are night blooming, gives out pineapple like fragrance in the afternoon and evening air with anticipation of the first night flower, huge and white. Based on climate, leaves and flowers both species are differentiated as:

Victoria amazonica	*Victoria cruziana*
It is a tropical species.	It grows in cooler and more southern areas of South America.
Leaves are larger with low rims and redder pads.	Leaves are smaller, greener, has higher upright rims and purple pads.
Plant is highly spiny and leaves are larger.	This plant is not as spiny as *Victoria amazonica*, the leaves remain smaller.
Flowers are white in colour on first night which becomes cotton candy pink to ruby red on the second night.	Flowers are delicately flushed pink.

6.0 Genetics and Breeding

6.1.0 Genetics

Diploid chromosome number of *Victoria amazonica* is 2n = 20-24 (Langlet and Soderberg, 1929; Okada and Tamura, 1981). Victoria has the largest chromosomes

Different Species of Victoria

V. cruziana
Bud

V. cruziana
First Day Flower

V. cruziana
Second Day Flower

V. amazonica
Bud

V. amazonica
First Day Flower

V. amazonica
Second Day Flower

Victoria amazonica and *Victoria cruziana* Flowers

Victoria 'Longwood' Flower

Source: Schneider, 1976.

Possibility of Hybridization in Two Species of Victoria

Source: Keith Hartley and Jeremie Fant of Chicago Botanic Garden.

(2-4 pm in length) among different genera under *Nymphaceae*. Gupta (1977) had also established that it is characterized by considerably long chromosomes varied between 3.08μ-5.39μ.

6.2.0 Breeding

Flowers are typically entomophilous and protogynous ensuring cross pollination. Morphology and behaviour of flowers favour geitonogamy and xenogamy. Individual flowers bloom for 2-3 days. On the first day flowers are functionally carpellate, on the second and subsequent days flowers are functionally staminate. Stigma is receptive on first day evening of flower opening ensuring the pollination on first day itself. Flower is pollinated by scarab beetles (Prance and Arius 1975; Prance 1980). Large, nocturnal flowers of *Victoria* are adapted for pollination by *Cyclocephala* beetles, as these are often attracted by the white colour and sweet pineapple-like fragrance of first-night flowers.

Manual pollination is also practiced in Northern countries, generally during June (Clifford, 2005). It is practiced during August to September in Longwood Gardens, Pennsylvania, United States. To prevent any chance of cross pollination, the flowers are covered with cheese cloth on first night. The pollen laden anthers of a second night flower are gently snipped off and laid in a tray. They are taken over to the first night flower on another plant. The petals are carefully opened and pollens are deposited. A soft brush is used to spread the pollen around the stigmatic bowl while pollinating the flower. After pollination the seed pod begins to swell and within about six weeks, it ruptures. To ensure that all the seeds are harvested, a muslin bag is tied over the pod once it looks ripe and up to 500 seeds can be produced per pod. These are covered in a glutinous white aril which needs to be washed off before the seeds are stored.

6.2.1 Hybridization

Patrick Nutt, at Longwood Gardens attempted the first successful hybridization between *Victoria amazonica* and *Victoria cruziana*, resulting in 'Longwood Hybrid' that first flowered during 1961. This was the only example of successful crossing among these two species with *V. cruziana* as pod parent and *V. amazonica* as pollen parent. The resulting progeny Victoria 'Longwood Hybrid' possesses the desirable characteristics of both parents viz, the extraordinary size and deep red colour on the undersides of the leaf of *Victoria amazonica*, and the large lip and relative cold hardiness of *Victoria cruziana* (Clifford, 2005). The giant waterlily, Victoria 'Longwood' is seasonally on display in the Princess of Wales Conservatory at Royal Botanic Garden England, Kew. In 1995, its leaves reached record-breaking dimensions when they grew to over 2.5 m in diameter and were registered in the Guinness Book of Records. During 1999, several other breeders independently attempted the reciprocal cross among the two species and named them as Victoria 'Adventure'. In general, the pads of both primary hybrids are larger than either parent and plants tend to bloom earlier, making them ideal display plants for botanic gardens around the world. Other Victoria hybrids, such as V. × 'Adventure', × 'Columbia', or × 'Atlantis', were also produced by Knotts (2004).

7.0 Propagation

7.1.0 Seed Propagation

Victoria is propagated mainly from pea-sized seeds. Since, seed viability is low, therefore, it is suggested to use fresh seeds for planting. Seeds must be rejuvenated every year and the plants must be regrown from those seeds. Victoria seeds are generally sown in February-March in Longwood Garden. Seeds of *Victoria amazonica* and Victoria 'Longwood hybrid' germinate best at water temperature of 26.5 to 27.5°C whereas, *Victoria cruziana* germinates slowly at lower temperature of 21 to 22°C. Victoria seeds must always be stored in distilled water or wet sand at 14°-15°C until their sowing in the following year. Drying out even for few minutes can kill the embryos. It has been reported that storing the varieties namely *Amazonica*, 'Adventure', 'Challenger' and 'Columbia', Fresh 'Longwood Hybrid', 'Atlantis', 'Discovery' and *cruziana* at cooler temperatures can greatly reduce viability and often require storage at 16°C to prevent unnecessary sprouting. Storage of seeds at temperature colder than 13°C can kill Victoria seeds. Due to impermeable hard seed coat, seeds are reported to be dormant. Nicking (careful removal of the operculum with a scalpel) is practiced to enhance seed germination rate usually in 3 – 12 days.

7.2.0 Micropropagation

Victoria Lily has high ornamental value, but is very difficult to initiate and multiply under *in-vitro* conditions (Bodhipadma *et al.*, 2011). Similar to other aquatic plants, the major challenge to the *in-vitro* propagation of Victoria lilies is high contamination rates. Experiments on *in vitro* establishment using seeds and ovule explants were conducted by Hoang (2016) to overcome these problems. Ovule and seed explants were successfully established under *in vitro* conditions, but their regeneration was poor. It was revealed that chemical inhibitors in the seed coats may inhibit hardy waterlily embryo germination in liquid medium. Victoria seedlings were successfully established by culturing isolated embryos. Double sterilization of seeds in a 3.3 per cent sodium hypochlorite solution for 30 minutes followed by embryo isolation and re-sterilizing in 0.6 per cent sodium hypochlorite solution for one minute were found effective in inducing higher germination rate (78 per cent) and lower contamination rate. The excised embryos were inoculated in half-strength MS medium solidified with 6 g/l agar and then ten ml liquid medium was dispensed into each tube after inoculation. This double-phase media allowed Victoria seedlings to be anchored inside culture vessels. Victoria seedlings were successfully established *in-vitro* by culturing of isolated embryos.

8.0 Soil and Climate

8.1.0 Soil

Loamy soil acts as best substrate as it is rich in nutrients. The roots are best anchored in organically rich loams at the bottom of the tank.

8.2.0 Climate

It is a tropical aquatic plant, requires warm water temperature and established ponds to grow. *Victoria* water lily is a perennial in tropics where the plant can survive for many years due to availability of uniform sunshine for more than 12 hours. Plants remain perennial as long as water temperatures remain at least 24 °C and thrive in calm shallow waters (<2 m), full sun, slightly alkaline to neutral pH (Lim, 2016). However, in temperate climate, plants can be maintained for many years with the aid of supplementary lighting. In higher latitudes, plants may be treated as annual due to lesser day length during prolonged winter.

9.0 Cultivation

Victoria cruziana performs better under greenhouse conditions compared to tropical Amazonian species. Seeds are sown at a depth of 2 to 3cm in soil mixture containing top soil and sand in 5:1 ratio. Cover the seeds with a layer of sand after sowing which keeps the soil in pots once the plant is submerged and also keeps water clean. Repotting is done after two true leaves appear. However, repotting may be continued until frost free weather to plant outside ponds. Potted Victoria lilies are planted in permanent concrete pots (45 cm deep and 1 m across), submerged in specially constructed large water pools or ponds.

In Royal Botanical Garden, Kew, England (RBGE), seeds are sown during late January, in 5cm clay pots filled with loamy soil and covered with crushed stone powder called chick grit. These pots are then placed to a depth of 100mm in water heated to 32°C. Germination usually occurs within 2-4 weeks and seedlings are potted into 10cm pots when the second set of true leaves has developed. Lily ponds constructed at RBGE, hold 82,000 litre of water and have surface area of 80 meters which enabled three plants to grow to full size. The water is kept at a constant 29°C and the air temperature in the glasshouse has a minimum day temperature of 24°C and night temperature of 19°C. However, in summer temperatures up to 30°C during the day is common. During March, young plants with true leaves measuring 30 cm diameter are transferred to water tubs. The tubs in which mature plants are grown are 60 cm deep and 900 mm in diameter and, when in place, the top of the tubs sit 400 mm below the surface of the water. Once the tubs are filled with loam, a 30mm deep layer of coarse grit or gravel is placed on top to ensure that the loam does not float out of the tub when the pond is refilled. One must be careful while placing plants into the tubs. Plants are transferred without any damage to fleshy roots and place plants firmly into the loam. Plants are fully grown with largest leaves of 2 m diameter, approximately by the end of June or early summer. The buds produced in the first month are cut off to encourage the plant to put all its energy into leaf production.

9.1.0 Manuring and Fertilization

Victoria lily is a heavy feeder. Fertilizer balls, approximately the size of an apple made of inorganic fertilizer containing NPK (2:1:1 ratio) and soil in 1:3 ratio is used to fertilize lilies. Weekly application of three balls to a depth of 30 cm near the end of rootzone of lilies is recommended by the Royal Botanical Garden, Kew, England. Later in the season, when roots have filled the tubs, the balls are pushed in wherever possible. At this depth the ball will break down effectively and the feed will percolate through the root system.

10.0 Diseases, Pests and Disorders

Victoria water lilies are not susceptible to many pests or diseases. However, in cooler climate, there is a problem of aphids, early in the growing season. The predatory midge *Aphidoletes aphidomyza* is very effective in controlling the aphid. Larvae of China Mark moth can also attack young pads but are rarely much of a problem with adults. Non-insect pest are turtles, ducks and crawfish can also damage emerging pads.

Excessive heat and/or humidity in greenhouse can damage the leaves creating light-colored patches in the case of high heat, dark patches if the leaves can't dry from humidity, both types of damage occur with high heat and humidity. Improving air circulation inside the greenhouse helps to alleviate these problems. Victoria lily can be reluctant to bloom, if its pads are damaged or removed prematurely, overlapping of pads results in premature death of plant.

11.0 Biotechnology

11.1.0 Molecular Markers

Molecular markers have become standard practice in the breeding process and for accurate variety identification. The first reliable and efficient method for isolating *Victoria amazonica* genomic DNA free from polysaccharides and polyphenols was developed by Puchooa and Khoyratty (2004). This protocol uses 1.5 M NaCl, 2 per cent polyvinylpyrrolidone (PVP), 5 per cent mercaptoethanol, 0.12 per cent sodium sulfite, and an incubation at 65°C for 4 h. The purity of isolated genomic DNA was confirmed by means of high-performance liquid chromatography (HPLC) profile and spectrophotometric analysis. With increasing knowledge about transposable elements in eukaryotic genomes, retrotransposon-based markers have been developed for various plant species and were sometimes found to be more informative than conventional marker methods. Due to the lack of information about retrotransposons in waterlilies, and to confirm the presence and analyze the diversity of the Ty1-copia retrotransposons 133 sequences of a conserved domain in the reverse transcriptase gene of Ty1-copia elements were isolated from 13 varieties of waterlilies and found to be homologous to either Ale, Angela, or TAR lineages. Sequences, homologous to the Ale lineage were found to be abundant and diverse. Those homologous to the Angela lineage were found to be very conserved, but scarce. Those homologous to the TAR lineage were relatively conserved and were found in ample amounts. Those homologous to the Angela or TAR lineages are more conserved to their counterpart in other plants than those homologous to the Ale lineage. From this study, 3 groups of Ty1-copia elements were found to be suitable candidates for development of retrotransposon-based markers for waterlilies in the future (Runch *et al.* 2018).

12.0 References

Allen, J. F. (1854) http://www.victoria-adventure.org/victoria_images/allen_sharp/victoria_regia 1.html.

Bodhipadma, K., Noichinda, S., Wachirabongkoth. P. *et al.* (2011) *J. App. Sci.*, **10**: 7-11.

Castner, J. L., Timme, S. L. and Duke, J. A. (1998) *A field guide to medicinal and useful plants of the upper Amazon*. Feline Press.

Clifford, P. (2005) *Sibbaldia: J Botanic Garden Horticulture*, **3**: 59-65.

Gupta, P. P. (1977) *Curr. Sci.*, **46**(16): 574-575.

Hoang, N.(2016) *PhD thesis*, University of Florida, pp. 233.

Knotts, K. (2020) www.victoria adventure.org/victoria/victoria_history.html

Langlet, O. and Soderberg, E. (1929) *Acta Horti. Bergiani*, **9**: 85-104.

Les, D. H. (2003) *Nymphaeales*. John Wiley and Sons Ltd, Chichester

Lim, T. K. (2016) *Edible Medicinal and Non-Medicinal Plants. Springer*, Cham. *pp*. 123-27.

Okada, H. and Tamura, M. (1981) *Journal of Japanese Botany*, **56**: 367-375.

Petko, L. and Holovko, T. (2020) *Berdyansk State Pedagogical University, pp*. 6–12.

Piedade, M. T. F. and Junk, W. J. (2000) *Concepts and Paradigms for Management of Ecosystem Resources, pp*. 721-729.

Prance, G. T. (1974) *Acta Amazonica*, **4**(3): 5-8.

Prance, G. T. (1980) *Brittonia*, **32**(4): 505–507.

Prance, G. T. and Arius, J. R. (1975) *Acta Amazonica*, **5**(2): 109–139.

Puchooa, D. and Khoyratty, S.U.S.S.(2004) *Plant Mol. Biol. Rep.*, **22**: 195–196.

Rosa-Osman S.M.D., Rodrigues, R., Mendonça, M.S.D. *et al.* (2011) *Acta Amazonica*, **41**(1): 21–28.

Runch T, Chatuporn K and Vipa H. (2018) *J. Hortic.*, **87** (4): 524–531.

Schneider, E. L. (1976) *J. Proc. Linn. Soc.*, **72**: 115–148.

Schneider, E.L. and Williamson, P.S. (1993) *Flowering Plants- Dicotyledons*. Springer, Berlin, Heidelberg, *pp*. 486-493.

Seaman, W. H. (1891) *In*: *Proc. Am. Soc. Microscopists, American Microscopical Society*, Wiley. **13**: 163-170.

Seymour, R.S. and Matthews, P. G. D. (2006) *Ann. Bot.*, **98**: 1129–1135.

Stephenson, M. (1885) Wisconsin State Horticultural Society, Oregon, USA, **XV**: 143-146.

17

KNIPHOFIA

V. Bhargav, B. Parvathi, T. Yatung, Kalkame Ch. Momin and Sunil Kumar

1.0 Introduction

Kniphofia is popularly known as 'Red hot poker'. The plant is easy to grow and deliver colorful, spiky, wand-like flowers. The genus *Kniphofia* was named in honor of Johannes Hieronymus Kniphof (1704-1763) who was a professor of medicine at Erfurt University in Germany. It belongs to family *Asphodelaceae*. Previously, the family Asphodelaceae was placed under the order Aspargales and considered to be a family of the lower asparagoids, which are characterized by simultaneous microsporogenesis, later *Kniphofia* was brought under the family *Asphodelaceae* (Chase *et al.*, 2000).

2.0 Importance and Uses

2.1.0 Aesthetic Uses

The genus *Kniphofia* is suitable for growing in home and botanical gardens. It is cultivated for indoor and outdoor gardening and for cut flower purpose. The plants have good ornamental characteristics (colour and shape), with prolonged flowering period (Li *et al.*, 2000). The flowers of few *Kniphofia* species are used as a minor food and apparently taste like honey. Naturally occurring species of *Kniphofia* are important source of pollen and nectar (Fichtl and Admassu Adi, 1994).

Traditionally, *K. parviflora* is used as a snake repellent; *K. rooperii* and *K. laxiflora* as a general medicine (Elovson, 1984). An infusion of roots is used to relieve or treat the symptoms of chest pain. The roots of *K. foliosa* are used in traditional Ethiopian medicine for the treatment of abdominal cramps (Abebe *et al.*, 2005). In addition, knipholone and related natural phenyl anthraquinones extracted from the roots of *K. foliosa* are among a new group of compound having anti-malarial property and anthraquinone aloe-emodin is known to exhibit anti-leukemic properties (Dagne and Steglich, 1984; Berhanu *et al.*, 1986; Bringmann *et al.*, 1999; Habtemariam, 2010). The roots of *K. foliosa* contains 10-hydroxy-chrysopanol-9-anthrone and chryslandicin (anthraquinone–anthrone dimmers) which show a high rate of inhibition on the growth of malaria parasite, *Plasmodium falciparum* with a very little host cell toxicity (Abebe *et al.*, 2005). Knipholone anthrone, a chemical present in kniphofia exhibits strong anti-HIV activity (Feilcke *et al.*, 2019).

3.0 Origin and Distribution

Kniphofia is native to Cape Province of South Africa and Madagascar. It was introduced to England during 1707 and was kept as greenhouse treasures until 1848. Later, outdoor cultivation of the plant was practiced. Majority of the species are found in South Africa, which is considered as the largest centre of endemism. It grows from costal area to the high altitude mountains of South Africa with majority of the species distributed in Afromontane grasslands (Ramdhani, 2008).

Uses of *Kniphofia*

As Specimen in Garden

In Pot

In Garden Bed

Kniphofia in Mixed Planting

Along Side of Waterbody

In Flower Vase

In Flower Arrangement

In Garden Border

4.0 Morphology, Species and Cultivars

4.1.0 Morphology

Kniphofia is an herbaceous perennial, growing up to 2 m in height. Most species are evergreen while few are deciduous and sprout in early summer. It forms large clumps of long arching leaves of about 90 cm length and 2.5 cm width having narrow and tapering ends. The leaves are non-succulent, having smooth to minutely serrulate leaf margin which distinguishes them from plant such as *Aloe cooperi*. Flowering stem emerges from the leaf axil, measures up to 90-120 cm height. Inflorescence is a raceme; 15 – 20 cm in length having pendulous tubular flowers. Buds and emerging flowers are red but mature to yellow, giving each spike a two-toned appearance. The small tubular flowers are produced in shades of red, orange, yellow and cream. The perianth is tubular, campanulate to cylindrical or slightly funnel-shaped and short, sub-equally lobed. The stamens are of same length or longer than the perianth and the ovary is sessile, ovoid, 3-locular consisting of many ovules in each locule. The fruits are globose to ovoid, often 3-angled with loculicidal dehiscence and seeds are somewhat flattened, acutely 3-angled or winged (Demissew and Nordal, 1997).

The underground part of the plant consists of a thick rhizome and fibrous, fleshy roots. In few species the rhizome divides forming group of stems, while in others the stems are more or less solitary (Elovson, 1984). The vast majority of *Kniphofia* species do not produce an aerial stem, but exceptions do occur in case of older plants of *K. caulescens* and *K. northiae* which can reach a height of 30 cm.

4.2.0 Species

The genus *Kniphofia* consists of 67 species with 13 varieties (Smith and van Wyk, 1998). Janaki Ammal (1950) confirmed 93 species, majority of them are distributed in Africa, with few exceptional species in Yemen. Codd (1968) indicated 70 species, of which 45 species occur in South Africa, 1 species in the Arab Republic of Yemen, 2 in Malagasy Republic and 23 in Tropical Africa, including seven species in Ethiopia. The seven species of the genus *Kniphofia* in Ethiopia were identified as *K. foliosa*, *K. hildebrandtii*, *K. insigns*, *K. isoetifolia*, *K. pumila*, *K. schimperi* and *K. thomsonii*. Of these, *K. foliosa*, *K. hildebrandtii*, *K. insignis*, *K. isoetifolia* and *K. schimperi* are endemic to Ethiopia whereas, *K. pumila* and *K. thomsoni* are widely distributed from West Africa to eastern and central Africa (Demeissew and Nordal, 1997). According to Marias (1973), *Kniphofia thomsoni* is common in Kenya, Uganda and Tanzania, in particular on Mount Kilimanjaro. Ramadhani (2008) identified 71 species, which are of African-Malagasy origin including *K. monticola* from Africa and one species from Yemen.

Kniphofia uvaria

It is popularly known as Red Hot Poker, Poker Plant, Common Torch Lily or Torch Lily. *Kniphofia uvaria* originated in Cape Province of South Africa and has been introduced into several parts of the world, such as North America, Australia, New Zealand, Patagonia and Europe as a garden plant, noted for attracting fauna.

K. uvaria is the progenitor of most of the garden forms. Plant is an evergreen perennial, growing up to 2 m height, a hardy plant and does not need much care. Flowering commences in the month of August and lasts up to September. The species bears hermaphrodite flowers has both male and female organs. It grows well in sunny locations, having sandy loam, moist soils with a pH ranging from acidic to alkaline nature (Klingaman, 2008).

Kniphofia foliosa

The species is distributed in the mountainous region of central and northern Ethiopia which produces flower from June to September or from December to January depending upon the availability of water (Maris, 1973). It is a robust, perennial plant forming dense clumps, growing up to a height of 175 cm, having thick and erect rhizome. The plant is usually stemless, but stems of 40 cm length are occasionally formed. It produces a basal rosette of sword-shaped leaves up of 100 cm length and 4 – 7 cm width. Flowers are yellow, orange or red (Marias, 1973; Demissew and Nordal, 1997). The plant is harvested from the wild by the locals for use as a medicine; rhizomes are edible and grown as an ornamental in Europe. The species grows best in well drained soil along the road sides, pasture land with scattered trees, hills and mountains ranging from 2500-4000 m above mean sea level (Demissew and Nordal, 1997).

Kniphofia hildebrandtii

This species is distributed in relatively smaller areas of Asgori, Ginch and Gedo, growing in wet grassland at an altitude of 2000-2450 m. Flowering occurs from June to August (Demissew and Nordal, 1997; Marias, 1973). It is a slender plant having only few shoots, with fibrous remains of leaves at the base. Leaves are broad at the base, long-linear, dark, slightly greyish green, keeled with smooth margins. Flowers are pale green or slightly greenish-white or yellowish, pendulous and usually found at one side. Bracts are ovate, acuminate or acute, slightly toothed or serrulate (Demissew and Nordal, 1997).

Kniphofia insignis

This species is distributed in Northern Shewa Zone of Oromiya Region at an altitude of 2550-3134 m (Teklehymanot, 2001). It is apparently solitary, slender, without fibrous remains of leaves at the base having fusiform roots. Leaves are wide at the base, linear, long and tapering with blunt tip, having minutely keeled margin. Peduncle 20-65 cm long, sometimes reaching up to 100 cm. Inflorescence is a raceme having fairly dense flowers. Flowers are white or largely pale pink in colour. The species grows in plain areas or water-logged conditions during the rainy season (Demissew and Nordal, 1997).

Kniphofia isoetifolia

This species is distributed in East Shewa Zone of Oromiya Region, North Shewa Zone of Amhara Region and West Arsi Zone of Oromiya Region at an altitude of 2050–3700 m above msl. This species thrives well under water logged meadows in the montane highlands to sloppy grounds (Teklehymanot, 2001). In

Species of *Kniphofia*

Kniphofia foliosa

Kniphofia uvaria

Kniphofia isoetifolia

Kniphofia thomsonii

Kniphofia rooperi

Kniphofia galpinii

Kniphofia reflexa

Kniphofia northiae

general, it is apparently solitary, slender, sometimes having five or six stems with little fine fibrous remains of leaves at the base. Roots are fusiform and leaves are wide, linear, blue–green, long and tapering with blunt tip which is soft and flaccid. Inflorescence has dense or sub-dense flowers, capitates, globose, short or medium length and are arranged centrifugally. Flowers are creamy, white, yellow, pale salmon, orange or red (Demissew and Nordal, 1997). The plants produces flowers from May to December, the flower opens from top to bottom in basipetal manner while in the rest of the *Kniphofia* species the flower opens from bottom to top, acropetally (Marias, 1973).

Kniphofia schimperi

The species is distributed in the highlands of Wollo Province at an attitude ranging from 2200-3500 m above msl (Teklehaymanot, 2001). The plant grows well on steep grassy or stony slopes, in rocky outcrops, and stony banks near streams (Demissew and Nordal, 1997). The plants are slender, with fibrous remains of leaves at the base, probably forming small clumps with a compact, slightly branched base. Leaves are broad, linear, dark-green, long-acuminate, soft and reflexed with almost smooth margins. Length of peduncle ranges from 43-130 cm. Flowers are yellow, orange or red with lanceolate, acuminate, scarious, white coloured bracts (Demissew and Nordal, 1997) and the flowering period is from June to January (Marias, 1973).

4.3.0 Cultivars

Red-hot pokers are grown in temperate gardens around the world, ranging in colours from red, orange through yellow to lime green and cream. Numerous cultivars and hybrids have been developed from species originated in South Africa.

Few important cultivars with their characteristics are summarized below.

Echo Mango: Inflorescence is apricot in colour. It blooms early with the first flower appearing at the end of April to the first part of May (Saul, 2011).

Echo Duo: It is an early flowering variety and flowers profusely throughout the summer. The variety has characteristic bicolour inflorescence with flowers having orange-red during emergence and later turns to creamy white colour at maturity (Saul, 2014).

Echo Rojo: The inflorescences are orange in colour initially and turn red as they mature. It is an early blooming variety with the first bloom appearing at the end of April to the first part of May and continues throughout the summer. It is a lodging resistant semi-evergreen plant with strong upright stem reaching up to 90 – 95 cm (Saul, 2012).

Creamsicle: It is having narrow and grassy leaves, compact plant habit with multiple crowns. The variety is characterized by very long flower stalks, blooms recurrently with creamy red orange flowers (Egger, 2012).

Papaya Popsicle and **Mango Popsicle**: These are vigorous, very early, recurrent blooming varieties having compact growth habit with multiple crowns, narrow, grassy leaves and red-orange flowers.

Cultivars of Kniphofia

Echo Mango

Echo Duo

Papaya Popsicle

Orange Vanilla Popsicle

Redhot Popsicle

Pineapple Popsicle

Pineapple Popsicle

Ember Glow

Fire Glow

Cultivars of Kniphofia

Poco Yellow

First Surprise

Flashpoint

Jackpot

Hot and Cold

Solar Flare

Lucky Lemons

Joker's Wild

High Roller

Orange Vanilla Popsicle: It is same as papaya popsicle with coral red to cream-coloured flowers.

Redhot Popsicle: It produces numerous spikes of coral red flowers during the first year. Flowering commences from June and continues till September.

Pineapple Popsicle: It is very early, recurrent bloomer which flowers from May to September.

Ember Glow and **Fire Glow**: These varieties grow vigorously and flower profusely from August to September, having compact growth habit with multiple crowns, narrow, grassy leaves, bicolor orange and yellow flowers.

Poco Orange: The variety is a recurrent bloomer, having very long flowering period from July to October.

Banana Popsicle: The plant is compact and vigorous with multiple crowns having narrow and grassy leaves. It profusely flowers from July to October producing numerous spikes of golden yellow flowers in the first year.

Poco Yellow: The plant forms multiple crowns very early and each crown produces one or two inflorescences in the first season, large inflorescences of yellow flowers are borne on short stems, having long flowering period of repeated blooming nature, from July to October.

Poco Sunset: The plant produces multiple crowns rapidly and each crown produces inflorescence with red orange and coral flowers on short stem. The variety produces flower from July and continues till October (Egger, 2017).

First Surprise: It is characterized by its relatively compact plant habit with upright and moderately strong flowering scapes. It is an early variety with free flowering habit; orange coloured flowers are arranged on racemes, suitable as a pot plant (Van Delft, 2005).

KNI609 and **KNI610:** These are characterized by upright plant habit having moderately vigorous growth, freely flowering; repeat flowering over a long period of time and produces pale yellow colour flowers with orange colour apices (Luo, 2014).

Flashpoint: It is a winter hardy and heat tolerant variety, having densely-growing habit. The variety is suitable for landscaping as a specimen, en masse, or as a cut flower.

Jackpot: It is a densely growing, winter-hardy, heat tolerant variety producing numerous scapes of vivid reddish-orange flowers on tall scapes, beginning from early summer and continuing till early October. This cultivar is suitable in landscaping as a specimen, mass planting or as cut flower.

Hot and Cold: It is a densely growing, winter-hardy, heat tolerant variety producing numerous scapes of vivid reddish buds and contrasting open flowers of pale greenish-yellow on tall scapes. Flowering starts from late spring and continues for six weeks. The cultivar is useful for landscaping, as a specimen, mass planting, or as cut flower (Hansen, 2019).

Other cultivars include 'Backdraft', 'Solar Flare', 'Lucky Lemons', 'Rockets red Glare', 'Poker Face', 'Gold Rush', 'Joker's Wild', 'Orange Blaze', 'High Roller' and 'Lady Love' (Hansen, 2020).

5.0 Genetics and Breeding

5.1.0 Genetics

The basic (haploid) chromosome number is six (6) and chromosome pairing and separation at meiosis is perfectly regular for most species. However, in few cases, pollen mother cells containing 12 or 24 chromosomes are found, which suggests that certain species of *Kniphofia* are tetraploid (Moffett, 1930). Webber (1932) examined the karyology of *K. aloides* (*i.e. K. uvaria*) in detail and found six pairs of chromosomes (2n= 12). 17 species of *Kniphofia* are diploid having somatic chromosome number 2n=12 but all the species showed little variation in chromosome morphology (De Wet, 1960). Nayak and Sen (1992) examined the karyology of *K. nelsonii* (*i.e. K. triangularis*) and *K. uvaria* clones and reported that most of the cells examined were diploid (2n = 12), while few were tetraploid. The species that are endemic to Ethiopia have diploid chromosome number 2n= 12 with a karyotypic formula of 1m + 3sm + 2st (Teklehaymanot, 2001). According to Teklehaymanot (2001), the evolutionary differentiation of the *Kniphofia* species in Ethiopia had occurred without involving either change in ploidy levels or karyotype differentiation. All the species of Ethiopia share a common ancestor, but show variation in floral and inflorescence character due to rapid evolution (Teklehaymanot *et al.*, 2004).

Morphological species of *Kniphofia* exhibited substantial leaf anatomical variation which is not influenced by environmental factors. However, it depends on genetic composition of the plant, since; hybridization plays a major role for genetic variation. The cpDNA and nuclear phylogenies for *Kniphofia* do not provide complete resolution and are not robust. Despite these limitations, a phylogenetic hypothesis for Kniphofia using molecular plastid and nuclear DNA sequence data were obtained (Ramdhani, 2006). In *Kniphofia,* knowledge of geographic patterns is essential in understanding its evolution and to interpret the DNA sequence data results. Despite horticultural popularity of *Kniphofia,* no general account of its biogeography is available. Geographic patterns combined with molecular data can provide important insights into evolutionary processes (Olsen and Schaal, 2000).

5.2.0 Breeding

A few species have been bred for commercial cultivation around the world, especially in the United Kingdom, Holland, the USA and Australia. Species and varieties of kniphofia were recently introduced in China, for use as indoor plant, landscape plant, and cut flowers. The plants have good ornamental characteristics (colour and shape), prolong flowering period, easily propagated and widely adaptable (Li *et al.*, 2000). The nectar-feeding birds are the main pollinators of *Kniphofia caulescens* (Brown *et al.*, 2009). It is apparent that many plant species, across several genera in southern Africa are adapted to pollination by short billed nectar feeding birds (Johnson *et al.*, 2006; Symes *et al.*, 2008). *Kniphofia leucocephala* is an extremely endangered species and exists only in Langepan, KwaZulu-Natal,

South Africa. Tissue culture is, therefore, a viable option to produce pollen under controlled conditions in order to increase the genetic diversity of the species, which can be used in breeding programme of this endangered species (McCartan and Van Staden, 2003).

6.0 Propagation

6.1.0 Seed Propagation

Kniphofia can be grown through direct sowing in the garden during summer, (also planted as a potted plant). It can be sown in late winter in glasshouse and can also be grown in cold frame followed by transplanting in the following spring. The seeds are sown thinly and evenly during last week of February-March on raised beds/protrays. The soil should be moist and soil temperature maintained at 21-24°C. Seedlings emerge in 20-30 days. After emergence, the seedlings should be exposed to bright sunlight during the day time. Seedlings can also be supplemented with artificial light, using fluorescent bulbs till the seedlings reach 8-10 cm high by turning on the lights for 16 hours per day and off for 8 hours at night. Water soluble fertilizer (19:19:19/20:20:20) is applied when the seedlings are 3-4 weeks old. The seedlings are ready for transplanting when they attain 2 – 3 pair of true leaves stage. Hardening of seedlings is done prior to transplanting. In areas experiencing low temperatures, seedlings are covered with polytunnels during night hours to prevent the plants from frost injury. Hardening process toughens the seedlings and protects from transplanting shock.

6.2.0 Vegetative Propagation

As flowering from seedling plants take long time, *Kniphofia* can be propagated by division of rhizomes. Division is the quickest method of propagation for the production of early flower. Large clumps are lifted and divided, which later can be replanted at new site. Irrigate the plants immediately after planting. The plants will take a year to settle down after being divided. In areas with cold winters, divide the plants in spring and protect the plant during winters, by covering the soil with mulch. In warmer regions, plants are divided after flowering.

6.3.0 Micropropagation

Large number of plantlets can be produced through micropropagation and also help in conservation of the species. Rooting and shooting in micropropagation is highly influenced by the concentration of benzyladenine (BA) in the culture media. The optimum concentration of BA for shooting is 2 mg/l (8.9 μM). After each cycle of 4 weeks interval, approximately 200 shoots can be obtained from a single shoot. The shoots were successfully rooted and acclimatized (McCartan and Van Staden, 2003).

7.0 Soil and Climate

7.1.0 Soil

Well-drained sandy loam soils with a neutral soil pH range between 6.6 – 7.0 located in an open sunny position or partial shade is suitable. Water logging leads

Propagation of *Kniphofia*

Seeds of *Kniphofia*

Propagation through Division

Rhizomes of *Kniphofia*

Micropropagation of *Kniphofia*

Cultivation of *Kniphofia*

Cultivation in Open

Full Bloom Stage

to rotting of rhizomes. In areas with clay soil, garden beds are amended with sand and organic matter to improve the drainage or may be planted on raised beds to avoid rhizome rot. It can also tolerate drought.

7.2.0 Climate

The plants grow best in full sun and most species tolerate frost but the winter-flowering species should be protected from scorching sunshine. In few summer-flowering species, leaves become dry and undergo dormancy during winters and emerge again in early summer.

8.0 Cultivation

8.1.0 Planting

The best time of planting in the garden is spring. The new growth commences every year from the fleshy rhizomes in the spring season. In mild Mediterranean climate, planting is done in late winter to take the advantage of rainy season. In colder areas, planting is done in the spring after ground thaws. The rhizomes are placed horizontally, at a depth of 10 cm and covered with the soil. The rhizomes are planted at a spacing of 45 x 45 cm or 60 x 60 cm. New growth emerges from the growing point called eyes along with the roots. Water the area immediately after planting to settle the soil. Unlike other bulbous crops, which needs thinning and division every 3-4 years, *kniphofia* should be left undisturbed once the roots are established for few years to get better flowering.

8.2.0 Irrigation

Torch lily requires plenty of water during growing season to thrive and flower well. In summer, irrigation should be given at weekly interval. Flower yield reduces under drought condition. Over irrigation or water logging should be avoided to protect from root rot. Once the plants are established, it can withstand drought, and can flower profusely with irrigation.

8.3.0 Manuring and Fertilization

Fertilizers should be applied at the time of emergence of new growth during spring season. To avoid foliar burn, granular fertilizers should be placed away from the plant crown and foliage. Use of slow releasing fertilizers at lower rates is more beneficial to the plants.

8.4.0 Mulching

Mulching helps to retain soil moisture, maintain soil temperature and control weeds. Mulching may be done using shredded bark or dried leaves. While mulching, care should be taken to place the mulches away from the plant to prevent rotting.

8.5.0 Weed Control

Weeds compete with plants for water, space and nutrients during the growing season. In the initial stage of plant growth, field should be kept weed free for better establishment of the plants. Therefore, weeds can be controlled by either cultivating the soil often or by using mulch. The frequency of weeding depends upon the season and weed infestation.

8.6.0 Other Cultural Practices

After the plants bloom, cut spent flower stalks to the base and the foliage back by half for a good appearance.

9.0 Diseases and Pests

9.1.0 Diseases

9.1.1 Alternaria Leaf Spot (*Alternaria* sp.)

Small, round reddish brown spots with white to gray center are observed on upper surface of the leaves and along the midrib. The lesions may encircle the stem and cause wilting. The disease incidence is severe under warm, wet or humid weather. To control the disease, wetting of foliage should be avoided. Remove infected plant parts and facilitate proper air circulation to avoid disease spread. Spraying of copper based fungicides Mancozeb or Dithane M-45 can effectively control the disease.

9.1.2 Botrytis (*Botrytis cinerea*)

The fungus causes grayish moldy growth on flowers, leaves, stems and buds. Infection is severe under cool and wet weather. As a control measure, remove affected plant parts, avoid watering at night and wetting of plants while watering. Spraying Benzamidazol @ 0.1 per cent can effectively control the disease.

9.1.3 Damping Off (*Rhizoctonia solani, Fusarium* spp.)

This is one of the most common problems in the nursery when the plants are raised from the seeds. The seedlings show sudden wilting. The disease is severe when the moisture content in the soil is high along with high soil and air temperature ($> 20^\circ$ C) or when the soil contains high amount of nitrogenous fertilizer. To avoid the infection, over irrigation and over application of nitrogenous fertilizers should be avoided, and maintain proper spacing for good air circulation.

9.1.4 Root Rot (*Sclerotinia sclerotiarum; Pythium* spp.)

Soil borne pathogen infects the roots of the plants. Symptoms include browning and rotting of the root system, resulting in wilting of the plants. In later stages, complete withering and drying off of the plants take place. To prevent the plants from rotting, proper drainage should be provided. Severely, infested plants should be removed and burnt. Drench the soil with fungicides like Ridomil or Metalaxyl @ 2g/l.

9.2.0 Pests

9.2.1 Aphids (*Myzus persicae; Aphis gossypii*)

Greenish peach colored sucking insects feed on the undersurface of the leaves. They excrete sticky honey dew which attracts ant. To control aphid infestation, introduce or attract natural predators such as ladybird beetles and wasps which feed on aphids. Aphids can also be washed off with a strong jet of water or by using an insecticidal soap such as neem or pongamia soap @ 2ml/l. Spraying of Imidacloprid @ 0.4 ml/l or Spirotetramat @ 0.75 ml/l can effectively control the aphid infestation.

9.2.2 Red Spider Mites (*Tetranychus urticae*)

Mite infestation is observed during dry spell. Mites suck the plant sap, resulting in speckling and discolouration of the foliage. The foliage turns yellow, becomes dry and stippled. Presence of web is the characteristic feature of the red spider mite infestation. Spraying of Diafenthiuron @ 0.8-1 ml/l or Abamectin @ 0.5 ml/l can effectively control the red spider infestation.

9.2.3 Thrips (*Haplothrips* spp.; *Frankliniella schultzei*)

Thrips are tiny insects which are black or straw in colour. They suck the plant sap by scrapping the plant tissue which results in discoloration, flecking or silvering of the infested surface. Proper cultural and sanitation measures help in avoiding the thrips infestation. Spraying of Fipronil @ 1.5 ml/l or Spinosad @ 0.3 ml/l is effective in controlling the thrips infestation.

9.2.4 Whitefly (*Bemisia tabaci*)

These are small tiny, white flying insects, infesting the plants mainly on cloudy weather. Nymphs and adults suck the plant sap from undersurface of the leaves resulting in discolouration, wilting and drying of leaves. In case of severe infestation, sooty mould may develop on the honeydew secreted by the nymphs. Spraying of neem oil or pongamia oil @ 5 ml/l or Acetamiprid @ 0.3 g/l can effectively control the white fly infestation.

9.2.5 Slugs

Slugs leave large holes on the foliage or completely defoliate the leaves. The pest can be traced by the slimy trails. They feed at night and are problematic mostly in moist weather conditions. Hand picking and destroying during the night is most effective control measure. Under severe infestation, dusting with Metaldehyde @ 25 kg/ha effectively controls the pest.

10.0 Harvesting and Postharvest Management

10.1.0 Harvesting

In kniphofia, flowering period commences from April to October. Flowers are harvested for cut flower purpose, when almost all florets show colour (Wilson, 2017). Scapes are cut at the base during early morning or late in the evening hours and placed immediately in a bucket of water. Later, the harvests are transferred to the packing house for pre-cooling and other postharvest activities.

10.2.0 Postharvest Management

10.2.1 Grading

There is no standard grading system available for the cut stems of kniphofia.

But in general, flowers having 60 cm stalk length are preferred for the flower arrangements or bouquets preparation.

10.2.2 Pulsing

Cut kniphofia flowers are subjected to geotropic bending and should be shipped in a vertical position. The gravitropism of kniphofia spikes can be overcome by pulsing treatment. The freshly harvested flowers are pulsed with 100ppm napthalene acetic acid (NAA) (Serek, 2000). Kniphofia flowers are ethylene sensitive. The sensitivity of ethylene can be reduced and vase life improved by pulsing with 20 per cent sucrose + 400 ppm silver thiosulphate (STS) solution (Prince and Tayama, 1988).

10.2.3 Storage

Cut flowers can be stored up to 6.5 days at 4° C by wrapping in wet newspapers and polythene which reduces the geotropic bending and maintains the keeping quality of the flowers. Flowers stored in wet and dry conditions have no significant difference on vase life. However, during storage, the weight loss is greater in flowers stored 'dry' compared to those stored 'wet' (Hettiarachchi and Balas, 2003).

10.2.4 Vase Life

Colour and visual appearance of the flowers plays an important role for consumer preference at the florist shop. Flower preservative solutions are composed of a mixture of chemicals that enhance the postharvest quality and longevity of cut stems. The solutions may contain carbohydrates (sucrose or glucose), biocides (8-HQS or 8-HQC), plant growth regulators (gibberellins, cytokynins), antiethylene compounds (silver thiosulphate) or chemicals known to aid water uptake and reduce water tension.

Vase life of kniphofia when tested using various vase solutions such as Chrysal (10 g/l), Flora (1 ml/l), Flower fresh (9 g/l), Biovin (1 ml/l) and 8-HQS (200 mg/l). Flora and Flower fresh were found to improve the vase life by 6.9 and 6.5 days respectively, than the flowers kept in normal tap water (Hettiarachchi and Balas, 2003). Use of floral preservatives is the most economical and practical method for extending the postharvest life of kniphofia cut flowers (Salunkhe *et al.*, 1990; Nowak and Rudnicki, 1990).

11.0 Biotechnology

11.1.0 Tissue Culture and *In vitro* Conservation

Kniphofia pauciflora is a threatened species found growing in marshy grass lands and is restricted to areas near Durban in Natal. For conservation of the species, tissue culture was initiated by using the apical regions of stolon. The explant is cultured in MS media supplemented with 100 mg/l myo-inositol, 3 per cent sucrose and 0.8 per cent agar. Shoot development occurs in presence of Kinetin (2 mg/l) and NAA (1 mg/l) in the culture media. Rooting is initiated by placing shoots in hormone free media. Hardening is done by placing the plantlets in the mist chamber for two weeks which is followed by transferring the plantlets in the greenhouse that are watered twice a week. In the species having problem of seed setting, *in vitro* culture is an excellent means of conserving and increasing the plant population (McAlister and Van Staden, 1996).

Under *in vitro* conditions, the carbohydrate source is essential for the induction of inflorescences in the endangered species like *Kniphofia leucocephala*. Taylor *et al.*, (2007) obtained high flowering percentage when media was supplemented with 60 g/l glucose and 10 μM BA, suggesting the direct influence of sugar over floral transition.

Organogenesis was studied in rhizome explants of *K. nelsonii* and *K. uvaria* grown on MS medium with added growth regulators. The best bud initiation (60 per cent of explants) was obtained with 1 mg/l BA for *K. nelsonii* and 1 mg BA + 8 mg 2,4-D per litre for *K. uvaria,* with induction periods of 8-11 and 40-46 days, respectively. Roots were induced within 4-5 days upon transfer to half-strength MS medium (Nayak and Sen, 1990). Cytokinins increase flowering in lower concentrations, best flowering response was induced when BA was used at 20 μM, however, it inhibits flowering at higher concentrations. Other cytokinins, like zeatin and 2iP failed to elicit the same flowering response in *K. leucocephala* species as observed with BA, indicating that both concentration and type of cytokinin are critical for flowering (Taylor *et al.* 2005).

12.0 References

Abebe, A.W., Bucer, F., Asres, K., Gibson, S., Rattray, L. and Croft, L.S. (2005) *Phytother. Res.*, **19**: 472- 476.

Berhanu, E., Fetene, M. and Dagne, E. (1986) *Phytochem.*, **25**: 847-850.

Botes, C., Johnson, S.D. and Cowling, R.M. (2008) *Oikos*, **117**: 875-882.

Botes, C., Johnson, S.D. and Cowling, R.M. (2009) *Int. J. Plant Sci.*, **170**(2): 151-156.

Bringmann, G., Menche, D., Bezabih, M., Molla, B.A. and Kaminsky, R. (1999) *Planta Med.*, **65**: 757-758.

Bringmann, G., Noll, T. and Rischer, H. (2002) *Plant Cell Rep.*, **21**: 125-129.

Brown, M., Downs, C.T. and Johnson, S.D. (2009) *S. Afr. J. Bot.*, **75**: 707-712.

Chase, M.W., Soltis, D.E., Soltis, P.S., Rudall, P. J., Fay, M.F., Hahn, W.H. and Pires, J. C. (2000) *Monocots: Syst. Evol.*, **7**: 3-16.

Codd, L.E. (1968) *Bothalia*, **9**: 363-513.

Dagne, E. and Steglich, W. (1984) *Phytochem.*, **23**: 1729-1731.

Dalia, E., Cipovistky, M. and Moshe, R. (2006) *Plant Cell Tiss. Organ Cult.*, **85**(1): 45-51.

Demissew, S. and Nordal, I. (1997) *Flora of Ethiopia and Eritrea* Vol.6. Hydrocharitaceae to Arecaceae. pp. 106-115.

Dimke, K.C., Still, S.K. and Gardner, D.S. (2008) *J. Environ. Hort.*, **26**: 222-226.

Egger, J.N. (2012) *U.S. Patent Application No. 13/066,143.*

Egger, J.N. (2017) *U.S. Patent Application No. 14/756,170.*

Elovson, S. (1984) *Wild Flowers S. Afr.*, 7th edn. Pretoria, South Africa.

Feilcke, R., Arnouk, G., Raphane, B., Richard, K., Tietjen, I., Marobela, A.K. and Fobofou, S.A. (2019) *J. Pharm. Biomed. Anal.*, **174**: 277-285.

Fichtl, R. and Admassu Adi. (1994) *Honeybee Flora of Ethiopia*. DED, Margeaf Verlang. pp. 510.

Habtemariam, S. (2010) *Fitoterapia*, **81**: 1013-1019.

Hansen, H.A. (2019) *U.S. Patent Application No. 16/350,526.*

Hansen, H.A. (2020) *U.S. Patent Application No. 16/350,527.*

Hettiarachchi, M.P. and Balas, J. (2003) *VIII Int. Symp. Posthar. Physio. Ornam. Plants,* 669: 359-366.

Janaki Ammal, E.K. (1950) *J. Roy. Hort. Soc.*, **75**: 23-30.

Johnson, S.D., Hargreaves, A.L. and Brown, M. (2006) *Ecol.*, **87**: 2709-2716.

Klingaman, G. (2008) Plant of the week: Red hot poker. *Division of Agriculture Research and Extension, University of Arkansas System.*

Li, Y., Lu, X. and Mao, Z. (2000) *J. Jiangsu Forestry Sci. Technol.*, **27**(6): 20-22.

Luo, S. (2014) *U.S. Patent Application No. 13/506,274.*

Marias, W. (1973) *Kew Bull.*, **28**: 465-483.

McAlister, B.G. and Van Staden, J. (1996) *S. Afr. J. Bot.*, **62**(4): 219-221.

Mc Cartan, S.A. and Van Staden, J. (2003) *In vitro Cellular Dev. Biol. Plant*, **39**(5): 496-499.

Moffett, A.A. (1930) *Merton,*: 315-337.

Nayak, S. and Sen, S. (1990) *Indian J Exp. Bio.*, **28**: 171-173.

Nayak, S. and Sen, S. (1992) *Biol. Plant.*, **34**: 135-141.

Nowak, J. and Rudnicki, R.M. (1990) *Postharvest handling and storage of cut flowers, florist greens and potted plants* (Ed. A.A. Duncan), Timber Press, Portland.

Olsen, K.M. and Schaal, B.A. (2000) *Procee. Nat. Acad. Sci., USA* **96**: 5586-5591.

Prince, T.A. and Tayama, H.K. (1988) *Bull. Ohio Flor. Assoc.*, **706**: 34.

Ramdhani, S. (2006) *Evolutionary and Biogeographic Studies in the genus Kniphofia Moench* (Asphodelaceae) Ph.D thesis, Botany Department, Rhodes University. pp. 410.

Ramdhani, S., Barker, N.P. and Baijnath, H. (2008) *J. Biogeog.*, **35**(12), 2258-2273.

Salunkhe, D.K., Bhat, N.R. and Desai, B.B. (1990) *Postharvest biotechnology of flowers and ornamental plants*. Springer-Verlag Berlin Heidelberg, pp. 6-24.

Saul, R.G. (2011) *U.S. Patent Application No. 12/653,154.*

Saul, R.G. (2012) *U.S. Patent Application No. 12/661,875.*

Saul, R.G.(2014) *U.S. Patent Application No. 13/507,434.*

Serek, M. (2000) *Plant Growth Regulators in Agriculture and Horticulture: Their Role and Commercial Uses*, 147.

Smith, G.F. and van Wyk, B.E. (1998) *The Families and Genera of Vascular Plants. III Flowering Plants; Monocotyledons. Lilianae* (except *Orchidaceae*): 130-140.

Symes, C.T., Nicolson, S.W. and McKechnie, A.E. (2008) *J. Ornithol.*, **149**: 13-22.

Taylor, N.J., Light, M.E. and Van Staden, J. (2007) *Plant Growth Regul.*, **52**: 73-79.

Taylor, N.J., Light, M.E. and Van Staden, J. (2005) *Plant Cell Tiss. Org. Cult.*, **83**: 327-333.

Teklehaymanot (2001) *Distribution, reproductive biology, demography and genetic diversity of endemic and indigenous Kniphofia species in Ethiopia.* Ph.D thesis, Addis Ababa University, Addis Ababa.: 119p.

Teklehaymanot, Endashaw, B., Sebsebe, D., and Nordal, I. (2004) *SINET. Ethiop. J. Sci.*, **27**(2): 143-152.

Van Delft, L. (2005) *Patent Application No. 10/968,804.*

Webber, J.M. (1932) *Amer. J. Bot.* 15: 411-422.

Wilson, C. L. (2017) *Postharvest Management of Horticultural Crops: Practices for Quality Preservation*, pp. 325.

Index